Plumbing:
Mechanical Services

PEARSON
Education

We work with leading authors to develop the strongest educational materials in technology, bringing cutting-edge thinking and best learning practice to a global market.

Under a range of well-known imprints, including Prentice Hall, we craft high quality print and electronic publications which help readers to understand and apply their content, whether studying or at work.

To find out more about the complete range of our publishing, please visit us on the World Wide Web at: www.pearsoned.co.uk

# Plumbing: Mechanical Services

# Book 2

*Fifth Edition*

**G.J. Blower**
Eng Tech (CEI), MIP, LCGI, Technical Teachers Cert.
Formerly Senior Lecturer, Plumbing Mechanical Services Section, College of North East London.
Currently an NVQ Visiting Assessor



PEARSON
Prentice Hall

Harlow, England • London • New York • Boston • San Francisco • Toronto • Sydney • Singapore • Hong Kong
Tokyo • Seoul • Taipei • New Delhi • Cape Town • Madrid • Mexico City • Amsterdam • Munich • Paris • Milan

**Pearson Education Limited**
Edinburgh Gate
Harlow
Essex CM20 2JE
England

and Associated Companies throughout the world

*Visit us on the World Wide Web at:*
www.pearsoned.co.uk

First published 1984
Second edition published 1989
Third edition published 1996
Fourth edition published 2002
**Fifth edition published 2007**



ISBN-13: 978-0-13-197621-4
ISBN-10: 0-13-197621-4

**British Library Cataloguing-in-Publication Data**
A catalogue record for this book is available from the British Library

**Library of Congress Cataloging-in-Publication Data**
A catalogue record for this book is available from the Library of Congress

Typeset in 10/12pt Times by 35
Printed and bound in Malaysia

*This publisher's policy is to use paper manufactured from sustainable forests.*

# Contents

# Foreword

It is said that every man is a builder by instinct and, while this may be true to a degree, the best way to become a craftsman is first to be an apprentice.

To be an apprentice or trainee means to learn a trade, to learn skills and have knowledge which will both be a means of earning a living and provide an invaluable expertise for life. But whereas this used to be the main purpose and advantage of serving an apprenticeship, nowadays many people in high positions in industry have made their way to the top after beginning with an apprenticeship.

The once leisurely pace of learning a trade has now been replaced by a much more concentrated period of learning because of the shortened length of training. In addition, there is an ever-increasing number of new materials and techniques being introduced which have to be understood and assimilated into the craftsman's daily workload. There is therefore a large and growing body of knowledge which will always be essential to the craftsman, and the aim of this craft series of books is to provide this fundamental knowledge in a manner which is simple, direct and easy to understand.

As most apprentices and trainees nowadays have the advantage of attending a college of further education to help them learn their craft, the publishers of these books have chosen their authors from experienced craftsmen who are also experienced teachers and who understand the requirements of craft training and education. The learning objectives and self-testing questions associated with each chapter will be most useful to students and also to college lecturers who may well wish to integrate the books into their teaching programme.

The needs have also to be kept in view of the increasing numbers of late entrants to the crafts who are entering a trade as adults, probably under a government-sponsored or other similar scheme. Such students will find the practical, down-to-earth style of these books to be an enormous help to them in reaching craftsman status.

*L. Jaques*
*Formerly Head of Department*
*Leeds College of Technology*

# Preface to the fifth edition

The contents of this volume, as with Book 1 of this series, relate mainly to the technology and practical activities necessary for the achievement of the technical and NVQ certificates, levels 2 and 3 in plumbing. Much of the text will also be helpful to students of associated crafts such as industrial and domestic heating, gas and oil work and those who wish to specialise in underground drainage and lead sheet weathering at an advanced level. It will also be helpful to those studying at professional levels in building subjects. This volume contains much of the necessary information relating to both the Water and Building Regulations, a good knowledge of which is essential for those installing and maintaining hot and cold water and heating schemes.

The current area of thought in the light of diminishing resources and changes in the environment is the government's commitment to energy saving. It is therefore important that plumbers and heating fitters understand the needs for energy conservation, because, due to the nature of their work, they can influence to a large degree the success or otherwise of energy savings.

I have endeavoured to implement some of the suggestions made by readers of the books in this series since the last edition was published and many of them have been embodied in these new editions of Books 1 and 2. I have, however, avoided detailed wiring plans of heating systems as this information is freely available in a clear and concise form direct from the manufacturers of heating controls. While the use of modern materials and techniques has simplified many of the traditional methods of installation, both the design of modern systems and fault diagnosis now form an important part of the work of plumbers and heating fitters. It is hoped that the contents of this volume will help the reader to achieve both the CGLI Technical Certificate and the plumbing NVQ's level 2/3. In conclusion, it should never be forgotten that a great deal of pride and satisfaction can be achieved in knowing that a job has been done well and carried out in a professional manner.

I would like to record my thanks to Mr D. Davis, the Technical Officer of the Institute of Plumbing and Heating, and to the staff of Essex County Library, Chelmsford, for their help in tracing and verifying many of the British and European Standards listed in this book. Also to Mr M. Mear for his permission to use illustrations of the Central Heating Calculator, The Institute of Plumbing for permission to use the tables and graph relating to pipe sizing from the design guide, The Energy Saving Trust and TACMA for their permission to reprint some of their illustrations and technical details. I should also thank all the other companies that have given me assistance and advice in the preparation of this book. I must also thank my wife Vilma for her help and support in the production of both Books 1 and 2 of this series.

# Acknowledgements

We are grateful to the following for permission to reproduce copyright material: Table 2.2 based on Table 3 in the *Essential Gas Safety* manual, p. 73 (The Council for Registered Gas Installers, 1998); Figures 5.35, 5.37 and 5.39 reproduced with permission from the *Domestic Central Heating Calculator* (M.H. Mear & Company Limited, 1994); Appendix C courtesy of The Association of Controls Manufacturers; Appendix D courtesy of Energy Savings Trust, Whole House Boiler Sizing Method for Houses and Flats – CE64, December 2003. For further or updated information: 0845 120 7799, bestpractice@est.org.uk or www.est.org.uk/housingbuildings. In some instances we have been unable to trace the owners of copyright material, and we would appreciate any information that would enable us to do so.

Extracts from British Standards are reproduced with the permission of BSI. Complete copies can be obtained by post from BSI Customer Services, 389 Chiswick High Road, London W4 4AL; Telephone: 020 8996 9001.

Copies of British Standards are also obtainable on loan from public libraries.

# 1 Welding and brazing processes

After completing this chapter the reader should be able to:

1. State the main safety precautions to be taken when working with compressed gases.
2. Identify the main causes of accidents in relation to gas heating equipment and describe how such accidents can be avoided.
3. Recognise the correct flame structure for various welding operations.
4. Select suitable methods of preparing sheet lead for welding.
5. Describe the principles and techniques of lead-welding processes.
6. Describe the techniques and underlying principles employed in bronze welding.
7. Identify the methods and techniques exployed with brazing processes for both sheet and pipes.

## Introduction

Welding and brazing and a knowledge of the high-pressure gas equipment used for performing these operations are an essential part of a plumber's skill. Welding and brazing have been in use almost since the dawn of time, certainly since metals were first smelted. Long before oxy-acetylene equipment became commonly used, blacksmiths used a process of welding by heating iron or steel to white heat in a forge and hammering them together. The term *brazing* is derived from brass, when common brass known as spelter, consisting of approximately 50 per cent copper and 50 per cent zinc, was used with a flux called *borax* to join iron and copper without actually melting the parent metals. In modern terminology *parent metal* relates to the metals to be joined, and the metal used to join them is called the *filler metal*.

One of the first welding processes to be used was lead burning, more correctly known as lead welding. The term *lead burning* probably relates to a very early form of welding this metal by joining the edges with molten lead. In 1837 a French engineer invented the first welding blowpipe using compressed air with hydrogen as a fuel gas; it was used with reasonable success for lead welding. It was quickly realised that the small concentrated flame required for welding could only be achieved with pure oxygen, and when compressed oxygen became readily available, it was used extensively with hydrogen or manufactured gas for lead welding. The most common combination of gases now in use for all gas welding and brazing processes is oxygen and acetylene, mainly because of their flexibility and high flame temperature.

## Welding safety

Compressed gases and fire are potentially dangerous and constant vigilance at all times is necessary when using welding equipment. There is an old saying 'familiarity breeds contempt', and it is a fact one can become so used to the equipment that the dangers involved are forgotten. Lack of care and attention can be the cause of serious accidents. The following safety precautions must be rigidly adhered to in order to ensure the safety of the operator and those working in the vicinity. Compressed gas cylinders are subject to statutory regulations and British Standards which are listed at the end of this chapter.

Correct identification of gases is the first rule of safety — do not use any gas cylinder which is not clearly identified by colour or labels. Cylinders should always be secured while being transported and should not be allowed to project beyond the sides or ends of the vehicle. They should not be rolled along the ground and the valves must be closed when they are moved. When cranes are used to lift them, chain slings or magnets must not be used, only approved webbing slings or cradles are permissible.

### *Storing cylinders safely*

All compressed gases should be stored in well-ventilated apartments or compounds. Fuel gases must be stored separately from gases which are not combustible, and empty cylinders must be marked 'MT'. 'Highly flammable' and 'No smoking' notices must be prominently displayed in areas where cylinders are stored. Acetylene and propane cylinders should always be stored in the upright position — if horizontal storage is used for other than fuel gas cylinders, they must be securely wedged and not stacked more than three cylinders high. Any store used for gas cylinders must be used for that purpose only, be securely locked when not in use, and not be allowed to become a dumping area for other materials. Oil, petrol or acids must never be stored with gas cylinders.

### *Avoiding dangers of fire or explosion*

After setting up welding equipment (described in Book 1, Chapter 3) check all joints for leaks using soapy water. Pay special attention to the hoses and ensure, by bending them, that no cracks appear. Take care when 'snifting' cylinder valves in confined places where there may be naked lights. Do not allow cylinders to become overheated by storing or using them near a heat source, and keep the burning blowpipe well away from them. If for any reason an acetylene cylinder becomes hot, the valve should be closed immediately and the cylinder removed to the open air where it should be cooled by being hosed down with cold water. The fire brigade should be contacted in all cases and the cylinder taken out of service. The supplier must be notified so that it can be tested to ensure it is safe for future use. The regulations governing acetylene cylinders are very strict indeed.

Take care not to allow sparks from welding or cutting operations to come into contact with cylinders or hoses as this can be a serious fire risk. If possible, do not allow hoses to trail over the floor where they are subject not only to falling fragments of red-hot metal but also to damage from machinery. A suitable fire extinguisher should always be at hand during any operation carried out using a flame. The dangers of allowing oil to come in contact with oxygen are mentioned in Book 1, Chapter 3, and should be carefully noted.

### *Cylinder valves*

Valves should be opened slowly to avoid a sudden surge of pressure on the equipment. When the valve is closed, never apply more pressure than necessary and do not extend the cylinder key in any way. When the equipment is out of use, close the valves. Never use broken or damaged cylinder keys.

### *Personal safety*

Do not wear overalls heavily contaminated with oil or grease. Where a welding operation causes sparks, leather aprons and gloves should be worn. Never use welding equipment without suitable eye protection. Many assume that because lead welding does not require high temperatures and subsequent glare, eye protection is not necessary; it is common sense always to protect the eyes, not only from glare but also from molten metal and sparks when welding or cutting steel. They are one of the most precious gifts humans have and it is very foolish to take risks with them. When welding steel or copper specially tinted goggles conforming to BS EN 7028:1999 must be used. Some welding processes employ the use of fluxes which, when heated, give off a bright glare causing severe eyestrain to the operator if the approved lenses are not used.

*Environmental safety*

Ensure there is adequate ventilation in any area in which a flame is used, as some oxygen is absorbed from the atmosphere and unventilated areas can be lacking in oxygen causing danger to the operator. Special care must be used when working in ducts or basements, and if possible operators should not work in such conditions on their own. If they are out of sight of the equipment, a responsible person should be stationed nearby to shut it down quickly in the event of an accident. Years ago when plumbers had to repair pumps in deep wells, they always took with them a lighted candle in a jar, the flame of which lengthened when the air became 'vitiated', meaning deficient in oxygen. Many plumbers do the same even today when working in confined spaces, as it serves as a warning when oxygen is lacking in the air.

Never be tempted to breathe in pure oxygen — it can result in pneumonia — and avoid the accidental enrichment of air by oxygen in a confined space which may lead to excessive fire risks. Material not normally combustible in the atmosphere will readily burn if it becomes enriched with oxygen.

When there are fume hazards, such as those encountered when working on painted surfaces or galvanised work, suitable respirators must be worn and the area must be well ventilated.

Do not use a flame on, or attempt to weld a tank or vessel which is suspected of containing (or having contained), flammable or explosive materials, unless one of the following treatments has been carried out:

(a) Boiling or steaming the vessel.
(b) Filling the vessel with water.
(c) Filling the vessel with a foam inert gas.

Treatments (b) and (c) ensure the exclusion of air, thus preventing the combustion of any traces of flammable materials which may be lingering in the vessel.

### The flame

The type of flame to be used varies, depending on the process for which it is required. For soldering or brazing, a large spreading flame is employed so that the whole area to be joined is at a uniform temperature enabling the filler metal to flow freely into the joint. If fusion welding is employed the edges of the parent metal are melted and fused together with the filler rod which is made of similar material to that of the parent metal. The flame used for fusion welding must be very hot and concentrated, allowing full control of the relatively small area of molten metal.

One of the most important prerequisites of successful welding is the ability of the operator to set the flame, having the correct proportion of gases and being of sufficient size to enable the weld pool to be retained under control. The weld pool is the area of the parent metal which is melted prior to adding the filler metal. If the area is too large, it will collapse leaving a large hole.

A welding flame can assume one of three forms: neutral, oxidising and carburising. The characteristics of these flames are shown pictorially in Fig. 1.1.

The neutral flame is the most useful for fusion welding of steel and lead — too much oxygen will result in the formation of an oxide film over the molten pool and prevent the filler metal merging with it. An excess of acetylene will result in a carburising flame, giving rise to a weld containing impurities and leaving a sooty deposit on the surface of the finished job. An oxidising flame is necessary for bronze welding as will be seen later, and even a carburising flame has its uses, one example being hard surfacing low-carbon steel, an engineering process known as *stelliting*. A slightly carburising flame is also recommended for brazing aluminium. A neutral flame is recognised by the small blue rounded cone seen on the end of the nozzle. This small cone is known as the area of unburned gas — complete combustion takes place at a point about 3 mm in front of it and this is the hottest part of the flame. Throughout all welding processes, the aim should be to keep the cone at about this distance above the weld pool. If it is allowed to fall into the pool, a small explosion takes place blowing small particles of molten metal in all directions. This is one of the reasons why it is absolutely necessary to use goggles. It is also one of the causes of 'backfiring'.

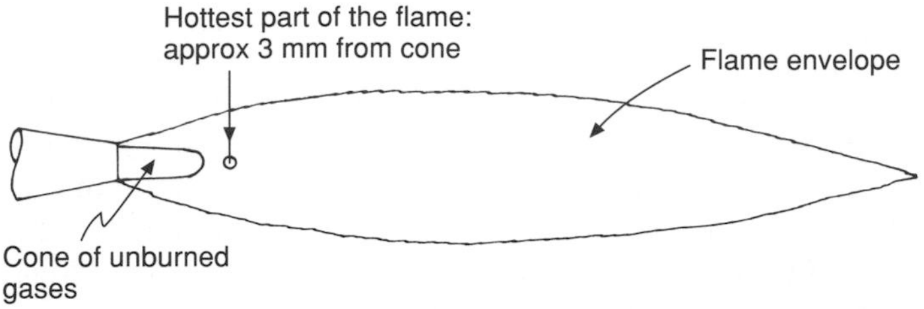


(a) Neutral flame
Used for most fusion-welding processes.

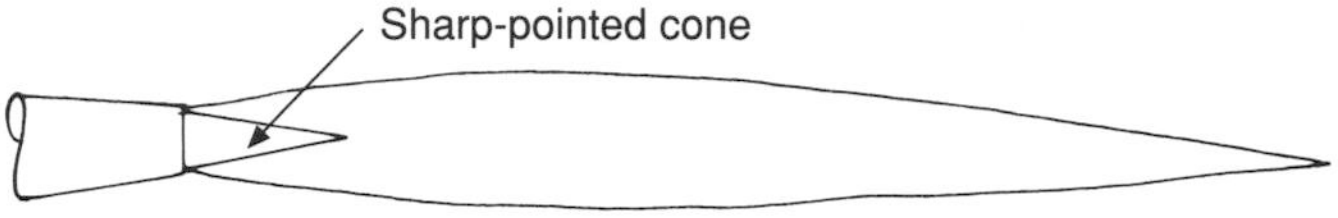


(b) Oxidising flame
Used for bronze welding copper, copper zinc and copper tin alloys.

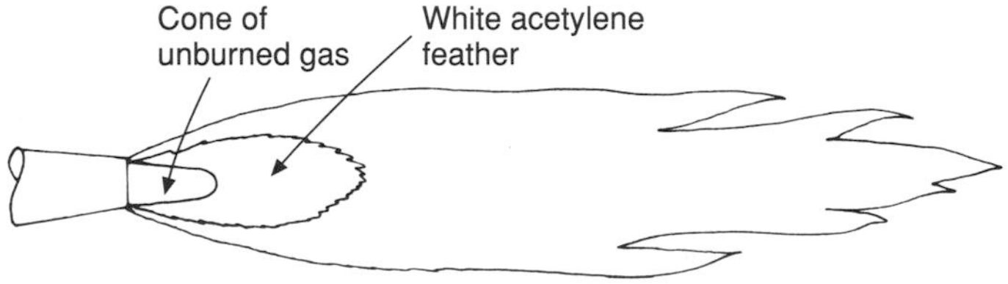


(c) Carburising or excess acetylene flame
Should not be used for welding copper, lead, iron or steel; mainly used in engineering for hard surfacing of steels.

**Fig. 1.1** Flame characteristics.

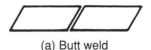

*The reducing flame*

The reader may be puzzled in that no flux is normally required for fusion welding (there are one or two exceptions). The combustion of oxy-acetylene produces a 'reducing' flame, which means the flame absorbs oxygen from the air surrounding it, causing the complete combustion of the hydrocarbons in the flame envelope which prevents the access of oxygen to the weld pool.

### Set-up of welded joints

All types of welded joints can be broken down into three main types called butt, lap and fillet welds. The set-up of these joints is illustrated in Fig. 1.2. Although the following text relates to lead, the joints used are common to all welding processes. In some instances, especially where

(a) Butt weld
Note: gap between sheet or plate will vary depending on the material. For lead no gap is necessary.

(b) Lap weld

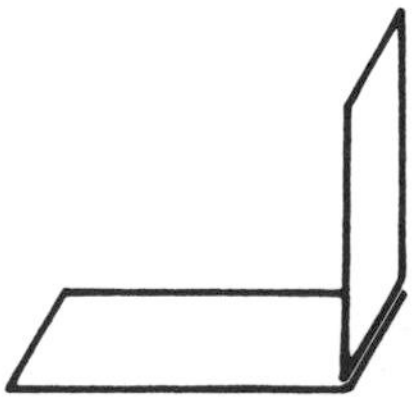

(c) Fillet weld (called 'angle' weld when on lead sheet)
May be 'set on' or 'set off' (see Fig. 1.22). Set on (no gap) is used for joining lead sheet.

**Fig. 1.2** Set-up of welded joints.

lead sheet is concerned, it is impossible to build up sufficient reinforcement in one run of welding and two or more depositions are laid to achieve a strong joint.

### Lead welding

It will be seen that the area in which fusion takes place varies with the type of joint (see Fig. 1.3). Note that the penetration does not occur right through the thickness of the undercloak with the lap joint. For this reason this joint is suitable for positional work without danger of the flame causing a fire in the surrounding timber work.

*Preparation for and welding a butt joint on lead*
It is essential that the surfaces to be joined are thoroughly clean and this should be achieved using a shave hook. Only the area covered by the weld should be shaved — a distance of about 6 mm on each edge is sufficient. If a larger area is shaved it not only spoils the appearance of the finished weld but also results in unnecessary thinning of the lead. Make sure that not only the top surface is clean but also the edges, and in the case of lap joints the underside too.

The lead should be assembled so that the edges to be joined are butted together and then tacked at 100–150 mm centres adding a little filler metal. This can be obtained in circular sections from lead manufacturers, although it is expensive to buy in this way if a lot of welding is to be done. It is usually more economical to cast one's own filler rods in specially prepared metal moulds of a type shown in Fig. 1.4. It is also possible to use strips of sheet lead cut from waste, but as these seldom exceed 2 mm in thickness (depending on the thickness of the lead from which they are cut) they are consumed too quickly. This results in a lot of stops and starts in a long run of welding. Do not forget to shave clean all filler rod before use to remove any oxide film from its surface.

Before actually commencing to weld, the operator should check that the correct nozzle is used, because if it is too small progress will be slow, and if too large it will be impossible to control the molten pool. Choice of nozzle size and correct adjustment of the flame are prerequisites to success in any welding operation; even an experienced welder will not be able to produce good work unless these conditions are right, much less one who is inexperienced.

The model 'O' blowpipe, illustrated in Book 1, Chapter 3, Fig. 3.5 has a range of five nozzle sizes most of which are suitable for lead welding. Table 1.1 shows suggested sizes for various thicknesses of lead, although this can vary slightly depending on the position of the work and the skill of the operator,

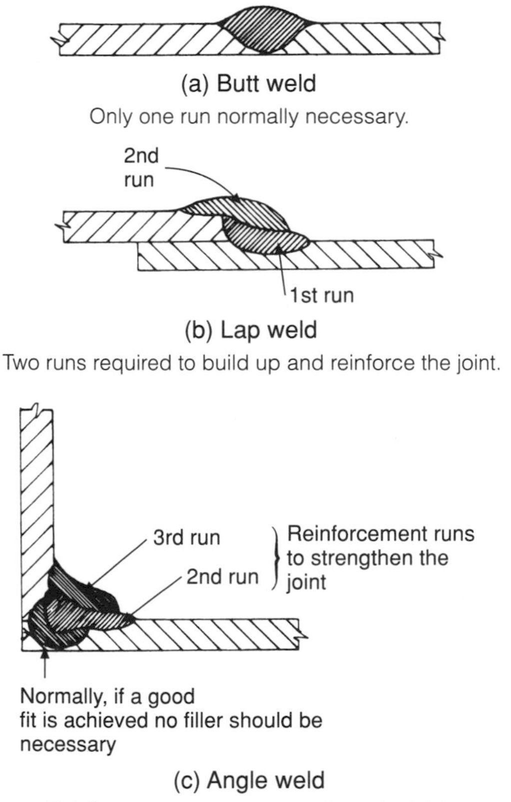
(a) Butt weld
Only one run normally necessary.
2nd run
1st run
(b) Lap weld
Two runs required to build up and reinforce the joint.
3rd run
2nd run
Reinforcement runs to strengthen the joint
Normally, if a good fit is achieved no filler should be necessary
(c) Angle weld
Reinforcement runs to strengthen the joint.

**Fig. 1.3** Sections through completed lead welds showing number of loadings recommended.

**Table 1.1** Nozzle sizes for lead welding.

| *Nozzle no.* | *Thickness of lead (BS code no.)* |
|---|---|
| 2–3 | 4–5 |
| 3–4 | 5–6–7 |
| 5 | Sand cast lead |

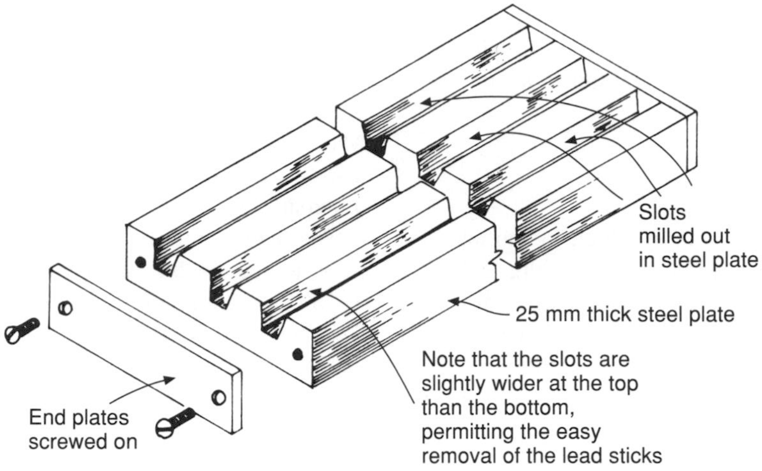
Slots
milled out
in steel plate
25 mm thick steel plate
Note that the slots are
slightly wider at the top
than the bottom,
permitting the easy
removal of the lead sticks
End plates
screwed on

**Fig. 1.4** Lead strip casting mould. Molten lead is poured into the grooves milled out in the steel where it quickly solidifies, producing a lead stick approximately 6 mm square which is ideal for most types of lead welding.

e.g. an experienced welder will use a larger nozzle size which will increase welding speed. It will be noted that no mention is made of the number 1 nozzle in the table because it is seldom used for lead welding. One should be aware, however, that blowpipes used for lead welding are also suitable for light welding work on other metals, in which case a very small nozzle is often necessary.

Prior to commencing any welding operation the cylinder regulators must be adjusted to give a pressure reading of approximately 0.33 bar (5 psi, 1 lb/in$^2$).

Having selected the correct nozzle for the work, the next step is to turn on the blowpipe acetylene valve and ignite the gas, its volume being adjusted so that the flame burns on the end of the nozzle and no soot is given off. The presence of soot indicates that the quantity and velocity of the gas is insufficient to draw in oxygen from the surrounding air. If, on the other hand, the valve is turned on too much the gas will burn in the air away from the end of the nozzle, a condition referred to in the next chapter as 'lift-off'. This must be corrected by reducing the volume of gas at the valve.

The oxygen valve is now turned on and adjusted until the white acetylene feather disappears. If it is found that the flame is too large and is overheating the work, it can be reduced by turning down the oxygen until the feather reappears, then shutting down the acetylene until the feather disappears into the area of the unburned gases.

The ability to adjust the blowpipe to give the correct flame for a specific work piece is something that will quickly be acquired by experience, but it is very important, and unless attention is paid to it successful welding will never be accomplished. If the flame is too large the molten pool will be difficult to control, and if it is not large enough full fusion of the sheets to be joined will not be achieved.

Once one is in a comfortable position and the body is relaxed, the welding operation can be commenced. The leftward method is generally used for lead welding — that is to say, for right-handed operators the filler rod is held in the left hand, the blowpipe in the right, with the welding operation proceeding from right to left. This can be confusing to the left-handed operator who will normally hold the blowpipe in the left hand and the filler rod in the right, the weld proceeding in a rightward direction. The thing to remember, however, when using the leftward technique is that the *filler rod always precedes the blowpipe*. The angles of the filler rod and blowpipe in relation to the work are shown in Fig. 1.5. A little leeway is permissible in the recommended angles, depending on the position in which one is working, but it will be found that the best results are obtained using these angles. In short, if it is found that the weld is not going as it should and the flame is correct, check the angles.

In any welding operation a molten pool must be established before the filler is added, as failure to

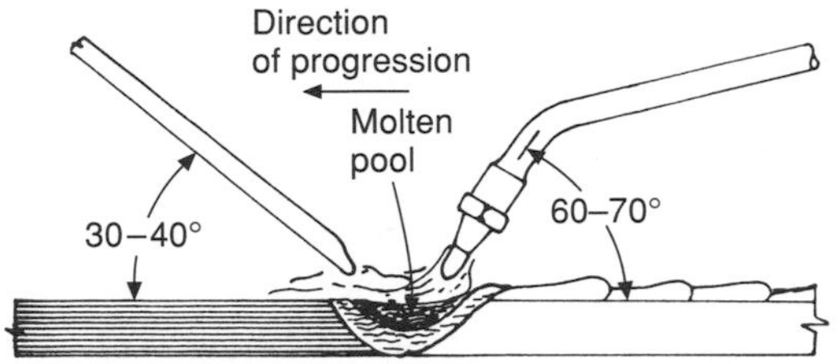


**Fig. 1.5** Angles of rod and blowpipe for leftward welding.

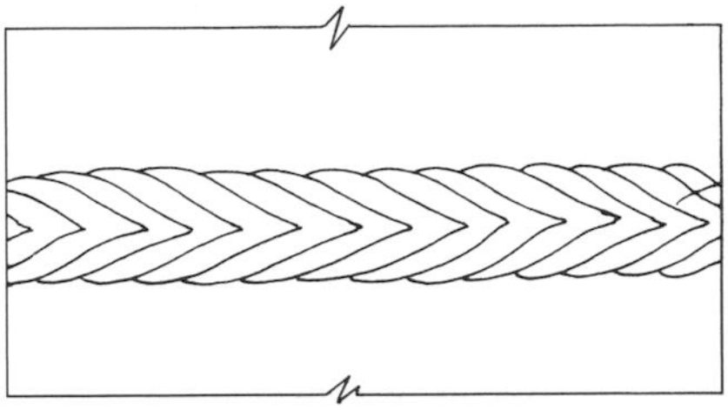

(a) Herringbone effect

Produced with a large flame and ideal for long flat runs.

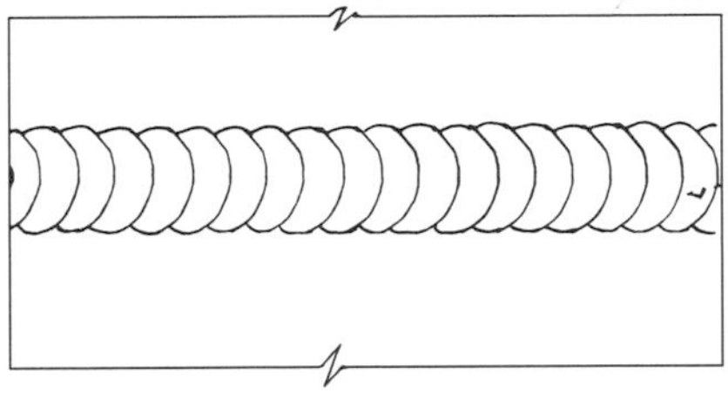

(b) Thumbnail effect

Produced with a smaller flame. Used in awkward corners and for intricate details.

**Fig. 1.6** Lead welding patterns.

observe this will result in unsightly blobs of metal in the work and incomplete fusion. The blowpipe is now raised slightly to melt a piece of the filler rod which is held in close proximity to the molten pool. As it melts and merges into the pool, the flame is brought down in a stroking action until the outer edges of the pool reach the edges of the shaved area where they cool, causing the distinctive ripple effect common to all welding operations. The filler rod and blowpipe are now moved towards the left, when the whole action is repeated. Each time a small portion of the filler metal is deposited in the molten pool a ripple is formed, the speed of the operator and the size of the flame determining the shape of the ripple. Bear in mind the control of the molten pool is achieved not only by the correct flame, but by the amount of filler metal deposited into it and the consequent cooling effect brought about by the speed of the forward movement.

It is points such as this that must be learned from experience by the welder until they are as natural as walking and breathing. A large flame and high welding speeds will produce the herringbone pattern that looks so effective on a long run of welding. The use of a smaller flame and lower welding speeds will produce a weld that looks very similar to a weld on steel, sometimes called by lead welders the *thumbnail* effect. While it does not look quite so effective, it must be remembered that if an intricate piece of work is attempted in an awkward corner, a smaller flame and lower welding speed are much more likely to lead to good results. A fast flame can lead to loss of control of the weld pool in such circumstances. Both types of finished weld appearance are illustrated in Fig. 1.6. The advantages and limitations of each should be understood so that the best technique can be selected for a given set of circumstances.

The term *penetration* has been used earlier; the other important term is *reinforcement*. This is applied to the build-up of metal over and above the surface of the parent metal and, as its name implies, its purpose is to strengthen or reinforce the area of the weld. It should not be excessive in height and the edge must merge into the parent metal smoothly.

### Defects in welds

Although at this stage we are considering lead welding, any defects are common to all types of welding. These defects are illustrated in Fig. 1.7, the most common being insufficient penetration or lack of fusion which will obviously result in a weak job, which when subjected to any stress will fall apart.

The other main weakness is undercutting and, as the illustration shows, this reduces the thickness of the parent metal adjacent to the weld. This defect is very common in fillet or angle welds when the molten metal on the vertical side of the weld tends to run down into the molten pool. It can be corrected by careful application of the filler rod to the upper edge of the weld pool. Generally, when

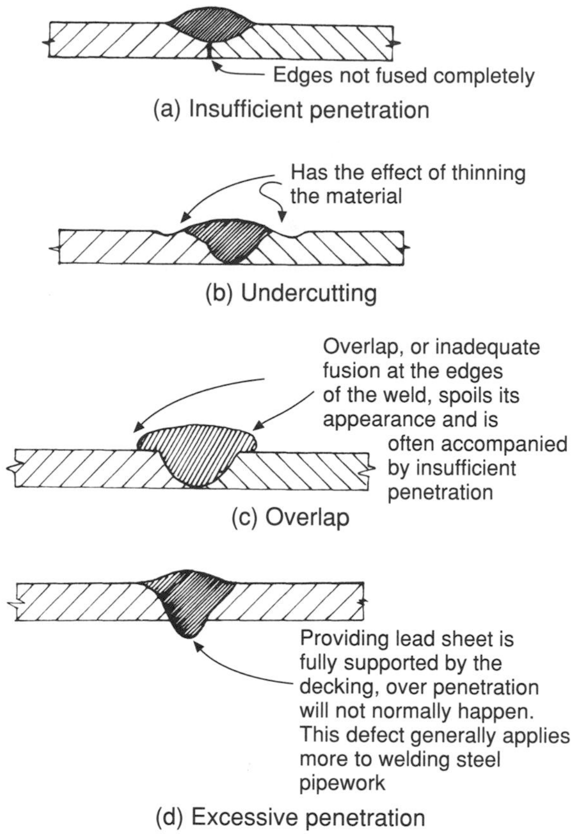
Edges not fused completely
(a) Insufficient penetration
Has the effect of thinning the material
(b) Undercutting
Overlap, or inadequate fusion at the edges of the weld, spoils its appearance and is often accompanied by insufficient penetration
(c) Overlap
Providing lead sheet is fully supported by the decking, over penetration will not normally happen. This defect generally applies more to welding steel pipework
(d) Excessive penetration

**Fig. 1.7** Common faults in welding.

undercutting takes place it is caused by too great a welding speed, addition of insufficient filler metal, incorrect welding angles or the use of a flame that is too large.

Overlap results in incomplete fusion of the edges of the filler metal with the parent metal, and is caused by an excessive build-up of the reinforcement.

Excessive penetration is a defect which can happen when welding all types of materials. It is the result of using too big a flame which causes the weld pool to enlarge, resulting in its collapse. Some metals, notably steel, can be welded without support providing careful control is maintained over the weld pool, but in the case of lead adequate support of the underside is essential at all times. Figure 1.11 illustrates a typical example.

The word *inclusion* is often employed when discussing defects in welds. This relates to impurities in the deposited weld metal. While it is not common in lead welds, the practice of cleaning lead for welding with steel wool can lead to the inclusion of small particles of steel in the weld, and for this reason it is not recommended. The other main point to bear in mind is the necessity to ensure that, when a weld is picked up or restarted after pausing for some reason, the weld pool merges into the metal already deposited. It is recommended that the restart is made 6–8 mm back into the previous deposit so that full fusion is maintained throughout the length of the weld. This also ensures that any oxides or scale contained in the weld pool, especially in the case of steel welding, are floated off to the surface and do not remain as an inclusion in the weld.

### Lap welds on lead sheet

This process is very similar to butt welding when this type of joint is made in the downhand (flat) position. The main differences are that the weld must not penetrate right through the underside of the lap or undercloak, and more than one run of filler metal is usually necessary to give adequate strength to the joint. About 25 mm should be allowed for the lap, and remember to shave all the jointing surfaces. The underside of the overcloak is often forgotten, resulting in loss of fusion causing a defective weld. The overcloak should be tacked in position and the first run made. This will almost certainly cause undercutting, and to produce a weld of adequate strength a second run must be added to provide effective reinforcement. A careful study of the section of a lap weld shown in Fig. 1.3(b) will clarify this.

This second run should be made so that the edge of the first remains exposed by about 3 mm. This ensures complete fusion to the parent metal without the overlap referred to in the section on weld defects.

*Lead welding in the vertical position*

The lead should be shaved and prepared in a similar way to that of an ordinary lap joint.
Two main techniques of welding on a vertical face are employed, one for welding at 90° and referred to as an upright joint, the other where the joint lies at an angle across the vertical face and is called an inclined seam. These joints are illustrated in Figs 1.8 and 1.9, and as a rule they do not require

the use of filler metal; the edge of the overcloak in each case is melted and fused to the undercloak.

To make these joints the blowpipe is held at about 90° to the face of the work, melting the undercloak to form a small pool while simultaneously describing an elliptical movement, melting the edge of the overcloak to flow into the molten pool. For control of the weld pool to be maintained when an upright joint is made, a step-by-step technique is employed. That is to say, as each weld bead is formed, the blowpipe is removed momentarily from the face of the work to allow the bead to congeal, forming a platform on which the next bead is deposited. This technique requires a lot of practice to perfect because if the molten edge of the overcloak is not carefully directed into the molten pool by the flame it will drop off. The only way to put this right is to make good the loss by using a very thin filler rod, preferably in this case cut from a strip of lead. This is easier said than done, and unless one is very skilled the appearance of the finished weld is spoiled. Careful selection of the welding nozzle and correct flame setting are essential in this method of welding.

Where possible, the inclined technique should be employed as the edge of the overcloak forms, to a greater degree, a support for the molten pool of lead and should control of the weld pool be lost, it is more easily rectified on an angled surface than on one at 90°. A stronger weld can be produced in this position by adapting the horizontal lap joint to the inclined position. The overcloak is turned out to a slight angle as shown in Fig. 1.10. This method

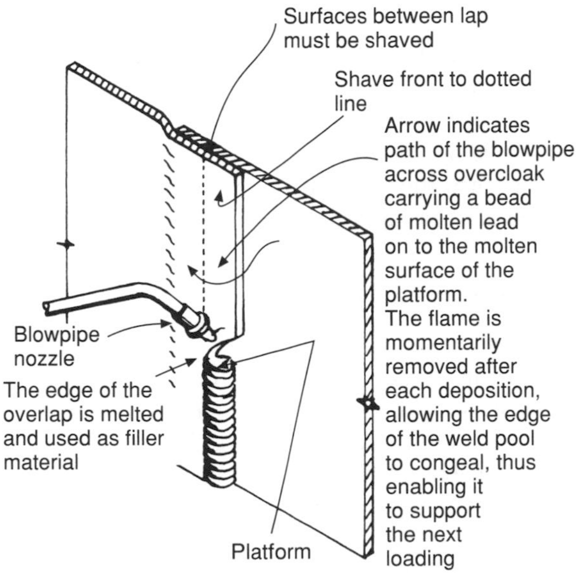


**Fig. 1.8** Vertical welding.

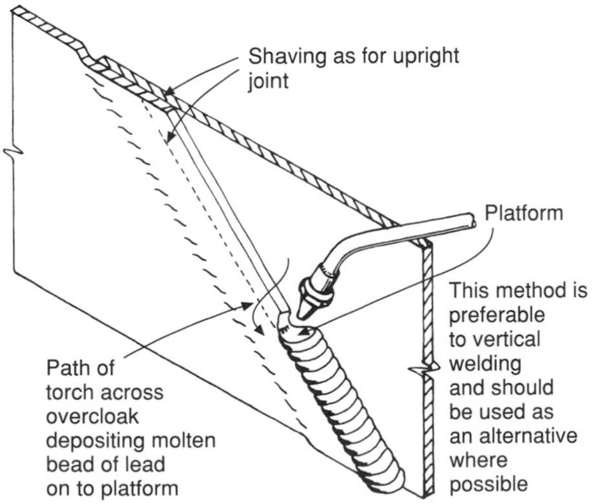


**Fig. 1.9** Inclined vertical weld.

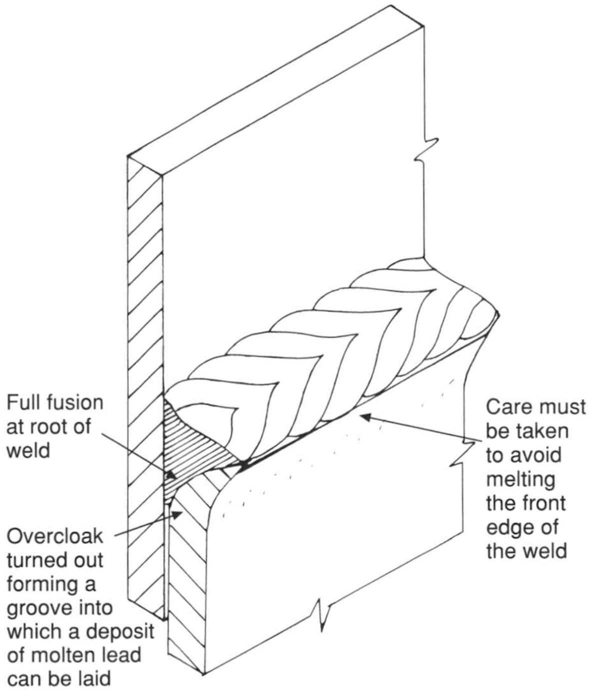


**Fig. 1.10** Horizontal lapped weld.

requires the use of the filler rod, but as a greater area of lead is exposed to fusion it results in a much stronger joint. It does not, however, produce quite such a neat effect as the former technique.

### Angle seams

If the work piece is relatively small and can be welded out of position, it can be set up on a wooden jig like that shown in Fig. 1.11 so that both faces are inclined. It is much easier to produce a good weld in this position than where one face is vertical.

The weld is accomplished by making three runs, the first simply fusing the two mating edges together, the second and third reinforcing the weld, the additional lead giving extra mechanical strength. It is on the first or 'burning in' run that accidents are likely to happen, such as burning away the edges, and it must be stressed that careful control of both the flame and the weld pool is essential.

It is more difficult to make an angle seam where one face is vertical: this is called *fillet welding* and the main difficulty encountered is undercutting the vertical sheet. The actual welding technique employed on lead is different from that for steel, but the problem of undercut is the same. It is essential to become skilled in welding angles in this position, however, as large pieces of lead are not easily handled and must often be welded *in situ*. Three runs are recommended as in the case of angle welds made in the inclined position, the last run being made taking great care to avoid undercutting the lead on the vertical surface. A thinner filler rod may be used with advantage here and should be held as close as possible to the vertical surface so that it melts at the same time as the surface of the lead, merging with it before it falls. Manipulation of the blowpipe from side to side as each bead is deposited, and avoiding the use of a fast flame, will also prevent undercutting.

The application of these basic techniques of lead welding are illustrated in Chapter 10 and with practice can show considerable savings in time over traditional bossing techniques.

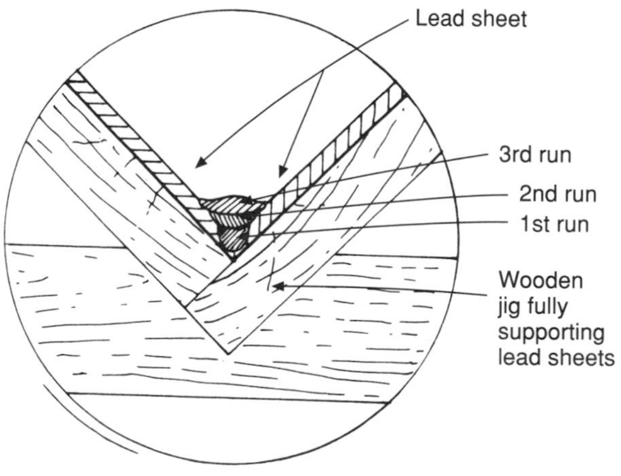


**Fig. 1.11** Welding angle seam on a work piece which is out of position. Suitable for bench fabricating small components such as lead outlets for flat roofs and catchpits.

### Lead-welded pipe joints

The use of lead pipe is now very limited and it is a fact that many plumbers will never come into contact with it. For this reason no specific information regarding this subject is given here. Most remarks relating to the welding of sheet, however, also apply to pipework, and with a little ingenuity the techniques previously described can be adapted for welding pipes should an occasion arise. In cases where help or advice is required on this subject it is recommended that contact is made with the Lead Sheet Association.

### Bronze welding

*The flame*

One of the most important factors to bear in mind when bronze welding is that, unlike fusion welding of lead and steel, a slightly oxidising flame must be used. The reason for this is that the zinc in the filler rod melts at a temperature of 410 °C and volatilises (gasifies) at higher temperatures. If the flame is incorrectly adjusted the gas will bubble through the molten filler metal as welding proceeds, leaving a series of blowholes which can result in a defective joint. The excess oxygen in the flame, however, combines with the zinc gas to form zinc oxide which melts at around 1,800 °C, a much higher temperature than either the copper or the filler rod, and in this way no loss of zinc occurs. Experience and practice soon enable one to identify the correct flame, but the following information will be useful. If the finished weld has a matt, yellow appearance, with some evidence of blowholes and poor weld ripple formation, the flame lacks oxygen. If, however, the flame is excessively oxidising it will cause the joint to blacken due to the oxide film and result in poor adhesion of the filler metal. This condition can be identified further by the exceptionally bright, almost golden colour of the deposited metal.

*Bronze-welded joints*

The basic concept in preparing joints for bronze welding is to allow sufficient space for a body of the filler metal to be built up, as this and its adhesion by 'wetting' the parent metal are the sources of its strength. Figure 1.12 illustrates a typical profile for bronze welding on metal thicknesses upwards of approximately 2.5 mm. This process was used extensively for jointing copper tubes for both water supply and sanitary pipework, but due to the fact that the filler metals used are subject to dezincification, jointing by this method is no longer acceptable and has been superseded by brazing techniques. It still has its uses in mechanical engineering services, especially for repair work on iron castings. Figures 1.13 and 1.14 illustrate typical repairs using the method. It is also suitable for light construction work such as hand railing or other light support structures as shown in Fig. 1.15, especially where galvanised steel tubes are used. A more satisfactory joint will be obtained using bronze-welding techniques on this material, causing less damage to the protective zinc coating, than by attempting to use fusion welds. It is very important to remember that welding galvanised work is very dangerous due to the fumes given off. Any such work must be undertaken in the open air or in a properly ventilated workshop, and a suitable respirator should be used.

It is essential that, prior to welding, the work is thoroughly clean. All joint surfaces should be scoured with card wire or a wire brush or prepared with a suitable mechanical grinder.

Any residues must be removed after welding because, like all fluxes, they are corrosive. One should be careful to use the right amount of flux as the residues set as a hard glass-like structure which is difficult to remove. Chipping with a chisel or screwdriver can cause damage to the work. The simplest way is to wash the joint with a weak solution of phosphoric acid. It is possible to obtain flux-impregnated filler rods at little extra cost (see

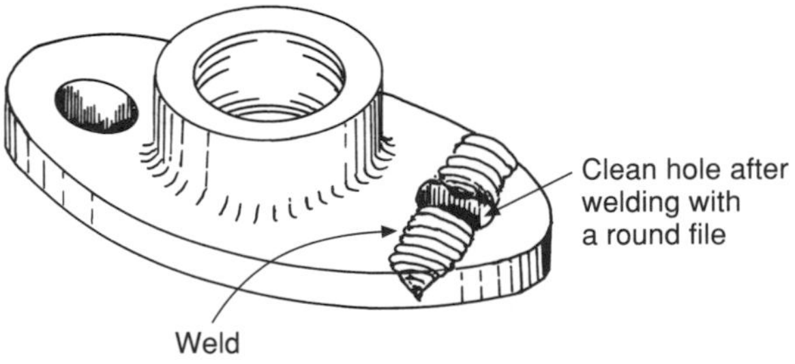


**Fig. 1.13** Repairing a broken cast iron flange.

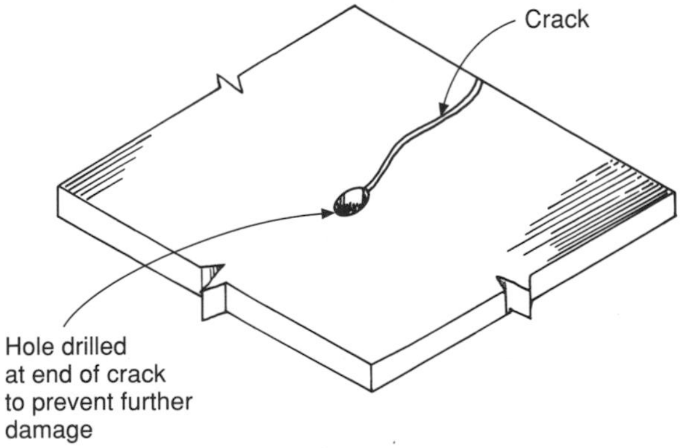


**Fig. 1.14** Bronze welding a cracked cast iron plate.

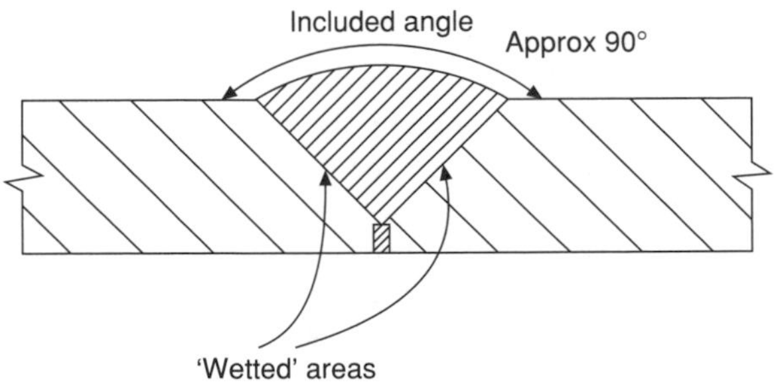


**Fig. 1.12** Preparation of steel or cast iron plate for bronze welding.

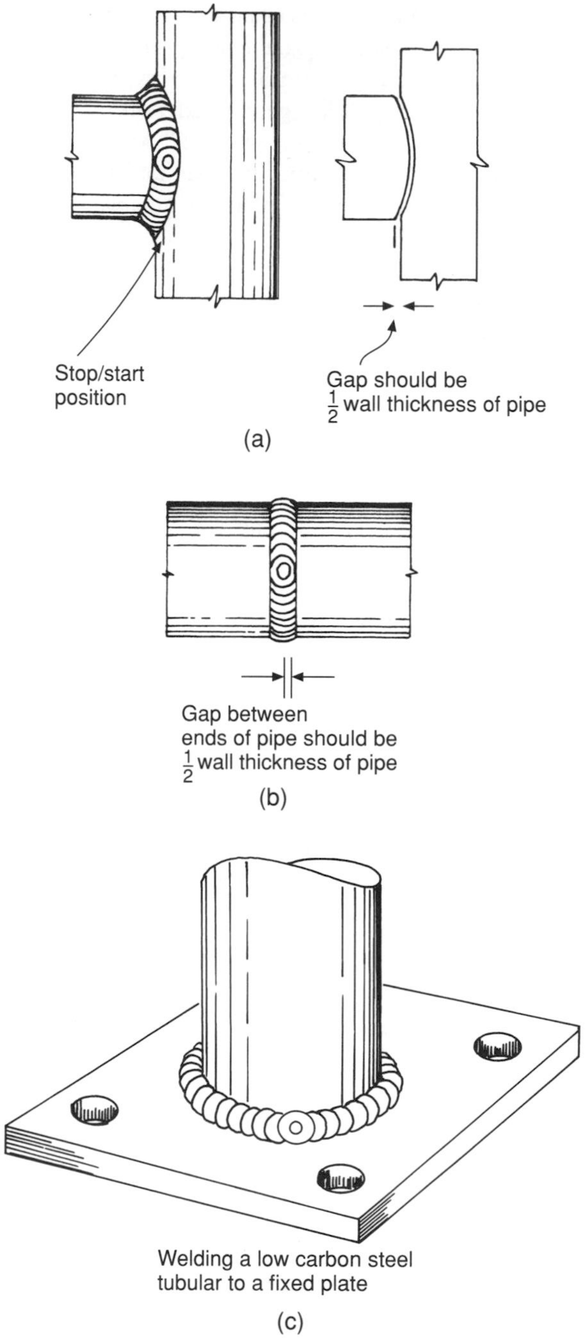
Stop/start
position
Gap should be
½ wall thickness of pipe
(a)
Gap between
ends of pipe should be
½ wall thickness of pipe
(b)
Welding a low carbon steel
tubular to a fixed plate
(c)

**Fig. 1.15** Butt-type bronze-welded joints for steel pipes. Mainly used for steel fabrications and general-purpose work where the use of high welding temperatures is undesirable, e.g. galvanised work. Note that although the finished appearance of these welds is identical to fusion welds on steel pipe, the gap for bronze welding is much less.

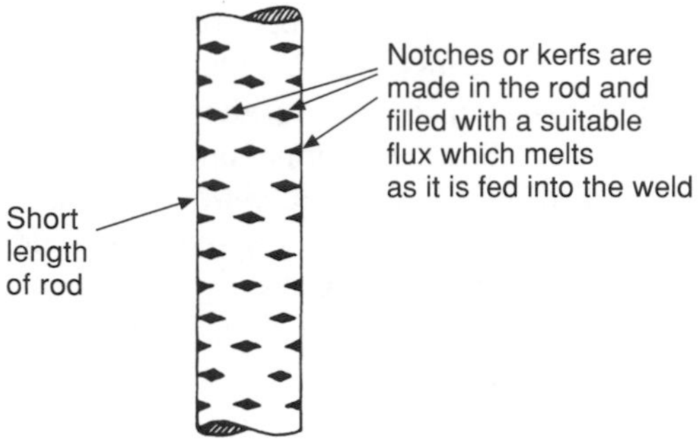
Notches or kerfs are
made in the rod and
filled with a suitable
flux which melts
as it is fed into the weld
Short
length
of rod

**Fig. 1.16** A flux-impregnated filler rod.

Fig. 1.16). The use of these ensures that only the right amount of flux is used and little trouble will be experienced in its subsequent removal. The rods must be kept dry to prevent deterioration of the flux content.

*Fluxes*

Special fluxes consisting mainly of borax and silicon are used for bronze welding. They may be applied to the joint as a paste by mixing the powder with water or by dipping the heated filler rod into the flux causing it to adhere to the rod. Table 1.2 shows the commonly used filler metals and their suitability for various types of parent metals.

*The welding process*

Most bronze-welded joints are normally made in the downhand (flat) position. Positional welding can be done using similar techniques to those used for lead, but it requires a lot of practice and is seldom necessary. After fluxing, the joint should be positioned and tacked if necessary. Heat the joint, keeping the flame moving from side to side to ensure the whole area is evenly heated. The filler rod should now be applied by means of a stroking movement until it is seen to run, wetting the edges of the joint and flowing into the root of the weld. The blowpipe should then be moved from side to side, or on smaller joints rotated so that the nozzle describes a series of circles as the weld proceeds in a forward direction round the joint, forming the characteristic weld ripples. Ensure that the filler rod at the stop position of the weld merges thoroughly with that at the start by remelting and overlapping it by at least 6 mm. To avoid a hollow at the stop/start

**Table 1.2** Bronze welding filler metals.

*Refs BS EN 1044–1999 CU 300 Series Brazing filler Metals*

| | Composition % | | | | | | Melting range | | |
|---|---|---|---|---|---|---|---|---|---|
| *Classification* | *Copper* | *Zinc* | *Silicon* | *Tin* | *Magnesium* | *Nickel* | *Solidus °C* | *Liquidus °C* | *Suitable parent metals* |
| CU 301 | 58.5 | Remainder | 0.2 | — | — | — | 875 | 895 | Brass, bronze, low carbon steel |
| CU 302 | 58.5 | Remainder | 0.2 | 0.2 | — | — | 875 | 895 | Carbon and galvanised steel |
| CU 305 | 46.0 | Remainder | 0.1 | 0.5 | — | 8.0 | 890 | 920 | As above slightly higher tensile strength |
| CU 306 | 56.0 | Remainder | 0.5 | 1.5 | 0.2 | 0.2 | 870 | 890 | Cast and malleable iron |

position of the weld due to contraction of the filler metal, the flame should be momentarily removed at this point. This allows the weld pool to cool slightly, permitting the edges to congeal before a little more filler metal is applied. The stop/start would then appear as in Fig. 1.15.

## Brazing processes

This is a method of making joints on metals using similar techniques to those employed with soft soldering; however, as the filler metals used have a higher tensile strength than soft solder, the depth of the sockets may be reduced. As with soft soldering, the filler metals used for brazing penetrate small gaps between the surfaces of the metals to be joined by capillary attraction. This means that some care is required to ensure the correct tolerances of the gap which will enable the filler metal to penetrate fully the surfaces of the joint. The considerable permissible variations in the gap depend on the type of filler alloy used. A branch joint, for instance, made with a hammer and bent bolt, might not be so accurately formed as one made with a purpose-made tool, and in such circumstances a gap-filling alloy would be used to make the joint.

### *Brazing alloys*

Those most commonly used in plumbing and associated crafts are based on silver–copper and copper–phosphorus alloys. Both these alloys melt at temperatures well below 800 °C, which is lower than those used for bronze welding. Copper–silver alloys nearly always contain zinc and cadmium in varying proportions, which have the effect of lowering the melting point and increasing their fluidity. The less silver the alloy contains the greater the temperature difference between the completely solid and the liquid state of the alloy. This gives it a long pasty range like that of plumbers' solder, and filler rods having this characteristic are capable of bridging and filling larger gaps than those having a higher silver content. Table 1.3 indicates the whole range of

**Table 1.3** Silver brazing alloys (reference BS EN 1044).

| *Type* | *Nominal composition* | | | | *Melting range (°C)* | | *Characteristics* |
|---|---|---|---|---|---|---|---|
| | *% Silver content* | *% Copper* | *% Zinc* | *% Tin* | *Solidus* | *Liquidus* | |
| Ag 1 | 56 | 22 | 17 | 5 | 618 | 652 | Free flowing |
| Ag 2 | 45 | 27.75 | 25 | 2.25 | 640 | 680 | Free flowing |
| Ag 3 | 40 | 30 | 28 | 2 | 650 | 710 | Gap filling |
| Ag 4 | 30 | 38 | 32 | X | 695 | 770 | Gap filling |

The alloys here are cadmium-free. The use of this constituent is no longer recommended due to its danger to health. These alloys are not recommended for hot and cold water services due to their zinc content, and the possibility of dezincification. For further details relating to all brazing alloys see the further reading section at the end of this chapter.

**Table 1.4** Copper–phosphorus brazing alloys (reference BS EN 1084 and BS 1854).

| *Type* | | *Nominal composition* | | | *Melting range (°C)* | | *Characteristics* |
|---|---|---|---|---|---|---|---|
| *BS EN 1084* | *BS 1854* | *% Silver* | *% Phosphorus* | *Copper* | *Solidus* | *Liquidus* | |
| CP 102 | CP1 | 15 | 5.0 | 80 | 645 | 700 | All types gap filling CP1/102 suitable for resisting torsional stresses, shock loads or flexing For copper tube/brazing |
| CP 105 | CP2 | — | 6.5 | 83.5 | 645 | 740 | |
| CP 202 | CP3 | 2 | 7.5 | 90.5 | 705 | 800 | |
| CP 104 | CP4 | 5 | 6.0 | 89 | 645 | 825 | |

Note that there are other alloys in this group. Those shown are common to the building services industry. Note also there are minor differences relating to the composition of these alloys depending upon to which standards they relate.

silver–copper alloys in general use. The letters Ag are the chemical symbol for silver and will indicate that the alloy contains this metal.

It should be noted that brazing alloys containing zinc and cadmium must be used in well-ventilated areas due to the dangerous fumes they give off. Copper–phosphorus alloys are more commonly used in plumbing as in most cases they contain little or no silver and are much cheaper than those having a high silver content. A further advantage of these alloys is that when brazing copper no flux is required as phosphorus is a deoxidising agent. While copper–silver alloys can be used to join both ferrous and most non-ferrous metals, copper–phosphorus alloys are more limited in their use, being confined to copper and copper–zinc or copper–tin alloys. When used on metals other than copper, a flux recommended by the manufacturer of the filler rod must be used. Table 1.4 lists the four main copper–phosphorus alloys in common use.

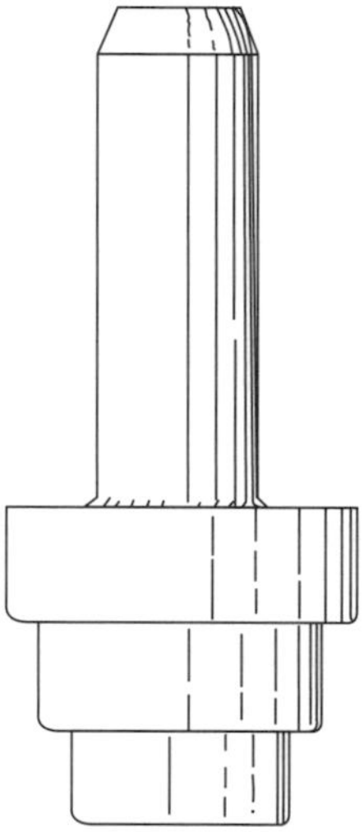

This tool is inserted into the annealed (softened) end of the tube and hammered home to form a socket


**Fig. 1.17** Multi-diameter steel mandrel for forming brazed socket joints.

*Joint design for brazing*

As a brazed joint is stronger than those made using soft solder as a filler, less surface contact between the metals to be joined is necessary. Where the depth of socket on a soft-soldered capillary joint is about 15–18 mm, the same joint using a brazing alloy will need a socket depth of only 6–8 mm. For making brazed socket joints on pipes, socket-forming mandrels can be used (see Fig. 1.17) or the special tools described in Book 1, Chapter 3. Branch joints can be formed with a hammer and bent bolt, but attention must be paid to ensuring an accurate fit between the mating surfaces due to the limitation of capillary attraction. When forming sockets or branch openings in copper pipes, they must always be worked in the annealed or softened state, the one exception being when using the special branch opening tools described in Book 1, Chapter 3. Similar methods of branch opening can be accomplished using power tools which enable branch holes to be opened very quickly. All mechanical methods have two disadvantages: they will only form 90° branches, and due to the size of the hole necessary to allow the entry of the forming tool, the flank of the branch is lower than the crotch. The branch must be profiled as shown in Fig. 1.18. Where oblique branches are required, traditional

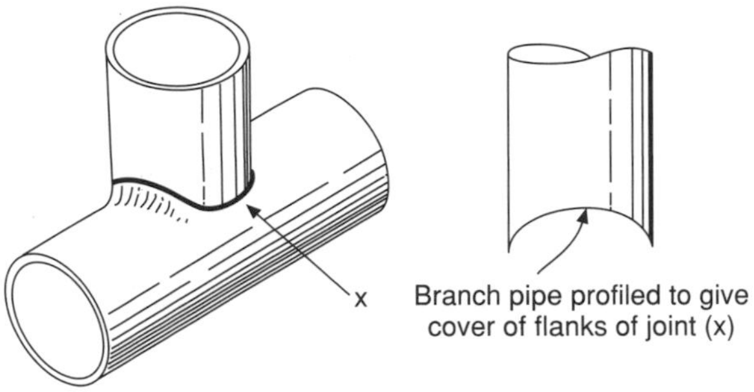


**Fig. 1.18** Profile of a branch formed with patent tools.

methods using a hammer and bent bolt are normally used. Figure 1.19 illustrates the techniques employed.

*Working up branch joints in copper tubes using hand tools*

To prepare a branch joint the first step is to mark its position on the main pipe. Two holes are then drilled, the edges of which (not the centres) should be 6–8 mm inside the confines of the marks indicating the branch position. After annealing, a slit is made between the holes with a strong knife such as a hacking knife. Take care not to nick the outer edges of the holes as this will cause them to split when the branch is opened. The branch can then be worked up with a bent bolt and hammer. The hole should then be rounded up with a piece of steel pipe or bar having as near as possible the same diameter as the copper branch. Finally, the top is filed level and the joint cleaned prior to welding.

Two outstanding points should be remembered:

(a) Always work the copper hot.
(b) Take care to 'lift' the area that will form the crotches of the branch as this will prevent the formation of a crease which would obstruct the waterway.

*Brazed joints for sheet metal*

Figure 1.20 shows typical types of joint design for sheet metal fabrications which the plumber may encounter during the course of this work. Notice that they are made in such a way that a suitable area of mating surfaces is achieved and, due to the turns made, distortion is minimised as this can present problems on light sheet metal work.

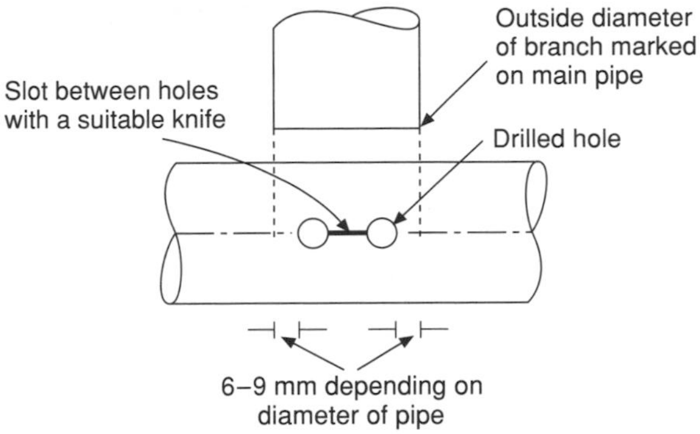

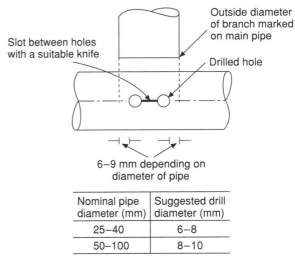

| Nominal pipe diameter (mm) | Suggested drill diameter (mm) |
|---|---|
| 25–40 | 6–8 |
| 50–100 | 8–10 |

(a)

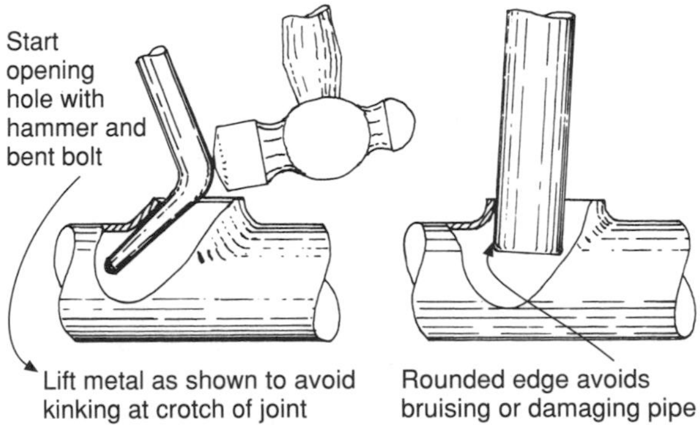


Round off hole with a short length of LCS pipe or bar. The final shape of the hole should be made using pipe or bar fractionally smaller than the branch diameter.

(b)

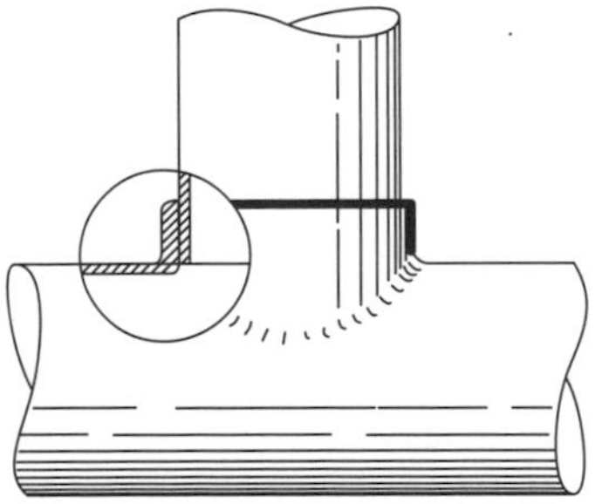

(c) Completed joint

**Fig. 1.19** Setting out and preparation of bronze-welded branch joints.

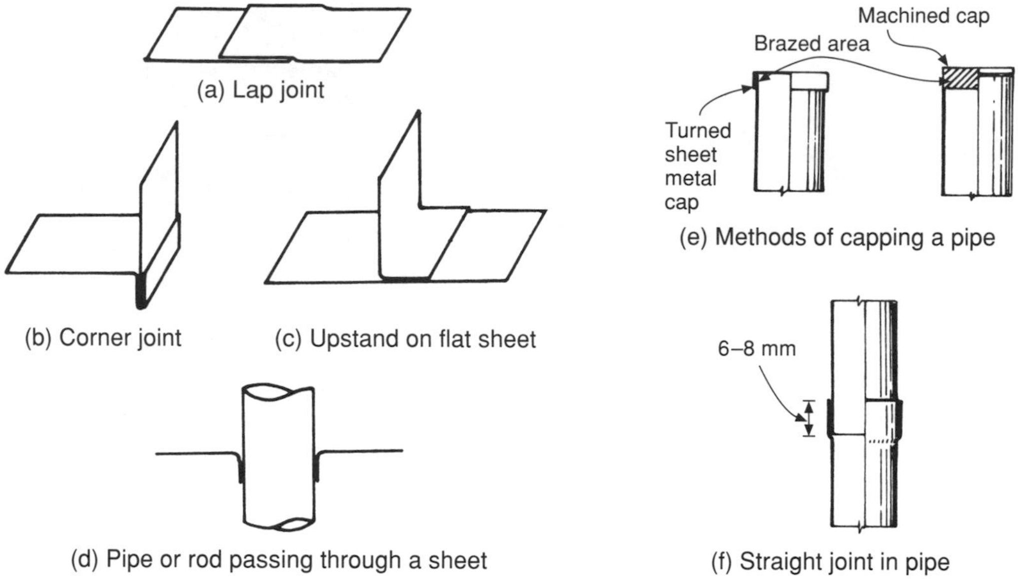


**Fig. 1.20** Joint designs for brazing sheet metal and pipe. Note that in each joint sufficient space is available to permit penetration of the filler material.

*Methods of heat application for brazing*

Welding requires a small concentrated flame, but a large spreading flame is necessary for brazing to ensure that the filler metal penetrates the full depth of the joint. Small-diameter pipes can be both fabricated into desired designs and brazed successfully on site using propane or butane blowlamps. For workshop use, natural gas/air brazing blowpipes are very effective, the air being pressurised by a small compressor. Oxy-acetylene equipment is also very useful — and essential where large-diameter pipes are to be brazed on site. Special nozzles having a series of holes similar to those in the top of a pepperpot are available which provide a more suitable flame structure for brazing than a normal welding nozzle.

One of the problems encountered with both welding and brazing copper is its high conductivity rate. In some cases heat losses due to conduction make it impossible to achieve the temperatures required and a secondary heat source, such as another blowlamp, is necessary to keep the work hot while the joint is made. While it is not often suitable for pipework, a brazing hearth can be constructed on site by placing bricks round the work. Heat absorbed by the bricks is reflected back by radiation enabling quite large pieces of work to be brazed with a relatively small heat source. A very important point to watch when brazing with a welding blowpipe is that one becomes so used to holding the flame close to the work that there is a tendency to do the same with brazing. This can lead to local overheating on one part of the joint and often leads to unnecessary spillage of the brazing rod over the edges of the joint, resulting in a very poor external appearance. The thing to remember is that brazing is a form of soldering, not welding, and if oxy-acetylene equipment is used with a welding nozzle the flame should be employed as illustrated in Fig. 1.21.

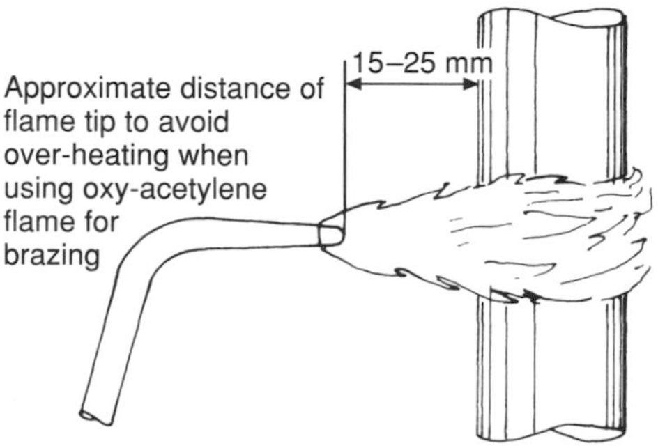


**Fig. 1.21** Using oxy-acetylene flame for brazing.

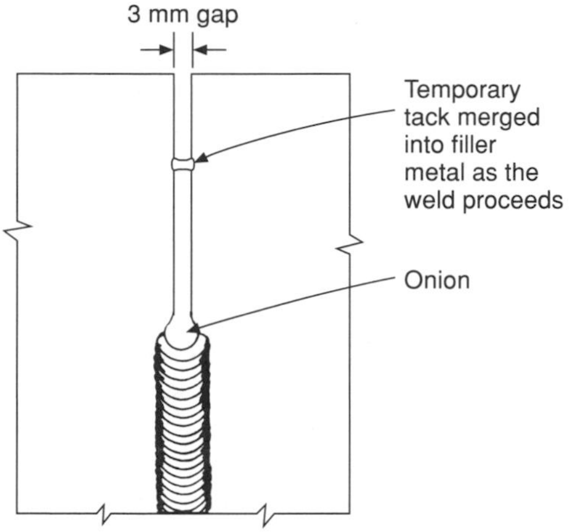


**Fig. 1.22** Fusion welding a low-carbon steel butt joint on plate.

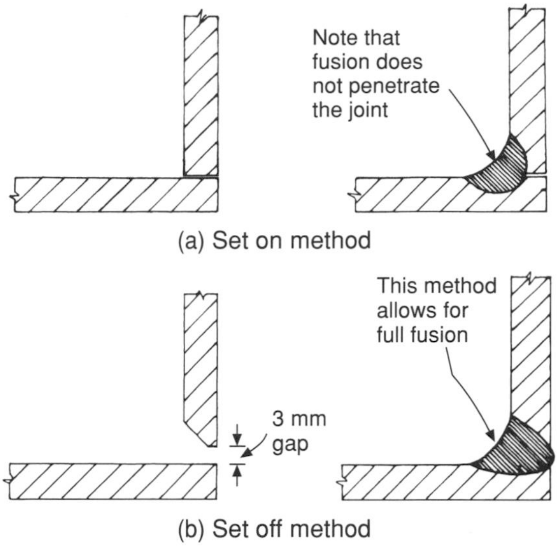


**Fig. 1.23** Fillet welds on low-carbon steel sheet or plate.

### Fusion welding low-carbon steel

At this level the trainee should be capable of effectively using the fusion-welding process for fabricating brackets and fixings for pipe clips and hangers. Most of the remarks about fusion welding of lead also apply to steel, especially those relating to penetration and full fusion of the metal. The most important point to bear in mind when butt welding two plates together is that a gap of about 3 mm must be left between them. This enables the edges of the metal to be melted, leaving what welders call an *onion*, or hole, which is filled, as the weld progresses, with a filler metal having roughly the same composition as the parent metal. Providing the onion is maintained, full fusion of the edges is assured (see Fig. 1.22).

Fillet welds can be made in two ways (see Fig. 1.23). For making brackets and hangers the 'set on' method is acceptable. For pipework joints the 'set off' method is necessary to achieve full fusion and is mandatory in any pipe-welding examinations.

A detailed study of either bronze welding or steel welding is outside the scope of this book and is indeed the subject of welding NVQs and special examinations set by the Joint Industry Board for Plumbing, and the Heating, Ventilation and Domestic Engineers' National Joint Industries Council. Plumbers wishing to qualify in this subject should contact these bodies or their local technical college for further information.

### Further reading

Much useful information can be obtained from the following sources.


'T' Drill copper pipe branch forming tools/'T' Drill Portable Tools, Baiford, 29 Thornycroft Lane, Downhead Park, Milton Keynes MK15 9BR Tel. 01908 667667.

REMS, 1a Greenleaf Rd, Walthamstow, London E17 6QQ Tel. 020 8521 9168.

The Lead Sheet Association, Hawkwell Business Centre, Maidstone Road, Pembury, Tunbridge Wells, Kent, TN2 4AH Tel. 01892 822773. www.leadsheetassociation.org.uk

Welding and Brazing Filler Metals and Fluxes/ Johnson Matthey Metal Joining Products, Unit C, Arundel Gate Court, 1 Frogget Lane, Sheffield S1 2NL Tel. 0114 241 9400. www.jm-metaljoining.com

Joining of Copper and ITS Alloys, Copper Development Association Publication No. 98. Copper Development Association, 1 Brunel Court, Corner Hall, Hemel Hempstead HP3 9XX Tel. 01442 275700. www.cda.org.uk

*Safe Under Pressure*. British Oxygen Company. Available free from HSE books (Tel. 01787 881165).

Welding of Carbon Steel Pipework TR/S Code of Practice. Recommended Practice and Welder Approval Tests HVCA, 34 Palace Court,

London W2 4JG Tel. 020 7313 4900. www.hvca.org.uk

BS 8451 Rubber hose for gas welding and allied processes.

BS 341 Part 1 Valve fittings for compressed gas cylinders.

BS EN 1089 Part 3 Identification of industrial cylinder containers.

### Self-testing questions

1. (a) Describe what action should be taken if it is discovered that an acetylene cylinder is hot to the touch.
   (b) List the main considerations for storage of compressed gas cylinders.
2. (a) Explain the safety hazards likely to be encountered when using oxy-acetylene welding equipment in basements and ducts.
   (b) What precautions must be taken prior to welding containers suspected of having contained flammable substances?
3. (a) Sketch a neutral flame and describe how it differs from oxidising and carburising flames.
   (b) State the type of flame that should be used for steel, lead and bronze welding.
4. State why no flux is necessary when fusion welding lead and low-carbon steel.
5. (a) Explain the two factors that influence control of the molten pool when welding sheet lead.
   (b) Describe two methods of welding lead on a 90° face.
6. List the common defects that can be found in welds produced by inexperienced operators.
7. Make a simple sketch of the set-up prior to the welding of a butt joint, a lap joint and a fillet joint in lead sheet.
8. Explain the differences between bronze welding, brazing and fusion welding.
9. (a) Sketch and describe the method used to open a branch hole for a brazed joint in copper pipe using hand tools.
   (b) State the constituents of two common brazing alloys used in plumbing.
10. (a) Explain why no flux is necessary when brazing copper pipe joints with copper–phosphorus filler alloys.
    (b) State why the socket depth for a brazed joint is less than that for a soft-soldered joint.

# 2 Gas installations

After completing this chapter the reader should be able to:

1. State the main characteristics of natural gas.
2. Outline the main regulations relating to gas installations and appliances.
3. Explain the need for pressure testing gas installations and equipment.
4. Identify the potential dangers of gas/air mixtures and explain the procedure for purging pipework and appliances. List the appliances that do not normally require a flue and explain the reasons for this.
5. Understand the necessity for the provision of an adequate air supply for the combustion of gas and identify the dangers of incomplete combustion.
6. State the working principles of a flue and sketch and describe the precautions to be taken when siting flue terminals. Explain the precautions to be taken where flues pass through combustible materials.
7. Indentify the causes of condensation in gas appliance flues and list the measures taken to minimise its effects.
8. Describe the main principles of balanced flue appliances and their advantages.
9. Explain the working principles of simple gas controls.
10. Identify the main methods of gas ignition on gas fires, water heaters and boilers.
11. Explain the working principles of simple flame failure devices.
12. Identify types of gas fires, and their installation and flueing requirements.

## Legislation

Due to the potential dangers of gas, suppliers, installers and users are subject to certain regulations. All competent gas installers must have a good working knowledge of regulations affecting their work. The following gives a broad outline of their requirements. Further details will be found in the Gas Safety (Installations and Use) Regulations. Due to the continuous development of gas appliances, changes often outstrip British Standards, and for this reason manufacturers' instructions always take precedence.

*Certification*

All installers of gas appliances must have a certificate of competence for each area of gas work they undertake. The validating certificates must be issued under one of the following schemes:

(a) Approved Code of Practice (ACOP).
(b) Nationally Accredited Certification Scheme (NACS).

NACS has superseded ACOP.

*The Council of Registered Gas Installers (CORGI)*

Although originally membership was voluntary, as a result of gas-related accidents it became mandatory in 1991 for all companies undertaking gas work using natural gas to be registered with a body approved by The Health and Safety Executive (HSE). CORGI is currently the organisation which maintains a register of all approved operators. It also carries out regular inspections of gas installations fitted by its members, maintains a public enquiry and complaints service and is responsible for publishing the importance of gas safety. Members of CORGI are kept up to date with changes in the law, technology and safe working practices. Gas installers who do not hold a certificate in energy efficiency for domestic heating are unlikely to be able to self-certify their work. This is also the date when gas installers must have gained the required level of electrical qualifications to self-certify their own installations. See reference to Building Regulations Part P in Chapter 11.

*The Gas Safety (Installation and Use) Regulations*

These are mandatory and deal with safe installation and maintenance practices in most types of building. The Regulations place responsibilities on installers, maintenance engineers, and suppliers and users of gas, including landlords of rented property. The Regulations are followed by guidance notes which (although not mandatory) will normally be sufficient to comply with the law if observed by an installer. This is an important document and all operators carrying out gas work should be familiar with it.

*The Gas Act 1995*

This updates previous legislation to include new licensing arrangements for Public Gas Transporters, permitting competition in the domestic gas market and allowing consumers to purchase gas from any supplier they wish to use. It also includes provision for safety regulation to be made in (a) the Gas Safety and (b) the Gas Safety (Rights of Entry) Regulations.

*Health and Safety at Work Act*

This has been dealt with in Book 1 of this series. It applies to everyone concerned with work activities, both employer and employees. It also includes provisions to protect the public from exposure to risks to health and safety. Failure to comply with the general requirements of the Act and those of other documents relating to this subject may result in legal proceedings.

*The Gas Safety (Management) Regulations*

These are designed to protect the public against dangers caused by failure to observe safe working practices when transporting or supplying gas. The HSE (Health and Safety Executive) is the enforcing body for gas safety and has legislation with which gas suppliers and transporters must comply. The main points of the legislation are as follows:

(a) Under these Regulations a gas supplier *must* produce a case showing systems and procedures that will be adopted to ensure a safe supply of gas. Subject to this being approved by HSE, permission to supply gas will be granted.
(b) The supplier must operate a full gas emergency service.
(c) The supplier must provide a gas incident service which reports gas explosions and cases of carbon monoxide poisoning.
(d) The supplier must operate an emergency telephone number for customers reporting gas leaks.

*The Gas Safety (Rights of Entry) Regulations*

These give authorised officials of gas transporters the right to enter a property and inspect any equipment connected to the gas supply. They have the power to disconnect any appliance they consider to pose an immediate danger. British Gas Transco is currently the main transporter and is responsible for dealing with all emergency calls. Lack of ventilation, seriously defective flues and gas leaks exceeding the permissible limits all constitute an immediate danger. In cases where the responsible person, e.g. owner or landlord, refuses to allow an appliance or supply of gas to be isolated, the National Gas Emergency Service Provider or the gas supplier must be notified immediately. Only they have the necessary powers of enforcement.

*Gas Appliances (Safety) Regulations (GASE)*

These implement an EC directive which requires appliances and fittings used for gas to conform with specified essential requirements, not the least being safety in use. Both the supply and installation of any gas appliance are prohibited unless it bears the **CE** (Conformité Européene) mark. The object of this legislation is to ensure all gas appliances for sale in the European market meet agreed safety standards and are designed to protect the consumer.

*The Building Regulations*
Compliance with these regulations is mandatory in England, Wales and Northern Ireland. Building Standards in Scotland have similar requirements. Approved Documents 'B' (fire safety), 'F' (ventilation) and 'J' (heat-producing appliances) give guidance on some of the specifications relating to the Gas Safety Regulations, and L1 which covers conservation of fuel. This document was the subject of a major review in 2005. In Northern Ireland the Approved Document part L (heat-producing appliances) applies. In Scotland, Technical Standards parts F and K of Building Standards are applicable. They relate to heat-producing installations, storage of liquid, gaseous fuel and ventilation. Approved Document part P relates to electrical works in buildings, and certain types of electrical installation work can now only be self-certified by a qualified person.

*The Electricity Supply Regulations*
These indicate the methods of supply that can be used to serve a specific property.

*The Electricity at Work Regulations*
These cover a wide area but relate here to the type of tests that must be carried out by installers, the test equipment used and its maintenance. See Chapter 11 for further details.

*The Water Act*
Since 1999 all water installation work is subject to the Water Supply (Water fittings) Regulations, which are mandatory in England and Wales. Northern Ireland and Scotland have their own Regional Regulations. Reference should be made to Chapter 3.

*Reporting of Injuries, Diseases and Dangerous Occurrence Regulations (RIDDOR)*
Reference should be made to Book 1 of this series where this legislation has been explained more fully. It is designed to allow the HSE to investigate dangerous situatations and accidents.

The responsibility for reporting dangerous gas fittings and installations lies with the service fitter, whether self-employed or of employed status. In the latter case it should be reported to the company for which the fitter works. Only after notifying the National Gas Emergency Services or the gas supplier should a report be made to the HSE. Some examples of what should be reported are listed as follows:

(a) Gas escapes outside the tolerances of soundness tests due to poor or unsatisfactory workmanship.
(b) Open-ended uncapped pipes, which may or may not be connected to a gas supply.
(c) Any evidence of 'spillage', the cause of which has not been rectified.
(d) Defective flues or chimneys not clearing the products of combustion. This includes appliances that should be flued but are not, and flues discharging into a roof space or excessively outside the parameters of the recommendations of BS 5440:2000 Part 1.
(e) Defective or insufficient ventilation of areas in which an appliance is fitted.
(f) Use of unsuitable materials for gas pipes.
(g) Faulty servicing making an appliance unsafe for use.

Having regard to the foregoing, if the installation/ appliance is immediately dangerous it must be disconnected and the supply capped off. If it is at 'risk' but not immediately dangerous it can simply be turned off. In both cases the responsible person, e.g. householder, landlord, etc., must be given an approved warning notice and a **DO NOT USE** notice attached to the appliance. The permission of the responsible person must be obtained prior to carrying out these procedures, but in the event of non-cooperation the National Gas Emergency Service must be contacted immediately. In all cases where an appliance or installation is considered **immediately** dangerous the HSE must be notified. If any appliance or installation does not comply with current regulations, standards or specifications but does not fall into the preceding categories, the responsible person should be informed orally. However, a record should be kept for reference at a later date.

*The Office of Gas Supply (OFGAS)*
This is a regulating body having powers under the Gas Act to:

(a) Issue licences to gas transporting and supply companies.
(b) Ensure the quality and calorific value of the gas.
(c) Appraise and certificate gas meters.
(d) Protect the interests of the consumer.

## Properties of natural gas

Natural gas is predominantly a mixture of hydrocarbons but, unlike the town gas used previously, contains no hydrogen. Its composition varies slightly according to its source, but a typical example of its constituents is shown in Table 2.1. Note that all these gases, except nitrogen, are combustible. Natural gas is non-toxic (which means it is not poisonous) as it contains no lethal gases such as carbon monoxide. It will, however, produce carbon monoxide if it is not completely burned, and for this reason careful attention must be given to flues and ventilation. In its natural state it has no smell, and to avoid possible dangers from leaks a chemical is added to give it a distinctive odour.

*Calorific values*
The calorific value of a fuel may be defined as the heat units it contains per unit volume. The volume by which gas is measured is the cubic metre and the unit of heat is the joule. One cubic metre of natural gas contains approximately 38 MJ (megajoules), but there are sometimes slight differences in this figure due to variations in the source of supply. The prefix *mega* means 1 million and enables one to deal with fewer digits, as 38,000,000 J (joules) is rather an unmanageable number when used for calculations.

**Table 2.1** Constituents of natural gas.

| *Constituents* | % |
|---|---|
| Methane | 93 |
| Ethane | 3 |
| Propane | 2 |
| Butane | 1 |
| Nitrogen | 1 |

*Specific gravity*
Natural gas is lighter than air. If air is taken to have a specific gravity of 1, then natural gas has a specific gravity of about 0.58. This is one of the reasons why higher pressures are required on the main and service pipes.

*Stoichiometric mixture*
This term relates to the quantity of air required to burn a fuel to achieve complete combustion. In the case of natural gas the figure is 10.57. In simple terms, 10.57 volumes of air are required to achieve the combustion of 1 volume of gas. Thus it will be appreciated how important it is to make sure that there is adequate ventilation to provide sufficient air for combustion. *Incomplete combustion will result in the production of carbon monoxide, a very dangerous toxic gas.*

## Combustion and gas burners

*Products of combustion*
When natural gas is completely consumed the products of combustion are harmless, being water vapour, carbon dioxide and the nitrogen originally contained in the air. Some gas appliances are flueless, typical examples being cookers and small water heaters, the products of combustion producing no ill-effects on the occupants of the rooms in which the appliances are installed providing they are correctly fitted and maintained. It is important, however, to ensure that there is adequate ventilation in such rooms, because, although these products are not toxic, it is possible that they could cause vitiated air (air from which the oxygen has been used) to recirculate in the appliance, thus producing carbon monoxide.

*Ignition temperature*
Natural gas requires a temperature of 700 °C to cause it to ignite; this is slightly higher than for town gas.

*Limits of flammability*
This relates to the amount of gas in air required to produce a flammable mixture, and is usually

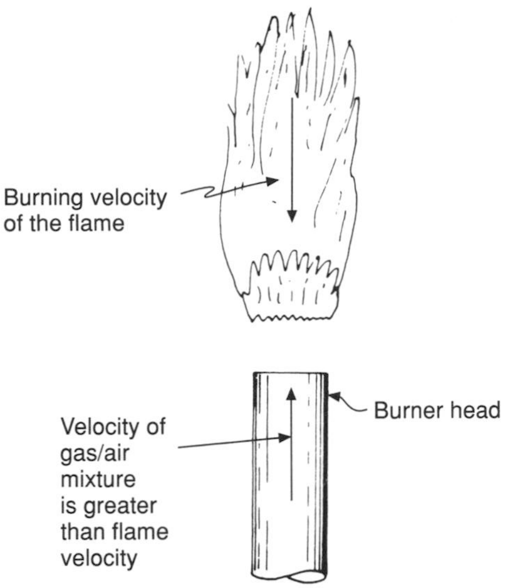


**Fig. 2.1** Lift-off. This very undesirable situation takes place when the velocity of the gas flow is greater than the burning velocity of the flame. Because the flame is not stabilised on the burner it can very easily be extinguished.

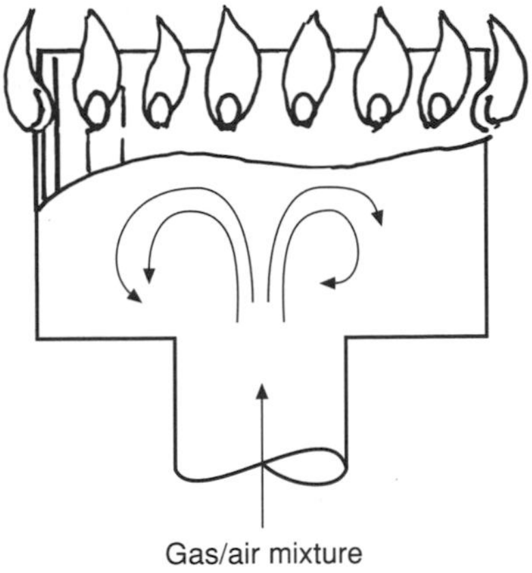


(a) Turbulence caused by the sudden enlargement of the burner head reduces the velocity of the gas/air mixture

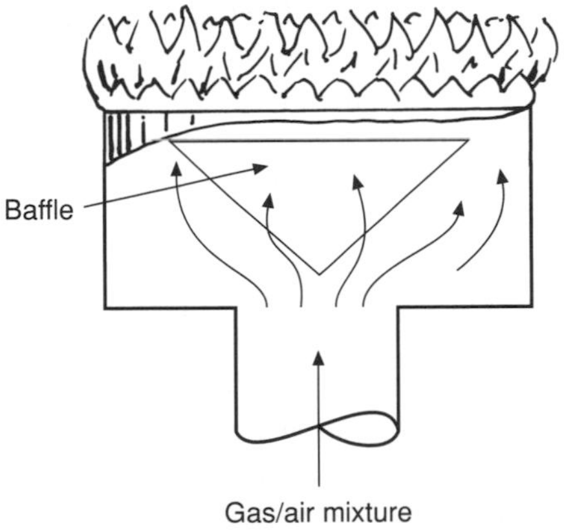


(b) Recirculating hot gases using a baffle to increase the effect of turbulence

**Fig. 2.2** Stabilisation of gas flames. Both types of burner head are designed to create turbulence.

expressed as a percentage. In the case of natural gas it is between 5 and 15 per cent. Anything less than 5 per cent will be too weak; if it is more than 15 per cent the mixture will be too rich.

*Burning velocity*

This term relates to the speed at which, on ignition, the flame spreads through the gas/air mixture. This speed is affected by the pressure of the gas which can accelerate its movement to such a degree that the gas/air mixture is unable to burn quickly enough. The result is an unstable flame burning in the atmosphere some distance away from the end of the burner as illustrated in Fig. 2.1. This is often referred to as *lift-off* and must not be tolerated, as the flame can easily be extinguished, allowing gas to enter the room with consequent danger of explosion. Burners are designed in such a way that the flame is stabilised as shown in Fig. 2.2.

The opposite of lift-off is *light back* which occurs when the velocity of the gas/air mixture is so low that the flame speed is greater and passes back through the burner to light on the injector as shown in Fig. 2.3. This is also dangerous as complete combustion is not achieved, giving rise to the production of carbon monoxide.

### Combustion air

When a fuel is burned it combines with the oxygen in the air to produce heat. From a purely academic view, a chemical change takes place during which oxygen in the air joins with the hydrocarbons in the gas, producing heat as it does so. The products of this chemical change are carbon dioxide and water vapour, which in this case are called products of

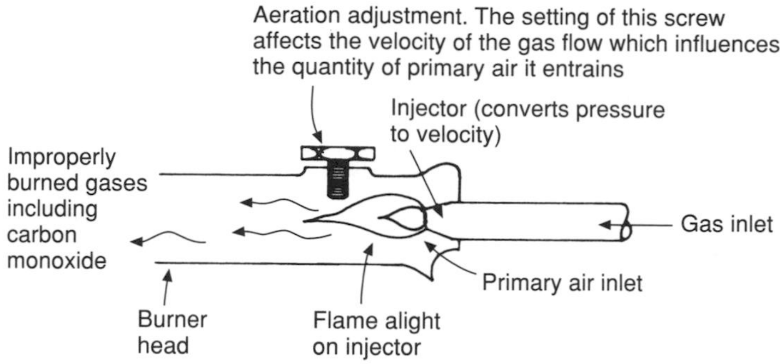
Aeration adjustment. The setting of this screw affects the velocity of the gas flow which influences the quantity of primary air it entrains
Injector (converts pressure to velocity)
Improperly burned gases including carbon monoxide
Gas inlet
Primary air inlet
Burner head
Flame alight on injector

**Fig. 2.3** Light back. This occurs when the speed of gas passing through the injector is reduced to such a degree that its velocity is less than the flame speed. This situation is potentially dangerous as it can result in the production of carbon monoxide.

combustion. The remainder of the air, consisting mainly of nitrogen, has no effect on the chemical change and simply passes out of the appliance with the products of combustion. Neither nitrogen, carbon dioxide nor water vapour are toxic, but it is essential that they are removed from the room in which the appliance is situated, as they can cause acute discomfort to the occupants. As stated previously, a build-up of these products will deter the entry of fresh air and may lead to an oxygen deficiency causing incomplete combustion in the appliance. This is very serious, because if the hydrocarbons in the gas are unable to combine with sufficient oxygen, carbon monoxide is produced. If this happens with a flueless appliance such as a cooker or small water heater or an appliance with a blocked or defective flue, it can result in the death of the occupants of the room.

All types of burners are classified as (a) pre-aerated and (b) post-aerated. Most modern appliances use pre-aerated burners where some air is mixed with the gas before it is burned. Some employ atmospheric or natural draught, others forced draught where the air for pre-mixing is supplied under pressure by a fan. Figure 2.4 illustrates a typical atmospheric burner correctly adjusted to give complete combustion, while Figs 2.5(a) and (b) illustrate examples of incomplete combustion. The physical appearance of the flame is a good guide to combustion fault finding. Defects due to inadequate supply of air are usually due to causes (a) to (c):

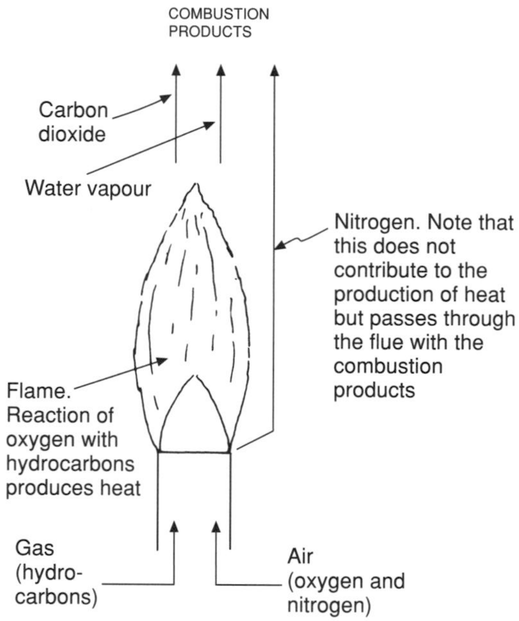
COMBUSTION PRODUCTS
Carbon dioxide
Water vapour
Nitrogen. Note that this does not contribute to the production of heat but passes through the flue with the combustion products
Flame. Reaction of oxygen with hydrocarbons produces heat
Gas (hydro-carbons)
Air (oxygen and nitrogen)

**Fig. 2.4** Combustion of gas.

(a) Blocked or undersized flues.
(b) Inadequate ventilation causing 'vitiation' or lack of oxygen.
(c) Under-aeration (insufficient air). The most common cause of this is blockage of the primary air ports of the burner by 'lint'; furnishing fibres and pet hairs are typical examples. Open-flued appliances are more prone to this than those that are room sealed.

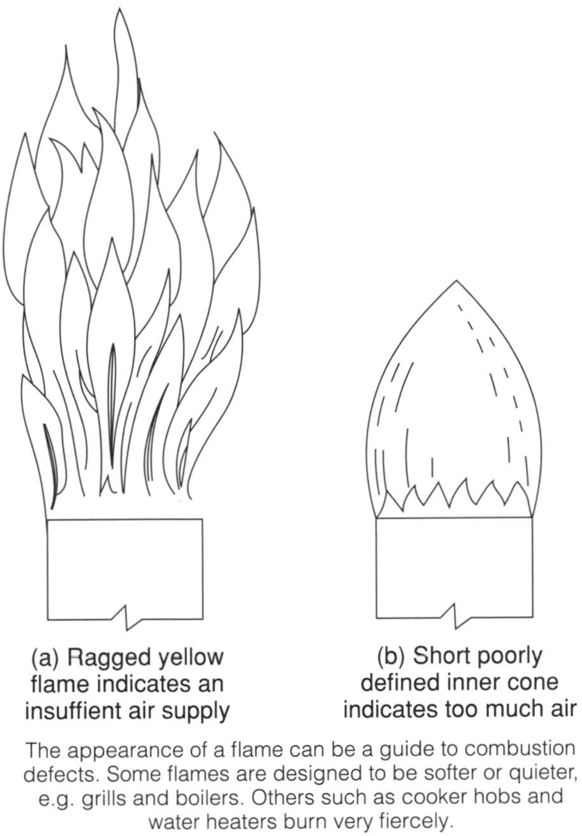

(a) Ragged yellow flame indicates an insuffient air supply
(b) Short poorly defined inner cone indicates too much air
The appearance of a flame can be a guide to combustion defects. Some flames are designed to be softer or quieter, e.g. grills and boilers. Others such as cooker hobs and water heaters burn very fiercely.


**Fig. 2.5** Examples of incomplete combustion.

(d) Over-gassing — this is usually due to an excessive gas rate or pressure.

### Installation of pipework

This is defined as any pipework fitted to the outlet side of the primary meter. To comply with the Gas Safety Regulations, all pipework and appliances must be installed using approved materials and recommended procedures. Copper (steel for industrial and commercial use) and plastic pipes specified as being suitable for gas work, also methods of jointing, are described more fully in Book 1 of this series. All open-ended pipes must be capped if left unattended. This avoids the risk to others, being unaware of any open ends, connecting or reconnecting the gas supply to the meter. If a blowlamp is used to repair or extend an existing supply, the meter outlet must be disconnected and both ends sealed to prevent a flashback. When using flux to make soldered joints, it is important that it is used sparingly to avoid corrosion inside the pipe. Although active fluxes are permissible, those of the non-active type are recommended for gas services. Pipework under screeded concrete floors and in wall chases should be protected against corrosion. Pipes having a factory-finished plastic sheath are recommended, but grease-impregnated tape or yellow-coloured wrapping tape are suitable. Compression fittings of any type must not be concealed or fitted in such a way as to preclude access to them. The main gas cock, filter, governor and meter must be installed in an accessible position to afford the consumer, installers and service engineers easy access to them. All domestic gas installations must have main equipotential bonding conforming to the IEE wiring regulations. It must be connected to the pipework within 600 mm of the meter outlet. Reference should be made to Chapter 11 for more detailed information on this subject. If a piece of pipe or a meter is to be removed, the electrical continuity of the installation must be maintained. If this is broken, even only temporarily, it could result in a fatal shock or an explosion due to a spark igniting a gas/air mixture.

To conform to the Gas Regulations a temporary bonding wire should be used as shown in Fig. 2.6.

All pipes and fittings used for a gas supply must be of suitable strength and comply to BS 6891. Materials used for making joints on threaded pipes

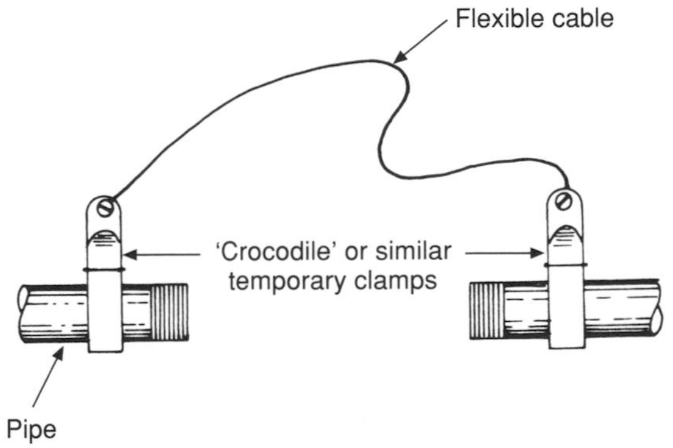


**Fig. 2.6** Temporary bonding wire. This must be used when a section of pipe or a meter is temporarily removed to ensure the continuity of electrical bonding.

and fittings must be suitable — not all jointing pastes meet this requirement and it is wise to check the label on the container before use. Polytetrafluoroethylene (PTFE) tape and most types of jointing media for oil pipelines are also suitable for natural gas. No pipework may be fitted in a cavity wall, and where it passes through such a wall it must be sleeved. The sleeve must have sealed ends to ensure that a leak in the pipe does not result in the cavity becoming filled with gas. Gas pipes must not be fitted in such a way that they are subjected to a compressive load. This means that they must not be installed under footings or load-bearing walls. When any new or additional work on an existing supply is carried out, the pipework and appliance (if applicable) should be air tested, using methods described in the section on testing.

Before fitting any gas appliance it is essential to ascertain that sufficient air is available for combustion, and should the appliance be fitted with a flue pipe it must be installed in such a way that the products of combustion are carried away to the external air.

All floor-standing appliances must be adjusted so that they are level, firm and stable. This is very important in the case of free-standing gas cookers or refrigerators. A very effective way of checking the level is to stand a pan of water on top of the appliance when any inaccuracy in level will be easily seen. In the case of free-standing boilers the floor or hearth must be of adequate strength and conform to the building regulations regarding its combustibility.

## Gas meters

The fixing of gas meters and the running of gas service pipes from the main is normally undertaken by the supplier's employees or one of their recognised subcontractors. The siting of meters is the subject of strict control relating to fire precautions and escape routes in the event of an emergency. If it is necessary to disconnect a meter in the course of one's work, the connections must be sealed to prevent any gas it contains becoming a fire or explosion risk, and when it is refitted any connection must be tested for leaks and purged (as described later) before any appliances are used.

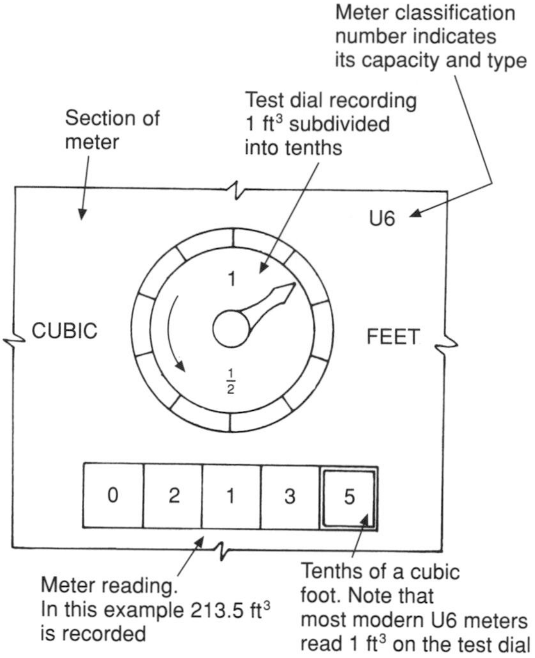


**Fig. 2.7** U6 gas meter dial.

Figure 2.7 illustrates the readings on a typical U6 meter.

*Conversion of meter reading to kilowatts*

U6 meters are calibrated to give readings in cubic feet ($ft^3$), and it may be necessary in some cases to convert this to cubic metres ($m^3$). In rounded-off figures it can be assumed that 1 $m^3$ = 35.336 $ft^3$. To convert cubic feet directly to kilowatts the following formula is used:

$$1 \text{ unit of gas } (100 \text{ ft}^3) = 2.83 \text{ m}^3$$

This is multiplied by the calorific value of the gas, which may vary slightly from area to area. The calorific value used here is 38.5 — the total thus produced is then divided by 3.6 which converts the original reading to kilowatts. Example: Assume a meter has passed 33,000 cubic feet — convert this to kilowatts:

$$33{,}000 \div 100 = 330 \text{ units}$$
$$330 \times 2.83 = 933.9 \text{ ft}^3 \text{ (cubic feet)}$$
$$933.9 \times 38.5 = 35{,}955.15 \text{ (total calorific value)}$$
$$35{,}955.15 \div 3.6 = 9{,}987.542 \text{ kW (kilowatts)}$$

*E6 meters (ultrasonic meter)*

These meters do not have a test dial but a liquid crystal display. They are calibrated in $m^3$ and the method of calculating gas flow rates is slightly different to that used with U6 meters. They do have several advantages over the U6 type: they are compact and unaffected by air mixture and temperature and also incorporate a fraud detection device. They also give a direct reading in cubic metres.

## Testing appliance pressures and gas rates

To check the working pressure at a meter, connect a pressure gauge to the meter test point. It should read 21 mbar plus or minus 1 mbar.

After an appliance has been fixed it must be checked to ensure it is working at the pressure recommended by the manufacturer and adjusted so that it consumes no more than the recommended quantity of gas. 'Over-gassing' may result in the production of carbon monoxide, flame lift-off and loss of efficiency. The procedures are as follows.

*Setting burner pressure*

A manometer should be fitted to the test point on the appliance. The pressure shown on the gauge should be as specified in the manufacturer's instructions. Any alteration necessary is made at the appropriate control, usually the governor.

*Setting the gas rate*

This is required by the Gas Safety Regulations. The object is to check the volume of gas used by the appliance, which will be found both on the badge affixed to it and in the manufacturer's instructions. As an example, a gas central heating boiler with an input rating of 52,000 Btu/h is to be checked. The procedure using a U6 meter which measures gas in cubic feet is as follows:

1. Check that the burner pressure is correctly set.
2. Turn off all other gas appliances in the premises.
3. Assume the calorific value of the gas to be 38 MJ/m (1,040 Btu/ft).
4. Turn the boiler on and ensure the burner is alight. Check the flame appearance for signs of poor combustion and allow approximately 10 minutes for it to reach operating temperature.
5. Watching the cubic foot dial on the meter, record the time in seconds for the pointer to rotate once; accurate timing is important. Assuming it takes 72 seconds to burn 1 cubic foot, the gas rate can be determined using the following simple formula:

$$\frac{\text{seconds in 1 hour} \times \text{calorific value Btu/h}}{\text{time in seconds to burn 1 ft}^3}$$

$$\therefore \frac{3{,}600 \times 1{,}040}{72} = 52{,}000 \text{ input rates in Btu/h}$$

$$3{,}421 \text{ Btu/h} = 1 \text{ kW}$$

$$\therefore \frac{52{,}000}{3{,}421} = 15.2 \text{ kW input}$$

To check the gas rate of an appliance, assuming the pressure is correct and all other appliances are turned off, the procedures are as follows:

1. Note the reading and time for 2 minutes.
2. Record the second reading.
3. Add the number of seconds until the next digit is shown.
4. Subtract the first reading from the second.
5. This gives the gas rate in $m^3$ over 2 minutes.
6. Convert these figures to kW using an E6 meter chart, or using the following formula:

$$\text{kW} = \frac{\text{number of seconds in 1 hour} \times \text{recorded volume in m}^3}{\text{120 seconds (2 mins)} + \text{number of seconds until next digit on test dial}}$$

Example: 1st reading 00836.322
2nd reading 00836.374 + 10 seconds
00836.374 – 00836.322 = 0.052
(recorded volume)

$$\frac{3{,}600 \times 0.052}{120 \text{ seconds} + 10} = 1.44 \text{ m}^3\text{/h}$$

Depending on the calorific value of the gas, 1 $m^3$ gives approximately 10.8 kWh
$\therefore$ input = 1.44 × 10.8 = 15.552, approximately 15.6 kWh

## Pressure

In practical terms the earth can be said to be surrounded by air to a height of approximately

7 miles and as a consequence its surface is subjected to a pressure equivalent to its mass, which is 101.3 kP/m or 1.013 bar. It will be seen that for all practical purposes atmospheric pressure can be expressed as 1 bar. Normally this can be ignored when pressure readings are taken because atmospheric pressure affects both sides of the gauge in the same way, any pressure recorded being technically called *gauge* pressure. When the pressure shown on a gauge is added to that of the atmosphere the combined pressures are called *absolute*. To give an example of absolute pressure, if 2 bar is recorded on a pressure gauge and added to that of the atmosphere, the result will be an absolute pressure of 3.013 bar (atmospheric pressure = 1.013 bar). The pressure in an installation when no appliances are in use is known as the standing pressure. Working pressures are those recorded when an appliance is in use. The causes of pressure loss are considered in the following text and must be taken into account when sizing pipes for both gas and water.

## Gas pipe sizing

Great care must be taken to ensure the pipe sizes for gas installations are adequate. Failure to do this may result in unsatisfactory performance of an appliance, e.g. low hot water temperature supplied from water heating equipment and poor performance of cooking appliances. In extreme cases an insufficient volume and pressure of gas at an appliance could cause the pilot and even the main burner flame to be extinguished. This is a dangerous situation which could lead to an explosion if the gas is reignited, and in some circumstances the production of carbon monoxide. The design of a gas piping installation should take into account the following. It should be capable of supplying sufficient gas at all the appliances connected to it and consideration should be given to possible future extensions. As with water supplies, allowances must be made for pressure loss caused by the frictional resistance of the pipe walls. The longer the pipe the greater this will be, and losses due to turbulence in tees and elbows must also be taken into account. It should be noted that pipes of various materials have differing carrying capacity and it is important that reference is made to the pipe sizing table which applies to the material being used. To give an example, a steel pipe having a nominal diameter of 15 mm and a length of 3 m will discharge 4.3 $m^3$ per hour. A copper pipe of the same nominal diameter will discharge only 2.9 $m^3$ per hour. Gas appliance manufacturers specify the recommended pipe diameter for their appliance only. Where more than one appliance is used it is necessary to determine a pipe diameter that will satisfy the total possible demand. **The permissible pressure drop on a gas pipeline during periods of maximum demand must not exceed 1 mbar.**

### *Example of pipe sizing*

Figure 2.8 shows a gas carcass supplying a variety of gas appliances. Copper tube as shown in Table 2.2 is used as the pipework material. Typical ratings of gas appliances are shown in Table 2.3. Bear in mind the two principal factors involved are the gas rate of each appliance and the frictional resistance of the pipework. A tabulation chart similar to that shown in Table 2.4 will be useful. The procedure is as follows:

(a) On the tabulation chart enter the appropriate section of pipe from the drawing.
(b) Enter the gas rate from Table 2.3 (in practice the gas rate will be taken from the manufacturer's instructions).

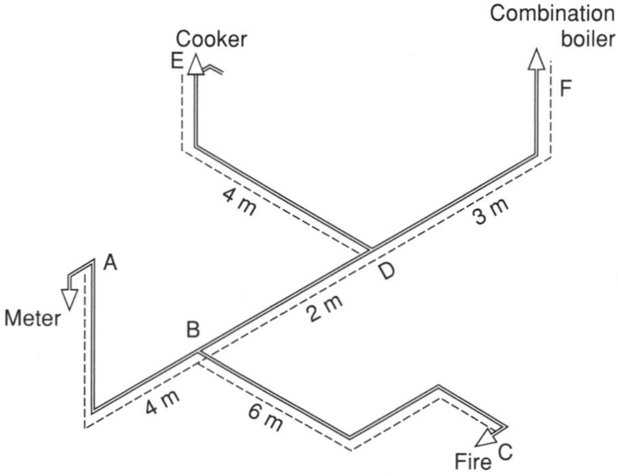


**Fig. 2.8** Pipe sizing for gas appliances.

**Table 2.2** Discharge in a straight horizontal copper or stainless steel tube with 1.0 mbar differential pressure between the ends, for gas of relative density 0.6 (air = 1). (Courtesy of CORGI)

| Copper piping in accordance with BS EN 1057 or corrugated stainless steel tube to BS 7838 | | | | | | | | |
|---|---|---|---|---|---|---|---|---|
| *Size of pipe* | *Length of pipe (m)* | | | | | | | |
| *mm* | *3* | *6* | *9* | *12* | *15* | *20* | *25* | *30* |
| | *Discharge ($m^3/h$)* | | | | | | | |
| 10 | 0.86 | 0.57 | 0.50 | 0.37 | 0.30 | 0.22 | 0.18 | 0.15 |
| 12 | 1.5 | 1.0 | 0.85 | 0.82 | 0.69 | 0.52 | 0.41 | 0.34 |
| 15 | 2.9 | 1.9 | 1.5 | 1.3 | 1.1 | 0.95 | 0.92 | 0.88 |
| 22 | 8.7 | 5.8 | 4.6 | 3.9 | 3.4 | 2.9 | 2.5 | 2.3 |
| 28 | 18 | 12 | 9.4 | 8.0 | 7.0 | 5.9 | 5.2 | 4.7 |

*Note*: When using this table to estimate the gas flow rate in pipework of a known length, this length should be increased by 0.5 m for each elbow and tee fitted, and by 0.3 m for each 90° bend fitted.

**Table 2.3** Typical gas ratings.

| *Appliance* | *Gas rate ($m^3/h$)* |
|---|---|
| Warm air heater | 1.0 |
| Multipoint water heater | 2.5 |
| Cooker | 1.0 |
| Gas fire | 0.5 |
| Central heating boiler | 1.5 |
| Combination boiler | 2.5 |

(c) Add the measured length of pipe to the allowances made for fittings. In this system of sizing a standard length of 0.5 m is allowed for both elbows and tees of all sizes (see Fig. 2.9).

Referring to Fig. 2.8 it will be seen that pipe A–B is supplying all the appliances in the installation, so the gas rate will be the sum of the following:

gas rate for fire = 0.5
gas rate for cooker = 1.0
gas rate for combination boiler = 2.5

Figure 2.8 shows the effective length of pipe A–B to be 4 m. Referring to Table 2.2 it will be seen that the pipe lengths are increased in multiples of 3 m. The nearest to 5.5 m is 6 m (third column from left): reading down this column it will be seen that a 6 m length of 22 mm pipe will pass 5.8 $m^3/h$ which is well within the limits required. The remaining pipes are sized in a similar way.

**Table 2.4** Tabulation of data.

| *Pipe section* | *Gas rate* | *Measured pipe length* | *Allowances for fittings* | | *Total length* | *Pipe diameter* |
|---|---|---|---|---|---|---|
| *ref Fig. 2.8* | *$m^3/h$* | *m* | *type* | *equivalent length* | *m* | *mm* |
| A–B | 4 | 4 | 3 elbows | 1.5 | 5.5 | 22 |
| B–D | 3.5 | 2 | — | — | 2 | 22 |
| B–C | 0.5 | 6 | 3 elbows, 1 tee | 2.0 | 8.0 | 12 |
| D–E | 1.0 | 4 | 3 elbows, 1 tee | 2.0 | 6 | 15 |
| D–F | 2.5 | 3 | 1 elbow | 0.5 | 3.5 | 22* |

* A 3 m length of 15 mm pipe will carry 2.9 $m^3/h$; 3 m is just short of 3.5 m so only a 0.500 m length would need to be 22 m diameter.

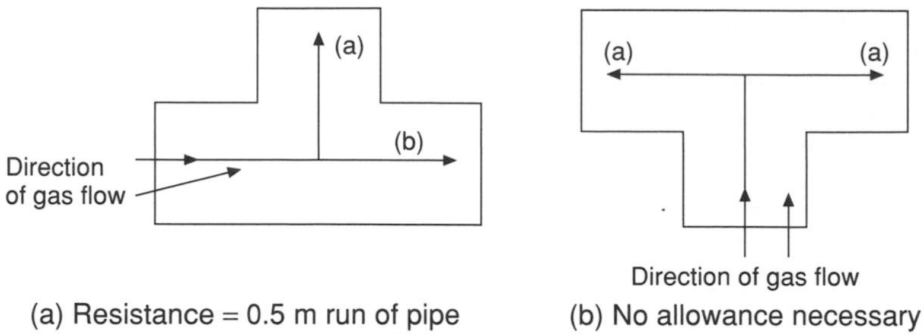


**Fig. 2.9** Tees showing where allowances for resistance to flow must be allowed for.

In situations where it is necessary to convert cubic feet to cubic metres the factor 0.0283 is used.

Example: convert 60 cubic feet to $m^3$

$$60 \times 0.0283 = 1.698, \text{ aproximately } 1.7 \text{ m}^3$$

## Testing for soundness

### *The U gauge, or manometer*

This instrument is used for both measuring the pressure and testing for leaks in gas installations. It is a very simple device consisting of a glass tube bent in the form of a letter U. A clear plastic tube is sometimes used and is less liable to breakage, but it has the disadvantage that it discolours after a period of time making it difficult to read. The tube is contained in a plastic or metal case which offers some degree of protection.

A graduated scale is attached to the two legs of the U, each large division on the scale representing 1 mbar, which is one-thousandth part of a full bar of pressure. It is important to note that low gas pressures are usually quoted in millibars, 1 mbar being in effect the pressure exerted by a head of water 9.8 mm high. For practical purposes a millibar is taken as 10 mm or 1 cm. Prior to metrication U gauges were graduated in inches and tenths of an inch, and where these gauges are still in use it is useful to remember that 1 inch water gauge is equivalent to 2.5 mbar.

U gauges may be of the direct reading or indirect reading type: both are shown in Fig. 2.10. In the case of Fig. 2.10(a), the direct reading type, it will be seen that although the scale is measured in 1 mbar divisions, each division is numbered to represent two. Providing the gauge has been zeroed properly, a direct reading can be obtained from one leg only. A reading on the indirect type is taken by adding together the number of millibars each side of the zero. Figure 2.10(b) shows how the same pressure is read on each type of gauge.

Careful scrutiny of Fig. 2.10(c) shows that the surface of the water in each leg of the tube is curved due to the adhesion of the water to the sides of the tube. The name given to the curve is the *meniscus* and it is important to make sure when taking a reading that it is taken from the bottom of each curve as shown.

Whenever a new gas installation is fitted, or additions are made to existing systems, they must be checked for leaks on completion of the work. This should be done before any pipework is covered over or any protective coating is applied. A soundness test must also be carried out when an appliance is serviced, if it has been fitted with new components or has been altered in any way.

### *Testing a new installation*

Many gas pipes have to be buried in walls and floors and must be fitted before the walls are plastered or any floor finish is applied. This pipework is often referred to as the *gas carcass* and the procedure for testing is as follows. All the open ends must be properly capped or plugged except one, to which is fitted a tee piece having a small test cock on one branch to allow air to be admitted, and a test nipple on the other end to which the U gauge, carefully zeroed, is attached by means of a rubber tube.

The set-up for testing is shown in Fig. 2.11. The U gauge must always be in an upright position when tests are made to ensure a correct reading.

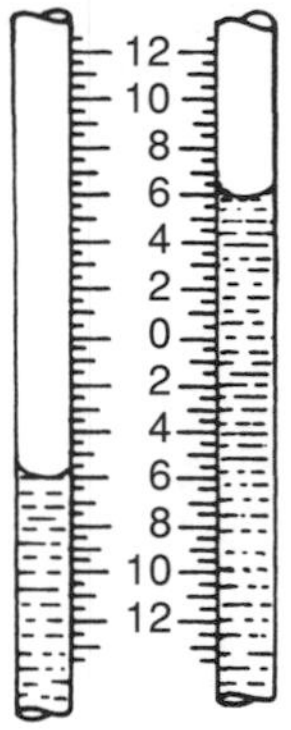


(a) Direct reading type

Sometimes known as a half-scale manometer. Each large division measures 1 mbar but is numbered in 2 mbar. When this gauge has been correctly zeroed it can be read directly. The reading shown here is 6 mbar.

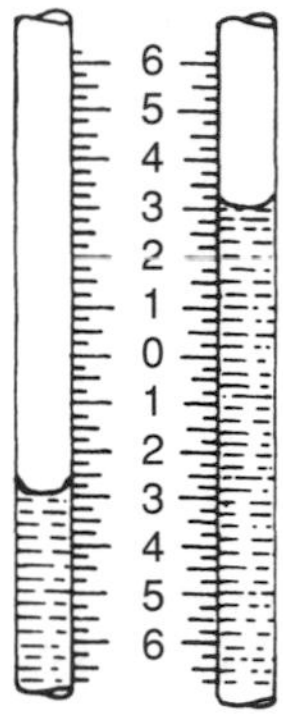


(b) Indirect reading type

With this type the number of millibars on either side of zero must be added. In this instance the water levels read 3 millibars on each leg. Thus 3 + 3 = 6 mbar. Note that in both cases the reading is taken from the lowest point of the meniscus.

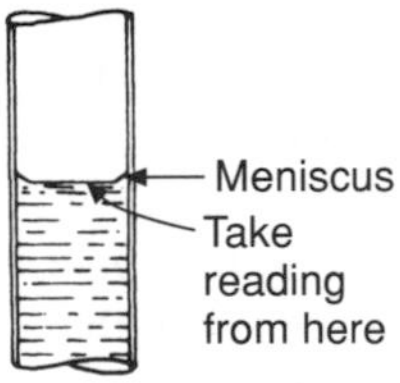


(c) Reading a U gauge

**Fig. 2.10** The U gauge (manometer).

The air in the system is pressurised by being pumped or blown through the test cock until a reading on the U gauge of 20 mbar is shown. The test cock is then shut and a period of time (usually 1 minute) is allowed for the air temperature to stabilise. This is important as the moisture content of warm air will condense when admitted to cold pipework and cause a reduction in its volume, resulting in a drop in the pressure showing on the U gauge. After a 1 minute stabilisation period has elapsed, the pressure should hold, without any further drop showing on the gauge, for a period of 2 minutes. If the distance between the water levels in each leg of the gauge lessens, a leak is indicated and must be found by painting the joints with leak detection fluid, any leaks being detected by the formation of bubbles at the source of leakage. Soap solutions are no longer recommended, as it has been found they have a slight corrosive effect on metals. It should also be noted that where possible, gas should be used for testing in preference to air.

*Testing extensions to existing pipework in domestic properties*

Before the extension is fitted or connected, the existing installation should be tested in the following way. The main gas valve on the inlet side of the meter and all appliance valves and pilot lights should be turned off, and the U gauge connected to the main test nipple, which is usually fitted on or adjacent to the meter. On most installations U gauge connections are fitted on the outlet side. The system should then be pressurised by turning on the main gas cock until a pressure of 20 mbar is recorded on the U gauge — the gas should now be turned off. After a 1 minute period for the temperature to stabilise, no loss of pressure should be recorded on the gauge for a period of 2 minutes. It is an accepted fact that many existing installations are not completely gas tight, e.g. minute leaks on appliance valves. A slight leakage may be acceptable on an appliance, providing it does not exceed that shown in Table 2.5 — but under no circumstances is it acceptable if the pipework is not sound. The procedure here is to isolate all appliances on the system and conduct a test on the pipework only. If it is defective, the leak must be found. However, if the test proves the pipework to be sound and no complaints are made about a smell of gas, there is no undue cause for alarm. Of course, if gas can be smelt the leak must be found and rectified. The differing pressure drops shown in the table relate

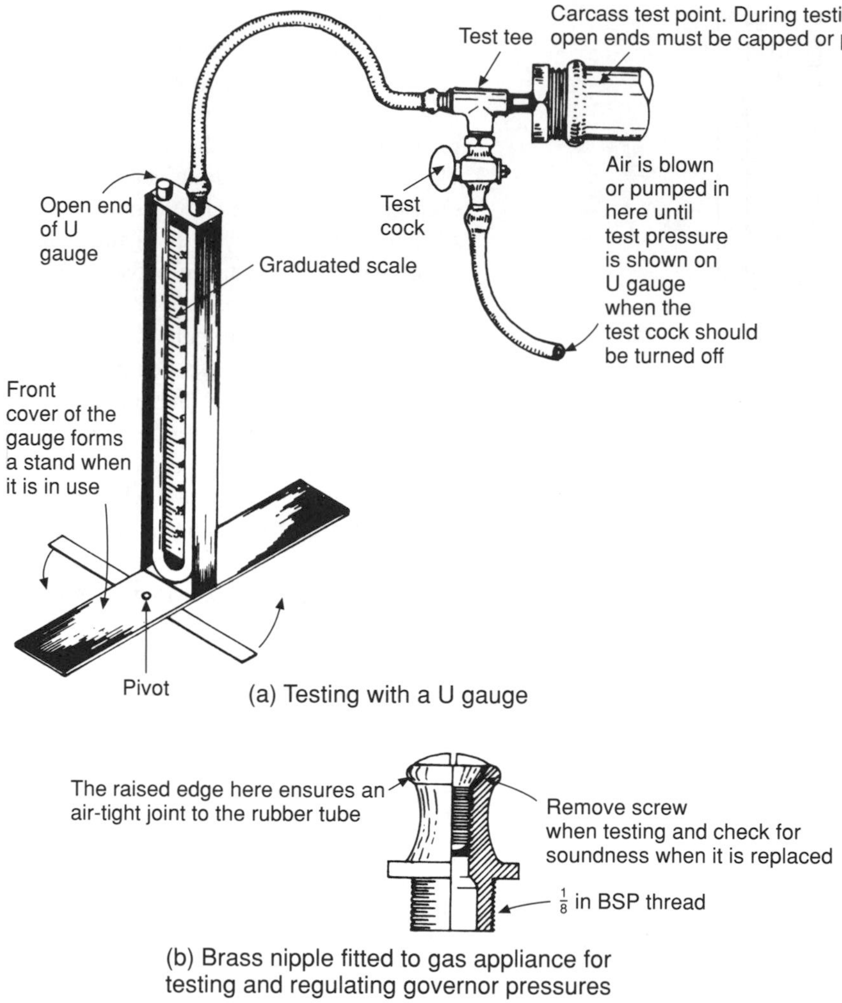
Carcass test point. During testing all other open ends must be capped or plugged
Test tee
Open end of U gauge
Test cock
Graduated scale
Air is blown or pumped in here until test pressure is shown on U gauge when the test cock should be turned off
Front cover of the gauge forms a stand when it is in use
Pivot
(a) Testing with a U gauge
The raised edge here ensures an air-tight joint to the rubber tube
Remove screw when testing and check for soundness when it is replaced
1/8 in BSP thread
(b) Brass nipple fitted to gas appliance for testing and regulating governor pressures

**Fig. 2.11** Use of a U gauge (manometer) for testing soundness of gas installations.

**Table 2.5** Permissible gas pressure drops.

| *Meter* | *Pressure drop (mbar)* | *Capacity in cubic feet per revolution* |
|---|---|---|
| E6 | 8.0 | N/A |
| U6 (D07) | 4.0 | 0.071 |
| P1 (D1) | 2.5 | 0.100 |
| P2 (D2) | 1.5 | 0.200 |
| P4 (D4) | 0.5 | 0.400 |

to the size of the installation and the number of appliances used, which is indicated by the type of meter. A U6 (DO7) meter is used for normal domestic installations whereas a P4 (D4) would be used in a commercial establishment with a larger pipe diameter. If it is understood that the pressure loss in a small pipe over a given period of time will be greater than that in a pipe having a larger diameter, it will be seen why there is a difference in the permissible drop in various installations. Although this difference exists, the ratio of the pressure loss has been calculated to be approximately the same for all types of installations. The rating of a meter relates to its capacity, i.e. the quantity of gas it will pass, and is indicated by a stamp on the front of the case.

One other very important point must be made here in relation to testing existing installations.

Never search for leaks with a match or naked light — it could result in a one-way ticket to the hereafter!

On existing installations any pressure test on existing pipework may be false if the main gas valve 'lets by'. After a period of time the lubricant used to enable it to turn freely and maintain a positive on/off action may dry out, and a small quantity of gas can seep past the valve. This can be checked by using a similar testing procedure to that used for existing installations, turning off the valve when a pressure reading of only 10 mbar is shown on the gauge. If the gauge indicates an increase in pressure the valve is defective and the fact must be reported to the gas supplier. This test must always be applied to existing installations after soundness tests are conducted.

### Purging domestic installations

One of the most important things that must be done when commissioning a new gas installation is to purge it of air, as if a mixture of gas and air occurs in the pipes, a blow back when lighting the appliance could cause an explosion. It is therefore essential to 'air test' the installation before purging is carried out.

The procedure for purging new installations is as follows. The main gas cock should be turned off before any of the appliances are turned on, preferably the one at the end of the main run of pipe. If the appliance is fitted with a flame failure device no gas will be able to pass to the burner, and the usual practice in such a case is to disconnect the burner union or remove the screw in the appliance test nipple. During the purging operation, it is essential that the area into which any gas/air mixture is discharged is well ventilated by opening any windows or doors. Avoid using any electrical switch which may cause a spark, ensure that there are no naked lights and, of course, smoking during the operation is not permitted. The main gas valve should be turned on until the meter is purged by passing a volume of gas not less than five times the capacity per revolution of the meter mechanism. This capacity is shown in the window housing the meter dials, and on a U6 meter is shown as 0.071 $ft^3$ per revolution, thus $5 \times 0.071 = 0.355$, just over a third of a cubic foot. Note that the volume of gas shown on the test dial of the meter varies. That shown in Fig. 2.7 will pass 1 $ft^3$ per revolution, thus to purge this meter a movement of the pointer through just under 3.5 divisions is required. This will ensure that no air remains in the meter or pipework. (Note that some U6 meters have a 2 $ft^3$ test dial.) The same volume of gas must pass an E6 meter when purging is carried out.

The main cock is now turned off and any connections that were broken for the purpose of purging should be tightened and retested for soundness. This is done in the usual way using leak detection fluid, after having re-established the gas supply by turning on the main cock. Each of the other appliances connected to the system should then be turned on until gas is smelt from the burner. As already stated, some appliances are equipped with flame failure devices and air in the pipes supplying them can only be purged as previously described. Only when all the air in the system has been removed should any attempt be made to light the appliances. If a new appliance or pipe run only has been fitted there is no need to purge the meter or the main pipe runs. It should only be necessary to disconnect the new appliance and proceed as previously described.

No appliance should be permanently connected to the gas supply until it has been commissioned, tested and purged.

### Ventilation

A good supply of air is necessary when all types of fuel are burnt and if an appliance is fitted in a confined space, e.g. a cupboard, a flow of air may be necessary to keep it cool. A supply of air is usually introduced into the building via vents built into the structure. The following relates to the air supply for open-flued appliances which take air for combustion from the area in which they are fitted. It is not exhaustive and for more detailed information reference should be made to CP 5440 Pt2.

Adventitious ventilation is the term applied to the natural ingress of air into a building via skirting boards and around windows and doorways, which occurs despite all efforts made to reduce it. For this reason an air vent is not normally necessary for

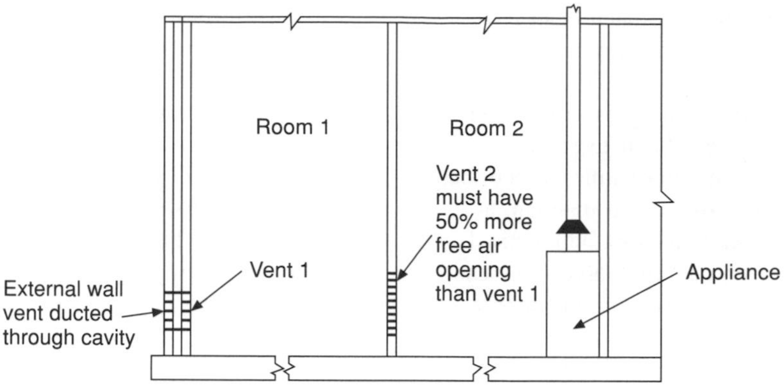


**Fig. 2.12** Air vents in series.

appliances rated up to 7 kW. This does not apply to Northern Ireland where a permanent ventilation opening of at least 4.5 $cm^3$ is required for any open-flued appliance, plus 4.5 $cm^3$ for every kW of input rating in excess of 8 kW. It must be noted that adventitious air is not taken into account when calculating the air requirements for certain fires, flueless appliances, and those fitted in compartments.

Care must be taken when fitting air vents. They should not be located where they would become blocked, flooded or allow the ingress of contaminated air. They should also be constructed to avoid allowing cold draughts into a room. In the past it has been the practice to take air for combustion from the ventilated area under a suspended floor. However, there are risks due to radon gas, which is produced naturally from the decay of uranium in rocks and soils. The level of concentration varies nationally, but in high risk areas any ventilation should be sited in, or ducted to, an external wall. If in doubt the local building control officer's advice should be sought.

Figure 2.12 illustrates the method of providing air for combustion to an open-flued appliance which is in a room with no access to an outside wall. It will be seen that air both for combustion and for ventilation of the room must pass through an adjoining room in which a vent can be fitted to the outside air. The important point here is the increase in size of the internal ventilation grills. Room-sealed appliances take air for combustion externally to the building, but in certain instances ventilation is necessary for cooling. Table 2.6 gives a guide based on the requirements of BS 5440 Part 2 for providing air for combustion and ventilation in both room-sealed and open-flued appliances.

### *Air bricks*

These are the usual method of providing ventilation in a building. They are usually made of terracotta, although cast iron may be used. A section of a terracotta air brick is shown in Fig. 2.13. Their size varies but is normally measured in millimetres as

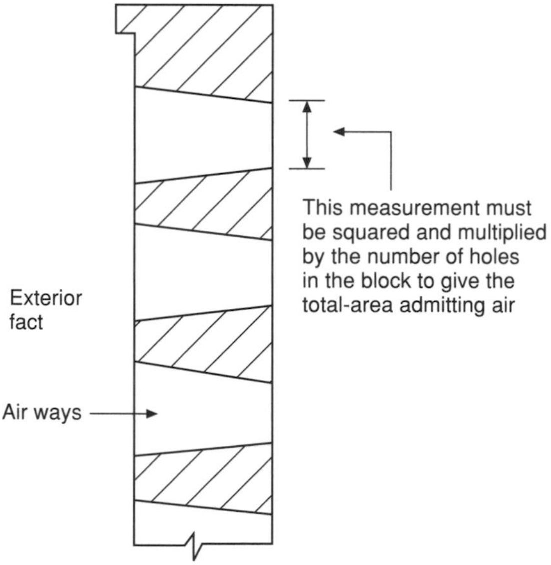


**Fig. 2.13** Section through terracotta ventilating brick.

**Table 2.6** Minimum effective air vents serving gas appliances up to 60 kW input.

| *Open-flued appliances* | *Position of vent levels* | *Free area of vent $cm^2$ per kW input* |
|---|---|---|
| In a room ventilated directly to the external air | High or Low | 4.5 in excess of 7 kW |
| In a compartment ventilated to the external air | High | 4.5 |
| | Low | 9.0 |
| In a compartment ventilated via an internal room or space | High | 9.0 |
| | Low | 18.0 |
| Room-sealed appliances | | |
| In a room | — | No vent necessary |
| In a compartment ventilated to the external air | High | 4.5 |
| | Low | 4.5 |
| In a compartment ventilated via an internal room or space | High | 9.5 |
| | Low | 9.5 |

*Notes*: A compartment is an enclosure designed or adapted to house a gas appliance. High- and low-level vents are necessary for cooling. Air for ventilation must always be taken from the same source, either from outside or inside the building. Vents must be fitted on the same wall of the compartment. All combustion air must be provided via low-level vents. The top of low-level vents should not be higher than 450 mm to avoid the passage of smoke in the event of a fire.

follows: 75 × 225, 150 × 225 and 225 × 225. A duct must always be provided through the wall connecting the ventilating grills. It will be seen that the holes taper and it is important that the free area is calculated by multiplying the area of one hole by the number of holes in the vent. Their sizes must be taken from the inside surface of the ventilator, which will be the effective area of each hole. Assuming the area of one hole is 64 mm and the vent contains 36 holes, the free area will be 2,304 mm or 23 cm.

Figure 2.14 illustrates an alternative to the bricks described. Because these ventilators are draught-proof the possibility of the occupant of the building blocking a draughty vent is avoided. They are circular in shape so that a round hole can be drilled using a core drill on the structure.

Note that the installer or service engineer is responsible for determining and checking ventilator sizes in both new and existing buildings. All manufacturers of gas appliances provide detailed instructions for their installations which include recommendations relating to the supply of fresh air. These instructions must be carefully studied and complied with to prevent the occurrence of accidents.

## Flues

### *Function of a flue*

The flue is an important part of the installation and must be sound, safe and efficient. Its main function is to remove the products of combustion from the appliance. In addition to this, an open flue acts as a ventilator, whether or not the appliance is in use, and this contributes to the physical comfort of the building's occupants and minimises the formation of condensation, especially in kitchens.

### *Flueless appliances*

Whether or not a gas appliance is fitted with a flue depends mainly on (a) the period of time during which it is in use and (b) the quantity of gas it consumes. Flueless appliances are permissible in rooms or spaces under the following conditions, providing adequate fixed ventilation is fitted. In most cases an openable window is also required.

(a) With low-rating appliances, e.g. refrigerators.
(b) When the period during which they are in use is of relatively short duration, e.g. a water heater serving a sink.
(c) When conditions associated with their operation render it likely that the user will

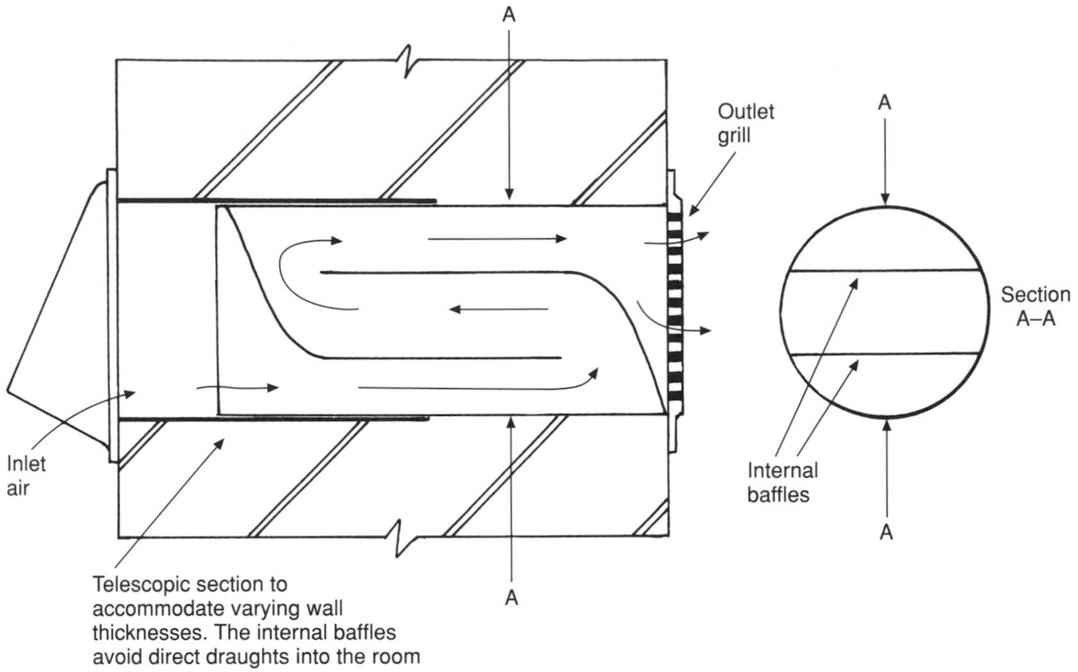
A
Outlet
grill
A
Section
A–A
Inlet
air
Internal
baffles
A
Telescopic section to
accommodate varying wall
thicknesses. The internal baffles
avoid direct draughts into the room
A

**Fig. 2.14** Draught-proof wall ventilator.

increase the ventilation of the room by opening the windows, e.g. cookers and hobs causing cooking smells and steam.

Gas appliances which fall into the categories listed above are: gas cookers; instantaneous water heaters up to 12 kW, providing they are not liable to prolonged use or fitted in a confined space (not less than 5 $m^3$); storage water heaters where the heat input is less than 3 kW, or 4.5 kW if the storage capacity is less than 45 litres; gas circulators, having a heat input of less than 3 kW, providing they are not installed in a bathroom, airing cupboard or badly ventilated area.

*Open flues (natural draught)*

A flue works on the principle of convection (see Fig. 2.15), a method of heat transfer which applies to fluids, i.e. liquids and gases. Convection has many applications in plumbers' work including the circulation of hot water and drain ventilation, the principle relating to the latter being exactly the same as that for a flue. In a natural draught flue the upward movement of the combustion products is

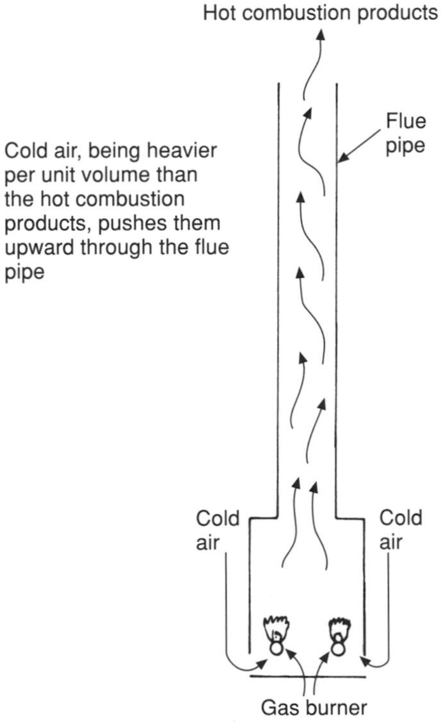
Hot combustion products
Flue
pipe
Cold air, being heavier
per unit volume than
the hot combustion
products, pushes them
upward through the flue
pipe
Cold
air
Cold
air
Gas burner

**Fig. 2.15** Operation of a flue.

brought about by the difference between their temperature and that of the air surrounding the flue.

*Flue gas temperatures* Generally, the updraught in a flue is improved as the temperature of the combustion products increases, but this tends to become proportionally less at higher ambient air temperatures.

### *Factors affecting the performance of flues*

*Height* An increase in the height of a short flue will raise its performance within certain limits. However, if it is too high the frictional resistance of the pipe walls will slow down the updraught and, more important, as the products of combustion cool, the convective current will become weaker. These factors do not apply to flues of less than 6–9 m in height, and longer flues are seldom encountered in domestic dwellings.

In commercial and industrial buildings and some domestic dwellings, a mechanical extractor may be necessary and specialist advice should be sought prior to its use. It is important to note that any appliance in which the products of combustion are removed by this means must be made in such a way that the valve admitting gas to the burner will not open unless the extractor is operating. Any condensing appliances will have lower flue temperatures and because they are fan assisted they do not rely on convection to remove the products of combustion. Stainless steel is now the most commonly used material for gas flues; it has been used as a flexible flue liner for many years and is also made in rigid form.

*Flue runs* Where possible, flues should be vertical. If bends are necessary those having an angle of 135° are preferred, as those having a more acute angle offer more resistance to the flow of combustion products. Bends of 90° and horizontal flue runs for natural draught appliances are no longer permissible. The diameter of the flue is dependent on the flue outlet spigot on the boiler and must never be reduced. The flue must always be fitted to the appliance in such a way that it can easily be disconnected; this can be accomplished with a sliding or split collar as shown in Fig. 2.16. The flue above the point of disconnection must be securely fixed to prevent it dropping down when the appliance is removed. Flue pipes should always be fitted with the socket upward, as this avoids any condensation running down the outside of the flue pipe leaving an unsightly stain.

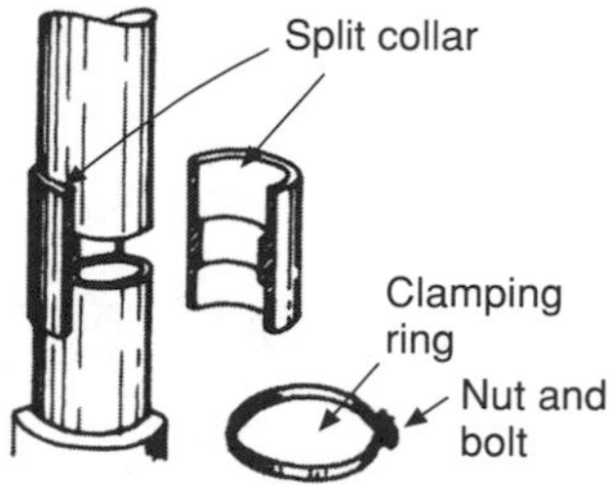


(a) The two halves of the split collar are secured with a clamping ring prior to sealing joints with fire cement

These must be fitted as closely as possible to the appliance.

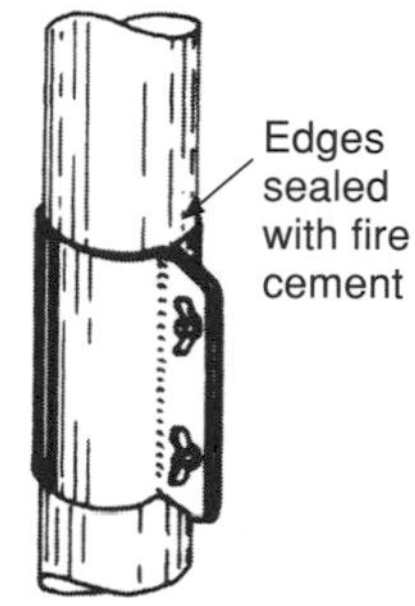


(b) A sheet metal clamp fixed round the pipe and secured by wing nuts

**Fig. 2.16** Methods of flue pipe disconnection for cleaning and maintenance.

### *Flue materials*

Stainless steel is now the most commonly used material for gas flues. It has been used as a flexible flue liner for many years, but it is also made in rigid form. A typical joint for this material is shown in Fig. 2.17. One end of each pipe is crimped and slightly tapered enabling a tight push-fit joint to be made into a plain end of pipe. Where necessary to give added support to the joints, self-tapping screws or pop rivets may be used to form a permanent fixing. Twin-wall insulated flue pipes may be used in situations where there is no existing brick flue.

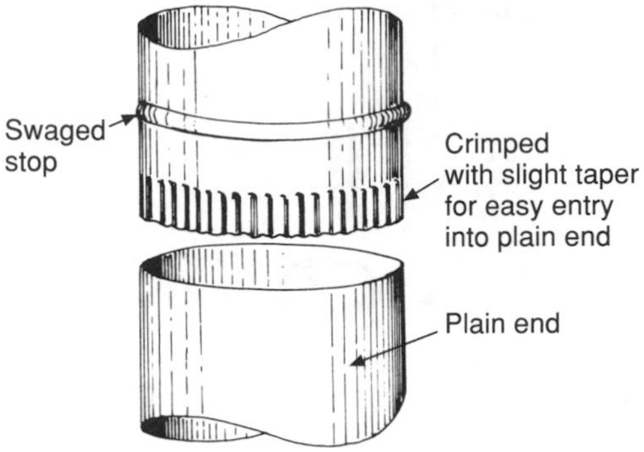


**Fig. 2.17** Push-fit joint for single-wall stainless steel flue pipe.

Unlike the previous materials discussed, which are normally only used in short lengths for connecting an appliance to an existing flue, twin-wall pipes provide a purpose-made insulated chimney, see Fig. 2.18. Their construction and the materials used vary slightly depending on the manufacturers, but the basic principles are the same. In some cases both the inner and the outer pipes are made of stainless steel, in others the outer casing may be galvanised steel or an alloy of zinc and aluminium. Some have push-fit joints, others employ a bayonet-type joint which locks the pipes together when given a slight twist. In both cases an external locking band is used to ensure rigidity of the joint. All manufacturers of this type of flue produce a complete range of components such as bends, appliance adaptors, varying pipe lengths and brackets. When ordering insulated flues of this type, remember it is not possible to cut and joint it on site, therefore the exact lengths should be obtained.

Enamelled steel socketless pipes are sometimes used to connect appliances to an existing flue or chimney, the joints being made with shaped rings and clamps as shown in Fig. 2.19. These pipes are made in a variety of sizes and lengths and can be cut with a hacksaw, but this is best avoided due to possible damage to the enamel. If it is necessary to make a cut apply a strip of masking tape around the point where the cut is to be made; this will prevent the hacksaw slipping. If careful measurements are taken the flue can, in most cases, be constructed using stock sizes of pipe. This particular material is normally only used to connect an appliance to the

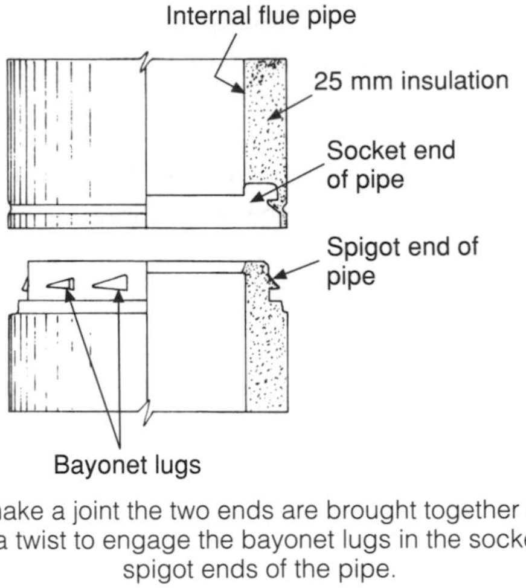

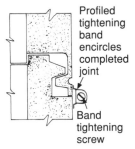

**Fig. 2.18** Steel twin-wall flue pipes.

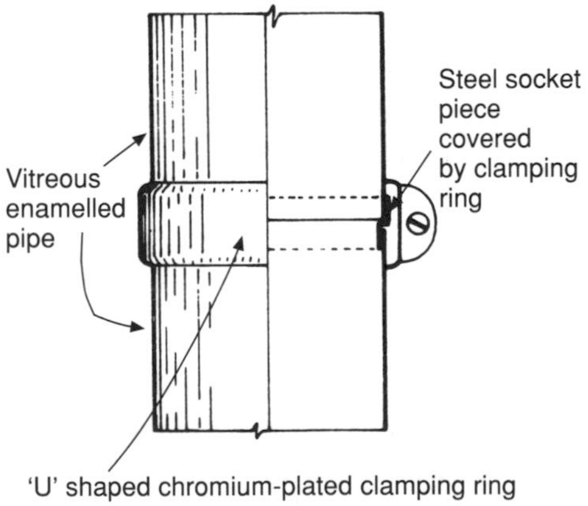


**Fig. 2.19** Vitreous enamelled sheet steel flue pipes. These pipes are obtainable in black or white and have a very pleasing appearance. They are socketless, being joined with a pressed steel socket piece. This is hidden underneath the chromium-plated steel clamping ring which is secured with a screw.

main flue or chimney. Flue lining is dealt with later in this chapter.

*Pre-cast flue blocks* These are blocks made of Portland or high-alumina cement conforming to BS EN 18581. They may be built into the building structure by the bricklayer or in a slightly different form for external use to construct what is best described as a traditional chimney. They incorporate an insulated ceramic flue liner, and although more expensive than a flue pipe, they do have a more aesthetic appearance. Always consult the manufacturer as to their suitability with, if necessary, condensing gas appliances.

*Flues for multi-storey buildings*

The two systems illustrated in Fig. 2.20 were developed by the gas industry and are extensively used in multi-storey buildings for the discharge of combustion products into a common flue. It must be stressed that normally it is not permissible to discharge more than one appliance into a flue. These systems are designed for this purpose. Special room-sealed appliances are required for use with the systems shown, and advice must always be sought from the appliance manufacturer prior to its installation in such schemes. Any replacement of the appliances must be like for like.

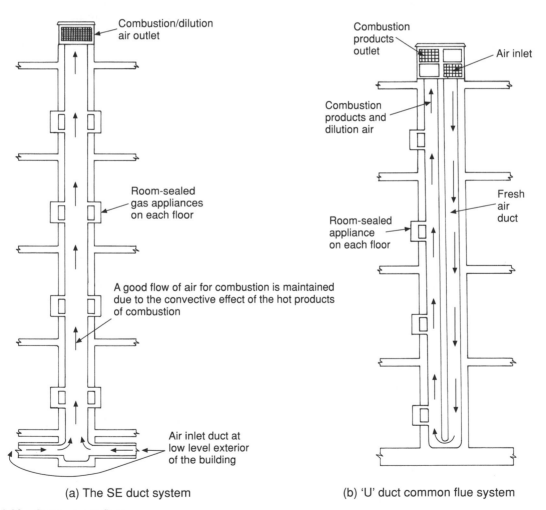

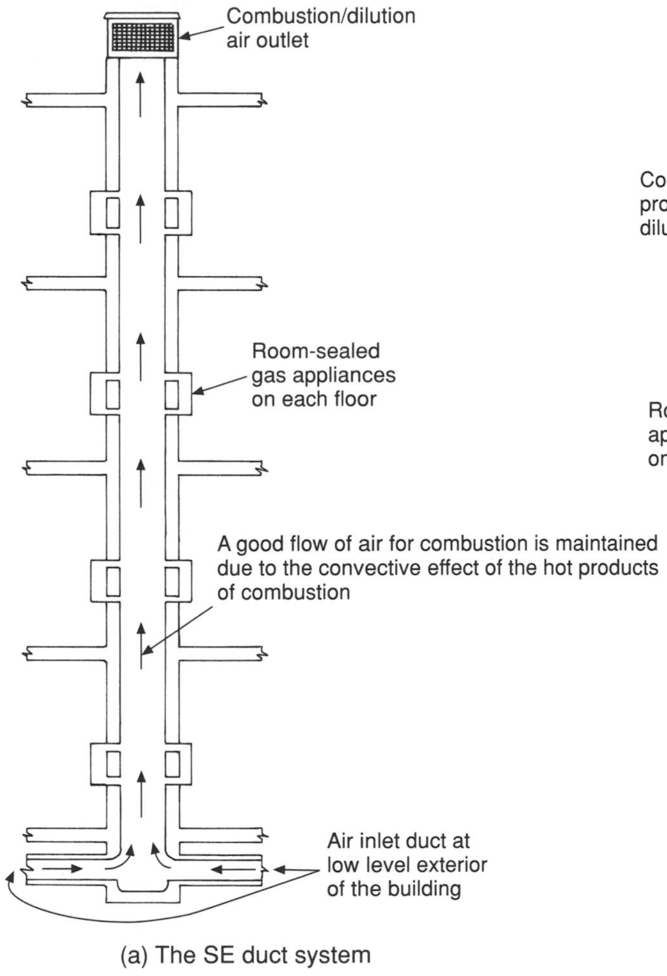


**Fig. 2.20** Common gas flues.

These systems were devised for water heaters or gas fires in multi-storey buildings. No individual forced draught appliances are permitted to discharge products of combustion into these ducts as they could become pressurised, resulting in a potentially dangerous situation.

The systems illustrated here are used in multi-storey buildings for the discharge of combustion products from balanced flue natural draught gas appliances. The manufacturer of any appliance used with common flues should confirm their suitability.

Permanent notices must be displayed adjacent to all air inlet and combustion outlets warning that they must not be obstructed in any way. Notices must also be displayed on all appliances warning that they are part of a shared flue installation.

*Flues passing through combustible materials* Although the flue temperature of most gas appliances is normally very low, accidents can happen, and if the risk of fire is to be avoided the flue must not be in direct contact with any combustible materials, e.g. wood floor or rafters. If it is within 50 mm of any such material it must pass through a sleeve of non-combustible material of sufficient width to form a space of 25 mm minimum between the sleeve and the flue pipe. Where the flue passes through a floor or ceiling, some method of preventing smoke or flame (in the event of a fire) from passing through the space around the pipe into the space above is necessary. The air space in these circumstances can be filled with a non-combustible insulation material such as fibreglass.

To comply rigidly with these fire regulations is not an easy task, but they are important. Figure 2.21 gives some idea of what is required, and it should be appreciated that as only a few special components are purpose-made for sleeving flues, in most cases it will be left to the plumber's ingenuity to devise something suitable. Where a flue pipe passes through a roof, it must be weathered with a pipe flashing as shown in Fig. 2.22.

*Fitting appliances to existing flues* It is often convenient to use an existing brick flue for a gas appliance such as a boiler or fire. If it has been used for an appliance burning solid fuel, it must first be properly swept. This will ensure that it is clean and is clear of birds' nests and other obstructions. The brickwork must be checked for soundness, and in old buildings it is recommended that a check is made to ascertain that the flue is serving only one appliance. It has been known for two fireplaces to be connected to one flue or chimney, and in the case of a gas appliance being fitted this could mean that combustion products are discharged into another room. It became mandatory around 1996

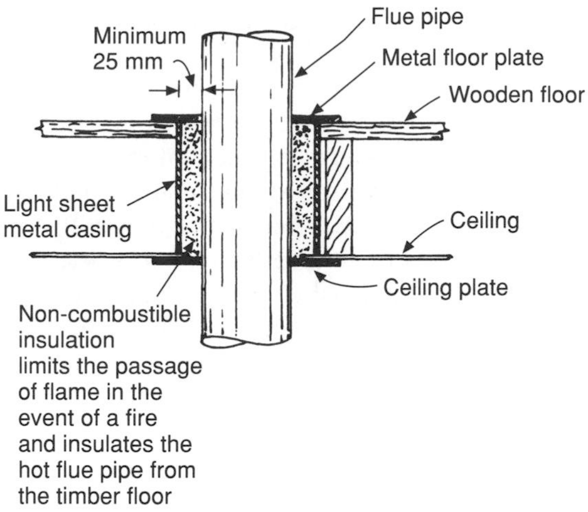


**Fig. 2.21** Passing a gas flue through combustible materials. This illustrates a flue passing through a wooden floor, but similar arrangements must be made where the flue passes through the roof if it is within 50 mm of any combustible material, e.g. rafters.

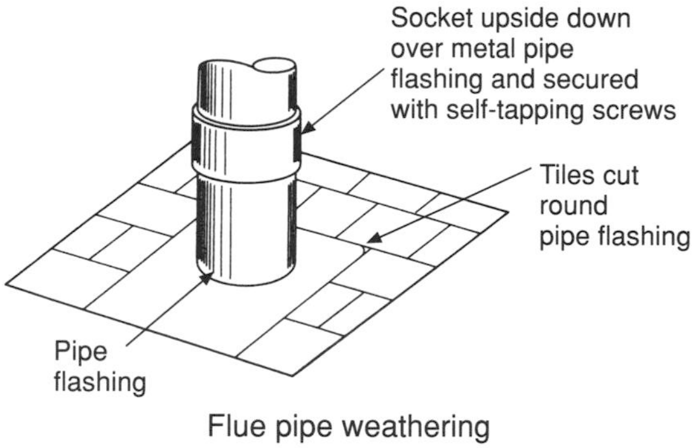


**Fig. 2.22** Weathering of flue pipes passing through a roof.

that all flues had to be lined. To comply with this the usual practice is to build in a ceramic lining (see Fig. 2.23). In buildings constructed prior to this date the flue is unlikely to be lined, and if a gas or oil appliance is to be installed it is usual to provide a liner. In the event of a replacement appliance being fitted, it is good practice to renew any existing metal lining. Even where the flue is to be lined this could mean that the insulation the lining provides is not as effective as it might be.

To test an existing flue for soundness, cap the top and warm it up by placing a blow lamp in the opening for about 10 minutes. Remove the lamp and light a smoke pellet prior to sealing the opening. This will verify whether the flue is sound and not interconnected with another. Do not confuse this with testing flues for updraught.

*Testing flues for updraught*

Before an appliance is fitted the flue must be tested to ensure it has a positive updraught. This is carried out using a smoke match enabling any sign of insufficient draught to be seen. Unless these tests are satisfactory, the appliance must not be fitted until the cause of any defect has been ascertained and corrected. Typical causes for adverse draught conditions are shown in Fig. 2.24.

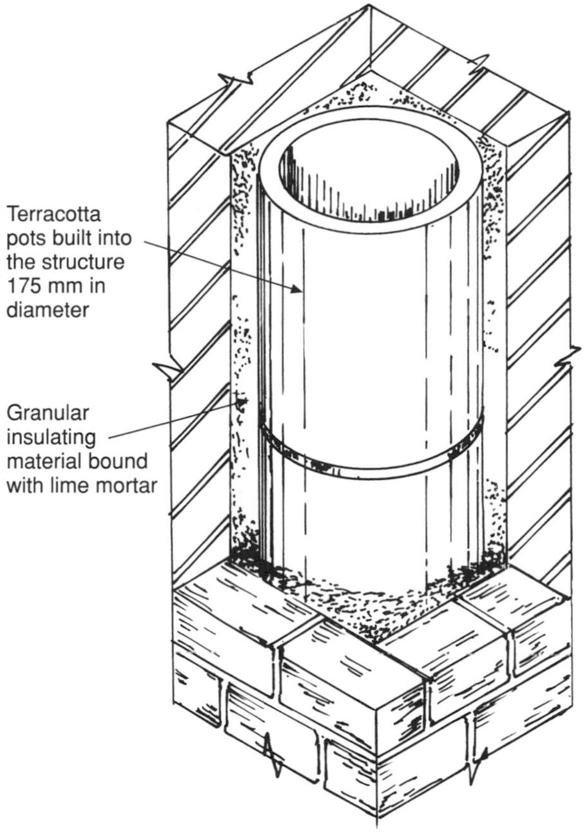


**Fig. 2.23** Built-in flue linings.

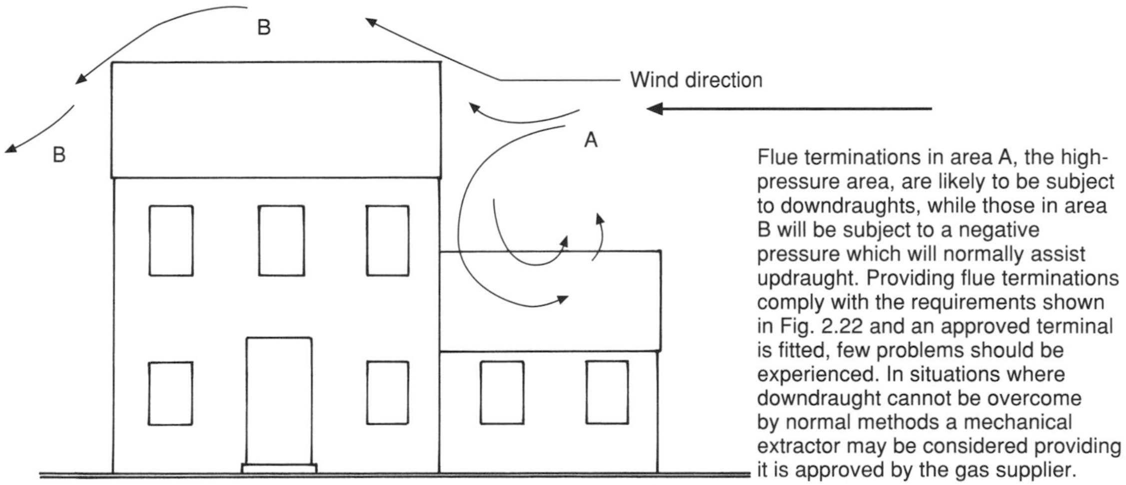

B
B
Wind direction
A
Flue terminations in area A, the high-pressure area, are likely to be subject to downdraughts, while those in area B will be subject to a negative pressure which will normally assist updraught. Providing flue terminations comply with the requirements shown in Fig. 2.22 and an approved terminal is fitted, few problems should be experienced. In situations where downdraught cannot be overcome by normal methods a mechanical extractor may be considered providing it is approved by the gas supplier.


**Fig. 2.24** Wind effect on natural draught (conventional) flues.

Smoke tests should be conducted under the worst possible conditions; in other words, all doors and windows through which air can be admitted should be shut. If the room contains an extractor fan, this must be working during the test. In the event of the draught being insufficient, under no circumstances should the appliance be fitted until the draught problem is solved. Any form of ventilation must be of a type which cannot be closed, to deter anyone, who may not be aware of the dangers, from sealing them. Ventilators of a type which avoids direct draughts into the room should be fitted; Fig. 2.14 illustrates a ventilator of this type. Spillage tests must be conducted when an appliance is commisioned or serviced and carried out as shown by the examples given in Figs 2.25(a), (b), (c) and (d). The term *spillage* relates to the products of combustion being discharged into a room or compartment due to downdraught. Spillage tests are conducted in a similar way to that described for initially testing for updraught, e.g. when the worst possible conditions are prevailing.

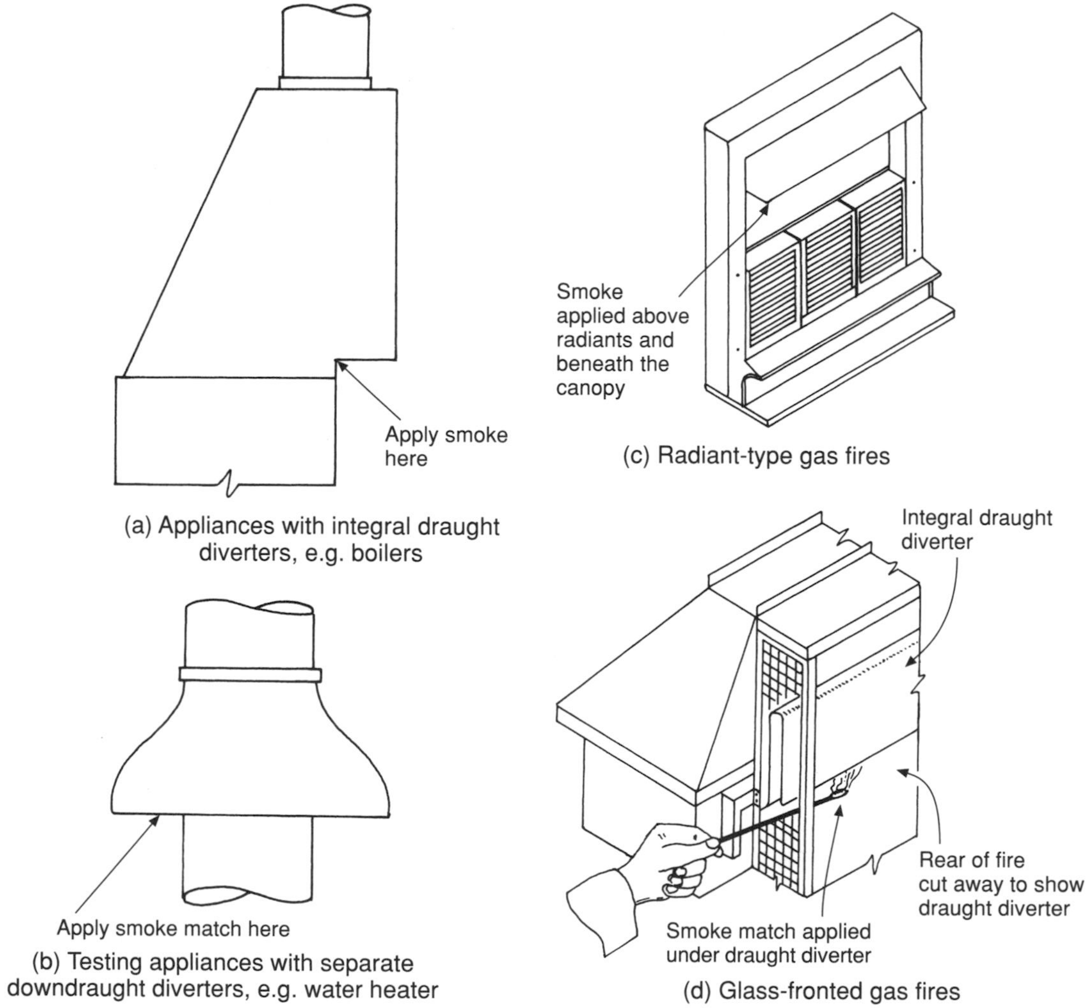


**Fig. 2.25** Testing gas appliances for spillage on natural draught appliances. Spillage is unlikely with fanned flued appliances.

*Condensation in flues*

The reader should note that this section relates mainly to flues in non-condensing appliances; boilers complying to SEDBUK Class A or B are designed for condensation to take place. Reference should be made to Fig. 5.8. It has already been stated that one of the principal products of combustion is water vapour and the reader will be aware that the effect of warm, moisture-laden air coming into contact with a cold glass window is to produce condensation. The same situation will occur if flue gases containing water vapour are discharged into a cold flue.

An explanation of condensation may be helpful here. The amount of vapour contained in the air varies according to its temperature, e.g. the higher the temperature, the greater the quantity of water vapour the air can absorb. There is a limit, however, and should a body of air at a specified temperature reach saturation point (i.e. it cannot absorb any more vapour at that temperature), any temperature drop will cause some of the water vapour to be precipitated, the temperature at which this takes place being called dew point. In the case of a gas flue two steps are taken to minimise the incidence of condensation: one is to keep it warm by various methods of insulation, the other is to make provision for what is called *dilution* of the combustion products. Dilution of the flue gases is achieved by allowing more air into the flue to absorb some of the water vapour. All products of combustion contain a certain amount of moisture, but its effect is aggravated in the case of gas or oil equipment as their high degree of efficiency means that the flue gas temperature is comparatively low in comparison with, for example, that of solid fuel, and therefore cannot absorb so much water vapour. Dilution air is admitted to the flue via the downdraught diverter or an open end in the boiler flue as shown in Fig. 2.26.

*Insulation of flues*

See Fig. 2.27. The usual method of avoiding excessive condensation in existing flues is to line them with a stainless steel flexible liner. This is marketed in coils in nominal sizes of 100 mm, 125 mm and 150 mm diameter. It can be cut with a fine tooth hacksaw, but beware — it is both springy

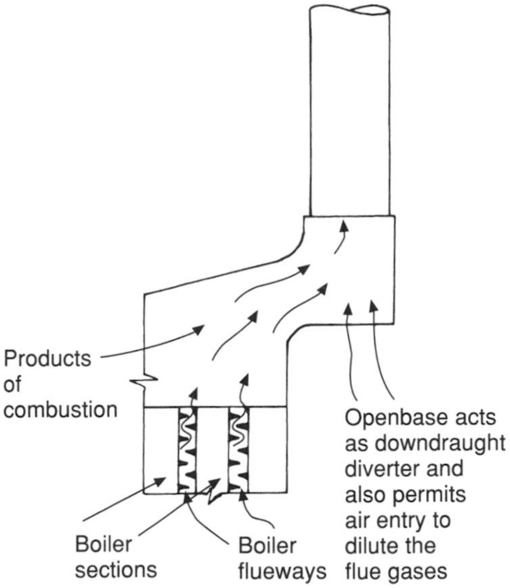


**Fig. 2.26** Admitting dilution air to appliance flues. If air is permitted to mix with the products of combustion it has the effect of absorbing some of the water vapour, thus reducing condensation problems in non-condensing appliances. A downdraught baffle (Fig. 2.29) is an alternative method of permitting flue gas dilution.

and sharp and has been the cause of many badly cut fingers and hands. A good idea is to tape the edges when it is cut as shown in Fig. 2.27(d). Figure 2.27 also illustrates the details relating to this method of lining. In some instances it will be necessary to remove an existing chimney pot and the cement mortar securing it. Extreme care must be taken if the flue is an old one with cracks and poorly jointed brickwork, as it may collapse. If in doubt, a proper scaffold must be used and the stack rebuilt. The fact that both the top and bottom of the flue are sealed prevents any movement of the air surrounding the liner, still air being an excellent insulating agent.

## Flue terminals

Typical examples are shown in Fig. 2.28 (part 1) and are constructed of metal or terracotta. A wide variety of types is available. The 'O H' terminal shown in Chapter 6 is also suitable for gas. All gas appliance flues must be provided with a terminal, except for certain types of open fire, and even here they are recommended. Their purpose is to prevent blockage by material such as leaves and

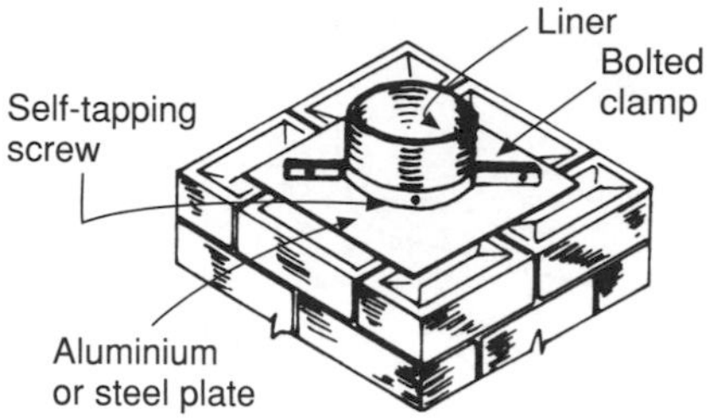


(a) Detail showing how the liner is secured at the top of the flue

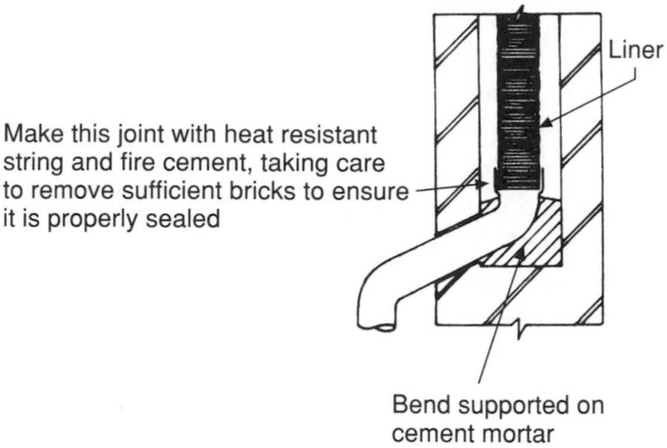


(c) Alternative method of connecting an appliance flue to the liner

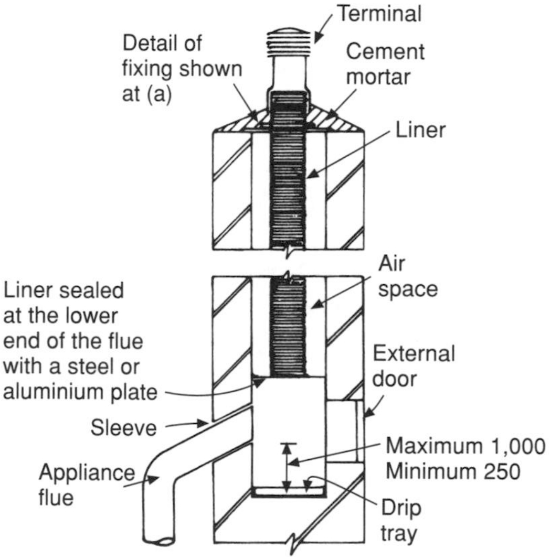


(b) Lining installed in flue

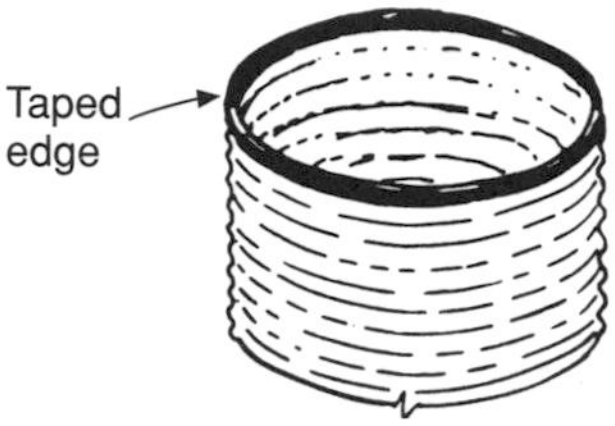


(d) To avoid accidents, seal the cut edges of the flue liner with suitable tape

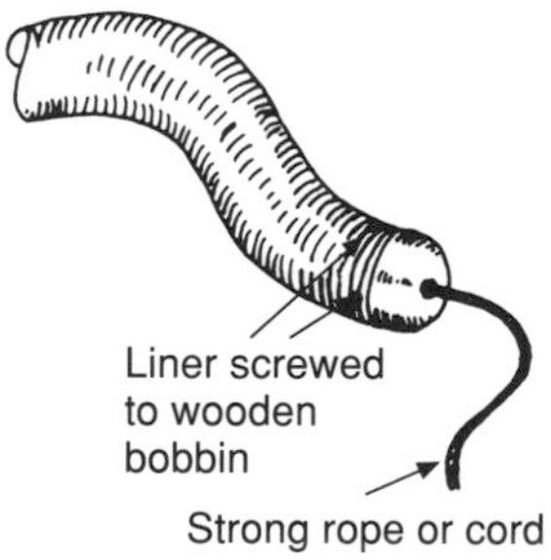


(e) Method of attaching rope to lining

The rounded bobbin, which can be hired or bought from the supplier of the lining, enables it to negotiate any bends in the brick flue.

**Fig. 2.27** Flue lining details with flexible stainless steel flue linings. These are the most convenient to use, especially in older buildings with previously unlined flues.

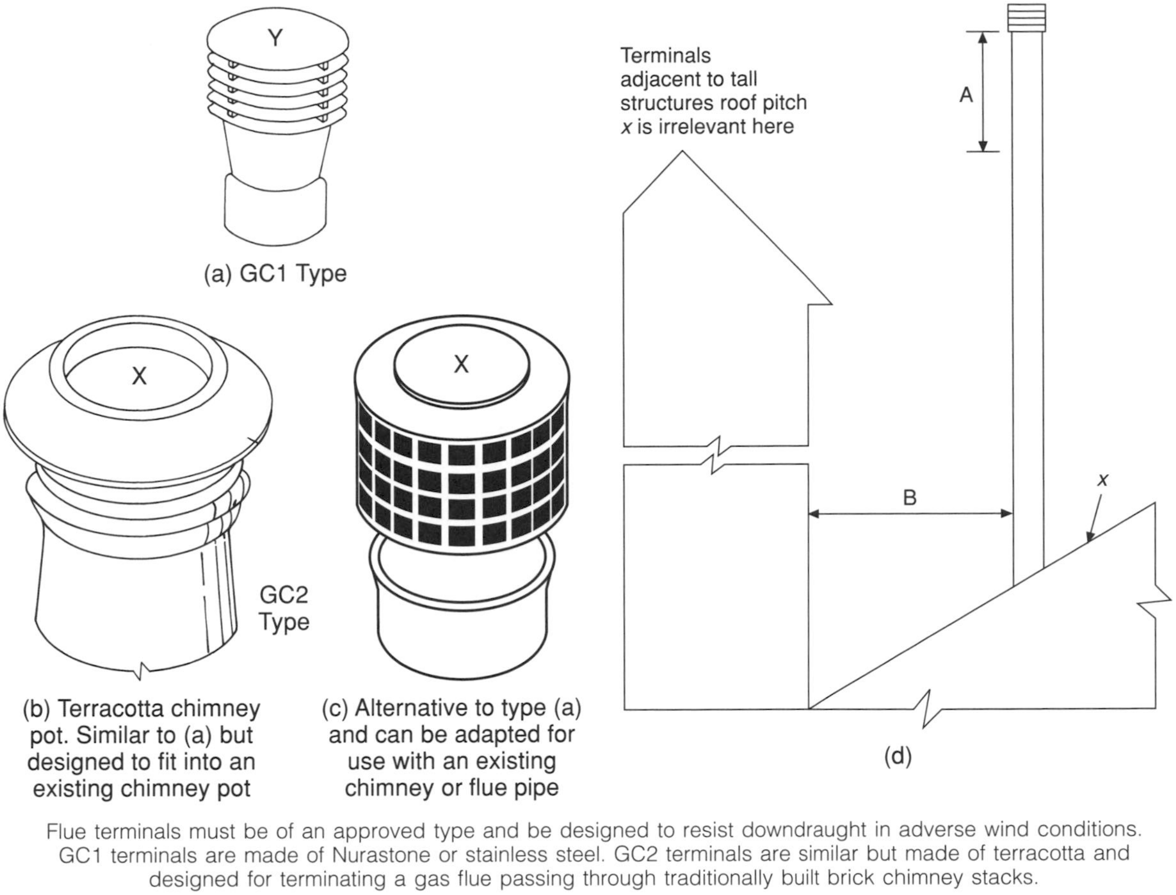
Y
(a) GC1 Type
X
GC2
Type
(b) Terracotta chimney pot. Similar to (a) but designed to fit into an existing chimney pot
X
(c) Alternative to type (a) and can be adapted for use with an existing chimney or flue pipe
Terminals adjacent to tall structures roof pitch x is irrelevant here
A
B
x
(d)
Flue terminals must be of an approved type and be designed to resist downdraught in adverse wind conditions. GC1 terminals are made of Nurastone or stainless steel. GC2 terminals are similar but made of terracotta and designed for terminating a gas flue passing through traditionally built brick chimney stacks.

**Fig. 2.28 (part 1)** Flue terminals.

birds' nests. Some types are made in such a way that they increase the flue updraught.

The illustrations shown in Fig. 2.28 (part 2) are based on the recommendations of BS 5440 Part 1:2000. They show most of the configurations likely to be encountered during the normal course of work when dealing with open-flued appliances. It will be seen the general requirements are such that (a) the terminal is high enough to avoid pressure zones and adverse draught conditions and (b) vitiated air is unlikely to enter the structure via windows and ventilation apertures. In some cases it may be necessary to provide for a fan-induced draught, but this should be avoided in domestic properties where possible.

*Downdraught diverters*

Most gas appliances having a conventional flue are fitted with a downdraught diverter. The main object of this component is to prevent any vitiated air from affecting the proper combustion of the gas at the burner. (Remember that the term *vitiated air* means air lacking in oxygen.) Downdraught diverters also prevent the flames lifting off the burner if the updraught is excessive. Figure 2.29 shows how a downdraught diverter functions. Providing the downdraught is not persistent, it will have no harmful effects because the products of complete combustion are harmless. They are, however, undesirable, and if the appliance is not burning correctly and is producing carbon monoxide the

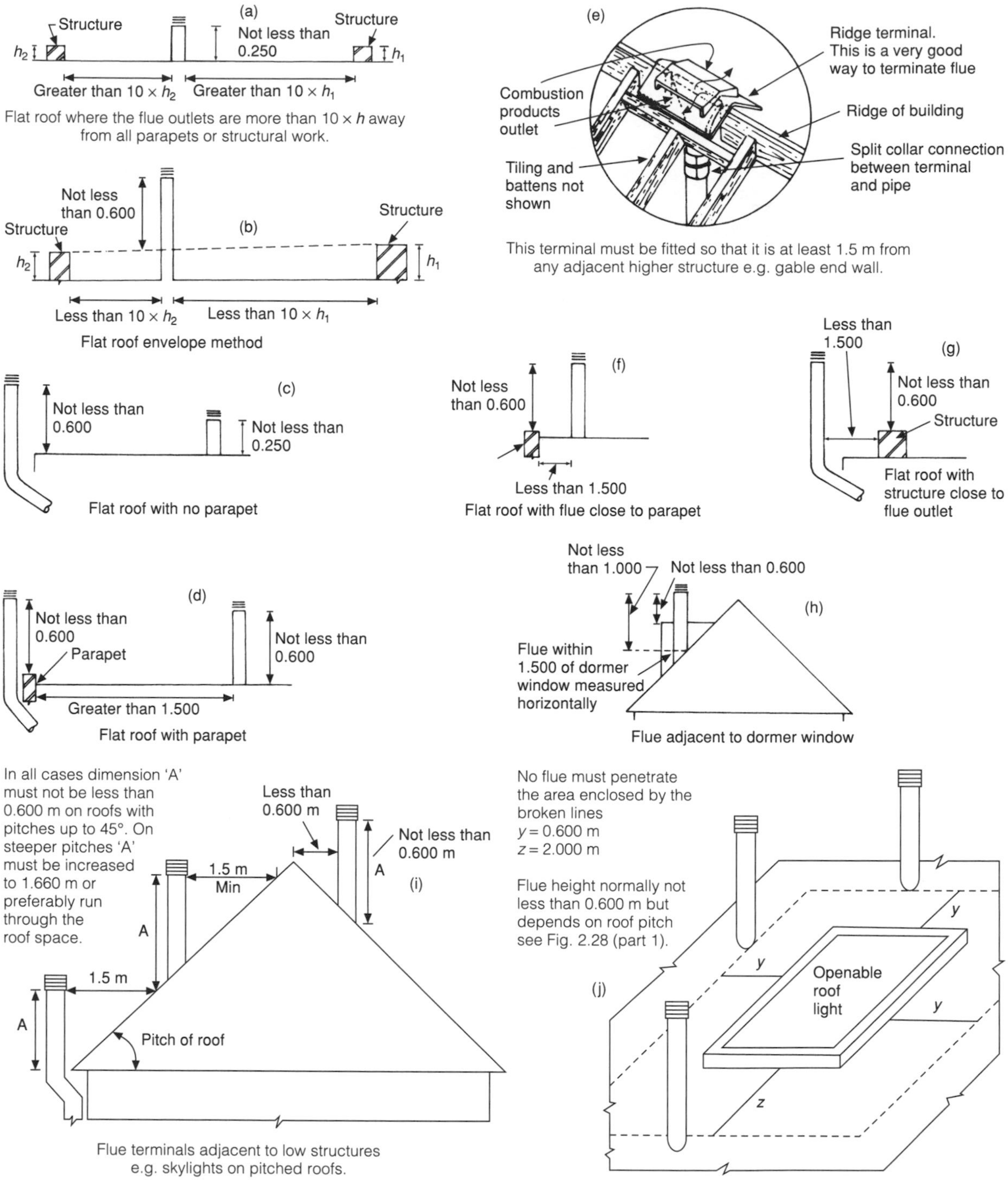


**Fig. 2.28 (part 2)** Approved positions for 'open' flue terminals. All the illustrations included in this group comply with BS 5440:Part 1:2000 and show the recommendations for fixing terminals to minimise the effects of downdraught in flues where they terminate in exposed positions. All dimensions are in metres. It should be noted that in most cases where the roof or a structure on the roof can affect the free discharge of combustion products, the minimum height of the flue above the structure is 0.600 m. The positions shown are all roof outlets and likely to comply with the requirements for condensing boilers.

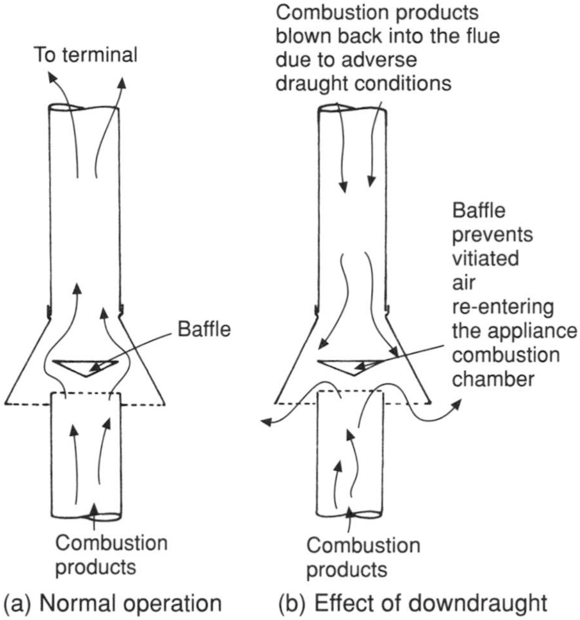


**Fig. 2.29** Downdraught diverter: its purpose is to prevent downdraught affecting complete combustion of the gas.

results could be fatal. Not all open-flued appliances are provided with downdraught diverters.

*Room-sealed appliances*

These do not require a conventional flue or chimney for the discharge of the combustion products, the flue being an integral part of the appliance. Those having open or conventional flues draw air for combustion through ventilators and openings in windows and doors, often causing uncomfortable draughts in the building; this does not apply to room-sealed appliances. Air for combustion is drawn in from outside the building via a wall-mounted terminal, the products of combustion being discharged through the same terminal in a separate duct. All the terminals for balanced flue appliances are made in such a way that there is no danger of vitiated air re-entering the appliance providing they are correctly installed.

Figure 2.30(a) illustrates the basic principles of a natural draught room-sealed appliance. Originally one of its limitations was that it had to be fixed on an outside wall. Modern appliances of this type have fan-assisted draught which enables flues and terminals to be made much smaller and longer, giving more flexibility to their installation. Figures 2.31(a–d) show some installations using these flues with modern fanned draught appliances. Other advantages of balanced flues are that there are no condensation problems and there is no possibility of combustion products entering the room. One disadvantage is that they are slightly less efficient than appliances with conventional flues due to the colder combustion air which is drawn from outside the building. This is, however, minimal with modern equipment having smaller flues and where the incoming air is warmed prior to entry into the combustion chamber. In a new building where the hot water and heating system is gas fired, a chimney or specially constructed flue will be unnecessary so there is a considerable saving. The flue terminals of any type of appliance must be carefully sited to avoid the possibility of combustion products entering the building through fresh air inlets or open windows. Table 2.7 and Fig. 2.32 show what are generally acceptable positions for non-condensing room-sealed appliance terminals (the corresponding minimum distances are given).

These terminals get very hot, however, and if touched could inflict serious burns, therefore a properly constructed guard must always be fitted where the terminal is less than 2 m from the ground level. High-level terminals can cause damage to plastic gutters and the approved methods of overcoming this are shown in Fig. 2.33. The terminal positions shown comply with the relevant British Standards and those of appliance manufacturers. Prior to fitting an appliance, however, the installer should conduct a survey to ensure that any products of combustion do not cause a nuisance to neighbours of adjacent properties. Thought should also be given to the possibility of future extensions to neighbouring property which may prevent the efficient function of the flue leading to a dangerous or high-risk situation.

## Gas controls

Modern gas appliances employ a wide diversity of control systems, most of which require the use of

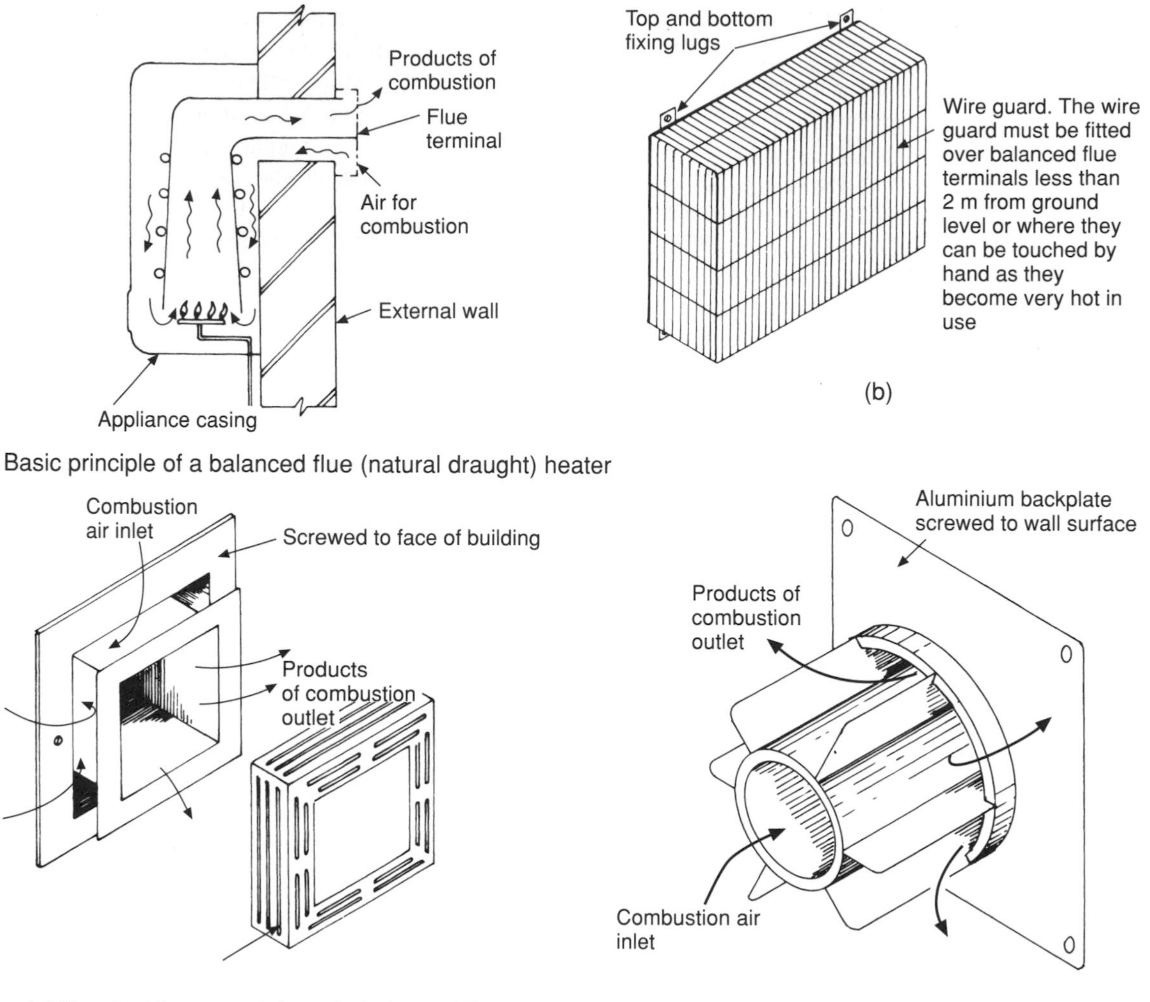


(a) Terminal for natural draught balanced flue non-condensing appliance

(c) Typical fanned flue terminal for balanced flue appliance

**Fig. 2.30** Principle of balanced flue room-sealed appliances and details of terminal with guard removed to show ducts. Room-sealed appliances draw no air for combustion from inside the room. Combustion air enters the appliance via a duct in the flue terminal, the products of combustion being discharged through a separate duct in the same terminal.

electricity. Originally all gas appliances, including gas boilers, had control systems operated by the gas pressure, and while they were very effective and did not require an undue amount of maintenance, their flexibility had some limitations. By today's standards some would not meet the requirements of the Gas Regulations as they did not shut down in a fail-safe position.

The most important controls from the viewpoint of the consumer are those which provide means of varying the temperature of the appliance and the periods when it is on or off. Controls must also be safe in use, and before approval they are examined and tested very thoroughly. It should be understood that although the controls dealt with in the following text are treated in isolation, it will be found that on most modern equipment they are fitted into what is commonly called a multifunctional, or composite control. If a defect occurs in a control, it can sometimes be repaired by

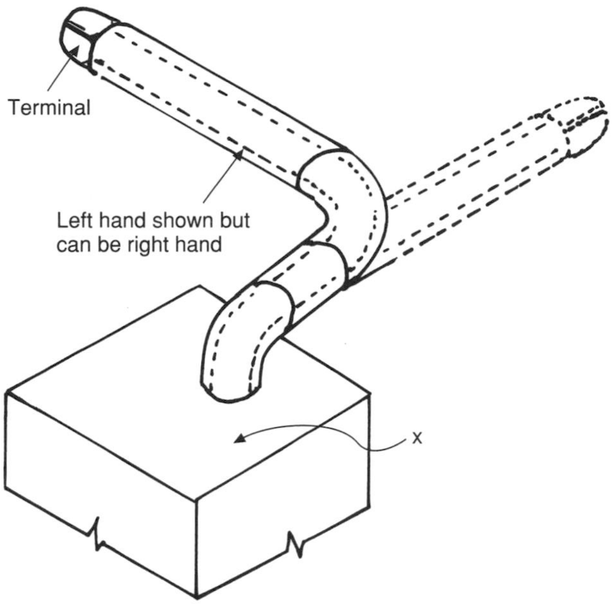

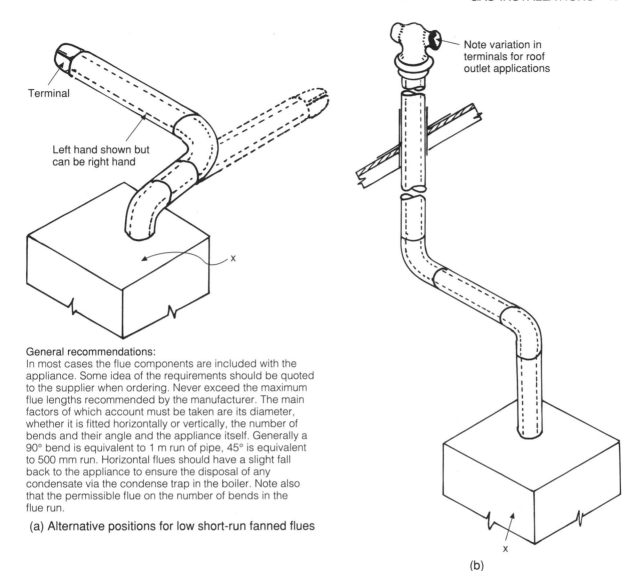

General recommendations:
In most cases the flue components are included with the appliance. Some idea of the requirements should be quoted to the supplier when ordering. Never exceed the maximum flue lengths recommended by the manufacturer. The main factors of which account must be taken are its diameter, whether it is fitted horizontally or vertically, the number of bends and their angle and the appliance itself. Generally a 90° bend is equivalent to 1 m run of pipe, 45° is equivalent to 500 mm run. Horizontal flues should have a slight fall back to the appliance to ensure the disposal of any condensate via the condense trap in the boiler. Note also that the permissible flue on the number of bends in the flue run.

(a) Alternative positions for low short-run fanned flues

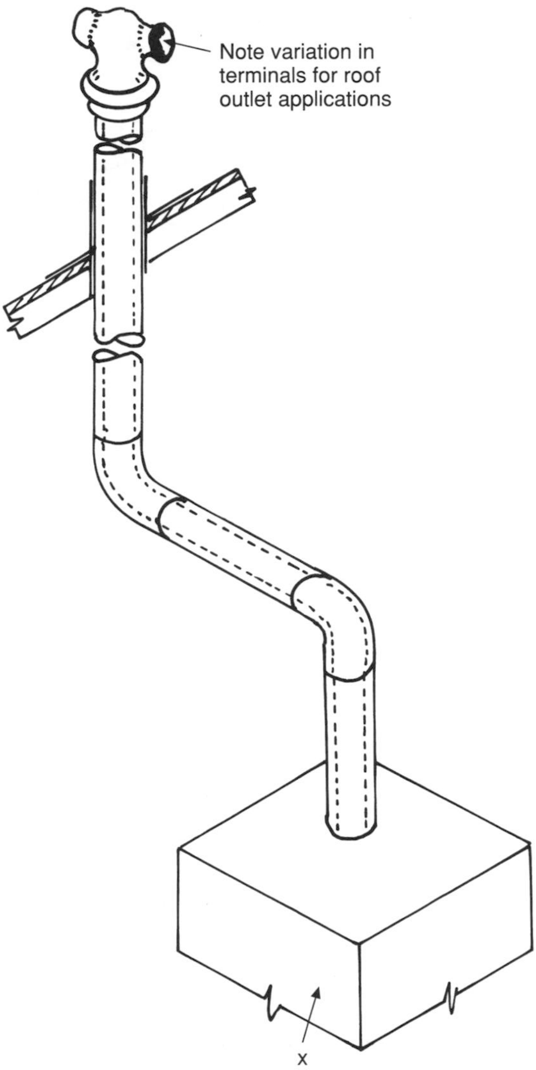


(b)

**Fig. 2.31** Twin-wall piped flue for fanned draught appliances (x indicates boiler).

someone who is qualified to do so, but it is often cheaper to replace it and many specialist suppliers offer replacements on a part-exchange basis. Should it be necessary to order new parts, do make sure the reference number of the control, and the appliance to which it is fitted, are quoted to the supplier as there are many variations.

*Main burner controls*

The controls employed will vary according to the appliance, but the following types are common to water heating equipment. Instantaneous heaters rely on the pressure of the water to lift the main gas valve, which ensures that the heater is charged with water before the main burner will fire. This type of valve is dealt with in Chapter 4.

*Solenoid valves*

As explained later in the section on thermoelectric valves, a solenoid, when energised, acts in the same way as an electromagnet and lifts a valve off its seating, permitting the passage of gas.

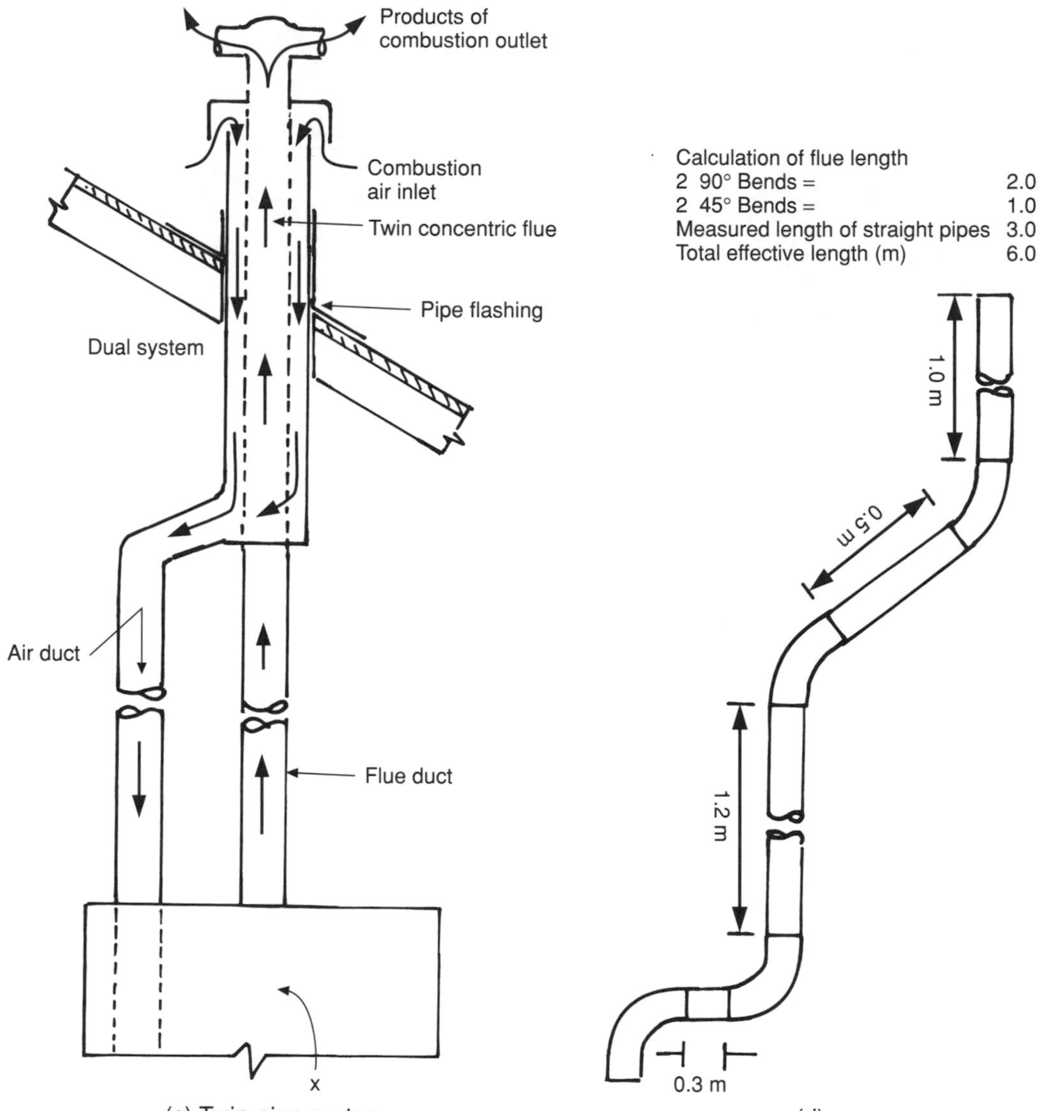


(c) Twin-pipe system (d)

**Fig. 2.31** (*cont'd*)

Most solenoid valves are operated by 240 V so it is very important that any electrical work conforms with the wiring regulations. Figure 2.34 shows a solenoid valve illustrating its main features.

*Thermostats*

The function of a thermostat is to cause the gas supply to the main burner to shut down when the appliance reaches a preset temperature. One exception to this is the older type cooker thermostat which allows a small quantity of gas to bypass the valve. Cookers have no permanent pilot, and should the thermostat shut down the gas completely, the oven burner would not be relit when the thermostat calls for heat.

One of the most common is the rod-type thermostat which has been fitted on cooking

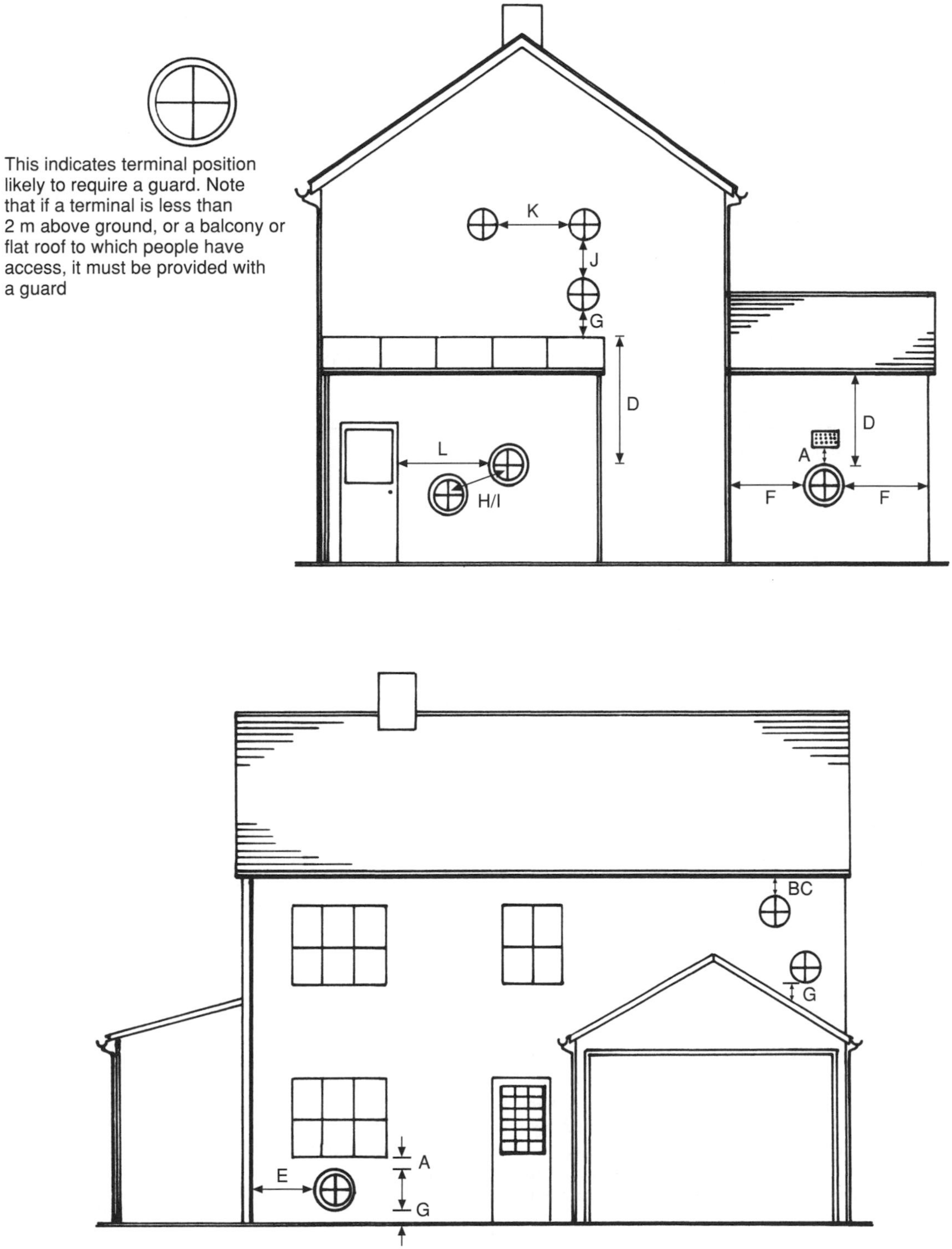
This indicates terminal position likely to require a guard. Note that if a terminal is less than 2 m above ground, or a balcony or flat roof to which people have access, it must be provided with a guard
K
J
G
D
L
H/I
D
A
F
F
BC
G
A
E
G

**Fig. 2.32** Specifications for room-sealed appliance terminals (ref. BS 5440 Part 1). Note that not all the terminal positions shown here are suitable for condensing appliances. Compliance with the manufacturer's instructions relating to flues is absolutely essential.

**Table 2.7** Suitable positions for room-sealed appliance terminals. Figures are permissible distances in mm.

| | *Terminal position (see Fig. 2.32)* | *Natural draught type* | *Fanned draught type* |
|---|---|---|---|
| A | Directly below an openable window or other opening, e.g. an air brick | 300 | 300 |
| B | Below gutters, soil pipes or drain pipes | 300 | 75 |
| C | Below eaves | 300 | 200 |
| D | Below balconies or car port roofs | 600 | 200* |
| E | From vertical drain pipes and soil pipes | 75 | 75 |
| F | From internal or external corners | 600 | 300 |
| G | Above ground, roof or balcony level | 300 | 300 |
| H | From a surface facing a terminal | 600 | 1,200 |
| I | From a terminal facing a terminal | 600 | 2,200 |
| J | Vertically from a terminal on the same wall | 1,500 | 1,500 |
| K | Horizontally from a terminal on the same wall | 300 | 300 |
| L | For an opening in a car port (e.g. door, window) into a dwelling | 1,200 | 1,200* |

* Not recommended for condensing boilers.

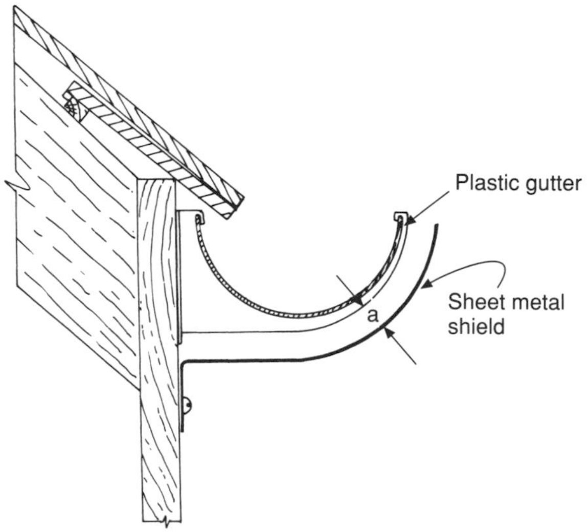


**Fig. 2.33** Protection of plastic gutters from hot products of combustion (ref. BS 5440 Part 1). Where a natural draught balanced flue terminal is fitted less than 1 m below a plastic gutter or less than 0.5 m below a painted surface, a sheet metal shield at least 1 m in length must be fitted as protection against the hot combustion products. The requirements for fanned flues may differ and the manufacturer's instructions must be complied with.

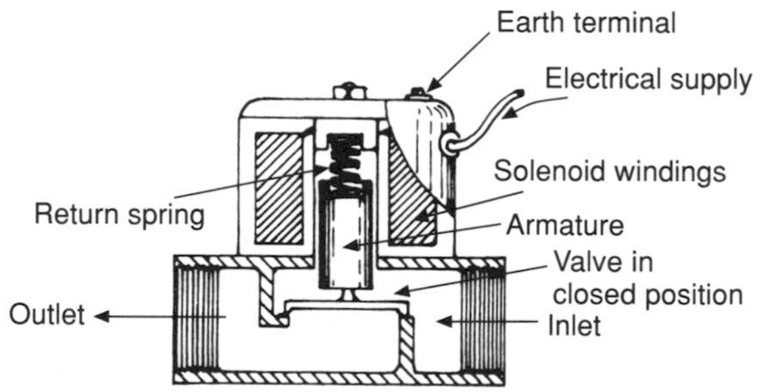


**Fig. 2.34** Solenoid (magnetic) valve. This illustrates the basic principle of magnetic valves. When the solenoid is energised it overcomes the pressure of the return spring and pulls the valve off its seating.

appliances for many years. The principle employed is the differential expansion rate of invar steel and brass, invar steel having a relatively low rate of expansion, while that of brass is comparatively high. Figure 2.35(a) shows the action of this simple valve. It will quickly be seen that it will close at one temperature only, the valve having no provision for adjustment.

In practice these thermostats are made as shown in Fig. 2.35(b). Their use for domestic appliances is very limited as most now have electrical control. They are still commonly used for gas cookers in large commercial establishments.

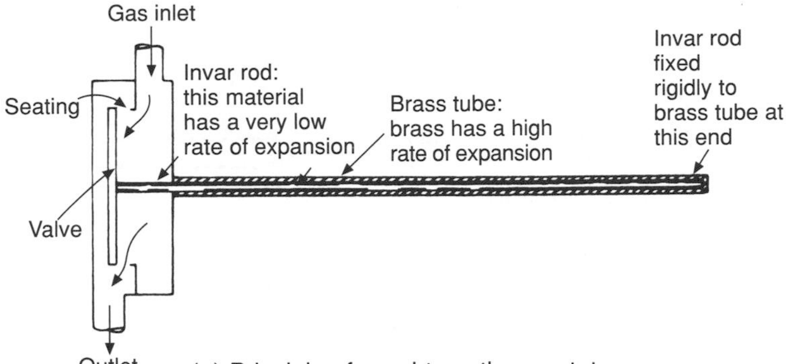
Gas inlet
Seating
Invar rod:
this material
has a very low
rate of expansion
Brass tube:
brass has a high
rate of expansion
Invar rod
fixed
rigidly to
brass tube at
this end
Valve
Outlet

(a) Principle of a rod-type thermostat

When the brass tube is heated it expands at a higher rate than the invar rod, thus pulling the valve on to the seating and closing off the gas supply. It will be seen, however, that no adjustment of temperature is possible.

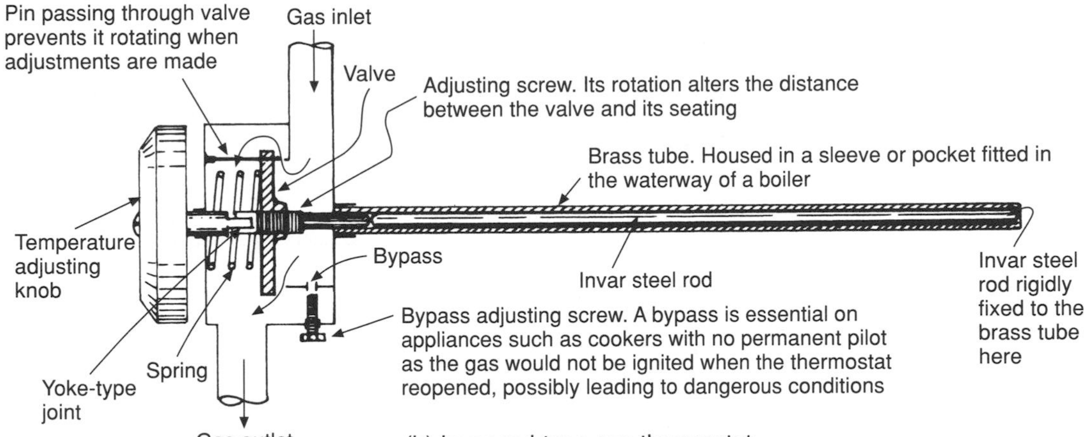
Pin passing through valve
prevents it rotating when
adjustments are made
Gas inlet
Valve
Adjusting screw. Its rotation alters the distance
between the valve and its seating
Brass tube. Housed in a sleeve or pocket fitted in
the waterway of a boiler
Temperature
adjusting
knob
Bypass
Invar steel rod
Invar steel
rod rigidly
fixed to the
brass tube
here
Bypass adjusting screw. A bypass is essential on
appliances such as cookers with no permanent pilot
as the gas would not be ignited when the thermostat
reopened, possibly leading to dangerous conditions
Yoke-type
joint
Spring
Gas outlet

(b) Invar rod-type gas thermostat

Here, the thermostat valve is fitted on a threaded spindle which is not actually joined to the invar rod. When the brass tube expands, carrying with it the invar rod, pressure exerted by the spring pushes the valve on to its seating; temperature variation is achieved by turning the adjusting knob which rotates the threaded spindle, moving the valve closer to or further away from its seating. The further away it is, the higher will be the temperature before the brass tube expands sufficiently for it to close.

**Fig. 2.35** Rod-type thermostats.

*Fluid expansion thermostats* These employ a sealed bulb or sensor and a bellows filled with a heat-sensitive fluid. The sensor is situated in a pocket in the appliance water ways, and on a rise in water temperature the fluid expands, extending the bellows. The illustration in Fig. 2.36(a) operates directly on a gas valve, causing it to open or close, depending on the temperature. The same principle can be employed to operate a microswitch, enabling it to be used for gas appliances controlled by electricity as shown in Fig. 2.36(b).

### *Gas governors*

As the name implies, these are devices for controlling the pressure and flow of gas to an installation or an individual appliance.

*Service governor* Gas pressure in the main distribution system is normally about 2 bar. This is reduced by the service governor fitted to the inlet side of the meter to give a working pressure at the consumer's appliance of 20 mbar. In the event of malfunction causing high-pressure gas to enter the

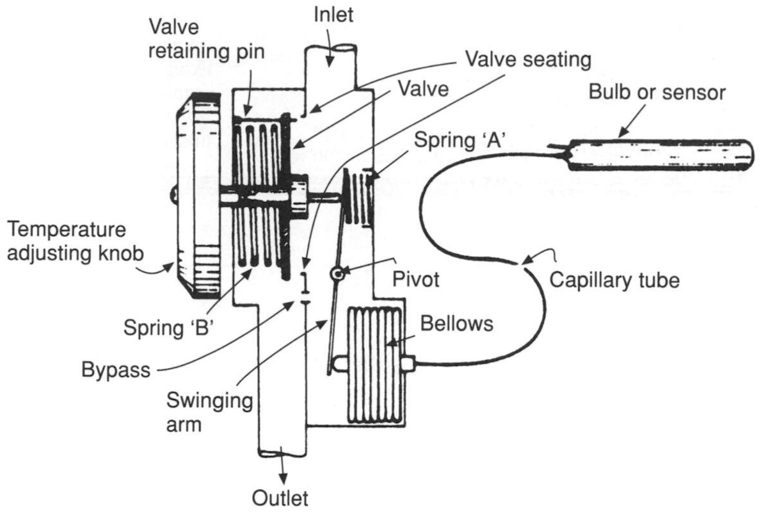


(a) Fluid expansion type thermostat – non-electric

When the sensor is subjected to heat, the heat-sensitive fluid it contains expands, causing the bellows to distend and apply pressure via the swinging arm on spring 'A', simultaneously allowing spring 'B' to close the valve. When the sensor cools and the fluid it contains contracts, the bellows returns to its normal position and pressure on spring A overcomes that of B, opening the valve.

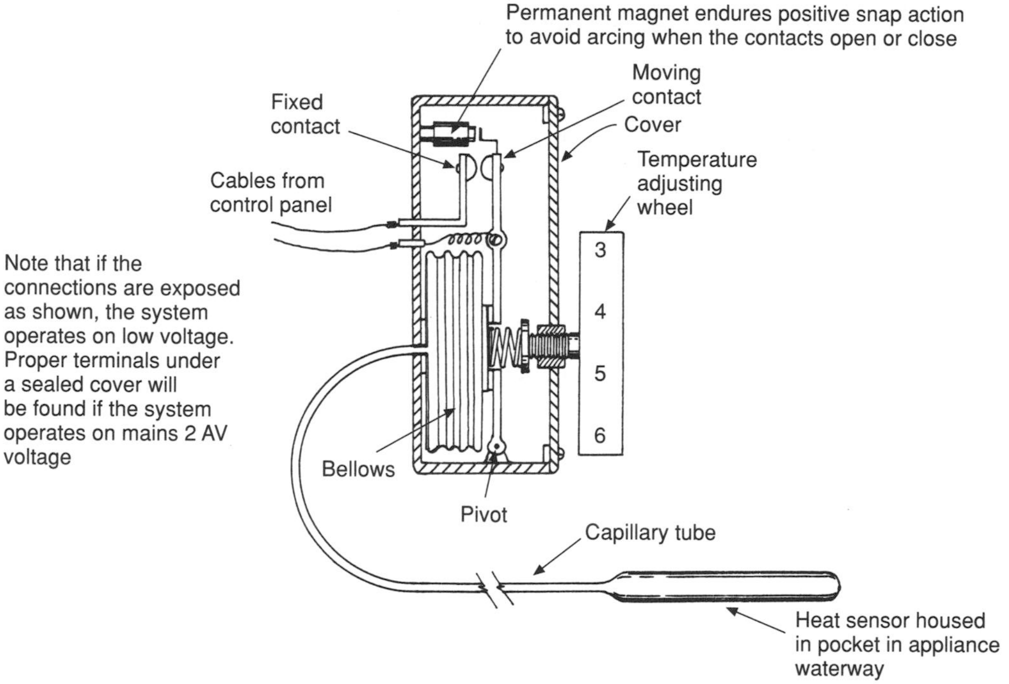


(b) Typical fluid-operated thermostat controlling a microswitch

Diagrammatic illustration. Operation: When the thermostat is at the correct temperature the contacts will be closed, but as the watch temperature increases the heat-sensitive fluid in the sensor expands causing the bellows to push the moving contact away from the fixed contact which has the effect of causing the main gas supply to the burner to be closed.

**Fig. 2.36** Fluid expansion thermostats.

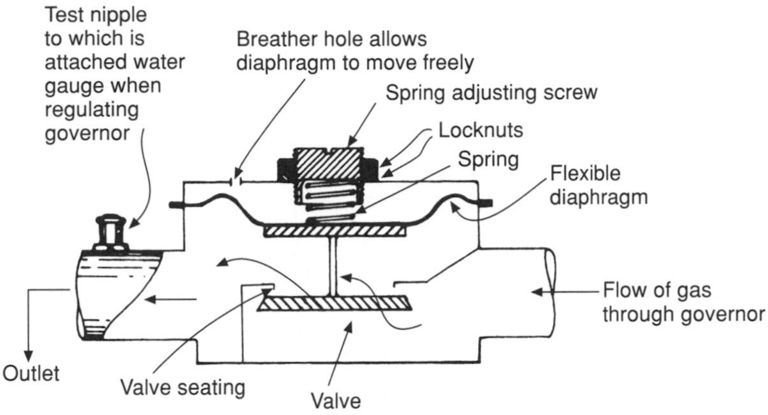


**Fig. 2.37** Simple gas appliance governor. A governor is designed to even out any variations of pressure that might occur in a gas installation, and to ensure a constant pressure at the appliance to which it is fitted.

gas services in the building with possible dangerous effects, it will automatically shut off the gas supply and it can be manually reset when the cause of the problem has been rectified. **Adjustments and repair of these regulators must not be undertaken by an installer** and any defects must be reported to the gas transporter or supplier.

*Constant pressure governors* Individual appliance governors are not now considered necessary except for gas boilers, enabling fine adjustments to be made to ensure maximum economy. Constant pressure governors are provided for this purpose. A simple governor of this type is illustrated in Fig. 2.37. It is really a variable restrictor in the gas supply. An increase of the inlet pressure will exert a greater force on the flexible diaphragm causing it to lift, carrying with it the valve. This has the effect of reducing the aperture through which the gas can pass to the outlet, causing a pressure drop. A lowering of the inlet pressure will result in less pressure on the diaphragm, thus allowing the valve to drop and thereby increasing the size of the outlet aperture. The movement of the valve enables a balance to be maintained between the inlet and outlet gas pressures.

Pressure can be increased by compressing the spring with the adjusting screw. The outlet pressure should be checked with a U gauge at the test nipple on the outlet side of the governor, and if it is found to be insufficient the diaphragm can be loaded by compressing the spring until the pressure recommended by the manufacturer is achieved. It should be noticed that the top of the governor has a small vent orifice which is open to the atmosphere, and if this becomes obstructed the governor will not function properly.

When a gas appliance is fitted it is important that it operates at the pressure recommended by the manufacturer. Most governors on domestic boilers are now integral with the multifunctional control. The procedures for setting gas pressures are described on page 27.

*Multifunctional controls*

The method of gas controls employing components as separate units is obsolete for domestic appliances. They are now housed in a unit called a multifunctional control, and are produced by various manufacturers, varying only slightly in detail. Some types employed with modern boilers, i.e. combis, incorporate a modulating valve, which enables the gas rate to meet the varying demands made on the heat-producing equipment. A typical example of the type used for domestic boilers is shown in Fig. 2.38. Their main advantages are the saving of space in boiler compartments; they are also more convenient for the full electrical control systems employed on modern appliances. A further advantage is that user controls can be conveniently situated near the front panel of the boiler. The flow of gas through the control should be carefully noted,

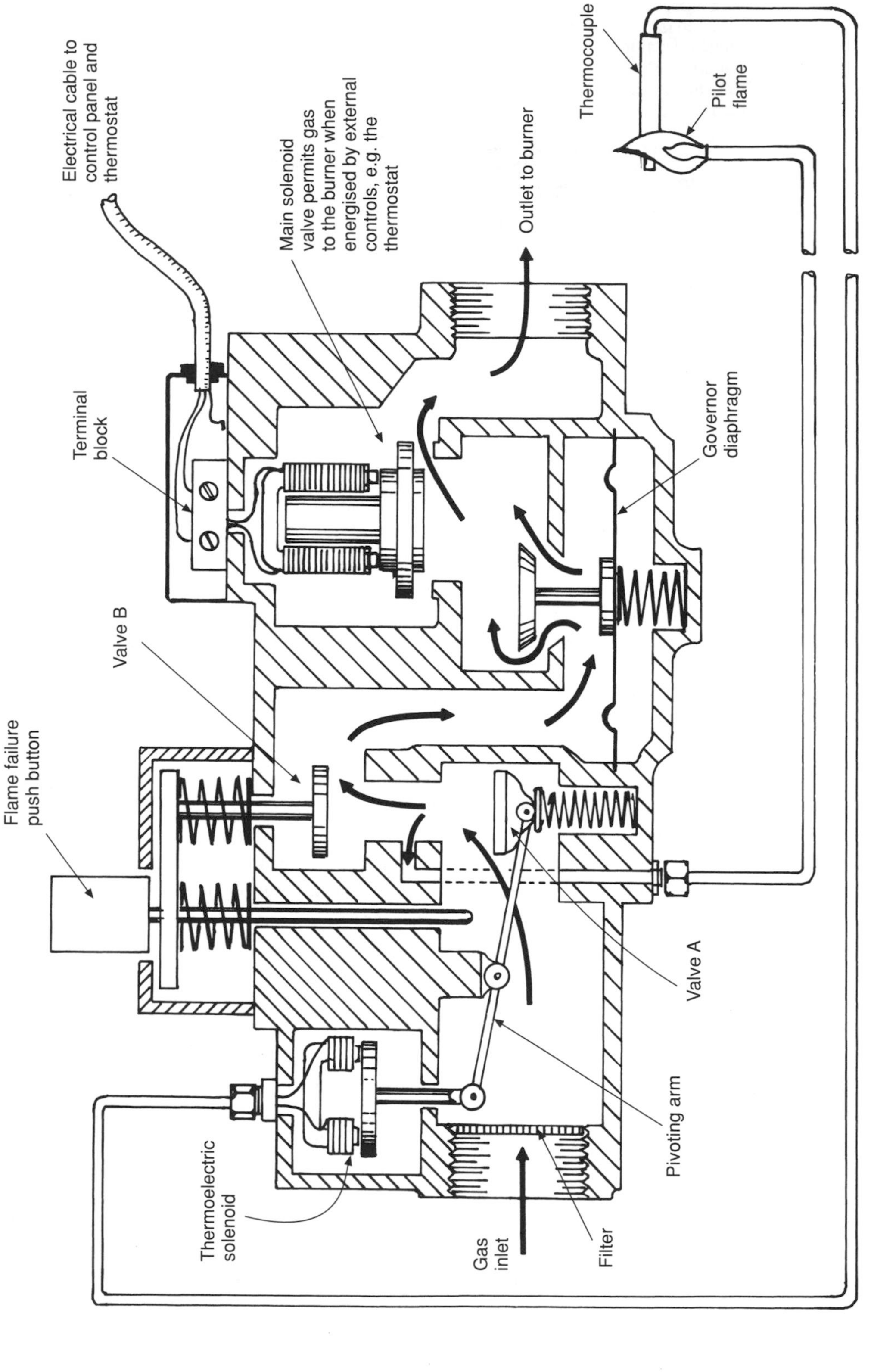


**Fig. 2.38** Simplified diagram of a typical multifunctional gas control.

and it will be seen how the individual components previously described are now combined into one integrated unit.

## Ignition devices

Some gas appliances employ a pilot flame, its function being to ignite the gas in the main burner when the controls are calling for heat. Others, such as gas fires and cookers, use a system whereby the main burner is lit directly by a spark or filament coil. Some pilot flames are permanent, irrespective of whether the main burner is on or off. Non-permanent pilots only ignite when a thermostat calls for heat, and some economy may be achieved using this system.

### *Manual ignition*

This system is mainly confined to older appliances where the pilot jet or main burner is easily accessible and can be lit with a match or taper. Manual ignition applies mainly to older types of gas appliances.

### *Spark ignition*

Most readers will be aware that in an internal combustion engine the fuel/air mixture is fired by the sparking plug. A similar arrangement is used to light either the pilot jet, or, in some cases, the main gas burner, by means of a spark caused by a high-voltage electric current arcing across a gap. The two main methods employed are (a) mains spark, or (b) piezoelectric ignition, the latter being illustrated in Fig. 2.39. The crystals are made of lead zirconate–titanate, which, when subjected to pressure, produce an electromotive force of approximately 6,000 V, which is transmitted via the metal pressure pad to the spark electrode. The illustration shows both a cam-operated ignitor, commonly employed with gas fires, and an impact type which works on a similar principle. Pressure, in this case, is applied by a blow from the hammer, which incorporates a trip mechanism and is activated by the operating knob.

The spark generator illustrated in Fig. 2.40, operates on the main electricity supply. Unlike the piezoelectric system, where the ignitor is operated manually and used with an appliance having a permanent pilot, spark generation is used with those employing non-permanent pilots. They are automatically operated by the control system when a signal from the thermostat indicates the appliance is required to fire. Some economy may be achieved using this system, as the pilot is only alight while the burner is firing. With all spark ignition devices one or two points should be noted. The electrodes must be clean and the gap between them must be maintained to the manufacturer's recommendations as the spark will not bridge it if it is too wide. The high-tension leads must be well insulated and preferably not be in contact with metal parts on the boiler if short-circuiting is to be avoided. All earth wires must be effectively connected to the appliance. In the event of spark failure these last points should be checked prior to looking further.

## Flame failure devices

The object of a 'fail-safe' device is to prevent gas reaching the main burner until a pilot flame has been established. If, for instance, the main burner is 'gassed' and some delay takes place before ignition, the ratio of gas to air on ignition could result in an explosion. The following devices are those most commonly found in domestic water heaters and boilers.

### *The bimetallic strip*

This device has been used on gas water heaters and boilers since they were first produced and has now been superseded by more modern controls. The main purpose for retaining the illustrations is to demonstrate a principle which has many other applications in the gas and electrical industry. See Fig. 2.41.

### *Thermoelectric valve*

This valve has been used on gas boilers for many years and is now commonly employed in many other gas appliances such as cookers and water heaters. As the name implies, it is an electrically operated valve, electricity being generated by the hot contact of the thermocouple when the pilot flame is ignited.

Figure 2.42 shows diagrammatically the principle upon which this valve works. Two

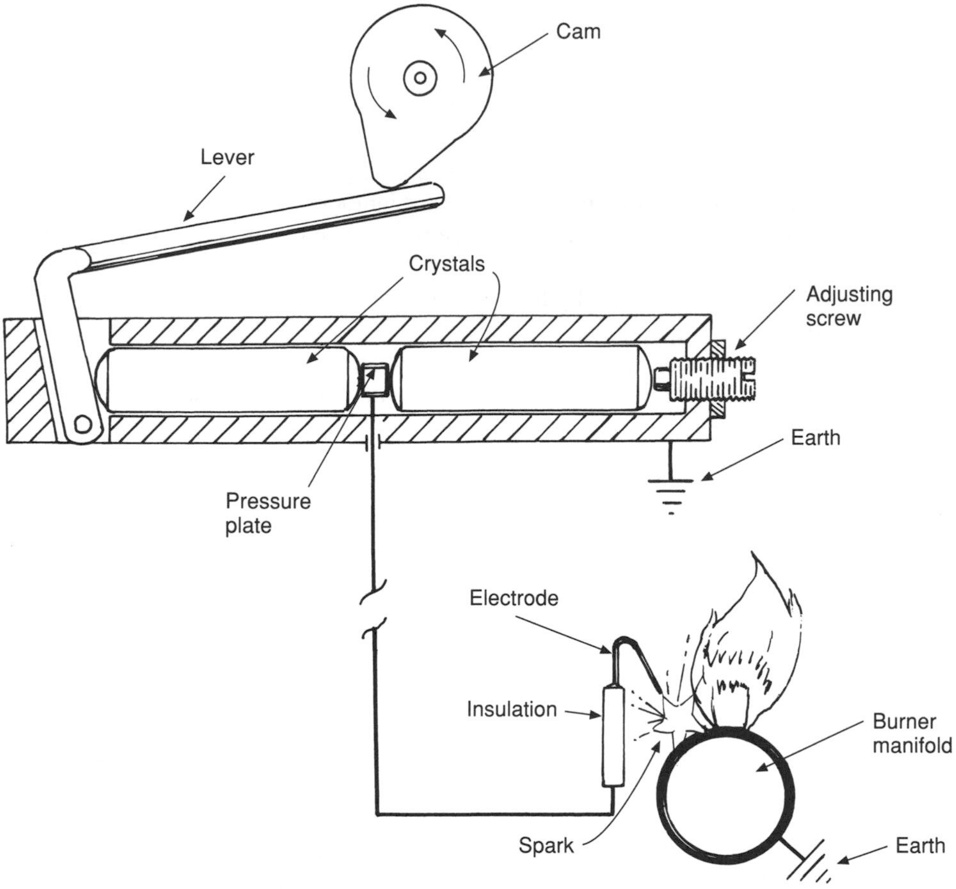


(a) The cam is operated when the gas control knob is rotated, simultaneously turning on the gas and exerting pressure on the crystals. This causes a voltage build-up causing a spark to arc across the gap between the electrode and burner igniting the gas. This arrangement is used to light the main burner directly and is commonly used in gas fires

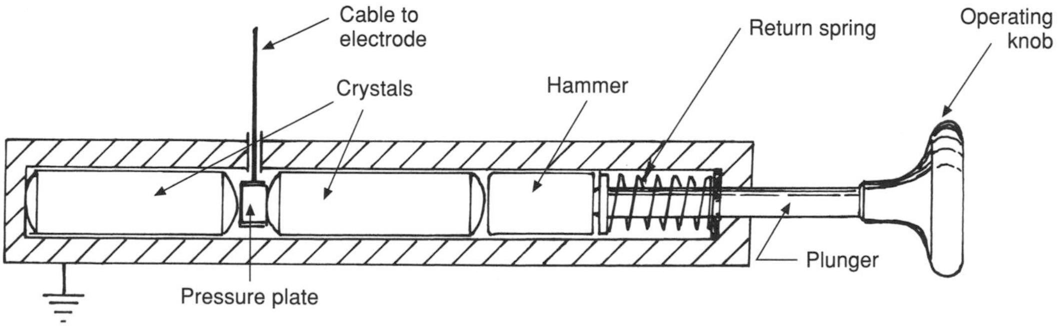


(b) Plunger type piezoelectric ignition. Pressure on the plunger exerts pressure on the hammer, causes it to strike a blow on the crystals which, as with (a), produces a spark

**Fig. 2.39** Piezoelectric ignition.

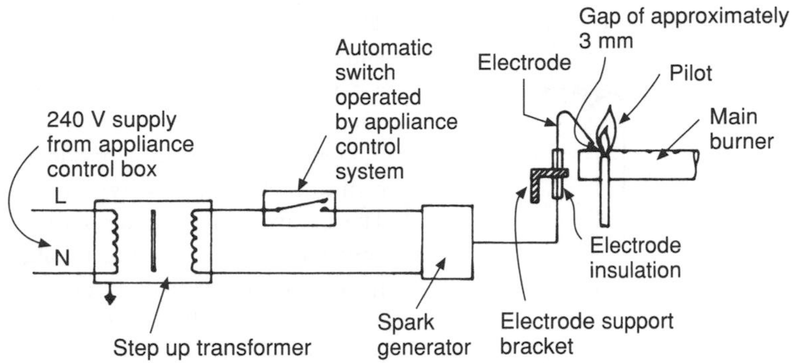
Gap of approximately 3 mm
Automatic switch operated by appliance control system
Electrode
Pilot
Main burner
240 V supply from appliance control box
L
N
Electrode insulation
Step up transformer
Spark generator
Electrode support bracket

**Fig. 2.40** Mains spark ignition. This system of ignition is usually employed with appliances having non-permanent pilots. Its action is normally fully automatic, being part of the control system of the appliance.

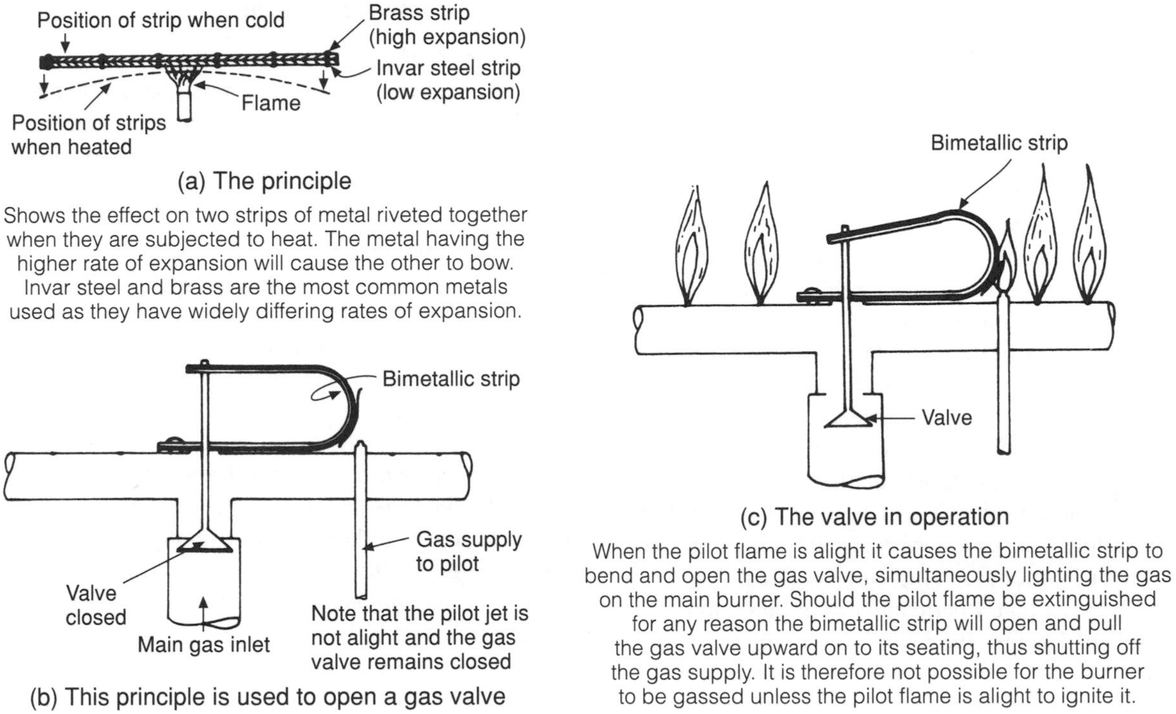
Position of strip when cold
Brass strip (high expansion)
Invar steel strip (low expansion)
Flame
Position of strips when heated
(a) The principle
Shows the effect on two strips of metal riveted together when they are subjected to heat. The metal having the higher rate of expansion will cause the other to bow. Invar steel and brass are the most common metals used as they have widely differing rates of expansion.
Bimetallic strip
Gas supply to pilot
Valve closed
Main gas inlet
Note that the pilot jet is not alight and the gas valve remains closed
(b) This principle is used to open a gas valve
Bimetallic strip
Valve
(c) The valve in operation
When the pilot flame is alight it causes the bimetallic strip to bend and open the gas valve, simultaneously lighting the gas on the main burner. Should the pilot flame be extinguished for any reason the bimetallic strip will open and pull the gas valve upward on to its seating, thus shutting off the gas supply. It is therefore not possible for the burner to be gassed unless the pilot flame is alight to ignite it.

**Fig. 2.41** Bimetallic strip flame failure device.

wires, one made of iron and the other of constantan, are joined together at one end and heated. If the free ends of the wires are connected to a galvanometer it will show that a small flow of electricity is generated due to the differential movement of the molecules of the two metals when subjected to heat. The thermoelectric valve utilises this small current of electricity to energise a solenoid in the thermoelectric valve to hold the gas valve open.

Figure 2.43 illustrates diagrammatically a thermoelectric valve. Figure 2.43(a) shows the valve in the closed position and it will be seen that gas cannot pass to either the main burner or the pilot. Figure 2.43(b) shows the position of the valves when the reset button is depressed. This has the

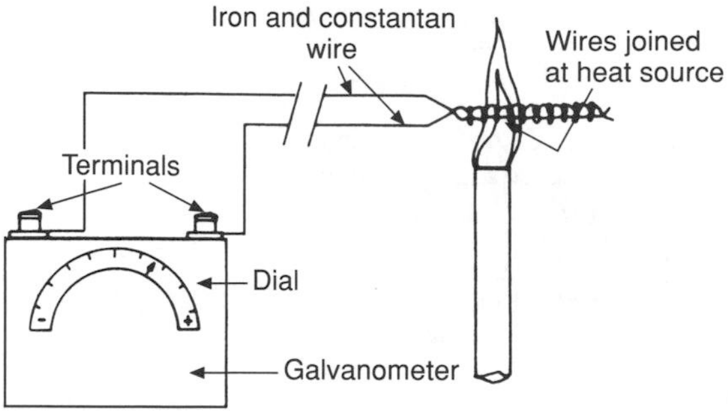


**Fig. 2.42** Illustration showing the principle of a thermoelectric valve. When heat is applied at the junction of the two wires an electromotive force is produced and is registered on the galvanometer.

effect of allowing gas to be admitted to the pilot jet while maintaining the outlet valve to the main gas burner in the closed position. At this stage it is possible to light the pilot flame, which impinges on the thermocouple, causing the generation of an electromotive force. This energises the solenoid and holds the main gas valve open when the reset button is released. The release of the reset button, shown in Fig. 2.43(c), due to the action of the integral springs, allows the lower valve to open and admit gas to the main burner. This will remain open until, for some reason, the pilot flame is extinguished, causing the solenoid to be de-energised and the main valve to be pushed back on to its seating, closing off supplies to the main burner and the pilot. It will be seen that, should the gas supply fail for any reason, both the pilot and the main burner are completely isolated. It will also be seen that a permanent pilot flame must be established before the main burner will function.

Electricity is conducted from the thermocouple to the solenoid via a mineral-insulated lead which looks like a small-diameter copper pipe. A section through the lead is shown in Fig. 2.44.

The cause of defects is the subject of more advanced study, but the most common faults affecting thermoelectric valves are:

(a) Pilot jet partly obstructed, preventing sufficient heat from reaching the thermocouple.
(b) Loose connection of the thermocouple lead to the valve.
(c) Damage to the lead due to careless handling.

*Mercury vapour flame safety valve*

Unlike the thermoelectric valve which requires a permanent pilot flame, the mercury vapour valve can be used where the pilot flame is ignited only when the appliance is in use, resulting in small savings on gas consumption. These valves are quicker acting than thermoelectric valves but more prone to breakdown. If the valve fails to open it is usually due to a damaged bellows unit which has allowed the vapour to escape, and a replacement valve will be required. These valves are now mainly used on gas cookers. Figure 2.45 illustrates the working principles of these valves.

### Flame conduction and rectification

This is an electronic flame protection device that was designed originally for commercial appliances, but due to the use of microelectronics on modern appliances, the components used with this type of flame-failure equipment can now be produced small enough to be used with domestic appliances such as cookers and boilers.

The basic principle of flame rectification is as shown in Fig. 2.46, based on the fact that when a substance burns, in this case gas, a chemical reaction or change is taking place. The flame we see when this happens also produces minute electrically charged particles called ions which can be made to pass between two conductors through the flame. In effect the flame is acting as a conductor to the flow of electricity. If a flame is not established, obviously there will be no flow of electrons and the control system of the appliance will not pass gas. A basic knowledge of electrical principles is required to fully understand the process, but the following information will be helpful. The flow of electrons through the flame produces an alternating current which means the flow is constantly reversing backwards and forwards, and to be effective for the purposes being considered, it must be changed or 'rectified' to 'direct' current. This type of current flows in one direction only. Rectification is achieved, in this case, by the electrodes which must be of a suitable type. The direct current thus produced is amplified and operates a relay which in turn operates the electrical controls on the main gas supply. An amplifier is an electronic device which makes it possible for a small current to operate a relay. A relay is basically a solenoid or electromagnet which enables one source of electricity to control another, in this case the solenoid valve controlling the gas inlet to the appliance.

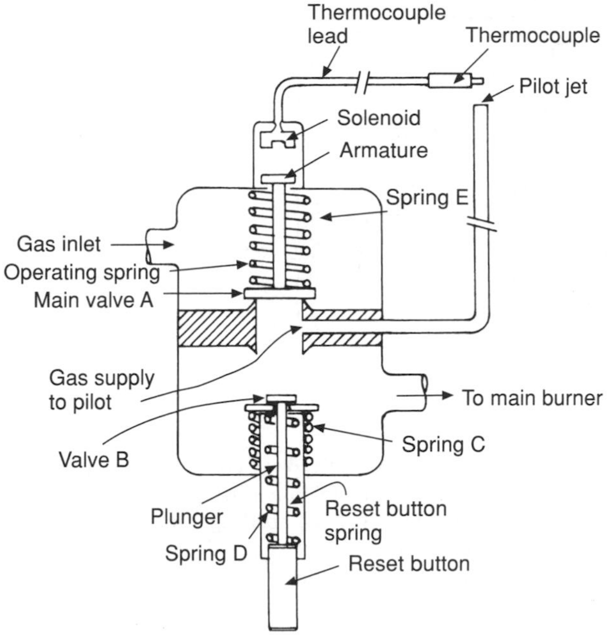


(a) Valve closed

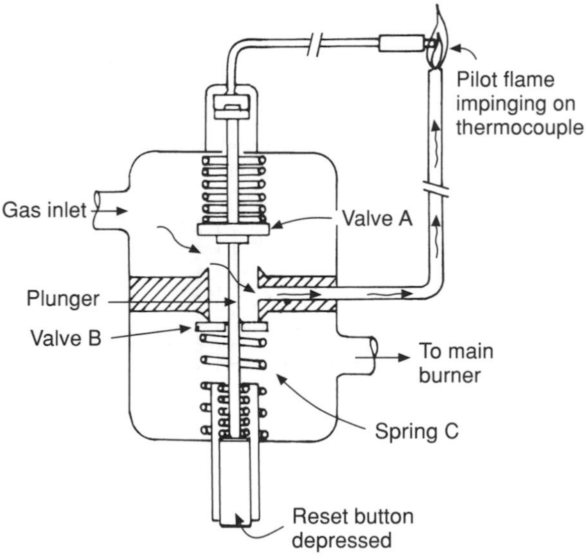


(b) Gas to pilot jet only

By depressing the reset button the plunger, passing through valve B, pushes valve A off its seating, permitting the passage of gas to the pilot jet which is ignited and heats the thermocouple. As valve B is retained in the closed position due to pressure from spring C, no gas can yet pass to the main burner.

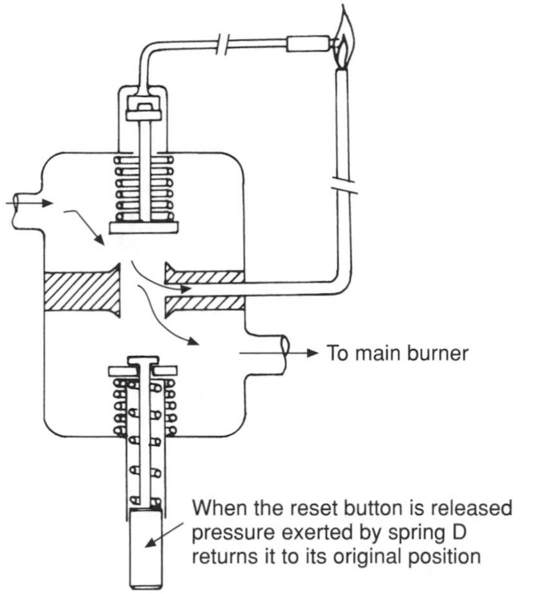


(c) Main gas valve open

After approximately 30 seconds sufficient electrical energy is developed by the action of the pilot flame on the thermocouple to energise the solenoid which holds the main gas valve A open. Both the reset button and valve B are now in their original positions due to the release of the reset button and pressure from spring D. Gas can pass to the main burner. If the pilot is extinguished the thermocouple will cease to produce electrical energy and the main valve will close due to pressure exerted by the operating spring E.

**Fig. 2.43** Operation of a thermoelectric valve.

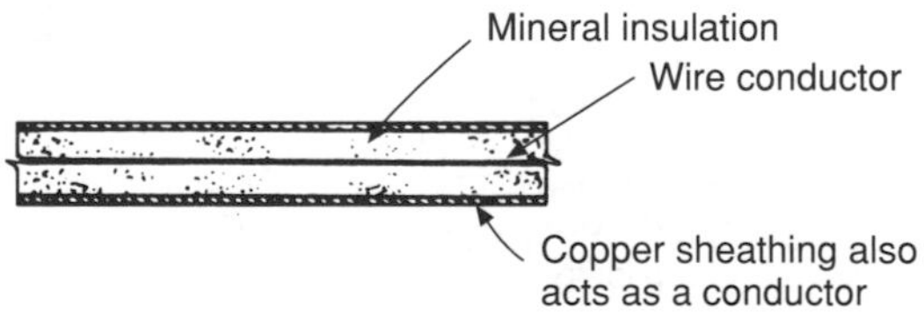


**Fig. 2.44** Section through thermocouple lead. This material is used extensively as an electrical conductor in situations where other conductors are unsuitable due to the relatively high temperatures, e.g. in boiler or water heater casings. When handling avoid very sharp bends and do not kink as this can result in short-circuiting the conductors.

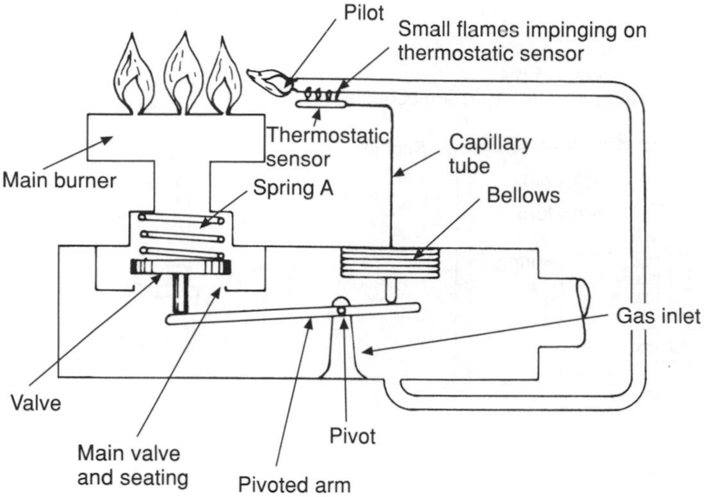


**Fig. 2.45** Mercury vapour flame safety device. These valves are used with a different type of control system from that used with a thermoelectric valve, the main gas control being a solenoid operated by an electrical thermostat. Upon a supply of gas reaching the mercury vapour valve the pilot is lit simultaneously by spark ignition and heats the thermostatic sensor. Expansion of the fluid opens the bellows, pushing down the pivot arm, thus opening the main valve admitting gas to the burner as shown. When the thermostat is satisfied the main solenoid will be de-energised, closing off the gas supply to both the pilot and the burner. On contraction of the bellows, spring A will push the valve back on to its seating. It is impossible for this valve to open unless a pilot flame has first been established.

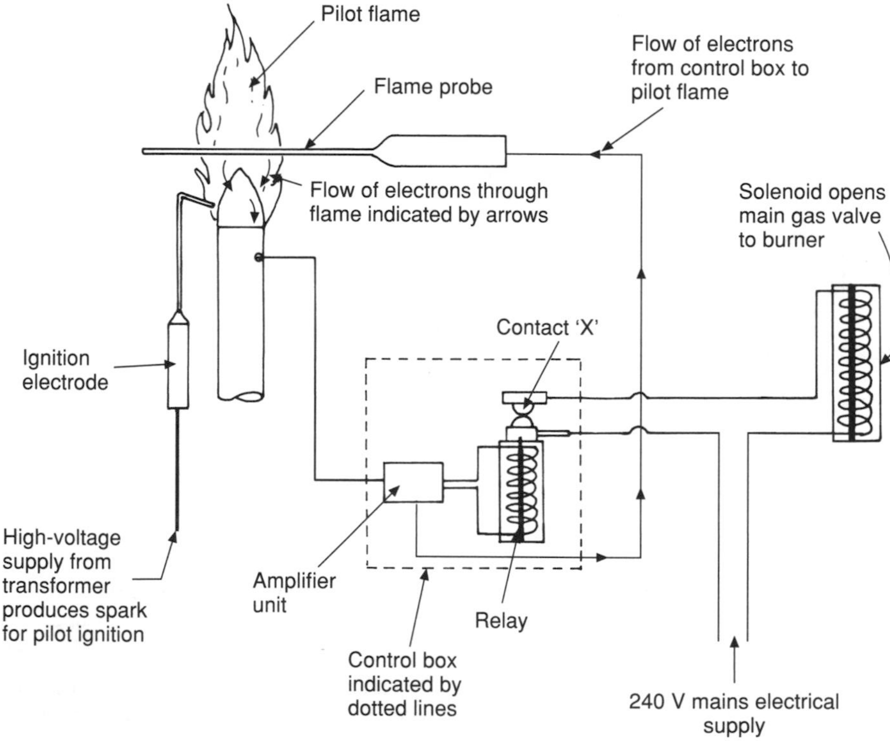


**Fig. 2.46** Flame conduction and rectification fail-safe device. The pilot flame must be established before a flow of electrons can pass through the flame to complete the electrical circuit which energises the relay. This causes contacts 'X' to close simultaneously, completing the circuit to the solenoid which opens the main valve admitting gas to the combustion chamber. Unless a pilot flame is established no gas can be passed through the main gas valve.

*Photo-electric flame failure components*
The use of an 'electronic eye' as a fail-safe device is well established in oil-firing practice and the development of similar burners for gas combustion has led to its use for the same purpose in some types of industrial gas appliances. The electronic eye, more correctly called a photo-electric resistor, looks very similar to an old-fashioned radio valve. It is housed in a suitable casing fitted into the blower tube of the burner so that its light-sensitive face can detect the ultraviolet rays in the flame. Electricity will only flow through the cell when it is exposed to these rays. The sequence of operations is as follows: on initial light-up the gas is ignited by a spark from the electrode which only operates for a limited period. If all is well and a flame is established, the electronic eye will 'see' the ultraviolet rays, permitting the flow of electricity to open the main gas solenoid valve. Should the gas not ignite, no ultraviolet rays will be produced and the electronic eye will be unable to permit the flow of current to the gas valve which will close. This is known as 'lock out' and is indicated visually by the appearance of a red light in the control box of the burner. The burner must then be reset manually, usually by depressing a button on the control box. If a fault exists with the burner or one if its components, it will again lock out and steps must be taken to rectify the fault. Typical examples may be:

(a) Gas turned off.
(b) Electrode gap incorrect.
(c) Breakdown of electrode insulation.
(d) Electronic eye requires cleaning or is not correctly fitted.

Figure 2.47(a) shows the cell and its casing while Fig. 2.47(b) illustrates its position in the blower tube of the gas burner — see also Fig. 6.24.

*Oxygen depletion valve*
This valve is an additional safety device incorporated into many modern gas appliances, especially fires. It is mandatory on open-flued appliances fitted in sleeping areas. It is designed to cut off the gas supply to the appliance if vitiation or lack of air for combustion causes oxygen depletion and the production of carbon monoxide. The pilot flame will be extinguished, deactivating the thermocouple. See Fig. 2.48(a).

*Thermistors*
These are very accurate non-metallic heat-sensing devices that will alter their ability to conduct a flow of electricity when subjected to changes in temperature. The higher the temperature to which they are subjected the greater will be the flow of electricity they will pass. Thermistors are used in many gas modulating controls where the supply of gas to a burner is variable. They are also found in programmable room thermostats and vitiation sensing safety devices as shown in Fig. 2.48(b).

## Gas fires and back boiler units

*General*
For many years now gas boilers, usually combined with a gas fire, have been used to replace the solid fuel appliances installed prior to the introduction of central heating. The production of gas appliances for this purpose has rapidly developed due to their efficiency, and possibly more so with the current trend to produce appliances using gas as a fuel to simulate solid fuel fires. The reader should note that not all these fires are suitable for use with gas back boilers and fall into three basic groups covered by BS 5871 entitled and listed as follows: *The Installation of Gas Fires, Convector Heaters, Fire Back Boilers and Decorative Fuel Effect Gas Appliances.*

Part 1 relates to gas fires, convector heaters and fire/back boilers 1st, 2nd and 3rd family gases. Part 2 concerns inset live fuel effect gas fires of heat input not exceeding 15 kW 2nd and 3rd family gases, and Part 3 decorative fuel effect gas appliances of heat input not exceeding 15 kW 2nd and 3rd family gases.

All this seems rather complicated, but is necessary because of the wide range of differences between the working principles of these fires and the methods of flueing, flue lining and hearth requirements relating to them.

It is important to note that appliance manufacturers' instructions take precedence over any British Standard and such instructions must be

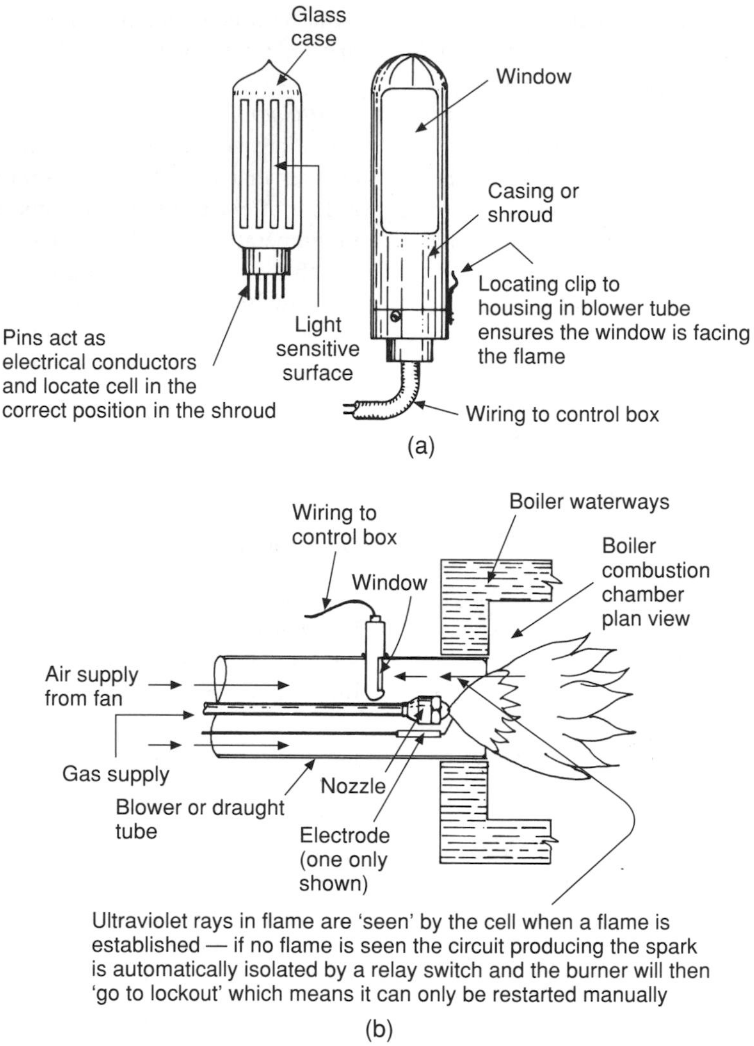
Glass case
Window
Casing or shroud
Locating clip to housing in blower tube ensures the window is facing the flame
Light sensitive surface
Pins act as electrical conductors and locate cell in the correct position in the shroud
Wiring to control box
(a)
Wiring to control box
Boiler waterways
Boiler combustion chamber plan view
Window
Air supply from fan
Gas supply
Nozzle
Blower or draught tube
Electrode (one only shown)
Ultraviolet rays in flame are 'seen' by the cell when a flame is established — if no flame is seen the circuit producing the spark is automatically isolated by a relay switch and the burner will then 'go to lockout' which means it can only be restarted manually
(b)

**Fig. 2.47** Photo-electric fail-safe device.

adhered to. To be safe with gas there are no short cuts! While this is true of all gas appliances it is especially true of gas fires, as it is very easy to go to sleep in front of the fire, and possibly never wake up if the fire is improperly fitted.

Although the installation requirements of modern gas fires vary widely, depending on their type, there are some features which are common to all and careful note should be made of the following. Before any such appliance is fitted the instructions given previously relating to the use of existing flues should be applied. Briefly they must be clean, have no obstruction and sufficient updraught. Any ventilation requirements and the suitability of existing hearths must be investigated. This is important as the recommendations vary widely depending on the type of fire used. Generally speaking, live fuel and decorative fuel fires are subject to the same hearth and flueing requirements as those for solid fuel appliances. This information can be found in the Building Regulations Part J.

*Gas back boilers with combined fires*

These boilers do not comply with the legislation relating to Part L, gas appliance efficiencies. However, should replacement become necessary,

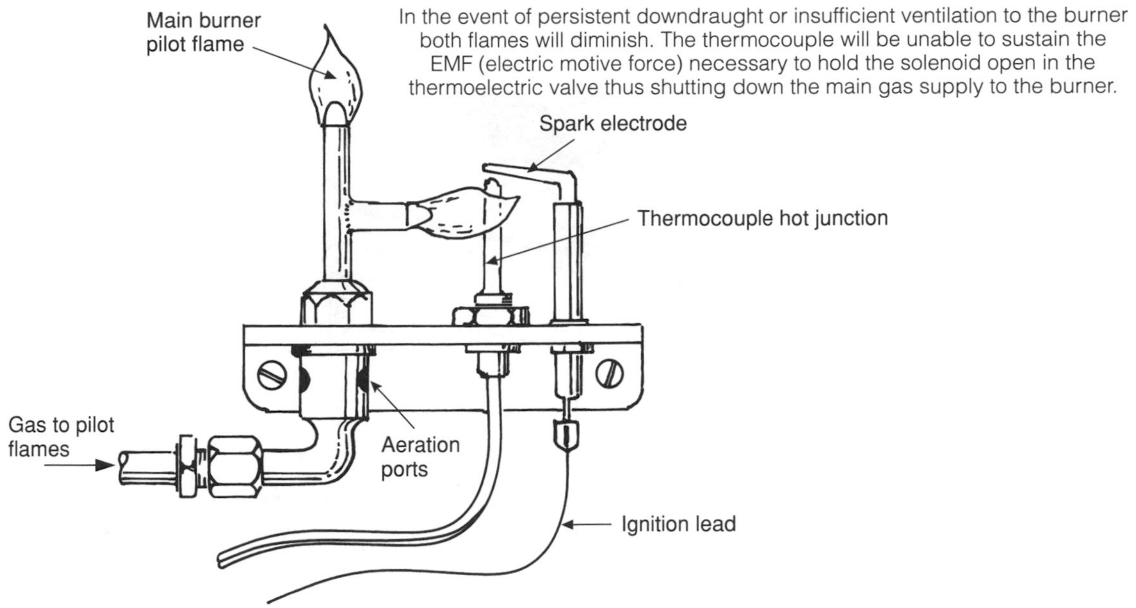
Main burner
pilot flame
In the event of persistent downdraught or insufficient ventilation to the burner both flames will diminish. The thermocouple will be unable to sustain the EMF (electric motive force) necessary to hold the solenoid open in the thermoelectric valve thus shutting down the main gas supply to the burner.
Spark electrode
Thermocouple hot junction
Gas to pilot
flames
Aeration
ports
Ignition lead

(a) Oxygen depletion sensor

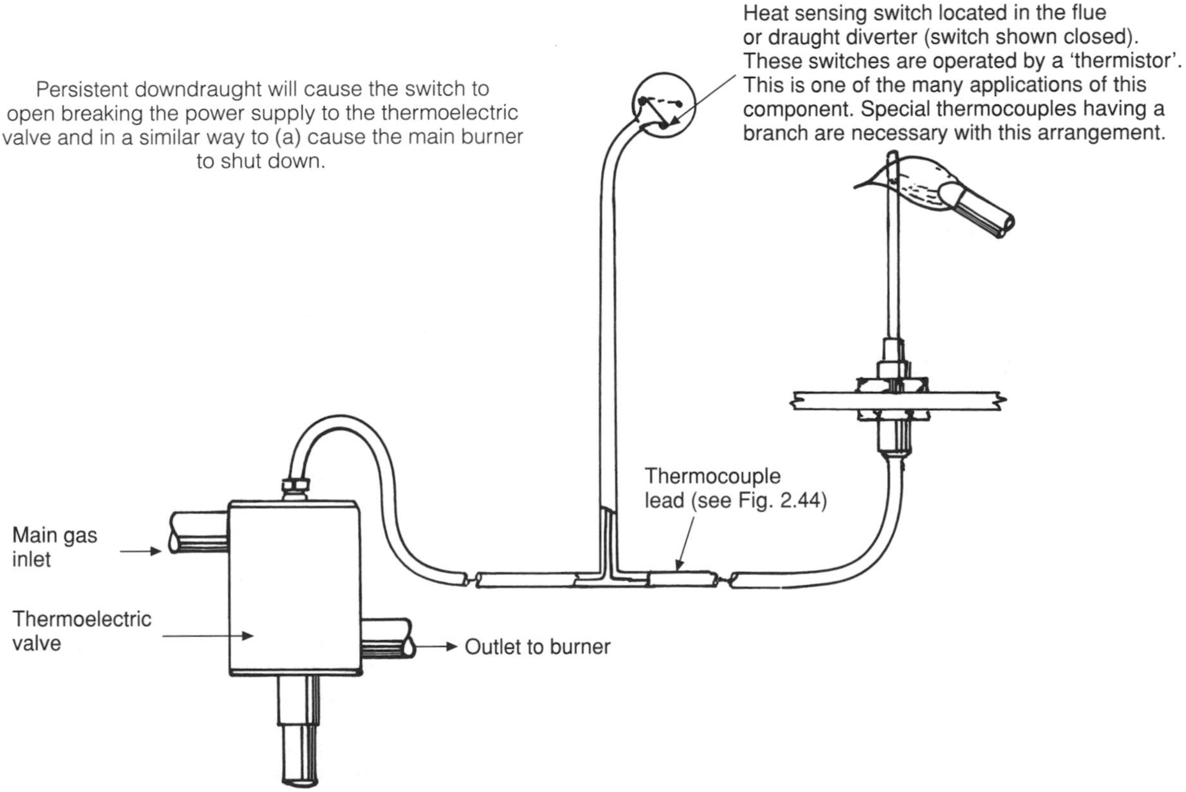
Heat sensing switch located in the flue or draught diverter (switch shown closed). These switches are operated by a 'thermistor'. This is one of the many applications of this component. Special thermocouples having a branch are necessary with this arrangement.
Persistent downdraught will cause the switch to open breaking the power supply to the thermoelectric valve and in a similar way to (a) cause the main burner to shut down.
Thermocouple
lead (see Fig. 2.44)
Main gas
inlet
Thermoelectric
valve
Outlet to burner

(b) Atmospheric sensing device

**Fig. 2.48** Vitiation sensing devices.

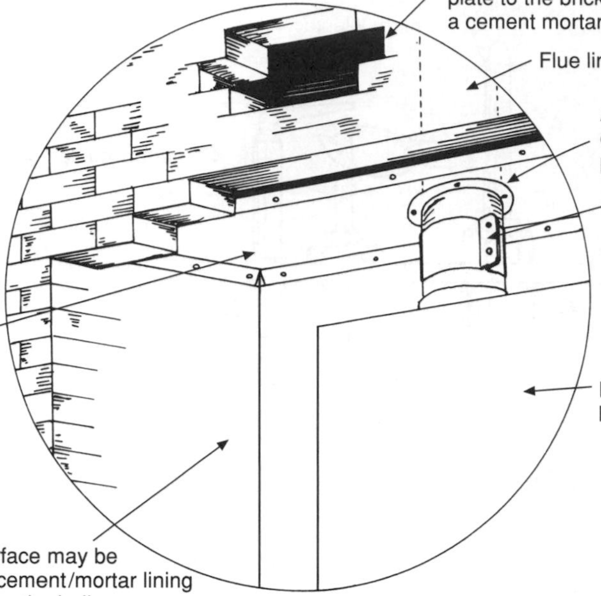


(a)

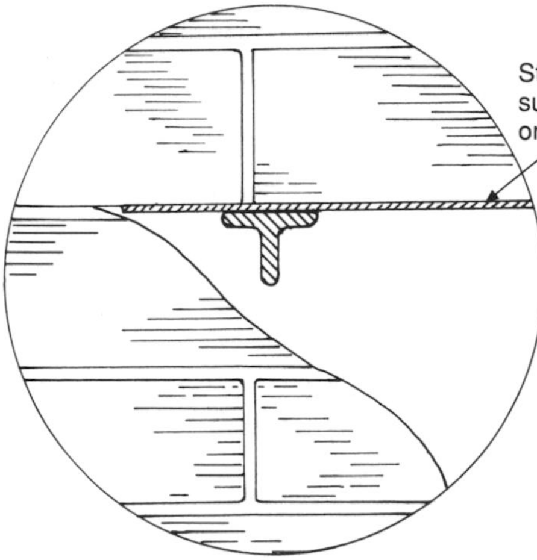

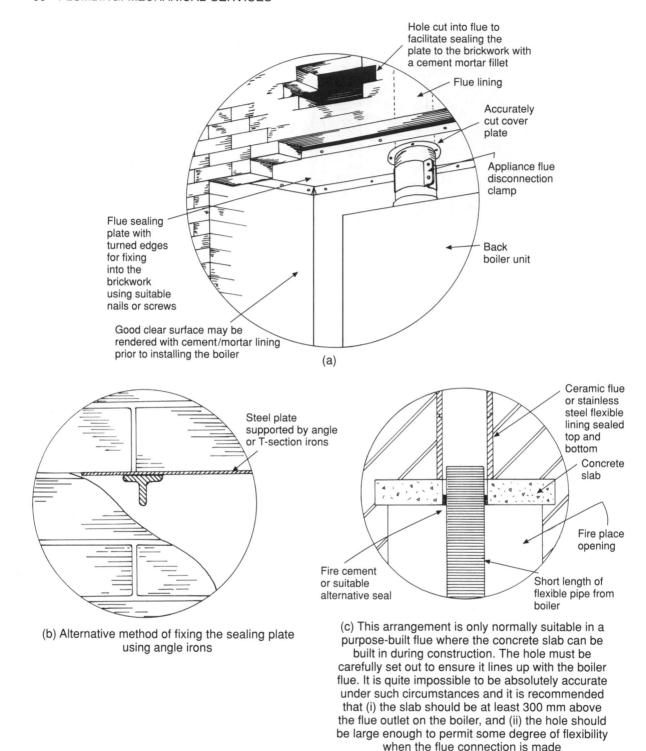

(b) Alternative method of fixing the sealing plate using angle irons

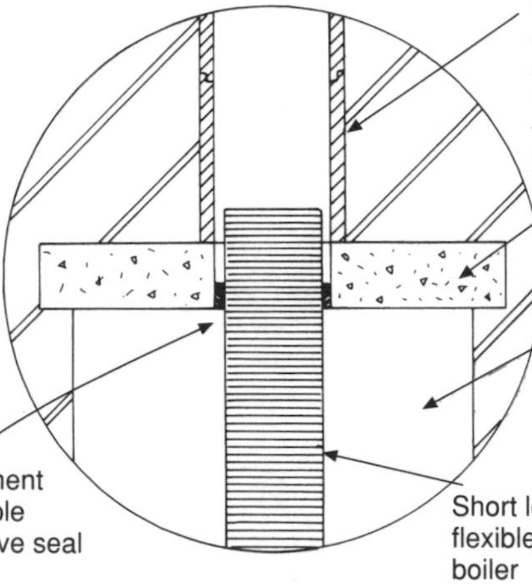


(c) This arrangement is only normally suitable in a purpose-built flue where the concrete slab can be built in during construction. The hole must be carefully set out to ensure it lines up with the boiler flue. It is quite impossible to be absolutely accurate under such circumstances and it is recommended that (i) the slab should be at least 300 mm above the flue outlet on the boiler, and (ii) the hole should be large enough to permit some degree of flexibility when the flue connection is made

**Fig. 2.49** Methods of sealing gas back boiler flues into chimneys.

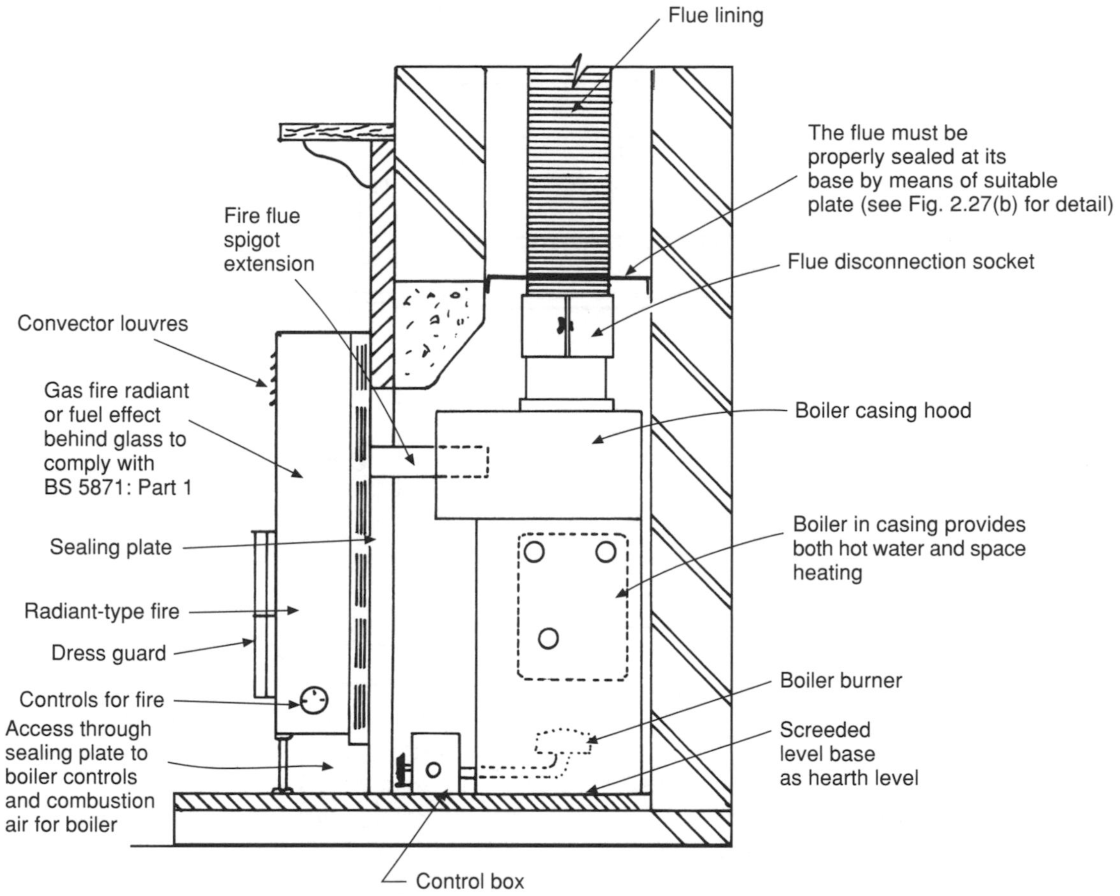


**Fig. 2.50** Section through gas back boiler and fire. These are supplied as a complete unit, the fire being made to be removed to allow the boiler to be serviced. It should be noted that adequate ventilation must be provided as the gas input for both the fire and boiler may range from a total of 8 kW to 30 kW.

this type of installation is likely to be classified for exemption due to excessive cost and other factors of making major changes to the heating system. Figures 2.49–51 illustrate the main requirement for this type of installation and the methods by which it can be achieved. Always ensure that any manufacturers' instructions are observed, especially for flues and hearths. In cases where the flue is very high a flue restrictor plate supplied by the manufacturer can be fitted to reduce the updraught.

*Gas fires — conventional type*

Figure 2.52 illustrates a section through an existing fireplace into which the fire is installed. In new buildings where the opening is designed for a gas fire the firebrick back is unnecessary, but the measurements shown always apply. Before the fire is fixed the closure or sealing plate, supplied with the fire, must be sealed with heat-resistant tape against the opening to which the fire is fitted. Its main purpose is to prevent warm air from the convector entering the flue. This plate has a pre-cut hole through which the flue spigot from the fire passes and a small aperture in its base which serves mainly to ventilate the flue, but also provides an element of flue gas dilution. As with gas back boilers, it may be necessary to fix the flue restrictor plate (supplied by the manufacturer) in cases where the updraught on the flue is excessive.

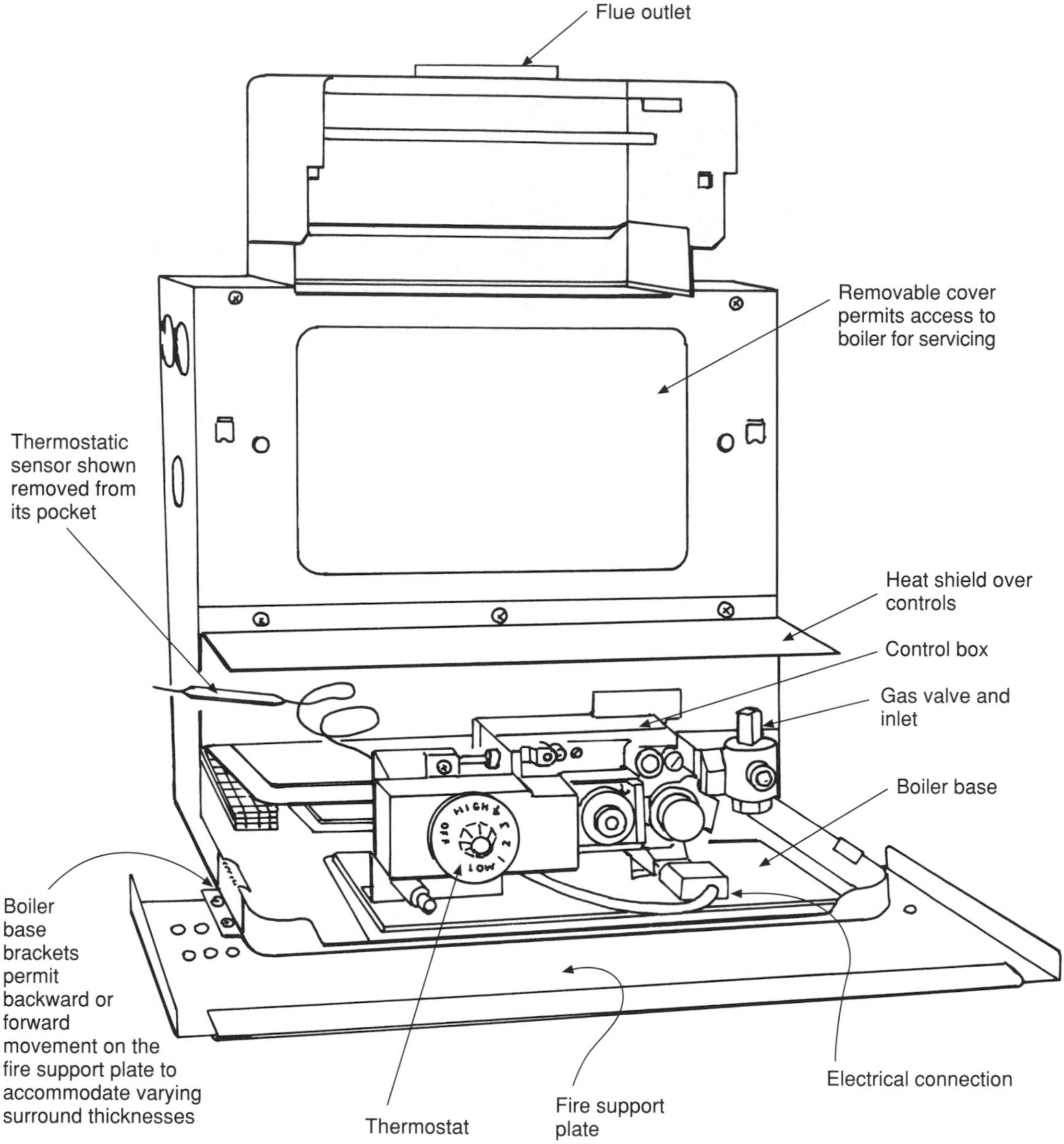


**Fig. 2.51** General arrangement of gas back boiler. These boilers are designed to be fitted into a standard builder's opening for a fireplace and are normally used in conjunction with a gas fire. Both the boiler and fire are marketed as a complete unit.

Any fire must be firmly fixed, as any movement may result in slackening the joints on the gas supply, thus causing a gas leak. Connection to the adjacent gas point is made using 8 or 10 mm copper pipe which may be bent using springs or one of the small bending machines marketed for this purpose. A suitable valve should also be provided to isolate the fire for servicing purposes.

### *Live fuel effect (LFE) fires*

Basically these are an open fire incorporating a form of convector, a section of which is illustrated

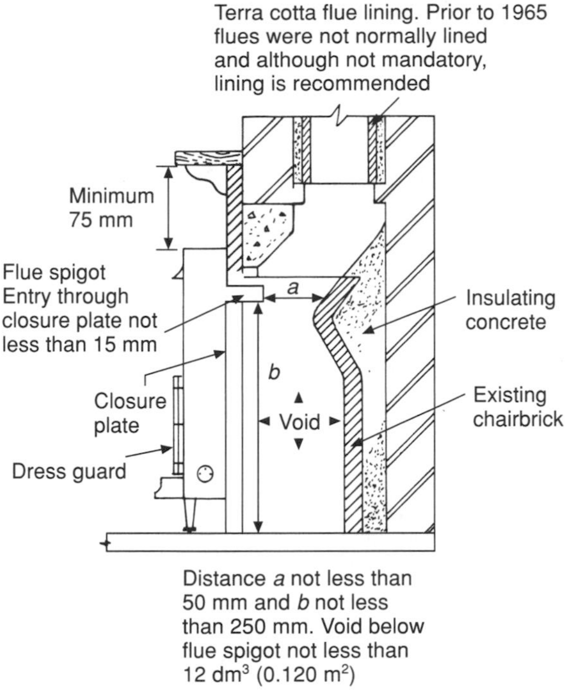


**Fig. 2.52** Arrangements for radiant or glass enclosed fuel effect fires complying with BS 5871 Part 1. Note that any damper or flue restrictor in an existing flue must be removed or, if this is not possible, fixed in the open position.

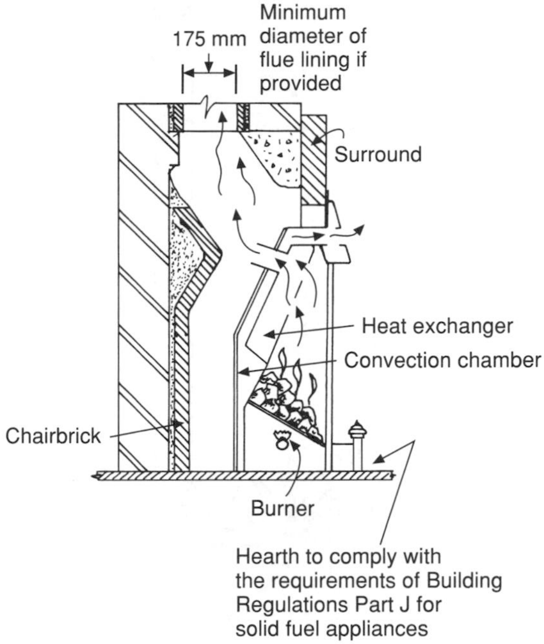


**Fig. 2.53** Section through a typical LFE gas fire constructed to the requirements of BS 7977-1. Fitting and installation must comply with BS 5871 Part 2. Supplementary ventilation is not normally required for fires of up to 7 kW input. Any flue lining for this type of fire must be of the same specification as those required for solid fuel. In buildings constructed prior to 1965 no linings were normally provided and while they are not considered essential they are preferred. The existing firebrick back (chairbrick) may sometimes have to be removed for some types of fire.

in Fig. 2.53. Efficiency varies depending on the type of fire, but generally they are less efficient than the fires previously described. As the products of combustion are of a higher temperature than conventional gas fires, flues must be the same in diameter or similar to those for solid fuel appliances. The same also applies to hearths, as fires of this type are not supplied with guards. Hearths must be at least 50 mm high to discourage the laying of carpets immediately adjacent to the fire, but the surface area may vary from fire to fire. With fires of this type having an input of up to 7 kW, no supplementary ventilation is normally necessary.

*Decorative fuel effect (DFE) fires* These are similar to LFE fires without a convector (see Fig. 2.54). They are made to simulate open fires and their effect is very realistic, but like the solid fuel appliances they are designed to replace, their efficiency is very low. Most DFE fires are designed to be installed in a normal fireplace having a chairbrick (firebrick) back, but basket types are also available for fitting into a large inglenook-type fireplace. These often require modification to gather in the flue (i.e. to make it smaller) so that the products of combustion are effectively removed. It must be stressed that the manufacturer's instructions relating to these fires must be strictly adhered to. It must also be borne in mind when any gas appliance is commissioned that a further spillage test will be necessary. Any fire not covered by BS 5258 must be subject to the Building Regulations Part J, relating to flues and hearths.

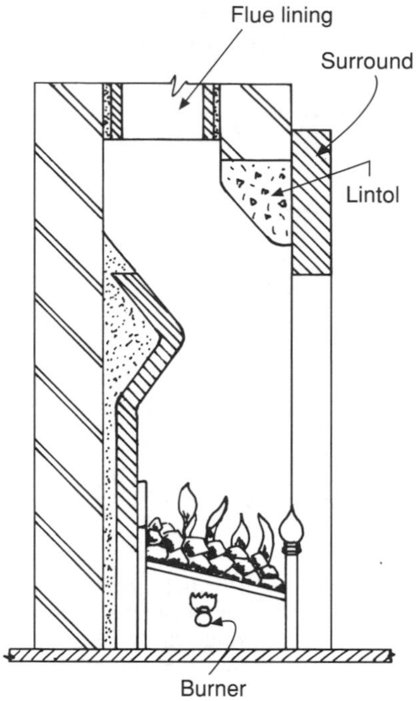


**Fig. 2.54** Section through a DFE fire constructed to the requirements of BS EN 509. Fitting and installation must comply with BS 5871 Part 3. The throat of the flue affects the efficiency of these fires and may require some modification in accordance with the manufacturer's instructions. These fires require a ventilator of at least 100 $cm^2$ for fires up to 15 kW input. Where this type of fire is fitted into a large open fireplace (e.g. inglenook) manufacturers may specify certain requirements to ensure a positive updraught in the flue. Flue requirements as for LFE fires.

*Room sealed gas fires*

These are very convenient in situations where no flue is available; they also have an advantage in that no additional ventilation is required in the room in which they are situated. Some types are in effect convector heaters, others have a sealed glass-fronted panel through which a solid fuel effect can be seen. It is important that the air inlet complies with the manufacturer's instructions in relation to safety and ensures an electrical supply is available if necessary.

*Flue boxes*

Figure 2.55 shows a typical insulated metal flue box which is primarily designed for installations where no flue is available. It can also be used in most existing fireplaces, providing a clean environment

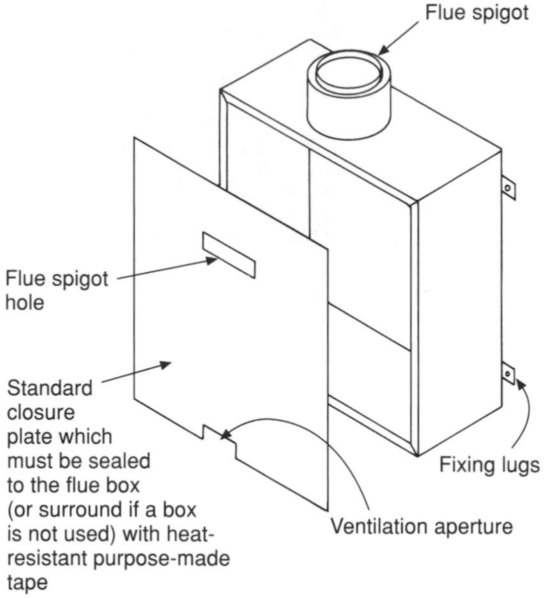


**Fig. 2.55** Flue box. A metal box developed for use with certain types of gas fires in effect to simulate a builder's opening. They are primarily designed for situations where a false chimney is constructed in existing properties using timber and plasterboard. They are also useful in conjunction with fire surrounds constructed of brick or stone to which it may be difficult to make an effective seal to the closure plate.

at the back of the fire, and although originally produced for conventional gas fires some companies are producing it for open flame effect fires. A check should be made on the suitability of any such fire for fitting into a fire-box of this type.

*Hearths for gas fires*

Hearths will vary as to the type of fire installed. Figures 2.56(a) and (b) illustrate the general requirements of BS 5871. These dimensions may vary, however, depending upon the manufacturer's requirements. Hearths for DFE and LFE fires will generally have to comply with the Building Regulations Part J. A hearth should be constructed having a fire-resistant surface of at least 12 mm and must be 50 mm above floor level unless a raised kerb or fender of this height is provided. This ensures that non-fire-resisting floor coverings are not laid over the hearth.

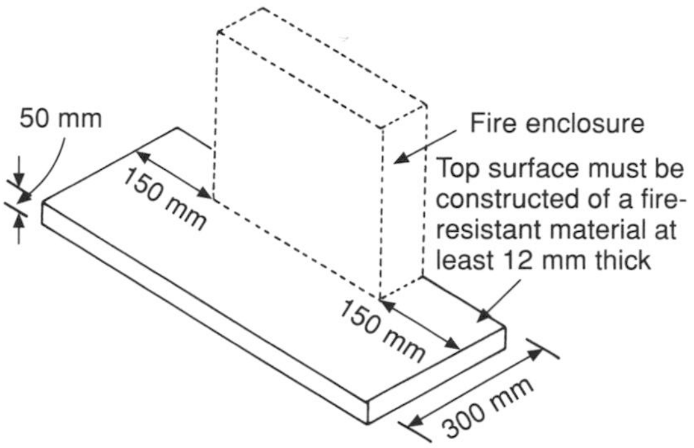


(a) Radiant or glass-fronted fires

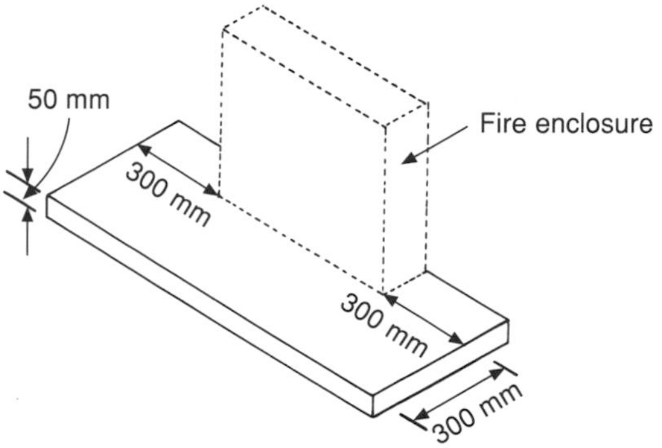


(b) Live and decorative fuel effect fires. Note the increase of hearth extension on both the front and sides of the fire. The dimensions shown are the minimum between any part of the incandescent fire bed

**Fig. 2.56** Hearths for gas fires. It should be noted the hearth dimensions shown are based on the relevant British Standards, but in some instances the manufacturer's instructions which must be complied with require an increase in these dimensions.

## Servicing gas appliances

Gas appliances should be serviced on an annual basis to ensure they are safe and working to maximum efficiency. However, because of the wide variety available it is very difficult to compile a complete list of servicing requirements for each appliance. This is one of the reasons why installation and maintenance manuals *must* be left with a responsible person on completion of a new installation. Plumbers or fitters should also be familiar with the servicing requirements of any appliance they are dealing with. There are, however, some servicing points which are common to all appliances listed as follows and can be used as a general guide.

1. Isolate any electrical connection if applicable.
2. Turn off any gas valve controlling the supply of gas to the burner.
3. Floor-standing boilers and fires tend to become very dusty and the interior of the appliance may need cleaning. An industrial vacuum cleaner will be very useful here.
4. Remove the burner and control assembly.
5. Inspect any flueing arrangement for damage, defective joints, and cleanliness. With fires ensure the void behind the sealing plate is clean.
6. Clean the flueways or heat exchanger with a bristle brush and clean the bottom of the combustion chamber, preferably using the vacuum cleaner.
7. Inspect and clean the burner and the gas injectors taking care not to damage them.
8. Clean the pilot light assembly.
9. Reassemble any dismantled parts, taking care to renew any seals that have been broken with proper replacements.
10. Reconnect the electrical and gas connections and check for soundness. Light the appliance and after approximately 10 minutes, with the thermostat set at its highest position, check the gas pressure at the test point on the burner.
11. If applicable check the electrical system for damaged cables and security of electrical connections.
12. It is wise to ensure visually that the appliance ventilation requirements have been complied with, especially if the appliance was originally fitted by another installer.
13. Finally, conduct a spillage test on the flue if applicable.

### *Tracing gas leaks*

This is sometimes a difficult job, especially in existing properties. The best procedure in such circumstances is as follows: ascertain from the customer

(a) Where the smell is more noticeable.
(b) Is it persistent or does it occur only when certain appliances are in use?
(c) Has it got worse over a period of time?

(d) Was the smell noticed before or after a new appliance was installed?

If the cause of the leak is not obvious check any valves, especially on gas cookers; the heat often dries out the grease on the plug. Valves on gas fires are another possible cause of leakage. Look for any mechanical damage on exposed soft copper pipes. It may be necessary to isolate the appliances one by one, testing the remainder of the installation each time. It may be possible to isolate individual branches using the same procedures. In the event of failure to find the leak the National Gas Provider or gas supplier should be contacted.

### Decommissioning

In the event of premises being unoccupied, depending on the circumstances, the gas transporter may cap off the main valve and remove the meter. Any pipework that is unlikely to be used again should be removed as far as is possible. All open ends must be securely plugged or capped where the pipework remains and an air test should be conducted to ensure this procedure is effective. It is an offence to leave any open ends on a gas supply pipe, even when it is no longer in use.

### Further reading

Much useful information can be obtained from the following sources:

The Gas Safety (Installation and Use) Regulations.
British Standards:
BS 5871 Installation of gas fires, convectors, and fire back boilers, Parts 1, 2, 3.
BS 5440 Part 1 Flues.
BS 5440 Part 2 Air supply.
BS 6798 Installation of gas-fired hot water boilers.
BS EN 1443 Chimneys. General requirements
'Corgi' Publications, Council for Registered Gas Installers, 1 Elmswood, Chineham Business Park, Crockford Lane, Basingstoke, Hants RG24 8WG Tel. 0870 401 2200. www.corgi-gas-safety.com

*Flues*
Security Chimneys UK Ltd, Dalilea House, St Mary's Road, Portishead, Bristol BS20 9QP Tel. 01275 847609.
Chimflue, Tel. 01707 266244, 01264 332878. www.chimflue.co.uk

*Testing equipment*
Testo Ltd, Newman Lane, Alton, Hampshire GU34 2QJ Tel. 01420 544433. www.testo.co.uk

*Fires*
Baxi Group, Wood Lane, Bromford, Erdington, Birmingham B24 9QP Tel. 0121 373 8111. www.firesandstoves.com

### Self-testing questions

1. State the physical properties of natural gas.
2. (a) Explain the term cross-bonding and state why it is necessary.
   (b) Describe the main cause of flame lift-off and explain why it must not be tolerated in gas burners.
3. (a) State the recommended pressure for testing a new gas installation and describe the procedures employed.
   (b) Why is it necessary to test the main gas cock before testing an existing installation? What action must be taken if it lets by?
4. (a) Describe how to purge gas installations.
   (b) Explain why it is necessary.
5. (a) Explain why a supply of air is essential for combustion and what is meant by vitiated air.
   (b) Sketch a downdraught diverter and describe its function.
6. (a) List the combustion products of gas.
   (b) Name the toxic gas which is present in the products of incomplete combustion.
7. (a) Make a simple sketch illustrating the principle of room-sealed, natural draught appliances.
   (b) Sketch and describe three methods of terminating a gas flue in a suitable position.

8. (a) Explain the working principle of an appliance governor and state its purpose.
   (b) Describe the procedure for regulating the working pressure of gas to an appliance.
9. (a) Describe the methods of igniting gas appliances.
   (b) List the essential controls for the correct functioning of a gas boiler.
10. (a) Explain the term *fail safe* in relation to gas appliances.
    (b) Make a simple sketch illustrating the working principles of a thermoelectric valve.
11. State the reason for sealing a flue lining at both the top and bottom.
12. A sealing plate must be used with gas fires complying with BS 5871 Part 1. State its purpose and why it is necessary to seal the edges to the fireplace opening.
13. From Table 2.2 determine the pipe diameter necessary to supply an appliance with an input rating of 1.5 m/h (pipe length is 9 m).
14. List the common procedures that must be carried out when commissioning a gas appliance.
15. Determine the area in a ventilating brick of free air necessary for combustion where an open-flued appliance rated at 16 kW is fitted in a room.
16. State the maximum permissible pressure drop between a meter and any appliance.
17. State the recommended procedure where, during a service call, an appliance is found to be 'at risk'.
18. State the minimum distance that the terminal on a room-sealed natural draught appliance must be from a corner of the building.
19. Describe the terms (a) the calorific value of a fuel and (b) the stoichiometric mixture.

# 3 Cold water supply

After completing this chapter the reader should be able to:

1. Explain the main purpose of regulations relating to water supply.
2. Identify water classifications and types of backflow prevention devices.
3. Identify the cause of noise in cold water systems and state the methods of prevention.
4. State the causes and means of prevention of contamination of water in domestic and industrial supplies.
5. Describe the methods of preventing frost damage to water pipes and fittings.
6. Describe the basic principles and operation of water treatment appliances.
7. Understand the basic principle of pump-boosted cold water supplies.
8. Describe the various water systems used for fire protection in buildings.
9. Understand the methods of sizing distribution pipework for hot and cold water services.

## The Water Regulations (1999)

These regulations superseded the 1986 Water By-laws in England and Wales. The water authorities in Scotland began to enforce new by-laws in April 2000 which mainly mirror the regulations for England and Wales. They have also been adopted by the Department of the Environment in Northern Ireland. The new regulations embody most of the requirements of the 1986 Water By-laws; the main differences relate to backflow protection, differences in the categories of water, and WC flushing. More emphasis is also placed on conservation.

As with previous by-laws, the main reasons for the need for regulations may be summarized as avoidance of: waste, contamination, misuse and undue consumption. Undue consumption may be interpreted as using more water than is actually required. It is also an offence to fit a supply of water to a metered premises in such a way that it does not pass through the meter.

Every practising plumber or fitter whose work entails the installation or repair of hot and cold water supply systems must be conversant with the regulations, as the penalty for their contravention is very severe.

*Restrictions on the use of water fittings*

The following is an abbreviated list of the principal requirements for any appliance or component used with water supplied by water authorities. For more detailed information reference should be made to the actual regulations, the Water Fittings and Materials Directory, and the Water Regulations Guide.

*Requirements of water fittings*

Every water fitting or appliance must comply with any relevant standard relating to quality and suitability for the purpose for which it is required. These include BS and EN harmonised standards or those having European technical approval which meet these requirements. Every water fitting must be installed, altered or repaired in a workmanlike manner and must comply with the relevant standards mentioned.

*Notification of work*

Except for minor extensions or alterations to a water system, permission must be applied for and granted in connection with the following:

1. The erection of a structure, not being a pond or swimming pool, requiring a supply of mains water.
2. Any alteration of, or extension to, any building except a dwelling.
3. A change of use of any premises.
4. Installation of:
   (a) Baths having a capacity of more than 230 litres.
   (b) Bidets with inlets below the spillover level or fitted with a flexible hose.
   (c) A single shower unit, not being a drench shower, which may consist of one or more shower heads (under review).
   (d) A pump or booster connected to a supply pipe drawing more than 12 litres per minute.
   (e) Any equipment which incorporates reverse osmosis processes.
   (f) A water treatment process which produces a waste water discharge or which requires the use of water for regeneration or cleaning.
   (g) Any mechanical device used for protection against a fluid category 4 or 5, e.g. an RPZ valve.
   (h) A garden watering system unless designed to be operated by hand or
   (i) Any water system or pipe laid externally of the building less than 750 mm or more than 1,350 mm below ground level.
5. Ponds or swimming pools with a capacity in excess of 10,000 litres which are designed to be replenished automatically and filled with water supplied by a water undertaker.

All notices to the water authority must include the following:

(a) The name and address of the person giving the notice.
(b) A description of the proposed work.
(c) Particulars of the location and premises to which the proposals relate and the intended use of the premises.
(d) Except in the case of work falling into category C.

There are exceptions to the foregoing: a plumber who is an 'approved contractor' will not require permission to install such appliances as those listed in 4(b) or 4(g).

*General requirements of Water Regulations*

The use of lead pipes and solders containing lead has been prohibited for use with potable water for many years. The only exception to this is in situations where lead services are still in use. Any repairs to such services, short of renewal, must be made using copper or a suitable plastic pipe and approved fittings.

Most water authorities also have reservations about the use of galvanised steel tubes, especially in areas where the action of the water on zinc causes dezincification. In most cases, where this material is permitted, its use is limited to distribution and hot water services only where the use of copper or plastic materials may be subject to damage.

The use of storage cisterns with purpose-made covers and overflow screens as illustrated in Book 1 is mandatory in all new buildings where any water is stored for drinking and domestic purposes, and any replacements to existing water-storage vessels must comply with the 1986 water by-laws. Any pipe-jointing compound used for making joints on pipework or storage vessels must be non-toxic and resistant to bacteriological growth.

Coal tar substances such as bitumen can no longer be used to protect the internal surfaces of pipes or cisterns against corrosion, but suitable anti-corrosion paints are available and a list of these may be obtained from the local water authority. It is not unknown for water to leak from the primary part of a central heating system into the secondary water, i.e. water drawn off via a hot tap. This can occur due to a leak in the connection or coil in the hot storage vessel. Any inhibitor used to protect the heating system from corrosion must therefore be of a non-toxic nature. It is not permissible to run underground service pipes in soil which may be contaminated by sewage or refuse of any description. While pipes made of plastic are very resistant to corrosion, they can be degraded and softened when subject to contact with petroleum products, oils and phenols, i.e. materials derived from coal tar such as bitumen and creosote. If there is any suspicion of such contamination the service may have to be rerouted or passed through a watertight duct. The local water authority should also be advised where any such doubts arise in relation to underground services.

## Types of cold water supply

Services may be supplied direct from the supply pipe and water main, and this is common in small domestic properties. The indirect type of supply is preferred, however, mainly because some storage is available and there is less risk of contamination. These two basic systems are described in Book 1 of this series. In the case of large commercial and industrial buildings, e.g. hotels, schools etc., an indirect supply is essential. In such cases the water suppliers should be consulted as to the capacity of any large storage cistern, as they will be aware of any difficulties in the volume and pressure of the main supply. Cisterns having a storage capacity in excess of 1,000 litres should be provided with a valved wash-out pipe at the lowest level. The wash-out pipe must not be connected directly to a drain except through a tun dish providing an AA-type air gap. When not in use the valve must be securely plugged. Table 3.1 shows the recommendations of BS 6700 for water storage in various types of building. Circumstances may vary but it provides a good general guide.

*Pumped systems cold water supply*

Many high-rise buildings are unable to be supplied with water direct from the main. The fact that main pressure varies (i.e. depending on whether the building is at the top or bottom of a hill) also has some influence on the pressure available. Daytime pressures are also lower than those at night due to the larger volume of water consumed. If for example a pressure of 350 kPa is available at the main (this is the equivalent of 35 m head) it will in theory serve a building 35 m higher than the main. After making an allowance for pressure loss, due to the resistance of the pipe and fittings, the effective pressure may only be approximately 32–33 m. For this reason water boosting is employed to supply cold water to upper drinking water draw-offs, storage cisterns and if necessary for fire-fighting in high-rise buildings. With the exception of installations where an interruption of the supply would not be serious, dual pumps are essential to allow for the possibility of mechanical failure or periodic maintenance. There are two main types of installation:

(a) Direct boosting, where the water is pumped directly from the main.
(b) Indirect systems, where the water is pumped from a break tank which is fed from the main via a float-operated valve.

Direct boosting is rarely permitted due to:

**Table 3.1** Water storage requirements in various buildings. Ref. BS 6700.

| *Type of building* | | *Storage (litres)* |
|---|---|---|
| Dwelling-houses and flats | (per resident) | 90 |
| Hostels | (per resident) | 90 |
| Hotels | (per resident) | 200 |
| Offices without canteens | (per head) | 40 |
| Offices with canteens | (per head) | 45 |
| Restaurants | (per head, per meal) | 7 |
| Nursing/convalescent homes | (per bed space) | 135 |
| Day schools — nursery–primary | (per pupil) | 15 |
| Day schools — secondary–technical | (per pupil) | 20 |
| Boarding schools | (per pupil) | 90 |
| Children's home or residential nursery | (per bed space) | 135 |
| Nurses' home | (per bed space) | 120 |
| Nursing or convalescent home | (per bed space) | 135 |

(a) The volume of water drawn from the main leading to the loss of supply, or at least lowering of pressure to other consumers.
(b) The possibility of backflow or cross-connection being much greater.

Under no circumstances is it permissible to connect a pump directly to a pipe connected to a service pipe without the written consent of the supplier, except if it draws less than 12 litres/minute.

*Indirect systems*
Indirect systems are so called because water is pumped from a break tank which is supplied from the main through a float-operated valve. They fall into two main types: those which pump to a drinking water header, shown in Fig. 3.1, and those which employ a low-level pressurised storage vessel, shown in Fig. 3.2.

In all cases of pumped systems, draw-offs within reach of the main pressure are connected directly to the main. This reduces the volume of water that has to be pumped, enabling the use of smaller pipes, pumps and, in some cases, storage vessels.

*Indirect boosting with a header*
The capacity of the break tank shown in Fig. 3.3 requires careful consideration. To avoid stagnation

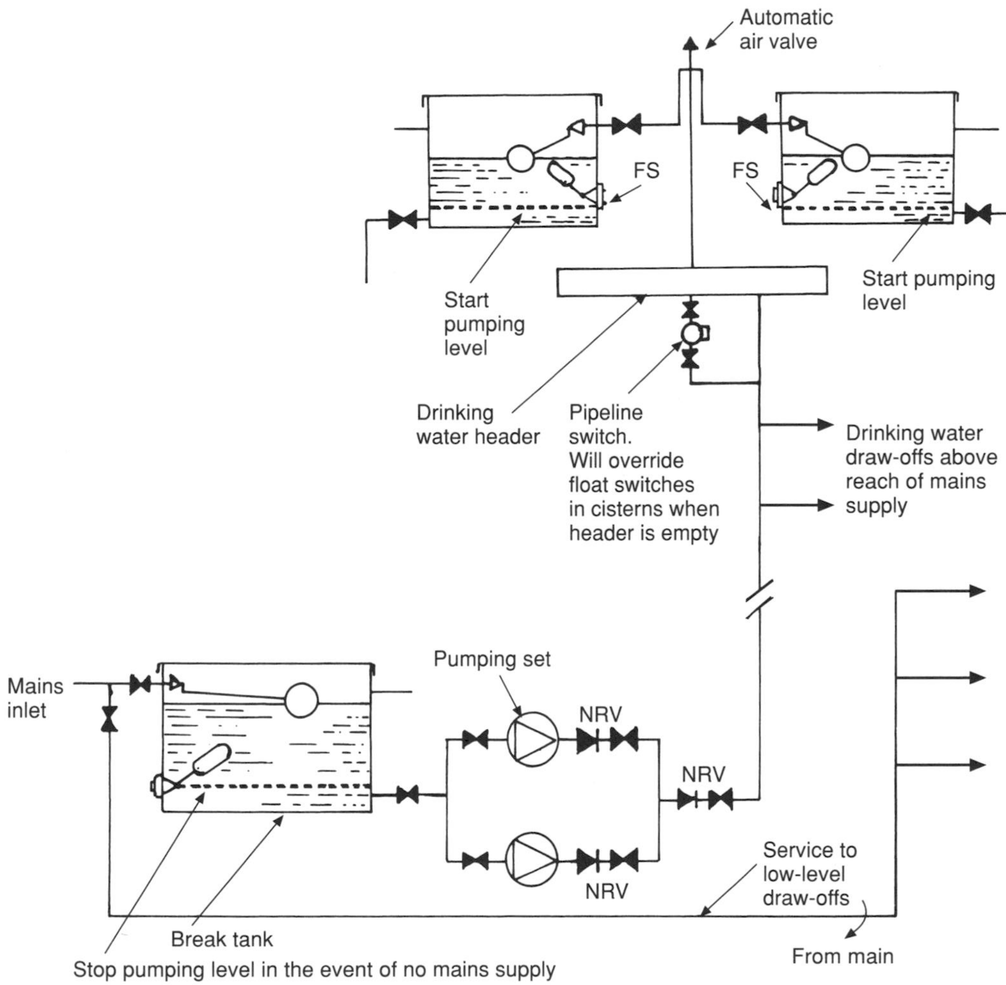


**Fig. 3.1** Indirect pump booster system with high-level storage cisterns and drinking water header.

break tanks are normally sized to provide 1 hour's supply, but conversely they should hold sufficient water to enable the pumps to function for 15 minutes before the low-level cut-out operates. In all cases of pumped supplies, the system must be designed to reduce the number of pump stop/starts to pump small quantities of water; failure to do this will shorten its working life. All pumping systems are therefore controlled in such a way that when the pump starts it will continue pumping for a preset period of time. To supply drinking water points above the reach of the main pressure using this system, a header is employed which is sized to provide 5–10 litres per day per dwelling served. When the pump is not running and drinking water is drawn off, the header begins to empty until the pipeline switch, illustrated in Fig. 3.4, activates the pump, and by means of a timing switch causes it to operate for a set period of time. If the water is replaced in the header before the pump timing cycle is incomplete, the excess water will be pumped into the storage cistern through the float-operated valve.

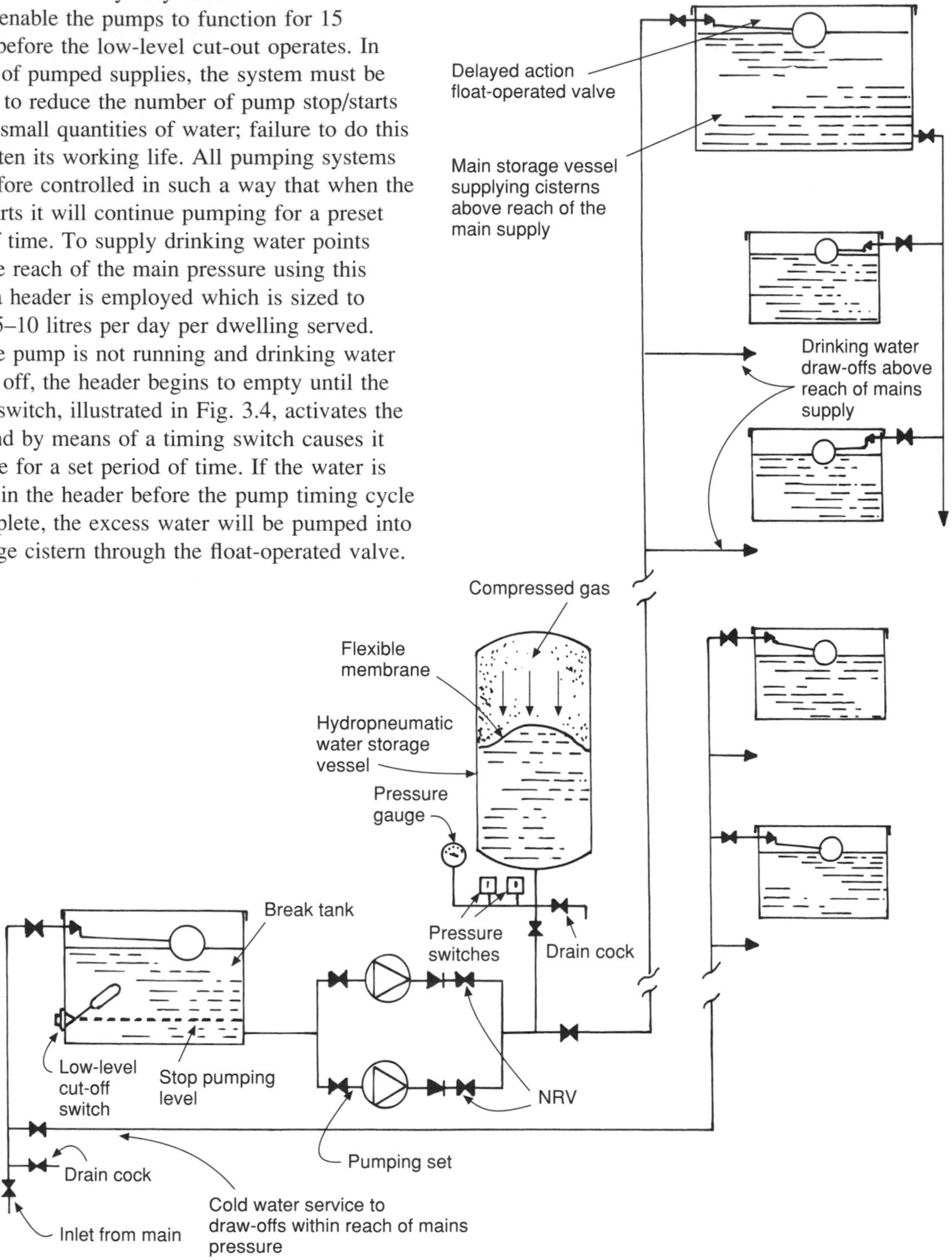


**Fig. 3.2** Hydropneumatic boosted cold water system.

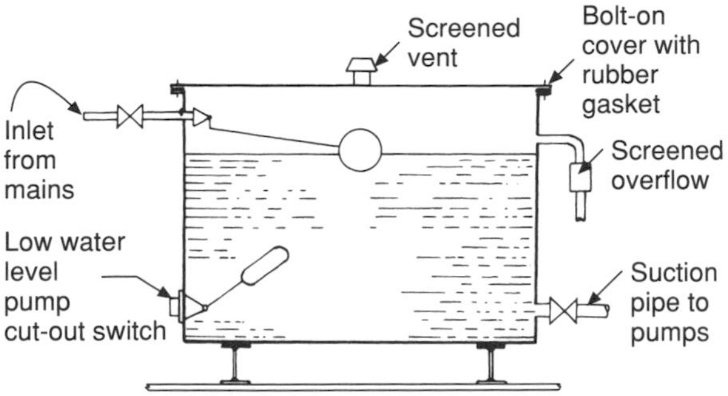


**Fig. 3.3** Detail of break tank.

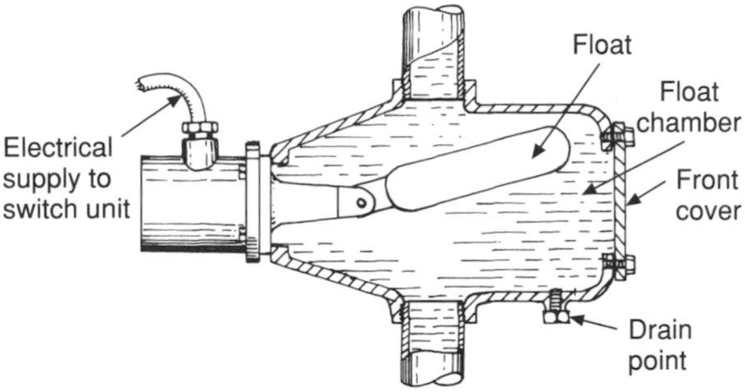


**Fig. 3.4** Pipeline switch. When the chamber empties the float will fall and operate the switch to activate the pump to refill the header.

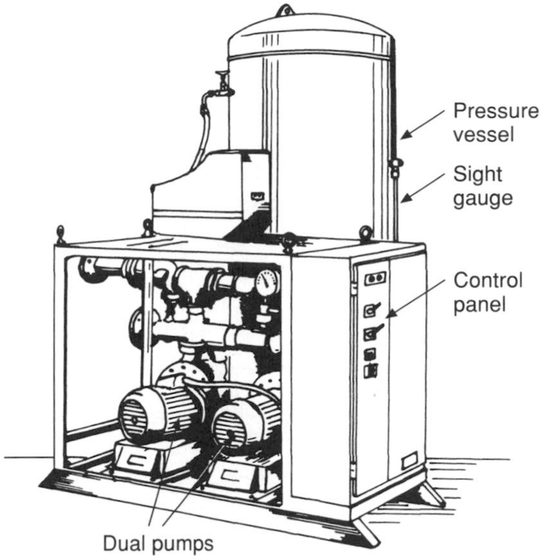


**Fig. 3.5** Typical hydropneumatic packaged pumping set.

Should this be closed while the pump is operating, no serious damage to the pump will take place, as being of the centrifugal type it can operate against a closed outlet for a limited period of time. If the pipeline switch is not activated, but the water level in the cistern falls to such a level that topping up is necessary, the float-operated switch will fall to the 'start pumping' level shown, restarting the pump. These switches are arranged to override the timing cycle, as the pump will run for a longer period to replenish the water in the storage cistern. From the foregoing it will be seen that the pump is activated in two ways: by a fall in the drinking water level, and by a fall in the water level in the storage cistern. An automatic air valve must always be provided to allow air into the drinking water header when water is drawn off, closing when it is full.

*Indirect hydropneumatic systems*

This is the most common type of installation now in use, its main advantage being that all the equipment and component parts are usually supplied as a complete package unit (see Fig. 3.5) for both large and small installations. It is also more convenient in situations where a number of storage vessels at different levels are to be served, which would be impracticable using several water-level-operated switches. The indirect hydropneumatic system operates on the principle of pumping water into a pressure vessel, causing the air it contains to be compressed. When a tap on the riser is opened, water is forced upward due to the pressure exerted by the air. Continued draw-off will lower the pressure in the vessel until it falls to a predetermined level, when the pump will restart on a signal from the pressure switch. Modern pumping units use a pressure vessel having a flexible membrane similar to those used with unvented hot water and sealed heating systems, its capacity depending on the size of the installation. When the system is serviced, the only maintenance requirement with this type of vessel is to check the pressure of the gas, which may be air or nitrogen.

*Delayed action float-operated valves* These are designed to ensure the valve is either fully open or closed, and while they may be used with advantage

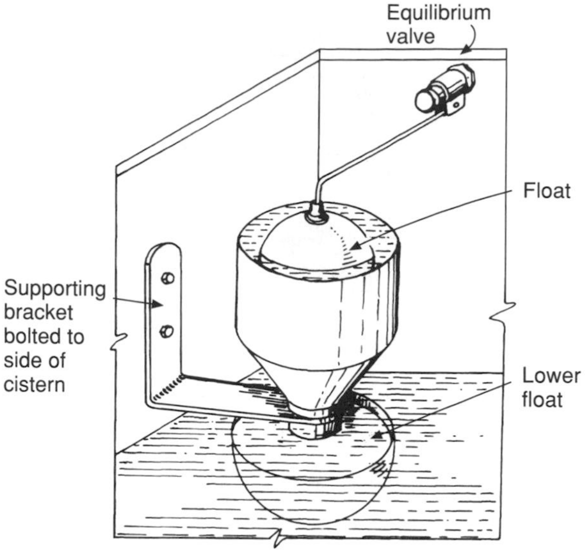


**Fig. 3.6** Pictorial illustration of a delayed action Portsmouth-type float-operated valve.

in many other situations, they are essential with hydropneumatic pumped systems, as the float does not fall until a large volume of water is required to replenish the contents of the cistern. This ensures the number of pump/stop starts are reduced to a minimum. Figure 3.6 illustrates a typical valve of this type and its operating cycle is illustrated in Fig. 3.7.

*The Aylesbury float-operated valves* These are quite unlike the traditional valves used by the industry for a long period of time. The rise and fall of the float is designed to partially rotate the activated tube, which operates a ceramic disc valve. The tube contains a rolling weight which moves in the tube to give a positive opening and closing of the valve. By adjusting the position of the floats and buoy the water levels can be adjusted. The valve shown in Fig. 3.8(a) can be used independently or as a pilot, opening and closing a main valve on the inlet to a large system having upwards a supply pipe of 50 mm nominal bore (nb) or more as shown in Fig. 3.8(b). The pilot valve and the inlet pipe are housed in a chamber above the cistern which is provided with a weir overflow, both of which are mandatory in circumstances where pollution of the main supply may be possible. The weir would only become operative if the overflow became blocked, which would be a rare event, but provision for the safe disposal of this water must be considered. The usual practice is to construct a metal safe in which the cistern is housed having a discharge pipe fitted to dispose of any water that may flow over the weir.

*Non-return valves* These must be fitted to prevent the possibility of back pressure caused by a moving column of water being suddenly halted. This happens when a pump stops and in some cases has been known to burst the pump casings. Pump manufacturers usually specify the use of a non-return valve on the riser and each individual pump. The normal disc or flap types of non-return valves are suitable where back pressure is unlikely to be excessive, but a spring-assisted recoil valve, shown in Fig. 3.9 is recommended where high pressures are anticipated. No provision is shown against backflow in Figs 3.1 and 3.2. A double-check valve would be necessary on each of the branches serving drinking water points and any other branches taken from the main riser. As a further precaution against contamination, a type BA or CA device may be necessary on the main inlet at the discretion of the water supplier; see Tables 3.2 and 3.3 below.

*Pumps, pump components and siting*

Pumps and the associated controls and components should be housed in a room as close as possible to the point where the main enters the building. The room should be dry, ventilated, protected against frost and flooding and of sufficient size to allow for maintenance and the replacement of component parts. Access should be restricted to authorised personnel only. In buildings which rely solely on the pumps for a supply of water, pumps must be duplicated. The previous comments relating to pump noise should be noted. Most packaged units are supplied with water hammer arrestors, but where a pumping set is constructed on site, provision must be made to include protection against water hammer, as the surge developed when a pump stops or starts can give rise to very high pressures. Reliable control systems are essential — pressure and float switches must be suitable and adequate for their purpose. Most pumping sets are electrically operated and should be controlled

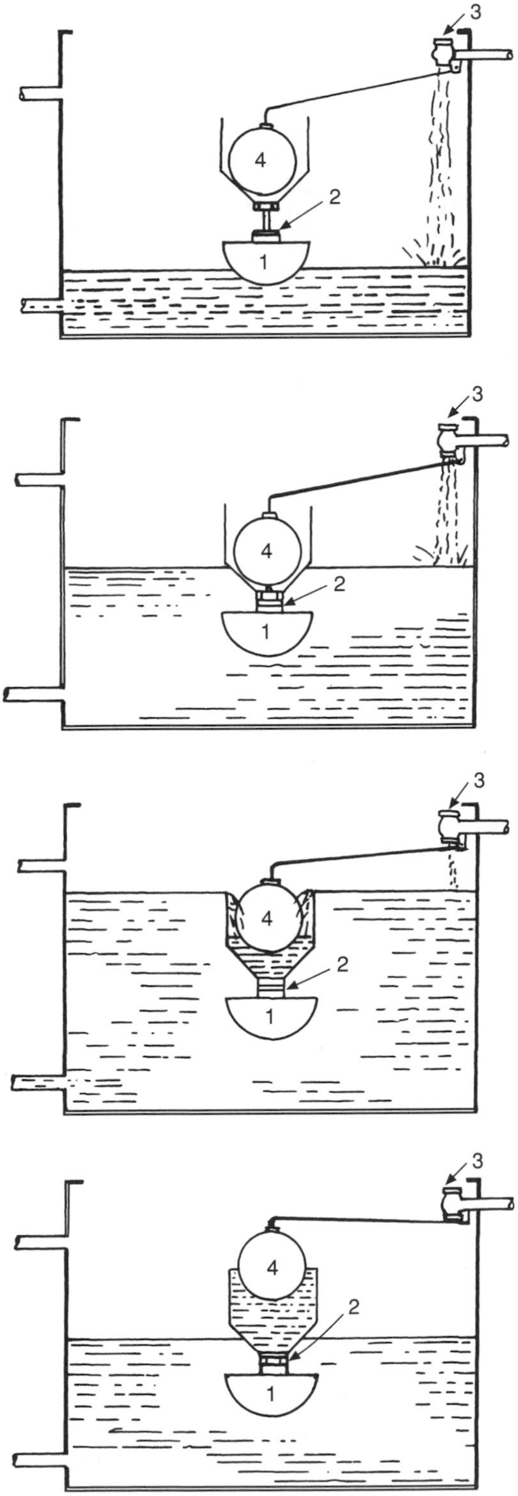


(a) The cistern is filling with both valve 2, which is carried by the semi-hemispherical float, and valve 3 open

(b) As the water level rises float 1 is lifted closing valve 2. Valve 3 is still fully open

(c) The cistern continues to fill until water flows over the edge of the tun-dish. This lifts the ball float 4 closing valve 3. As the tun-dish holds little water the closure is very sudden and for this reason an equilibrium float-operated valve and the use of water hammer arrestors is essential to reduce noise

(d) As the cistern empties valve 3 remains closed until the water level falls to a level where it no longer supports float 1. This causes it to fall and open valve 2 allowing the tun-dish to empty. Float 4 will now fall opening valve 2, restarting the sequence as shown in (a)

**Fig. 3.7** Delayed action float-operated valve sequence of operation.

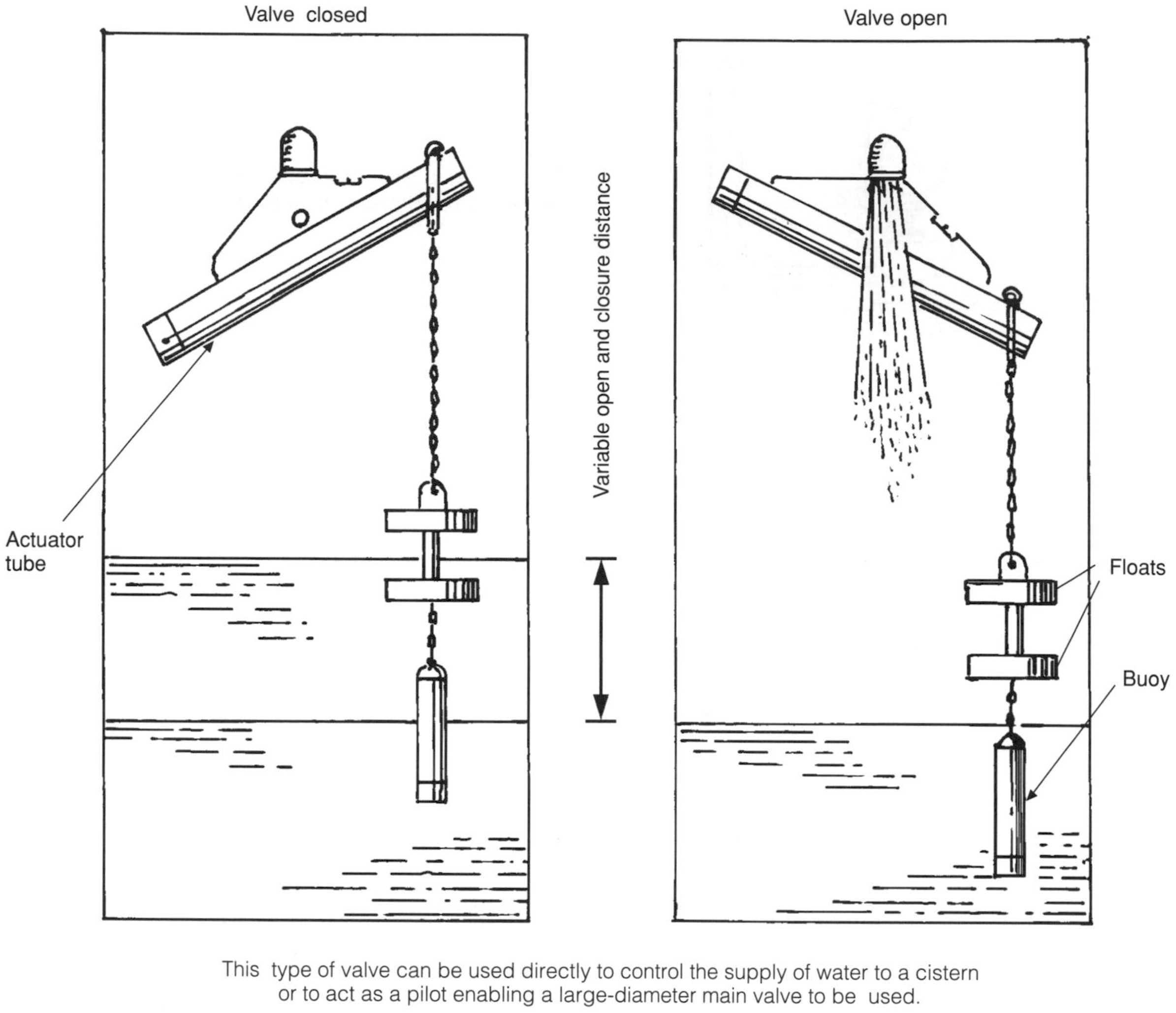
Valve closed
Valve open
Variable open and closure distance
Actuator tube
Floats
Buoy
This type of valve can be used directly to control the supply of water to a cistern or to act as a pilot enabling a large-diameter main valve to be used.

**Fig. 3.8a** Aylesbury KB float-operated valve.

by a pump selector switch so the pumps can be operated alternately. On most modern equipment switching arrangements are fully automatic. Where the failure of the main power supply could be serious, back-up plant such as a generator set must be provided.

*Maintenance and inspection*

The user should make arrangements for servicing the plant at regular intervals. Any work carried out and the dates of inspection should be recorded in a log book, which is usually retained in the plant room. As the equipment is almost entirely automatic, maintenance consists mainly of keeping the components clean and checking that the controls are functioning correctly.

*Noise from pumps or boosters*

All pumps make some degree of noise due both to the motor which 'hums' and to the rotation of the impeller which causes vibration. This vibration is transmitted to the floor supporting the pump and its pipework connections. By introducing flexible pipe joints and pump fixings as shown in Fig. 3.10, this type of noise is almost eliminated. Manufacturers produce these pumps having regard for the noise

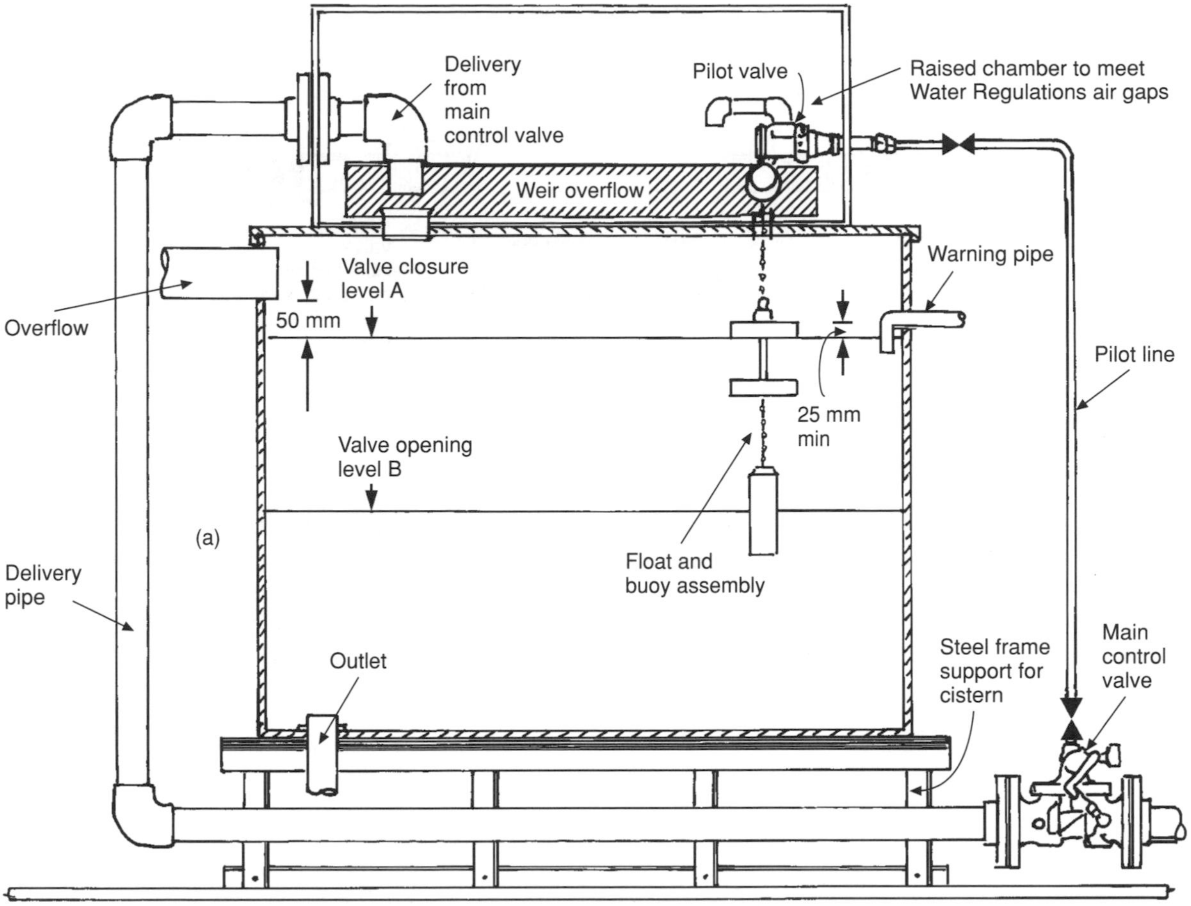


**Fig. 3.8b** The Aylesbury KB valve acting as a pilot for the main inlet control valve.

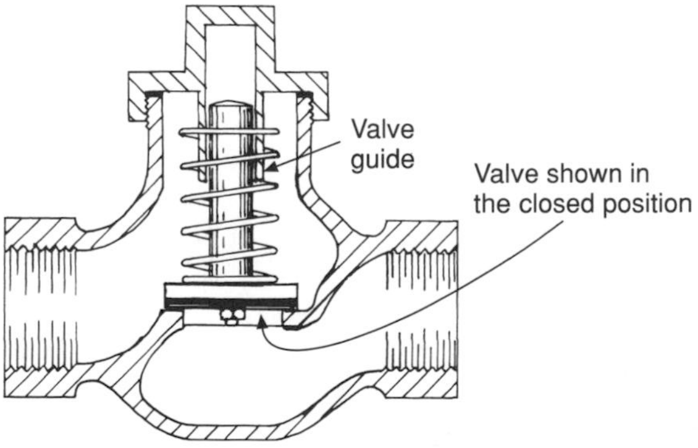


**Fig. 3.9** Spring-loaded non-return valve. Used as an alternative to the normal flap or disc NRV on boosted cold water systems where a quick-acting valve is necessary to avoid back pressure on pumping equipment.

factor, and motor speeds are kept to a minimum in relation to the output of the pump. To avoid the 'hum' generated by electric motors becoming a nuisance, pumps should be situated as far as possible from living apartments — in some cases they are contained in a soundproof room.

## Large water storage cisterns and associated pipework

In buildings where large quantities of water are consumed it may be necessary to use two or more cisterns either to ensure a more equal distribution of the mass of water over a greater load-bearing surface or because a cistern large enough to meet

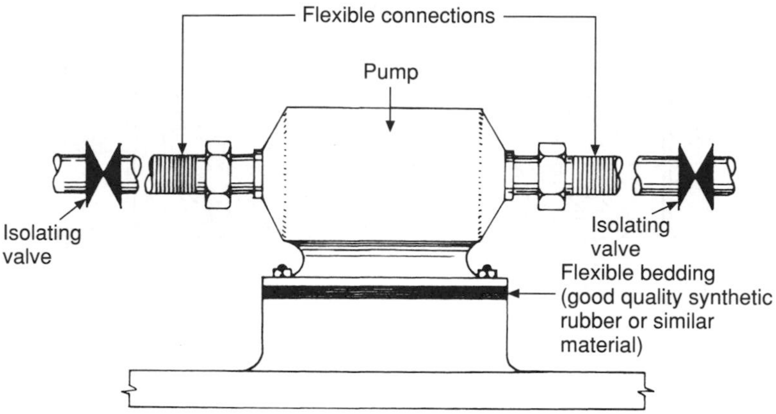


**Fig. 3.10** Noise reduction in pumping equipment.

the requirements is not available. In some instances additional storage may be required in an existing building; Fig. 3.11 shows two arrangements for coupling two or more storage cisterns together. To avoid stagnation, it is essential to ensure that the water flows through both cisterns (as though they were a single cistern) and that the outlet connection is on the opposite side to that of the inlet.

Where it is practicable at least one outlet from the cistern should be connected in the bottom to

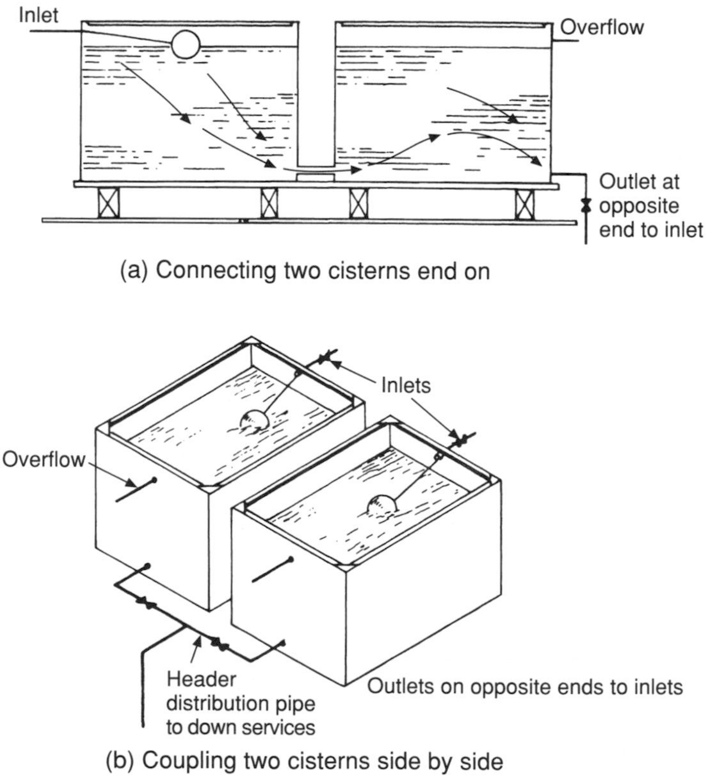


**Fig. 3.11** Coupling large water storage cisterns to avoid stagnation.

avoid the build-up of sediment, which can allow bacteria and other harmful organisms to multiply. Where large storage accommodation is necessary it should be broken down into two or more cisterns in such a way that each can be isolated for maintenance to be carried out without an interruption of the supply. Cisterns having a capacity of more than 1,000 litres must be provided with a wash-out pipe with a full way valve permanently capped when not in use.

*Storage cistern overflows*

Book 1 of this series deals with the size and siting of overflow pipes for cisterns in domestic properties and those having capacities of up to 1,000 litres. It is recommended in the Water Regulations that cisterns having an actual capacity greater than this are fitted with both an overflow and a warning pipe; see Fig. 3.12(a). This is mandatory where the capacity is in excess of 5,000 litres, the invert of the overflow pipe being 50 mm above the normal shut-off level. The warning pipe may be omitted if a level indicator is fitted which provides either visual or audible warning if the water exceeds the normal cut-off level. Figure 3.12(b) illustrates a typical method of achieving this. It is important that the relationship between the inlet valve and overflow pipe in a cistern complies with the regulations against backflow, see Fig. 3.12(c) and Tables 3.2 and 3.3 below.

*Ventilation pipe terminations*

Any ventilating pipe terminating in a cistern must comply with the requirements of Fig. 3.12(c).

## Noise in cold water systems

It is unavoidable that a certain amount of noise is created by plumbing systems, either by the flow of water through the pipes, or by actuating the fittings, a typical example of the latter being the use of a flushing cistern and its subsequent refilling. For some people a certain amount of noise is acceptable, but to others the same noise level would be intolerable, and every plumbing system should be as noise-free as possible. The following text will indicate the more common causes of noise in cold water supplies and its prevention.

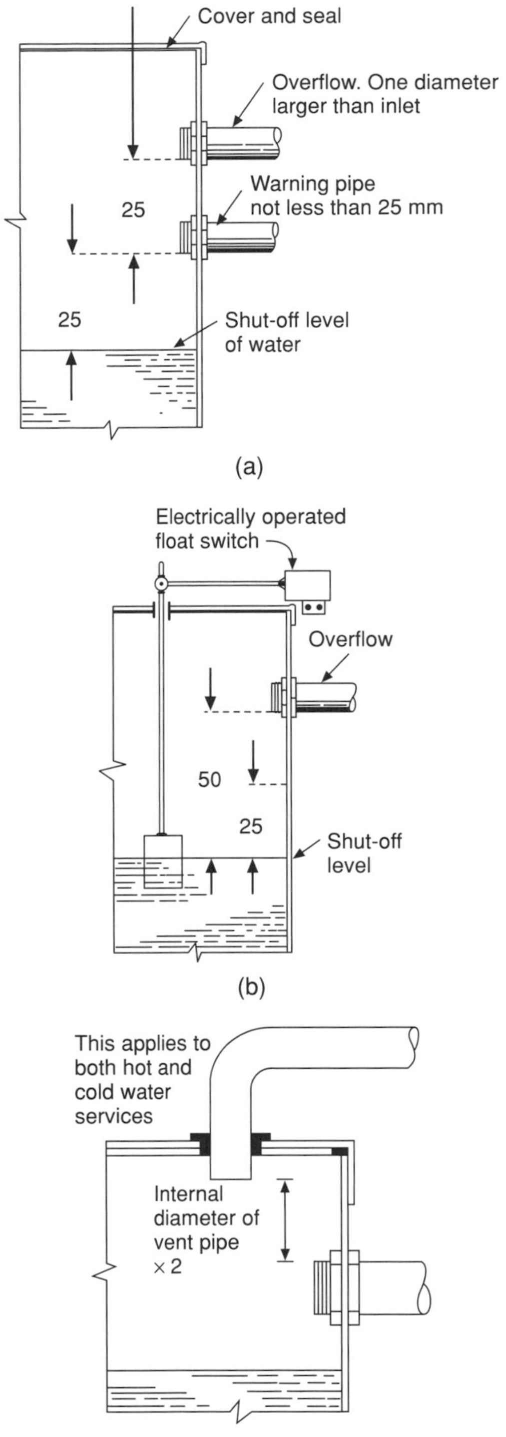


(c) Termination of vent pipes in cisterns

**Fig. 3.12** Positioning overflow and warning pipes in large storage cisterns (distances in mm).

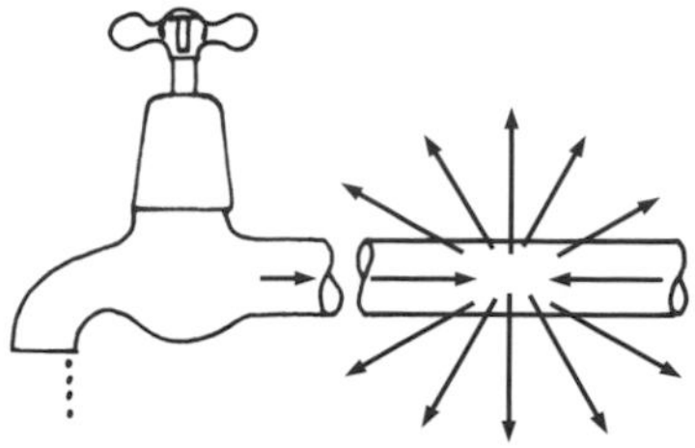

**Fig. 3.13** Water hammer. Sudden closure of tap causes opposing pressure to that of the incoming water resulting in a loud bang in the pipework.

*Water hammer*

This not only is undesirable from the point of view of noise, but in some forms can cause damage to plumbing systems. It is always associated with high-pressure supplies, whether direct from the main or boosted by pumps, and in some cases pressure-reducing valves may have to be employed to lower the pressure in certain sections of a building. Water hammer usually occurs when a high-pressure flow of water is suddenly arrested, as in the case of a tap that is turned off quickly (see Fig. 3.13). This has the effect of causing a loud bang or series of bangs throughout the pipework, and momentarily subjecting the whole system to a pressure almost double that of the incoming water. If this is allowed to persist the excessive pressure can cause a pipe, perhaps already weakened by frost, to start leaking.

The plug cock, classified as a quarter-turn tap, was the first type of tap used in plumbing systems, its history going back many hundreds of years. However, due to the increase in piped water supplies in the early nineteenth century and subsequently higher water pressures, it was found, because of their quick closure, these taps were a major cause of water hammer. This prompted water authorities to insist, prior to the 1986 by-laws, on screw-down taps (which are designed to effect gradual closure) to be fitted on main water supplies. Quarter-turn taps are now permissible for use as servicing valves on such appliances as storage cisterns, flushing cisterns and clothes- or dish-washing machines, and as they are used infrequently they are unlikely to cause persistent water hammer. Comparatively recently ceramic-disc taps have become more common and like the plug cock they are quarter-turn taps which in high-water-pressure areas can give rise to water hammer. Screw-down taps, when the spindle thread and gland become worn, can cause a humming or singing sound as the tap spindle rapidly moves up and down as the flow of water at high pressure passes through it. In such circumstances, although the gland may be repacked as a temporary measure, it is better to replace the tap.

Some appliances, such as washing machines and dishwashers, employ electrically operated valves which are also noisy in operation. Only those having the approval of the local water authority should be used, as apart from the noise they cause they may not conform to the back siphonage regulations.

Special measures are required to limit the effects of water hammer in larger buildings or a series of dwellings supplied by the same service pipe.

In such instances where there are long pipe runs of high-pressure water, it may be necessary to fit an air vessel similar to that shown in Fig. 3.14. These vessels operate on the principle that gases are easily compressible. This enables any momentarily high pressure caused by a sudden closure such as a draw-off to be absorbed.

Float-operated valves are often responsible for noise in plumbing systems due to high-velocity supplies. This is caused by the following:

(a) The water passing through the valve orifice.
(b) Splashing as it falls into the cistern.
(c) Creating ripples or waves on the surface of the water, causing the float to bounce.

When the valve is nearly closed, these waves cause it to open partly and close very quickly, giving rise to a persistent banging, often terminating in a shrill whine just prior to fully closing. While it is not possible entirely to eliminate noise caused by float-operated valves, it can be minimised by the following methods:

(a) The use of velocity restrictors shown in Fig. 3.15.
(b) The use of spray-type outlets in the valve shown in Fig. 3.16.
(c) This is a development of the silencing pipe principle and is shown in Fig. 3.17.
(d) An equilibrium float-operated valve is a very effective method of preventing noise

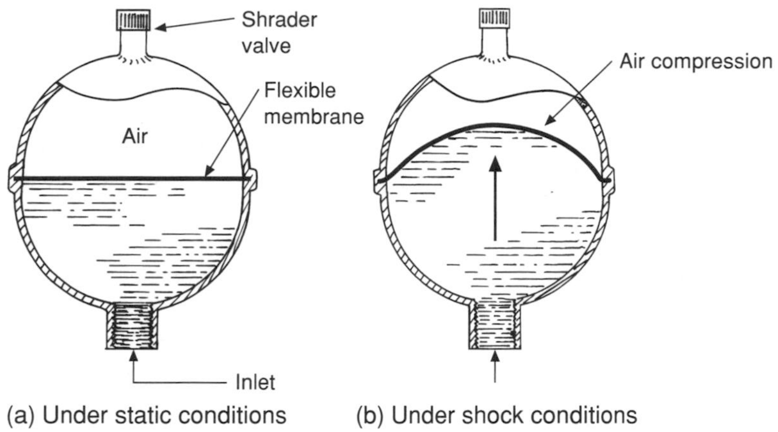


**Fig. 3.14** Water hammer arrester.

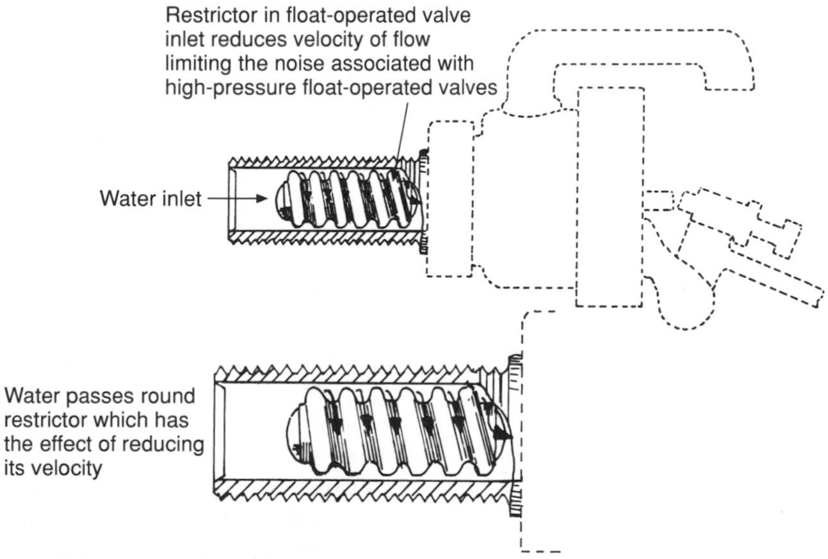


**Fig. 3.15** Reducing velocity of water flow to float-operated valves.

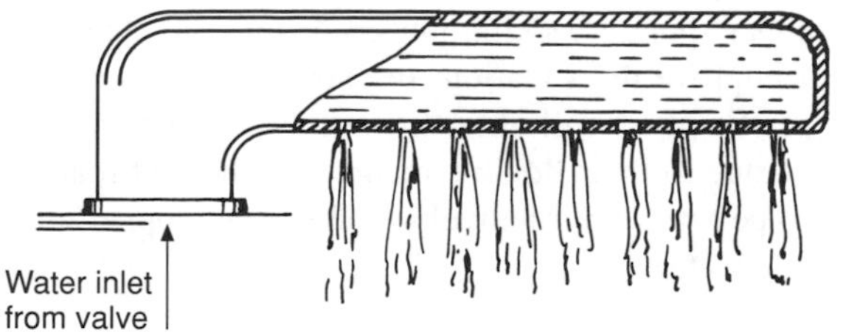


**Fig. 3.16** This attachment has the effect of breaking up the water into a number of finely divided jets which reduce the velocity of the incoming water and the associated noise.

provided suitable provision is made for an appropriate air gap, as most of these valves are of the Portsmouth pattern with a bottom outlet.

The arrangement shown in Fig. 3.12 complies with the Water Regulations provided that a suitable air gap is arranged.

It has already been stated that water hammer is always due to high pressure and velocity. In some cases a reduction of noise can be achieved by partly turning down the outside stop valve so that it passes just sufficient water for the householders'

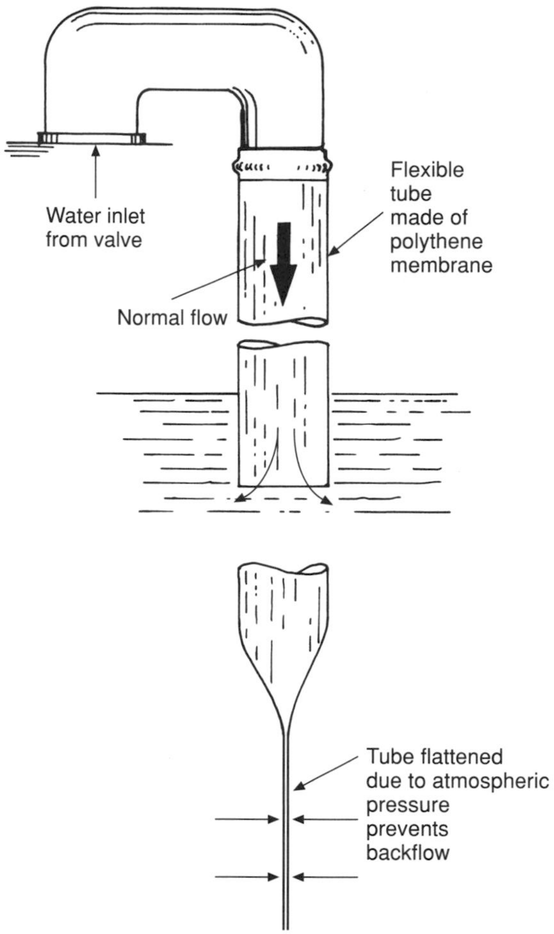


**Fig. 3.17** This arrangement permits the flow of water to discharge below the surface of the water in a manner similar to the original rigid silencing pipes. In the event of backflow causing a negative pressure, atmospheric pressure will flatten the tube as shown in the inset, thus preventing possibly polluted water entering the main. It should be noted that some authorities may not permit this arrangement on water supplied from the main and it should not be used in industrial premises.

requirements. This does not have, as many think, the effect of reducing the pressure, but it does reduce the velocity, and in some instances can cure the problem.

*Pressure-reducing valves*

These valves may be fitted in areas having very high water pressures. Those produced for domestic purposes are not expensive and their use solves many of the noise problems caused by modern taps of the quarter-turn pattern. They are also effective in preventing splashing due to the reduction in pressure when the water is discharged into a sink or wash basin. A simple pressure-reducing valve is shown in Fig. 3.18 where it will be seen that its working principles are similar to those of other pressure-regulating devices with which a plumber is familiar, such as gas governors and regulators for high-pressure gases. As water flows through the inlet, pressure is exerted on the diaphragm which lifts the valve, via the stirrup, closer to its seating. The outlet pressure is varied by adjusting the tension of the large spring. The more the spring is compressed the greater will be the outlet pressure. The purpose of the spring under the valve assists in evening out any pressure fluctuations. It is important to note that the correct pressure adjusting spring is supplied with this valve and the supplier should be informed of the pressure range over which it is required to work. A pressure gauge fitted at a convenient point on the outlet side of the installation will enable fine adjustments to be made when commissioning.

*Noise due to high-velocity flow*

The flow of high-velocity water through water pipes can also be responsible for unnecessary noise, especially pipes of copper or stainless steel. These two materials, being thin walled and rigid, tend to vibrate with the passage of water, and if the pipes are fixed on a hollow surface, such as a partition wall constructed of plasterboard and timber, the wall reacts as a sound box, magnifying the vibration. If fixing pipes to such surfaces cannot be avoided, enough clips or brackets must be used to avoid too much pipe movement, and in some cases a rubber sponge backing placed behind the clips will improve the situation. In extreme cases the use of plastic pipes will almost certainly prevent any noise due to their flexible nature and the fact that the pipe walls are relatively thicker than copper and stainless steel.

## Pollution of water supplies

When water has been discharged from the main water pipe into a cistern or sanitary appliance of

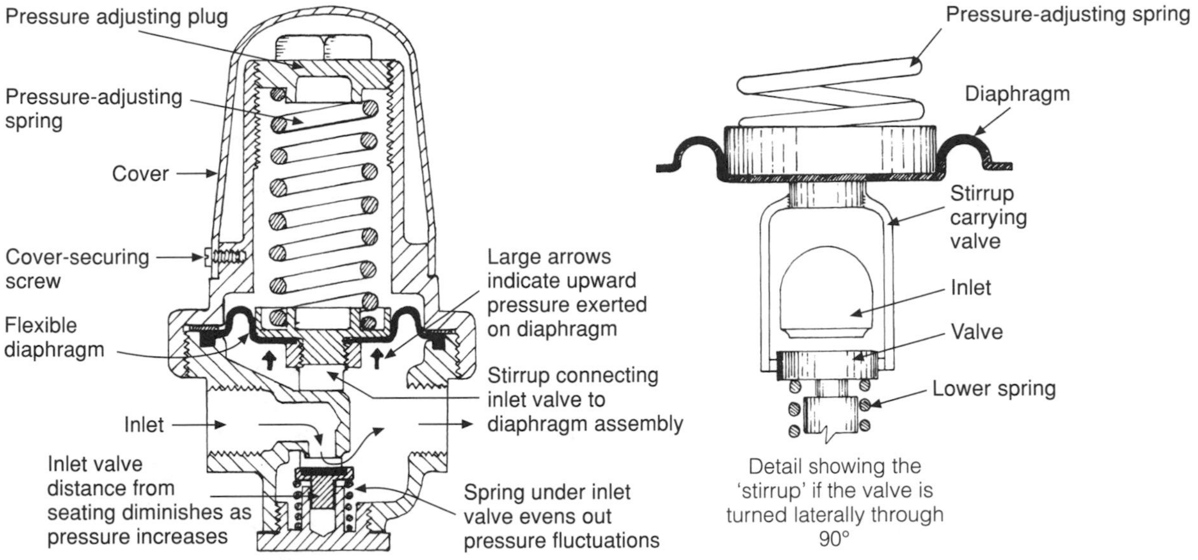


**Fig. 3.18** Pressure-reducing valve.

any sort, it can be assumed that it may be polluted (i.e. has been in contact with possibly harmful bacteria) and must in no circumstances come into contact with water in the main or service pipe. The only exceptions to this are pumped schemes for the supply of drinking water in high buildings, and in such cases there are special requirements to prevent the water becoming contaminated. There are many examples on record of water contamination, all of which were a possible danger to health, not only to the occupants of the premises concerned, but to the community as a whole.

One of the most common causes of contamination occurs where water for drinking and culinary purposes is drawn from uncovered cisterns. It is well known that birds, rodents and insects have been found in such cisterns, quite apart from impurities such as dust which settles on the bottom forming an unpleasant sludge. A fine wire mesh screen must be fitted in the overflow and a sealed cover made of the same material as the cistern must be fitted. It is not necessary or advisable that it should be airtight, simply that it prevents the ingress of debris. Covers made of wood or hardboard must not be used as these materials, being organic, produce a mould in damp conditions which can fall into the water and in itself cause contamination.

The 1999 Water Regulations list five water categories in ascending order of risk to the consumer. They also specify the minimum acceptable methods of preventing contamination of water supplies, which usually occurs due to 'backflow'. The following text is based on the recommendations of the Department of the Environment, Transport and Regions (DETR).

*Water Categories 1–5*

Previous by-laws categorised water in three groups according to risk. The Water Regulations now contain five groups as listed:

1. Water supplies direct from a main supply, generally without being stored before use. One exception to this is the water in a break tank fitted with pumped systems.
2. Water which is pure and wholesome like that of category 1, except it has undergone a change in taste, smell, temperature or appearance, none of which are considered to be a hazard to health. Typical examples are domestic hot water, a mixture of category 1 and 2 water discharged from mixer taps or showers, and water softened in appliances using salt during the regeneration process. In connection with water softeners a drinking water tap should be supplied directly

from the main, especially if used by people with certain medical conditions.

3. This category may pose a slight health hazard and is not suitable for drinking and culinary purposes. Primary water in indirect heating systems, and water used for ablutionary purposes, clothes or dishwashing machines are typical examples.
4. Water falling into this category is not suitable for drinking or culinary purposes and may contain substances causing cancer, micro-organisms, bacteria and viruses, all of which constitute a positive danger to human health. Typical waters classified as category 4 are listed as follows:
   (a) Primary water in commercial or industrial heating systems whether or not any treatment using additives has taken place.
   (b) Water treatment processes using materials other than salt.
   (c) Domestic irrigation systems using perforated hoses or sprinklers fitted less than 150 mm above ground level.
   (d) Commercial clothes and dishwashing machines.
5. This is the most dangerous of all hazard categories and human exposure to water in this category is very serious indeed. Pathogenic organisms are a general term given to the many types of bacteria, viruses and parasites capable of causing serious illness; salmonella and cholera are two of the most commonly known but there are many others. The types of water falling into this category are in many cases similar to those listed in category 4, but they are classified as category 5 if the period of exposure to the risk is longer or if the concentration of the toxic substances in the water is higher. Discharged water from sanitary appliances, both domestic and commercial, is always classified as category 5, together with what is termed 'greywater'. This is the discharged water from ablutionary appliances which instead of passing into the foul drain is stored, and after treatment may be used for flushing WCs. Water collected from roofs or paved surfaces and used for the same purpose is also classified as category 5.

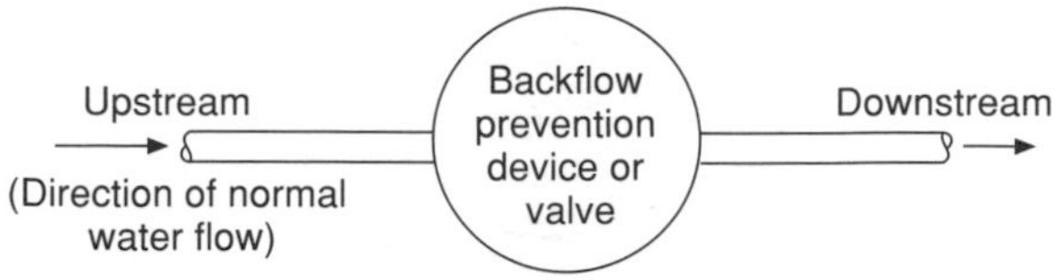


**Fig. 3.19** Illustration of the terms 'upstream' and 'downstream'.

*Backflow*

Water supplied through the 'undertaker's' main is pure and wholesome and it is important that it is not contaminated in any way. Schedule 2 section 6–2 of the Water Regulations deals specifically with backflow and its prevention. Backflow may be defined as the reversal of the normal flow direction in supply pipes and water mains. Two terms are commonly used in connection with the flow of water in pipes: (a) upstream and (b) downstream, both of which are illustrated in Fig. 3.19. Backflow may occur due to either back pressure or back siphonage. If the pressure of the water upstream of a valve or fitting falls below that of the water downstream, backflow may take place. A typical example of back pressure is where a conventional hot storage vessel is fitted, and due to the expansion of the water it is forced back into the supply pipe. It is uncommon in domestic premises and is far more likely to happen in an industrial environment. It has been known to happen in unvented hot water systems and combination boilers provided with storage vessels, but the volume of water involved in these cases is small and is unlikely to exceed a category 2 risk. Provided the temperature of the water is not in excess of 25 °C this risk can be ignored. Back siphonage is far more common and will take place if a negative pressure occurs in the water service or mains pipework. This is by no means unusual and can happen for example if a washout on the main is opened for cleaning, or the main is shut down and drained for repair work. In the event of a major fire, the connection of fire brigade pumps to a hydrant can also cause an appreciable drop in main pressure. A classic instance of backflow due to back siphonage is shown in Fig. 3.20(a) and can still happen if the suitable safeguards are not applied.

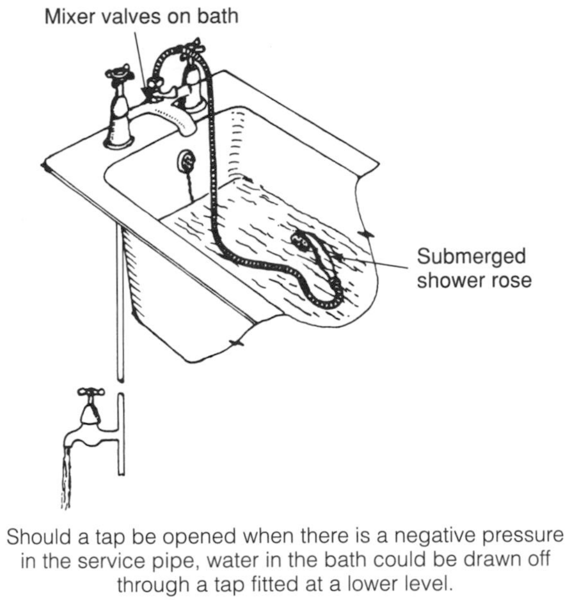
Mixer valves on bath
Submerged shower rose
Should a tap be opened when there is a negative pressure in the service pipe, water in the bath could be drawn off through a tap fitted at a lower level.

**Fig. 3.20(a)** Back siphonage due to a submerged inlet.

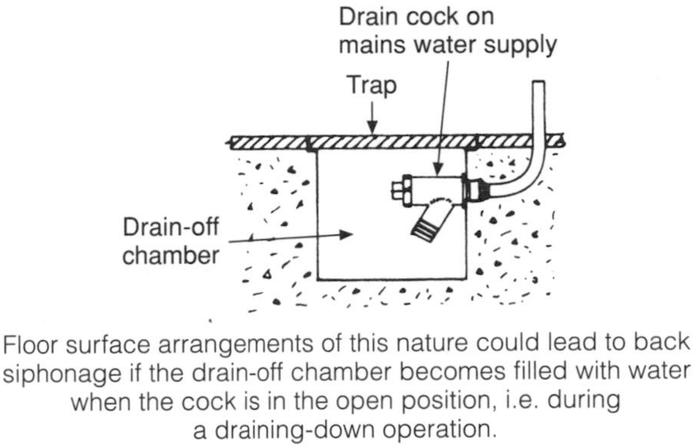
Drain cock on mains water supply
Trap
Drain-off chamber
Floor surface arrangements of this nature could lead to back siphonage if the drain-off chamber becomes filled with water when the cock is in the open position, i.e. during a draining-down operation.

**Fig. 3.20(b)** Back siphonage due to submerged drain-off. This is not permissible due to the reasons described.

Yet another cause of contamination can arise where a drain-off cock is fitted in a floor well (see Fig. 3.20(b)). When the valve is opened and becomes submerged, should a negative pressure occur in the main at that moment, back siphonage can cause the contaminated water to enter the main supply. The seriousness of this problem is further illustrated by the fact that situations have arisen whereby it was possible for water from gutters, and in one known instance a urinal, to enter the water supply system.

Hose-union taps have always been a possible source of contamination in both domestic and industrial situations. The end of the hose can be left in vessels that contain or have contained unpleasant and dangerous substances. They can be left in garden ponds during refilling or put into inspection chambers to wash out a system of drains, these being only a few of the possible risks of contamination due to hose pipes. To illustrate this point further consider the use of hose pipes which are often used with attachments containing solutions for washing down cars or chemicals used in the garden (see Fig. 3.21). These are perfectly safe while water is discharged through the hose, but if a negative pressure occurs on the main, all sorts of dubious substances could find their way into both the service pipe and water main with possibly very serious results. It must also be borne in mind that the number of hot water systems supplied directly from the main will increase in the future. This

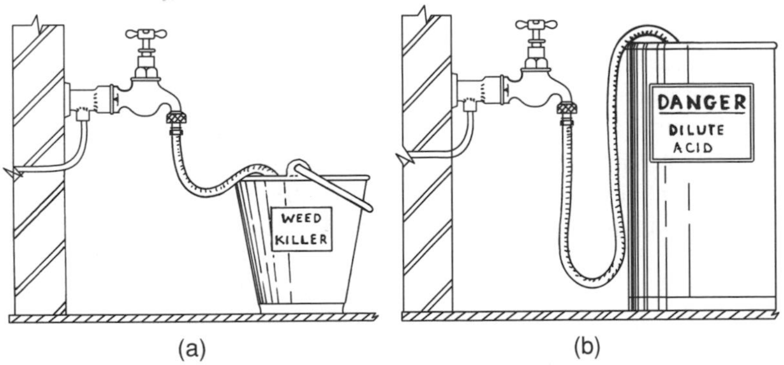
WEED KILLER
DANGER
DILUTE ACID
(a)
(b)

**Fig. 3.21** Illustrating the danger of using hoses connected to taps with inadequate or incorrect backflow protection.

**Table 3.2** Schedule of non-mechanical backflow prevention arrangements and the maximum permissible fluid category for which they are acceptable. (Crown Copyright 1999 Reprinted with permission of HMSO)

| *Type* | *Description of backflow prevention arrangements and devices* | *Suitable for protection against fluid category* | |
|---|---|---|---|
| | | *Back pressure* | *Back siphonage* |
| AA | Air gap with unrestricted discharge above spillover level | 5 | 5 |
| AB | Air gap with weir overflow | 5 | 5 |
| AC | Air gap with vented submerged inlet | 3 | 3 |
| AD | Air gap with injector | 5 | 5 |
| AF | Air gap with circular overflow | 4 | 4 |
| AG | Air gap with minimum size circular overflow determined by measure or vacuum test | 3 | 3 |
| AUK1 | Air gap with interposed cistern (for example, a WC suite) | 3 | 5 |
| AUK2 | Air gaps for taps and combination fittings (tap gaps) discharging over domestic sanitary appliances, such as a wash basin, bidet, bath or shower tray shall not be less than the following:<br>Size of tap or combination fitting — Vertical distance of bottom of tap outlet above spillover level of receiving appliance<br>Not exceeding $G^1/_2''$ — 20 mm<br>Exceeding $G^1/_2''$ but not exceeding $G^3/_4''$ — 25 mm<br>Exceeding $G^3/_4''$ — 70 mm | X | 3 |
| AUK3 | Air gaps for taps or combination fittings (tap gaps) discharging over any higher risk domestic sanitary appliances where a fluid category 4 or 5 is present, such as:<br>a. any domestic or non-domestic sink or other appliance; or<br>b. any appliances in premises where a higher level of protection is required, such as some appliances in hospitals or other health care premises,<br>shall be not less than 20 mm or twice the diameter of the inlet pipe to the fitting, whichever is the greater. | X | 5 |
| DC | Pipe interrupter with permanent atmospheric vent | X | 5 |

*Notes*: X indicates that the backflow prevention arrangement or device is not applicable or not acceptable for protection against back pressure for any fluid category within water installations in the UK.
Equivalent to Table S6.1 in DETR Guidance document.

means care must be taken to prevent any possible pollution, via not only cold water supplies but also those of hot water too. It cannot be stressed enough that compliance with the Water Regulations must be strictly observed to ensure that any danger to public health by water supplies is minimized.

## Backflow prevention

There are many devices listed to prevent backflow. Some are mechanical, typical examples being check valves. Others are classified as non-mechanical, the traditional air gap being one of the most commonly known. They are all classified as being suitable for certain types of fluid category. Tables 3.2 and 3.3 show the main types of backflow prevention methods used against the relevant water category. They are not exhaustive but carry most of the common risk levels encountered for domestic and commercial applications in the UK. Reference to these tables will be necessary when reading the following text to determine the degree of protection

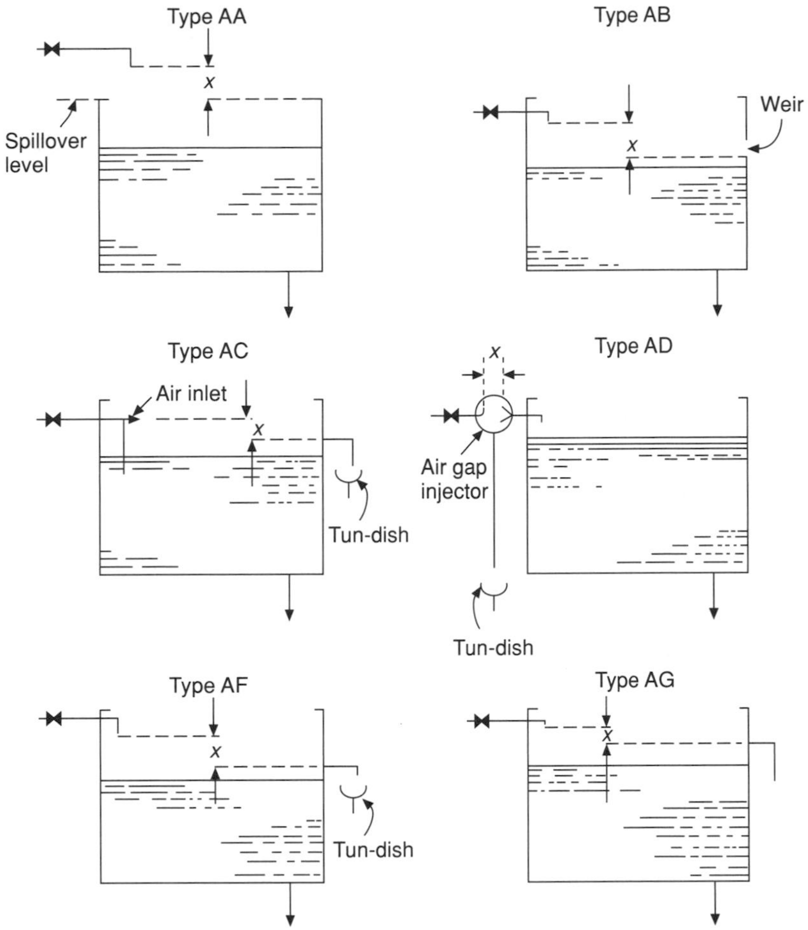


Reference must be made to Table 3.2, which shows the types of risk for which the air gaps are suitable. Some are used only for certain industrial processes. The air gaps for systems with tun-dishes must comply with Table 3.2.

**Fig. 3.22** Types of air gap. In all illustrations, the air gap is indicated by x.

offered by each method of backflow device shown against back pressure and back siphonage. The fluid catergories for which they can be safely used should also be noted. Note that many of the devices listed relate to situations that are unusual in the normal course of work.

It is the responsibility of the plumber or fitter to ensure the correct type of device is fitted to satisfy the requirements of the Water Regulations. If any doubt exists the water supplier must be consulted.

## Non-mechanical backflow devices

*Air gaps*

An air gap must be clearly visible and constitutes a positive break between the level of water in a cistern or appliance and the lowest level of discharge from the inlet. It must not be less than 20 mm or twice the internal diameter of the inlet, whichever is the greater. The angle of flow must not be more than 15° from the vertical centre line of the water inlet. Figures 3.22 and 3.23 illustrate

**Table 3.3** Schedule of mechanical backflow prevention arrangements and the maximum permissible fluid category for which they are acceptable. (Crown Copyright 1999 Reprinted with permission of HMSO)

| *Type* | *Description of backflow prevention arrangements and devices* | *Suitable for protection against fluid category* | |
|---|---|---|---|
| | | *Back pressure* | *Back siphonage* |
| BA | Verifiable backflow preventer with reduced pressure zone (RPZ valve) | 4 | 4 |
| CA | Non-verifiable disconnector with difference between pressure zones not greater than 10% | 3 | 3 |
| DA | Anti-vacuum valve (or vacuum breaker) | X | 3 |
| DB | Pipe interrupter with atmospheric vent and moving element | X | 4 |
| DUK1 | Anti-vacuum valve combined with a single check valve | 2 | 3 |
| EA | Verifiable single check valve | 2 | 2 |
| EB | Non-verifiable single check valve | 2 | 2 |
| EC | Verifiable double check valve | 3 | 3 |
| ED | Non-verifiable double check valve | 3 | 3 |
| HA | Hose union backflow preventer. Only permitted for use on existing hose union taps in house installations | 2 | 3 |
| HC | Diverter with automatic return (normally integral with some domestic appliance applications only) | X | 3 |
| HUK1 | Hose union tap which incorporates a double check valve. Only permitted for replacement of existing hose union taps in house installations | 3 | 3 |
| LA | Pressurised air inlet valve | X | 2 |
| LB | Pressurised air inlet valve combined with a check valve downstream | 2 | 3 |

*Notes*:

1 X indicates that the backflow prevention device is not acceptable for protection against back pressure for any fluid category within water installations in the UK.

2 Arrangements incorporating a Type DB device shall have no control valves on the outlet of the device. The device shall be fitted not less than 300 mm above the spillover level of an appliance and discharge vertically downwards.

3 Types DA and DUK1 shall have no control valves on the outlet of the device and be fitted on a 300 mm minimum Type A upstand.

4 Relief outlet ports from Types BA and CA backflow prevention devices shall terminate with an air gap, the dimension of which should satisfy a Type AA air gap.

Equivalent of Table S6.2 in DETR Guidance document.

diagrammatically various configurations of the air gap principle. Reference must be made between the illustrations and Tables 3.2 and 3.3 so that a clear picture of the requirements of the various categories of water can be seen. The degree of protection for both back pressure and back siphonage is also shown in the tables.

It will be seen that types AA and AB give good protection against both types of backflow. Such protection is necessary when installing cattle drinking troughs, supplies to a sewage treatment plant, or vats containing toxic chemicals. Types AC, AD and AF are again mainly required in industrial or commercial premises. It will be seen that some types of protection require the overflow to discharge into a tun-dish. The reason for this is because the overflow may discharge into, say, a contaminated pond or river. If backflow via the overflow took place from such a source, it could contaminate the contents of the cistern. The AG air gap is suitable for lower risks and it will be seen from the table that it offers protection up to category 3

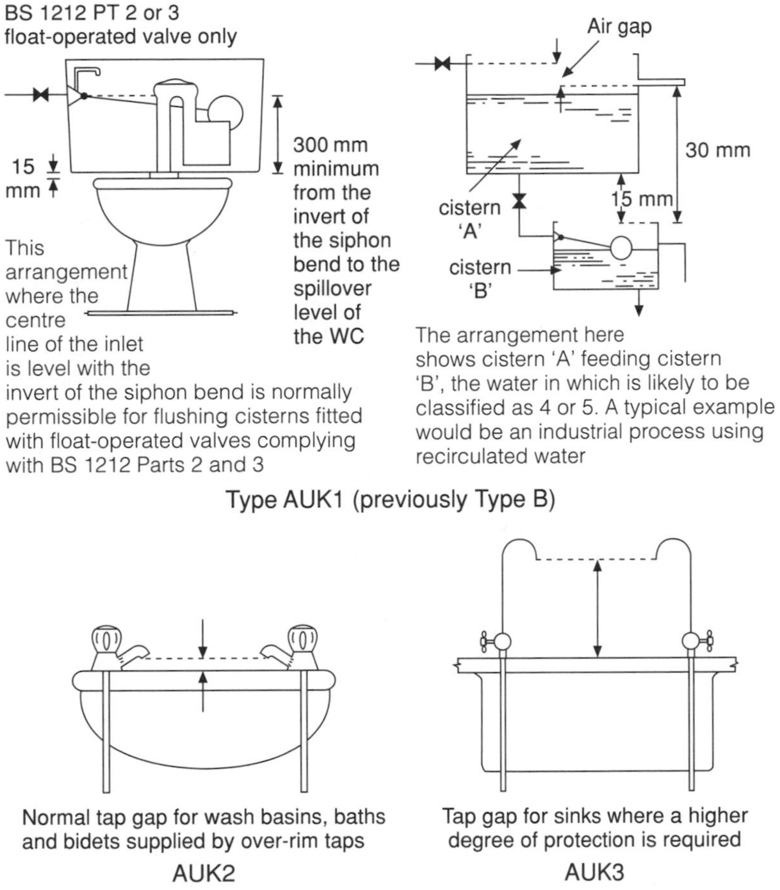


**Fig. 3.23** AUK air gaps. Most of these are 'built-in' by the manufacturer. Refer also to Tables 3.2 and 3.3.

for both types of backflow. It complies with the requirements of BS:6281 Part 2 and was formerly known as the class B air gap, which satisfied the requirements of previous by-laws in relation to flushing cisterns fitted with BS:1212 Part 2 or 3 float-operated valves.

In most cases air gaps complying to type AUK 1,2,3 are built into an appliance, e.g. wash basins and baths, and a plumber will usually be familiar with these. Care must be taken, however, in relation to type AUK 1 and close-coupled WC suites with flushing cisterns fitted with valves. If the internal overflow pipe is shortened in any way, a check must be made to ensure the minimum distance of 300 mm is maintained between the spillover level of the WC and the overflow weir or invert. If a siphon is used as shown in Fig. 3.23 this measurement is usually built in, but it is recommended that a measurement check is made to ensure compliance with the regulations.

*Pipe interrupter*

This is a device with no moving parts and is shown in Fig. 3.24. Its working principle is very simple: should a negative pressure take place on the inlet (upstream) of the valve, air is drawn in via the air ports. This has exactly the same effect as when air is admitted to a siphon bend: it stops siphonic action. This type of arrangement can be made as an integral component of such equipment as washing machines or dishwashers. It can also be used when controlled by an upstream valve for delivering water to or cleaning vehicles, ships or large cisterns where the supply is operated by a solenoid valve. Where

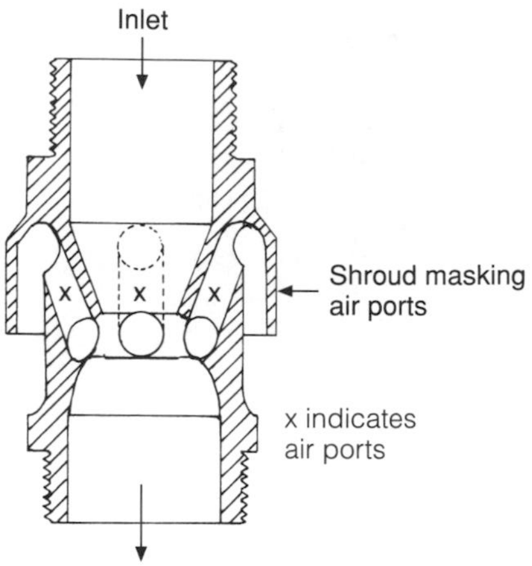
Inlet
Shroud masking air ports
x indicates air ports
x
x
x

**Fig. 3.24** Pipe interrupter type DC. This device is non-mechanical and has no moving parts. It is fitted in line with the pipe run where there are no control valves downstream of the outlet.

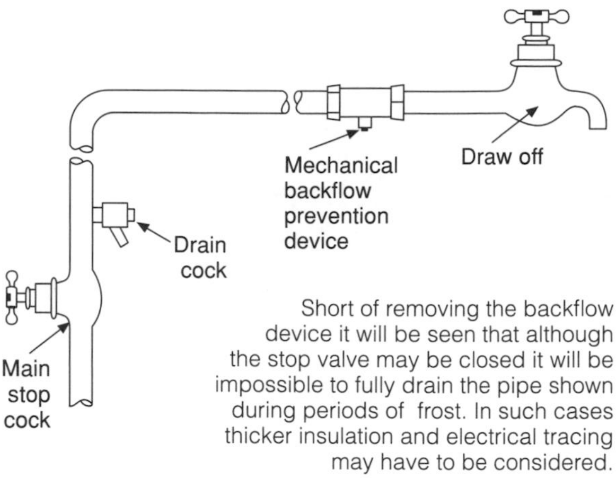
Mechanical backflow prevention device
Draw off
Drain cock
Main stop cock
Short of removing the backflow device it will be seen that although the stop valve may be closed it will be impossible to fully drain the pipe shown during periods of frost. In such cases thicker insulation and electrical tracing may have to be considered.

**Fig. 3.25** Draining down water services.

flushing valves are permitted, they must be provided with a pipe interrupter, often built into the outlet of the valve itself. They are suitable for category 5 risks against backflow.

### Mechanical backflow protection

In most cases this type of backflow prevention has some moving parts, usually spring-loaded valves. Some are classed as 'verifiable', which means if necessary they can be checked for effectiveness. To test a verifiable double check valve, for example, turn off the main supply and open the valve or test screw between the two internal valves. If only a small quantity of water emerges it can be assumed that all is well. If a larger quantity is discharged it can be assumed that one of the valves is letting by and needs either servicing or replacing. It should be noted that it is not possible to verify the first valve embodied in a double check valve. One other point that must always be taken into account with mechanical backflow valves is the fact that it may be impossible, in periods of frost, to completely drain a system (see Fig. 3.25). Exposure to very cold weather in some industrial and commercial installations may require some form of tracing, and frost protection in such situations must be seriously considered. All mechanical backflow devices must be accessible for inspection, maintenance and testing. Except for double check valves fitted to hose union bib taps, all such devices must be fitted within the premises. Strainers must be fitted immediately upstream of any backflow prevention device used to protect an installation against category 4 or 5 risks. This is necessary to avoid any debris becoming deposited on the valve seats preventing their closure. Service valves must also be provided both upstream and downstream of the device. Any relief outlets must be provided with a type AA air gap not less than 300 mm above floor level. Table 3.3 lists most of the mechanical backflow devices currently available and the following text and illustrations describe those most commonly used in the UK. **The water supplier must be informed if it is proposed to install any mechanical device to prevent backflow risks from category 4 or 5.**

#### *The reduced pressure zone (RPZ) backflow prevention valve*

These valves are currently the only mechanical device suitable for both types of backflow against category 4 risks and can be used for both whole site protection or point of use application. They are rarely used or indeed necessary in domestic dwellings as the contamination risks do not warrant it. Figure 3.26 illustrates diagrammatically an RPZ valve, and it will be seen that it has similar

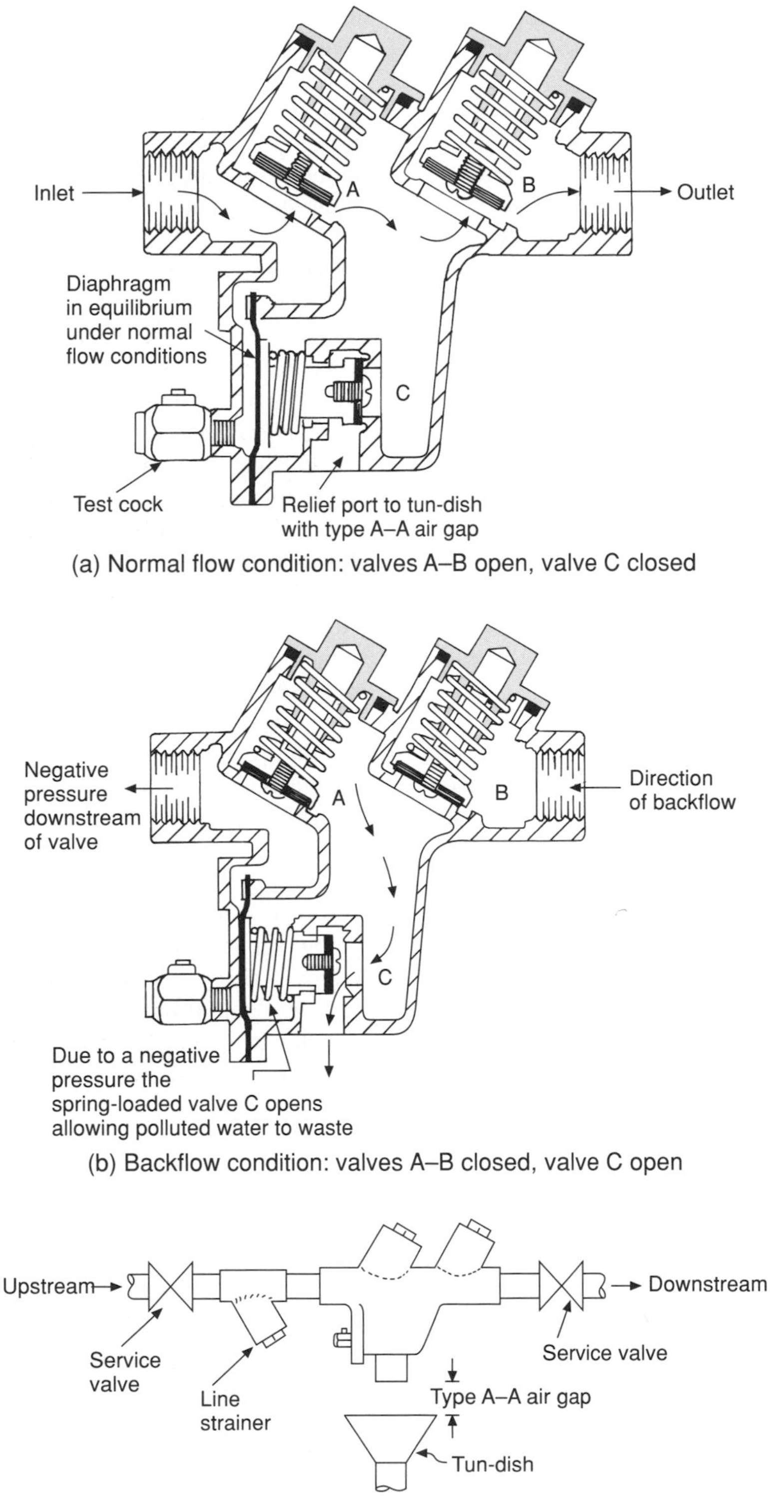


(c) Installation details of RPZ valve

**Fig. 3.26** RPZ valve (verifiable backflow preventer with reduced pressure zone).

characteristics to those of a double check valve. Both check valves are designed to progressively reduce the water pressure in two stages so that under normal flow conditions pressure is always less on the downstream side than it is upstream, hence the name 'reduced pressure zone valve'. This pressure imbalance and the check valves themselves act to prevent backflow. The other main characteristic of the RPZ valve is that if a negative pressure occurs upstream, even if one of the check valves A or B is not operating correctly, it will open to the atmosphere via the relief valve C creating an air gap. This will prevent water which may have become contaminated entering the upstream part of the system. It will be seen that valve C is normally closed, as the upstream pressure of water acting on the diaphragm overcomes the spring and keeps the valve closed. Only when the upstream pressure falls and backflow occurs will the relief valve open, creating an air gap due to the action of the spring and a lowering of the pressure on the diaphragm.

*Installation* Manufacturers of RPZ valves specify the installation requirements necessary for servicing and testing. Generally they should be fitted no less than 300 mm from floor level with a minimum clearance behind the assembly of 100 mm. Suitable methods of discharging waste water must be provided for the air gap to always be maintained. The usual method of achieving this is to provide a tun-dish as shown in Fig. 3.26(c).

*Maintenance* These valves must be inspected annually to ensure they are functioning correctly. A differential pressure gauge is connected to each of the three test points in the valve body (not shown in the illustration). The pressure shown on the gauge will indicate whether all the valves are functioning correctly. Most manufacturers of these valves recommend that maintenance is carried out by plumbers or fitters who have undertaken an approved course and are suitably qualified.

*Non-verifiable disconnector with different pressure zones (type CA)*

This is similar to the RPZ valve in that it provides a positive disconnection area between the upstream and the downstream flow of water. In a similar way to the RPZ valve, the area between the two main valves will open to the atmosphere in the event of an upstream pressure drop. Any water discharging through the outlet is allowed to run to waste through a tun-dish. It is only suitable for use with water up to a category 3 risk.

*Check valves*

These are basically a spring-loaded valve which will close due to pressure of the spring if the upstream pressure falls below that of the downstream.

*Single check valves* Type EA is a single verifiable check valve, the term 'verifiable' meaning that all valves of this type can be checked for effective functioning. Type EB valves are non-verifiable. However, both types are suitable for category 2 risks. They will permit water to flow upstream but will close against downstream pressure. They are suitable for use against both back siphonage and back pressure. They are mainly used with mixing valves supplied with hot and cold water at differing pressures, and are often built in by the manufacturer. Figure 3.27 shows a typical non-verifiable valve of this type.

*Double check valves* These are similar in construction to single check valves, but because they incorporate two valves, in the event of one failing to close under backflow conditions, the other will still afford the necessary protection. The classification EC indicates the valve is verifiable and is shown in Fig. 3.28 Non-verifiable

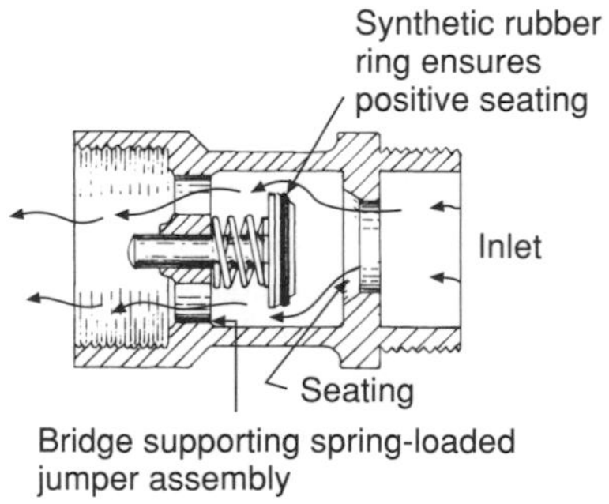


**Fig. 3.27** Non-verifiable single check valve type EB.

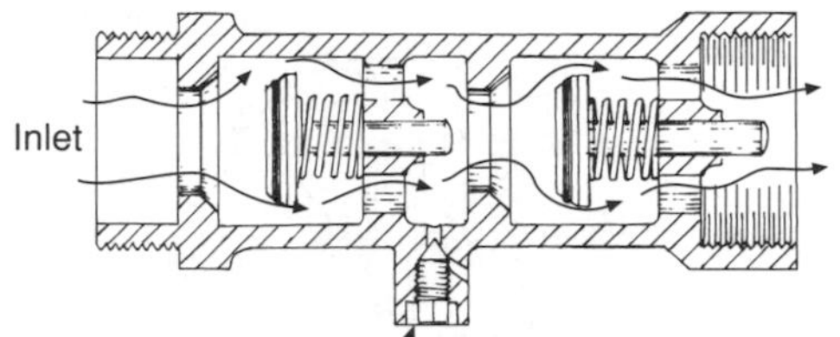

Inlet
Test cock. With no water flowing through the valve both jumpers should be seated. If a persistent flow of water can be seen when the test cock is opened, one or both valves are defective. The cause of the trouble should be investigated or the complete unit replaced


**Fig. 3.28** Double check valve type EC.

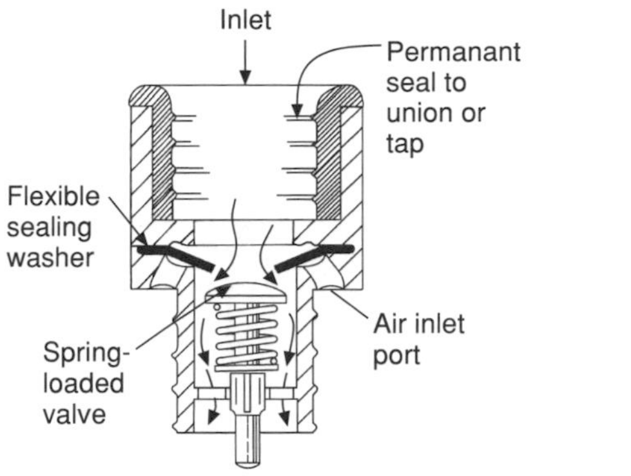


(a) Hose union backflow preventer type HA

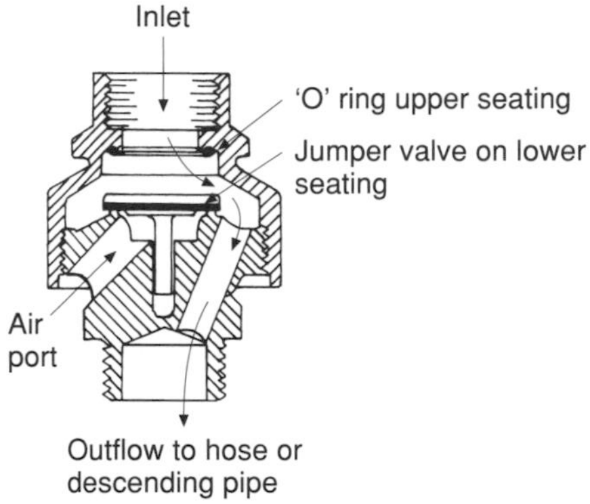


(b) Check and vacuum breaker type HA

Both valves shown work on the same principle except that (a) is spring-loaded. In the event of a negative inlet pressure the valves close and admit air to the outlet allowing any water in the hose to drain.

**Fig. 3.29** Combination check and vacuum breakers.

types without a test cock are classified as the ED type. Both are suitable for protection against category 3 contamination risks.

Prior to commissioning a system it should always be flushed out, but it is especially important if any mechanical devices such as check valves are fitted. Flushing out should be undertaken with the valves out of position, as any detritus such as filings or swarf could lodge on the valve seatings and prevent them closing. Verifiable valves must always be checked during commissioning to ensure they are working correctly. Taking as an example the verifiable double check valve, it will be seen that it has a test cock between the two valves. By shutting off the water supply and opening the test cock, little or no water should appear. This indicates the downstream valve is closed and in working order. If it is not then the complete unit should be removed for repair or replacement, because, although the valve upstream may be holding, there is no way of checking it unless the valve incorporates two test cocks.

*Hose union taps – backflow protection*

Because the category of risk will depend upon the circumstances for which they are used, the type of backflow protection will vary. For domestic properties using hand-held hoses, category 3 backflow protection devices are normally acceptable. For new build work a double check valve situated inside the building must be fitted to the supply. Book 1 of this series illustrates hose union taps type HUK1 which embody two check valves and can be used as a replacement only in existing properties. Another acceptable method of protection, again only for existing properties, is the use of a combined check and anti-vacuum valve (type HA in Table 3.3). Two very similar types are shown in Figs 3.29(a) and (b) and are designed to screw onto the tap outlet. The reason that HUK1 and HA types should not be used on new work is because of the difficulty of draining down in frosty weather, making them very prone to damage.

*Pipe interrupter with vent and moving element*

Classified as DB this is currently an unfamiliar device in the UK. Figure 3.30 illustrates a typical valve of this type. The flexible membrane closes off

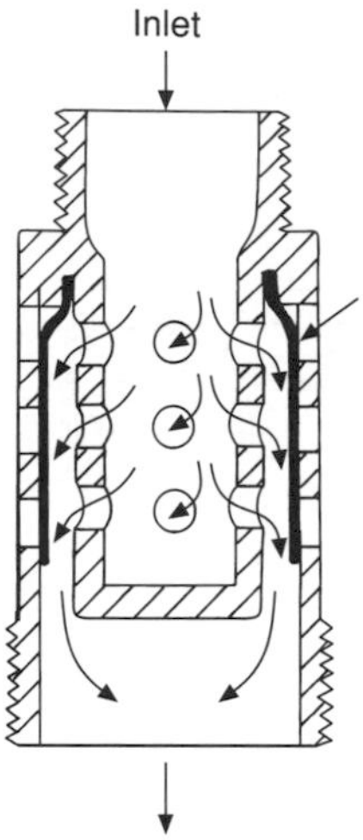


**Fig. 3.30** Pipe interrupter with vents and moving element, type DB.

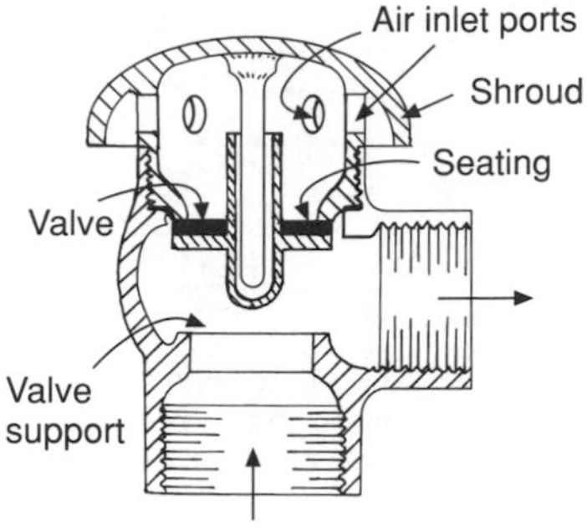


A negative pressure upstream allows the valve to drop maintaining equilibrium in the pipe work system.

**Fig. 3.31** DA type anti-vacuum valve.

the air ports when in normal use, but in the event of a negative pressure in the supply the membrane will be drawn towards the inlet ports, closing them and thus preventing any backflow and the possibility of contaminated water entering the supply. It must be installed in a vertical position as shown with no valves or restriction on its outlet and fitted no less than 300 mm above the spillover level of the appliance it supplies. It will be seen that it has similar characteristics to a pipe interrupter and in some circumstances it is used as an alternative.

*Diverter with automatic return*

This device is usually built into combined bath–shower mixers and is classified in Table 3.3 as the HC type. When the taps are opened the valve is lifted manually to allow a mixed supply of water to the shower head, the flow of water keeping it open while the shower is operating. It is designed to drop under its own weight if the supply is interrupted.

*Anti-vacuum valves (vacuum breaker)*

A typical valve of this type is shown in Fig. 3.31. Under normal conditions the valve is held in the closed position by the water pressure. If the pressure upstream falls to or below that downstream of the valve, it will fall open under its own weight to prevent backflow. When the valve is used on its own it is listed in Table 3.3 as type DA. It is often used in this form in conjunction with unvented hot water cylinders to prevent their collapse. When it is used as a backflow device it is more likely to be used in conjunction with a single check valve. The various configurations shown diagrammatically in Fig. 3.32 illustrate this. These are quite often built into domestic washing machines and dishwashers, and into many types of commercial appliances with similar applications. They are also used in connection with large-scale irrigation systems. As it is not unknown for leakage to occur due to failure of these valves to re-seat properly after operating, it is not advisable that they are used in situations where this is likely to cause serious damage.

### Application of backflow-prevention devices in domestic premises

Figure 3.33 illustrates a typical arrangement of both hot and cold water services in a dwelling. It will be seen that it does not vary a great deal with traditional practice. All draw-off taps are protected by a type AA air gap apart from the external hose union tap which must be provided with either a check and anti-vacuum valve at the hose connection, or a double check valve on the supply pipe. Float-operated valves should be of the diaphragm type complying to BS 1212 Part 2 or 3. The storage cisterns must be provided with a type AG air gap and flushing cisterns with type AUK1. The gap between the float-operated valve and overflow is the same with both types, but the AUK1

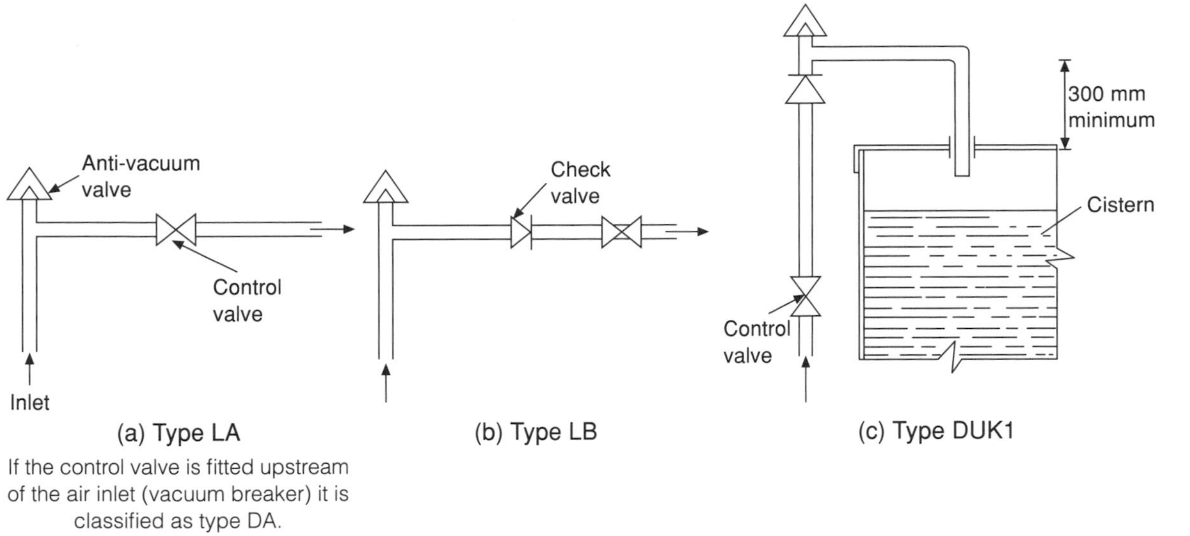


**Fig. 3.32** Backflow configurations with air admittance and check valves.

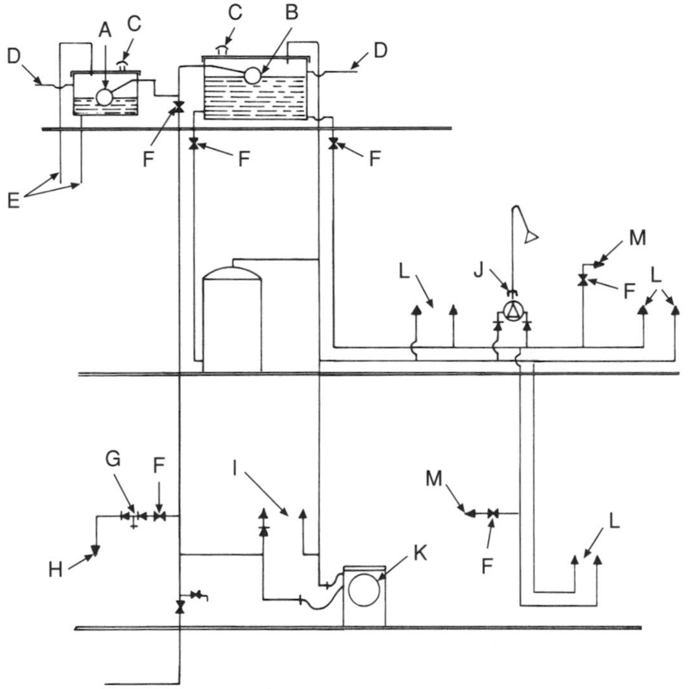


**Fig. 3.33** Water supply for domestic premises complying with the Water Regulations. Both storage cisterns (A) and (B) have type AG air gaps; (C) cistern vents; (D) screened overflow; (E) feed and vent pipe to heating system; (F) service valves; (G) double check valve; (H) outside tap; (I) sink mixer of the biflow type with a single check valve on the cold supply; (J) bath–shower mixer tap with single check valves on inlets and flexible hose; (K) washing machine with integral AG air gap; (L) all draw-offs including the over-rim bidet have type AUK air gaps. (M) Flushing cisterns incorporate type AUK1 air gaps.

type also incorporates a vertical gap not less than 300 mm from any contaminated water, in this case the WC. Service valves must be fitted so that float-operated valves can be repaired or replaced without shutting down supplies to other appliances. These service valves may be of the quarter-turn ball type, stop- or gate valves, bearing in mind the latter are not suitable for main water supply. Where a sink mixing tap is subject to unequal pressures, unless it is of the biflow type it should be fitted with a single check valve to prevent possible backflow from the hot water supply into the cold service pipe. If, for example, a hose was connected to the outlet of the mixer, backflow could take place if the main supply, or the water main, were to be subjected to a negative pressure. Such hoses are available for users of washing machines not permanently connected to the hot and cold supplies. Permanently plumbed-in clothes- or dish-washing machines should comply with the Water Research Centre's requirements and BS 6614 which specifies the requirements for connection of these appliances to water supplies. If this is the case they will be provided with a type A6 air gap or pipe interrupter made integrally in the machine. Although it is not mandatory at present, some manufacturers of these machines may provide single check valves as an integral part of the hoses.

Showers having flexible hoses do present a serious risk of contamination, which can be addressed as follows:

(a) The flexible hose may be attached in such a way that the shower rose is prevented by a restraining ring from becoming submerged in the appliance it serves, thus maintaining an air gap above the flood level of the appliance.
(b) The mixer is provided with single check valves to both the hot and cold supplies. A further check valve is fitted in the flexible hose outlet. All these valves are normally built into the mixer unit and give the equivalent protection of double check valves on each supply. A further form of protection is the use of a hose restraining ring which prevents the shower rose falling below the flood level of the appliance.

It may be asked why, in the system shown, it is necessary to go to these lengths as many are in current use without any backflow protection. The reason for this is that at some future date the premises might be fitted with an unvented hot water system with both hot and cold supplies taken from the main. It will be seen that in such circumstances a very real danger of contamination could exist. Reputable manufacturers are aware of this and produce taps and valves with built-in backflow protection where possible.

*Appliances with submerged inlets*

The only type of appliance normally falling into this category is a bidet, which due to the very nature of its use poses a serious risk of contamination. In previous by-laws these appliances have never been permitted to be directly connected to a main supply pipe and the same prohibition is maintained in the Water Regulations. When the cold supply to a bidet is provided by a storage vessel, a separate distribution pipe from that supply to other draw-offs is necessary, except that where applicable the same pipe can be used to supply a WC or urinal. The methods of supplying hot water may vary. A separate water heater supplying the bidet only is the ideal solution. The heater would have to be a low-pressure type as it would not be permissible to connect it to the main supply. Figure 3.34(a) illustrates an acceptable solution. It is not mentioned in the Water Regulations but was accepted by the 1996 Water By-laws. Yet another alternative is shown in Fig. 3.34(b) where both the hot and cold supply are connected to a mixer valve. This is the most suitable arrangement and is that used in hospitals and care homes where thermostatic control is essential. Although several types of backflow devices are listed as being suitable for this arrangement, the pipe interrupter is probably the most suitable, but remember there must be no restriction to flow downstream of the water. From the foregoing text it will be seen that to meet the requirements of the Water Regulations, any fitting having an inlet below flood level is both difficult and expensive to install, and the use of bidets having over-rim supplies is becoming normal practice in domestic properties. As these comply with the requirements of a type AUK2 air gap they are the only type of bidet suitable for use with unvented hot water systems.

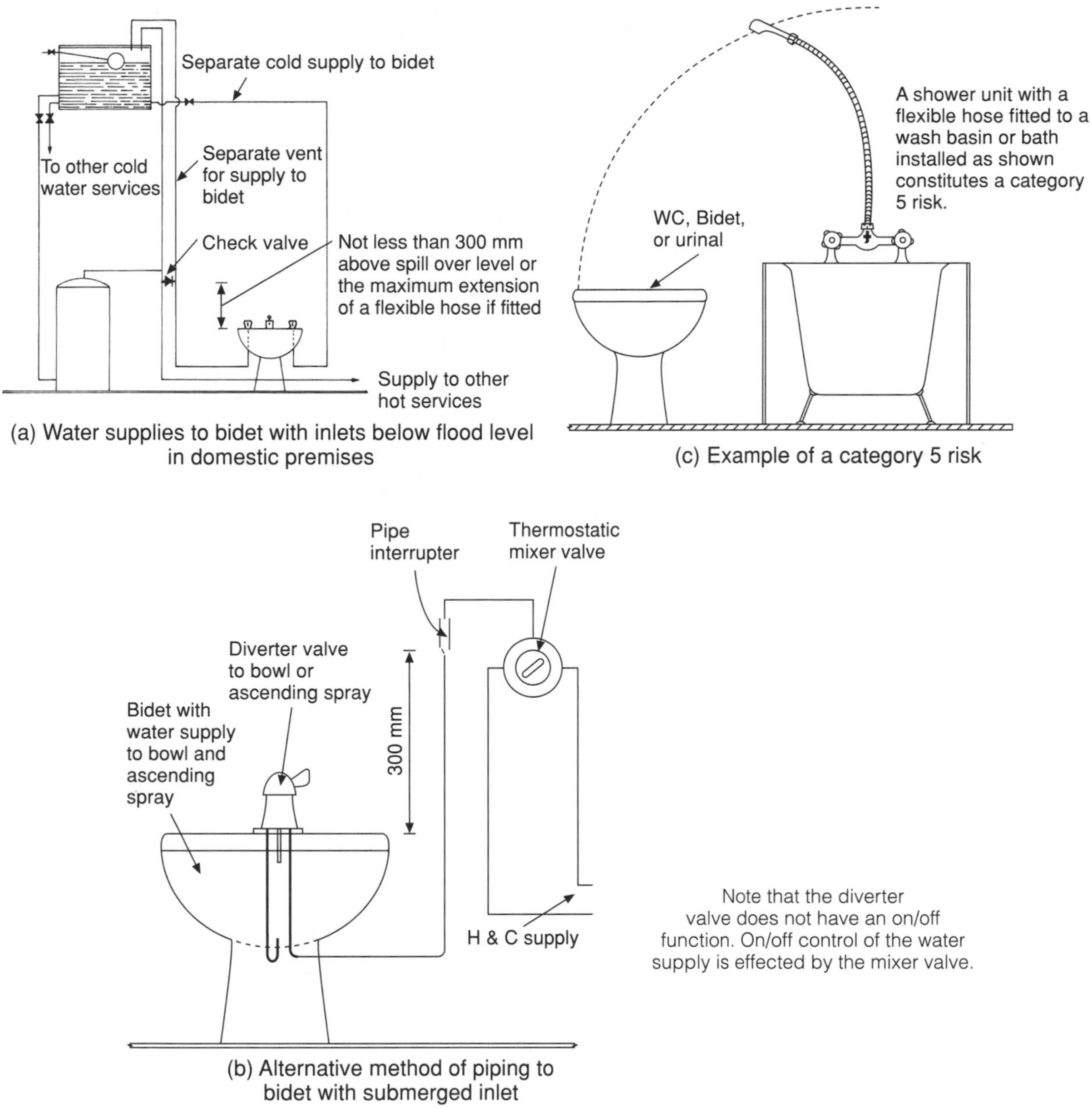


**Fig. 3.34** Appliances with submerged inlets, and an example of a category 5 risk.

*Flexible hoses*

The use of flexible hoses attached to sanitary appliances has always constituted a serious backflow risk. A typical example is shown in Fig. 3.19, which would be considered a category 3 risk. Figure 3.34(c) illustrates an even greater danger where the hose may fall into water classified as category 5. A hose restraining ring would definitely not be suitable here as it can be detached; a fixed shower head would be more practical, indeed the only option if it is connected to a main water supply. Even if the hose is connected to a distribution pipe serving other appliances the danger is still very real, and a type AA air gap would be the only solution complying with the Water Regulations.

### Backflow protection in multi-storey public and commercial buildings

This is a very wide area and to list every instance of possible contamination cannot be done within the limitation of this volume. The following text, however, illustrates some examples of good practice so that the reader will be able to recognise installations that do not comply with the Water Regulations. One of the most common instances of water contamination that may occur in industrial and agricultural premises, apart from back siphonage, is a cross-connection between water from the main and water from some other source. Surprising as it may seem, main water pipework has been found to be connected to supplies, the original source of which was subsequently discovered to be from wells, and in one case, a pond. Possibly even greater dangers exist in industrial premises dealing with dangerous chemicals and a good plumber must always investigate an existing system very carefully before making any connection to a main water supply. The illustrations in Fig. 3.35 are mainly concerned with the prevention of back siphonage. Figure 3.35(a) shows how water from another source is prevented from contaminating mains supplies, and Fig. 3.35(b) shows an arrangement where water is used for cooling in industrial premises. Typical examples include cooling for plastic moulding machines, milk-cooling and air-conditioning plant. Note that in both cases the type AA air gap *must* be used. Figure 3.35(c) shows how water is supplied to cisterns and tanks to which toxic chemicals may be added, i.e. laboratory mixing tanks, etc. The risks are so great here that a tun-dish, or a service cistern as shown is fitted.

Standpipes in industrial premises are subject to strict control and should be supplied with water from a cistern. In certain circumstances where there is little risk of contamination, it may be possible, with the prior agreement of the water authority, to supply such taps direct from the main, providing a double check valve is fitted in the pipeline. Where bidets of the submerged inlet type or those having

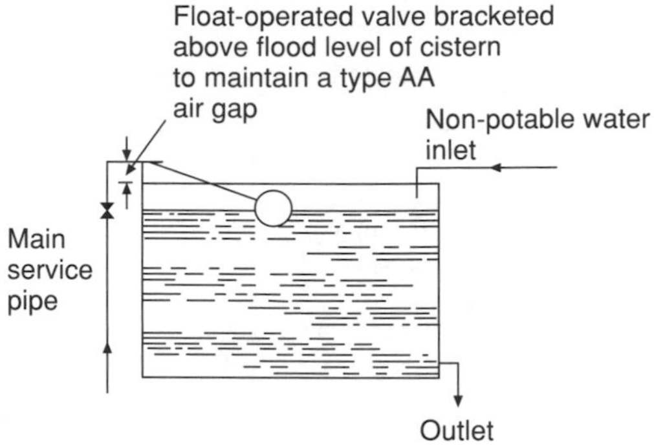


(a) A type AA air gap must be provided where any supply of non-potable water is used in conjunction with water supplied by a water authority

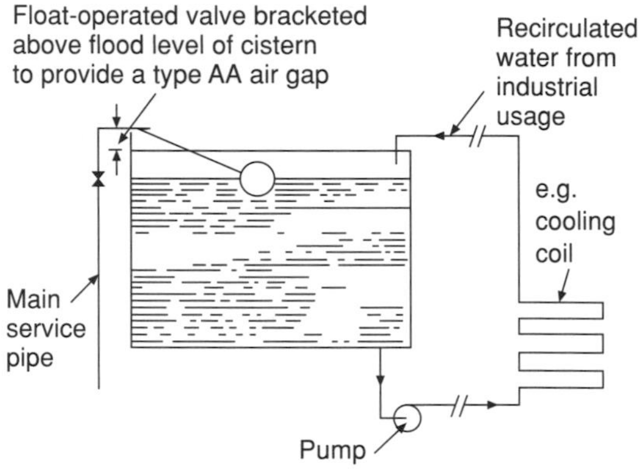


(b)

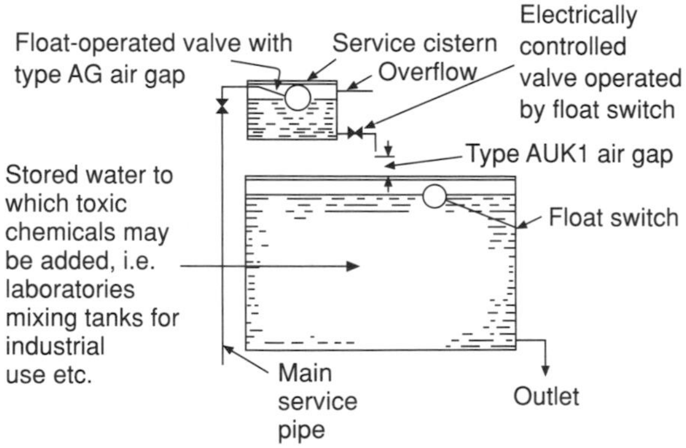


(c) As an alternative to the arrangement shown the outlet pipe could discharge into a tun-dish fitted in such a way that the air gap is maintained. Water level control in this case would be via a solenoid valve operated by a float switch

**Fig. 3.35** Backflow protection in public and commercial buildings.

flexible hoses are fitted in hotels, hostels or nursing homes having several floors, the arrangement shown in Fig. 3.36 is an acceptable method of installation. Both the hot and cold water services must be completely separate from any other water supplies. The pipework shown is designed to prevent siphonage from the bidets on upper floors from contaminating the supplies from those fitted below. The single check valves and 300 mm upstands provide primary backflow protection here, the vents on the distribution pipes secondary protection.

*Secondary backflow protection*

This term relates to any device fitted to supplement primary or point of use protection such as an air gap on each tap or drawoff. Figures 3.37(a–c) show some typical methods of providing secondary backflow protection in multi-storey domestic

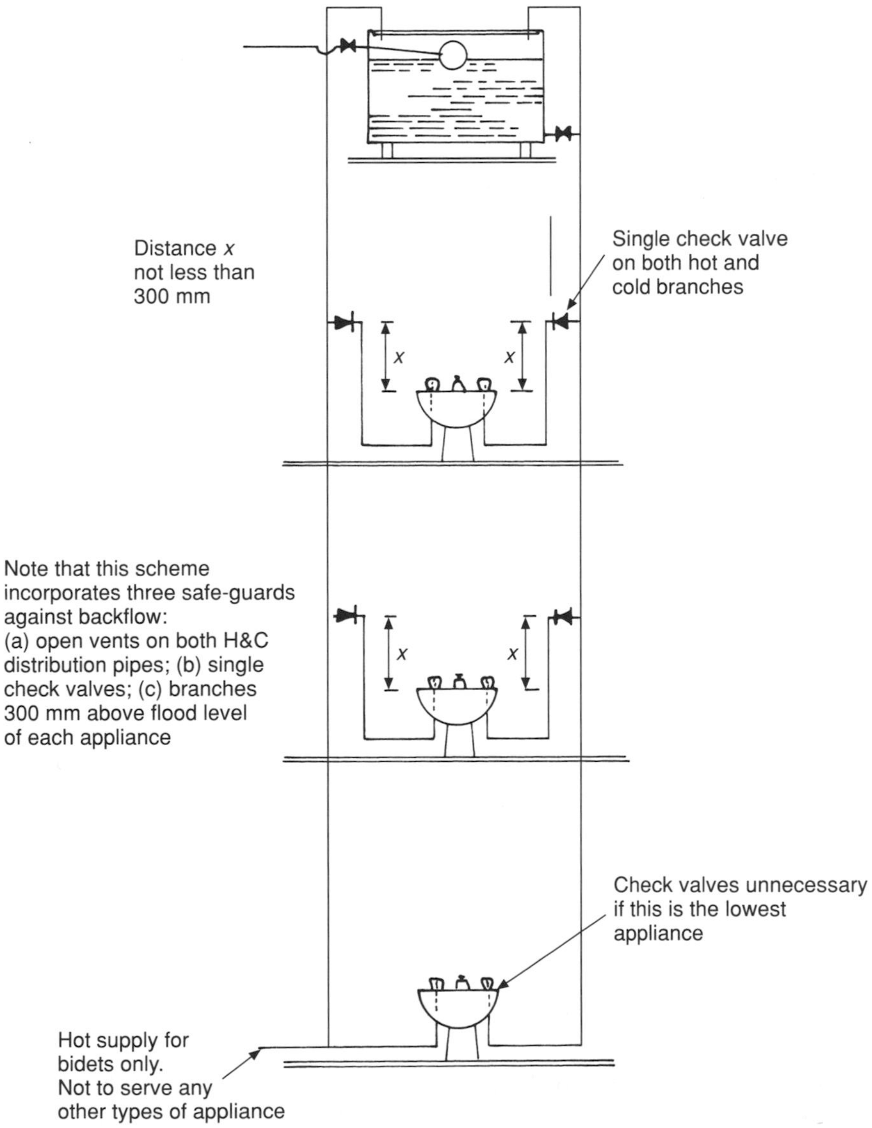
Distance *x* not less than 300 mm
Single check valve on both hot and cold branches
x
x
Note that this scheme incorporates three safe-guards against backflow: (a) open vents on both H&C distribution pipes; (b) single check valves; (c) branches 300 mm above flood level of each appliance
Check valves unnecessary if this is the lowest appliance
Hot supply for bidets only. Not to serve any other types of appliance

**Fig. 3.36** Backflow protection for appliances, e.g. bidets and bed pan washers in hospitals, with inlets below flood level in a public or commercial building.

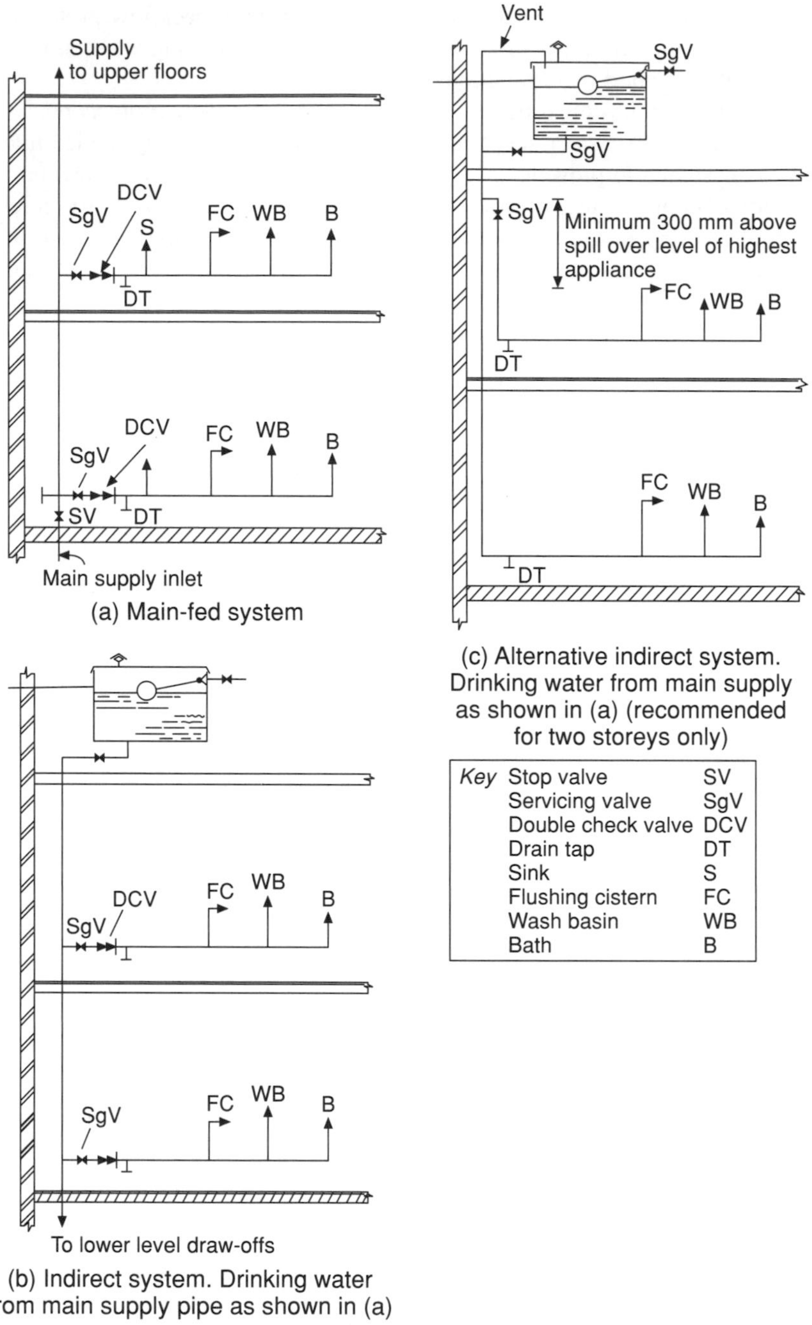


**Fig. 3.37** Cold water supply systems in multi-storey dwellings with secondary backflow protection.

dwellings where risks in excess of category 3 are unlikely to be encountered. In Figs 3.37(a) and (b) a verifiable double check valve is fitted on each branch downstream of the stop valves. An alternative to this is shown in Fig. 3.37(c); 300 mm upstands and a vented distribution pipe provide the secondary backflow protection here. The Water Regulations Guide suggests that ventilated distribution pipes are only suitable for two-storey buildings. Care must be exercised with pipe sizing to ensure adequate protection of the lower floor appliances. Two other provisions are also necessary: (a) the branch pipe must not run above the level of its junction to the distribution pipe at any point, and (b) a minimum of 300 mm must be maintained between the branch and the flood level of the highest appliance. One of the reasons why the water suppliers must be informed prior to undertaking certain types of work is to enable them to assess the possible backflow risks. In some situations, especially in commercial and industrial buildings, category 4 or 5 risks may exist and the supplier will make the necessary recommendations. Generally an RPZ valve will satisfy category 4 risks. No direct connection to a supply pipe is permissible in the case of a category 5 risk. Where such risks exist water must be supplied via a storage cistern, the inlet having an AA, AB or AD type air gap. Suitable protection must be provided with sprinkler systems and is dealt with in the section on fire-fighting.

### Sterilisation of cold water systems

Where large water installations are carried out, often over a period of many months or even years, it is possible for storage vessels and pipework to become polluted. In such cases the system must be sterilised to ensure that all harmful bacteria are destroyed before it is commissioned.

This procedure is usually carried out by water authority employees or it can be done by the plumbing contractor. The procedure is as follows. The cistern is cleared of all debris, after which the system is thoroughly flushed out. It must then be refilled and at the same time a sterilising chemical containing chlorine is added, care being taken to ensure that it is thoroughly mixed with the water. The recommended dosage is 50 parts chlorine to 1 million parts water. If other chemicals are used the manufacturer's instructions must be rigidly adhered to regarding the quantity to be used with a given volume of water. When the system has been completely filled the supply of water must be shut off and the taps on the distributing pipe opened, those nearest the cistern first. When the water from each tap begins to smell of chlorine they should be closed. The cistern should then be topped up with water, the topping-up water also containing the same proportion of chemical. The whole system is then allowed to stand for a period of 3 hours, after which a test is made by smell for residual chlorine. If none is found the whole procedure must be repeated. When the treatment is concluded the system should be emptied and flushed out with clean water to remove any taste or smell of the chlorine.

### Provision for fire-fighting in buildings

The reader should note that water supply for fire-fighting and the associated equipment is very specialised, and it is not possible to deal with this subject fully within the confines of this book. The aim here is to deal with the basic principles only.

In the event of a fire in low-rise dwellings and small commercial premises, the fire service uses the fire hydrants connected directly to the water main for a supply of water for fire-fighting. In larger and multi-storey buildings, special arrangements are necessary for fire protection, and are mandatory in many cases. Methods of fire protection fall into two main categories:

(a) Equipment that can be used by the layman, such as portable fire extinguishers or hose reels.
(b) Systems of pipework which can only be used by the fire service.

#### *Hose reels*

Figure 3.38 illustrates a typical hose reel and its associated components. They are designed as a first-aid measure, and in the event of a fire can be operated by the occupants of the building. They should be sited in such a way that the user can escape if the fire becomes out of control.

Suitable siting positions are near fire exits, adjacent to landings or along recognised escape routes. The following lists the essential requirements of hose reel installations:

(a) When the hose is fully unrolled, the nozzle must not be more than 6 metres from any part of the floor area it covers.
(b) When the two highest or most remote hose reels are in simultaneous use, the minimum

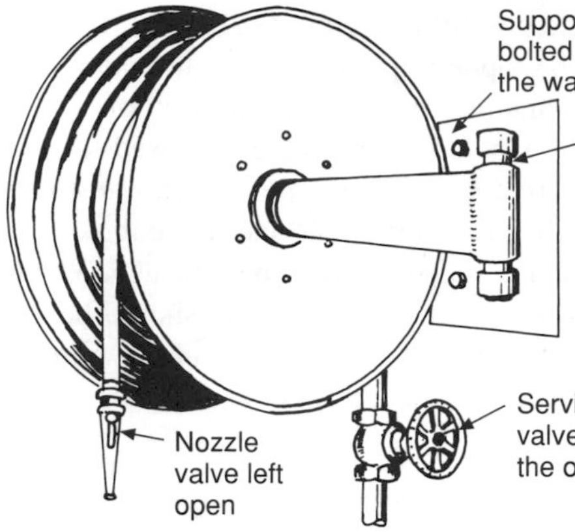


**Fig. 3.38** Hose reel. The type shown is designed for housing into a wall or cupboard. When in use it can be pulled through 90° on the pivot. Most hose reels are designed to open an integral valve when the hose is pulled out. Lengths of hose available: 18 m, 24 m, 30 m, 36 m. Diameters 20 or 25 mm.

flow rate through each reel must be 0.4 litres/second. A water pressure of 2.5 bar at the nozzle is necessary to give the water jet a range of 8 m when used horizontally. This distance will be less when used in the vertical position.

(c) The diameter for a pipe serving a single hose reel is normally 25 mm, depending on its length and the number of bends on the pipe run, and in some cases a larger pipe diameter may be necessary. In buildings having several storeys, the fire main servicing the hose reels should not be less than 50 mm in buildings up to 15 m in height. If this height is exceeded, the main must be increased to 64 mm. Branch pipes serving a hose reel must not be smaller in diameter than the hose on the reel.

*Water supply to hose reels* In small low-rise installations, hose reels may be connected directly to the main, subject to the pressure being sufficient to meet the foregoing requirements. Water in hose reels supplied direct from the main is unlikely to constitute more than a category 2 risk in the event of backflow. This is due to the possibility of the stagnant water in the hose reel re-entering the supply pipe or main.

A typical installation of this type is shown in Fig. 3.39. In multistorey buildings, fire mains to hose reels must be boosted in a similar way to that used for cold water supplies. Figure 3.40 shows a typical pumped hose-reel installation with a break tank. In some circumstances, water authorities will permit direct boosting from the main for fire-fighting purposes. In such cases the requirements of the system will be the same, except that no break tank will be necessary. If the capacity of the high-level storage vessel meets the requirements of the hose-reel installation, water can be pumped downward as the alternative shows. Pumped hose-reel installations must be fully automatic, so that when a hose reel is operated the pressure on the pipeline is lowered, causing a pressure switch to activate the pump. The purpose of the small pressure vessel shown is to prevent any pressure loss, due, for example, to small leaks starting the pump when a hose reel is not in use. An alternative to a pressure switch is a flow switch which will detect a flow of water when a hose reel is operated.

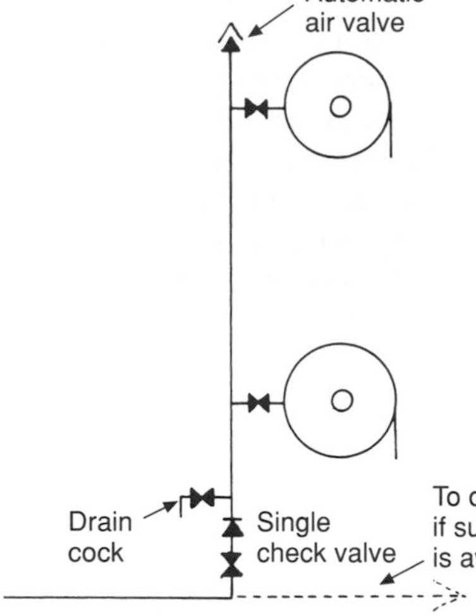


**Fig. 3.39** Hose reel connected directly to the main. Subject to the permission of the local water authority where the main pressure and the service pipe size is adequate, hose reels may be connected directly to the main or possibly an existing service pipe. This arrangement may be possible for small low-rise buildings.

*Fire protection systems for use by fire service personnel*

In buildings of up to 61 m in height, dry risers are normally provided — except in very special

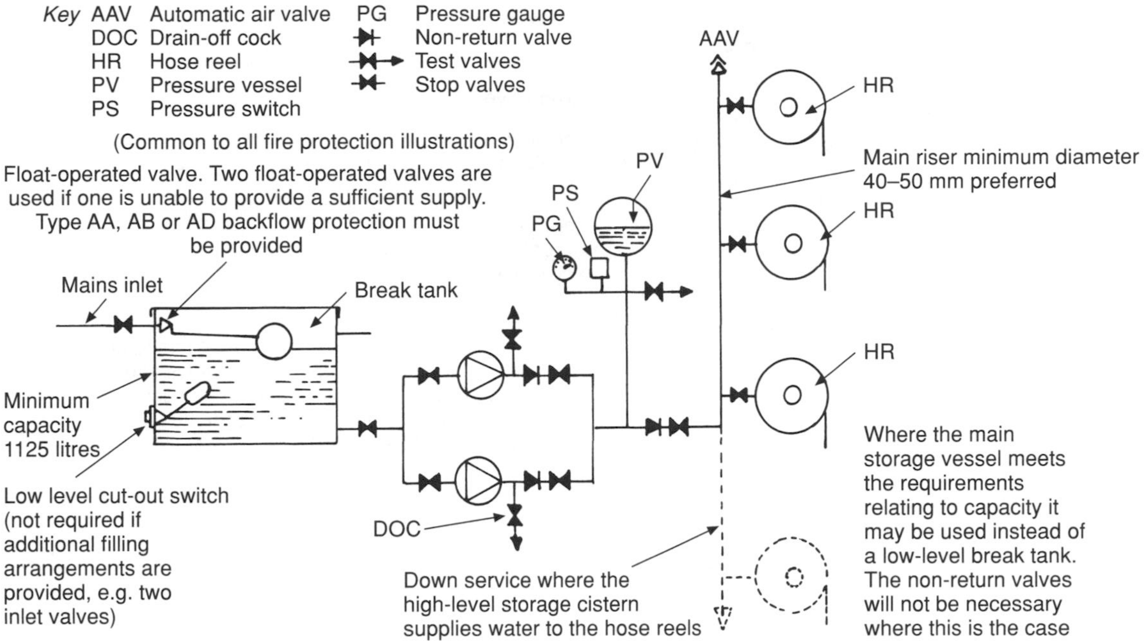


**Fig. 3.40** Wet riser serving hose reels.

circumstances. They are usually not charged with water and are intended for fire service use only. Their purpose is to avoid running out long lengths of fire service hose up stairways or passageways, thereby obstructing what may be a means of escape for the occupants of the building. The lower end of the riser terminates external to the building in a purpose-made box containing the inlet breeching. The box is usually provided with a wire-reinforced glass door, which can be opened from the inside when the glass is smashed in the event of an emergency. A breeching piece serving a 100 mm riser should have two connections for fire service pumps (see Fig. 3.41), a 150 mm riser should have four. They are fixed with their centre line being 760 mm above ground level. As with any fire-fighting installation, where any part of the pipework extends above roof level, it must be provided with a lightning conductor. Figure 3.42 illustrates a typical dry riser installation.

*Wet risers* A typical diagrammatic installation is shown in Fig. 3.43. These are similar to dry riser systems, but are permanently charged with water.

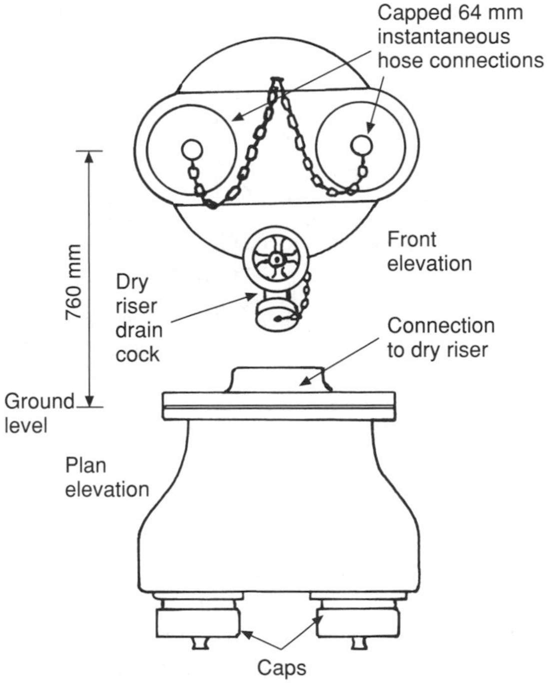


**Fig. 3.41** Breeching piece. To ensure an adequate supply of water to the dry riser provision is made to connect two fire service hoses from the fire engine pumps.

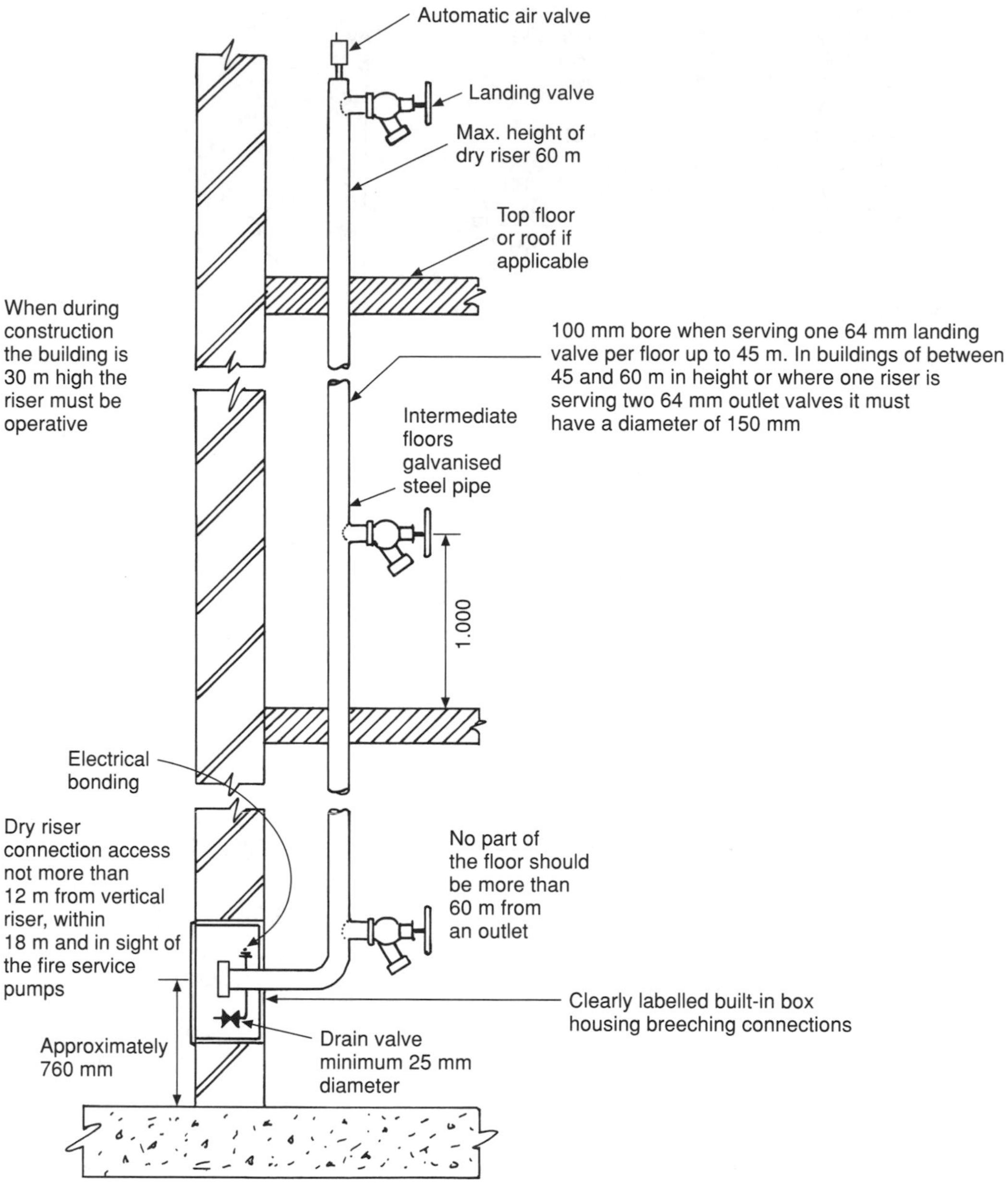
Automatic air valve
Landing valve
Max. height of dry riser 60 m
Top floor or roof if applicable
When during construction the building is 30 m high the riser must be operative
100 mm bore when serving one 64 mm landing valve per floor up to 45 m. In buildings of between 45 and 60 m in height or where one riser is serving two 64 mm outlet valves it must have a diameter of 150 mm
Intermediate floors galvanised steel pipe
1.000
Electrical bonding
Dry riser connection access not more than 12 m from vertical riser, within 18 m and in sight of the fire service pumps
No part of the floor should be more than 60 m from an outlet
Clearly labelled built-in box housing breeching connections
Approximately 760 mm
Drain valve minimum 25 mm diameter

**Fig. 3.42** Dry riser installation.

They are normally only provided in buildings higher than 61 m. As with dry risers, there are some standard regulations relating to wet risers, but any installation will be subject to the approval of the local authority fire control officer, who should be consulted regarding the suitability of any system. The main requirements are listed as follows:

(a) The installation should be capable of maintaining a pressure at the top outlet of 4 bar with a flow rate of 22.7 litres per second.
(b) The maximum working pressure when only one outlet is in use is 5 bar.
(c) Because of the very high pressure involved, and to prevent damage to the water main,

pumping directly from the town supplies is not permissible. Therefore water for wet risers must be pumped from a break tank having a minimum capacity of 45.5 $m^3$ (45,500 litres), and the valve feeding the tank must be capable of delivering water at a minimum of 7.6 litres per second.

(d) The requirements for landing valves relating to their fixing heights, and the areas protected, are the same as those for dry risers. They are, however, provided with an adjustable regulator which limits the pressure to 4.5 bar. This avoids the possibility of bursting the fire service canvas hoses. Additional protection is provided by means of a pressure relief valve in the outlet of the landing valve, which will open at 6.5 bar. A piped supply from these valves is returned to the break tank as shown.

*Sprinkler installations*

There are two basic systems of sprinkler installation:

(a) The wet system, which is permanently charged with water and most frequently used.
(b) The alternative wet and dry system, which is used in unheated buildings, or where the heating system only operates when the premises are occupied.

Sprinkler systems are fully automatic and usually fitted in large public and commercial premises such as cinemas, department stores and warehouses, where a fire, and sometimes the method used to

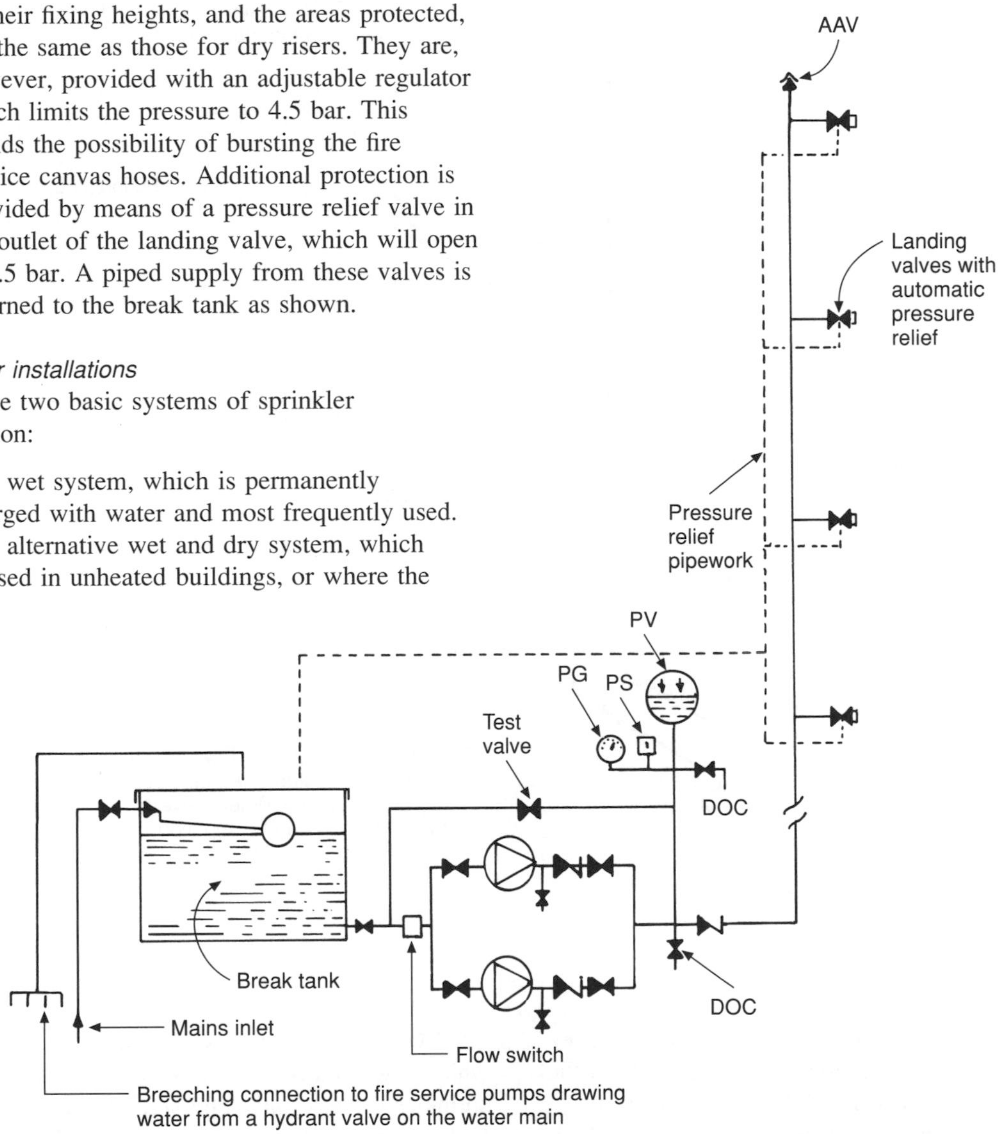


**Fig. 3.43** Wet riser pumping system. In schemes of this type where total reliance for fire-fighting depends on internal pumping equipment, provision must be made for an alternative supply of electricity such as a standby generator due to the possibility of power failure. Due to the very high pressures developed in this type of system water can only be pumped from the break tank. To ensure sufficient water is available for fire-fighting, two float-operated valves taken from the mains supply are often provided in addition to the fire service connection. (See Fig. 3.40 for key to abbreviations.)

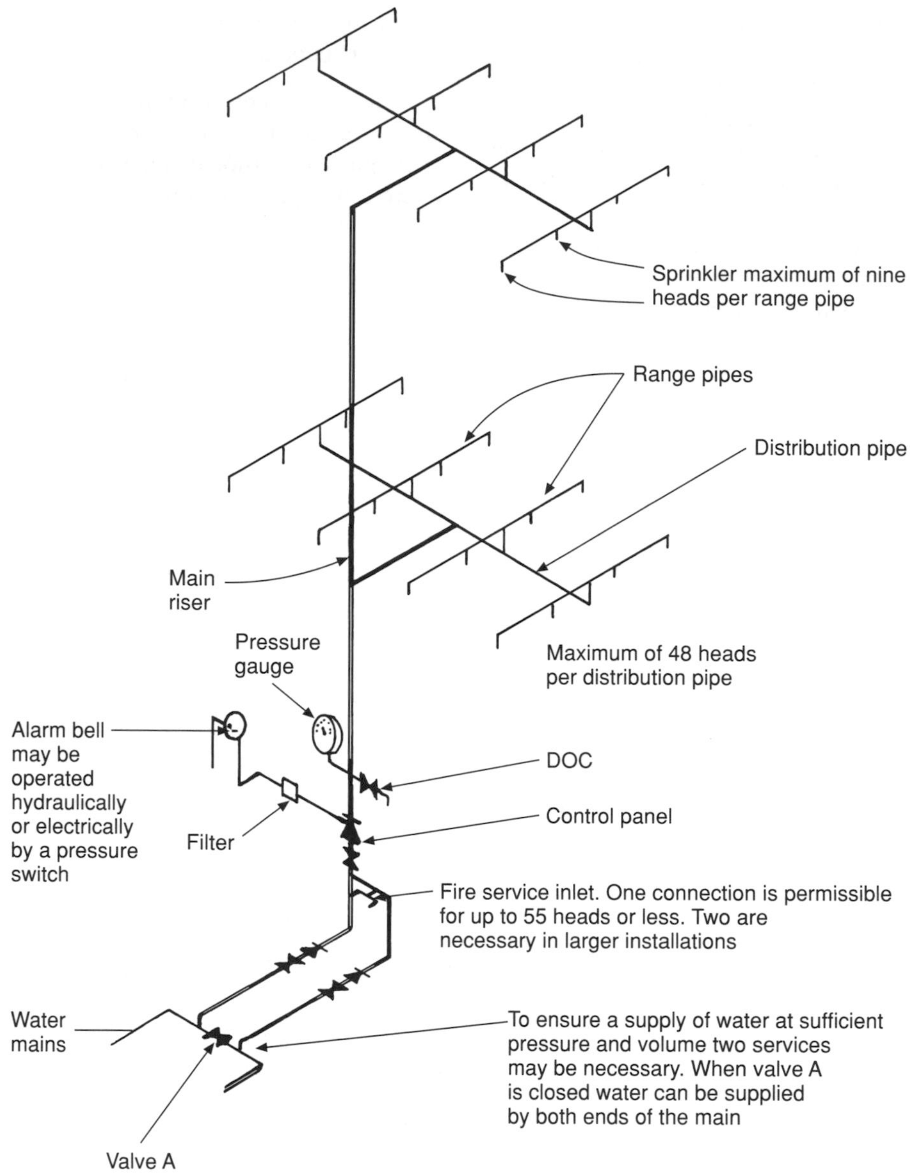


**Fig. 3.44** Sprinkler system.

extinguish it, could cause a great deal of damage. The purpose of the sprinkler system is to limit the extent of the fire and possibly to extinguish it before it gets out of control. Because this method of fire protection is so effective, its installation in commercial buildings reduces the cost of fire insurance premiums. A typical wet system is shown in Fig. 3.44. It consists of a system of pipework fitted with sprinkler heads, shown in Fig. 3.45, which react to a rise in temperature, causing a sprinkler valve to open and spray the area it covers with water. The type of head illustrated employs a glass bulb, which bursts due to the expansion of the fluid it contains when it detects an increase in temperature. These bulbs are manufactured covering a wide temperature range, which allows

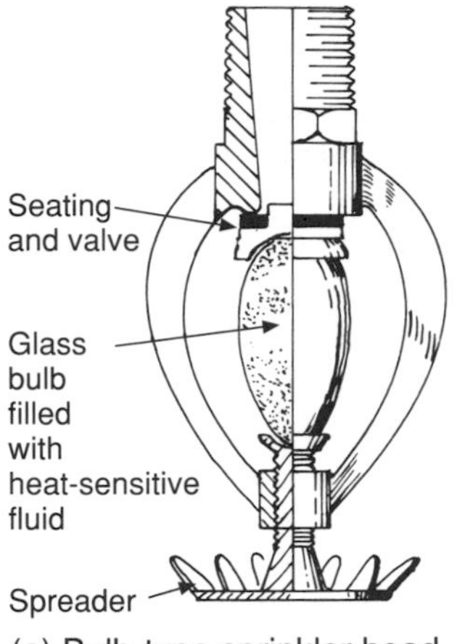


(a) Bulb-type sprinkler head

When the fluid in the bulb senses a temperature rise it expands and breaks the glass, releasing the valve. The ensuing flow of water is diverted outward covering a specified floor area.

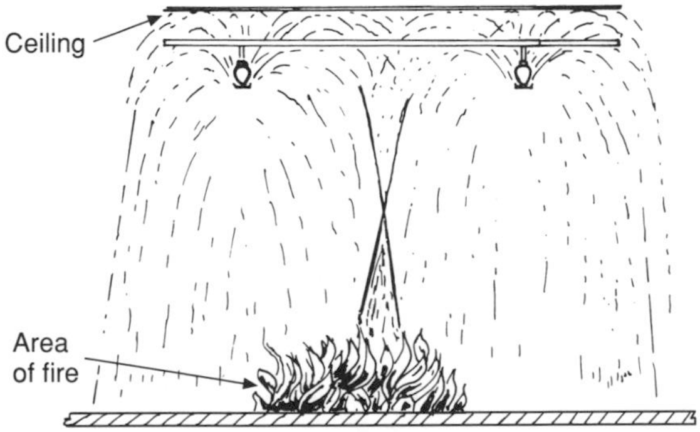


(b) The area covered by sprinklers and the overlap

Sprinkler head should preferably be within 75 and 150 mm of ceiling, but not more than 300 mm below combustible ceilings or roofs.

**Fig. 3.45** Bulb-head sprinklers.

for variations in the ambient temperature in both public and industrial buildings. Seven variations are listed covering temperatures from approximately 57 to 290 °C, each variation being identified by its colour.

Another type of sprinkler head employs a series of metal plates joined together with low-melting-point solders. These plates hold the valve in the closed position, and in the event of a fire the solder melts at a specified temperature. Variation in the temperature at which the sprinkler head operates is achieved in this case by using solders having

**Table 3.4** Maximum area covered by a sprinkler.

| *Hazard class* | *General ($m^2$)* | *Special risk areas or storage tanks ($m^2$)* |
|---|---|---|
| Very light hazard | 21 | 9 |
| Ordinary hazard | 12 | 9 |
| Extra high hazard | 9 | 7.5–10 |

differing melting points. The number of sprinkler heads, their diameters and the area each covers depend on the fire hazard classification shown in Table 3.4.

Such buildings as offices and libraries would be classified as light hazards. Ordinary hazards would cover industrial and commercial premises such as shops and warehouses containing or handling combustible materials unlikely to burn intensely during the early stages of a fire. Industrial or commercial buildings having a very high fire hazard would include premises having high piles of combustible stocks or those handling very flammable materials.

When a sprinkler system is installed it normally covers the whole building. There are exceptions, due to low fire risk, where this rule may be waived. For example, rooms having fire-resistant walls and doors which are separated from the area covered by sprinklers.

*Wet and dry sprinkler installations* This type of system is filled with water during the summer months only. When periods of frost are expected, the system is emptied and charged with compressed air at a greater pressure than that of the water supply. In the event of a fire, the sprinkler head will operate releasing the air in the system, allowing it to become charged with water. An alternative to the wet and dry system is to charge it with a solution of anti-freeze fluid. If this is the case it is considered to be a category 4 backflow risk which would necessitate the use of a type BA device such as a RPZ valve. In some instances a sprinkler system may be supplied by a storage cistern in a similar way to that shown in Fig. 3.40. The inlet of all cisterns supplying water to fire-fighting equipment must be provided with a type AA, AB or AD air gap. A single check valve must also be fitted in the supply pipe.

All sprinkler systems should comply with the rules of the Fire Officers Committee (FOC). The purpose of this body is to control the standards of fire protection equipment, installation and practice. Insurance companies responsible for insuring against fire risks will insist that the FOC recommendations and rules are complied with. It is therefore prudent to contact this body prior to the installation of any fire protection systems.

### Water treatment

The causes of hardness of water and the treatment of hard water by public authorities have been dealt with at some length in Book 1. While it is true that water authorities make considerable efforts to minimise the total hardness of the water they supply, it would be too expensive in those parts of the country where the water is initially very hard to reduce the hardness to the levels required by the consumer.

Conversely, in areas where the water is very soft, water authorities increase the hardness content of the water they supply. This is due to the fact that soft water, possibly slightly acidic in nature, can, when conveyed in lead pipes, take lead into solution which could result in lead poisoning in extreme cases. Although lead pipe for the conveyance of water has been discontinued for many years and its use for this purpose is now forbidden by the water by-laws, many old properties are supplied with water through pipes made of lead. As the renewal of all existing lead pipes would be very expensive, most water authorities in areas where the water is naturally soft increase its total hardness content, substantially reducing any risk of lead poisoning.

There are many advantages in the use of soft water. Greater economy is obtained from the use of soap and detergents for ablutionary and laundering processes. There is a reduction of scale in kettles and culinary utensils and an absence of unpleasant scum which forms in ablutionary appliances due to the reaction between soap and the hardening salts in the water. Largely due to the foregoing, water softeners and scale reducers are being fitted in increasing numbers and the following text deals with the equipment in common use for domestic and small industrial premises.

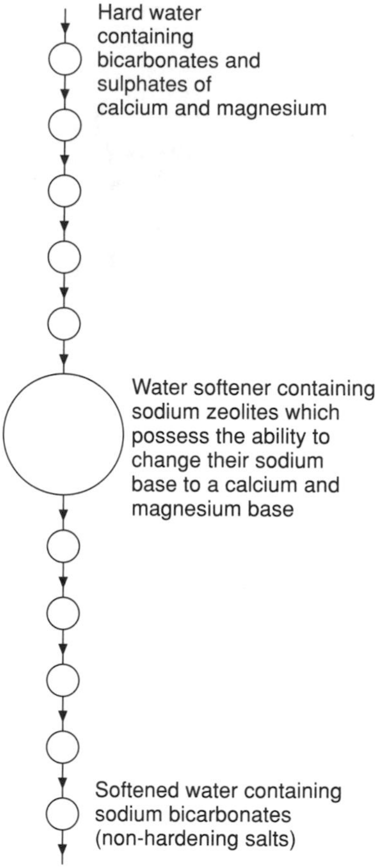


**Fig. 3.46** Working principles of a water softener.

### Base exchange water softening

The principal method of softening water for both industrial and domestic use in most cases is that known as the *base exchange process*. It employs the use of zeolites and sodium chloride (common salt) in the water. The principles of base exchange, to be understood fully, require some knowledge of chemistry, but it is sufficient to say at this stage that the sodium zeolite base is exchanged for one of calcium as shown in Fig. 3.46.

A domestic simple water softener is shown in Fig. 3.47. Hard water enters at the top, passes

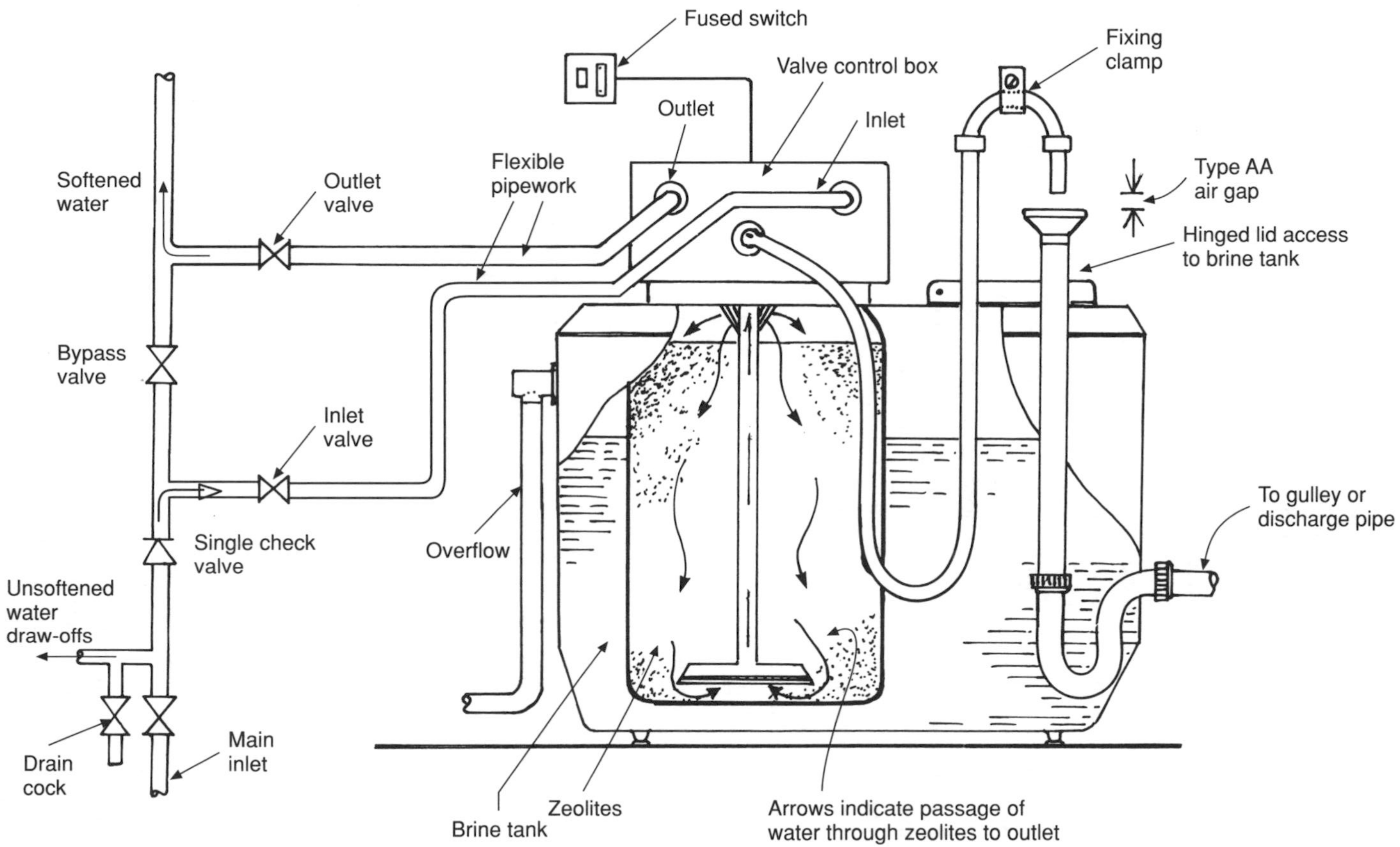


**Fig. 3.47** Domestic water softener.

through the zeolite bed and emerges through the outlet with the hardening salts completely removed. After a period of time, depending on the capacity of the softener, the zeolite becomes exhausted or so saturated with calcium that it is no longer capable of softening the water. The softener must then be regenerated, first by back washing, i.e. reversal of the flow of water through the softener prior to the addition of salt. This is accomplished by a control system of automatic valves which permit the flow of water through the softener to be reversed, washing out the hardness salts and simultaneously drawing a measured quantity of brine from the brine reservoir. This in effect regenerates the zeolites enabling another softening cycle to be commenced. The softener shown here is designed to be plumbed into the cold water supply and is automatic in operation. The control system, incorporating an electrical timer, operates the valves which activate the regeneration process including the addition of brine. The time switch should be set so that the regeneration process takes place when no water is being used, usually at night. Time control only is the simplest form of automation, the problem being that it does not take into account the volume of water that has passed through the softener. This means that regeneration takes place whether or not it is necessary and leads to waste of salt. More sophisticated systems of control are based on measuring the volume of water used or monitoring the hardness of the water. The initial cost of softeners incorporating these systems of control is, however, more expensive than for those with simple time control.

The risk of water contamination due to the use of water softeners is not considered to be high and as such a single check valve, usually supplied on the inlet to the softener, is all that is normally necessary. It is important, however, that a tun-dish providing a type AA air gap is provided at the waste connection to prevent any possible contamination from the foul water drain. The removal of the hardening salts may leave the water soft and acidic. The usual practice is to mix some hard water with the softened water to give a blended supply. At least one drinking water point should be provided directly from the main, as in most cases it is more pleasant to the taste than soft water. To reduce running costs, supplies to standpipes and WCs should bypass the softener.

*Scale reducers*

These are not water softeners in the true sense of the word, and although the technology relating to their working principles is not new, it is only recently they have been produced for domestic use. It was discovered that passing water containing calcium crystals (hardening salts) through a magnetic field causes a change in the shape of the crystals, an increase in their size and a decrease in their ability to dissolve in water. An increase in the size of the crystals has two beneficial effects. First, they lose their ability to 'coagulate' or join together as easily as those of a smaller size thus preventing the build-up of rock-like scale in the plumbing installation. Secondly, the presence of these larger crystals disrupts the equilibrium between the water, and tends, in very simple terms, to dissolve any existing scale. Unlike the base exchange process of water softening, the calcium remains in the water, but instead of forming scale or fur it is drawn off via the taps or deposited in the form of sludge in the base of boilers or hot storage vessels where it can easily be washed out. Figure 3.48 shows in diagrammatic form the basic principles of scale reducers, the heart of which is a permanent magnet made mainly of iron with additions of cobalt, nickel, aluminium and copper. The resulting alloy is coated with PTFE which protects it from corrosion. The magnet is housed in a metal case which is designed to increase the turbulence of the water as it flows round the magnet.

*Electronic water conditioners*

These produce similar physical changes to calcium crystals to those produced by magnetic-type scale reducers. In simple terms the calcium loses its solubility and ability to coagulate to form the hard, rock-like fur on the inside of hot water pipes and storage vessels. The principle of electronic conditioners differs from the magnetic type in that they employ magnetic hydrodynamics (MHD), which is effected by passing a form of radio signal through a coil wrapped around the pipe to which it is fitted. The signal is inaudible and sets up a dynamic field around and through the coil, pipe

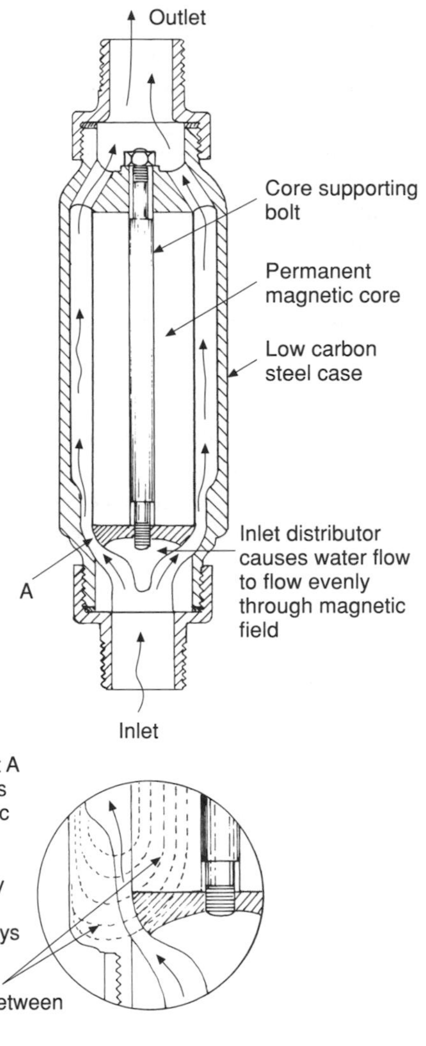


**Fig. 3.48** Scale reducer.

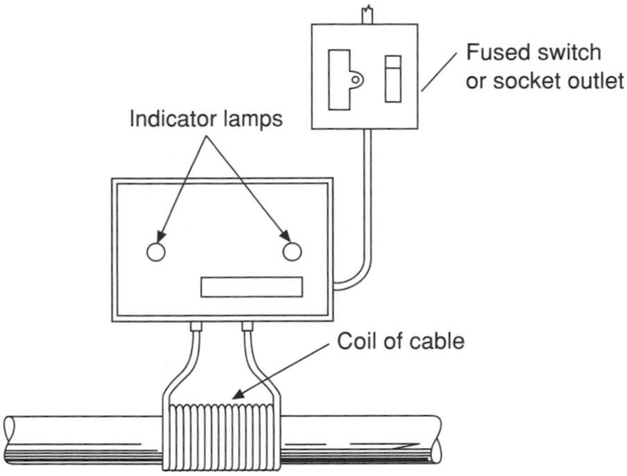


**Fig. 3.49** Electronic water conditioner, normally fitted to the incoming main water supply.

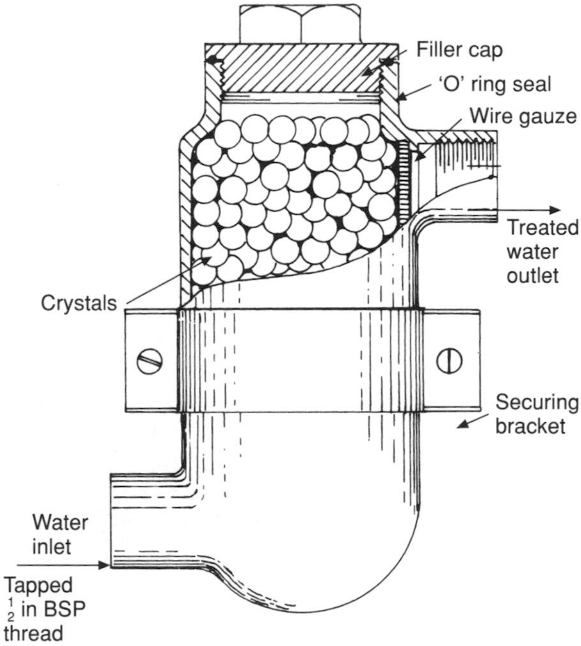


**Fig. 3.50** Scale inhibitor.

and water. Because the signal field changes at high audio frequencies the scale-forming crystals are changed but remain in the water as a fine sediment which is washed away with the flow of water. An illustration of a typical unit of this type is shown in Fig. 3.49, and it will be seen that an electrical supply via a fused switch or socket outlet is necessary. The control panel contains a step-down transformer giving an output, in most cases, of 6 volts. Unlike water softeners these units do not produce immediate results; usually a period of between 3 and 6 months must elapse before any significant change can be seen. This also applies to scale reducers of the static magnetic type.

*Chemical methods of scale prevention*

As with scale reducers this method of treating water to protect plumbing installations against the build-up of scale is not new, and similar equipment to that shown in Fig. 3.50 may be found on the inlet of gas water heaters fitted in areas of water supply known to have a high hardness content. Because of their relatively low cost, these scale inhibitors, as they are called, are sometimes used as an alternative to water softeners and are ideal for use in individual appliances such as drink-vending equipment, washing machines and electric instantaneous water heaters. It must be emphasised that like scale reduction chemical treatment of water is not 'softening' in the true sense. Water to be treated passes through a non-ferrous metal dispenser containing the chemical crystals which have the effect of 'inhibiting' or suspending in the water the hardness salts responsible for scale formation. This prevents the hard scale building up in pipes and fittings. The working life of the chemicals varies with the volume of water used. In a domestic system it can be anything from 9 to 15 months. Replacement of the crystals is a simple operation and can be carried out by the householder by unscrewing the top of the container and topping up as necessary. It must of course be fitted in a reasonably accessible position. The suppliers of the chemical crystals state there are no harmful effects from their use and as the natural minerals are not removed from the water, its beneficial qualities are not impaired.

### Hydraulic pressure testing

It is often necessary on large installations to test sections of the work while the job is progressing, especially if it is likely to be difficult to gain access to the work on completion. Typical examples occur when pipes are fitted under floors or in wall ducts where they will be covered when the job is finally commissioned. Pressure testing is carried out with the equipment illustrated in Fig. 3.51 on both hot

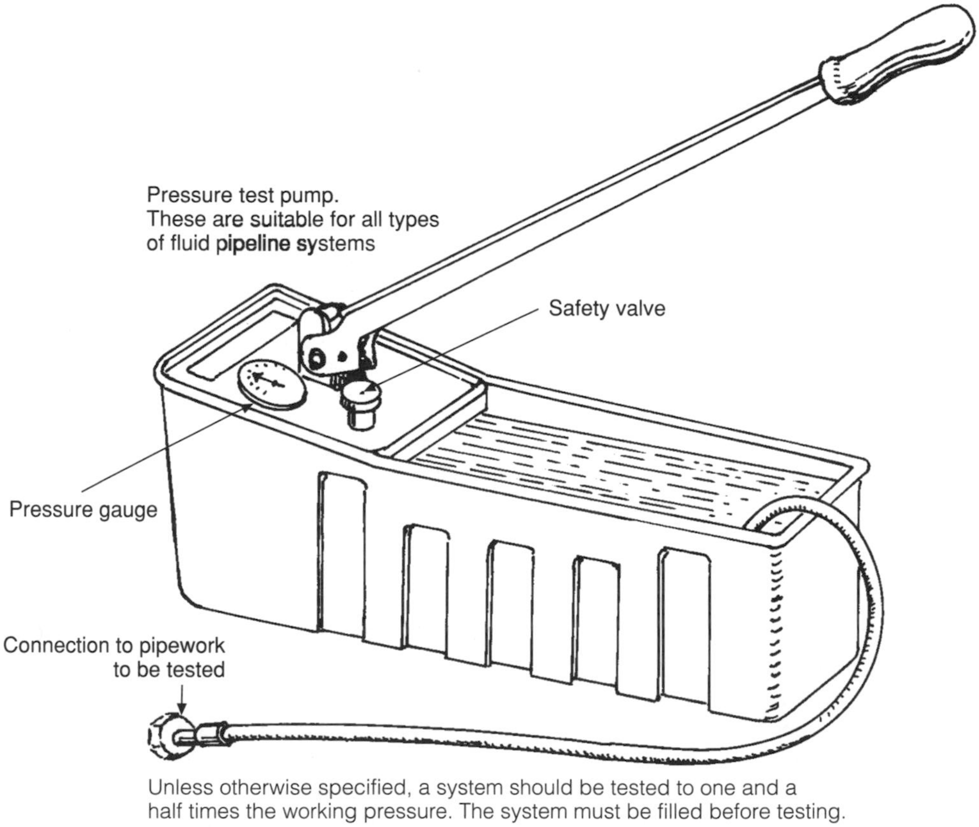


**Fig. 3.51** Typical force pump used to pressure test water services.

and cold water services. Installations are normally tested to at least one and a half times the normal working pressure, and in cases where access is extremely difficult, twice the working pressure. In all cases the pressure should be maintained for at least 30 minutes.

### *Testing plastic pipes*

The method employed differs from that used when testing rigid pipe systems. Due to elasticity the diameter of the plastic pipe will expand slightly under pressure. The Water Advisory Service recommends the following two methods of testing plastic pipe:

(a) The test pressure is maintained by periodic pumping over a period of 30 minutes. The pressure is then reduced by a third and no drop in pressure should occur over a period of 90 minutes.
(b) As with (a) the test pressure is maintained for 30 minutes after which pumping stops and the pressure is noted. A pressure drop of less than 0.6 bar is permissible for a period of the next 30 minutes, then a further 0.2 bar during the next 2 hours.

Provided the test meets the specifications of (a) or (b) and there is no visible leakage, it is satisfactory.

### *Inspection, commissioning and testing*

Reference should be made to BS 6700 which details the procedures that must be followed in relation to the water supplies in large buildings. The following text lists the main points to which attention must be given:

(a) The materials, equipment and workmanship must comply to relevant standards and job specifications.
(b) The installation should comply with the current Water Regulations and any other appropriate regulations. This includes the Health and Safety at Work Act, which applies not only to those installing the system but also to those using it.
(c) The installation must be visually inspected on completion for damage to pipework and protective coatings, and for security of fixings.
(d) Storage vessels must be inspected to ensure they are clean and securely supported. Cistern covers must be correctly fitted.
(e) The system should be slowly filled with water, with the highest draw-off open allowing any air to escape. After inspecting for leaks, the installation should be tested as previously described (see Hydraulic pressure testing p. 118).
(f) Each draw-off, shower unit and float-operated valve should be checked for delivery of the specified flow rate when the system is permanently connected to the main supply, see Fig. 3.52.
(g) Float-operated valves must be correctly adjusted to achieve the correct water level in cisterns.
(h) Any backflow devices fitted in the installation must be checked for efficient working.
(i) Any insulation specified should be fitted at this stage, and where necessary colour coded to BS 1710. All pipeline control valves should be labelled showing their purpose.

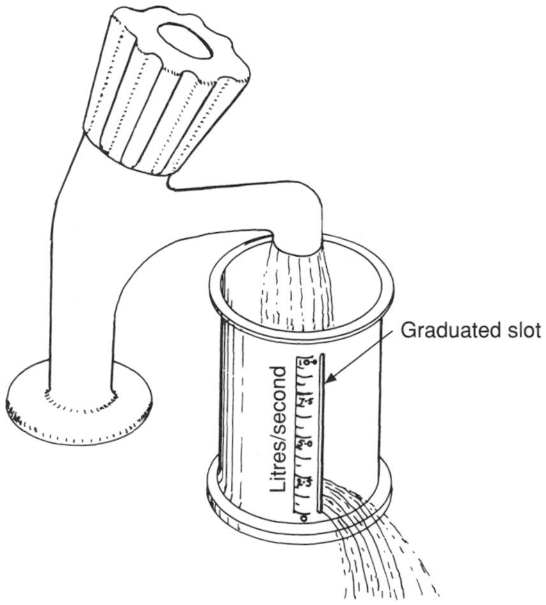


**Fig. 3.52** Testing the flow rate at draw-offs. A reading is taken at the highest point of water flow in the slot. In the example a discharge rate of 2.3 litres per second is shown. Gauges are also available with graduations from 2.5 to 20 litres per second.

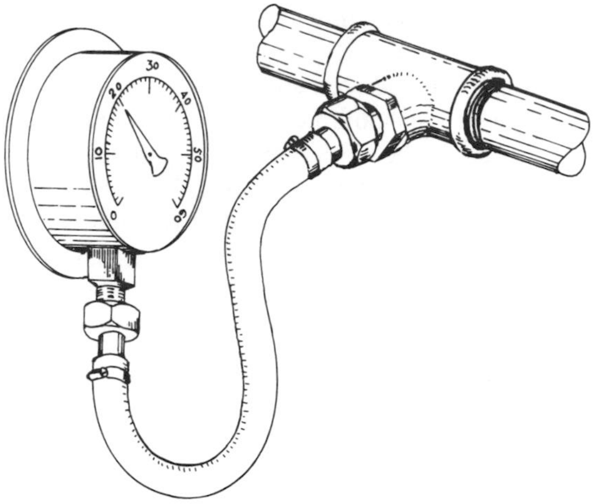


**Fig. 3.53** Pressure testing equipment. The water pressure in any pipeline may be tested by using a Bourdon-type gauge as shown or by using a special adaptor which can be connected directly to a draw-off tap. Most pressure gauges give alternative readings in both psi and bars.

(j) On completion all relevant information, such as drawings, test results where applicable and manufacturers' installation and maintenance documents, should be given to the building owner for safe keeping.

*Pressure testing*

It is sometimes necessary to test the pressure on existing supplies, for example where it is proposed to install an unvented hot water system or to make checks on the pressures of fire services.

Figure 3.53 shows a portable pressure gauge which may be connected to a test point as shown, or connected directly to a draw-off by means of a rubber adaptor.

## Decommissioning

When properties are left unoccupied for long periods, all water systems must be valved off at the mains and drained down when not in use. Failure to observe this procedure may result in frost damage in winter, with the possibility of an undetected water leakage over a long period of time. All systems, including hot water and heating, must be clearly labelled 'EMPTY'. Any electrical or gas services related to such systems must also be effectively isolated or capped off.

## Distribution pipe sizing in hot and cold water services

In most domestic plumbing systems the plumber uses tried and tested 'rule of thumb' methods of sizing pipework, and if there is any doubt, the next pipe size larger is selected. In commercial, industrial and multi-storey properties where there are many draw-off points, a greater degree of accuracy is necessary to ensure there is a sufficient volume of water supplying each appliance. To make provision for all the draw-offs on a system to be opened simultaneously is unnecessary and would result in uneconomic oversized pipework.

A system called 'the probability theory' is based on the likelihood of a given number of draw-offs being open at any given time, and has been found to be reliable in that, to the best of the writer's knowledge, it has never led to underassessment of simultaneous demand calculations. Frequency of use is the time between each use of an appliance and whether it is classified as 'low' for dwellings, or 'medium' where a larger group of people are using the appliance on a random basis, such as a sanitary annex in an office block or hotel. 'High' use would be in a situation where there is only five minutes between each use of the appliance by a large number of people in a short period of time. Buildings such as theatres, concert halls and cinemas would fall into this category. A system of loading 'units' has been developed which takes into account the type of sanitary appliance used, its flow rate, capacity, period of time and frequency of use. These are applicable to the various types of appliances shown in Table 3.5 and are converted to flow rates using Fig. 3.54 (read from right to left).

Figure 3.55 illustrates the details of the sanitary appliances in a small commercial hotel in which the pipe sizes are to be determined; the loading units and frequency of use are classified as 'medium' in this example. The drawing shows the measurements of pipe runs, the number, type and position of all the appliances, and any valves or fittings which may result in head loss. To simplify this exercise it can be assumed that all control valves in this example

**Table 3.5** Loading units. (Courtesy of the IPHE)

| *Type of appliance* | *Frequency of use* | | |
|---|---|---|---|
| | *Low* | *Med* | *High* |
| Basin, 15 mm sep. taps | 1 | 2 | 4 |
| Basin, 2 × 8 mm mix tap | 1 | 1 | 2 |
| Sink, 15 mm sep/mix tap | 2 | 5 | 10 |
| Sink, 20 mm sep/mix tap | — | 7 | — |
| Bath, 15 mm sep/mix tap | 4 | 8 | 16 |
| Bath, 20 mm sep/mix tap | — | 11 | — |
| WC suite, 6 litre cistern | 1 | 2 | 5 |
| Shower, 15 mm head | 2 | 3 | 6 |
| Urinal, single bowl/stall | — | 1 | — |
| Bidet, 15 mm mix tap | 1 | 1 | — |
| Hand spray, 15 mm | — | 1 | — |
| Bucket sink, 15 mm taps | — | 1 | — |
| Slop hopper, cistern only | — | 3 | — |
| Slop hopper, cistern/taps | — | 5 | — |
| Clothes washing m/c, dom. | 2 | — | — |
| Dishwasher m/c domestic | 2 | — | — |

**Table 3.6** Equivalent pipe lengths (copper, plastics and stainless steel). (Courtesy of the IPHE)

| *Bore of pipe (mm)* | *Equivalent pipe length* | | | |
|---|---|---|---|---|
| | *Elbow (m)* | *Tee (m)* | *Stopvalve (m)* | *Check valve (m)* |
| 12 | 0.5 | 0.6 | 4.0 | 2.5 |
| 20 | 0.8 | 1.0 | 7.0 | 4.3 |
| 25 | 1.0 | 1.5 | 10.0 | 5.6 |
| 32 | 1.4 | 2.0 | 13.0 | 6.0 |
| 40 | 1.7 | 2.5 | 16.0 | 7.9 |
| 50 | 2.3 | 3.5 | 22.0 | 11.5 |
| 65 | 3.0 | 4.5 | — | — |
| 73 | 3.4 | 5.8 | 34.0 | — |

*Note*: The losses through tees are taken to occur on a change of direction only, losses through fully open gate valves may be ignored. See also Fig. 2.9.

are of the gate or ball full-way pattern through which any pressure loss can be ignored. Table 3.6 shows the head loss through fittings of various sizes as equivalent pipe lengths. For the purpose of calculating the total head loss, these allowances must be added to the measured pipe lengths. Figure 3.56 gives an example of how the total head loss due to frictional resistance is calculated on a run of 28 mm diameter copper tube.

The next step is to make a tabulation schedule on which all the necessary data can be transferred from the drawing. As this is a comparatively small job only the essential information is shown, but the knowledge gained can be extended with some experience to large multi-storey and commercial premises. From the drawing identify each pipe A to F which is entered in column 1 of the schedule in Table 3.7. The loading units for each appliance supplied by each pipe can be identified and tabulated in column 2. Those relating to pipe A have been shown as a worked example and are shown in Table 3.8. The flow rates in column 3 are determined from Fig. 3.54 by following the horizontal line from right to left. As an example pipe A shows 146 loading units and taking a line from right to left from between 100 and 200 a flow rate of approximately 1.5 litres per second is shown. Note the pipe size through which this line passes; it should be entered on the schedule under column 4 as the estimated pipe size. The velocity and head loss can also be read directly from Fig. 3.54 and is determined by taking a vertical line from the estimated pipe size. Following the heavy broken lines upwards will show the velocity for pipe A to be approximately 1.0 m/s with a head loss of 0.02 m per m run (see bottom line of figure). This information should be entered into column 5 and 6 of the schedule. The effective pipe lengths are the total measured length plus the allowance for loss of head through the pipe fittings and valves etc. as shown in Fig. 3.55 and entered into column 7. The total frictional loss on each pipe run is found by multiplying the loss per metre run by the total pipe length (columns 6 and 7). The result should be entered in column 8. The actual head available is taken from the measurements shown on the drawing and should be entered in column 10. 'The progressive head' which should be entered in column 9 takes into account the increase in pressure by the head of water available to each pipe. It is calculated by simply adding columns 8 and 9 as shown. The first entry in column 9 will be the same as that in column 8. To give an example, pipe B will have a progressive head of 0.15 + 0.75 + 0.9. The purpose of this is to ensure that the head losses are never greater than the available head shown in column 10. Column 11 is used to confirm the

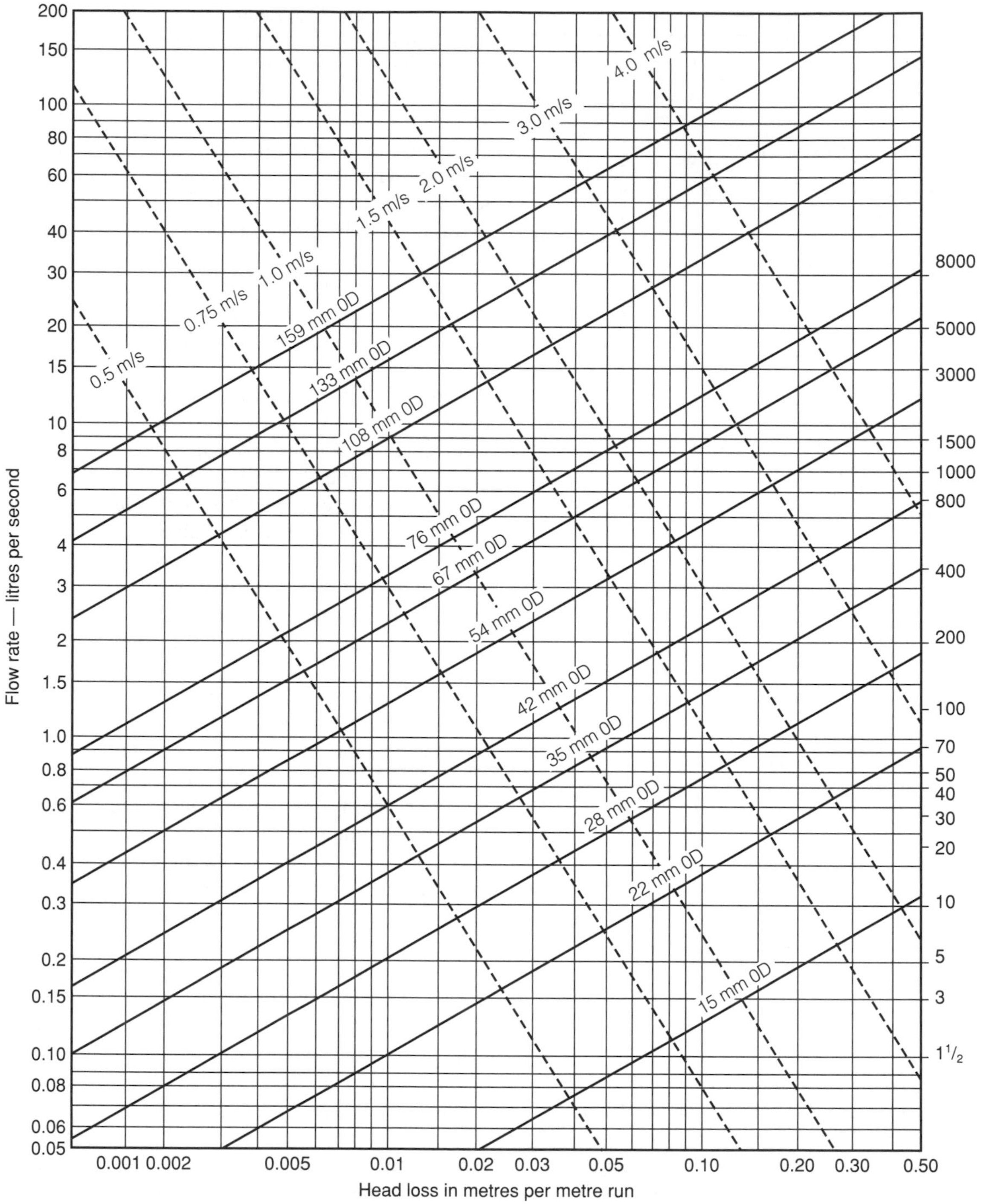


**Fig. 3.54** Pipe sizing chart — copper and stainless steel. (From the Plumbing Engineering Services Design Guide. Reproduced by permission of the Institute of Plumbing and Heating Engineering)

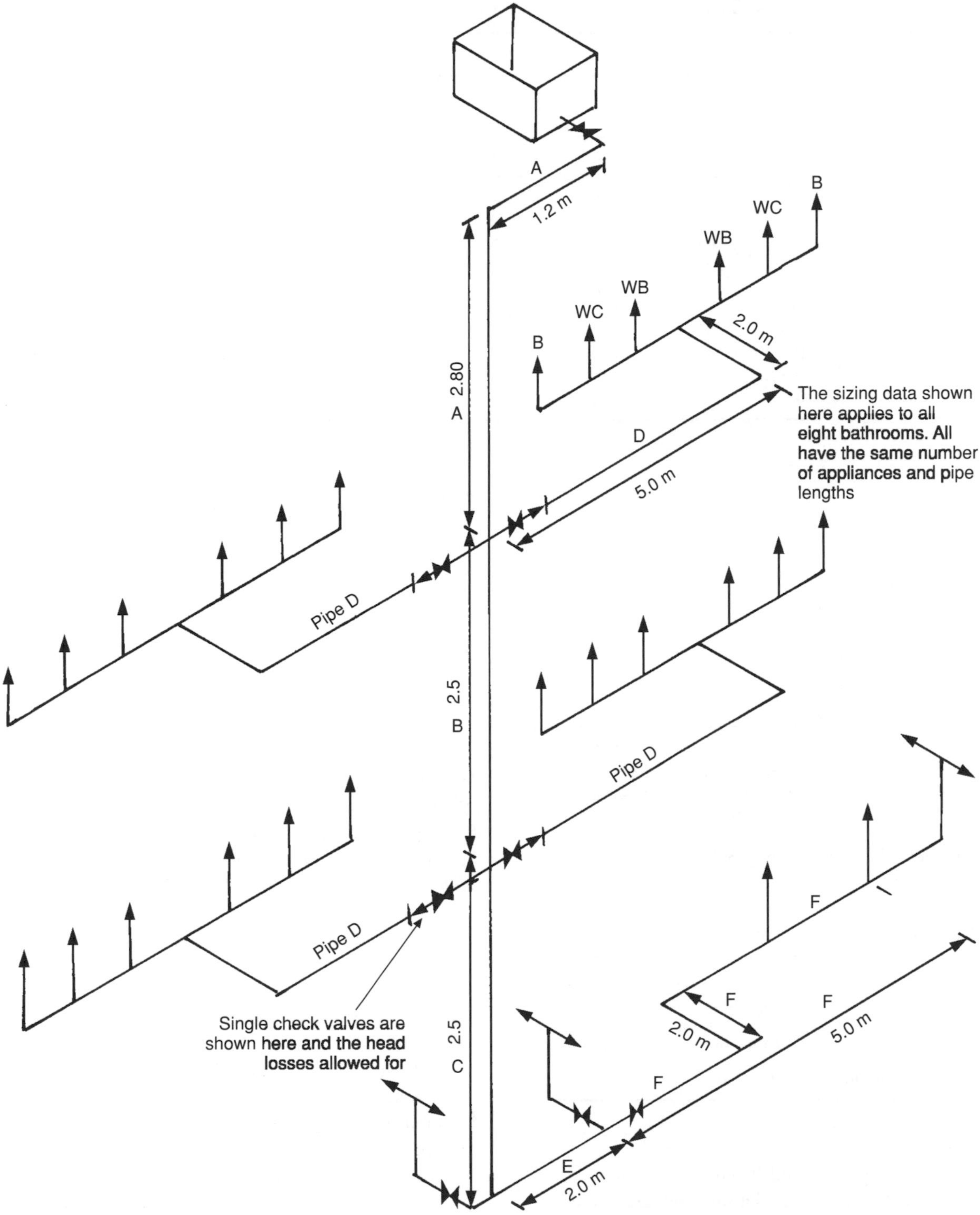


**Fig. 3.55** Distribution pipe sizes for the cold water services in a small commercial hotel (not to scale). Pipes are sized using medium-use loading units (see Table 3.5).

**Table 3.7** Pipe sizing tabulation schedule.

| 1 | 2 | 3 | 4 | 5 | 6 | 7 | 8 | 9 | 10 | 11 |
|---|---|---|---|---|---|---|---|---|----|----|
| Pipe section | Loading units | Flow rate (l/s) | Estimated pipe size | Velocity (m/s) | Loss of head (m per m run) | Effective pipe length | Loss of head due to friction | Progressive head | Actual head available | Confirmation of estimated pipe size |
| A | 146 | 1.5 | 42 | 1.0 | 0.02 | 7.4 | 0.15 | 0.15 | 2.8 | 42 |
| B | 58 | 0.8 | 35 | 1.0 | 0.03 | 2.5 | 0.75 | 0.9 | 5.8 | 35 |
| C | 26 | 0.5 | 28 | 0.75 | 0.05 | 2.5 | 0.8 | 1.7 | 7.8 | 28 |
| D | 24 | 0.5 | 28 | 1.0 | 0.04 | 16.2 | 0.65 | 2.35 | 2.8 | 28 |
| E | 22 | 0.5 | 28 | 1.0 | 0.05 | 2.6 | 0.13 | 2.48 | 7.8 | 28 |
| F | 18 | 0.4 | 22 | 0.75 | 0.04 | 13.6 | 0.5 | 3.00 | 7.8 | 22 |

Measured length. Total 7 m + 2 m
and 1.5 for two elbows and one
tee = total effective length 10.5 m

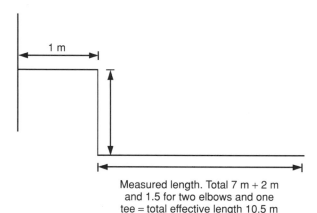

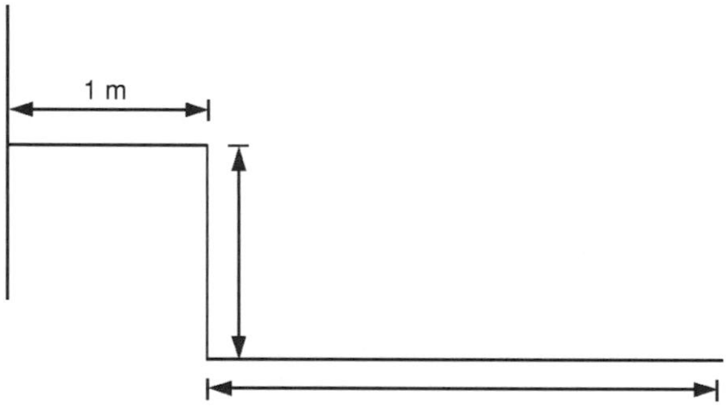

**Fig. 3.56** Calculating the effective pipe length.

**Table 3.8** Pipe A calculating loading units.

Wash basins $11 \times 2 = 22$ loading units
WC $\quad\quad 11 \times 2 = 22$ loading units
Baths $\quad\quad 8 \times 11 = 88$ (20 mm taps) loading units
Sinks $\quad\quad 2 \times 7 = 14$ (20 mm taps) loading units

Total loading units $= 146$

estimated pipe size and it will be seen how all the factors listed in the tabulation schedule come together to give an accurate calculation of the final pipe size.

## Further reading

BS 6700: Design, installation, testing and maintenance of services supplying water for domestic use with buildings and their curtilages.
BS 5306: Part 1. Hydrant systems, hose reels and foam inlets.
BS 5306: Part 2. Specifications for sprinkler systems.

*The Building Regulations*
BS EN 806 Parts 1 and 2.
Parts 3–5 Plumbing installations in building (currently in preparation).
*Water Regulations Guide*, Water Regulations Advisory Scheme, Fern Close, Pen-y-fan Industrial Estate, Oakdale, Gwent, NP11 3EH Tel. 01495 248454. www.wras.co.uk

*Water treatment and conditioning*
Hydropath UK Ltd, Unit F, Acorn Park, Redfield Rd, Nottingham, NG7 2TR Tel. 0115 986 9966. www.hydroflow.force9.co.uk
Scalemaster, Unit 6, Emerald Way, Stone Business Park, Stone, Staffs ST15 OSR Tel. 01785 811636. www.scalemaster.co.uk
Fast Systems Ltd, Dalton House, Newtown Road, Henley-on-Thames, Oxfordshire RG9 1HG Tel. 01491 491200. www.scalewatcher.co.uk

Tap Works, Mill Rd, Stokenchurch, High Wycombe, Bucks HP14 3TP
Tel. 01494 480621. www.tapworks.co.uk

*Hose reels*
Norsen Ltd, Unit 17, Airport Industrial Estate, Kenton, Newcastle-Upon-Tyne NE3 2EF
Tel. 0191 2866167.

*Water and fire pumps*
Armstrong Holden Brooke Pullen Ltd, Wenlock Way, Manchester MI2 5JL Tel. 0161 220 9660.
www.holdenbrookepullen.com

*Delayed action and equilibrium float valves*
Keraflo Ltd, Griffin Lane, Aylesbury, Bucks HP19 8BF Tel. 01296 435785.
www.keraflo.co.uk
H Warner and Son Ltd, Arclion House, Hadleigh Rd, Ipswich 1P2 OEQ
Tel. 01473 253702. www.hwarner.com

*Water controls*
Reliance Water Controls Ltd, Worcester Rd, Evesham, Worcestershire WR11 4RA
Tel. 01386 47148. www.relianceworldwide.com

### Self-testing questions

1. Define the difference between indirect and direct cold water systems.
2. Describe how stagnation is avoided in cold water supplies.
3. List the causes of noise that can occur in high-pressure cold water supply.
4. (a) Explain the term backflow in relation to cold water services and mains.
   (b) List three possible causes of backflow.
5. Describe the difference between a type AUK2 and AUK3 air gap.
6. (a) State the water category for which the RPZ valve is suitable.
   (b) Explain why a tun-dish is necessary on the discharge pipe.
7. (a) Explain why it is essential that potable water does not come into contact with water of category 5.
   (b) State the requirements regarding a supply of water to a bottle washing plant which is a category 5 risk.
8. Describe the procedures and equipment used to pressure-test a cold water system where large sections of pipework are to be concealed in floor ducts and suspended ceilings.
9. Describe the special arrangement that must be made regarding overflow and warning pipes in cisterns containing more than 5,000 litres of water.
10. (a) Define the term secondary backflow protection and the type of building in which it would be required.
    (b) Describe the methods and components necessary for secondary backflow protection.
11. Explain the action of zeolites used in the water softening process.
12. State the maximum height of a dry riser and its minimum diameter.
13. Sketch and describe a suitable system of water supply for a hose reel installation where the main supply is of insufficient pressure.
14. State the minimum pressure necessary for the effective operation of hose reels.
15. Evaluate the advantages and disadvantages of the various methods of treating hard water.
16. State the vertical distance from the spillover level of an appliance and the outlet of a $\frac{3}{4}$ inch tap in a domestic property.
17. State the type of backflow device that must be fitted to the outlet of pressure flushing valves.
18. List the procedures which must be carried out when commissioning cold water systems.
19. Explain the relationship between loading units and flow rates in connection to pipe sizing.
20. State the effect of excessive velocities in pipework systems.

# 4 Hot water supply

After completing this chapter the reader should be able to:

1. Identify the main cause of lime scale formation and corrosion in hot water systems
2. Understand the working principles and advantages of unvented hot water systems and the associated operational and safety controls.
3. Recognise and state the working principles of pumped hot and cold water supplies.
4. Calculate the boiler power required to heat a given quantity of water.
5. Understand the principles and limitations of circulating pressure in connection with gravity hot water systems.
6. Describe and sketch the methods of supporting and making connections to cylindrical vessels fitted in a horizontal position.
7. Select the methods of connecting towel rails and space-heating equipment in various circumstances and to different types of systems.
8. Identify the systems, applications, advantages and principles used in relation to gas and electric water heaters.

## Corrosion and scale formation in hot water systems

It is assumed that the reader is conversant with the basic principles and design factors relating to small hot water systems which are fully dealt with in Book 1 of this series. These also apply to more complex systems, and if they are fully understood no difficulty will be experienced in the study of more advanced work.

One of the main factors which the plumber has to bear in mind in relation to all water supply work is corrosion, to which is added in the case of hot water supply the problems associated with temporary hardness. Both temporary and permanent hardness are undesirable in water supplies, and water having a high temporary hardness content will cause scale or fur to form on the internal surfaces of the boiler and circulation pipes. The effect will be to lower the efficiency of the boiler and will sometimes result in noise in the circulating pipes.

Temporary hardness in water is produced when water having a high carbon dioxide gas content comes into contact with carbonate rocks. These carbonates are only soluble in water due to the presence of carbon dioxide which enables the water to take them into solution. It is a physical fact that when water is heated to temperatures of approximately 65–70 °C, all traces of gas including the carbon dioxide are given off, and as the water is then no longer able to contain the carbonates in solution, they are deposited in the boiler and circulating pipes as fur or scale. The fur builds up in the pipes and can over a period of time cause a serious obstruction.

In areas where soft water is encountered, corrosion problems are more common than furring and scaling, due mainly to oxidation and electrolysis.

*Oxidation*

This is caused by the oxygen contained in the water attacking unprotected ferrous metal surfaces. In most hot water systems this will apply to the boiler, which is generally made of cast iron or low-carbon steel. Oxidation brings about the formation of red rust or black magnetic oxide of iron, both of which can be the cause of discoloured water becoming discharged from the hot draw-offs. In combined hot water and heating schemes employing sheet steel radiators for space heating it can also, in conjunction with other forms of corrosion, result in their complete destruction.

*Electrolysis*

Due to the fact that soft water is capable to varying degrees of dissolving all metals, when such water is conveyed, for example, through copper pipes, it can take into solution a small percentage of this metal. When water containing dissolved copper comes into contact with zinc, which is sometimes present in old hot water schemes in the form of a galvanised coating on cisterns and cylinders, the zinc will be destroyed by electrolysis leaving the steel surfaces unprotected against further attack by the water, resulting in rapid corrosion of the steel. Some degree of immunity can be achieved by 'cathodic protection' and it is quite common to fix a sacrificial anode in any hot storage vessel made of galvanised steel. It consists of a block or rod of magnesium or aluminium hung or bolted inside the vessel, and these metals, being lower on the electrochemical scale than zinc, are attacked in preference to the galvanised coating.

The best protection against electrolysis is to use only one metal in hot water schemes, but this is usually economically impossible. For instance, in the case of cylindrical hot storage vessels, while it is common to use copper for the smaller sizes, to make larger ones of this material would be too expensive. This is due to both the quantity and thickness of the copper sheet necessary to cope with the higher internal pressure to which these vessels would be subjected. Cylinders or hot water storage having capacities in excess of 240 litres are almost invariably made of galvanised steel plate, or in the case of gas heated vessels, vitreous enamelled steel.

In combined systems of hot water supply and heating it is almost impossible to avoid mixtures of metals. For example, in most cases copper pipes are used in conjunction with thin sheet steel radiators in heating installations, and unless some preventive measures are taken serious corrosion will take place in the radiators.

It has been found that corrosion and scaling problems are more common in direct schemes of hot water than in the indirect type. This is due to the fact that the water in a direct system, including that in the boiler and circulating pipes, is constantly changed, thus introducing more corrosive or hard water (depending on the locality) into the system. This gives rise to continual corrosion attack on any ferrous metal components it may contain. Further reference is made to this subject in Chapter 5 (page 199).

### Supplementary storage systems

When a system of hot water supply is considered in tall buildings with the main storage vessel at low level, it is sometimes necessary to install a supplementary storage vessel at high level, as shown in Fig. 4.1. This is really an enlargement of the pipe and ensures an adequate supply of hot water to draw-offs on the upper floors in the building which might otherwise be starved of water when those at lower levels are in use. One could, of course, increase the pipe size to ensure that an adequate supply of water is available at all draw-off points, but this would increase the heat losses. A little thought will show that the larger the pipe diameter, the greater will be its surface area capable of dissipating heat, which in most cases would be in excess of that presented by the surfaces of the supplementary storage vessel. It is important that this vessel is not too big, causing unnecessary heat loss. The maximum capacity should not be more than one-fifth of the total storage content, i.e. if the total storage is 1,000 litres, not more than 200 litres should be stored in the supplementary storage vessel. It is necessary to ensure that both hot storage vessels and the secondary circulation are well insulated to avoid excessive operating costs.

### High-level flow systems

The following relates to the use of solid fuel boilers where a gravity circulation, usually the primary flow and return to the hot water storage vessel, must be

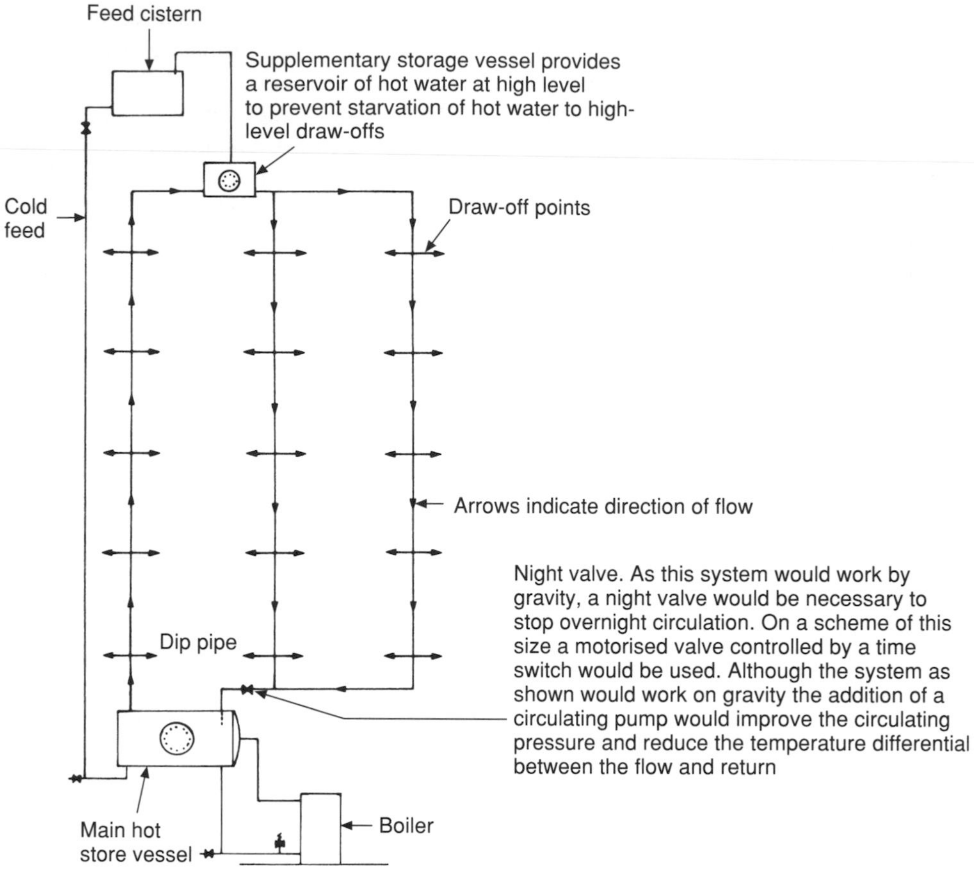


**Fig. 4.1** Supplementary storage system.

provided to avoid overheating. This does not apply to fully automatic gas- or oil-fired boilers where fully pumped systems can be employed. There are situations where a door or window interrupts the most suitable way of running the flow and return.

A simple but effective method of overcoming this problem is by using what is called the *high-level flow system* illustrated in Fig. 4.2. The hot storage vessel must not be fitted too low or reverse circulation may take place for the following reason. If the water temperature in the boiler falls below that of the hot storage vessel (this could happen if the fire dies down overnight) a circulation could take place from the hot storage vessel to the boiler. Not only would this cause unacceptable overnight heat losses, but due to reversed circulation the system will be unacceptably noisy when the fire is relit and normal direction of circulation is resumed. To ensure an acceptable circulation with this system, the horizontal runs should be as short as possible and the $\frac{4}{5}:\frac{1}{5}$ ratio from the centre of the boiler (shown in Fig. 4.2) should be adhered to. Failure to do so may result in very sluggish circulation.

It is always necessary to provide what is called a *heat leak* with solid fuel boilers. This means that any heat produced by the mass of fuel in the boiler, even when the thermostat is closed, must be accommodated by a gravity circulation. Failure to do this will result in overheating and possible boiling in the system.

## Dual boiler systems

Some systems employ two boilers, a typical example being where a traditional range is employed for cooking and possibly domestic hot

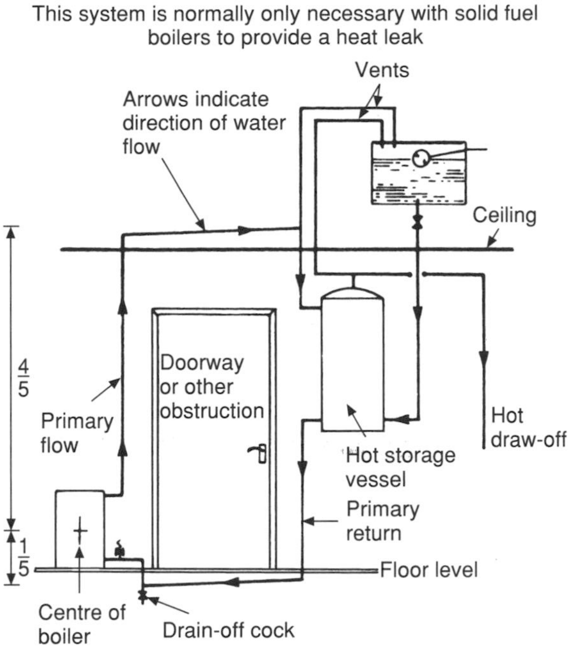


**Fig. 4.2** High-level flow system used to overcome obstacles which would preclude a normal flow and return pipe run. Providing the ratio of $\frac{4}{5}$ above, $\frac{1}{5}$ below, the centre line of the boiler is observed and the horizontal runs are not excessive, an acceptable circulation can be expected.

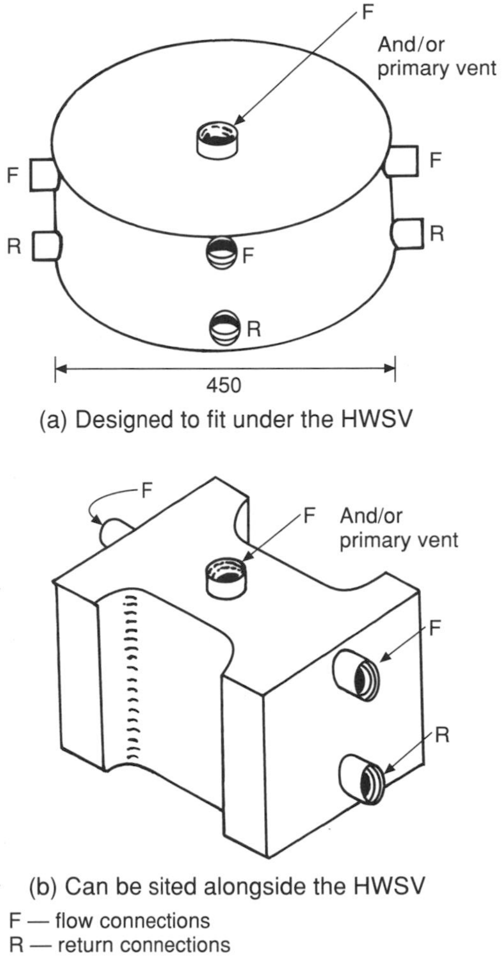


**Fig. 4.3** Neutralisers for dual boiler installations.

water supply, and is interconnected with a boiler for space heating. Unless the system is very carefully installed, water may circulate through the boiler not in use, and the updraught through the flue will rapidly dissipate any heat into the open air — not a very economic proposition. Although the output from range boilers is usually very low and their recovery rates are slow, they can be used to augment a space heating appliance. This may be a wood or solid fuel appliance if a plentiful supply of such fuel is available, but many customers like them for their aesthetic appeal. The type of fuel used with dual boiler schemes is not important, except in the case of solid fuel where a suitable heat leak must be provided. The main problem encountered with this type of installation occurs when only one boiler is operating. This system employs a component called a circuit neutraliser (see Fig. 4.3) which, in effect, provides an area in which there are no positive or negative pressures. It acts in a similar way to a boiler (except those with a low water content) in a combined hot water and heating system, where the boiler is the neutral point. The neutraliser serves the same purpose and is fitted where the circulating pipes from the two boilers interconnect. It may be fitted in the cupboard housing the hot water store vessel (HWSV) if it is of sufficient size, but it can be installed at any convenient point, providing the base is at least 300 mm above the highest boiler. Figure 4.4 illustrates a typical layout using in this example two solid fuel boilers.

In order to preserve the 'heat leak' when using solid fuel boilers, it is generally an accepted practice that no electrical controls are fitted in such a way that the heat leak is closed. The one possible exception is hopper-fed boilers, which usually incorporate forced draught. These are normally thermostatically controlled, and it is possible using

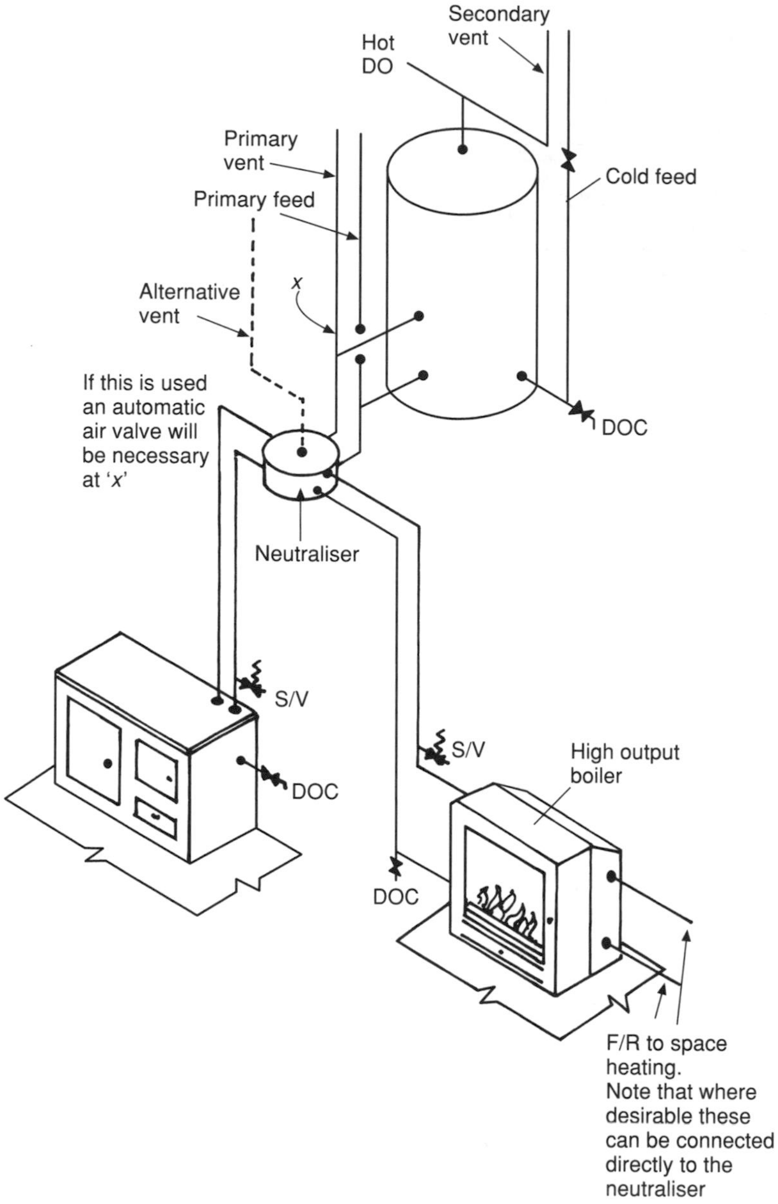


**Fig. 4.4** Interconnecting two boilers. This system can be adapted for fully pumped schemes with gas or oil boilers.

the system shown in Fig. 4.5 to exercise a degree of control over the stored water. Although the original concept of the neutraliser was to intercouple solid fuel appliances, it has been developed for fully automatic gas and oil boilers and can be usefully employed in any situation where two or more boilers are interconnected. It is recommended that the manufacturer of the components used should be contacted for more detailed information if a system of this type is contemplated.

### Unvented hot water systems

Traditionally, hot water storage systems in the United Kingdom have been of the low-pressure type, water being supplied to the storage vessel

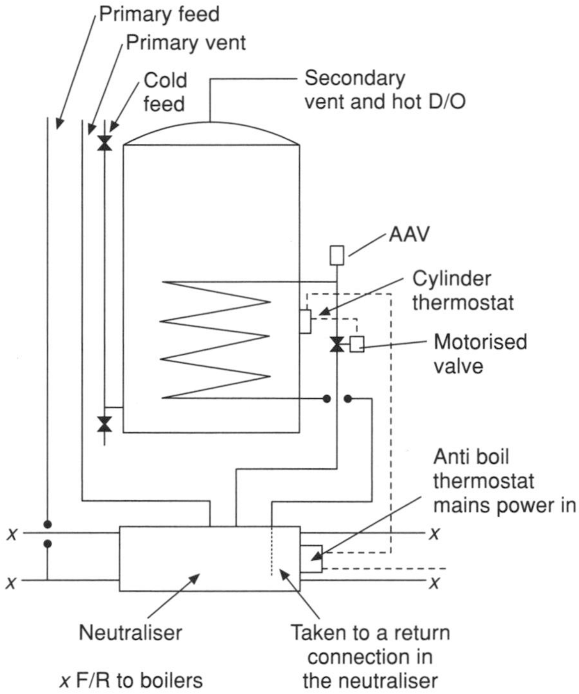


**Fig. 4.5** Electrical control for dual boiler schemes. This form of control depends on the effectiveness of the boiler thermostats, but it is designed to provide a heat leak in the event of a pump or power failure.

from a feed cistern usually situated in the roof space. Until the revision of the Model Water By-laws 1986 all previous legislation by the water authorities precluded main feed storage systems having storage capacities of more than 15 litres, their main objection being:

(a) Greater possibility of contamination.
(b) The fact that in some areas the distribution system might not be able to meet the demand of all draw-offs, both hot and cold being taken from the main.
(c) The possible dangers due to explosion in such systems.

Most of the foregoing problems have been resolved, although the installer of unvented systems must be sure that the main supply is of sufficient pressure and volume to meet the draw-off demands, as an unvented hot water system does not possess any magical qualities for improving a poor water supply. Most readers will be aware that for a long period of time main pressure hot water equipment has been available mainly in the form of single- or multi-point gas water heaters. While in the right circumstances such heaters are perfectly satisfactory, they suffer from the disadvantage of having a low flow rate in comparison with that of storage systems.

Unvented hot water systems have many advantages and some disadvantages and these should be considered very carefully before a decision is made on which type of system is best for a specific installation. It is a fact that the elimination of a traditional feed cistern and the pipework necessary for its installation saves both material and on-site labour costs, as most unvented systems and their necessary controls are supplied as a packaged unit by the manufacturers. Such systems eliminate the need for water storage and its associated pipework in the roof space. This is a very important advantage. Due to the requirements for ventilation and ceiling insulation in roof spaces, they have become very cold areas indeed. It should not be forgotten, however, that if the premises are heated by a traditional system having a feed and expansion cistern, it will normally remain in the roof space. Unvented hot water systems give greater flexibility to the design of taps, mixers and shower-heads, some of the latter being dependent for their satisfactory operation on higher pressure than those normally associated with traditional systems. Many modern tap designs incorporate mixing devices and due to the requirements of the water by-laws prior to 1986 it was not permissible to connect such mixers to the main supply, biflow types being the only exception. Due to the more stringent requirements to prevent backflow and water wastage, non-biflow mixers are now acceptable on main-fed supplies and this should result in lowering costs for the installation of mixers for all applications. Savings on pipework and fittings may be made because, due to the higher pressures involved, smaller pipes can be used.

The disadvantages for unvented storage systems are few but somewhat formidable. The storage vessel, which is generally a packaged unit (this is to say all the necessary controls are already fitted by the manufacturer), tends to be expensive in comparison with the components of a traditional vented system. Such a system, if properly installed, normally lasts for a long period of time, often the lifetime of the installation, requiring little or no maintenance. Unvented systems rely mainly on automatic controls which require periodic maintenance and if they do become defective they must be replaced. **No attempt should be made to repair them** as they must be regulated and checked under factory conditions. Generally speaking unvented systems are most likely to cost more to maintain. Any savings made by the installation of unvented systems are offset by the cost of stronger storage vessels and ancillary controls that must be provided. Another important consideration is the lack of water storage and should the main be shut down for any reason, no water will be available until the supply is restored. Such circumstances are fortunately rare, but to be in a situation where no water is available at all is, to say the least, very inconvenient. Some authorities have suggested that the possibility of water contamination is greater with the use of unvented systems as there are more draw-off points liable to backflow. It must be pointed out, however, that in new premises with backflow protection devices conforming to the 1999 Water Supply (Water Fittings) Regulations this danger should be minimal. The greatest danger exists where unvented systems are installed in existing properties in which backflow prevention devices are not fitted. In such cases the installers must satisfy themselves that no danger from contamination exists before the installation is completed. Most existing systems would, for example, have to be fitted with check valves on such fittings as bath and sink mixers in order to meet the requirements of the Water Regulations.

*System layout*

Figure 4.6 shows the main components required for an unvented hot water system. In practice most

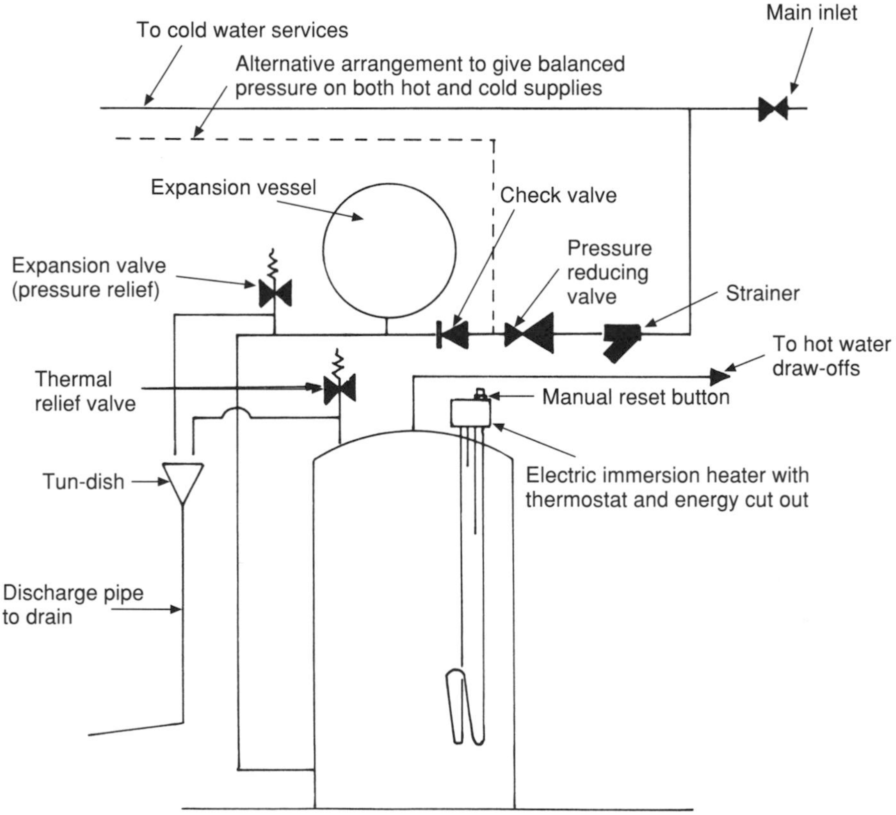


**Fig. 4.6** Diagrammatic layout of operating and safety controls for unvented hot water storage systems.

of them are fitted to the storage vessel by the manufacturer, all the plumber has to do on site is to make the necessary connections. Do make sure when installing the storage vessel that sufficient space is available to carry out maintenance work. These systems should be serviced annually and it is a messy and complicated job if the unit has to be completely removed for this purpose.

The water may be heated by an electric immersion water heater or the storage vessel may be of the indirect type, the secondary water being heated by a boiler in the traditional way. Whatever method of heating is used the equipment must be provided with a safety cut-out device which isolates the heat source in the event of failure of the thermostat.

### *Operating components*

*Strainers* These are essential to prevent any debris such as silt or shrimps (microscopic organisms) passing into the system possibly causing problems with the valves downstream. A suitable strainer is shown in Fig. 4.7. When the system is serviced the strainer should be inspected and washed in clean water.

*Pressure-reducing valves* The working principle of these valves has been described in the previous chapter. They are fitted to reduce the inlet pressure to the working pressure of the equipment used and to maintain a constant flow rate to the draw-off points. It is currently recommended that the pressure on these systems is limited to 2 or 4 bar, depending on the type of material, i.e. steel or copper, from which the storage vessel is made. Only steel vessels should be used for the higher pressures. The working pressure of the system is usually two-thirds of the test pressure of the storage vessel, and the pressure relief, or expansion valve (described later), is usually set at this pressure. The outlet pressure of the reducing valve must therefore be closely related to the pressure at which the expansion valve will open. All the valves affected by pressure, when obtained with a packaged unit, are factory set, thus ensuring a close relationship with each other.

*Check valve* These are single check only and are provided to prevent backflow of hot water into the cold water services, but they also prevent 'implosion', a term used to describe what the plumber understands as cylinder collapse. A study of the system will show that it is possible for the water in the storage vessel to be siphoned out by opening a cold water tap if the main stop valve is shut down.

*Anti-vacuum valve* As a further safeguard against implosion an anti-vacuum valve is provided to admit air to the storage vessel should the pressure inside fall below that of the atmosphere. Figure 4.8

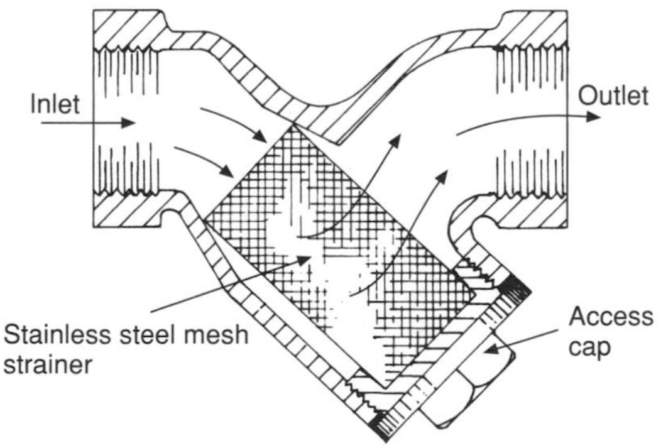
Inlet
Outlet
Stainless steel mesh strainer
Access cap

**Fig. 4.7** Strainer.

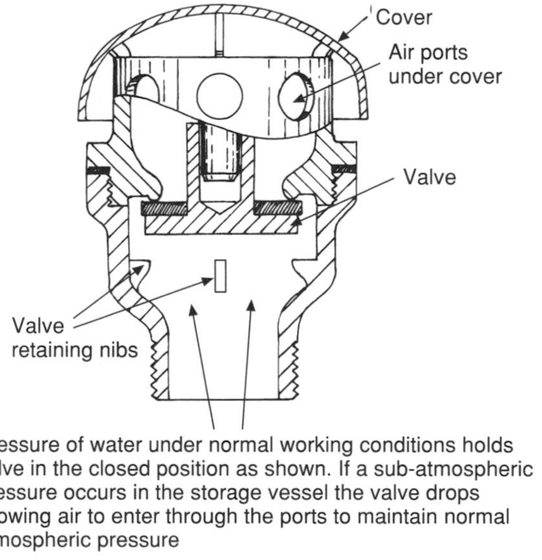
Cover
Air ports under cover
Valve
Valve retaining nibs
Pressure of water under normal working conditions holds valve in the closed position as shown. If a sub-atmospheric pressure occurs in the storage vessel the valve drops allowing air to enter through the ports to maintain normal atmospheric pressure

**Fig. 4.8** Anti-vacuum valve.

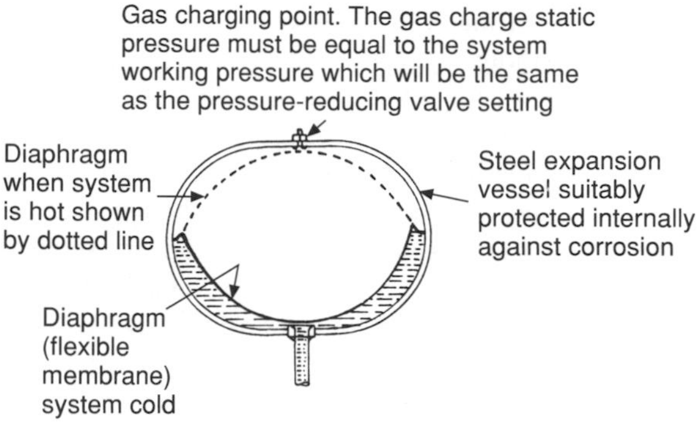
Gas charging point. The gas charge static pressure must be equal to the system working pressure which will be the same as the pressure-reducing valve setting
Diaphragm when system is hot shown by dotted line
Steel expansion vessel suitably protected internally against corrosion
Diaphragm (flexible membrane) system cold

**Fig. 4.9** Section through expansion vessel. See also Fig. 5.22.

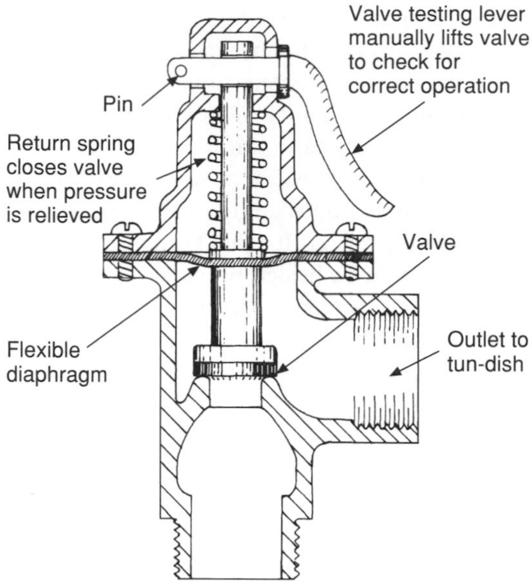
Valve testing lever manually lifts valve to check for correct operation
Pin
Return spring closes valve when pressure is relieved
Valve
Flexible diaphragm
Outlet to tun-dish

**Fig. 4.10** Expansion valve (pressure relief valve). This valve is designed to protect the storage vessel from bursting. It is designed to open only when provision for expansion has failed or the pressure-reducing valve is malfunctioning and the system is subjected to excessive pressure.

illustrates a typical valve of this type, although in some cases they are made as an integral part of the thermal relief valve.

*Expansion vessel* This is designed to accommodate the expansion of the water in the system when it is heated and to prevent any operation of the expansion valve which could lead to wastage of water. It is important that it is correctly sized so that it can absorb the required volume of expansion. Figure 4.9 shows a section through a typical expansion vessel and it will be seen that as the water in the system expands it causes the diaphragm to distend, compressing the gas, usually nitrogen or air. If due to a fracture in the rubber diaphragm the gas escapes, the vessel will become waterlogged and because liquids are for practical purposes incompressible, expansion of the water when it is heated will cause the expansion valve to open.

*Expansion valve* As previously explained this valve is used to relieve the pressure in the system due to the expansion of the water if for any reason it is not accommodated in the expansion vessel. Figure 4.10 illustrates its working features. As it is fitted to the cold inlet side of unvented hot water systems, it is not considered as a safety valve and should not be referred to as such. Any discharge through these relief valves must be passed through a tun-dish to provide a type AUK 3 air gap, thus avoiding contamination via the drain. A typical tun-dish is shown in Fig. 4.11.

*Safety controls* Possibly the greatest disadvantage of unvented hot water systems is the possibility of serious explosion. It must be pointed out, however, that such a catastrophe is unlikely due to the safety systems built into the installation. The real danger lies not with the system but with the possibility of untrained people interfering with the factory-set controls or incorrectly installing a system. Unfortunately the legislation governing installation of these systems is very difficult to enforce except in new properties which are normally inspected by a building control officer and a water board official. The reader should be aware that at atmospheric pressure water boils at a temperature of 100 °C and is converted to steam. If, however, the water is in a sealed container its temperature can be increased without its conversion to steam. Should the storage vessel burst, the water, if at a temperature in excess of 100 °C when escaping into the atmosphere, will immediately be converted to steam, which because it occupies a much greater space than water will cause an explosion. It is perhaps worth noting that

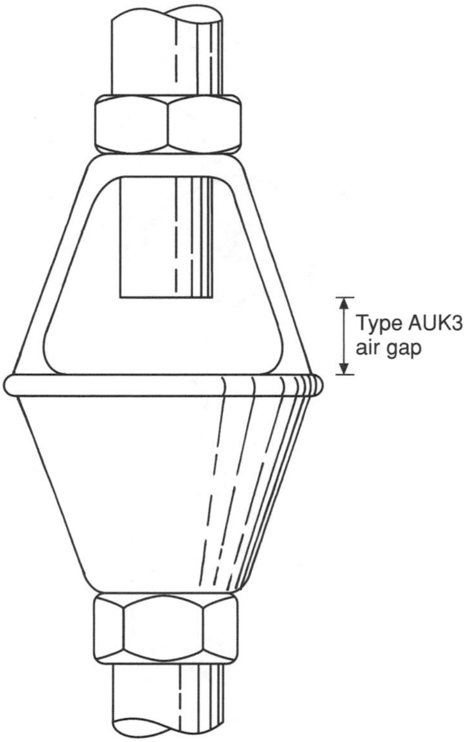


**Fig. 4.11** Detail of tun-dish.

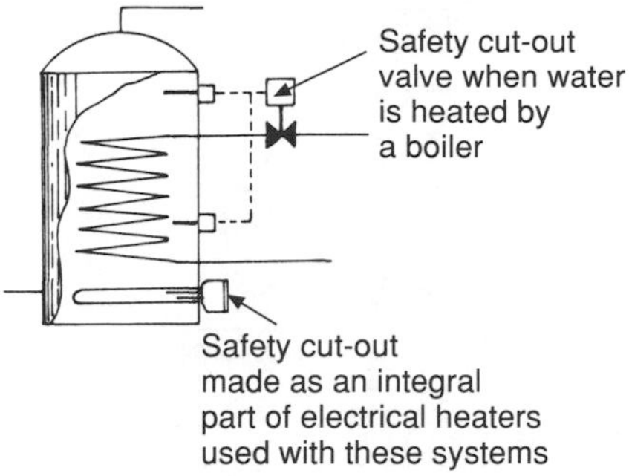


**Fig. 4.12** Safety controls for unvented systems.

water in a sealed container at 100 °C, if released to the atmosphere, would occupy 1,600 times its volume when it changes to steam. The safety devices built into unvented hot water systems include three lines of defence against such a disaster, all of which are designed to prevent the water achieving temperatures of 100 °C. These safety controls are designed to act in sequence as the temperature rises and are listed as follows.

*Thermostats* These are common to all types of heating equipment, their object being to permit varying operating temperatures of appliances. Domestic hot water temperatures are normally between 60 and 65 °C, as higher temperatures may cause scalding, increase of heat loss of the stored water and the formation of lime scale. The use of invar rod, fluid-expansion, and bi-metallic strip types of thermostats are suitable with these systems, the working principles of which are described in Chapter 2.

*Temperature-operated cut-outs* These usually take the form of a second thermostat, factory set at approximately 85 °C, and should the appliance thermostat fail, will limit the water to this temperature and automatically cut off the source of heat. All cut-outs of this type are fitted with a reset button which must be operated manually to restore the source of heat. If manual operation of the cut-out is continual the consumer will be made aware that all is not well with the thermostat. Where the water is heated electrically the cut-out is usually located near or in the heater cap. When a boiler is employed a thermostatically operated motorised valve may be used as shown in Fig. 4.12. Both the thermostat and the cut-out sensors must be located at a point in the storage vessel where they are in contact with the hottest water. Please refer to the Further Reading section of this chapter for relevant British Standard codes.

*Temperature relief valves* These must comply with BS EN 1490 and BS EN 1491 and only valves conforming to these specifications must be used. Figure 4.13 shows the operating principles of this component. It has a spring-loaded valve which is opened at a temperature of 90–95 °C due to the expansion of temperature-sensitive fluid acting on a push rod which moves the valve upward, causing it to open. The temperature relief valve is designed to operate only after the failure of both the thermostat and the temperature-operated cut-out. The type shown illustrates only the basic principles of these valves. They are obtainable having built-in vacuum relief valves and a pressure relief device that opens if the discharge pipework is obstructed. Some

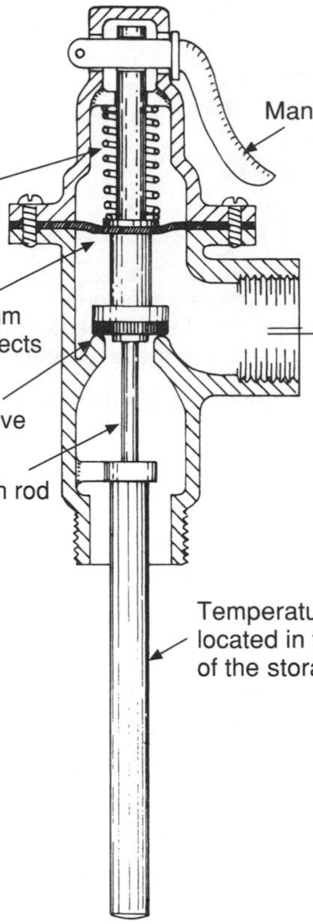


**Fig. 4.13** Combined temperature and pressure relief valve. This valve will open when (a) the design pressure in the system is exceeded and (b) when the temperature exceeds 90–95 °C due to the failure of the thermostat and safety cut-out. The type shown doubles as an expansion relief valve, which saves the cost of using two valves.

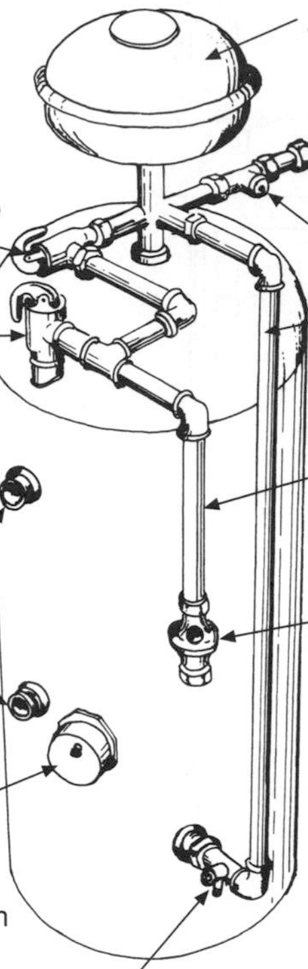


**Fig. 4.14** Unvented hot water storage vessel showing arrangements of controls and safety components.

valves of this type also incorporate expansion relief which saves the cost of two valves. As with expansion valves a discharge pipe is necessary and must conform to current legislation. Both expansion relief and temperature relief valves must be capable of automatic closure and be watertight when closed. This should be checked periodically by operating the levers attached to these valves, which will ensure that they function correctly. Figure 4.14 illustrates pictorially a typical unvented hot storage vessel showing the relative portions of all the necessary components.

*Discharge pipes (see Fig. 4.15)*

In the event of a pressure relief or expansion valve opening, water at possibly 100 °C will be

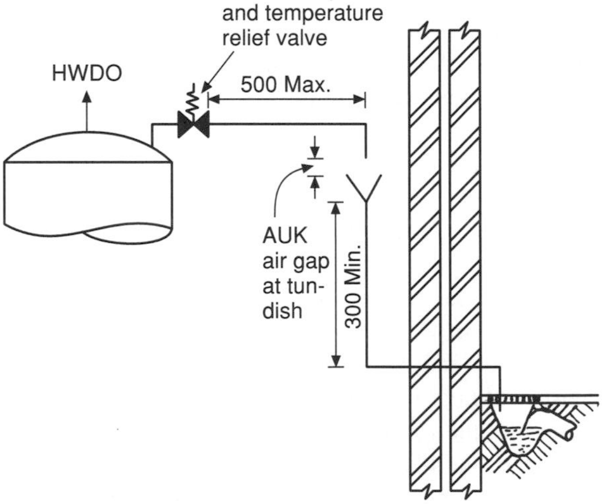


**Fig. 4.15** Discharge pipe details (unvented hot water systems).

discharged. The safety requirements relating to its disposal are quite specific and are listed as follows:

(a) The pipe must be made of metal with metal fixings.
(b) It should not be longer than 9 m with no more than three easy-radius bends, and should have a continuous fall to the point of termination. Sharp bends should be avoided, but if this is not possible the necessary allowance for resistance must be made (see Table x.x, p. xx). Its diameter must be enlarged if it is over 9 m long.
(c) Its diameter must not be less than that of the valve outlet, and one size larger than the tun-dish.
(d) Pipes fitted externally of the building require frost protection (a slight let-by may result in freezing and consequent blockage by ice). The discharge point must be visible, but not so that it could cause scalding. The distance between the point of outlet and the gulley grating should not exceed 100 mm, and if in reach of very young children, it should be guarded with a suitable mesh cover.
(e) Any pipe termination points at high level may discharge onto a roof (providing the covering will not be damaged by high temperature) or into a metal hopper and pipe. The termination point must always be visible.

*Legislation relating to unvented hot water systems*
The Water Regulations specify the requirements of expansion and temperature relief valves in the context of prevention of water waste. The Building Regulations specify the requirements of these systems in terms of safety and are found in the approved document G3. This states that the components should be supplied in the form of a unit or package conforming to the British Board of Agrément.

A unit is defined as an appliance to which both the operating and safety components are fitted by the manufacturer, a package being supplied only with the safety devices fitted, the operating components being supplied separately and fitted by the installer. In both cases this ensures that the safety devices are 'factory set' and should not be tampered with. The approved document G3 also requires that the installation of these systems is carried out only by an installer who has undertaken an approved course of training. The requirements for the discharge pipe on both expansion and temperature relief valves and also specified in G3 have been dealt with in the previous text.

*Methods of heating for unvented hot water systems*
The diagrammatic system shown in Fig. 4.14 is heated by means of an electric immersion heater, but a boiler can be connected in the same way as a traditional vented system providing it is fitted with a suitable thermal cut-off arrangement should the thermostat fail. If the primary system is of the open vented type having a feed and expansion cistern the advantage of avoiding pipes and cisterns in the roof space will be lost, and serious consideration should be given to the installation of an unvented heating system which is described in the following chapter.

### Combination boilers

Combination boilers, commonly called combis, incorporate what is, in effect, a central heating low-water-content boiler, and a secondary heat exchanger which indirectly heats a mains water supply for domestic hot water. These boilers combine the basic principles of both gas instantaneous heaters and, generally speaking, sealed heating systems, although some types may be fitted with an open vented heating system. Because they are relatively small and light in weight, most of them are designed to be wall hung. All boilers now except for a few exceptions must be SEDBUCK 'A' or 'B' rated, both of which meet the requirements of Part L of the Building Regulations. Details of condensing boilers are given in Book 1 of this series and a typical condensing wall-hung combination boiler is shown in Fig. 5.8. Where forced draught is provided fans are built into the boiler unit in such a way that the burner will not fire until it is operating.

Combination boilers have all the advantages of both sealed heating and unvented hot water systems in that no storage vessels are necessary, and as the hot water supply is at mains pressure they are a positive advantage for showering. The electrical control system is incorporated in the boiler unit itself, thus external wiring is normally limited to the requirements of the room thermostat and time switch, if externally fitted. The one main disadvantage with most combination boilers is their flow rate in comparison with storage systems of hot water. This is not quite so true as was once the case, as due to improvements of design, manufacturers have been able to increase the flow rates without reducing the water temperature. Some of the larger types of these boilers have a flow rate of 14.5 litres per minute at a temperature rise of approximately 35 °C, which compares reasonably with the requirements of BS 6700 which specifies 18 litres per minute for a $\frac{3}{4}$ in bath tap. Obviously if another tap is open on the system the flow rate will diminish, but in practice it has been found that combination boilers of the larger type normally satisfy the demands of an average domestic property having only one bathroom. The heat exchanger shown in Fig. 4.16 in these boilers is made to allow the incoming water to be heated to a high enough temperature for domestic hot water.

The actual installation of combination boilers should present no difficulties to a competent plumber. Having ensured the unit is correctly fixed on a surface which is suitable for its support and installation of the flue, it is simply a matter of connecting the water, heating and gas services. The manufacturer's instructions, and both gas and electrical regulations, must be complied with.

Servicing, maintenance and fault finding on combination boilers is slightly more complicated than most other appliances, and for this reason it is recommended that any repairs and maintenance be carried out by someone experienced in this work. Most manufacturers run special courses relating to their equipment at a very low cost for those who wish to specialise in this area of work.

Although combination boilers are designed to limit operating temperatures to a maximum of 65–70 °C, most manufacturers recommend fitting a scale reducer where the water has a high temporary hardness content. If a water softener is already installed this, of course, will not be necessary.

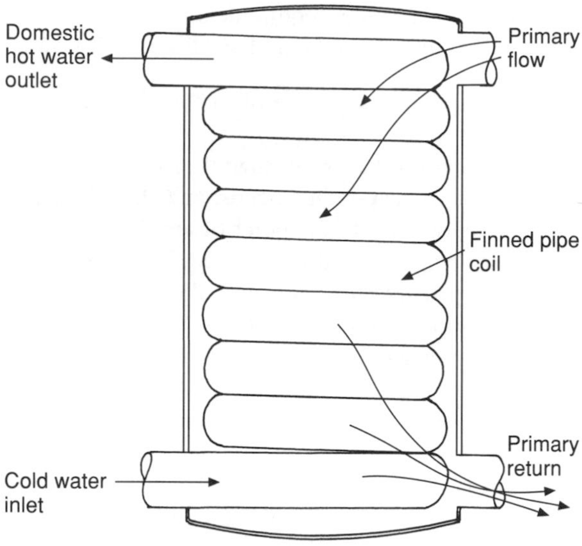


(a)

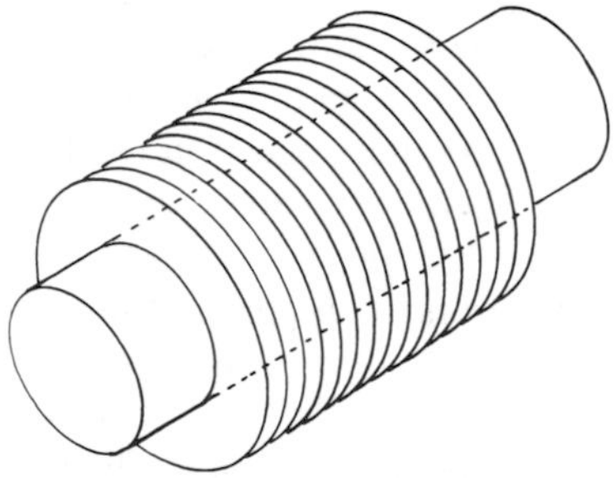

(b) Section of finned flexible pipe used to conduct the maximum quantity of heat from the primary water to that of the secondary serving the hot draw-off

**Fig. 4.16** Water-to-water heat exchanger as used in combination boilers to heat domestic hot water.

To avoid damage to the pump a bypass circuit must be provided in a similar way as that shown in Fig. 5.14. Some combination boilers incorporate this bypass circuit integrally, others do not, so always check the manufacturer's instructions on this point.

*Thermal storage systems* Figure 4.17 illustrates a typical system of this type and it will be seen that it is designed like the unvented hot water systems described previously to enable hot water to be delivered direct from the mains. It works in a similar way as an instantaneous heater, but in this

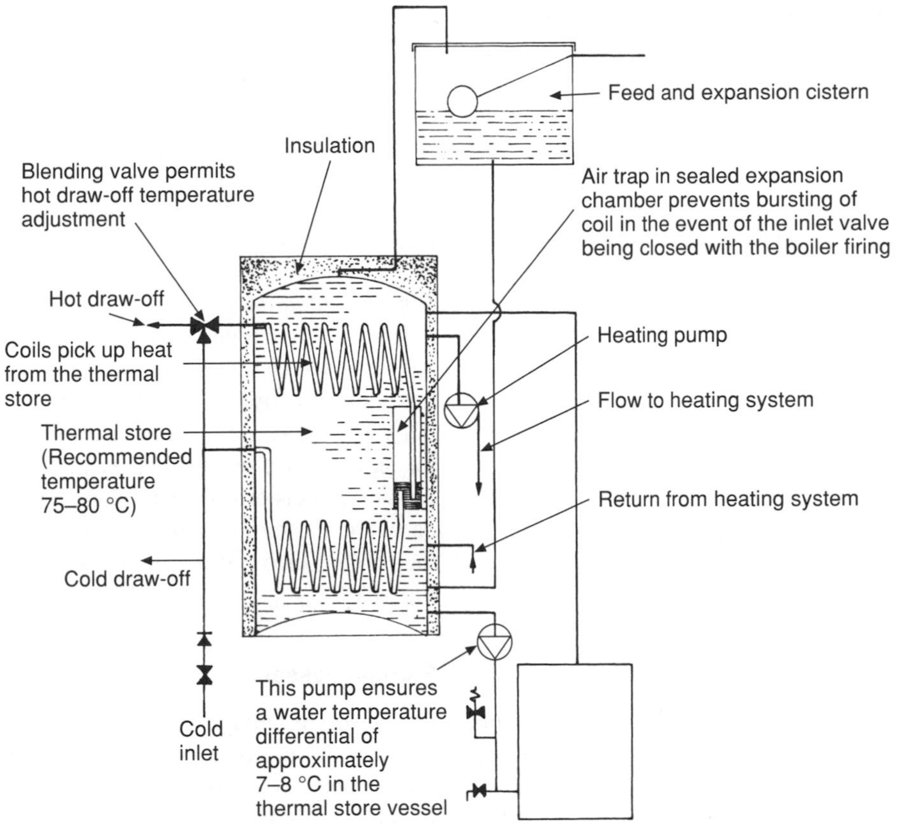


**Fig. 4.17** Thermal store system. These systems have the advantage of main flow hot water without the use of the high pressures that are used in unvented systems. The type shown has an independent boiler and feed cistern. Other models, incorporating both a boiler and feed cistern as a unit, are available for small domestic properties.

case the carefully designed heat exchanger operates on a water-to-water basis, being completely immersed in the 'thermal store'. As can be seen from the illustration, a well-insulated body of water is maintained at a temperature of 80–85 °C by the boiler. The volume of domestic hot water delivered at usable temperatures is an improvement on most combination units, and generally for a temperature rise of 45 °C with an inlet pressure of 2 bar from the mains, they will deliver 12–24 litres per minute, depending on the type used. If the mains pressure exceeds 5 bar a pressure-reducing valve is recommended, because, like combination units, if the water pressure is excessive, water will pass through the heat exchanger too quickly and will not pick up sufficient heat for domestic purposes. It should also be noted that the water in the top part of the thermal store must be maintained at 80–85 °C to give the quoted temperature rise for domestic hot water. As the stored water is only subject to pressure from the feed and expansion cistern there is no danger of explosion so the control system is very simple. These units must be installed by a competent person who is qualified in gas and electrical installation work. All work must comply with both the Water and Building Regulations. Many of the smaller units are supplied as a complete unit including the boiler, others where higher outputs are required can be fitted with a boiler in the usual way. As with unvented systems pipework and cisterns in the roof space can be eliminated, as the feed and expansion vessel supplying the primary water to the radiators, thermal store and boiler is in most cases made as an integral part of a unit. Space heating can be controlled by a room thermostat although better and more economic control would be achieved by using thermostatic radiator valves.

## Pumped systems

For many years equipment has been available to 'boost' or pump shower fittings in situations where the static head in the cistern supplying both hot and cold supplies is unable to provide sufficient pressure. This arrangement can be extended to enable all the cistern-fed supplies to be delivered at higher pressures. There are many permutations with this arrangement, dependent upon the type of system employed. Figure 4.18 shows two arrangements:

(a) When the cold water supply is of the direct type where all cold water draw-offs are connected to the main service.
(b) The indirect type where all but one drinking water tap is supplied from a cistern.

Suitable pumps are available to enable the owners of existing premises to boost the supply of both hot and cold services to permit the use of many of the special mixer fittings and shower heads now available. Unlike combi and thermal storage appliances which are limited to the domestic market, boosted systems can be used in all types of buildings. Like other equipment currently developed for modern plumbing systems, the pump, storage vessels and necessary controls may be obtained as a complete unit or, alternatively, suitable pumps are available for fitting into existing systems. See also pages 245–247.

### *Surrey vented flange*

It is sometimes found that even if the pump duty has been selected carefully, air may be drawn down the vent pipe giving an unsatisfactory supply of hot water. This may be because the cold feed is of inadequate size, or more likely the pressure exerted by the feed cistern is insufficient. The problem may be overcome by making an additional connection to the HWSV or by fitting the surrey flange shown in Fig. 4.19.

## Storage vessels

### *Capacities of storage vessels*

In industrial or commercial buildings the capacity of hot storage vessels is based on several factors which include the number of persons to be accommodated,

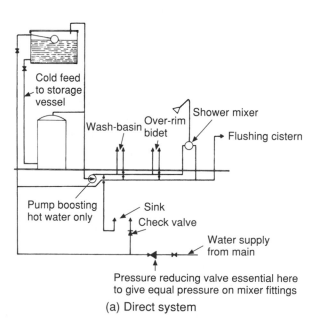

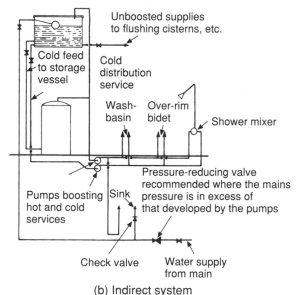

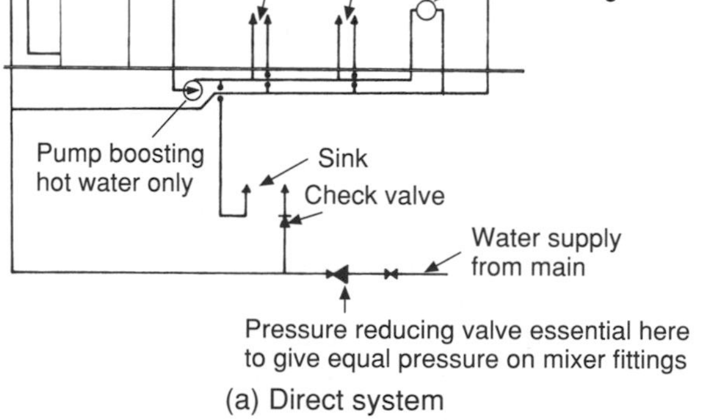


**Fig. 4.18** Pumped supplies for multi-appliance installations.

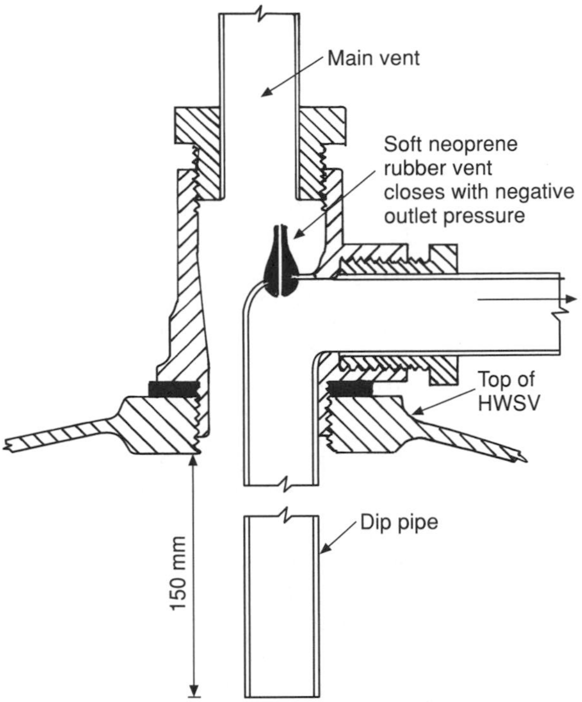


**Fig. 4.19** Surrey flange. Prevents ingress of air via the vent pipe into pumped hot water draw-offs.

the type of building, the incidence of usage and any peak demand the storage vessel has to meet. In small domestic properties the storage capacity is based on the number of bedrooms the building contains as this is generally an indication of the number of people in occupation. Table 8.4 in Book 1 lists the capacity of hot storage vessels for small dwellings heated by a boiler. If electricity is used to heat the water and full advantage is to be taken of cheap night rates, larger storage vessels than those listed are recommended. It should be noted that the capacities are for copper direct cylinders only; those for indirect cylinders will be slightly less due to the space occupied by the coil or annulus. The boiler power, unless otherwise specified, should be capable of raising the temperature of the stored water through 50 °C in 2 hours and 30 minutes. The following illustrates a typical example. To determine the boiler power required (in kilowatts) to heat 120 litres of water through 50 °C in 1 hour the following formula should be used:

$$\text{No. of kilowatts} = \frac{\text{Quantity of water in litres} \times \text{Temperature rise in °C} \times \text{Specific heat of water}}{\text{No. of seconds in 1 hour}}$$

$$\therefore \text{kW} = \frac{120 \times 50 \times 4.2}{3{,}600}$$

$$= \frac{12 \times 5 \times 4.2}{36}$$

$$= \frac{5 \times 4.2}{3}$$

$$= \frac{21}{3} = 7$$

To heat 120 litres in 1 hour requires 7 kW, but a firing period of 2.5 hours is allowed:

$$\frac{7}{2.5} = 2.8 \text{ kW (in round figures 3 kW)}$$

In fact, 2.8 kW would be required to heat the water in the time given.

The recovery chart in Table 4.1 is not comprehensive, but it may be useful to give a rough guide for sizing smaller heaters or boilers.

To use the chart as a check to the foregoing calculation, draw a horizontal line beneath the 3 kW loading. Then look at the numbers of litres of water being considered on the top row. Unfortunately, 120 litres is not shown so look at the two columns 100 and 150; where your horizontal line from 3 kW cuts beneath these columns, observe the numbers given, i.e. 115 and 173. The average of these two numbers is 144 which is the number of minutes taken to heat the water through 50 °C. Our example of 120 litres would fall approximately at 144 minutes, or 2 hours 24 minutes, which is roughly equal to the time calculated.

### *Hot water requirements for domestic appliances*

Table 4.2 will enable the reader to recognise the approximate volume and temperature of water required for domestic appliances.

### *Stratification*

This relates to the temperature differences of the water at the top and bottom of a hot storage vessel. The term *stratification* is used because the water

**Table 4.1** Recovery chart: approximate time in minutes to heat water.

| *Loading (kW)* | *Litres heated through 50 °C* | | | | | | | | | | | | | | | |
|---|---|---|---|---|---|---|---|---|---|---|---|---|---|---|---|---|
| | *5* | *8* | *10* | *15* | *30* | *60* | *80* | *100* | *150* | *200* | *250* | *300* | *400* | *600* | *800* | *1000* |
| 1.0 | 18 | 28 | 35 | 53 | 105 | 210 | 280 | | | | | | | | | |
| 2.0 | 9 | 14 | 18 | 26 | 53 | 105 | 140 | 175 | 263 | | | | | | | |
| 3.0 | 6 | 9 | 12 | 17 | 35 | 70 | 92 | 115 | 173 | 230 | 288 | | | | | |
| 4.0 | 5 | 7 | 9 | 14 | 27 | 54 | 72 | 90 | 135 | 180 | 225 | 270 | | | | |
| 6.0 | 3 | 5 | 6 | 9 | 18 | 36 | 48 | 60 | 90 | 120 | 150 | 180 | 240 | 360 | 480 | 600 |
| 8.0 | 2 | 4 | 5 | 7 | 14 | 27 | 36 | 45 | 68 | 90 | 113 | 135 | 180 | 270 | 360 | 450 |
| 9.0 | 2 | 3 | 4 | 6 | 12 | 24 | 32 | 40 | 60 | 80 | 100 | 120 | 160 | 240 | 320 | 400 |
| 12.0 | 2 | 3 | 3 | 5 | 9 | 18 | 24 | 30 | 45 | 60 | 75 | 90 | 120 | 180 | 240 | 300 |
| 15.0 | 1 | 2 | 3 | 4 | 8 | 15 | 20 | 25 | 38 | 50 | 63 | 75 | 100 | 150 | 200 | 250 |
| 18.0 | 1 | 2 | 2 | 3 | 6 | 12 | 16 | 20 | 30 | 40 | 50 | 60 | 80 | 120 | 160 | 200 |
| 24.0 | 1 | 2 | 2 | 2 | 5 | 9 | 12 | 15 | 23 | 30 | 38 | 45 | 60 | 90 | 120 | 150 |
| 36.0 | 1 | 1 | 1 | 2 | 3 | 6 | 8 | 10 | 15 | 20 | 25 | 30 | 40 | 60 | 80 | 100 |

**Table 4.2** Approximate volume and temperature of water required for domestic appliances.

| *Appliance* | *Approximate quantity of water in litres* | *Suitable temperature (°C)* |
|---|---|---|
| Bath | 114 | 43 |
| Wash basin | 4.5–6.0 | 43 |
| Sink | 4.5–9.0 | 60 |
| Shower | 4.5–9.0 | 40–43 |

forms layers or strata of different temperatures, the hottest at the top and the coolest at the bottom. When a storage vessel has cooled overnight, the layers are more distinct and can be felt by running a hand down the side of the vessel before any water is drawn off. This is a physical fact and it is significant in that if the temperature of the stored water were the same throughout the storage vessel, no circulation would take place. Quite apart from this, every time a hot draw-off tap is opened the water that is drawn off is replaced by cold water from the feed cistern which has the effect of reducing the water temperature in the lower part of the storage vessel.

The effects of stratification on the temperature of the hot draw-off are not so noticeable with vertically fitted cylinders as those lying in a horizontal position. For this reason, where possible, cylindrical vessels should be fitted in an upright position, although this is not always practicable. A rough estimate of the average temperature of water stored may be made by adding the temperature of water at the top of the vessel to that at the bottom and dividing the result by two. A more accurate method would be to find the average of a series of temperature readings taken over the height of the vessel.

### Support and connections to horizontal hot storage vessels

When very large hot storage cylinders are necessary to meet the hot water requirements of a building, it may not be possible to position them vertically due to their length and the need to provide sufficient circulating pressure by gravity (see Fig. 4.20). By fitting the cylinder in a horizontal position the circulating pressure can be increased without the use of a pump. In large commercial buildings greater use is made of pumps, which despite their initial and running costs have many advantages. Their use provides much greater circulating pressures enabling smaller circulation pipes to be used, better control systems and, possibly the most important, greater flexibility in system design. This means the relative positions of the boiler and hot store vessel are not important, nor the position of a hot store vessel in relation to a secondary circulation. A typical example is shown in Book 1, Fig. 8.39. Such a system would not circulate by natural means.

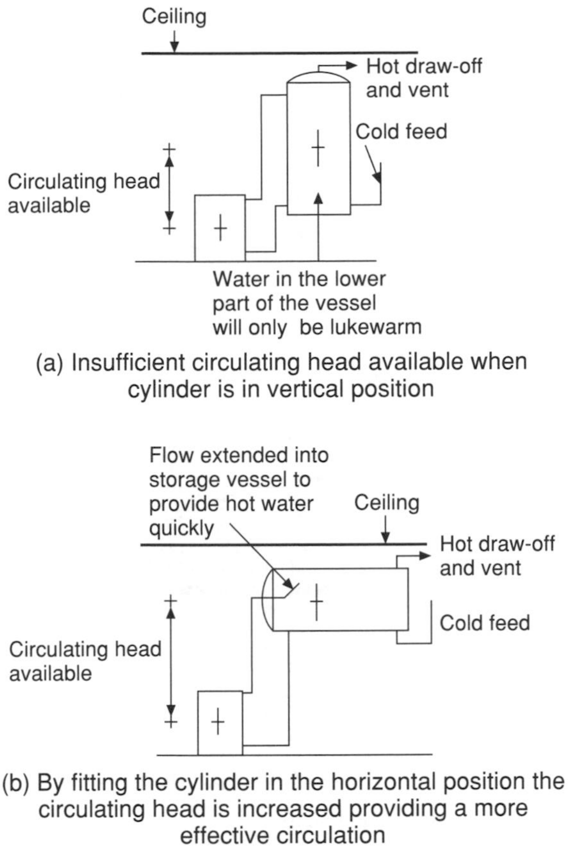


**Fig. 4.20** Circulating head can be increased by fitting the hot storage vessel in a horizontal position.

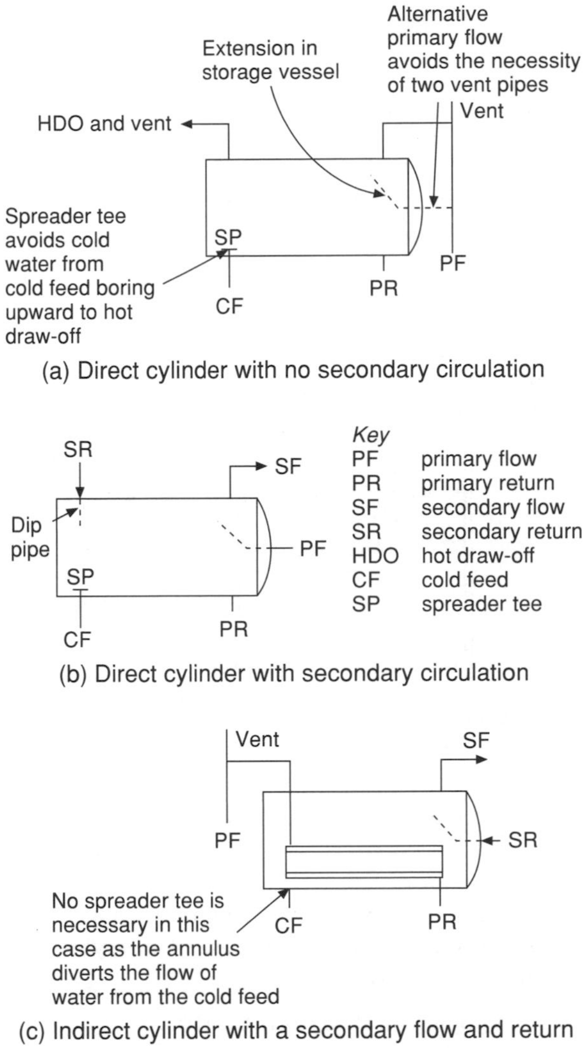


**Fig. 4.21** Positions of connections for cylinders fitted in the horizontal position.

Figure 4.21 shows the connections to both direct and indirect cylinders fitted in a horizontal position. The reader should assume that the vessels shown are made of galvanished steel, as those made of copper lack the strength to withstand the stresses to which they would be exposed in this position, and indirect copper cylinders employ coils as heat exchanges which would result in the coil becoming airlocked. Small cylinders of limited capacity are sometimes fixed on specially shaped brackets cantilevered into the wall, but the safest method is to provide support from the floor such as brick piers or a suitably constructed cradle. Figure 4.22 shows alternative methods of support for horizontally fitted cylinders. Care must be taken to ensure that maximum advantage is taken of the standard tappings.

It is sometimes necessary to fit a cylindrical vessel in a position, possibly a narrow airing cupboard, where its diameter may be limited, but where there is no restriction on its height. If its required capacity and diameter are known its height can be calculated. First ascertain how much water will be contained in a cylinder of such a diameter having a height of 1 m. The capacity required is then divided by this result, which will give the height of the cylinder. For example, assuming that the diameter of the vessel is to be 0.6 m and the required capacity 425 litres, the volume can be

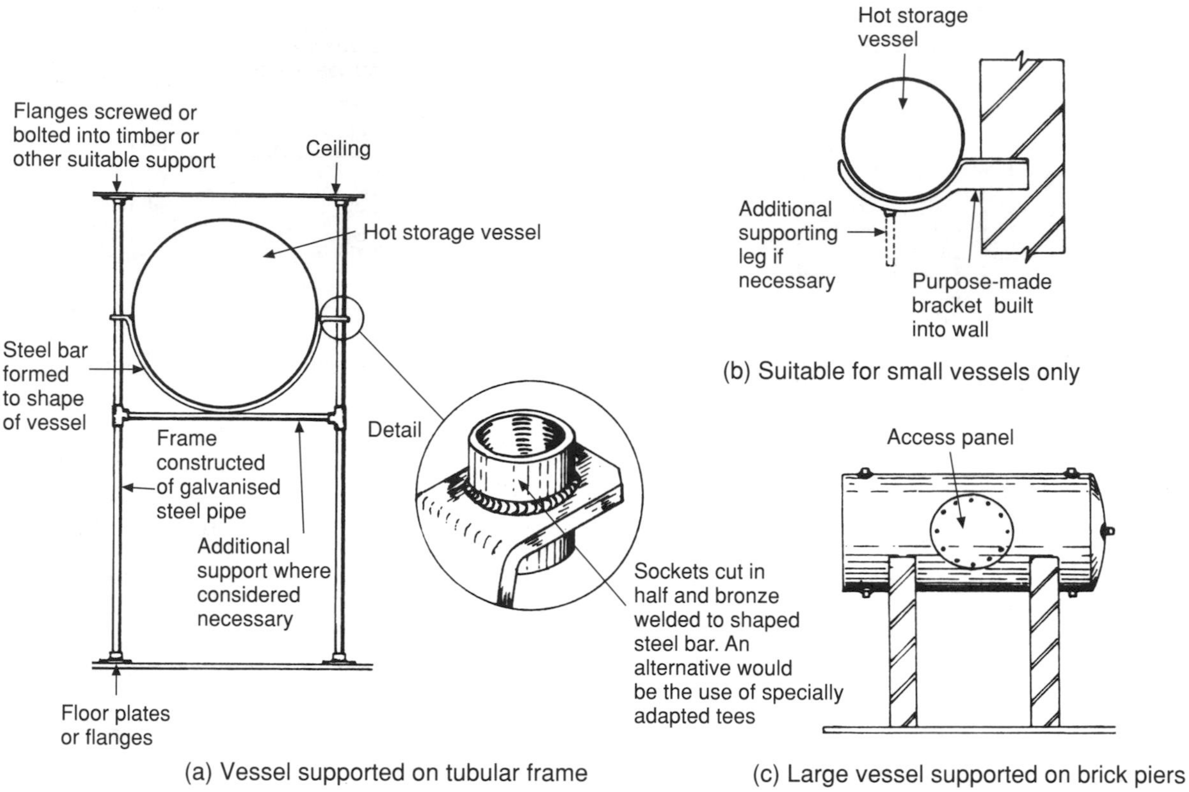


**Fig. 4.22** Alternative methods of supporting horizontally fitted hot water storage cylinders.

found in the following way. Using the formula $A = \pi r^2$ to find the area of the base of the cylinder (where radius = 0.6 m ÷ 2 = 0.3 m):

$$\text{Base area} = 3.142 \times 0.3 \times 0.3$$
$$= 0.283\ \text{m}^2 \text{ approx.}$$

The volume of a cylinder with this base area and height of 1 m = 0.283 $m^3$.

There are 1,000 litres in 1 $m^3$, therefore volume = 283 litres. The required capacity (425 litres) is now divided by 283 to find the required height:

$$\text{Height} = 425 \div 283 = 1.502$$

Therefore the height of the cylinder will be approximately 1.5 m.

### Towel rails

Before central heating became commonplace it was unusual to find any form of heating in the bathrooms of small dwellings. In buildings with a secondary circulation of hot water supply it was realised that a heated rail for towel airing could be fitted into it very easily. As central heating came within the reach of more property owners, both towel rails and the method of heating them changed.

In domestic properties where central heating is fitted, it is more economical to fit a radiator rather than a heated towel rail, due to the fact that a radiator has a greater heating surface. A non-heated rail can be fitted above the radiator. Chromium-plated copper towel rails having inset radiators are an alternative but are very expensive. In commercial hotels or hostels where a secondary circulation of hot water is available, it is still usual to fit the towel rail on the secondary circulation. In this way any bathroom radiators are on quite separate circuits and can be shut down during the summer months. A typical hot water scheme in a small boarding house or hotel where the towel rails or airers are fitted to a secondary circulation is illustrated in Fig. 4.23. A minimum temperature of 60–65 °C is normally necessary for the towel rails/airers to be effective, and in such premises thermostatic control would be essential for water used with ablutionary appliances. In a very large building of this type a dedicated system serving the towel rails only might be a more viable proposition. Figure 4.24 shows a towel rail connected to the central heating system in a small dwelling. It should be fitted in such a way that it is

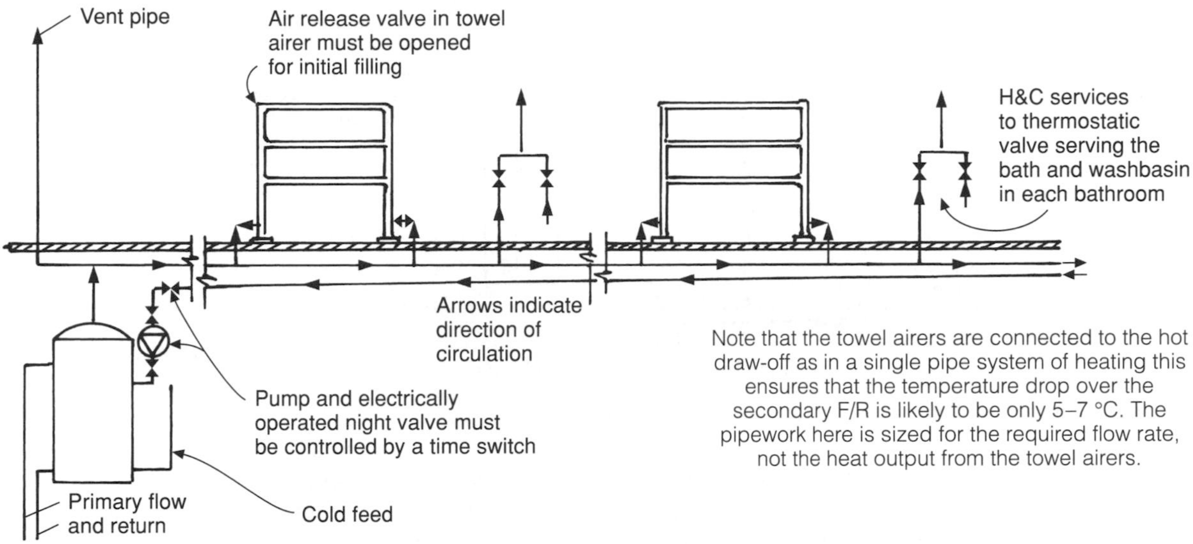


**Fig. 4.23** Pipework system for towel airers in a small commercial hotel.

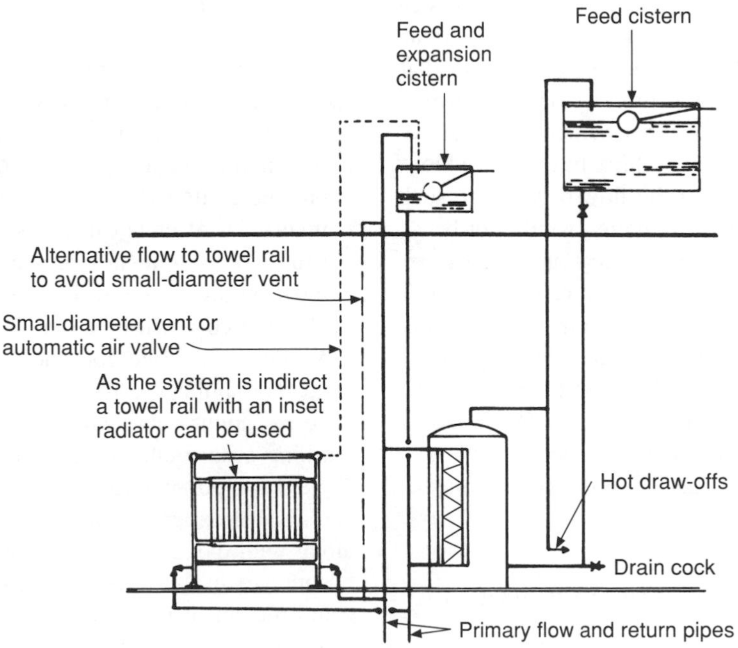


**Fig. 4.24** Towel airer connected to an indirect system of hot water supply in a small dwelling.

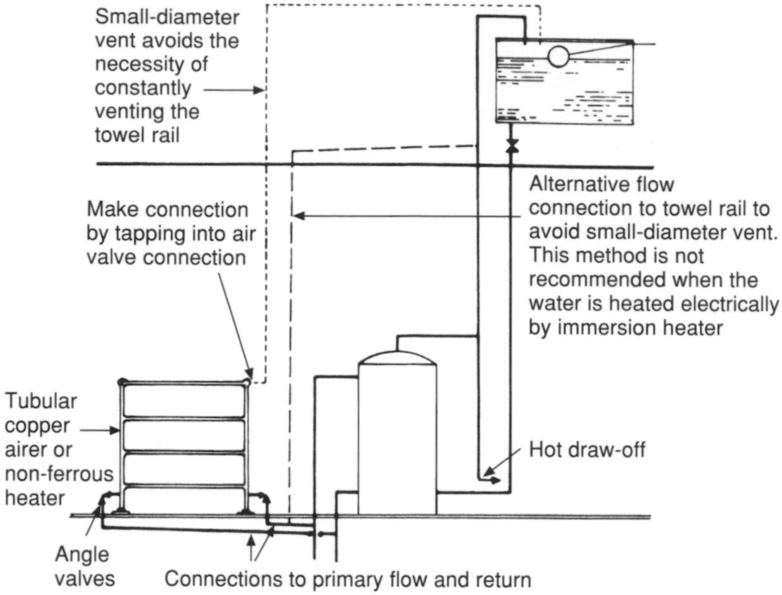


**Fig. 4.25** Single towel and airer fitted to a domestic direct system.

independent of the main central heating system as it can be used during the summer months when the main heating system is shut down. Figure 4.25 shows a similar pipe layout that is connected to a direct system of hot water supply. This method would be unlikely now, unless an alternative system of space heating is used, e.g. warm air or electrical heating. It is important that any towel rail/airer that works on a gravity system employs only full-way radiator valves on any of the schemes shown as many cheaper valves have a high head loss, which means they would restrict the flow of water to such an extent that a gravity circulation would not be effective. Where fully pumped heating systems are employed it is considered undesirable to connect any form of space heating on the pipes serving the hot storage vessel. With a system of this type the bathroom radiator must be fitted on the pipework serving the other space-heating equipment in the building. When the bathroom only needs to be heated, all the other radiators on the system must be shut off.

### Fuels used for hot water and central heating

The choice of a fuel is often dependent on its availability, as not all parts of the country are within reach, for example, of a natural gas supply. There are many other considerations such as cleanliness, convenience and both initial and running costs. Standing charges, as well as maintenance costs, apply to both gas and electrical supplies. Apart from this, the actual heat energy possessed by a fuel must be taken into account when a comparison is made between the cost of fuels. For example, certain types of solid fuel may be cheaper weight for weight, but the cheaper fuel may contain a lower heat content, which is usually expressed as its *calorific value* and is stated as the number of heat units (in joules or kilojoules) per mass or weight (in grams or kilograms). By comparing this value in different fuels some idea of the true cost will be ascertained.

While the average householder can maintain a solid fuel appliance, a specialist is required, usually once per annum, to clean out and check for good working order both gas and oil appliances. The cost of new components for burners using these fuels can also be expensive and must be considered along with the running costs. Electricity has many advantages in that it is considered to be 100 per cent efficient, it is clean, it requires no storage and little maintenance is necessary. It is, unfortunately, expensive, although savings can be made if off-peak electricity is used. This is usually referred to as Economy Seven heating, where during periods when the demand for electricity is low, e.g. overnight, its cost per unit is reduced considerably. Its use for space heating is limited because it is expensive at the normal tariff rates, and night storage heaters, which build up a store of heat during off-peak periods, are difficult to control effectively. Whether or not heating will be needed on the following day has to be decided the night before, and some reliance on the weather forecast, which may not be entirely accurate, is essential.

## Water heaters

### *Instantaneous water heaters*

These appliances are fuelled by either gas or electricity and, as the name implies, produce an instant supply of hot water but have no storage capacity. Instantaneous gas heaters have been used for many years and are well developed. Those using electricity as a fuel are comparatively new and have become very popular due to their use in instantaneous showers.

The advantages of instantaneous water heaters are as follows: they only function when a supply of hot water is required; a supply of hot water is immediately available; there is no limit to the supply of heated water; and, having no storage, heaters of this type have little heat loss. The main disadvantage with all instantaneous heaters is that the quantity of water delivered at a suitable temperature is considerably reduced if more than one draw-off is in use at the same time. They are, however, especially useful where a supply of hot water is desired in an isolated position which might need an excessively long draw-off if fed from a central source. A typical instance might be a supply of water for hand washing in a toilet situated a long distance from the main hot water services.

### *Storage water heaters*

These are also fuelled by gas or electricity, possibly the most common being an electric immersion heater which is fitted into the main hot water storage vessel to heat the water when one does not want to operate the boiler. They are especially useful when the normal method of heating the water is by a solid fuel boiler, as the latter would tend to overheat the room in which it is situated, especially in hot summer weather.

Storage vessels are purpose-made for both gas and electric heaters, many of which are very economical to run due to the high standards of thermal insulation used. The main disadvantage of storage heating equipment is the limitation imposed by its capacity on the quantity of hot water available, which, when exhausted, takes a period of time to replenish. The temperature of the water is controlled by a thermostat, unlike that of instantaneous heaters where the temperature is governed by the volume of water passing through it. There are two other terms which the reader should understand relating to water heaters whether they are of the storage or the instantaneous type. *Single point* heaters are those serving only one point and are usually fitted with a swivel spout, while *multi-points* serve several draw-offs.

It must be stressed that when gas or electric water heating appliances are to be installed the relevant regulations must be complied with. All manufacturers provide detailed fitting and fixing instructions with their products and these should be studied carefully before work is commenced. In all cases where multi-point heaters are installed long dead legs should be avoided, which means careful planning to ensure grouping of the appliances to which hot water is to be supplied.

One important point should be considered when the running and installation costs are compared between gas and electrical appliances. Although the running costs of gas are less than those of electricity, the installation costs are usually higher due to the need for both flueing and ventilation of the space in which the appliance is fitted. Servicing charges are also more expensive.

### *Gas instantaneous water heaters*

Although the market for these types of heaters has diminished, mainly due to the increase of combined space and hot water systems in private dwellings, they do have many applications. Typical examples are small dwellings with alternative space-heating systems, and small industrial and commercial premises for ablutionary purposes. Most modern heaters of this type are multi-point, e.g. serving more than one outlet. They are very economical in use as there is little or no heat loss and any pipe runs should be as short as possible. If possible they should be fitted near to the draw-off most used, usually the sink. They do have the disadvantage of being capable of supplying water to only one tap at a time, e.g. if another tap on the system is opened the volume of water delivered will be halved. These heaters are normally designed to be connected to the main cold water supply, but some with modified pressure ratings can be used with a low-pressure cistern-fed supply. Most modern heaters are room-sealed, although open-flue types are still available. Room-sealed appliances must be used if installed in bathrooms, bedrooms and garages. Figure 4.26 shows the basic working principles of a typical heater of this type. The essential controls they embody are very similar to those used when they were first produced.

*Gas storage heaters*

This term includes both boilers and circulators which are fitted independently of the water storage vessel and those which are purpose-made vessels with a small gas burner fitted directly underneath. Gas circulators are really small boilers and are quite often fitted in the airing cupboard close to the hot storage vessel. They can be used to augment another source of water heating, although they are often employed as the sole source. Although they are very effective and economical to run, the main disadvantage with this form of heater is the relatively high cost of both the heater and its installation. Due to this and the availability of alternative and more economic methods of achieving the same purpose, the demand for these heaters is very small.

A purpose-made gas storage heater is shown in Fig. 4.27. The burner is incorporated in the base of the storage vessel and is thermostatically controlled. These units are made with a great variation in water capacity, ranging from 75 to 285 litres. They can be used as the sole water-heating appliance in both domestic and industrial buildings where gas is used in preference to electricity for water heating. They are available for both unvented and open vented systems and are very economic in use. Many modern heaters of this type are room-sealed.

### Electric water heating

The main features of instantaneous electric water heaters are discussed and illustrated in Chapter 7 where their application to shower heating is dealt with. Apart from this, their use is at present confined to small single-point heaters serving spray taps for hand washing. Storage heating by electricity has been well established for a long period of time. The methods used range from the installation of electric immersion heaters in a normal hot storage vessel to purpose-made units of varying capacities. The main advantage of these latter vessels is the high-quality insulating jacket with which they are provided, permitting only minimal heat loss.

*Under-sink electric water heaters*

These are small, single-point storage heaters, usually of approximately 10 litre capacity with options of 3 or 1.2 kW element ratings. The 3 kW provides very fast recovery from cold. They must be fitted with a special tap which allows for expansion of the water during heating, but isolates the inlet until the tap is turned on (see Fig. 4.28). These heaters are a very convenient alternative to a wall-hung heater with a swinging arm outlet.

*Electric immersion heaters*

All methods of using electricity for heating stored water employ an immersion heater, one type of which is illustrated in Fig. 4.29. The type shown is available in various lengths and ratings and is suitable for installation in an existing storage vessel. Immersion heaters are often described as 100 per cent efficient as all the heat they generate must pass into the water, but as electricity is expensive it is important that any vessel into which they are fitted is well insulated. Some of these heaters are fitted with two elements, a short one heating the top third of the vessel's contents, which is normally sufficient for general use, and a long one used only when the entire contents of the storage vessel are required, each element being separately switched. The better

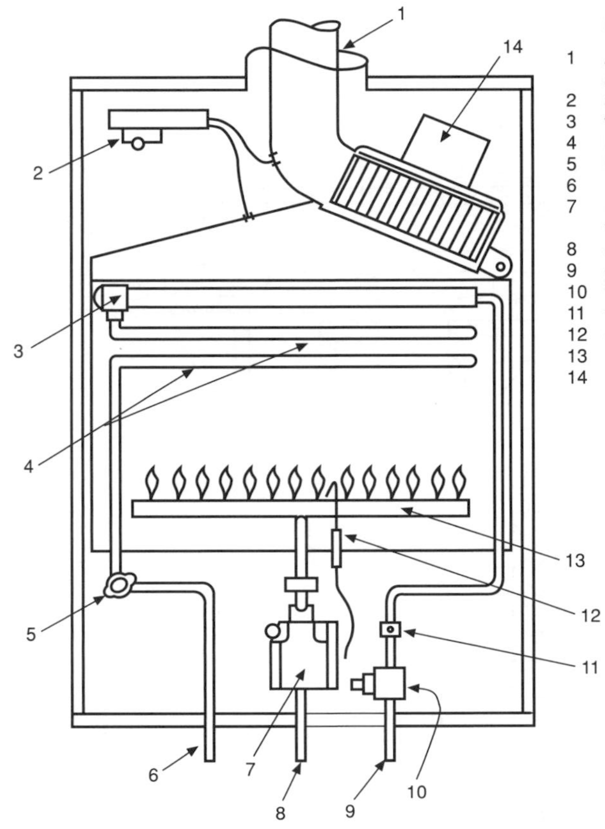
1
2
3
4
5
6
7
8
9
10
11
12
13
14

Key

1 Annular coaxial flue and air inlet
2 air pressure switch
3 thermistor
4 heat exchanger
5 high-limit thermostat
6 hot water outlet
7 main gas modulating valve and control unit
8 gas inlet
9 cold water inlet
10 flow switch
11 water flow restricter
12 spark electrode
13 main burner
14 extractor fan

(a) Typical modern instantaneous water heater

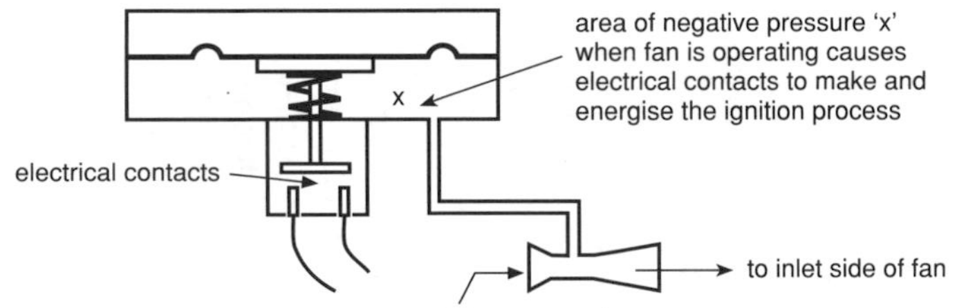
area of negative pressure 'x' when fan is operating causes electrical contacts to make and energise the ignition process
x
electrical contacts
to inlet side of fan
inlet to venturi from flue outlet (positive pressure)

Note that a venturi tube due to its profile has the effect of converting pressure head to velocity head thus causing a negative pressure at 'x'

(b) Diagrammatic illustration showing the working principles of the air pressure switch

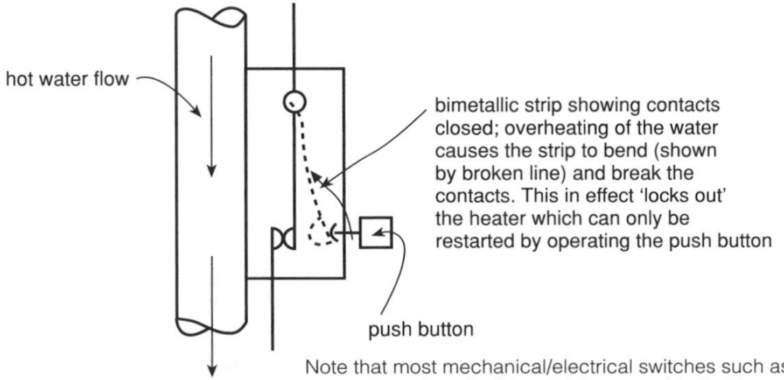
hot water flow
bimetallic strip showing contacts closed; overheating of the water causes the strip to bend (shown by broken line) and break the contacts. This in effect 'locks out' the heater which can only be restarted by operating the push button
push button

Note that most mechanical/electrical switches such as those shown here in (b) and (c) are provided with a permanent magnet to ensure a positive on/off to avoid arcing

(c) Working principle of the high-limit thermostat

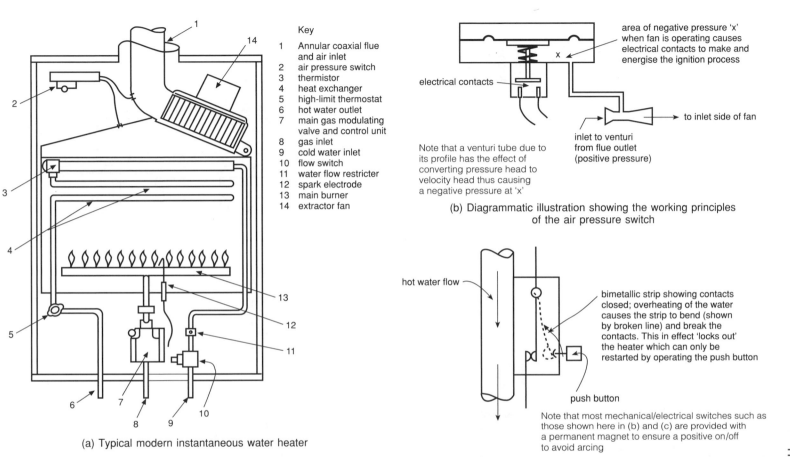

**Fig. 4.26** (a) is based on the Vokera instantaneous water heater and is typical of the type now used. Like most of these heaters, it is electrically controlled to enable a positive flow of air through the combustion chamber by means of a fan extractor — unlike its open and balanced flue predecessors which operated on natural draught. Prior to installation both the water pressure and flow rates must be checked to ensure they are sufficient for the correct operation of the appliance. Its operation is very simple — when a hot tap is opened the flow switch senses a drop in pressure and energises the fan. This causes a positive pressure in the flue, operating a pressure switch which incorporates a venturi tube. A diagrammatic illustration of this switch is shown in Fig. 4.26(b). When the contacts are made the ignition process commences via the spark electrode. When the water has attained the required temperature, which can be varied, the thermistor (see page 63), working in conjunction with the modulating gas valve, maintains a constant water temperature at the outlets. In the event of overheating, the high-temperature cut-out isolates the supply of electricity. A manual reset button will restore the power. If the operation of the cut-out persists the cause must be investigated. The basic principle of the temperature cut-out is shown in Fig. 4.26(c).

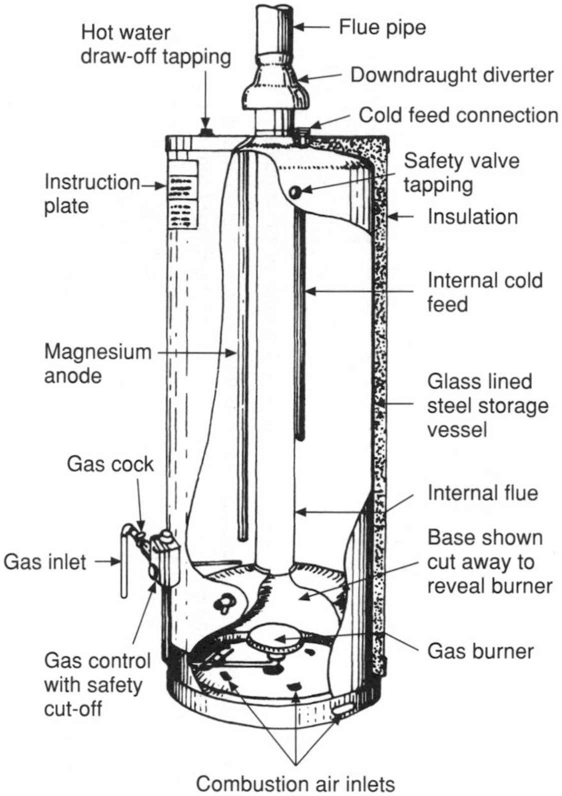


**Fig. 4.27** Gas-fired hot water storage heater.

types of heater have separate thermostatic controls for each element, others use one to control the temperature of both. While the latter are cheaper they are often not so effective in controlling the longer element, often causing it to switch off when only about half the contents of the storage vessel have achieved the desired temperature.

Thermostats controlling electric immersion heaters usually work on the invar rod principle. A diagrammic illustration is shown in Fig. 4.30. In hard water areas the thermostat setting should not be more than 60–65 °C, otherwise scaling of the element will take place causing it to overheat and burn out. Due to the possibility of serious scalding, since April 2004 all immersion heaters have been required to comply with EN60335-2. This standard requires that an overheat thermostat is provided which operates independently of the water thermostat. If it is actuated it must be reset manually; continous need to do this will indicate a faulty thermostat. This is now a legal requirement due to a fatality caused by the collapse of an improperly fitted plastic water cistern when the cistern immersion thermostat failed allowing near boiling water to be discharged by the primary vent pipe.

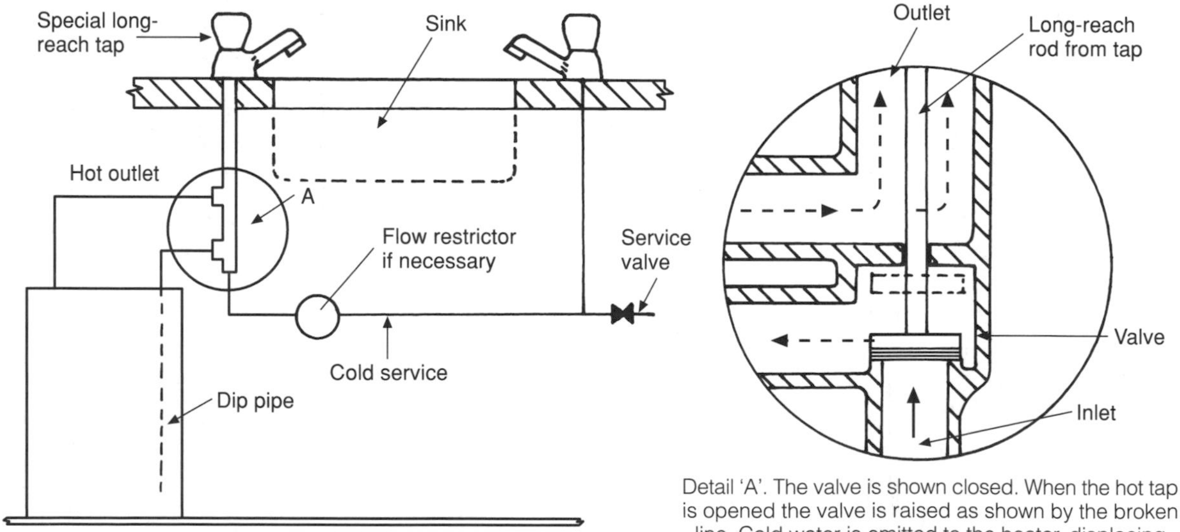


Detail 'A'. The valve is shown closed. When the hot tap is opened the valve is raised as shown by the broken line. Cold water is emitted to the heater, displacing the stored hot water which issues from the tap.

**Fig. 4.28** Under-sink electric water heater.

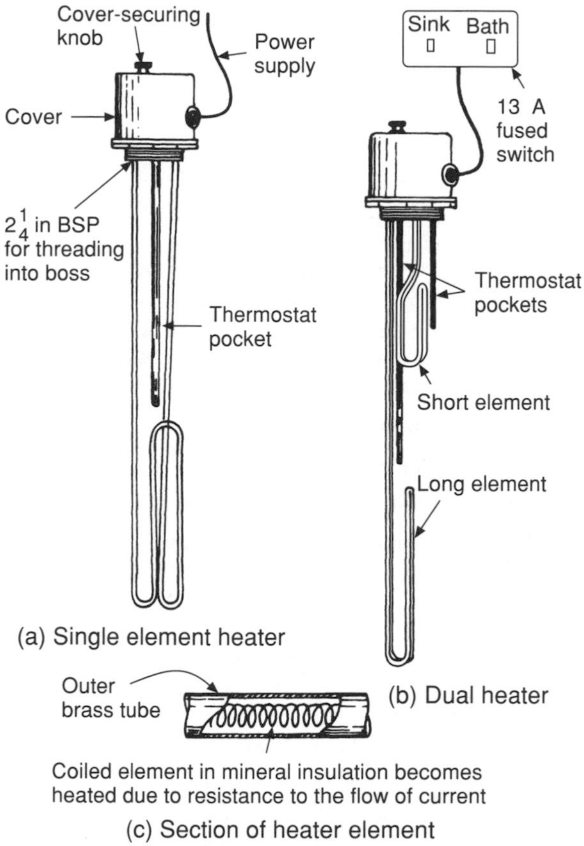


Dual heaters have two elements and better types have two thermostats. Each element is separately switched in the heater cover or, as shown, a wall-mounted switch; when the 'sink' switch is thrown the short element is energised and will heat only the upper part of the hot-store vessel. The 'bath' switch controls the long element which heats the entire contents of the cylinder.

**Fig. 4.29** Electric immersion heaters.

Most modern cylinders are supplied with a boss tapped with a $2\frac{1}{4}$ in BSP thread into which the immersion heater is fitted. If the storage vessel has no boss or if a bottom entry heater is used, special flanges can be used which are made and fitted in such a way that they can be installed without access being gained to the inside of the vessel. They are suitable for both copper and galvanised storage vessels but, when ordering, the surface on which they are to be fitted must be specified, i.e. whether it is flat or circular. A boss of this type is illustrated in Fig. 4.31.

*Immersion heater arrangements* Figure 4.32 shows arrangements for fitting immersion heaters for various purposes. That shown in Fig. 4.32(a) is a single, top entry type which will heat the entire contents of the vessel. It is useful as a supplementary form of heating for summer use, but is not as economical as the dual heater previously described. Figure 4.32(b) shows an arrangement using two short heaters and is very useful when electricity is the only source of heating as, by increasing the size of the storage vessel, full advantage can be taken of Economy Seven rates of electricity. The lower heater which heats all the water in the vessel is connected to the cheap rate supply, while the one at the top is used only for top-up purposes using electricity at normal tariff rates.

*Fitting and removing immersion heaters* Specially made ring spanners are available for the installation or removal of immersion heaters. They form a good fit on the octagonal flange of the heater and cause less damage to the brasswork than large wrenches. When installing an immersion heater make sure the element is not in contact with any part of the surface of the storage vessel (or the coil or annulus if an indirect cylinder is used) as this may cause the element to overheat and burn out. If a heater is fitted at low level in the storage vessel it should be at least 50 mm above the base to prevent movement of the circulating water disturbing any sediment that may be deposited there. It is sometimes difficult to unscrew a defective heater from its boss, and damage to the vessel will be avoided if the heater is eased or unscrewed slightly before draining down when the vessel is more rigid due to its water content. This applies especially to vessels made of copper.

### Introduction to solar heating

This form of heating is not new and has been commonly employed in countries nearer to the equator for many years. With certain exceptions it has not become popular in the UK, mainly due to the climate and the fact that the pay-back period on the initial cost is said to be 10–15 years. However, the economic cost of solar heating is

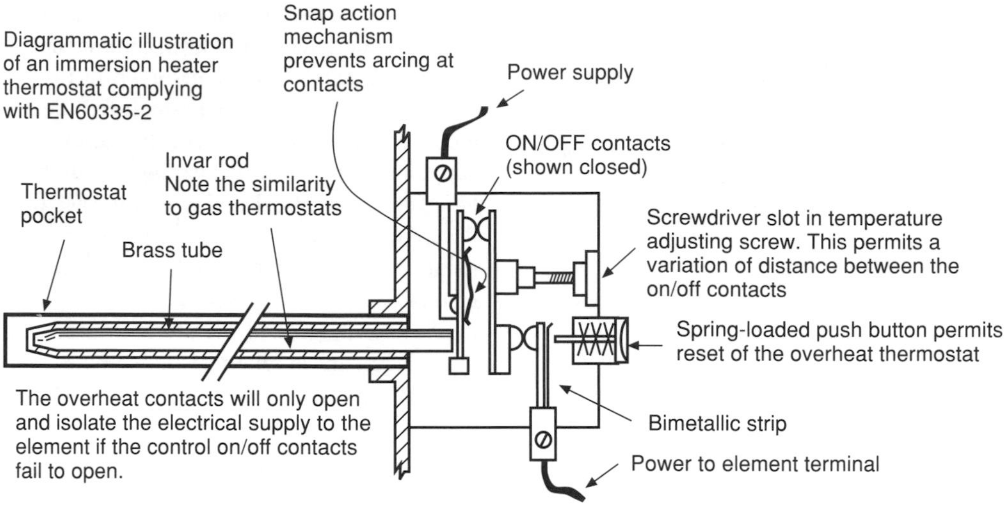


**Fig. 4.30** Diagrammatic section of immersion heater thermostat. The permanent magnet ensures positive making and breaking of the contacts and avoids arcing with consequent burned contacts and possible fire risk.

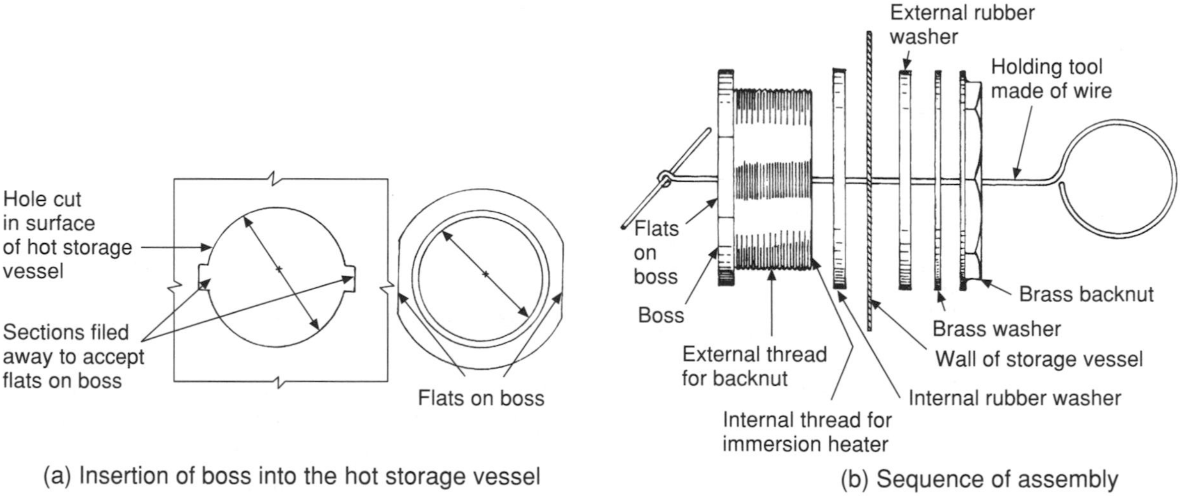


**Fig. 4.31** Patent boss for fitting immersion heaters. This is one of several types of boss manufactured to enable connections to be made in vessels accessible from one side only. A hole is first cut in the vessel to dimension $x$, then two sections are filed away to admit the flats on the boss which is then turned through 90° so that the flats do not coincide with the filed area. A rubber washer is then passed through the hole and over the external thread of the boss which is now held firmly in position with the wire holding tool. The external rubber and brass washers are then fitted over the threaded end of the boss protruding from the vessel, prior to tightening the backnut which secures the complete assembly. The holding tool is then removed and the immersion heater can be fitted in the usual way, but care must be taken not to overtighten.

becoming more realistic due to (a) the apparent increase in global warming, (b) the incidence of long, hot summer periods and (c) the gradual increase in the price of fuel. To encourage energy saving, VAT is reduced to 5 per cent on labour and parts for the installation of solar heating systems at the time of writing.

The basic principles are very simple. A collector panel or series of panels are fixed to the roof of the building in the most favourable position for

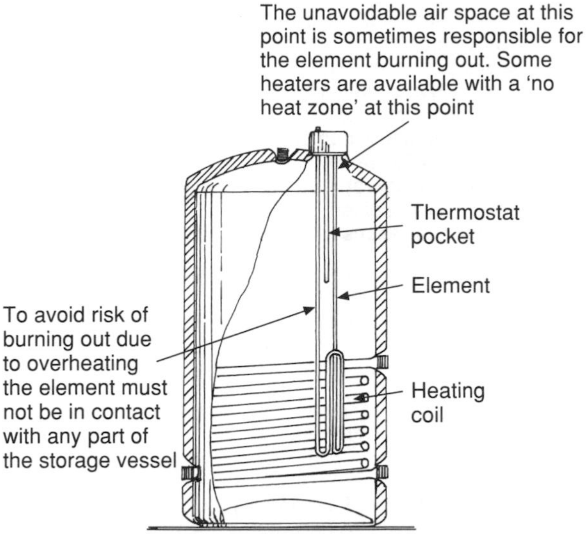
The unavoidable air space at this point is sometimes responsible for the element burning out. Some heaters are available with a 'no heat zone' at this point
Thermostat pocket
Element
To avoid risk of burning out due to overheating the element must not be in contact with any part of the storage vessel
Heating coil

(a) Single top entry immersion heater

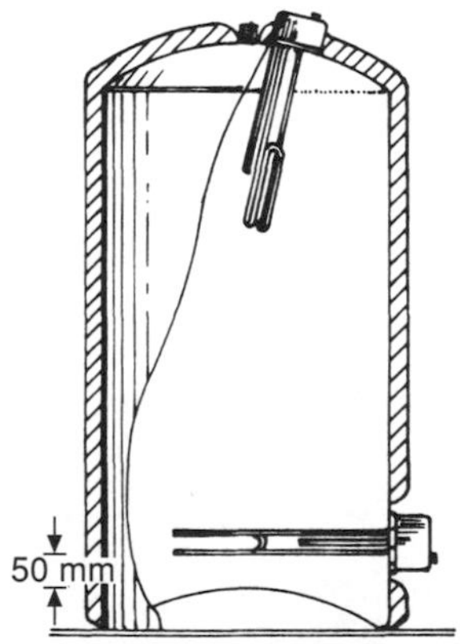
50 mm

(b) Immersion heater using two short heaters

This arrangement is convenient and economical when electricity is the only source of water heating. The heater in the top of the vessel operates on electricity at the normal rate. The lower one heats the entire contents of the vessel and is normally supplied by cheap-rate electricity.

**Fig. 4.32** Immersion heater arrangements.

maximum exposure to the sun. These panels are designed to absorb as much of the sun's heat as possible and transfer it to a fluid which is pumped through a coil in a storage vessel. The storage vessel may be of the vented or the unvented type and is often interconnected to a sealed or open system of space heating. Three types of solar heat collection are employed. In some instances the collecting panel may be fitted below the storage vessel using gravity convection currents to heat the water. This is not always convenient as the most advantageous point for solar heat collection is usually above the hot store vessel. The two most common systems used are shown in Figs 4.33(a) and (b), both of which require a pump to circulate the heat transfer fluid, usually water, through the collector. Figure 4.33(a) is a sealed system, which as it is permanently charged requires protection against frost damage. This is achieved using a suitable anti-freeze fluid in the circulation system. The drain-back system (Fig. 4.33(b)) is designed so that when the pump is inoperative the collector is not charged and freezing is not a problem. To avoid the possibility of overheating the water in the hot storage vessel (outside temperatures can exceed 35 °C on very hot days) a thermostat is fitted in the storage vessel to stop the pump when the water achieves the temperature of approximately 60–65 °C. However, it is usual to match the panel size to the capacity of the storage vessel, largely overcoming the problem.

*Controls*

The principal function in a basic system is to activate the pump when there is sufficient heat available for collection. This information is obtained by the two temperature sensors, one on the solar panel, the other on the storage vessel. For example, if the storage water is at a higher temperature than that in the panel, heat will be lost — not gained — if the pump is operating. These two thermostats are interwired with the controller to ensure the pump only operates when heat can be gained. It will also be seen that the hot store vessel could act as a boiler, heating the solar panel by gravity when the pump is not operating. This reversed circulation on smaller schemes is prevented by a non-return valve or a two-port motorised valve governed by the controls.

*Solar panels*

These are the source of heat collection and vary considerably in cost and efficiency. The simplest type is a plate collector but there are other more complex types with increased efficiency; for details of these the Solar Trade Association should be contacted.

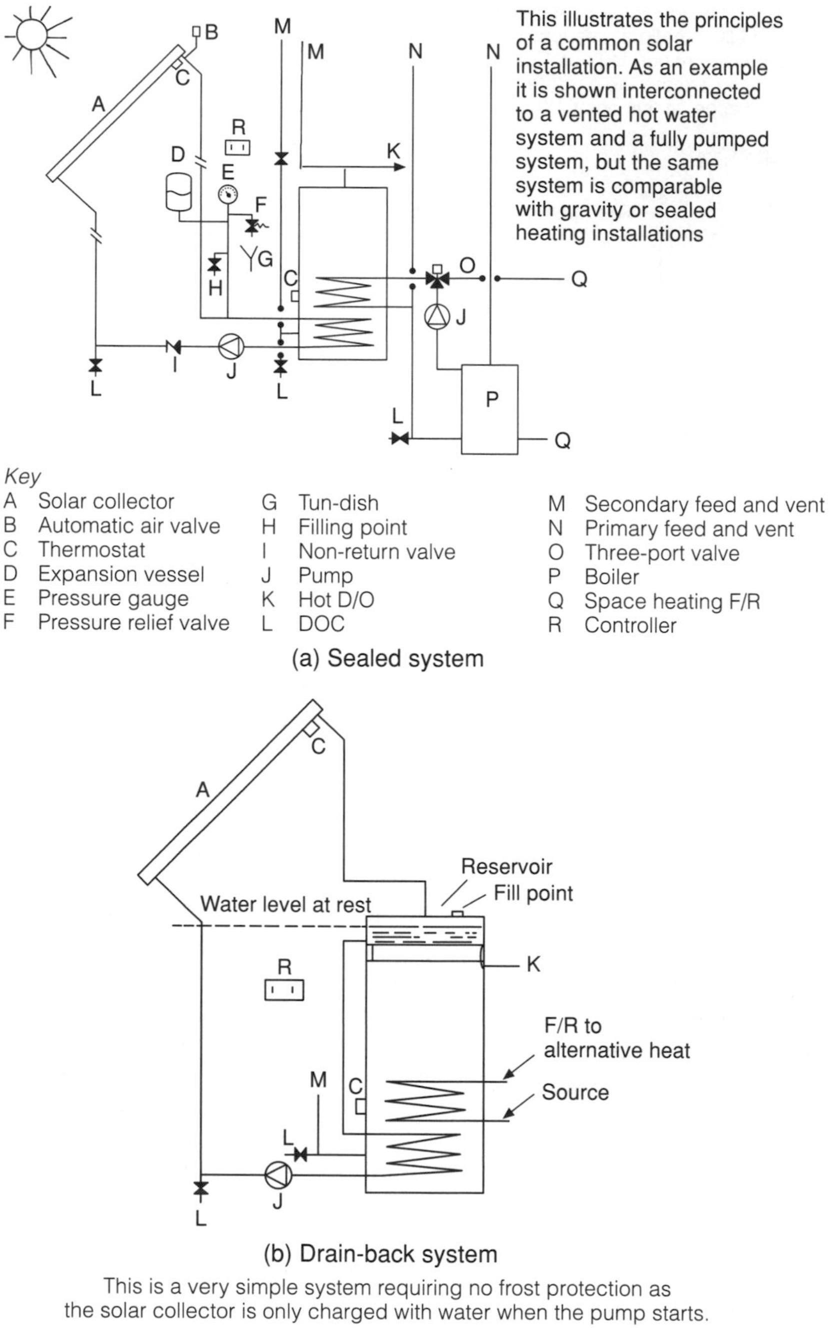

B
M
M
N
N
This illustrates the principles of a common solar installation. As an example it is shown interconnected to a vented hot water system and a fully pumped system, but the same system is comparable with gravity or sealed heating installations
C
A
R
D
E
K
F
G
H
C
O
Q
J
I
J
L
L
L
P
Q
Key
A Solar collector
B Automatic air valve
C Thermostat
D Expansion vessel
E Pressure gauge
F Pressure relief valve
G Tun-dish
H Filling point
I Non-return valve
J Pump
K Hot D/O
L DOC
M Secondary feed and vent
N Primary feed and vent
O Three-port valve
P Boiler
Q Space heating F/R
R Controller
(a) Sealed system
C
A
Reservoir
Fill point
Water level at rest
K
R
F/R to
alternative heat
Source
M
C
L
J
L
(b) Drain-back system
This is a very simple system requiring no frost protection as the solar collector is only charged with water when the pump starts.


**Fig. 4.33** Solar heating systems.

This section on solar heating deals mainly with water heating for domestic use. It can, however, be extended to heating systems, especially those of the underfloor radiant type, and can also be used for heating (or partially heating) the water in small swimming pools. Many manufacturers of solar heating market their products in package form and will give advice and supply all the necessary components required for any installation, providing they are given all the relevant details. This avoids time wasted in obtaining them from a variety of suppliers who are unlikely to keep them as stock items. As the traditional sources of heat and power dry up or become too expensive, solar heating and solar power will certainly become more frequently used as a source of renewable energy.

## Heat pumps

The government is actively seeking more energy efficiency and carbon reduction from fuel in both commercial and residential buildings. Heat pump technology, like solar heating, is a sustainable and self-renewing form of energy. Unfortunately, at the moment the equipment it requires is very expensive. Heat pumps are designed to extract heat from low-temperature sources such as the ambient air, water and heat stored in the earth's crust and raise it to a higher, more useful temperature. There are several types of heat pump: (a) water to water — this is where heat is taken from groundwater and used to heat water for heating and hot water supply. (b) Air to water — probably the most suitable for heating outside swimming pools providing the ambient air is 20 °C or more. This arrangement can give a high level of efficiency, but as the ambient temperature of the air falls, so will the efficiency. (c) Air to air — this type of system can be employed with advantage when the air used to heat a building can be recirculated with little or no heat loss. (d) Water to air — as with (a) this is one of the most effective systems. The evaporator can be constructed or buried over a large area enabling it to pick up large quantities of low-grade heat.

### *Working principles of heat pumps*

The basic working principle of most modern heat pumps is based on either vapour compression or on an absorption cycle, the vapour system being most commonly used. The main components in this type of system are the *compressor*, *an expansion valve* and *two heat exchangers* — called *an evaporator* and *a condenser*, which are connected together to form a closed circuit as shown in Fig. 4.34. A volatile liquid which can be easily compressed, called the refrigerant, circulates throughout the complete unit. The temperature of the refrigerant is kept lower than that of the heat source, allowing it to pick up low-grade heat and subsequently evaporate to become a vapour. It is then compressed to a higher pressure causing it to increase in temperature. This can be proved in a practical way if, when pumping up a bicycle or car tyre, the connector which is under pressure feels warm. The hot compressed vapour is then passed through

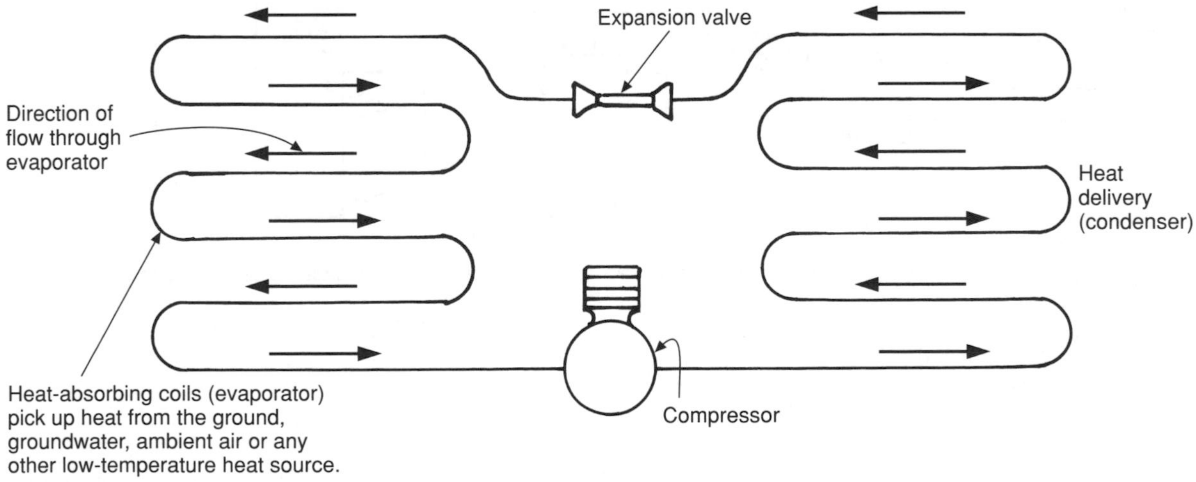


**Fig. 4.34** Basic principle of the heat pump. The heat pump converts low-temperature heat into useful (higher temperature) heat.

the condenser, where it condenses and gives up its heat for use in one of the alternative systems previously described. As the vapour is now condensed to a fluid, when it leaves the condenser it passes through the expansion valve where it reverts to its original state and re-enters the evaporator to commence the next cycle. Figure 4.35 shows a heating and hot water scheme using a heat pump to augment the temperature of water heated by a boiler. In most heat pumps the compressor is driven by an electric motor and it is this energy requirement that must be taken into account as a measure of heat pump efficiency. This is given by its coefficient of performance (COP). This is the ratio of the useful heat output to the electrical input to run the compressor. To give a typical example, if the thermal output is 12 kW and the electricity used to run the compressor is 3 kW the COP can be calculated as follows:

$$\text{COP} = \frac{12}{3} = 4$$

For the cost of 1 kW we are gaining 4.

This does not take into account the capital or running costs of the system. If the COP falls below 3 it may not be financially viable. The purpose of a heat pump may be reversed and used for cooling; this is one of its advantages, especially in the commercial sector. Unfortunately, in this mode it uses more energy than it produces. For more details, The Heat Pump Association will provide further information on the subject.

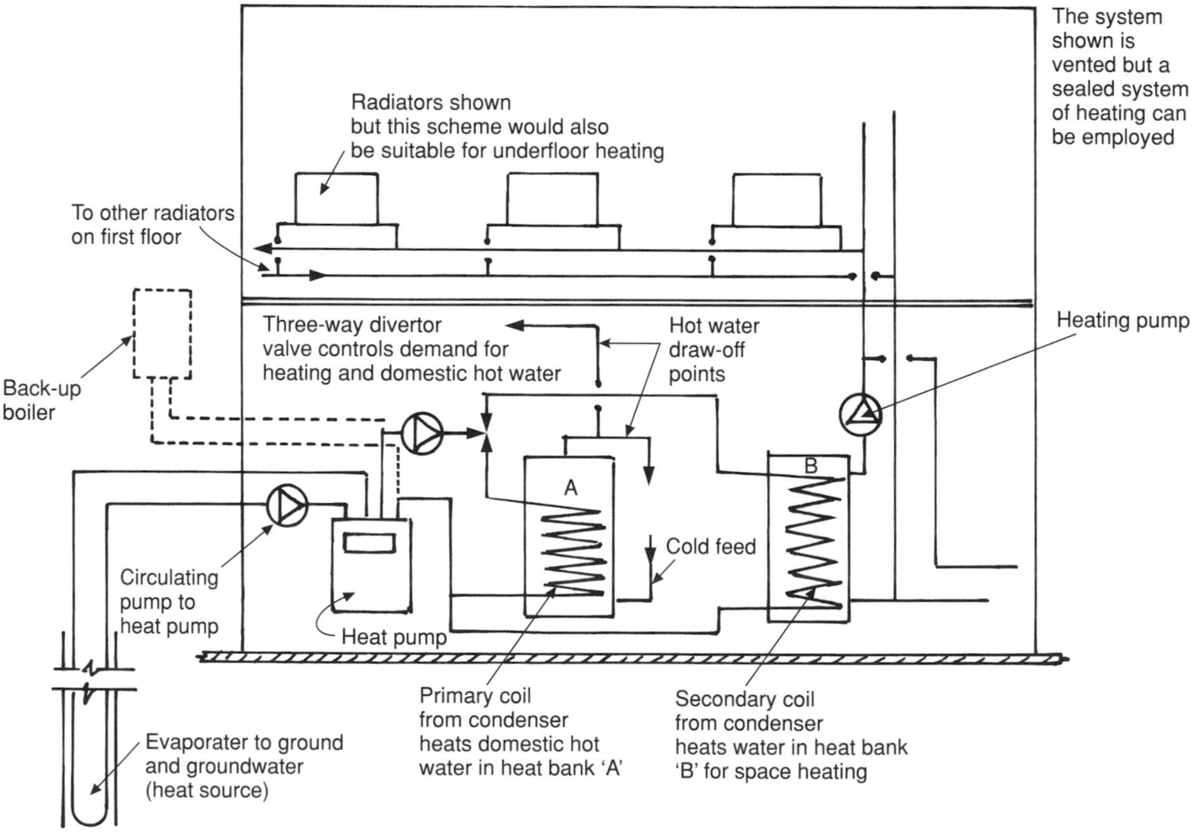


**Fig. 4.35** Heat pump installation. This scheme is broadly based on the recommendations of 'Viessmann'. Note that no valves or electrical controls are shown and that the evaporator is diagrammatic only. In practice it would need a much larger system of buried pipes.

### Micro combined heat and power (micro CHP)

All condensing boilers are designed in such a way that the maximum heat is extracted from the fuel resulting in much reduced flue gas temperatures. These gases are still at temperatures over 50 °C and as such still contain heat energy, which can be used to power a hot air engine, which in turn can be used to generate electricity. These engines, commonly called 'Stirling' engines after their inventor, were first produced in 1816. Although there has never been a mass market for them, they are currently used for some specialist applications such as auxiliary power generators in, for example, submarines or yachts. One of their characteristics is their quietness when in operation; for instance, like a steam engine no explosion takes place in the cylinder. The Stirling engine uses an external heat source, in this case the flue gases which drive a generator and produce electricity which can be used in the home or in commercial buildings. It has been suggested that any surplus electricity produced in this way could be sold back to the electricity supplier. There are other methods of energy conservation such as photovoltaic cells, wind turbines, tidal and wave sources, to name but a few, all of which produce electricity. Although they are not directly connected to plumbing or heating, it is important that the reader is aware of alternative sources of energy.

### Legionnaires' disease

This is a potentially fatal form of pneumonia which was first identified following an outbreak of the disease among people who attended an American Legion convention in America in 1976. It is normally contracted by inhalation. Legionella bacteria are common and low numbers are found naturally in water sources such as lakes, rivers and reservoirs. They can survive in water temperatures of between 6 °C and 60 °C, but water temperatures of between 25 °C and 45 °C appear to favour growth. The presence of sediment, sludge, scale and bio-films (a thin layer of micro-organisms forming a slime on the surface of stagnant water) provides favourable conditions in which Legionella bacteria can multiply. From a practical point of view, all crafts working in the mechanical service industry should be aware of the following, especially those working on the maintenance of commercial and industrial systems.

Cisterns and storage vessels connected with cooling and air-conditioning plants are known sources of infection, bearing in mind the critical temperatures of 25–45 °C. Careful installation of cold water storage cisterns is necessary to avoid stagnation (see Fig. 3.11). Cold water services running adjacent to hot water or heating pipes must be sufficiently well insulated to avoid any appreciable temperature rise. Pipes carrying mixed hot and cold supplies at temperatures of approximately 40 °C to showers or other sanitary fittings should be as short as possible; the Department of Health recommends a maximum of 2 m. From the foregoing it will be seen that some degree of thought must be given to any situation in the working environment in which Legionella bacteria could proliferate. As the Health and Safety at Work Act may be invoked here, it may be necessary for employers to carry out a COSHH risk assessment and provide suitable measures, including information and instruction, to protect their employees. Where considered necessary the building owner and HSE should be notified of any possible danger to health.

### Further reading

BS 6283: Parts 2 and 3 Safety devices for use in hot water systems.

BS 6144 Specification for expansion vessels using an internal diaphragm for unvented hot water systems.

BS EN 60730-2-9: Particular requirements for temperature-sensing controls.

BRS Digest 308 *Unvented domestic hot water systems*, BRE Bookshop, 151 Rosebery Avenue, Farringdon, London EC1R 4GB Tel. 020 7505 6622. www.brebookshap.com

*Heating appliances and hot water systems*

Solid Fuel Association, Swanwick Crt, Alfreton, Derbyshire DE55 7AS Tel. 0845 601 4406. www.solidfuel.co.uk

*Instantaneous water heaters* Baxi (Potterton), Brownedge Road, Bamber Bridge, Preston PR5 6UP Tel. 0870 606 0780. www.baxipotterton.co.uk

Johnson and Starley Ltd, Rhosili Road, Brackmills, Northampton NN47 7LZ Tel. 01604 762881. www.johnsonandstarleyltd.co.uk

Heatrae Sadia Heating, Hurricane Way, Norwich NR6 6EA Tel. 01603 420100. www.heatraesadia.com

Gledhill Water Storage Ltd, Sycamore Estate, Squires Gate, Blackpool, Lancs FY4 3RL Tel. 01253 474444. www.gledhill.net

Reliance Water Controls Ltd, Worcester Road, Evesham, Worcestshire WR11 4RA Tel. 01386 47148. www.rwc.co.uk

*Pumped systems* Harton Heating Appliances Ltd, Unit 6, Thistlebrook Industrial Estate, Eynsham Drive, Abbeywood, London SE2 9RB Tel. 020 8310 0421. www.hartons.co.uk

BRS Digest 254 *Reliability and performance of solar heating systems.*

*Legionnaires' Disease. Advice to employers.* HSE Books, PO Box 1999 Sudbury, Suffolk CO10 2WA. www.hsebooks.com

*Dunsley Baker Neutralizer System* Dunsley Heat Ltd, Bridge Mills, Huddersfield Road, Holmfirth, West Yorkshire HD9 3TW Tel. 01484 682 635. www.dunsleyheat.co.uk

*Andrews Boilers 2 Water Heaters* Wednesbury One, Black Country New Rd, Wednesbury, West Midlands WS10 7NZ Tel. 0121 506 7400. www.andrewswaterheaters.co.uk

*Solar heating — heat pumps* Viessmann Ltd, UK Office, Hortonwood 30, Telford, Shropshire TF1 7YP Tel. 01952 675000 www.viessmann.co.uk

Solar Trade Association, The National Energy Centre, Davy Avenue, Knowlhill, Milton Keynes, MK5 8NG Tel. 01908 442290. www.greenenergy.org.uk

### Self-testing questions

1. In what circumstances would it be necessary to install a supplementary system of hot water supply?
2. (a) Make a sketch of an indirect cylindrical hot water storage vessel fitted in the horizontal position showing all the connections including those for a secondary circulation.
   (b) State why it is sometimes necessary to install a cylinder in this position.
3. (a) List three advantages of using instantaneous water heaters for domestic supplies.
   (b) State the main disadvantages of these heaters.
4. List and describe the essential operating and safety controls necessary for the safe and efficient functioning of unvented hot water systems.
5. (a) Explain what is meant by the term *stratification* in a hot water storage vessel.
   (b) Assuming the temperature of the water at the top of the vessel is 65 °C, and at the bottom 45 °C, state the mean water temperature in the vessel.
6. State the most likely cause of persistent discharge of an expansion valve in an unvented hot water installation.
7. Describe three appliances or methods of providing main-fed hot water systems that can be fitted by a non-BBA-approved installer.
8. State the reason for fitting a single check valve on the mains inlet to an unvented hot water system.
9. State the advantages and disadvantages of combination boilers.
10. (a) State the possible causes of burning out the element on an electric immersion heater.
    (b) Sketch the fixing position of the heater for (i) a small quantity of water heated in a short period of time, (ii) a larger quantity heated over a period of 2 or 3 hours.
11. List the essential safety requirements for discharge pipes from pressure and temperature relief valves on unvented hot water storage vessels.
12. State the purpose of sacrificial anodes.
13. Specify the type of towel rail for use on a secondary circulation.
14. State how the relative height between a boiler and hot water storage vessel affects gravity circulation.

# 5 Hot water heating systems

After completing this chapter the reader should be able to:

1. Recognise and evaluate heat emission appliances.
2. Recognise the basic pipework systems common to wet central heating.
3. Understand the purpose and working principles of appliances and components used in hot water heating schemes.
4. Identify the main types of heating systems and their advantages for specific applications.
5. Identify control systems essential for economy.
6. Identify methods of energy conservation in housing.
7. Understand the basic principles of designing domestic heating systems.

### Space-heating systems

Some form of space heating has become a prerequisite for modern comfort and it is a recognised part of the work of a plumber to install and maintain hot water heating systems. This subject is wide and is impossible to cover completely in the limited space available. Modern schemes include small-bore heating, schemes using very small pipes having diameters of only 8 or 10 mm, called mini-bore, and fully pumped schemes where the circulation of both the domestic hot water and heating water depends on the use of a pump. The type of boiler and the fuel used also influence the type of system employed. It is important to ensure that all components and appliances using modern heating systems carry the energy efficiency logo and the scheme complies with the requirements of the Building Regulations.

The term *wet central heating* is derived from the provision of heat by hot water from a central source as distinct from a series of unconnected heaters. Central heating systems have been in use for many years, some installed with little concern for design or the conservation of fuel. Most of the early systems worked on a gravity basis, e.g. a cold column of water displacing a hot column.

Due to the low circulating pressures available when used to heat radiators long distances from the boiler, gravity circulation has some limitations. The lower circulating pressures require larger pipe diameters and large-radius bends to ensure that minimum resistance is offered to the flow of water. Larger pipes are often very difficult to conceal and installation costs are higher.

For these reasons, during the early 1950s, the British Coal Utilisation Research Association introduced a system of heating using silent, sealed rotor pumps to circulate the water, thereby enabling smaller pipes to be used, the pump head or pressure easily overcoming the frictional resistance of the smaller pipes. The system was designed to enable owners of small- and medium-sized dwellings to enjoy the benefits of low-cost central heating. Although this system was designed initially for solid fuel installations, it is readily adaptable for use with gas- or oil-fired boilers. A further advantage of using pumps and small-diameter pipes is the degree

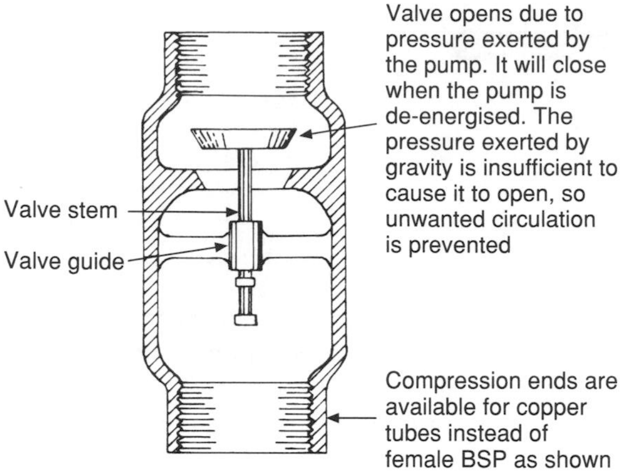

Valve opens due to pressure exerted by the pump. It will close when the pump is de-energised. The pressure exerted by gravity is insufficient to cause it to open, so unwanted circulation is prevented
Valve stem
Valve guide
Compression ends are available for copper tubes instead of female BSP as shown


**Fig. 5.1** Anti-gravity valve.

of flexibility and control that can be obtained. The small volume of water in the system circulates quickly and less time is necessary for it to heat up or cool down. When the pump is stopped, in a well-designed system, circulation to the heat emitters should also cease.

Despite the resistance of the small pipes used, some systems tend to work on gravity when the pump is not functioning, especially radiators fitted on upper floors. This is very unlikely to occur with fully pumped systems — it is one of their advantages. It is likely to happen when the domestic hot water is heated by a gravity flow and return, as may be the case if a solid fuel boiler is used. To avoid this an anti-gravity valve may be fitted in a vertical position on the space heating flow pipe. Figure 5.1 illustrates such a valve, which is normally closed and will only open due to pressure exerted by the pump.

### Heat emitters

There are two main types of heater used in hot water heating systems to heat the rooms in which they are situated. One is called a *radiator*, the other a *convector*.

The object of a radiator is to expose a hot surface to the air in a room. The air passes over the radiator, being warmed as it does so, and a convection current is created, causing a circulation of air in the room. Despite its name, approximately 90 per cent of the heat emission from a radiator is by convection; the remaining 10 per cent only is by direct radiation although this does vary according to the type of radiator.

A convector heater relies for its heat output on a relatively small area of heating surface over which air flows through a series of fins. It is essential that the water temperature in this type of heater is in excess of 80 °C if this form of heating is to be effective, and for this reason high-pressure sealed systems are sometimes employed with heating schemes using convector heaters. It is debatable how effective this form of heating will be with the lower F/R temperatures used with condensing boilers.

#### *Radiators*

Most modern radiators are made of pressed steel sheets of welded construction, although originally they were made of cast iron. Aluminium alloy, although expensive, is also commonly used as a material for modern column-type radiators. Although they are lighter and easier to handle, steel radiators are more prone to corrosion than those made of cast iron or aluminium alloy. It is recommended that heating systems are treated with an inhibitor, not only to prevent corrosion but also to avoid pump seizure. Radiators may be of the column, panel or hospital type (see Fig. 5.2).

Column radiators are the most compact for any given heating surface. The number of columns in each section varies and a radiator of this type is named by the number in each section. Hospital radiators are very similar, but each section consists of only one column, which reduces their overall heating surface. The name 'hospital' derives from the fact that these radiators are easier to clean than column radiators, an important detail where cleanliness is essential. The most popular and effective radiator in use is the panel radiator, despite the fact that it occupies more wall space than other types. Its popularity is mainly due to its higher heat output in relation to its area in comparison with other radiators. These, too, were originally made of cast iron, but they are now easily and quickly produced by passing a sheet of steel through a set of rollers which press out the waterways. A continuous welding process is used to join the top and bottom of the two sheets, after

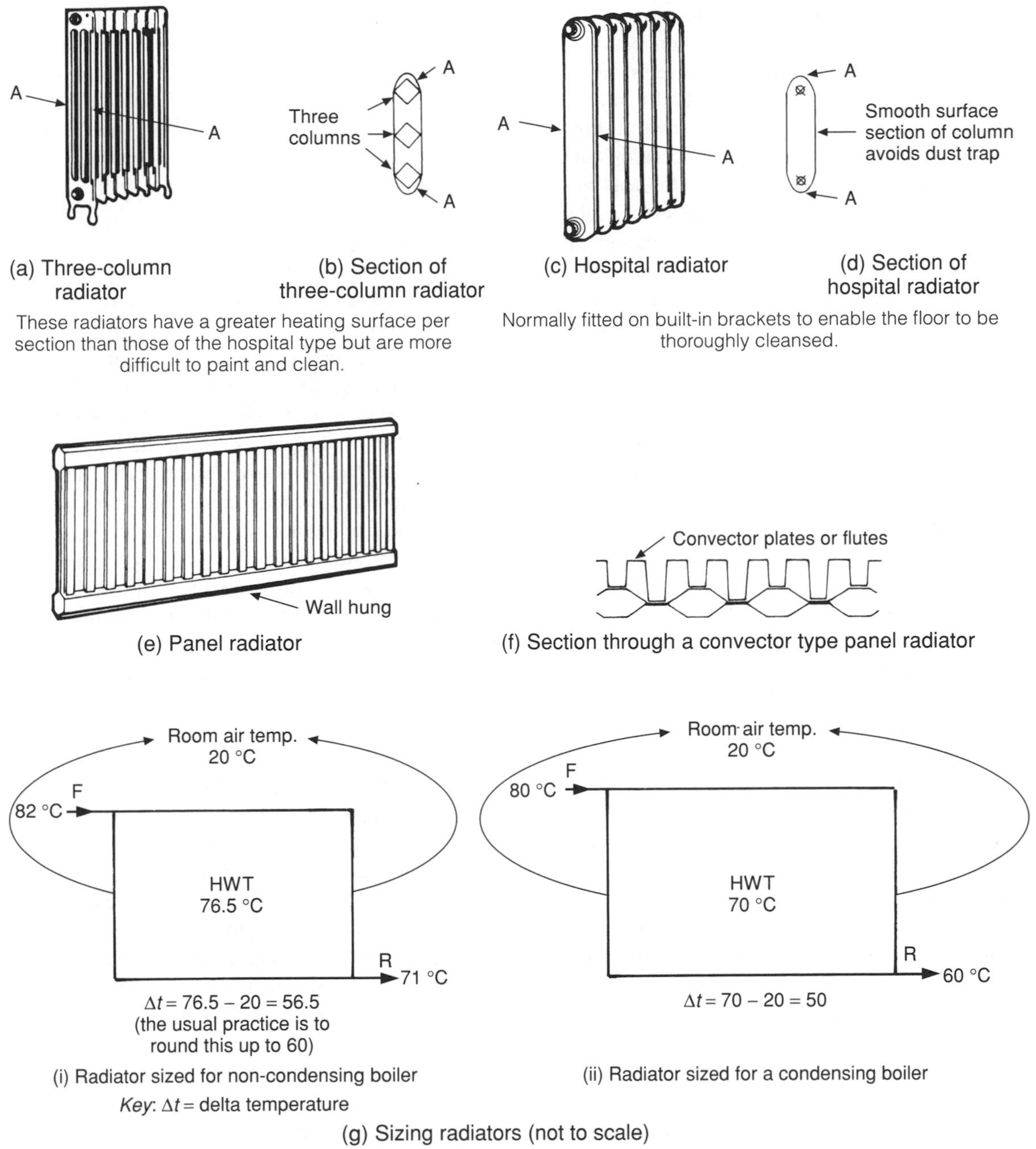


It is important to understand the relationship between the mean water temperature (MWT) in the radiator and the ambient air surrounding it, both of which affect the heat output of a room heater. Radiator tables give these outputs in watts at a variety of delta temperatures. Taking as an example a heater which emits 1000 watts at a $\Delta t$ of 60, the quantity of heat given off will increase if the room temperature drops, but if it rises it will be reduced.

**Fig. 5.2** Radiators.

which they are cut to the required length prior to sealing the ends and welding in the connections. Sheet steel panel radiators can be curved or angled to order, enabling a radiator to be fitted into a shaped bay window. The radiators shown in Fig. 5.2 are basic types, but there are many variations of each which are designed to increase the heat output and improve the appearance. It is possible to buy a top and side casing for some panel radiators, the top having a grilled outlet to permit air movement. Some panel radiators have a series of flutes or convector plates (see Fig. 5.2(f)) welded on to the back which improves the convected effect thus increasing the output. When fitted on the inside surface of double panel radiators these flutes reduce the radiant heat loss from one panel to the other. The heat output from radiators is based on the relationship of the MWT (Mean Water Temperatures) of the F/R. Heating schemes using non-condensing boilers have traditionally operated with flow temperatures of 82 °C and a return temperature of 71 °C, the mean or average water temperature being 76.5 °C. If the desired room temperature is 20 °C then what is sometimes called 'delta' temperature ($\Delta t$) would be 56.5 °C. Most radiator tables round this up to 60 °C as it is this temperature on which the radiator heat output is based. Figure 5.2(g) illustrates the foregoing pictorially. Because condensing boilers are designed to work on lower temperatures, e.g. F/R 80–60 °C, this gives an MWT of 70 °C. At a room temperature of 20 °C the $\Delta t$ will be $70 - 20 = 50$. It will be seen in theory that a larger radiator will be necessary to match the output of a radiator sized for $\Delta t = 60$. It has been suggested that because heat emitters have traditionally been oversized, in many cases it may not be necessary to replace them if a condensing boiler is fitted. It is advisable however that the client should be made aware that larger radiators may be necessary. It is generally recommended to work with a temperature difference of 20 °C between the flow and return. This will enable energy efficient boilers to operate in condensing mode for a longer period of time. This will vary depending upon the type of radiator used — those of the panel type are generally more efficient than those of the column type for example, due to the slightly lower level of radiant heat they emit.

It is recommended that heat emitters are fixed under windows for the following reasons: much of the incoming cold air enters the room via a window; condensation that forms on the glass is evaporated; and as furnishings are seldom situated under windows, useful wall space is not taken up.

### *Pattern staining*

Where it is not possible or desirable to fit a radiator under a window (this is obviously impossible in a room with patio doors), it is important that a suitable shelf is provided over the radiator. This prevents what is called 'pattern staining', which is caused by dust in the air, due to its movement by convection, being deposited on the walls and ceilings immediately above the radiator. The shelf has the effect of diverting the air out into the room. Most radiator manufacturers produce shelves of varying lengths made of metal and it is also possible to make or purchase them in polished hardwood. Any shelf should be fitted with a minimum clearance of 70–75 mm above the radiator to allow the convected air to circulate freely.

### *Convector heaters*

These fall into two main groups, those which rely on natural air movement and those which are fan-assisted (Fig. 5.3). A typical example of the former type is called skirting heating. See Figs 5.3(a) and (b). The heating unit is encased in a moulded steel case and is fitted instead of a normal skirting. Specially shaped angles are produced, enabling changes of direction to be made, and also valve boxes with hinged doors allowing the valves to be concealed. All larger wall-hung heaters have similar features to skirting heater as shown in Fig. 5.3(c). Those heaters which incorporate a fan are much more compact and often occupy less space than a radiator. They are not completely silent in operation due to the rapid movement of the air caused by the fan. Most are provided with two-speed fans, the greater speed being used for a quick heat-up of the room, while the lesser (and quieter) speed can be used for normal running. Figure 5.3(d) illustrates a typical heater of the type.

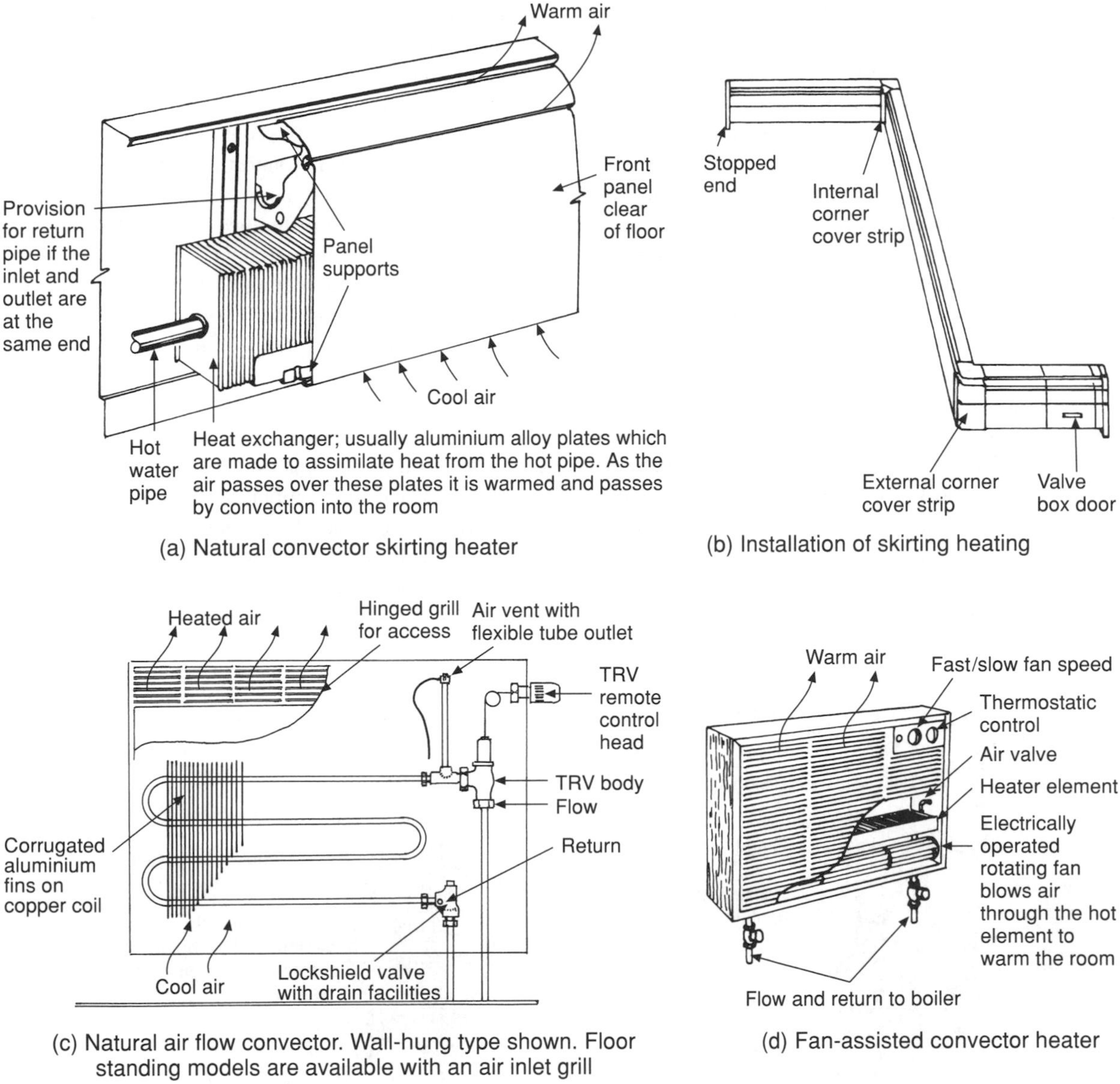


**Fig. 5.3** Convector heaters.

## Central heating components

*Radiator valves*

Both angle and straight patterns are made, but the angle type is the more popular due to its convenience of connection and neatness in appearance. All straight valves are of the gate pattern and are therefore full way, but some angle valves cause considerable head or pressure loss to the flow of water. This has little effect on a two-pipe scheme, but if for any reason a one-pipe system is used where the water gravitates to the radiator from a pumped circuit only, the use of a better quality, more expensive angle valve is essential. In most cases they are specified as full way.

The glands of radiator valves are prone to leakage, especially when they are continually being turned on and off. The leak is often very slight and often imperceptible, especially in carpeted rooms where the carpet soaks up the water. It is important that when servicing heating systems the valves are inspected for leakage as it may avoid serious damage to valuable household furnishings. The glands of some types of valves may be resealed without draining the system or freezing the pipes locally. Both valves shown in Fig. 5.4 are of this type, although it is still necessary to take extreme care when carrying out this operation. It is recommended that a sheet of plastic material is laid under the valve, and always have an absorbent cloth handy for mopping up if necessary.

Each radiator must be fitted with two valves, one for on/off operation, the other to control the flow of water. The latter must be of the lockshield pattern. This prevents short-circuiting, to which two-pipe systems are especially prone, and ensures an equal distribution of water to each radiator. The lockshield valve is regulated by the plumber when the system is commissioned. This is called 'balancing' and is carried out by adjusting the lockshield valve and noting the temperature at both the inlet and outlet of each radiator on the system, until a difference of approximately 11 °C is achieved. Providing the pipes have been correctly sized this should not be too difficult, but if it is found that the difference is more than 11 °C it may be necessary to increase the pump pressure. If the difference is less than 11 °C the pump pressure can be reduced. Electronic methods of balancing heating systems are available and the results they give are very accurate. An alternative is to use two rotary clip-on thermometers.

*Pumps*

These provide the pressure causing the water to circulate through the system. Most modern pumps are made so that their output can be regulated to accommodate a variety of requirements. It is obvious that the frictional resistance on a run of pipe 100 m long will be less than that of a run of, say, 150 m, and the pressure necessary to overcome

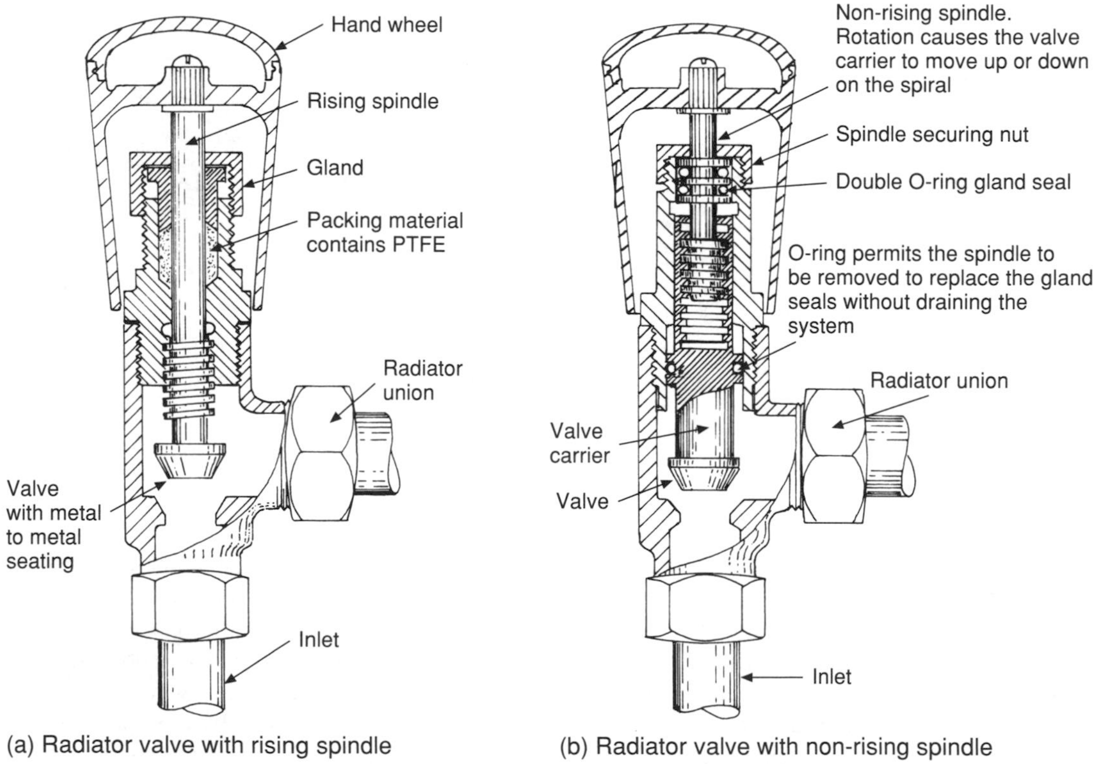


**Fig. 5.4** Radiator valves.

this resistance will be greater. By regulating the speed, a pump can be made to cover a wide range of pressures. If the pressure is too high it will result in high water speed (high velocity) causing noise to occur in the system. The recommended velocity for small-bore schemes should not exceed 1 m per second. For mini-bore schemes 1.5 m per second is recommended to overcome the frictional resistance of the smaller pipes used.

Pumps should be valved so that they can be removed without draining down the system in the event of a breakdown. In order to avoid undue wear they should be fitted as shown in Figs 5.5(a), (b) and (c) with the bearing in the horizontal position. Most pumps are now more energy efficient and consume a varying quantity of electricity, depending on the heating load at any given time. To ensure a long working life, the pump must be fitted to comply with the manufacturer's instructions.

The system must be washed out to remove any debris such as traces of wire wool or grit before the pump is fitted. It must also be purged of air before it is commissioned, because, if the impeller runs dry, the bearing will be damaged. Normal lubrication is provided by the water in the system only, and no oil or grease should be used on the bearing at any time.

The pump must be fitted into the system in what is termed the *neutral point* to avoid water being pumped into the expansion cistern. Figure 5.6 shows in a simple way how this can happen. It is much less effort for the pump to draw water from the expansion cistern than to overcome the frictional resistance of the radiator pipework.

In the early period of forced circulation systems of heating it was recommended that the pump should be fitted on the return. It was found, however, that this causes a negative pressure on most systems which allows air to be drawn in through leaks which are normally unseen as they are too small to admit the passage of water. This is due to the fact that the cohesion between the molecules of a gas is less than that between those of a liquid. This can be proved quite simply with a fine-meshed wire strainer. It will be found that water will not pass through it. However, if a candle is lit and air is blown through the strainer, it will be found possible to extinguish the flame. Air can often be drawn into a heating system via valve glands and defective joints, and also through an open vent and, due to its oxygen content, it can be the source of many corrosion problems. It is now common practice for the pump to be fitted on the flow pipe, thus pressurising most of the system, which avoids the problems mentioned. The pipework adjacent to the pump should be supported

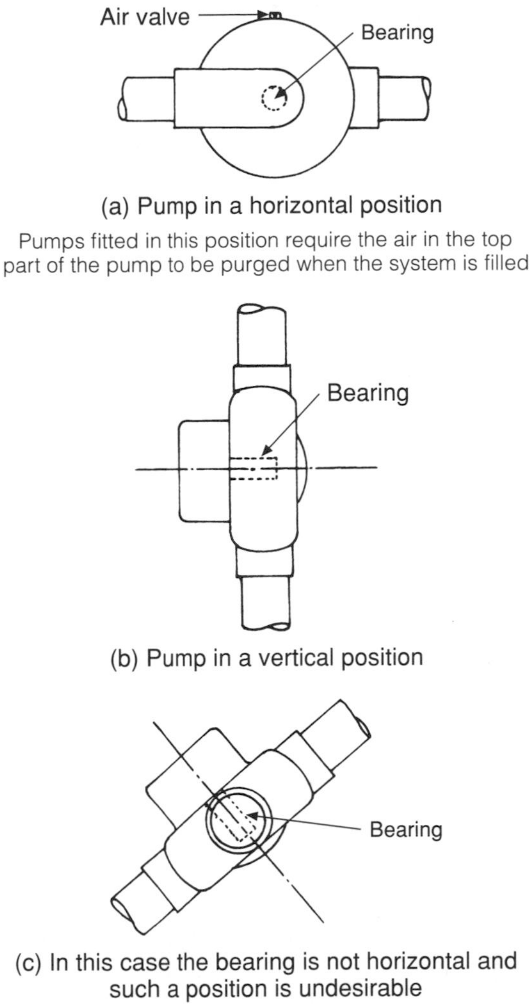
Air valve
Bearing
(a) Pump in a horizontal position
Pumps fitted in this position require the air in the top part of the pump to be purged when the system is filled
Bearing
(b) Pump in a vertical position
Bearing
(c) In this case the bearing is not horizontal and such a position is undesirable

**Fig. 5.5** Positioning central heating pumps. To avoid excessive wear on the bearing on which the impeller rotates it must always be in a horizontal plane as shown in (a) and (b). If fitted as (c) the pump is likely to fail after a very short period of time. Note that pumps should be fitted with isolating valves (not shown here) so that they can be removed for maintenance.

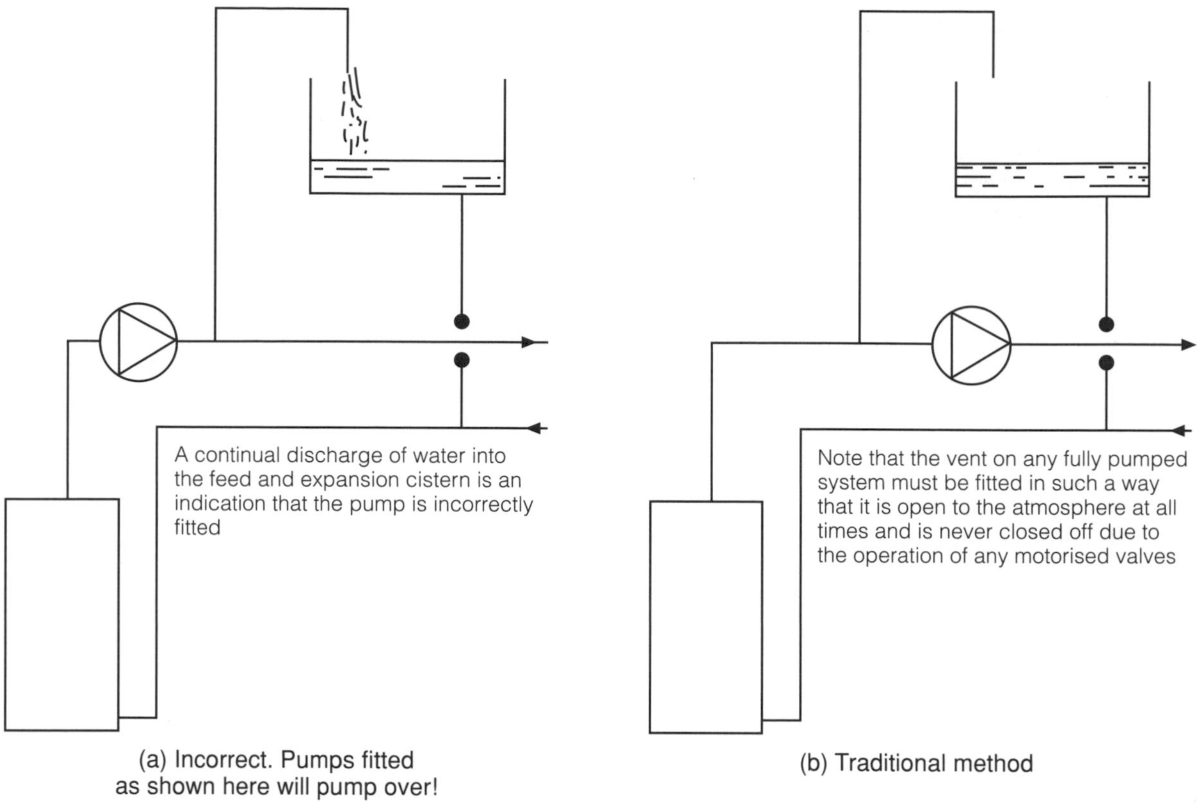
A continual discharge of water into the feed and expansion cistern is an indication that the pump is incorrectly fitted
(a) Incorrect. Pumps fitted as shown here will pump over!
Note that the vent on any fully pumped system must be fitted in such a way that it is open to the atmosphere at all times and is never closed off due to the operation of any motorised valves
(b) Traditional method

**Fig. 5.6** Position of heating pump in the system.

by brackets and lined with a resilient material to reduce pump noise.

*Air eliminator*

Figure 5.7(a) shows a simple device for ejecting any air or other gas contained or picked up by water in low-pressure heating systems. Water containing air can cause many problems such as air locks, corrosion due to oxidation and noise, and their use prevents water being pumped through the vent where the static head on the system is very low. The associated illustration Fig. 5.7(b) shows its application in a heating installation. It is equally effective with fully pumped or pumped heating/gravity hot water schemes.

## Boilers

General details of boilers are dealt with in Book 1 of this series. More detailed information on gas and oil appliances is covered in Chapters 2 and 6 of this volume. Low-water-content and condensing boilers are dealt with in the following text.

*Low-water-content boilers*

These are mainly gas-fired condensing, but some using oil as a fuel are available. Care must be exercised when these boilers are used, as failure to follow any instructions in fitting and connecting such a boiler can result in serious damage to the heat exchanger. These boilers must only be used on fully pumped schemes. They rely for their efficiency upon circulating a small amount of water very quickly. It is essential, however, in order to avoid overheating, that a movement of water occurs at all times through the heat exchanger. To this end these boilers will only operate when the pump is functioning; a delay switch is incorporated which ensures that the pump continues to run for a short

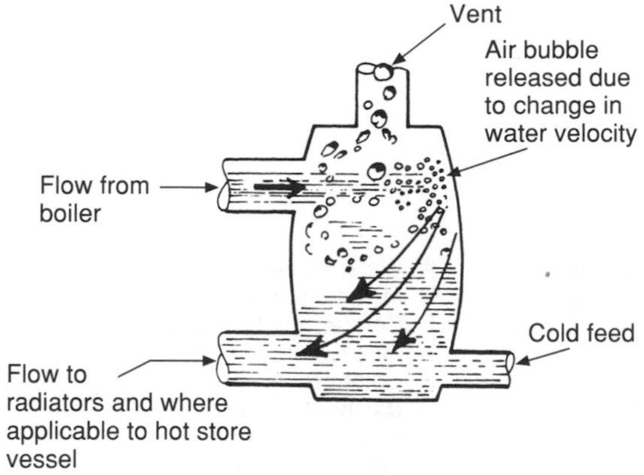


(a)

This very simple device is designed to disperse and eliminate air bubbles in a low-pressure heating system. As the water from the boiler impinges on the back of the chamber its velocity is suddenly arrested causing it to release any air it contains, the air passing into the vent. These fittings are not essential in boilers having a high water capacity, as any air contained in the water can easily escape via the vent. However, they can be used with advantage on schemes using boilers having a high-pressure drop across the flow and return.

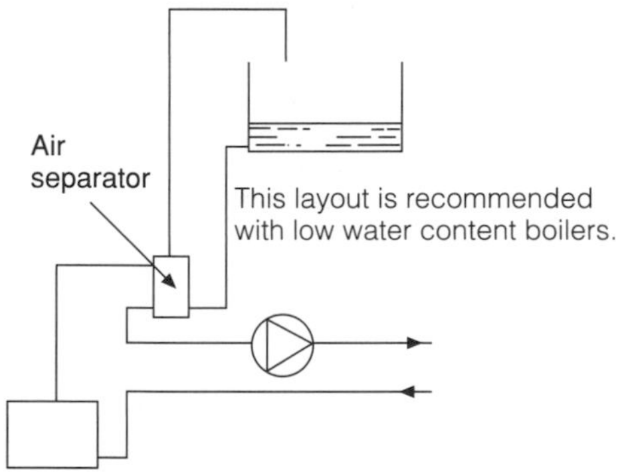


(b) Application of an air eliminator

**Fig. 5.7** Air eliminator.

period of time after the main boiler shuts down. See also bypass valve, page 185.

*Condensing boilers*

The 2005 Building Regulations made it mandatory that, unless for a variety of reasons, any new or replacement boilers must be of the condensing type. The basic principles and the reasons for the use of condensing boilers are fully explained in Book 1 of this series. Their main advantages are: (a) their energy efficiency with consequently lower running costs and (b) the reduction of nitrous oxide, which is responsible for acid rain, and carbon dioxide, two of the main gases responsible for global warming. It is important to understand that the efficiency of condensing boilers depends mainly on two things. With a non-condensing boiler having an efficiency of 75–78% with a flow temperature of 82 °C, the flue gases are at a temperature of approximately 200 °C. By lowering the flow temperature and passing the flue gases over a secondary heat exchanger, yet more heat can be extracted from the fuel. Further to this, if the temperature of the flue gases is lowered to what is called 'dew point', at a temperature of approximately 55–60 °C more heat will be released as the flue gas changes its state and condenses to water. The water produced by condensation in the boiler is collected in the base and discharged into a drain, preferably via a gully where the acidic nature of the condensate will be diluted. An alternative is a special soakaway as shown in Book 1, page 296. One further point is worth noting: the relatively small quantity of water produced when the boiler is in operation would, in frosty weather, cause the discharge pipe to freeze. To prevent this a device similar to that employed in automatic flushing cisterns is used which allows a body of condensed water to build up in the base of the boiler until it operates what the plumber would call an automatic siphon. The relatively large body of water thus discharged is then unlikely to freeze. Although the condensate is acidic in nature containing traces of nitric and sulphurous acids, tests have been conducted on materials from which domestic drains are constructed and the following conclusions reached. Drains made of plastic material and clayware showed insignificant damage while cast iron is likely to be affected in the long term and gives rise to staining. Cement and concrete products appeared to be affected more seriously than other materials. This could lead to problems in older properties having salt-glazed drain pipes with cement joints. In practice, however, the adverse effects due to condensation are unlikely to be serious as it will be appreciated that it will be diluted very quickly by the discharges from sanitary appliances.

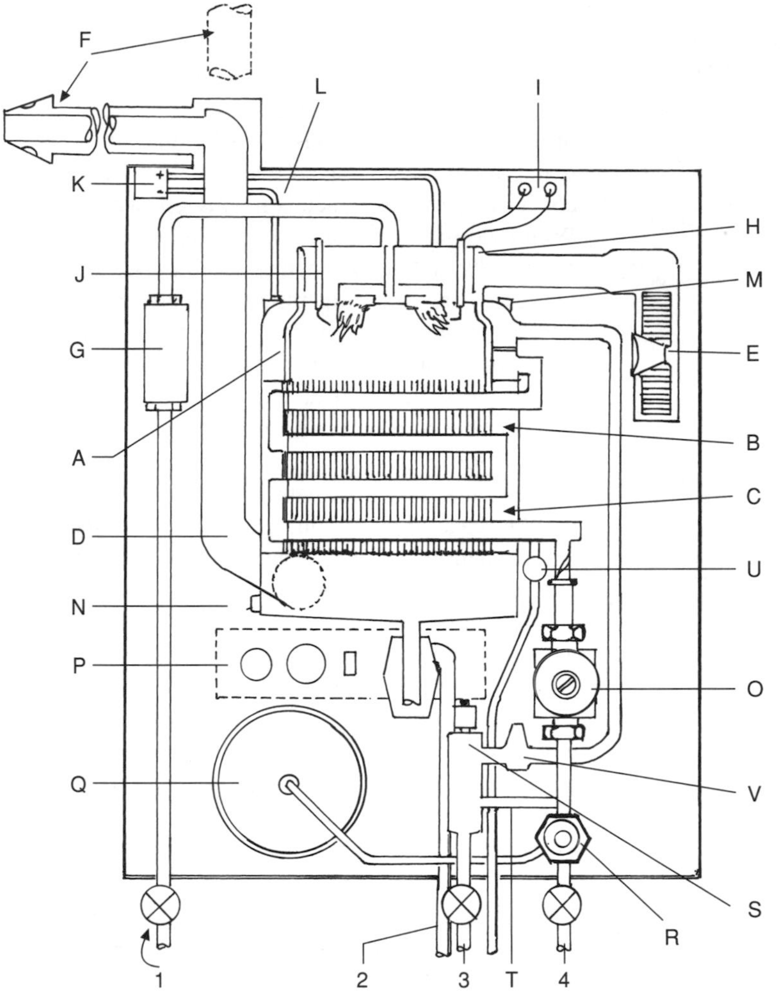


**Fig. 5.8** Typical wall-hung domestic condensing boiler.

Figure 5.8 shows a typical modern condensing boiler made for domestic use; wall-hung or floor-standing models are also available. The burner shown is for pressurised air to be mixed with gas prior to ignition. When alight the flame is established on a stainless steel grill. Automatic spark ignition is normal with these boilers as it is more economic and easier to control electrically. The associated list of parts A–V is very similar to a standard combination boiler. The main heat exchanger is an aluminium casting which combines both the main and condensing secondary heat exchanger and has a water-cooled combustion chamber. As with most boilers of this type, the fan is controlled by an air pressure switch which will isolate the boiler should the fan stop working. The water and flue gas overheat thermostats are safety devices, both of which prevent the boiler operating if they are activated. The main gas valve is of the modulating type which matches the flow of gas to the boiler firing requirements. Should the flame fail for any reason an ionisation probe will prevent any gas reaching the burner. It acts in a similar way to the thermoelectric valve shown in Chapter 2. Heat is extracted from the combustion products in two ways:

(a) In the form of sensible heat, i.e. the transmittance of heat from a hotter medium (in this case the combustion products) to the cooler return.
(b) By the latent heat of evaporation.

The reader should understand that the water vapour in the combustion products reverts to a liquid when condensed, giving up the latent heat that caused it to become a vapour. These two sources of heat can be usefully employed in properly designed boilers to dramatically increase their efficiency, even when the electrical energy necessary to operate the forced draught fan is taken into account.

A further study of the illustration will show that the lower the temperature of the return water in relation to that of the products of combustion, the greater will be the heat transference to the water. Some authorities have suggested that by using much larger radiators to permit lower operating temperatures with consequently lower boiler return temperatures, greater efficiency can be achieved. In practice, however, such a design concept would only fractionally increase operating efficiency and any saving would be offset when compared with the initial cost of installing much larger heat-emission equipment such as radiators. A more realistic approach would be to increase the heating flow and return differential. This would have the effect of reducing the average or mean temperature of the heat emitters, but if the radiators are accurately sized this might not be a practical proposition. It is, however, worth checking on a refurbishment job where a condensing boiler is replacing an older model, as it is quite common for existing radiators to be oversized.

The initial cost of condensing boilers is more than that of conventional types, and although the 92–94 per cent efficiency claimed by manufacturers is attractive, as already explained this depends largely on operating temperatures as illustrated by the graph in Fig. 5.9. Dewpoint is the term used to signify the temperature at which the water vapour condenses and reverts to water. The ideal temperature at which this occurs is 59 °C when the air-to-gas ratio is just sufficient to cause complete combustion of the gas. In practice, however, to ensure safe working conditions, a quantity of air in excess of the ideal requirements must be provided which has the effect of lowering the combustion temperature and the dewpoint of the combustion products. Despite this it will be seen that with a dewpoint of approximately 53–54 °C very high efficiency can be obtained with low return-water temperatures. Condensing boilers are an economic alternative to a conventional boiler in terms of running costs. Radiant heating schemes employing a system of pipework embedded in walls, floors or ceilings, operating at lower temperatures and heated by a condensing boiler, show significant savings on heating costs.

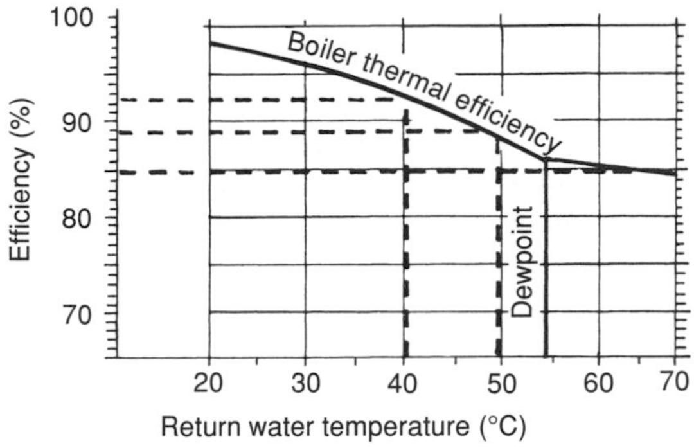


| Return water temp (°C) | Efficiency (%) |
|---|---|
| 70 | 85.9 |
| 48 | 89 |
| 40 | 91 |

**Fig. 5.9** The relationship between boiler efficiency and the return-water temperature.

When upgrading an existing boiler one of the practical aspects that must be considered is the position of the flue outlet. Due to their efficiency the flue gases from condensing boilers tend to produce a greater degree of 'pluming' than those of the traditional type. Flue termination under windows, or adjacent to doors, car ports or opposing walls, must be avoided.

## Pipework systems used for heating

There are basically only two main pipework installations used for heating, the one-pipe and two-pipe systems. There are many variations of these, but they can all be classified under one of the two main headings. Both systems are well established for heating by gravity, and are just as effective for use with forced circulations. The reader should note that the following illustrations are diagrammatic only and do not include details of valves, pumps, etc.

*The one-pipe system*

This consists of a single pipe to which both the radiator connections are made and it is illustrated in Fig. 5.10. The main advantage of this system is that only one pipe is necessary to convey hot water to the radiators; the main disadvantage is that hot water passing through radiator no. 1 is cooled and so, when it returns to the main pipe and then supplies the next radiator, it has the effect of producing a lower temperature in radiator no. 2 than in radiator no. 1. As this process is repeated at each radiator, the water becomes progressively cooler and the temperature of the last radiator on the circuit is noticeably lower than that of the first. This can be overcome, to some extent, by careful regulation of the lockshield valves on each radiator and by limiting the number of radiators on each single pipe circuit to three, or at most four. There is a further factor to take into account: with the one-pipe system, hot water is pumped round the main circuits only, the radiators being heated by the convection currents which occur between the cool water in the radiator and the hot water in the supply pipe beneath it. As most modern radiators have only $\frac{1}{2}$in BSP tappings, these may not be large enough to convey sufficient water to the radiators at the low flow rates produced by gravity, resulting in an unacceptable temperature difference between the top and bottom of the radiators. This and the fact that full-way radiator valves are required limit the application of this scheme when modern heating components are used.

*The two-pipe system*

This is shown in Fig. 5.11. It has quite a number of advantages over the one-pipe system and, in most cases, should be used in preference to it. Its main advantage is that the hot water is pumped through each heat exchanger or radiator, which gives a quick heat-up and a more positive flow. It is for this reason that when convector heaters are used, this system of pipework is essential. The main disadvantage is that the first radiator on the circuit tends to short-circuit the system, as water will take the easiest path and it will be seen from the diagram that there will be less resistance to the flow of water through the first radiator on the system than to that through the last. This can be overcome in most cases by careful regulation of the lockshield valves. Reference should be made to balancing on page 155.

A modification of the two-pipe system is the three-pipe system, or reversed return system, shown in Fig. 5.12. It is expensive to install due to the extra quantity of pipe necessary, but close inspection of the diagram will show that the length of pipe to each radiator is much the same so that

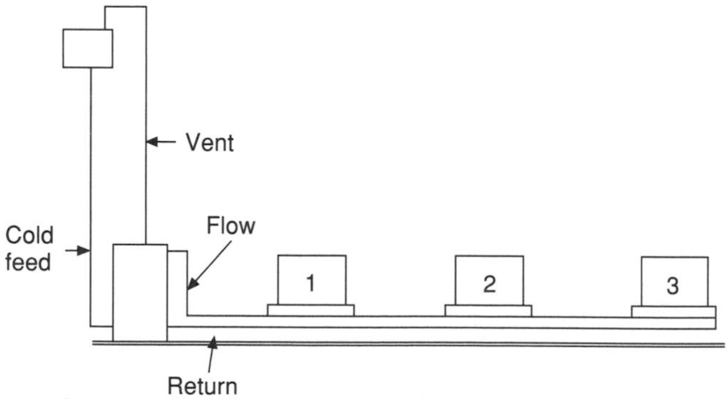


**Fig. 5.10** One-pipe system. The main feature of this system is that both flow and return connections to each radiator are connected to a single loop of pipe.

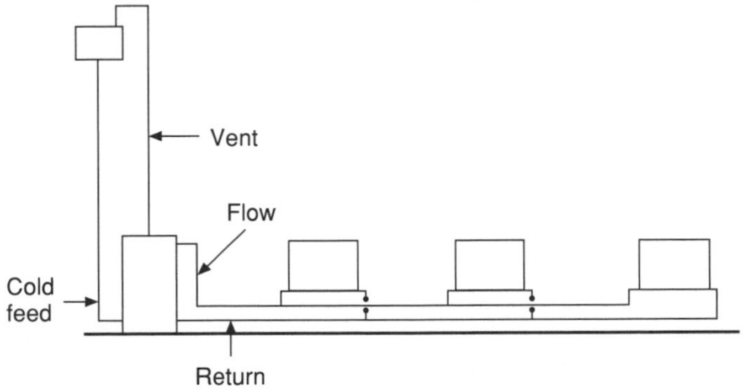


**Fig. 5.11** Two-pipe system. The inlet of each radiator in this installation is connected directly to the boiler flow which ensures an even distribution of hot water to each radiator.

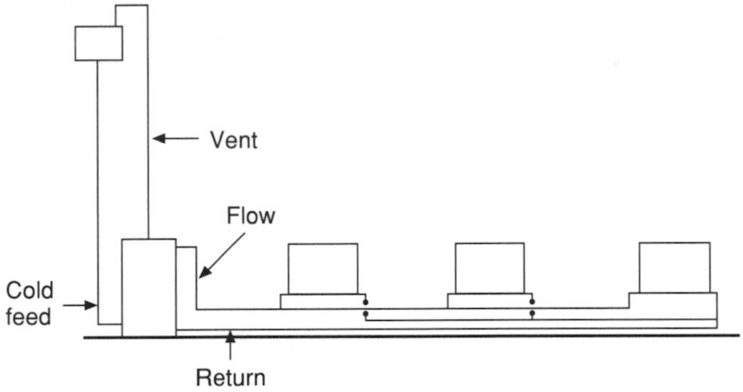


**Fig. 5.12** Two-pipe reversed return system, sometimes called a three-pipe system. This is very similar to a two-pipe scheme but, due to the fact that the pipe run to and from each radiator is of the same length, this system is easier to balance.

the flow of water to each is subjected to the same frictional resistance. This makes the balancing of the system a very simple operation.

### Heating systems

*Pumped heating, gravity hot water systems*
Heating systems using small-bore pipes and a pump to circulate the water round the system have been used for many years. The original arrangement was for the heating circuit to be pumped, the domestic hot water operating on a gravity circulation. Such a system is shown in Fig. 5.13 and is still commonly used and recommended for schemes employing solid fuel boilers, where a heat leak is essential. It should be noted that normally no valve or temperature control is fitted on a gravity circulation that provides a heat leak. The one possible exception to this is boilers fitted with fan-assisted draught or when the systems shown in Figs 4.4 and 4.5 are employed. However, a note of caution! In the case of solid fuel appliances always check with both the neutraliser and boiler manufacturer that the proposed scheme meets their approval. The main disadvantage of systems where the domestic hot water is heated by a gravity circulation is the slow recovery rate of the domestic hot water. It is also sometimes difficult to arrange a gravity circulation between the boiler and hot storage vessel, especially in bungalows, and this is another reason for fully pumped systems as they have a greater degree of flexibility.

*Fully pumped systems*
This arrangement of heating is a development of the small-bore principle where not only is the water

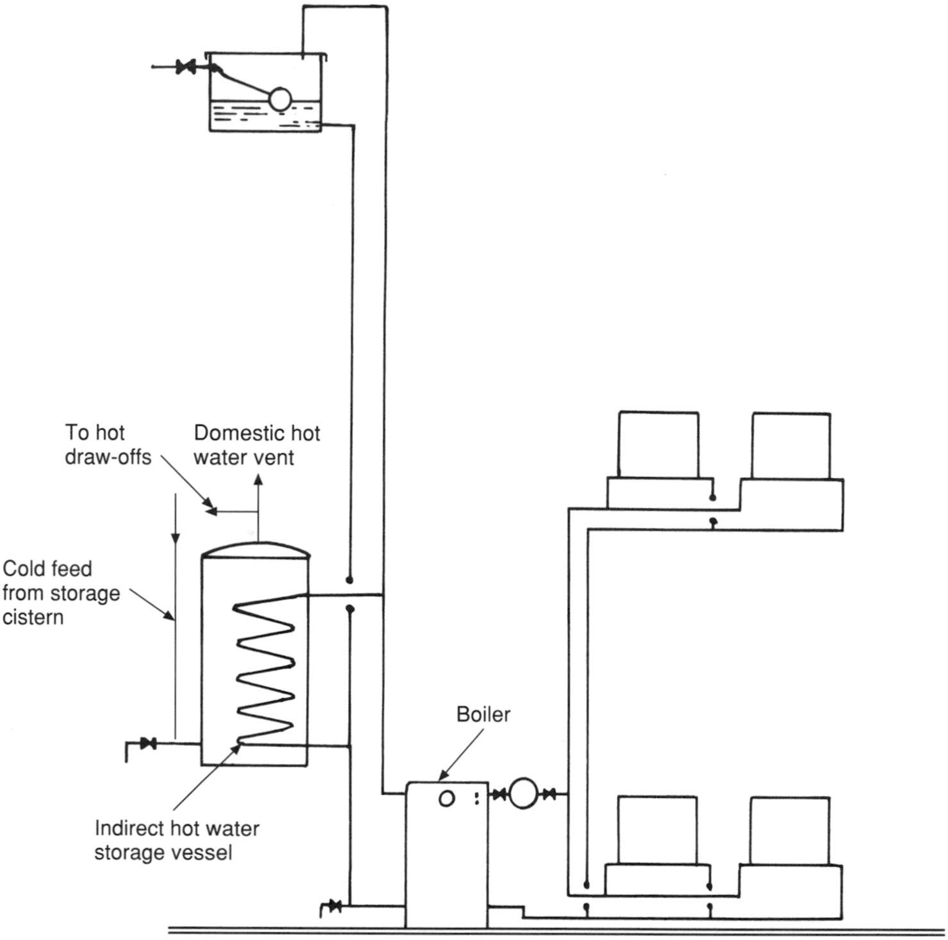


**Fig. 5.13** Small-bore system. This is a traditional system with the domestic hot water heated by a gravity circulation. Note that no electrical controls are shown. This type of system is not now recommended for fully automated systems with gas/oil burners as it can result in boiler cycling.

to the radiators pumped, but so is the primary flow and return to the heating element in the hot store vessel. Figure 5.14 illustrates the installation details and it will be seen that the heart of the system is the three-way motorised valve shown in Fig. 5.15 which is controlled by both a room thermostat and a cylinder thermostat. The original concept of this system was to give priority to the hot water storage vessel should the thermostat be calling for heat. This has the effect of causing the heating port in the motorised valve to be closed, allowing all the hot water produced by the boiler to be diverted to the storage vessel. This continues until the thermostat signals that the water is hot enough and only then will the water be diverted to the space-heating system. This ensures a constant hot water supply, and because the recovery rate is so rapid (approximately 120 litres of water can be raised 40–50 °C in 15–20 minutes) any slight cooling of the radiators during this period is unnoticed. Only when both requirements of domestic hot water and space heating are met will the boiler shut down. Three-way valves of this type are provided with a lever, which, when set in the manual position, has the effect of allowing both outlet ports to be held in the open position while the system is being filled, and should the motor be defective, enabling both hot water and space-heating demands to be met.

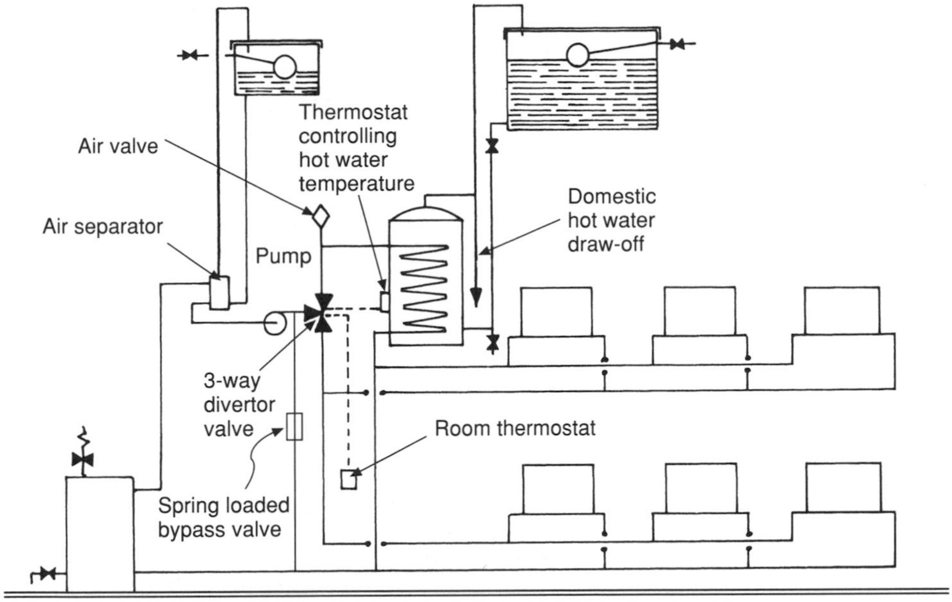


**Fig. 5.14** Fully pumped system. General layout for a fully pumped small-bore system. The room and hot water thermostats control the three-port valve to divert hot water from the boiler to the radiators or heaters in the domestic hot water storage vessel. If the valve is closed against the hot water circulation and all the radiators are shut off water can still circulate via the bypass valve which opens the bypass circuit due to pump pressure.

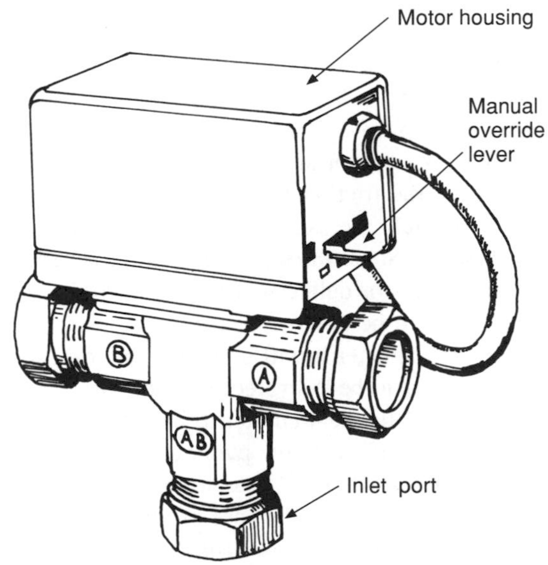


**Fig. 5.15** Three-port motorised valve used with fully pumped heating systems. The valves must be fitted in such a way that the vent and feed pipe to the system are not closed off.

This is useful, because, although it cancels the control system, both hot water and heating will function until the defective valve is serviced. It is perhaps worth noting that manufacturers of these valves make them in such a way that the motor can be replaced without removing the valve from the system. Similar valves called 'mid' position valves are also available which permit both space heating and domestic hot water to be heated simultaneously. A similar system of control can also be achieved by using two two-port motorised valves, one serving the heating circuit, the other the primary circuit to the hot storage vessel. With all systems of this type the recommendations of the boiler manufacturers relating to the pipework details and the electrical wiring system must be observed. The advantages of fully pumped systems can be summarised as follows:

(a) Rapid recovery period for hot water supply even after a heavy draw-off such as running a bath.
(b) Greater circulating pressure by the pump avoids the limitations of a gravity primary circulation

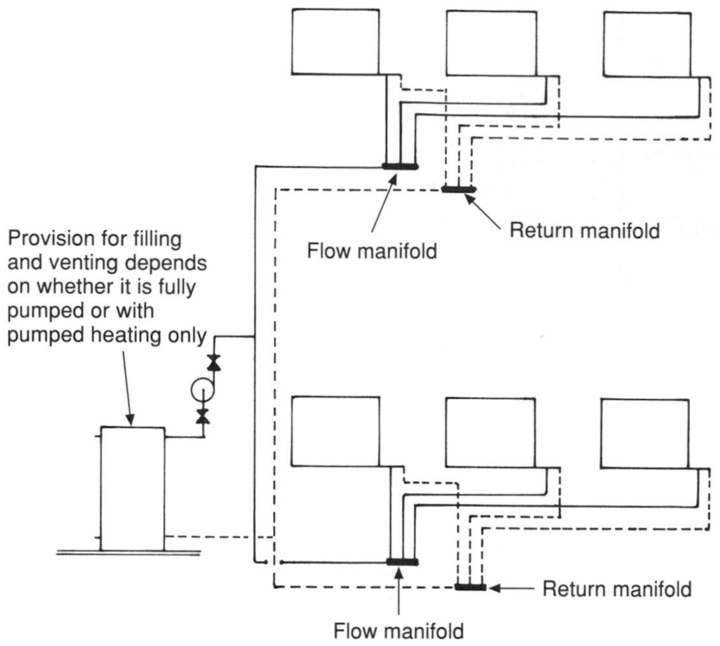


**Fig. 5.16** Mini-bore system. This system differs from that of small bore in that each radiator has its own flow and return pipes which are connected through a manifold to the main flow and return. The system shown can be used with sealed or vented fully pumped systems.

to the hot storage vessel in single-storey buildings.

(c) Reduction of the effects of stratification; this means the temperature difference between the water at the top and bottom of the storage vessel is much less, resulting in a greater quantity of water at a higher temperature.

(d) Much more positive control is exerted over a system which is electrically operated and unwanted circulation by gravity is non-existent in a well-designed scheme.

(e) The circulating pipes to the hot store vessel can be reduced in diameter enabling costs of pipework and fittings to be reduced. This must of course be offset against the cost of the motorised valves.

### Mini-bore systems

This system shown in Fig. 5.16 is really a development of the small-bore principle using smaller diameter soft temper copper pipes which are obtainable with a PVC sleeve which serves to both protect and insulate the pipes. The main difference is that the flow and return to each heat emitter is connected to the main flow and return through a manifold, two typical examples being shown in Fig. 5.17. If the manifold can be situated in such a way that the branch flow and return to each radiator are approximately the same length, the frictional resistance will also be approximately the same, making the system self-balancing. Some thought must also be given to the flow rates or delivery of hot water to the radiators. It will be obvious that if the pipes are too small in diameter it will be impossible for sufficient water to be delivered to enable the mean output of the heat emitter to be achieved without the use of excessive pump pressure.
Of course these facts must be considered in all types of installations. In the case of mini-bore systems the soft temper copper pipes used are made having 8, 10 and 12 mm bores. It is found in practice, however, that for domestic work 8 and 10 mm diameters are satisfactory for most purposes and the use of only two sizes reduces the number of

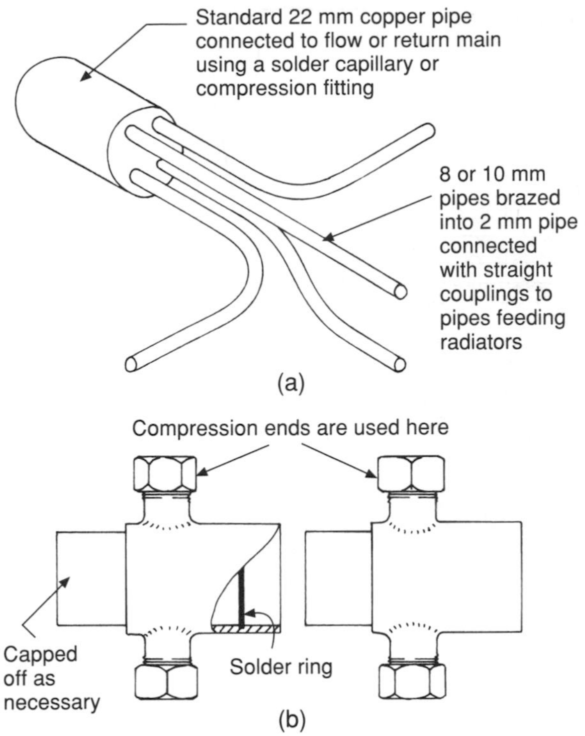
Standard 22 mm copper pipe connected to flow or return main using a solder capillary or compression fitting
8 or 10 mm pipes brazed into 2 mm pipe connected with straight couplings to pipes feeding radiators
(a)
Compression ends are used here
Capped off as necessary
Solder ring
(b)
These four-way tees are used together to build up manifolds as required. Reducers are used in the branches to accommodate the appropriate pipe size.

**Fig. 5.17** Manifolds for mini-bore heating.

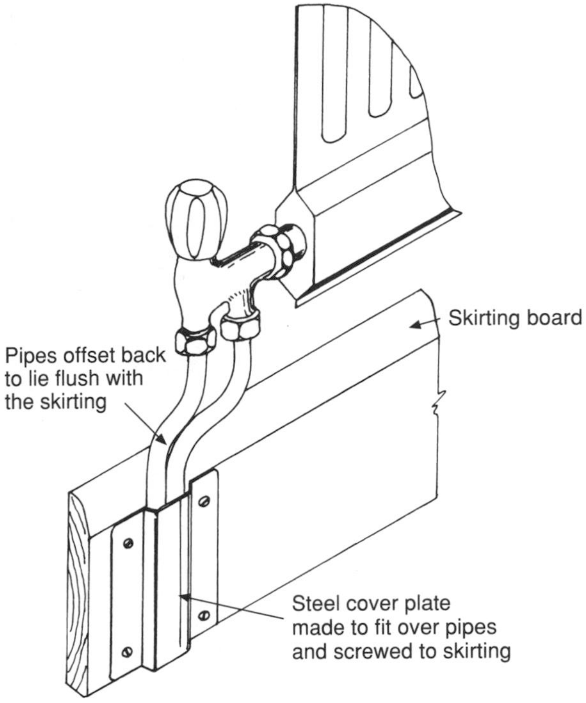
Skirting board
Pipes offset back to lie flush with the skirting
Steel cover plate made to fit over pipes and screwed to skirting

**Fig. 5.18** Protecting soft copper or plastic pipes at skirting level.

adaptors for valves and manifolds that have to be carried. All diameters of this pipe may be obtained in coils of 7.5, 15 and 30 m in length. Being soft temper they are prone to damage where they are exposed, especially at floor level, and Fig. 5.18 shows a method of protecting them at this point. These small pipes are easily concealed and can be run behind skirtings with a minimum of trouble, or in shallow ducts in a floor screed. Where they are passed through or over joists in suspended floors the holes or notches necessary are smaller than required for larger pipes, therefore there is less possibility of weakening the joists. The non-rigid nature of the pipe, being dead soft temper, reduces the noise factor sometimes associated with more rigid pipes. The longer lengths require fewer joints and the lower water content in the system makes it very responsive to temperature change and therefore more efficient. Because of the smaller pipe diameters the pump pressure must be raised to overcome the increase in frictional resistance, the recommended flow rate being 1.5 m per second. The use of soft copper pipe may lead to traps of air in the pipe runs, but the higher pump pressure used (unlike the traditional small-bore scheme) is normally capable of shifting the air. Mini-bore schemes are suitable for use with all types of system layouts, including solid fuel, providing provision for a heat leak is made. The radiator valves are usually of the annular type (see Fig. 5.19) which enables the flow and return connections to be made on one end of the radiator only. No provision for balancing is made on the valve shown, but some do have this facility. In a well-designed system balancing should not be necessary as the aim is to keep all pipe runs to heat emitters at, as near as possible, the same lengths.

## Sealed heating systems

The basic principles of sealed heating systems are similar to those of unvented hot water systems, but

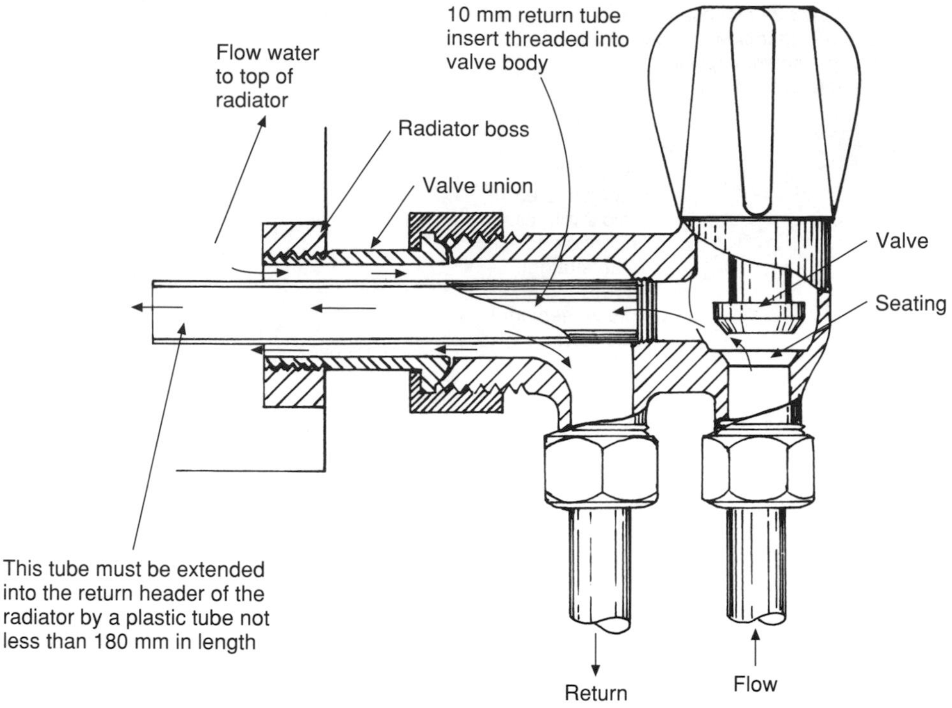


**Fig. 5.19** Double entry radiator valve. They may be manually or thermostatically controlled and are specially designed for use with small pipes and permit both the flow and return to the heater to be connected at one end. When used with double panel radiators a special flexible insert is required.

while these have only recently been permitted by water authorities, a sealed system of heating in its present form has been used for a long period of time and historically it is almost as old as low-pressure heating. These systems have many advantages, possibly the most important being as follows: pipework and cisterns at high level or in a roof space are unnecessary and the system can be operated at higher working temperatures if required. If radiators are used the maximum temperature now recommended is the same as for low-pressure schemes. For the really effective use of convector heaters higher temperatures are necessary, and if this type of heating is specified serious consideration should be given to a sealed system of heating. The reader should be aware, however, that to achieve higher flue temperatures a condensing boiler will not be working at its maximum efficiency. Figure 5.20 illustrates a typical scheme combined with an unvented hot water system. As with unvented hot water systems they are usually provided by the manufacturer in the form of a package. It should be noted that both ventilated and unventilated hot water systems can be used with all types of heating schemes, whether they are small-bore, mini-bore or fully pumped systems.

*System components*

The system is filled by one of two methods, either manually via a small top-up vessel at the highest point of the system or by direct connection to the main, providing the following requirements of the water by-laws are met. *The connection between the mains supply and the heating system must be temporary only, to be used when filling or topping up; after use it must be disconnected.* A backflow device such as a check and anti-vacuum valve or double check valve must be permanently fitted at the inlet point. This ensures there is no possibility of contamination of the mains supply by water in the heating system, which may contain a corrosion inhibitor or other unpleasant additives. To avoid

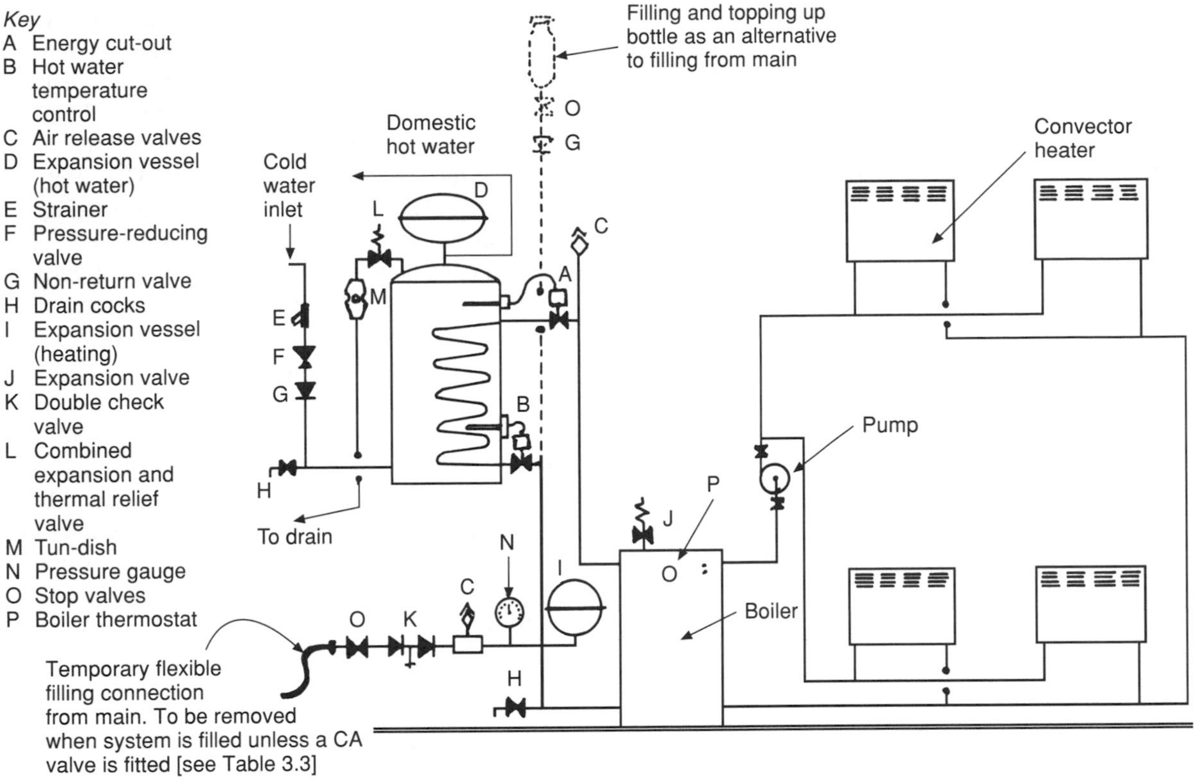


**Fig. 5.20** Sealed system of heating with an unvented domestic hot water system.

continually disconnecting the hose a CA valve may be fitted on domestic boilers (see Table 3.3).

*Air purger* This is necessary only when filling takes place from the main. Its purpose is to remove the air bubbles from the water as it enters the system. The air is expelled through an automatic air valve which is sometimes made as an integral part of the air purger. Both these components are illustrated in Fig. 5.21.

*Expansion vessel* This takes the place of the feed and expansion cistern used in a low-pressure heating system to accommodate the expansion of the water when it is heated. The capacity is important and it must be capable of absorbing the expansion of water when it is raised through 100 °C. Manufacturers of these vessels will advise on their capacity for a specific installation. The factors involved are the static pressure on the system, the maximum working pressure which will be the same as the relief valve setting and the volume of the water in the system. An approximation of the water content can be made working on a basis of 14 litres per kW of boiler power. The static pressure is calculated by measuring the highest point in the system to the expansion vessel. The situation of the expansion vessel in the system is not critical, but it is likely that the life of the flexible membrane will be extended if it is not subjected to water at very high temperatures. For this reason it is usually fitted to the system return via a non-circulating pipe as shown in the illustration. It is also suggested that the pump exerts a negative pressure rather than positive pressure on the membrane, as this reduces the pressure to which it is subjected. The gas charge in the vessel can be air or nitrogen, but the latter is considered to be the better, because if the gas escapes into the system, due to a fractured membrane, air, containing oxygen, is likely to cause corrosion. Figure 5.22(a) shows an expansion vessel

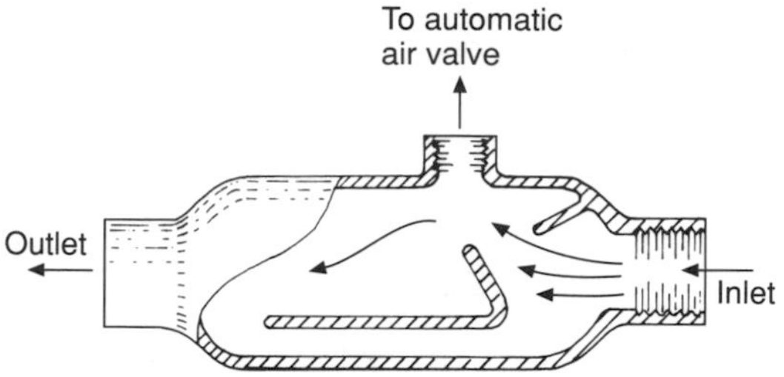


(a) Air purger: sealed heating systems

This works in a similar way to that of an air eliminator for low-pressure heating. The turbulence caused by the internal plates causes the minute air bubbles to merge together where they are easily dispersed through the automatic air valve.

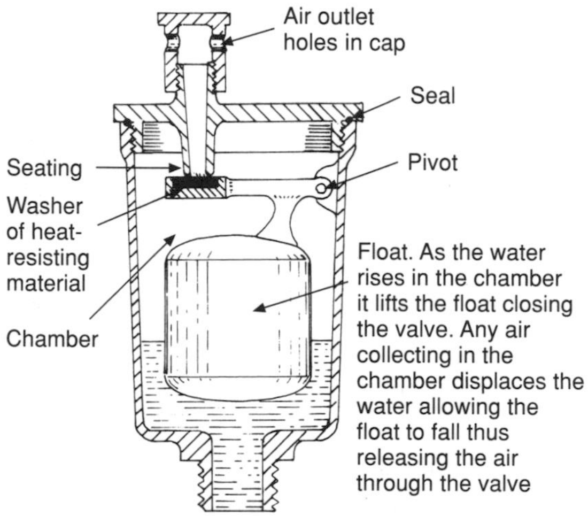


(b) Automatic air valve or eliminator

These valves are used in combination with an air purger in sealed systems of heating, but they have many other applications in low-pressure systems where the use of an open vent is undesirable, i.e. high points on a heating system or on radiators that persistently collect air. It must be stressed, however, that the use of these valves in low-pressure systems does not mean that an open vent at some point in the system is unnecessary.

**Fig. 5.21** Air elimination in sealed heating systems.

when the system is cold, Fig. 5.22(c) when the normal operating temperature is achieved.

*Safety devices*

As with unvented hot water systems there are three safety devices to prevent the build-up of excess pressure. The first is the thermostat permitting variation of temperature of the appliance. Should this fail, a temperature cut-out comes into operation in the same way as an unvented hot water system. Thermal relief valves are not normally fitted, but a pressure relief valve is of course essential. They are very similar to the expansion relief valves used with unvented hot water systems. The safety valves traditionally used for low-pressure heating are unsuitable for this type of work as they are often made of poor-quality brass which is affected by dezincification, and the valves often become sealed on to the seating due to corrosion, or in some cases, limescale. Safety devices for sealed systems should be made of bronze and any seals, washers or O-rings made of heat-resistant synthetic rubber or plastic. A pressure gauge, often combined with a rotary-type thermostat, should be provided to enable the user and service engineer to monitor the working characteristics of the system. Working pressures of these systems should not exceed 3 bar and the water temperature should not exceed 99 °C.

*Under-floor systems of heating*

Both installation and design procedures are dealt with later in this chapter.

### Heating controls

Controls for automatic shut-down of the system and temperature are absolutely necessary for both hot water and heating installations to comply with the current Building Regulations. The 2005 Part L Regulations currently in force specify the following:

(a) Time control for all wet central heating systems except natural draught solid fuel boilers.
(b) Zone control to provide separate temperature control in areas with different requirements.
(c) Cylinder heat exchangers (the internal coil) must comply to at least BS 1566. This is to prevent wastage of fuel due to boiler cycling. The cylinder must also be insulated to a minimum standard and the first metre length of any pipe connected to it must also be insulated.
(d) Temperature control of the stored water is also mandatory.
(e) The thermostats on gas- and oil-fired boilers used for wet central heating systems must be

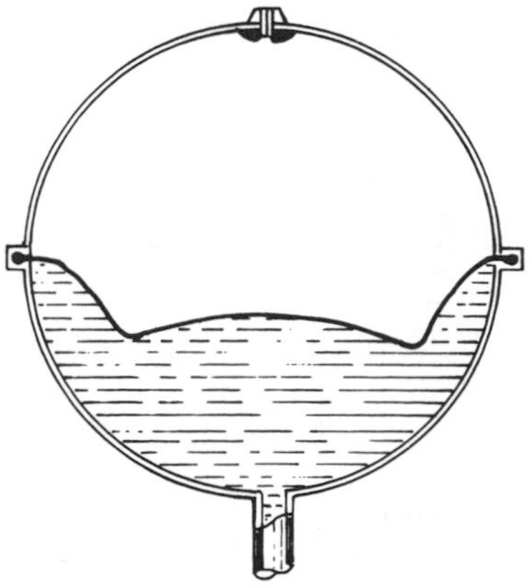

(a) The system cold with the vessel charged with gas

The gas charge is equal to the static head exerted on the membrane + a margin. Normally this is 3 bar.

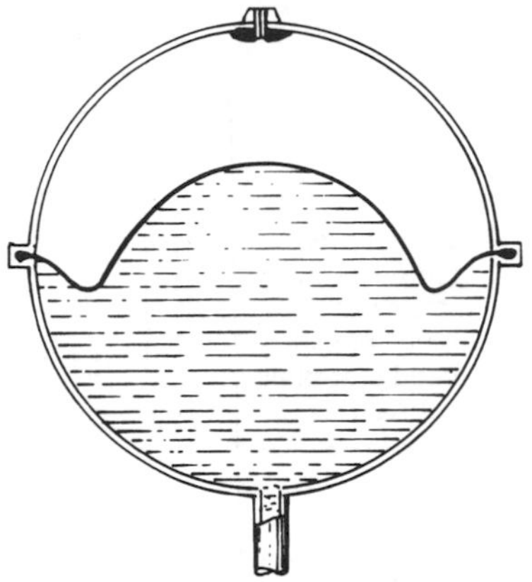

(b) The water is heated and the gas is compressed due to expansion of the water

Note that expansion vessels used for sealed heating systems are not suitable for unvented hot water.

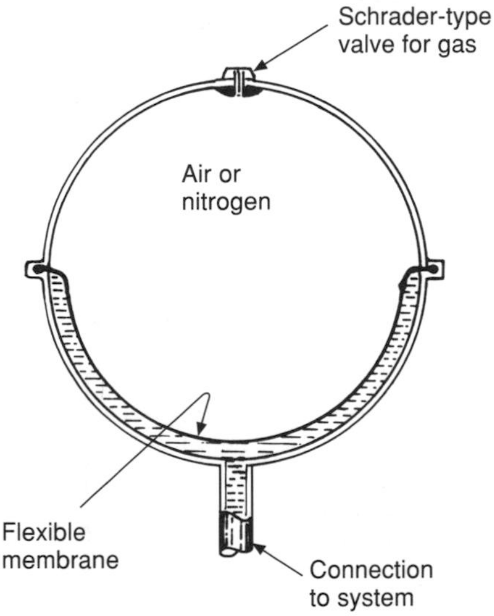


(c) Operating temperature is achieved

**Fig. 5.22** Expansion vessel. The gas, air or nitrogen, in the vessel is precharged to a predetermined pressure related to the static head on the system. A gas charge of 1 bar will support approximately a 10 m head of water. If there were no charge, water would fill the vessel leaving no space for the expansion of the water when it is heated.

interlocked with the thermostats controlling space and hot water storage heating. This ensures that the boiler will only fire when there is a call for heat, not simply when the water in the boiler cools. See page 182 and Figure 5.28.

*Boiler thermostats*

These are usually an integral part of the boiler. Those used for modern gas- and oil-fired boilers work on the principle of expansion, either of heat-sensitive fluids or metals which operate an electric switch. Typical examples are shown in Chapter 2. The boiler thermostat controls the temperature of both the space-heating and domestic hot water equipment.

*Time control*

Some provision for automatically switching on and off the boiler and pump is essential, and this is achieved by means of a time switch, or a programmer.

*Time switches* A simple one-circuit control is suitable for combination boilers and should be selected so that it is easy to use and understand.

*Programmers* (1) A mini-programmer allows the space heating and hot water to be on together, or hot water alone. (2) A standard programmer uses the same settings for both space and hot water heating. (3) A full programmer allows the time settings for space and hot water to be fully independent. All the foregoing may be of the electric mechanical type, in which case they operate

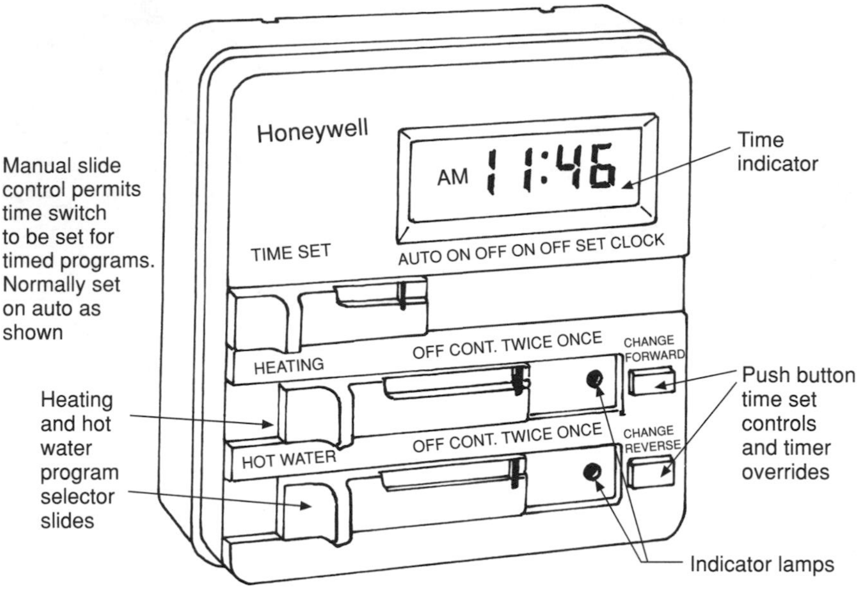


**Fig. 5.23** Programmer.

over a 24 hour or 12 hour period of time. Those of the electronic multi-channel type are more flexible and can be programmed for up to seven days with independent switching of heating and hot water circuits. They are normally provided with battery back-up so that in the event of a power cut they will continue to operate. A basic programmer is shown in Fig. 5.23.

*Cylinder thermostats*

The average or mean design temperature on which most domestic heating schemes operate can overheat the domestic hot water, the maximum temperature of which should not exceed approximately 65 °C. To control the hot water temperature a cylinder thermostat controlling a motorised valve is the usual method employed. A cylinder thermostat and its application is shown in Fig. 5.24.

The position of the thermostat on the vessel is important, and should be fitted as shown to prevent the boiler 'cycling'. The motorised valve will remain closed preventing wasteful circulation until a call for heat is made by the cylinder thermostat.

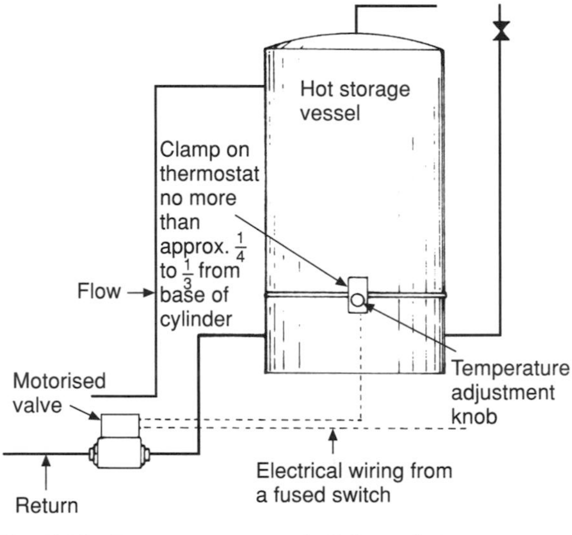


**Fig. 5.24** Temperature control of domestic hot water with a motorised valve. This is an electrical system of control which is suitable for both domestic and industrial use. The thermostat controls the motorised valve causing it to open when the water temperature in the vessel drops, and close when the predetermined temperature is achieved.

*Space-heating control*

*Room thermostats* The boiler is normally set for the temperature required for heating, usually about 70–80 °C, and unless some form of control is used, this can lead to unnecessarily high space-heating temperatures. The heating system seldom operates at its designed temperature, except for a few weeks in the year when the weather is very cold. Overall control of space heating is achieved by a room thermostat which is normally set at a temperature of 20–22 °C. When the temperature falls the contacts close and start the pump. In heating systems fitted with thermostatic radiator valves, room thermostats provide overall control of the space-heating system and should therefore be fitted in an area of the building requiring a lower temperature than that of living rooms and bedrooms. The room thermostat shown in Fig. 5.25 is a basic type. Some now available can be programmed to provide both timing and variations of temperature at different periods in a 24 hour or 7 day cycle. The latter is very useful when the pattern of building occupation varies during the period of a week. A further development in control systems is the use of radio waves, which makes conventional wiring unnecessary. It is important to ensure that any radio control has a satisfactory level of immunity to 'blocking' by other radio transmissions. Products bearing the new 'radio mark' symbol are certified to meet the requirements relating to quality and fitness for purpose. Night set-back features are also available where a minimum temperature can be maintained during the night. This can be included in some types of room thermostat.

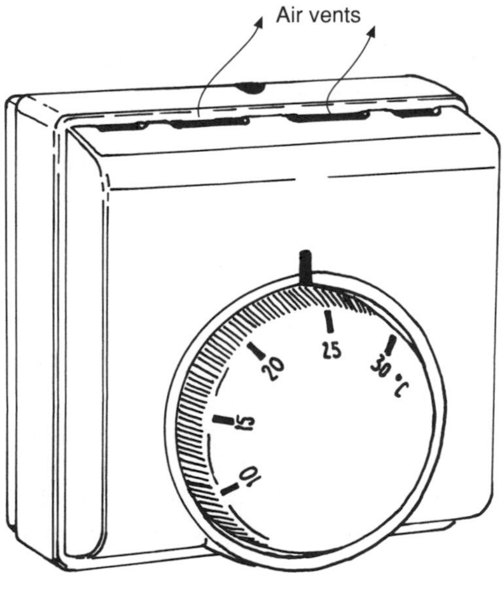


**Fig. 5.25** Room thermostat. These operate by a flow of air passing over a bimetallic coil or bellows causing it to expand or contract making or breaking a microswitch.

*Zone control* This is now mandatory in all buildings. In some types of commercial and larger dwellings, energy savings are made by breaking down the space-heating system into zones, e.g. living spaces and bedrooms. This allows different parts of the building to be either isolated when not in use or heated to different temperatures. The use of thermostatic radiator valves accomplishes this and is considered satisfactory for small dwellings. In the case of larger buildings considerable savings can be made by controlling the heating circuits to various zones using two-port motorised valves and programmable room thermostats. These control not only the zone temperature but also the on/off periods when the heating is required. Figure 5.26 illustrates a typical system having separate zones.

*Delayed optimum start feature* This type of control has been used in commercial premises for many years to save energy. It is essentially a room thermostat that embodies an energy-saving delayed start feature, which depends on how close to the set point the room temperature is at the time the programmer is scheduled to start the heating system. In simple terms it reduces the time on a daily basis during which the system operates without causing any loss of comfort. See Fig. 5.27. These controls are increasingly used in domestic premises for the same purpose. The delayed start will overide the programmer/time switch, both of which operate at a preset time.

*Boiler energy manager* This is an electronic control that embodies weather and load compensation by measuring and coordinating both the outside air temperature and boiler return temperature. Its use ensures the system produces

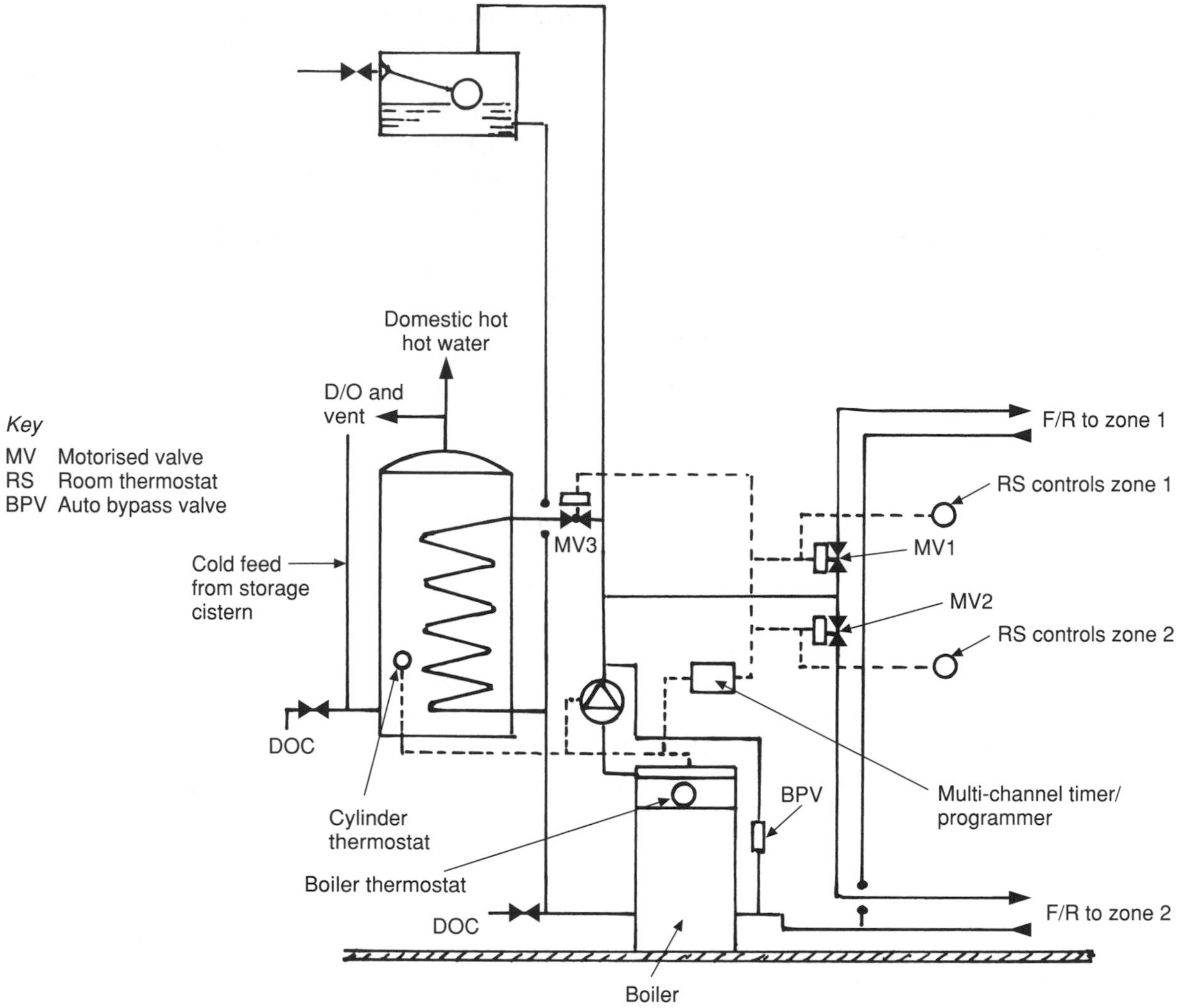


Zone control must now be incorporated into schemes having floor areas in excess of 150 m². It enables areas in a house or building to be isolated when not in use. Multi-channel timers/programmers are considered best practice for this type of scheme. See text on programmers on pages 179–80.

**Fig. 5.26** Zoned system of heating.

the necessary heat in relation to the current weather conditions and the comfort of the user. This type of control is electronic in operation and can be fitted to an existing heating scheme without the necessity of draining it down, making it an ideal system upgrade option.

*Temperature control interlock* To ensure maximum economy and efficiency, heating control systems must be fully interlocked, which means the boiler will not fire until the room or cylinder thermostat calls for heat. Figure 5.28 shows in schematic form how the controls are wired to achieve this. With older systems the boiler thermostat was controlled only by the programmer, which allowed the boiler to fire every time its water content cooled. This resulted in wasteful cycling when there was no call for heat from the room or boiler thermostat.

*Weather-compensating controls* These have been fitted for many years in commercial buildings and

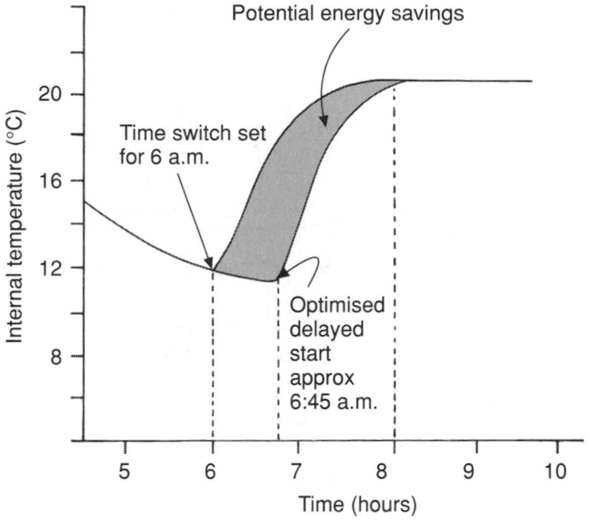


**Fig. 5.27** Delayed optimum start graph. The warmer the internal temperature, the later can be the start-up of the scheme.

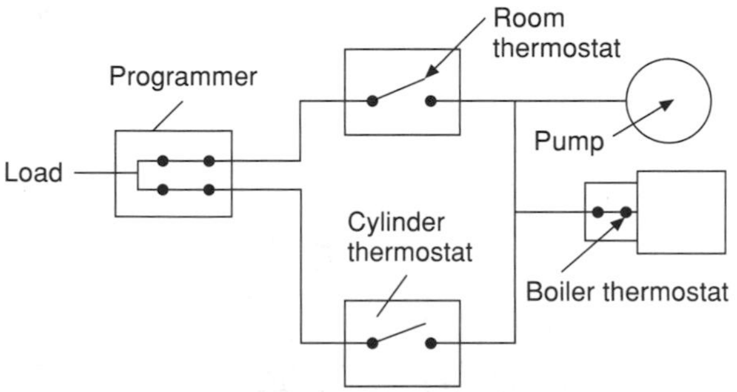


**Fig. 5.28** Interlocking of controls. Although the programmer switches and the boiler thermostat are closed, the boiler will not fire until the room or cylinder thermostats close on a call for heat.

may be used with advantage in larger domestic properties. They ensure that energy is not wasted in milder weather when the design flow temperature for space heating is unnecessary. A simple valve of this type is shown diagrammatically in Fig. 5.29. Some of the larger types used in industrial and commercial buildings are electrically operated, but the basic principles are the same as shown.

*Thermostatic radiator valves (TRVs)* These are now mandatory for space-heating control and work in conjunction with the room thermostat. Details of these valves are shown in Fig. 5.30. They are

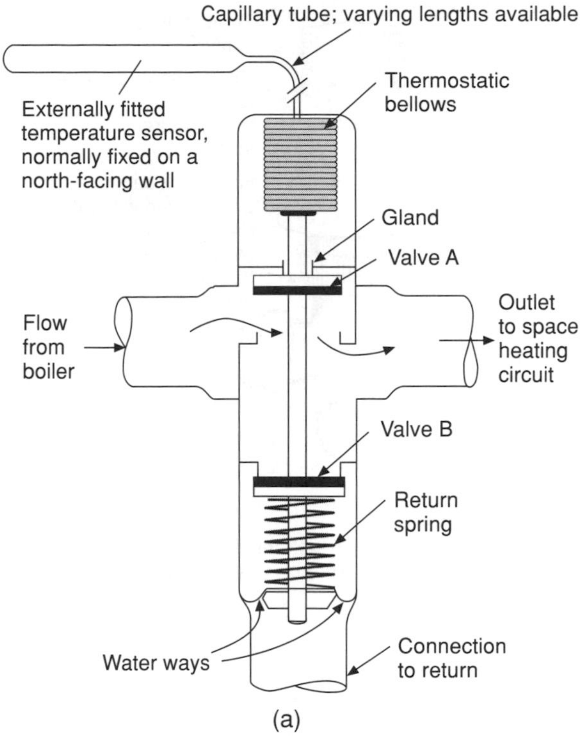


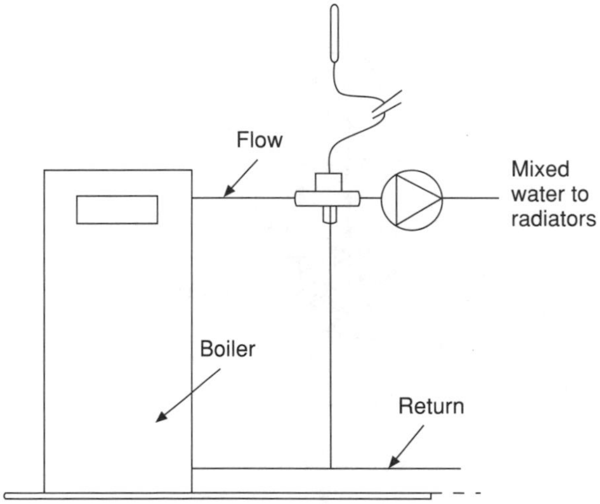


(b) Position of weather-compensating control in heating circuit

**Fig. 5.29** Weather-compensating control. The valve is shown with valve A open, and in this position water at boiler temperature is flowing to the radiators. An outside temperature rise will be transmitted via the sensor to the bellows, partially closing valve A and opening valve B. This allows water from the heating return to mix with the water from the boiler, effectively lowering the temperature of the water circulated to the radiators.

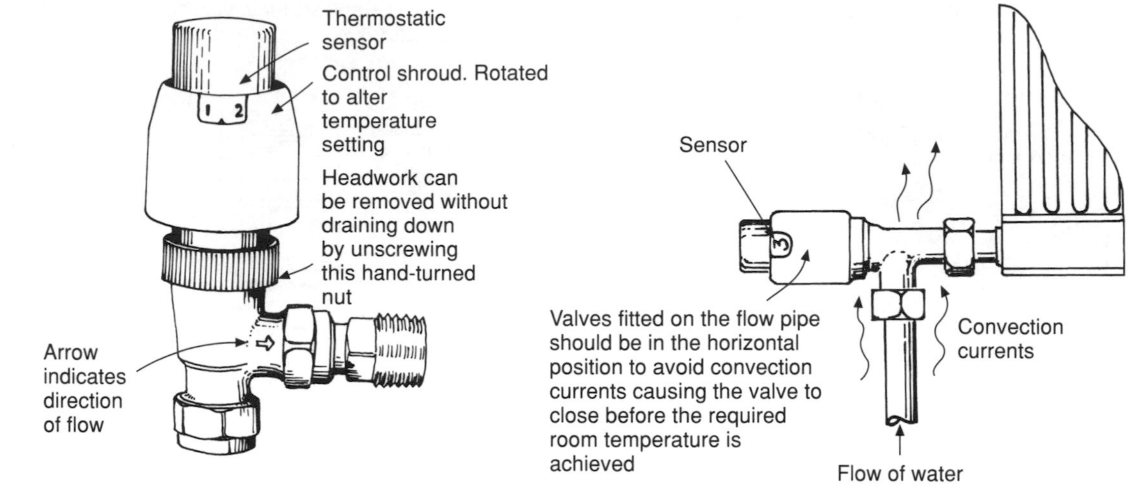

Thermostatic sensor
Control shroud. Rotated to alter temperature setting
Headwork can be removed without draining down by unscrewing this hand-turned nut
Arrow indicates direction of flow
Sensor
Valves fitted on the flow pipe should be in the horizontal position to avoid convection currents causing the valve to close before the required room temperature is achieved
Convection currents
Flow of water

(a) Thermostatic radiator valve

Compliance with the manufacturer's instructions is essential if the valve is to operate effectively. Generally, the best position is on the flow with the sensor in the horizontal plane.

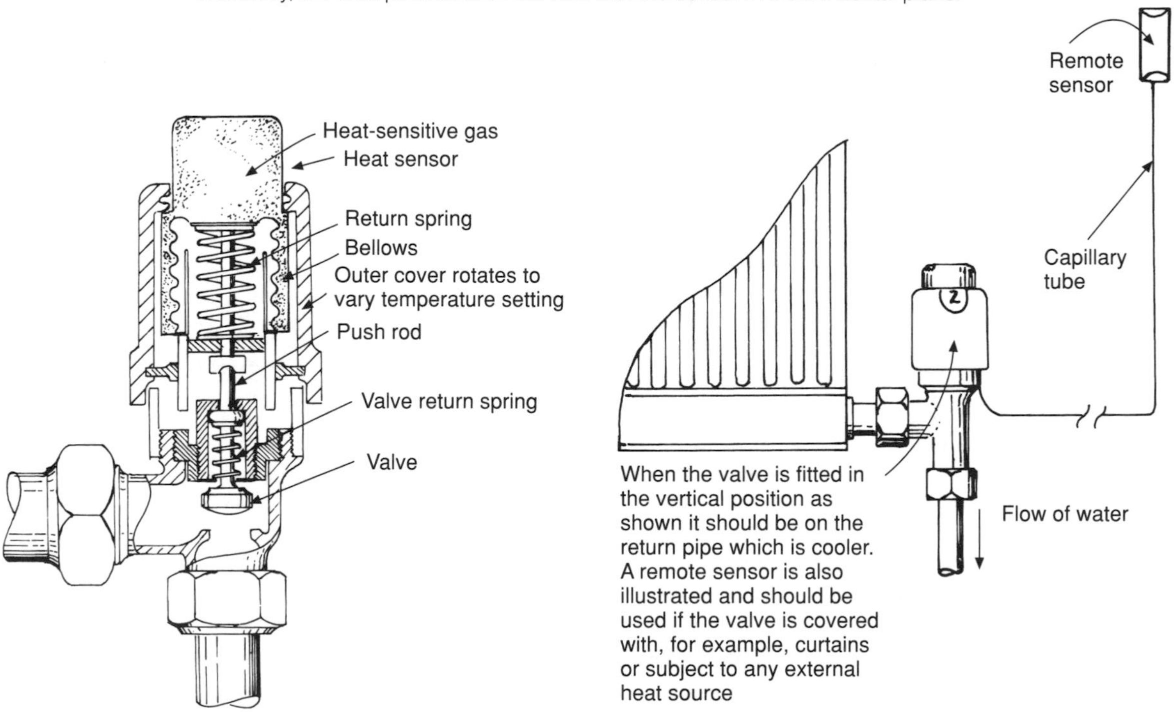

Heat-sensitive gas
Heat sensor
Return spring
Bellows
Outer cover rotates to vary temperature setting
Push rod
Valve return spring
Valve
Remote sensor
Capillary tube
Flow of water
When the valve is fitted in the vertical position as shown it should be on the return pipe which is cooler. A remote sensor is also illustrated and should be used if the valve is covered with, for example, curtains or subject to any external heat source

(b) Operational details of a thermostatic valve

(c) Fitting thermostatic radiator valves

On a rise in room temperature the heat-sensitive gas expands causing the bellows to push the valve on to its seating. Both the valve and bellows are spring loaded to ensure a positive opening of the valve on a drop in temperature.

**Fig. 5.30** Thermostatic radiator valves.

fitted to each radiator and control the temperature of each room independently of any other. The fitting instructions supplied by the manufacturers of these valves must be carefully observed to ensure their full effectiveness. They should not be positioned where they can be affected by heat from the sun as this can result in their closure before the desired room temperature is achieved.

TRVs embody a sensor containing a heat-sensitive fluid or wax which expands when the required air temperature in the room is achieved, closing the valve, and contracts when the temperature in the room falls below that required, thus opening the valve.

TRVs have a low temperature setting so that if the room temperature falls below approximately 4 °C, they will automatically open. This means that if in very cold weather the heating system is turned on in empty premises, these valves provide some degree of protection in the event of a frost. To prevent possible damage to both low-water-content boilers and pumps it was common practice to leave one open circuit. This is not now recommended due to energy loss which can be avoided by the use of an automatic spring-loaded bypass valve shown in Fig. 5.31. It must be fitted on a bypass pipe between the outlet side of the pump and the main heating return. It is designed to open when there is a build-up of pressure in the system, e.g. if all valves are closed. When a heating circuit opens the valve closes, allowing the full output from the pump to circulate through the system. It must be manually adjusted to meet the requirements of the system.

*Frost thermostats* As with room thermostats, frost thermostats sense the temperature of the air and, except for working over a lower temperature scale, they are almost identical. They are essential if the premises are left empty for a long time or during periods of frost or if the boiler is fitted outside the building. They are wired into the control scheme in such a way that the time switch or programmer, when in the off position, is overridden when there is a sharp drop in temperature. They are sometimes installed with a limit thermostat as shown in Fig. 5.32. The foregoing controls are generally applicable to oil- and gas-fired schemes. Due to the necessity of providing a heat leak with solid

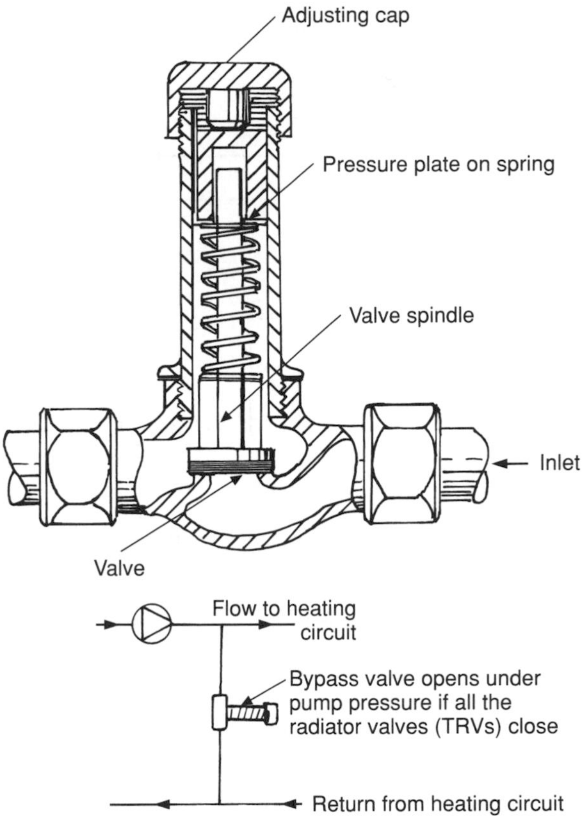


**Fig. 5.31** Bypass valve.

fuel appliances, it is not recommended that any automatic control is fitted to the domestic hot water circuit.

*Wiring electrical control systems* Most manufacturers of control systems publish suitable information for installers who undertake this work, and also run short courses on this subject. It is essential that all electrical work complies with the IEE wiring regulations and Part P of the Building Regulations and no short cuts should be taken. Failure to comply could not only be dangerous but also make it very difficult to trace a fault should one occur at a later date. In short, the electrical installation must be carried out strictly according to the diagrams supplied with the components, and the wiring colour code used throughout, if applicable. See also text relating to Part P of the Building Regulations on pages 345–6.

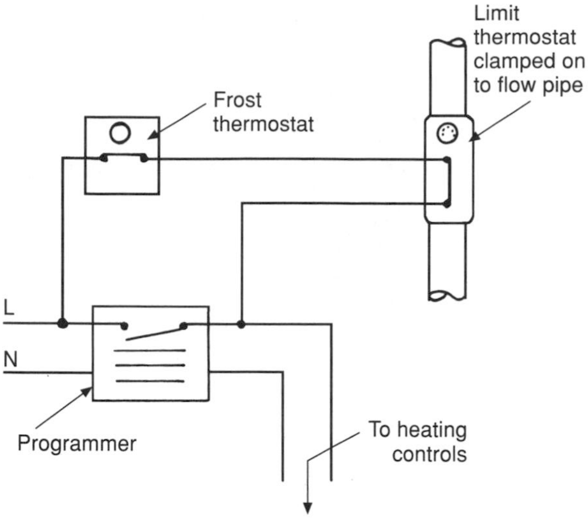


**Fig. 5.32** Wiring to frost thermostats. The illustration shows a schematic wiring diagram for frost protection. The time switch is open and the system is shut down. If the air temperature falls below the thermostat setting (2–3 °C) the contacts close, overriding the time switch. The limit thermostat, shown strapped to a flow pipe adjacent to the boiler, will normally be closed when the system is cold. Under these conditions the system will operate until the limit thermostat opens. It is generally set at 10–15 °C which prevents the system operating at design temperatures when, but for very cold weather, it would normally be shut down.

## Central heating design

It should be noted that the requirements of the Building Regulations relating to the heat loss in buildings are ahead of the relevant British Standards. Some parts of BS 5449 and BS EN 12828 are still valid, but until this code is updated any design calculations should be based on figures quoted by the Energy Saving Trust, and when appropriate the equipment manufacturer's publications.

### *Heat loss*

Modern, efficient central heating systems are designed on the basis of calculated heat loss, the main sources being loss of heat through the building fabric by conduction, and ventilation or air change. While bearing in mind that some ventilation of rooms is necessary for fresh air, odour removal, prevention of condensation and combustion air for heating appliances, excessive air changes are both unnecessary and costly.

Prior to designing and installing a heating system, a simple survey should be made to investigate and rectify unnecessary heat loss, especially in older buildings. Heat loss can usually be minimised quite simply without too much cost, being identified as follows:

(a) Is the roof space sufficiently well insulated? A minimum thickness of 100 mm of insulation blanket is essential, but a thickness of 200 mm is now recommended. Heat loss in this area can be further minimised by a layer of aluminium foil between the ceiling joists.
(b) Non-insulated cavity walls can be filled with insulating material by companies which specialise in this work. New buildings must comply with the requirements of the Building Regulations, and external walls must be constructed to specific standards to reduce thermal transmittance. Prior to 1965 many buildings were constructed with solid brickwork 225 mm thick, the heat loss through this form of structure being very high by today's standards. The cost of dry lining external walls in such buildings is worth consideration.
(c) Double glazing does reduce heat loss through windows by approximately 50 per cent, but is very expensive, and because window areas are relatively small, any savings on heating costs would take a long time to pay for its installation. The main advantages of double glazing are the reduction of condensation on inside windows, and the minimisation of noise levels from external sources.
(d) The closure of all apertures causing draughts is also worth investigating. Gaps around doors and windows can easily be weatherstripped. Shrinkage of floors and skirting board on suspended floors can also present a problem unless the room is carpeted. In rooms without carpet, gaps can be sealed using a suitable filler, or fixing a small wooden moulding around the base of the skirting. Open fires in a room should be sealed if no longer in use, as they are very wasteful due to the convective effect of the flue, but should a client wish

**Table 5.1** Room temperature and ventilation rates.

| *Room* | *Temperature (°C)* | *Air changes per hour* |
|---|---|---|
| Living | 21 | 1.5 |
| Dining | 21 | 1.5 |
| Bed/sitting | 21 | 1.5 |
| Bedroom | 18 | 1.0 |
| Hall/landing | 18 | 1.5 |
| Bathroom | 22 | 2.0 |
| Toilet | 18 | 2.0 |
| Kitchen | 18 | 2.0 |

to retain a fire, a sealed solid fuel or gas appliance will result in lower heat losses than an open fire.

*Heating requirements*

Calculated heat loss is normally based on an outside temperature of −1 °C. Table 5.1 gives the inside temperature recommended by BS 5449:1990.

Design flow temperatures for systems where the heat-emitting surfaces are exposed, e.g. radiators, should not exceed 82 °C to avoid the possibility of burns. The recommended return temperature is 71 °C for pumped schemes giving an 11 °C temperature drop. The F/R temperatures here are quoted from BS EN 12828 but in the case of condensing boilers these temperatures would be lower. By circulating the water more quickly the temperature drop could be lowered, but the higher water velocity necessary could result in pipework noise. Where heating systems operate intermittently, i.e. where they shut down overnight, an allowance should be added to the calculated heat loss of between 10 and 15 per cent. This has the effect of oversizing the heat emitters and allows for a quicker heat-up. The higher percentage is normally used for very well-insulated buildings only, because of the low design heat requirements.

*Boiler output rating*

The boiler should be capable of meeting the sum total of the design heat requirements and the emission of the system pipework. An allowance of 2–3 kW is normally made for boilers supplying both hot water and heating.

*Thermal transmittance (conduction)*

This term relates to the flow of heat through the building fabric. The first step in heating design is to determine the heat loss. There are three main factors to consider here:

(a) The building fabric (some types resisting the rate of flow better than others).
(b) The superficial area of walls, ceilings and floors.
(c) The temperature difference between the inside and outside of the building. The greater the difference, the greater will be the heat flow.

These factors are all taken into account by the use of *U*-values, which have been devised by experiment and calculation and are published in tables covering most forms of construction. **A *U*-value may be defined as the rate of heat flow through a structure in watts/m$^2$ per hour, per 1 °C temperature difference** and is illustrated in Fig. 5.33(a). Window glass has a very high *U*-value of 5 W/m$^2$ because of its comparatively thin section. To carry this one stage further, assuming a 22 °C temperature rise is required as shown in Fig. 5.33(b), the heat loss will be very much greater, and is calculated as follows: Area of glass m$^2$ × temperature rise × *U*-value, thus:

$$1 \times 22 \times 5 = 110 \text{ W}$$

The result is a heat loss of 110 W through a single-glazed window.

It will be seen that this is a considerable loss of heat, and fortunately most other construction materials have a lower conductivity rate. In order to save energy, it should be noted that current Building Regulations specify that the maximum permitted *U*-value for domestic buildings is:

0.25 W/m$^2$ °C for roofs
0.45 W/m$^2$ °C for external walls and floors.

The reader should bear in mind that *U*-values relate to the heat transference factor of various combinations of building materials: the lower its numerical value, the less will be the heat loss.

The following text shows how heat loss is calculated (a) through the building fabric and (b) due to air changes. Figure 5.34 illustrates a typical room of which the heat loss is to be determined,

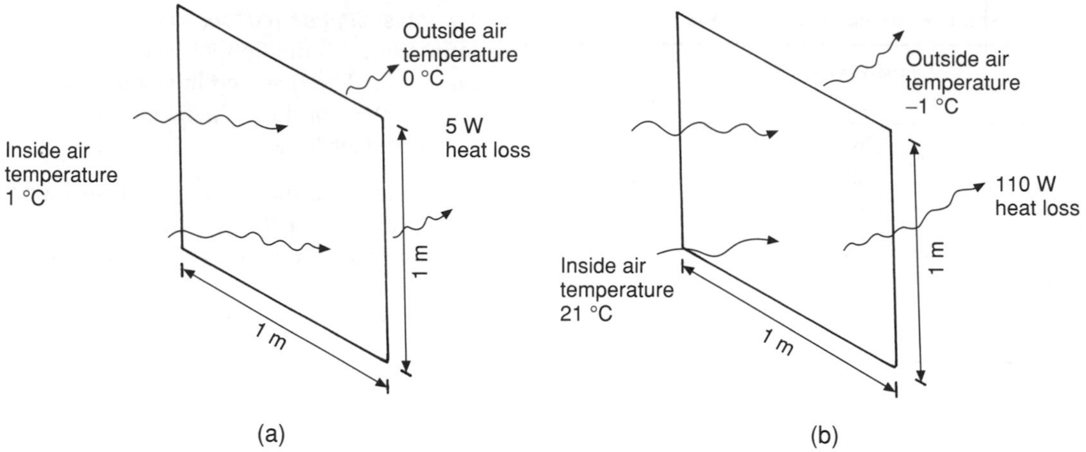


**Fig. 5.33** Heat losses through glass.

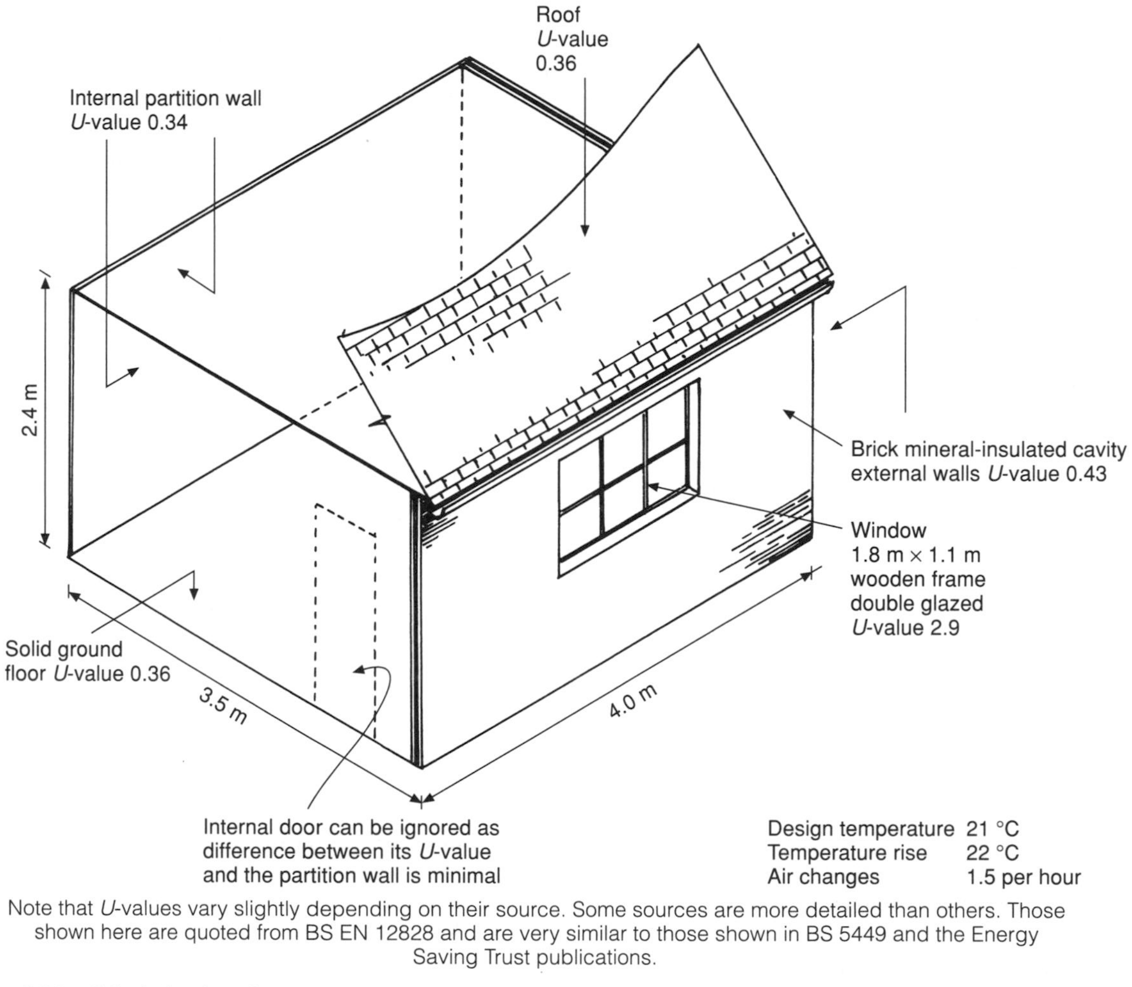


**Fig. 5.34** Calculating heat losses.

**Table 5.2** Tabulating heat requirements.

| *Component* | *Area ($m^2$)* | *U-value* | *Temperature rise* | *Watts* |
|---|---|---|---|---|
| Floor | 14.0 | 0.36 | 22 | 110.88 |
| External walls | 16.02 | 0.43 | 22 | 151.60 |
| Internal walls | 18.0 | 0.34 | 5 | 30.60 |
| Window | 1.98 | 2.9 | 22 | 126.32 |
| Roof | 14.0 | 0.36 | 22 | 110.88 |
| Heat loss through the building fabric (total) | | | | 530.28 |
| + Heat loss due to air change 1.5 watts per hour | | | | 365.90 |
| Total heat loss in room | | | | 896.18 |
| + 15% for intermittent firing | | | | 134.40 |
| Heat requirements (watts) | | | | 1030.60 |

and shows the relevant *U*-value of its components, the heat loss of these being tabulated in Table 5.2. To illustrate how this is done the floor is dealt with first thus. Measurements:

$$3.5 \times 4.0 = 14 \text{ m}^2 \text{ total floor area}$$

If the loss through 1 $m^2$ of floor area is 0.36, the loss through the whole floor will be

$$14.0 \times 0.36 = 5.04 \text{ W}$$

This will be the heat loss for 1 °C temperature difference. A temperature difference of 22 °C is, however, required:

$$5.04 \times 22 = 110.88 \text{ W}$$

This will be the heat loss per hour through the floor. The other components are calculated in the same way: Total area of two external walls:

$$7.5 \times 2.4 = 18 \text{ m}^2$$

Less window area:

$$\begin{aligned} 1.98 \text{ m}^2 \quad &= 16.02 \text{ m}^2 \\ 16.02 \times 0.43 \times 22 &= 151.60 \text{ W} \end{aligned}$$

The remaining components have been calculated and entered into Table 5.2 using the same method.

To ascertain the heat loss due to air change, the cubic contents of the room are calculated and multiplied by the temperature rise, the number of air changes and 0.33, this figure being a constant relating to the specific heat of air:

$$\begin{aligned} 2.4 \times 3.5 \times 4.0 &= 33.6 \text{ m}^3 \\ 33.6 \times 1.5 &= 50.4 \\ 50.4 \times 0.33 &= 16.63 \\ 16.63 \times 22 &= 365.9 \end{aligned}$$

Thus 365.9 W must be added to the fabric losses in the example, and by adding 15 per cent for intermittent firing, the total heat loss in watts can be seen.

## Domestic heating calculators

A typical calculator is shown in Fig. 5.35 and has been used by heating installers for many years to determine the design requirements of domestic properties without resorting to lengthy calculations. It should be noted that their use is generally confined to domestic buildings with standard methods of construction, using traditional materials. It has been found that their use may slightly oversize heat emission surfaces, but because TRVs are now mandatory this can be ignored for all practical purposes. Due to the improvement of insulation in buildings, to ensure accuracy in assessing heat losses the appropriate calculator for the type of building should be used. It is basically a series of rotating scales, which, when lined up, provide the information necessary to design the

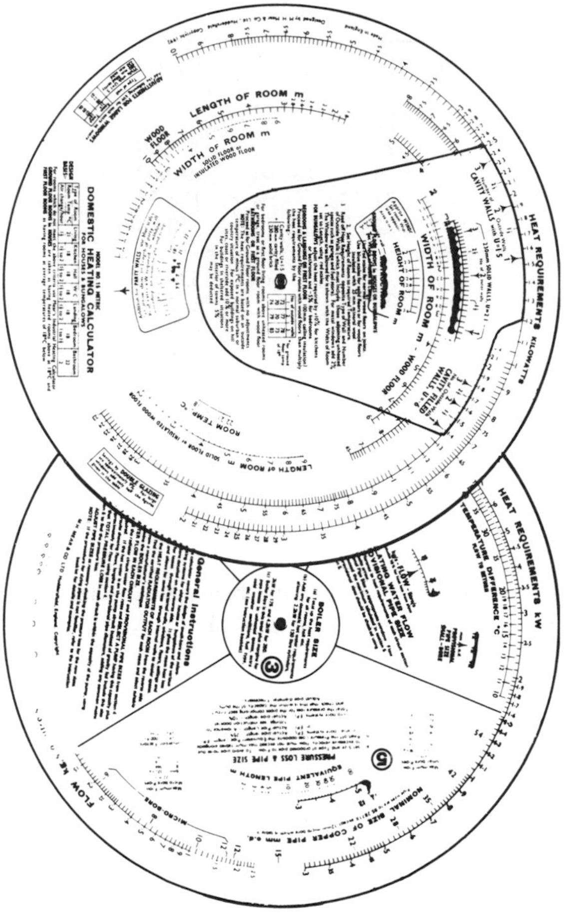


**Fig. 5.35** Illustrates the two sides of a domestic central heating calculator. (Courtesy of M.H. Mear & Company Limited)

complete system. The calculator used here is for general purposes. All the information necessary for its use is shown on the calculator itself, including the *U*-values used in this example. For buildings complying to the latest requirements of the Building Regulations, a calculator should be used which takes into account the permissible heat transference rate of energy-saving buildings. This is necessary, as the heat losses in such buildings will be less. All calculators take into account any necessary allowances, e.g. the percentage increase of heat emission where the heating system is used intermittently.

Figures 5.36(a) and (b) show the principal dimensions of a small two-bedroomed house, and

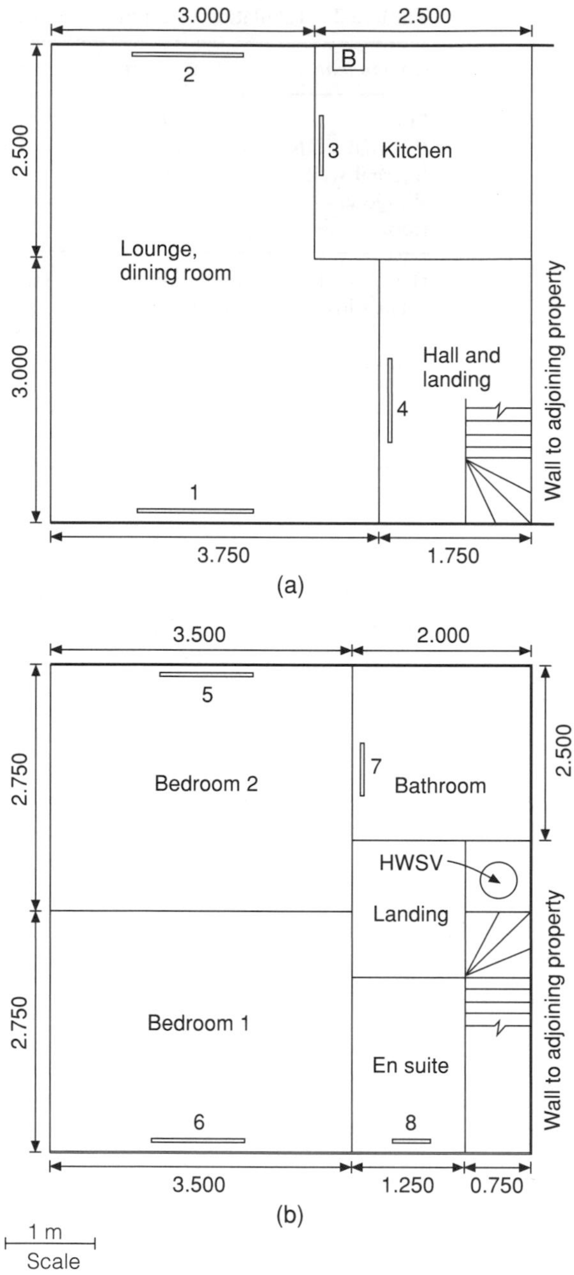


*Notes*
1 Solid ground floor
2 Wood floor on joists first floor
3 280 mm external cavity walls
4 All room heights 2.400

**Fig. 5.36** Plan of small two-bedroomed semi-detached house: (a) ground floor showing measurements of rooms, radiator and boiler positions; (b) first floor plan. Note that the dining and lounge areas have been calculated as two separate rooms.

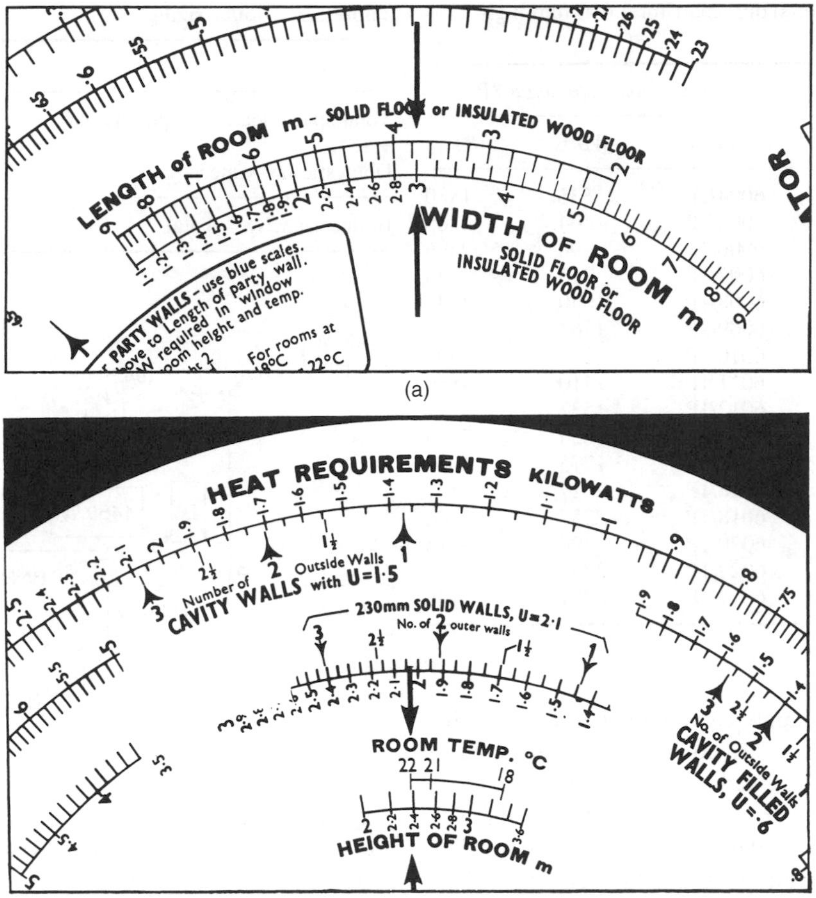


(a)

(b)

**Fig. 5.37** The arrows indicate the position of the calculator dials to assess the heat requirements of the lounge which are shown as 1.8 kW.

Figures 5.37(a) and (b) show how the calculator is used to determine the heat requirements of the lounge area. This is demonstrated by illustrations of the relevant scales using a step-by-step approach.

Set the width of the room by its length — in this case for solid floors (a separate scale is provided for suspended floors).

Set the height of the room to the required temperature rise, e.g. 22 °C.

Read off the heat requirements against the number of outside walls and its construction. In this case they will be kilowatts.

The calculator can be used to indicate the area of heat emitter surface required, but it is often more convenient to read this information directly from the manufacturer's information sheets, a section of which is shown in Table 5.3. All radiators are of the single panel convector type, 600 m high. A radiator schedule should be prepared at this stage as shown in Table 5.4. The boiler capacity may now be selected by totalling the space-heating requirements of each room and making an allowance of 2.5 kW for domestic hot water as shown in Table 5.5. Do not forget the increase of 15 per cent to allow for a quick heat-up period which is shown for systems operating intermittently.

The next step is to determine the pipe sizes that will provide sufficient hot water to supply the radiators and the main circuits. It can generally be assumed that for small-bore schemes 15 mm is of

**Table 5.3** Radiator sizing chart for a 600 mm high radiator.

| *Nominal length* | | *Single convector type 1P* | | |
|---|---|---|---|---|
| *mm* | *in* | *Order no.* | *Watts* | *BTUs* |
| 400 | 15.7 | 6004/1P | 536 | 1830 |
| 500 | 19.7 | 6005/1P | 664 | 2265 |
| 600 | 23.6 | 6006/1P | 790 | 2696 |
| 700 | 27.6 | 6007/1P | 915 | 3123 |
| 800 | 31.5 | 6008/1P | 1040 | 3548 |
| 900 | 35.4 | 6009/1P | 1164 | 3971 |
| 1000 | 39.4 | 6010/1P | 1287 | 4392 |
| 1100 | 43.3 | 6011/1P | 1410 | 4810 |
| 1200 | 47.2 | 6012/1P | 1532 | 5227 |
| 1300 | 51.2 | 6013/1P | 1654 | 5643 |
| 1400 | 55.1 | 6014/1P | 1775 | 6057 |
| 1600 | 63.0 | 6016/1P | 2017 | 6881 |
| 1800 | 70.9 | 6018/1P | 2257 | 7700 |
| 2000 | 78.7 | 6020/1P | 2496 | 8516 |
| 2200 | 86.6 | 6022/1P | 2734 | 9328 |
| 2400 | 94.5 | 6024/1P | 2791 | 10136 |

sufficient size to supply each individual radiator. If, for example, there is any doubt about a very long circuit serving a large radiator, the pipe size can be ascertained in a similar way to that of sizing a circuit. To determine the circuit loadings, a drawing illustrating the circuit layout for the building under consideration is given in Fig. 5.38.

Figure 5.39(a) shows how the calculator is used to determine the flow rates and pipe diameters. The method of determining the flow pressure drop

**Table 5.5** Boiler sizing.

| | *kW* |
|---|---|
| Total space heating requirements | 9.06 |
| + 15% | 1.36 |
| Domestic hot water | 2.5 |
| Boiler capacity = Total | 12.95 |

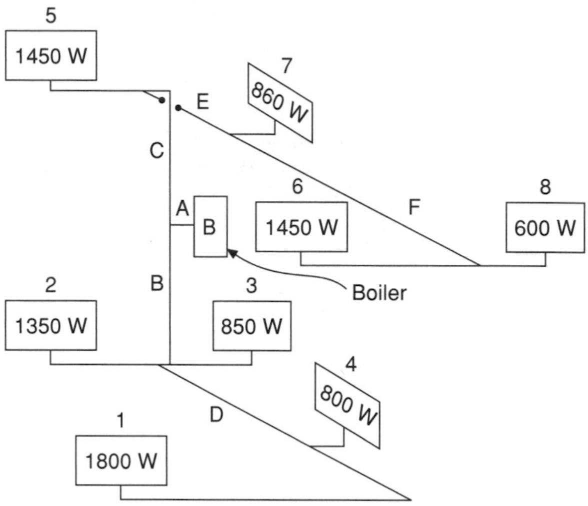


**Fig. 5.38** Diagrammatic layout of pipe runs showing how they are broken down into individual circuits feeding groups of radiators, so that their heating load and pipe diameters can be established. Each radiator is identified by its number on the radiator schedule. It will be seen that circuit A, although very short, is carrying the whole circuit load. Circuit B is feeding the ground floor; circuit C the first floor; E, F, D, are sub-circuits of C and B.

**Table 5.4** Radiator schedule.

| *Radiator no.* | *Location* | *Design temperature (°C)* | *Heat emission (kW)* | *Radiator size (mm)* |
|---|---|---|---|---|
| 1 | Lounge area | 21 | 1.70 | 600 × 1400 |
| 2 | Dining area | 21 | 1.35 | 600 × 1100 |
| 3 | Kitchen | 18 | 0.85 | 600 × 700 |
| 4 | Hall/landing | 18 | 0.80 | 600 × 700 |
| 5 | Bedroom 1 | 18 | 1.45 | 600 × 1200 |
| 6 | Bedroom 2 | 18 | 1.45 | 600 × 1200 |
| 7 | Bathroom | 22 | 0.86 | 600 × 700 |
| 8 | En suite | 22 | 0.60 | 600 × 500 |

*Note*: Although the design temperature is shown, the actual temperature rise is 1 °C in each case. This must be taken into account when using the calculator, e.g. for a design temperature of 21 °C the actual temperature rise will be 22 °C. The radiator sizes are read from Table 5.3, which is an extract from a manufacturer's list.

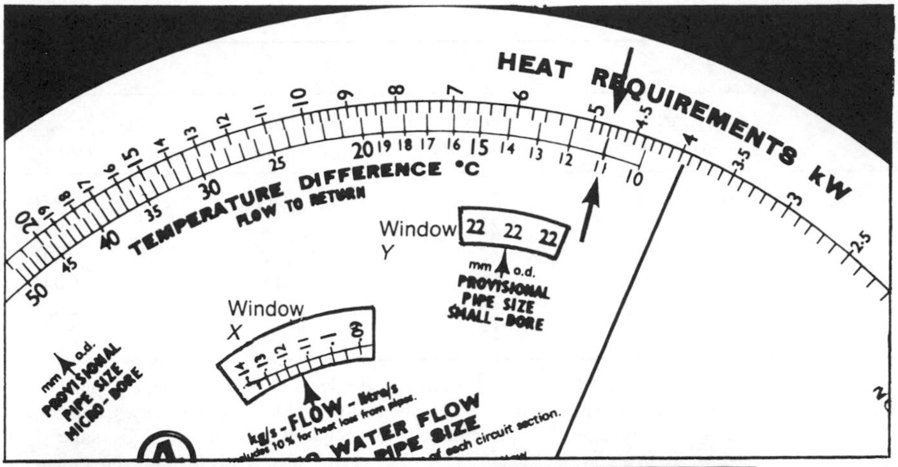


(a) Determining the water flow and provisional pipe size

Section B has a total heating load of 470 kW, and the scale is rotated until the load lines up with the temperature drop (11 °C) shown arrowed. The flow rate can then be read in window *X* as just under 0.12 kg/s (litres per second), and a provisional pipe size of 22 mm is shown in window *Y*.

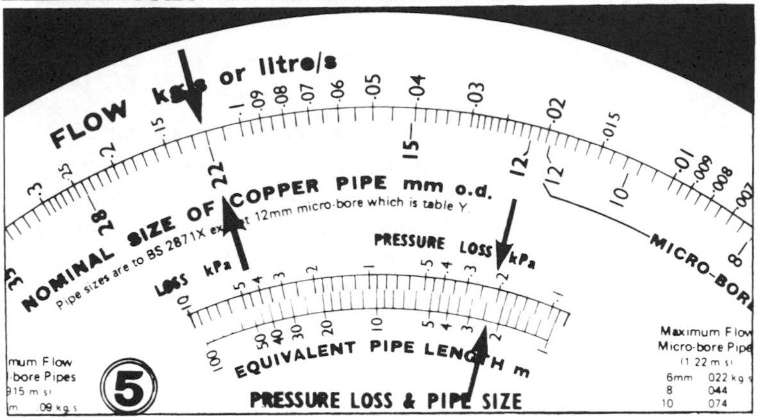


(b) Determining the pressure drop on circuit B

From Fig. 5.37 it has been established that the circuit length is 2.4 m, with a flow rate of 0.12 l/s. The pipe size (22 mm) is set against the water flow scale (top arrows). The pressure loss can now be read off through the window and will be seen to be approximately 2.5 mbar (bottom arrows).

**Fig. 5.39** Determining flow rates and pipe diameters.

for section B is shown in Fig. 5.39(b). The circuit lengths are taken from the plans (Figs 5.36(a) and (b)), the usual practice being to take the measurements from the centre of each radiator. Note also that the circuit length must be doubled for two-pipe schemes. All the foregoing details for each pipe section must be recorded as shown in Table 5.6.

The final step is to determine the 'index circuit'. This is the circuit which is the most difficult to feed by virtue of its heating load and pressure drop. It will be seen that the scheme has two main circuits serving the ground and first floors. Dealing with the ground floor first, it will be seen that it comprises pipe sections A, B, D. The total water flow for these sections is obtained from Table 5.6 and found to be 0.42 litres per second. Also from Table 5.6 it will be seen that the total pressure loss is 85 mbar, which divided by 10 gives 8.5 kN (kPa). The first floor circuit is dealt with in the same way, and from Table 5.6 it will be seen that the volume of water to be circulated is 0.48 litres per second against 95 mbar. This will be the index circuit upon which the pump loading is determined. Reference should be made to Fig. 5.40 which shows a typical graph for setting the pump to give the required outputs. The broken lines show the correlation of the water volume and the pressure. It will be seen that they intersect in the area where a pump setting of (2) is shown to be adequate.

**Table 5.6** Circuit sizing.

| *Col. 1* | *2* | *3* | *4* | *5* | *6* |
|---|---|---|---|---|---|
| *Pipe section* | *Heat emission per section k/W* | *Flow rate (l/s)* | *Pipe diameter (mm)* | *Circuit length (m)* | *Pressure drop (mb)* |
| A | 9.06 | 0.23 | 22 | 0.5 | 20.0 |
| B | 4.70 | 0.12 | 22 | 2.4 | 2.5 |
| C | 4.36 | 0.11 | 22 | 2.4 | 2.0 |
| D | 2.50 | 0.07 | 15 | 18 | 60.0 |
| E | 2.91 | 0.08 | 15 | 16 | 20.0 |
| F | 2.05 | 0.06 | 15 | 15 | 35.0 |

*Notes*: Column 1 lists the pipe section taken from Fig. 5.38. Column 2 is the total heating load carried by each section, e.g. pipe section B is carrying the load for all the ground floor, section D is carrying the load for the lounge and hall radiator only. Columns 3–6 list the values obtained from the calculator; for example, Fig. 5.39 shows calculations for section B. The calculator gives the maximum flow rates for 15 mm pipe as 0.13 litre and for 22 mm pipe as 0.29 litre at a velocity of 1 m/s. It can be seen above that these velocities are not exceeded and that the original pipe sizes selected are suitable.

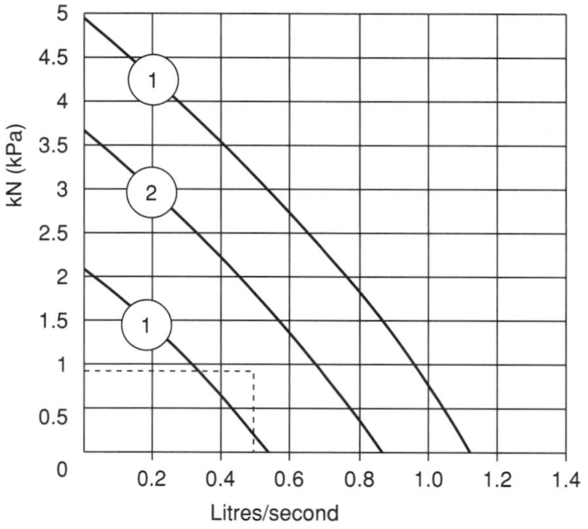


**Fig. 5.40** Pump duty selection graph. The pressure 0.85 kN and volume of water to be pumped, 0.48 litres per second, are identified by the horizontal and vertical broken lines. It will be seen that they fall into area 2 on the graph which corresponds to setting no. 2 on the pump.

### Under-floor central heating

This type of system employs, in effect, a heating pipe coil embedded or laid in a floor, which heats the space above it mainly by radiant heat. As radiant heat requires no medium through which to travel, floor coverings of whatever type used have little effect on the heat output. It is not entirely new: radiant heating using copper tubes has been fitted for many years but has never been used extensively for domestic work. The use of modern flexible plastic pipes, e.g. made from cross-linked polythene, and more sophisticated methods of control have made this system more viable, certainly for new-build work. It has many advantages, not the least being that the design temperature is much less than that for radiators, and at 45–60 °C is ideal for the use of condensing, energy-saving boilers. While most people do not object to the appearance of radiators, they do occupy wall space and might in some cases limit the interior design, and in very old historic properties radiators are hardly authentic.

Figure 5.41 illustrates the layout of the pipe coil for each room or space and Fig. 5.42 shows a section of a solid floor installation. The methods of pipe support vary if suspended or floating floors are heated (the term 'floating floor' relates to an installation fitted over an existing floor). The flow and return from each coil are taken to a distribution panel housed in a box and connected to the main manifolds from the boiler. This box must be situated in a convenient position in the building and also houses the on/off and lockshield valves to each circuit. As with most modern systems a fully pumped scheme is recommended. Individual control of each room or area can be achieved by

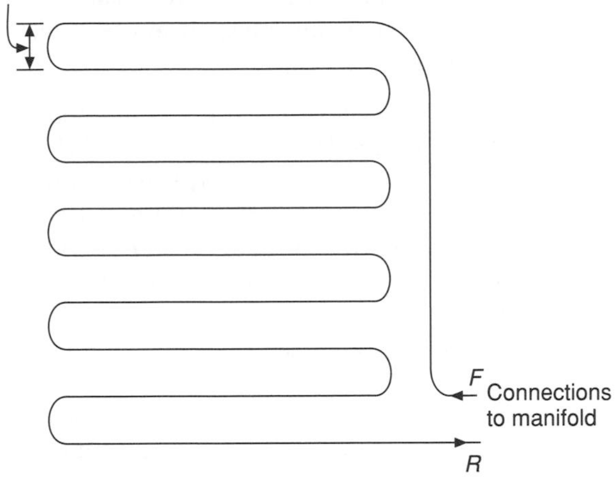


**Fig. 5.41** Pipe coil for under-floor heating.

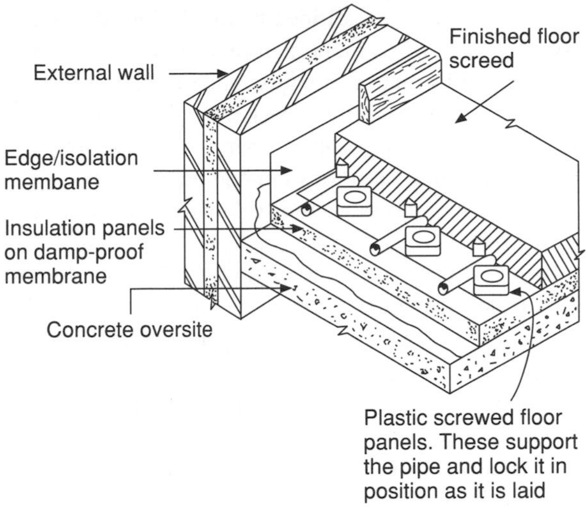


**Fig. 5.42** Section of under-floor heating in solid floors. Refer to manufacturers for details of suspended and floating floors.

using TRV with a remote sensor. For systems with zone control the use of two-port motorised valves controlled by a room thermostat for each circuit is another alternative.

Because the flow temperature is lower than that necessary for hot water and radiator schemes, where a single boiler is used for both space heating and hot water, the flow temperature to the under-floor circuits must be reduced. This is accomplished by a blending valve very similar to a three-way thermostatically controlled mixing valve, which allows water from the return manifold to mix with that from the boiler flow, thereby lowering the water temperature in the under-floor coils. The blending valve is normally manually operated to give the prescribed temperatures. Figure 5.43 shows this valve and its position in relation to the distribution panel.

*System design*

The following procedures are common to most methods of installation and the following text will

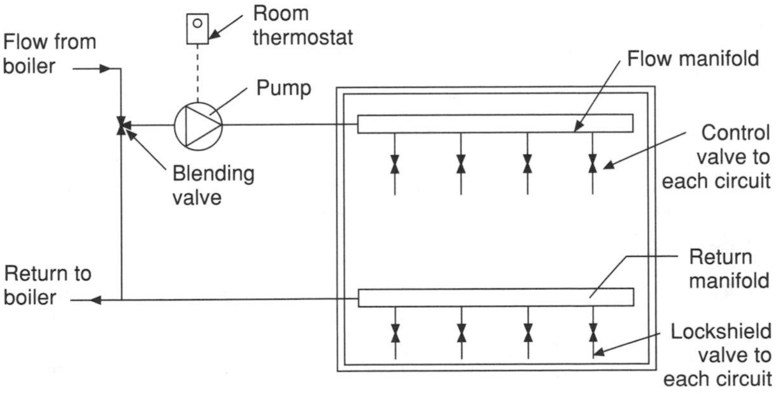


**Fig. 5.43** Detail of manifolds and control box.

give the reader a broad idea of the design of under-floor heating. Most manufacturers supply on request sufficient information for this type of heating to be carried out successfully. The heat emission will depend upon the pipe length per $m^2$ of floor area. There are some variations depending upon the type of floor construction — the following relates to solid floors.

It is first necessary to calculate the heat losses for each room in the usual way. If a calculator is used it should be of a type specifically for under-floor heating. Heat loss through the floor on which the coil is fixed can be ignored as it is laid on a well-insulated base. The next step is to divide the calculated heat losses by the floor area, which will give the heat requirements in $W/m^2$. To give an example, assume the heating requirements of a room with a floor area of 14 $m^2$ are 1.120 watts and the design temperature is 21 °C:

$$\therefore \frac{1{,}120}{14} = 80$$

Thus, each $m^2$ must produce 80 watts.

Table 5.7 shows output per $m^2$ at various flow temperatures and pipe centres. It will be seen that for pipes fitted at 100 mm centres, the higher the flow temperature, the greater will be the output per $m^2$ of floor area. From the table it will be seen that to give 80 $W/m^2$ there are two alternatives. Pipes fixed at 100 mm centres at 50 °C flow temperature gives 82 $W/m^2$. If the flow temperature is raised to 55 °C the pipe centres can be 200 mm which gives 90 $W/m^2$. Generally the temperature drop over the flow and return on the system is approximately 10 °C. The following points must always be taken into consideration:

(a) The maximum heat output from a solid-floor installation with a design temperature of 21 °C and a floor temperature of approximately 30 °C is 99 $W/m^2$. If this is insufficient it may be necessary to fix two separate coils or provide heat back-up using a radiator or other heat source.

(b) The maximum recommended pipe length in one coil is 100 metres. Pipes fixed at 100 mm centres will require 8.2 metres of pipe per $m^2$ of floor area, giving a total coverage of floor of approximately 11 metres. If the pipes are fixed at 200 mm centres, each $m^2$ will require 4.5 metres of pipe with a maximum floor coverage of 21 $m^2$.

### *Commissioning and testing under-floor heating systems*

The system should be pressure tested to 6 bar before the floor screed is laid. A constant pressure of 3 bar must be maintained throughout the period of screeding and curing. The optimum thickness of screeds is 50 mm from the top of the pipes as this gives the best performance. When the under-floor heating is initially turned on it must be allowed to warm up gently for a few days. Adjust the flow temperature at the manifold and each floor area to the design temperatures using a suitable

**Table 5.7** Design criteria for under-floor heating (solid floors).

| *Design* (room) *temperature* | *Pipe* centres *(mm)* | *Watts/m²* *Flow temperature at 45 °C* | *Flow temperature at 50 °C* | *Flow temperature at 55 °C* |
|---|---|---|---|---|
| 18 °C | 100 | 75 | 92 | 109 |
| | 200 | 62 | 75 | 89 |
| 21 °C | 100 | 65 | 82 | 99 |
| | 200 | 53 | 67 | 90 |
| 22 °C | 100 | 62 | 79 | 95 |
| | 200 | 51 | 64 | 78 |

Reproduced Courtesy of Polypipe Ltd.

thermometer in a similar way to when balancing radiators. It is recommended that reference is made to BS EN 1264 2001 Part 4 Installation of Floor Heating Systems and Components. It deals with floor screeds, testing pipework and initial heat-up period. It also recommends that plastic pipes used for under-floor heating should have an oxygen barrier layer to avoid oxygen permeating the pipe walls and causing corrosion in the metal parts of the system. Generally, because the floor emits radiant heat there is little or no resistance to the heat output into the room. It has been found, however, that carpets having a high tog number can have a limiting effect on heat emission.

## Commissioning and maintaining central heating systems

On completing the installation it should be washed out prior to refilling, venting the radiators, pump and all high points on the system. Check for leaks, fire the boiler and run until operating temperature is achieved — then examine the system for leaks again. Switch off the boiler — pump and drain the system while it is still hot. This will flush out any remaining debris or chemical deposits. Refill the system, and if specified add a suitable inhibitor to the proportions recommended by the manufacturer. A label stating the type of inhibitor used, and the date of the application, should be exhibited in a prominent position on the system. Check and adjust the float-operated valve in the feed and expansion cistern to the correct level. After a final examination for leaks any insulation necessary should be fitted. The boiler should be commissioned to the manufacturer's instructions, and the pump adjusted to the required setting to provide the designed flow rate. The radiators should be balanced to ensure an even distribution of hot water. To maintain the system in good working order the following checks should be made on an annual basis:

1. Service boiler as specified by the boiler manufacturer.
2. Examine the system for leaks, paying special attention to the radiator valve glands. Ensure the spindles of any stop- or gate valves are not seized and can be operated if necessary.
3. Examine all pipe fixings and insulation for damage and rectify if necessary.
4. Check the water level in the feed and expansion cistern (it can evaporate over a period of time). Also ensure the float-operated valve is in working order.
5. Vent the radiators and any high points on the system.
6. Fire the system and verify the correct operation of the controls. If the service is carried out at the beginning of the heating season, ensure the pump is operational.
7. Notify the client in writing that the service has been carried out, carefully noting any factors which may require attention but are not covered under servicing arrangements.
8. If possible, check if or when a suitable inhibitor was last added. If in doubt advise the client that the system should be properly cleaned and inhibitor added. This information must always be entered in the boiler log book if one is available.

### *Useful maintenance tools for central heating systems*

It is often necessary to replace radiator valves if they are beyond repair. This would normally necessitate draining and refilling, possibly admitting air to the system in the process, causing an air lock. The freezing equipment shown in Book 1 is one method of overcoming this problem, but if this is not available, the tapering plugs shown in Fig. 5.44 can be used temporarily to plug the feed and vent pipes connected to the primary part of the system.

The theory is that if no air can obtain access to the system, little or no water will escape if a radiator is disconnected. This method is not advocated where there is any possible chance that extensive damage may occur should something go wrong, but apart from the loss of a little water, which is easily mopped up, the use of these plugs can save time. This method of avoiding drain-downs should not be used for the domestic hot water system — its use has been known to result in a collapsed cylinder!

A useful adaptor shown in Fig. 5.45 can be used to avoid draining a radiator, thus preventing the

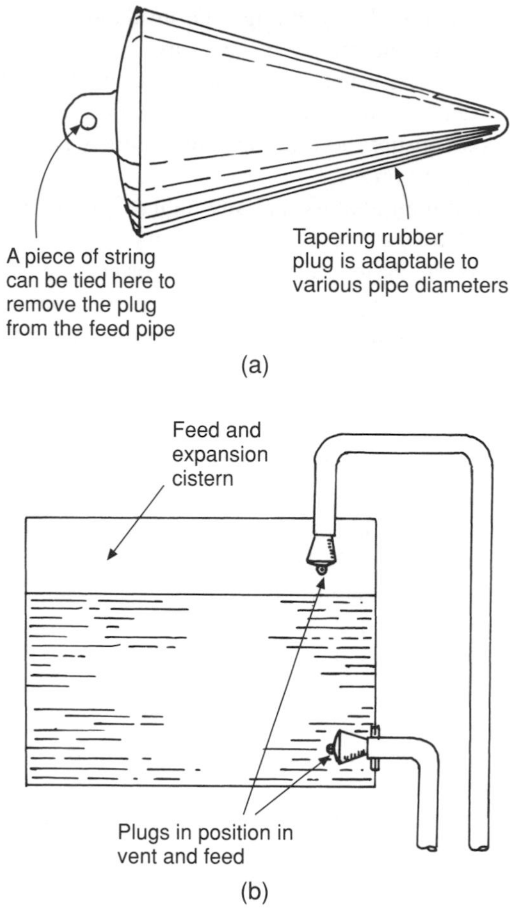
A piece of string can be tied here to remove the plug from the feed pipe
Tapering rubber plug is adaptable to various pipe diameters
(a)
Feed and expansion cistern
Plugs in position in vent and feed
(b)

**Fig. 5.44** Use of plugs to avoid draining heating systems for maintenance.

loss of water which may have been treated with an inhibitor. Using this adaptor enables the water to be pumped back into the feed and expansion system. To use, first close both radiator valves and remove the air valve, substituting it with the adaptor, which is connected to a cycle or foot pump. Open one of the radiator valves and commence pumping until a gurgling sound is heard, indicating the radiator is emptied. Reclose the valve and the radiator may be removed.

If a radiator has to be drained, Fig. 5.46 shows a radiator union incorporating a drain cock, enabling it to be drained easily if the valves are closed and the air valve is open.

## Noise in hot water supply and heating systems

Most noise in hot water and heating systems is caused by expansion and contraction due to temperature changes in pipes and components such as radiators and boilers. One of the most objectionable noises encountered with hot water systems occurs when a pipe has insufficient freedom of movement where it is notched into a joist or passes through a floor board. While it is not necessary to cut massive notches in the joists, or overlarge holes in flooring, pipework must have freedom of movement in such situations. The foregoing also applies to clips and brackets securing hot water pipes, especially if long straight runs are considered. Air trapped in a system can also give rise to noise, especially if it is present in circulating pipes. Hot water pipes should always be fitted in such a way that any air or gases in the system are automatically passed into the atmosphere via a vent or air valve at the highest point on the system. The remarks relating to pumps in the cold water supply relate also to those used with hot water if pump noise is encountered. It is rare, however, as the velocities are usually lower, and the pumps used for hot water supply and heating are seldom as powerful as those used for cold water supplies in multi-storey and industrial buildings.

Boiler noise is not uncommon and may be due to a restriction of the waterways, scale formation or the movement of rust deposits and debris in the base of the boiler when a violent circulation of water takes place. A 'singing' noise is sometimes caused by the explosion of small bubbles of steam on the interior surface of the boiler when heat is transferred quickly through the boiler walls causing the water in immediate contact with them to boil. This results in the rapid formation of minute bubbles of steam which burst on contact with the cooler water. Noise due to this cause may cease as the water temperature increases in excess of 80 °C. If the boiler is firing a combined hot water and heating scheme, an increase in pump velocity may improve the situation. Where the boiler is fired by gas, a check must be made to make sure the blue oxygen core of the flame is not impinging on the

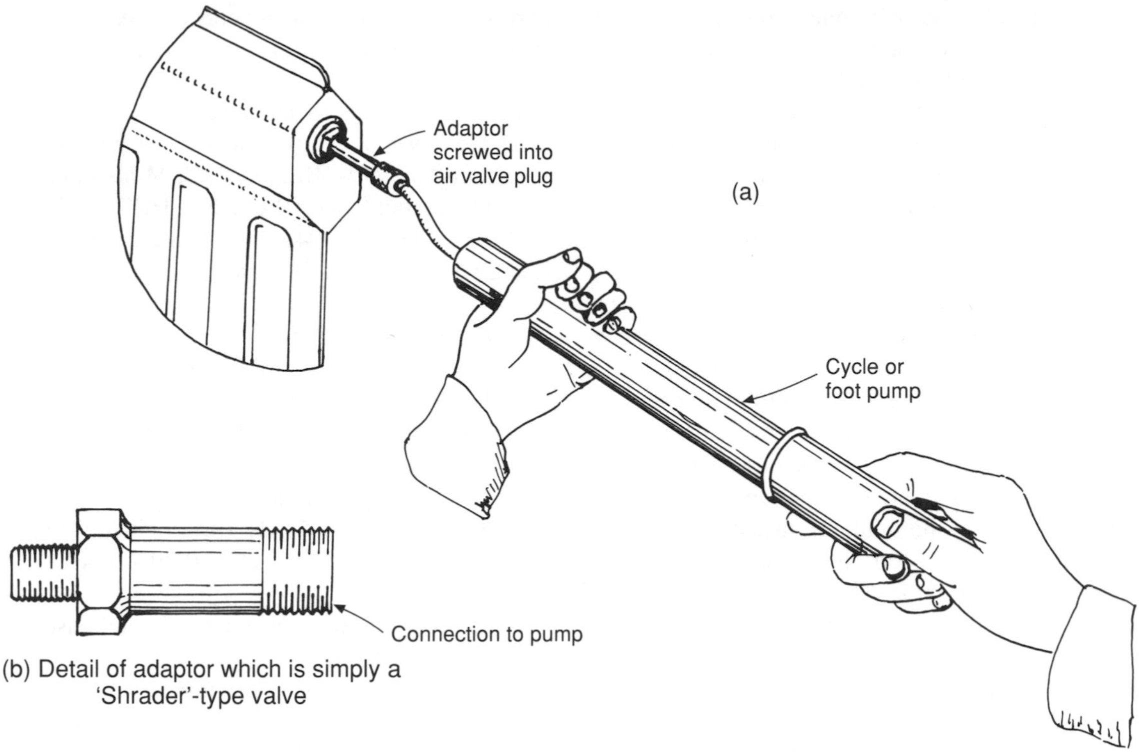


**Fig. 5.45** Emptying radiators by air pressure. By pumping air into the radiator the water is pressurised and forced back into the feed and expansion cistern. Note that if the cycle pump is inadequate, a foot pump would exert greater pressure.

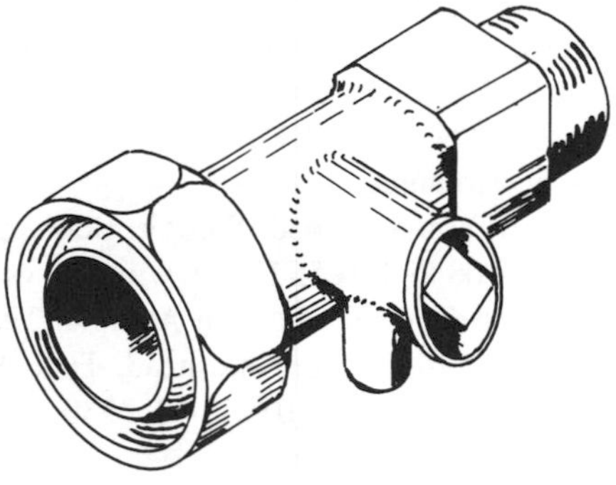

**Fig. 5.46** Radiator union with integral drain cock. The type shown has a miniature drain cock moulded into an extended union. Some have a screw which if removed serves the same purpose.

boiler as local overheating may be the cause of noise. The addition of a suitable chemical inhibitor usually solves the problem, but it cannot be guaranteed to do so.

### Corrosion in heating systems and its prevention

There were very few instances of corrosion problems in heating systems where the radiators were made of cast iron and the pipework of low-carbon steel. Both these materials are ferrous metals and any electrolytic corrosion between them would be minimal, and because of their physical thickness any chemical attack would have little effect. The use of systems with radiators made of thin sheet steel and copper pipe coupled with the use of acidic fluxes have brought about very serious problems, one of the most common being 'pin-holing' in the radiators. The effect produced is a series of very small holes on the external surface of the radiator with subsequent leakage of water. The adverse effects of air containing oxygen in the system have been mentioned in several instances in this chapter and modern systems should be designed to exclude air where possible. Electrolysis takes place between dissimilar metals, and in simple terms, minute local corrosion cells are formed giving off hydrogen, which, like air, collects at the top of the radiator making continual venting necessary. The worst effects of corrosion can be overcome, however, by the use of a chemical inhibitor, the word *inhibit* meaning 'to slow down or prevent' the process of chemical change that produces corrosion. When a new system is commissioned it must be thoroughly flushed through to remove any debris or residue in the system. This also applies if major alterations, e.g. a boiler change, are carried out. Failure to do this may result in nullifying any guarantee. Inhibitor manufacturers recommend a cleansing additive to assist the process. The system is heated for approximately 1 hour and then flushed out prior to adding the inhibitor, usually via the feed and expansion cistern, or in the case of sealed systems through the filler bottle. Always follow the instructions on the vessel containing the inhibitor regarding the quantity to be used, and in the case of cleansing additives always observe the rules on health and safety, as these additives often contain strong acids. For existing systems which may be severely scaled and corroded, special treatment may be necessary and advice should be sought on treatment procedures from a suitable builders' or plumber's merchant. See also Book 1, page 71.

### Further reading

BS 5449 Specification for forced circulation in hot water central heating systems for domestic premises.

BS 2767 Specification for valves and unions for hot water radiators.

BS 4814 Specification for expansion vessels using an internal diaphragm for sealed hot water heating systems.

BRE Digest 108 *U values*. BRE Bookshop, 151 Rosebery Avenue, Farringdon, London EC1R 4GB Tel. 020 7505 6622. www.brebookshop.com.

BS EN 1264 Floor Heating

Part 1. Systems and components

Part 2. Systems and components determining thermal output

Part 3. Systems and components dimensions

Part 4. Systems and components installation

*Heating design and recommendations*

M H Mear & Co. Ltd (Calculator Designers), Ramsden Mills, Britannia Rd, Milnsbridge, Huddersfield, West Yorkshire HD3 4QG Tel. 01484 485404. www.mhmear.com

Energy Saving Trust literature.

*Heating controls and systems*

The Association of Control Manufacturers (TACMA), Westminster Tower, 3 Albert Embankment, London SE1 7SL Tel. 020 7793 3008. www.heatingcontrols.org.uk

Energy Saving Trust, 11–12 Buckingham Gate, London SW1E 6LB Tel. 020 7931 8401. www.est.org.uk

Polyplumb (Underfloor Heating), Broomhouse Lane, Edlington, Doncaster DN12 1ES Tel. 01709 710000. www.inspiredheating.co.uk

Honeywell Ltd, Arlington Business Park, Bracknell, Berkshire RG12 1EB Tel. 01344 656000. www.content.honeywell.com

Danfoss Randall Ltd, Ampthill Road, Bedford MK42 9ER Tel. 01234 364621. www.danfoss-randall.co.uk

Horstmann Controls Ltd, A division of Horstmann Group Ltd, Roman Farm Road, Bristol BS4 1UP Tel. 0117 9788700. www.horstmann.co.uk

Invensys, Q4 Farnham Road, Slough, Berks SL1 4UH Tel. 0845 130 5522. www.invensys.com

Myson Heating Controls, Eastern Avenue, Team Valley Trading Estate, Gateshead, Tyne & Wear NE11 0PG Tel. 0191 491 7530. www.mysoncontrols.co.uk

Pegler Limited, St Catherines Avenue, Doncaster, South Yorkshire, DN4 8DF. www.pegler.co.uk

Siemens Building Technologies Ltd, Building Automation/HVAC Products, Hawthorne Road, Staines, Middlesex TW18 3AY Tel. 01784 461616. www.landisstaefa.co.uk

Sunvic Controls, Bellshill Rd, Uddingston, Glasgow G71 6NP Tel. 01698 812 944. www.sunvic.co.uk

*Radiators*

Caradon Stelrad Ltd, Stelrad House, Marriott Road, Mexborough, Rotherham, South Yorkshire S64 8BN Tel. 08708 498056. www.stelrad.com

Barlow Radiators, Barlow House, Spinning Jenny Lane, Leigh, Lancashire WN7 4PE Tel. 01942 261291.

Clyde Combustions Ltd, Unit 10, Lion Park Avenue, Chessington, Surrey KT9 1ST Tel. 020 8391 2020. www.clyde4heat.co.uk

Zehnder Ltd, B15 Armstrong Mall, Southwood Business Pk, Farnborough, Hants GU14 0NR Tel. 01252 515151. www.zehnder.co.uk

*Heating pumps*

Grundfos Pumps Ltd, Grovebury Rd, Leighton Buzzard LU7 8TL Tel. 01525 850000. www.grundfos.com

*Boilers*

The Little Blue Book of Boilers, Free Phone 0800 512 012

Worcester Bosch Ltd, Cotswold Way, Warndon, Worcester WR4 9SW Tel. 01905 754624. www.worcester-bosch.co.uk

### Self-testing questions

1. Name three types of heat emission devices for space heating.
2. State the differences between, and advantages and disadvantages of, one-pipe and two-pipe systems of heating.
3. List three advantages of using an air eliminator in a low-pressure heating system.
4. Explain the working principles of a gas condensing boiler.
5. Explain the need for balancing a system and state how it is carried out.
6. List the advantages of fully pumped systems.
7. Explain the main difference between mini-bore and small-bore heating systems.
8. Describe the method and equipment used for filling a sealed heating system directly from the main in compliance with the Water Regulations.
9. Make a list of the essential controls necessary for a domestic heating system and state their purpose.
10. Describe how $U$-values are used to ascertain heat losses through the fabric of a building.
11. List the factors that must be taken into account when determining a boiler output rating.
12. Explain the term boiler cycling, its effects and how to prevent it.
13. List any observations that you would take into account relating to heat loss in an existing building prior to installing a central heating system.
14. During a survey of an existing building prior to installing a central heating system, list the points you would make to the client relating to improvements in the structure to reduce heat loss.

# 6 Oil firing

After completing this chapter the reader should be able to:

1. State the recommendations and principal regulations to oil storage vessels.
2. Identify the components used in connection with oil storage and know their requirements.
3. Understand the principles of burning oil.
4. Know the functions and working principles of the components of an atomising burner.
5. Understand testing, commissioning and service procedures.

## Fuel oils

The concept of heating using oil as a fuel goes back for many years and was pioneered in America *c.* 1920 due mainly to the fact that a plentiful supply of cheaply produced oil was available. Oil-fired boilers were also available in this country during the 1930s, but were mainly confined to large commercial installations. It was not until the 1950s, when large quantities of oil began to be refined in this country, that it became a serious competitor to solid fuel and gas for domestic use. Very briefly, crude oil, as it is called when abstracted from the earth, is a mixture of gases, light and heavy oils and bituminous residues. Before it can be used commercially it must be refined or broken down into its constituent parts. The two basic methods used to refine oil are a chemical process called 'cracking', and 'distillation', where the lighter constituents are separated by heating the oil. After these have been extracted, a mixture of fuel and heavy (thick) bituminous oils remains which must be further refined to separate them for commercial use.

Oil is generally classified by its viscosity or fluidity, which is determined by the length of time in seconds it takes to measure a given quantity as it flows through a standard orifice, see Fig. 6.1. The main oils used for domestic and some industrial burners are C2 (28) second oil and class D (35) second oil. The heavier or thicker oils of 200, 960 and 3,500 seconds viscosity respectively, although much cheaper, are unsuitable for domestic use as the preheating equipment necessary to keep the oil fluid, and the special burners required, would make it uneconomic to both purchase and maintain. Even class D 35-second oil can thicken and become waxy at temperatures just below freezing point. The use of these heavy oils is therefore confined to industrial premises where its low cost makes it economic to use. Table 6.1 shows the main characteristics of light fuel oils.

Oil with a viscosity of (class D) 35 seconds has been and still is used for domestic work. It was cheaper than C2 oil but the difference in price is now minimal and most modern atomising/pressure jet burners for non-industrial use burn (C2) 28-second oil. Its advantages are its very low sulphur content, and its solidification temperature is well below that normally experienced in the United Kingdom.

## Oil-fired boilers

These are similar to gas boilers and when oil firing became popular for the domestic market *c.* 1955 gas

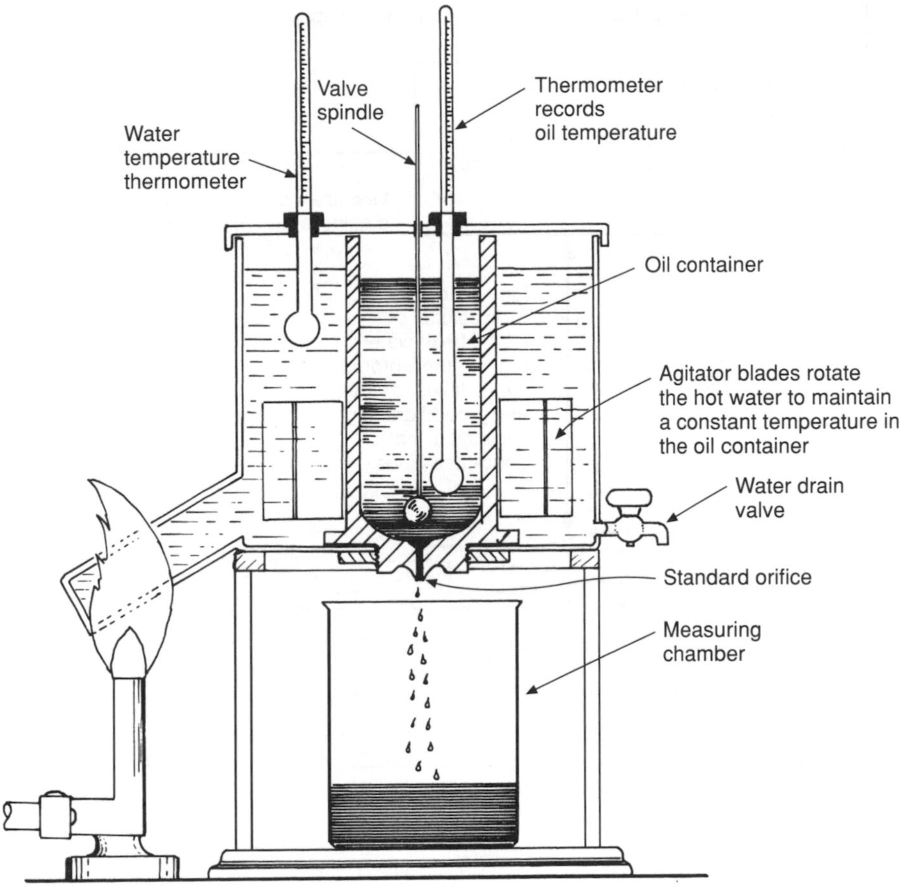


**Fig. 6.1** Diagrammatic viscosometer. This apparatus is used for testing the viscosity of light oils.

**Table 6.1** Approximate characteristics of light fuel oils.

| *Characteristic* | *Class* | |
|---|---|---|
| | *C2.28s* | *D.35s* |
| Specific gravity at 15 °C | 0.79 | 0.83 |
| Viscosity. Redwood No. 1 scale (oil at 37.8 °C) (seconds) | 28 | 35 |
| Calorific value (mJ per kg) | 46.6 | 45.5 |
| Flash point | 38 °C | 55 °C |
| Sediment | — | 0.01 |
| Solidification | −40 °C | 10 °C |
| Ash (% by mass) | — | 0.01 |
| Sulphur (% by mass) | 0.2 | 1.0 |
| Water content (% by mass) | — | 0.05 |

boilers of that period were easily converted to oil firing with few modifications. Like gas, oil is a relatively clean fuel and providing the boilers are properly maintained will give little trouble. As of 2007, oil boilers must be of the condensing type to comply with the Building Regulations. Most modern oil burners are of the atomising type, and although it is not possible to modulate the burner to meet varying heating loads, they are very efficient and compare favourable with gas boilers. The heat exchangers are usually constructed of cast iron or welded steel plates and most domestic oil boiler flues are, like gas, room sealed and fan assisted.

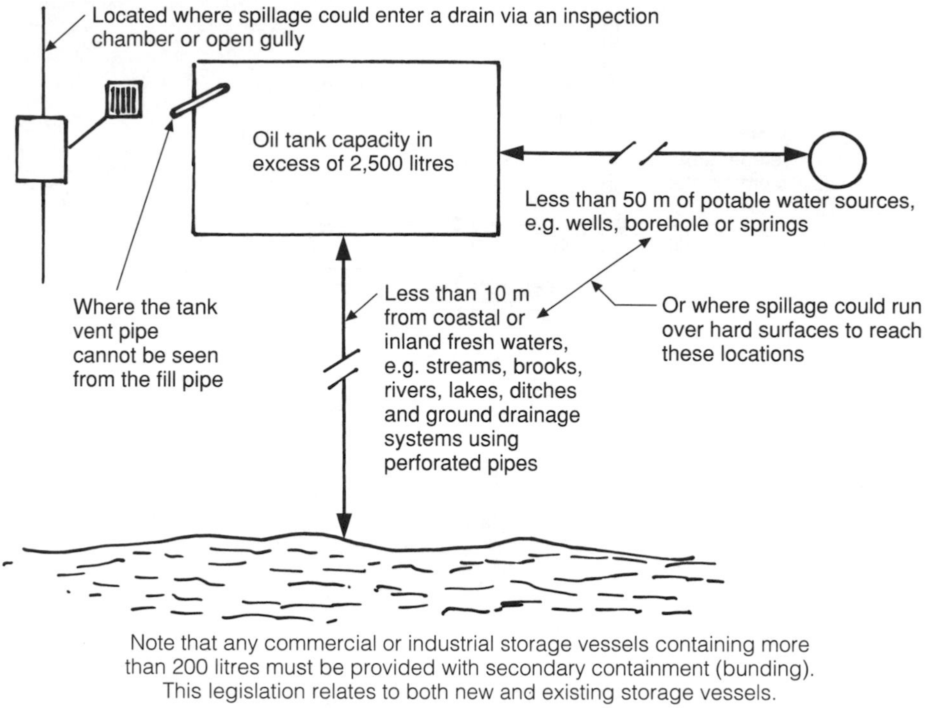
Located where spillage could enter a drain via an inspection chamber or open gully
Oil tank capacity in excess of 2,500 litres
Less than 50 m of potable water sources, e.g. wells, borehole or springs
Where the tank vent pipe cannot be seen from the fill pipe
Less than 10 m from coastal or inland fresh waters, e.g. streams, brooks, rivers, lakes, ditches and ground drainage systems using perforated pipes
Or where spillage could run over hard surfaces to reach these locations
Note that any commercial or industrial storage vessels containing more than 200 litres must be provided with secondary containment (bunding). This legislation relates to both new and existing storage vessels.

**Fig. 6.2** Control of oil pollution. Oil Storage (England and Wales) Building Regulations J5/6. Location of *domestic* oil tanks where spillage or leakage would cause environmental damage. Any storage vessel falling into the categories shown **must** be provided with secondary containment.

*Environmental pollution*

Oil storage in England and Wales must comply with the Control of Pollution Regulations which relate to storage requirements. Until recently, bunding of vessels for domestic use has not been obligatory, but new regulations make this necessary where spillage or leakage of oil could gain access to water courses, cause land pollution or a nuisance to neighbouring premises, see Fig. 6.2. Bunding or secondary containment of oil not only prevents environmental pollution, but also contains oil leaks and spills preventing oil from spreading and vaporizing. All bunding and secondary containment must be capable of containing the contents of the tank when full + 10 per cent. Plastic tanks having an integral bund are becoming increasingly used as they require little maintenance. Steel tanks are still used and can be made like those of plastic with an integral bund. It is mandatory for storage vessels of over 200 litres capacity in industrial and commercial premises for both new and existing installations. Where oil leakage from the tank cannot be seen, a sensing device is available which is fitted in the bunded area so that periodic checks can be made to ensure there are no leaks in the main storage vessel, see Fig. 6.3. In domestic dwellings storage without bunding is permissible in tanks up to 3,500 litres capacity unless the locations shown in Fig. 6.4 apply. Regulations relating to oil storage in public buildings are usually subject to local fire authority control.

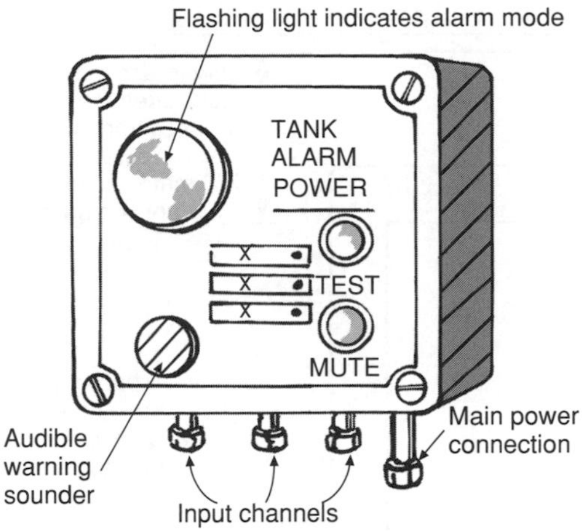


Flashing light indicates alarm mode

**Fig. 6.3** Tank alarm unit. This has three input channels and is designed to detect overfill, low oil levels or oil leaking into the secondary containment of the tank. Main unit shown but battery operation can be used on some models. The panels marked ‘x’ can be used to indicate which channel is operative if in alarm mode. This alarm should be housed in a water proof lockable box adjacent to the storage vessel.

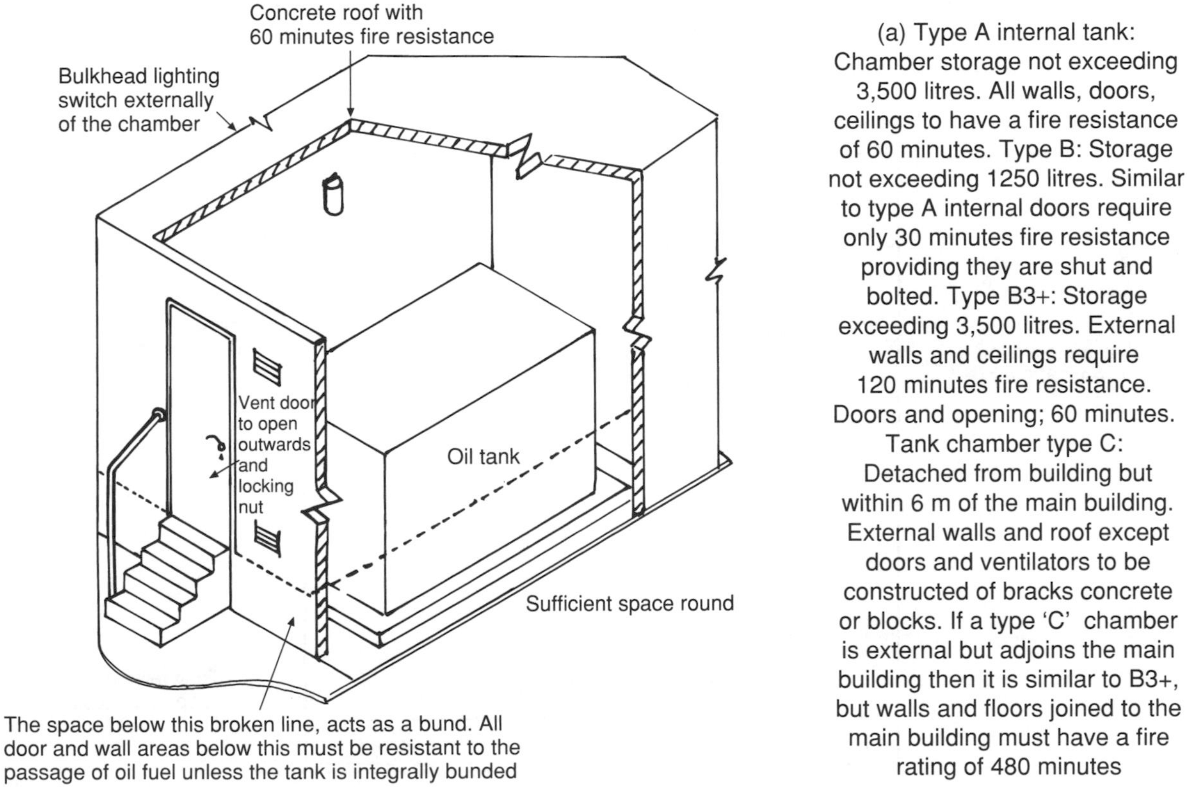

Concrete roof with
60 minutes fire resistance
Bulkhead lighting
switch externally
of the chamber
Vent door
to open
outwards
and
locking
nut
Oil tank
Sufficient space round
The space below this broken line, acts as a bund. All
door and wall areas below this must be resistant to the
passage of oil fuel unless the tank is integrally bunded
(a) Type A internal tank:
Chamber storage not exceeding
3,500 litres. All walls, doors,
ceilings to have a fire resistance
of 60 minutes. Type B: Storage
not exceeding 1250 litres. Similar
to type A internal doors require
only 30 minutes fire resistance
providing they are shut and
bolted. Type B3+: Storage
exceeding 3,500 litres. External
walls and ceilings require
120 minutes fire resistance.
Doors and opening; 60 minutes.
Tank chamber type C:
Detached from building but
within 6 m of the main building.
External walls and roof except
doors and ventilators to be
constructed of bracks concrete
or blocks. If a type ‘C’ chamber
is external but adjoins the main
building then it is similar to B3+,
but walls and floors joined to the
main building must have a fire
rating of 480 minutes


**Fig. 6.4** Oil storage requirements relating to fire risks.

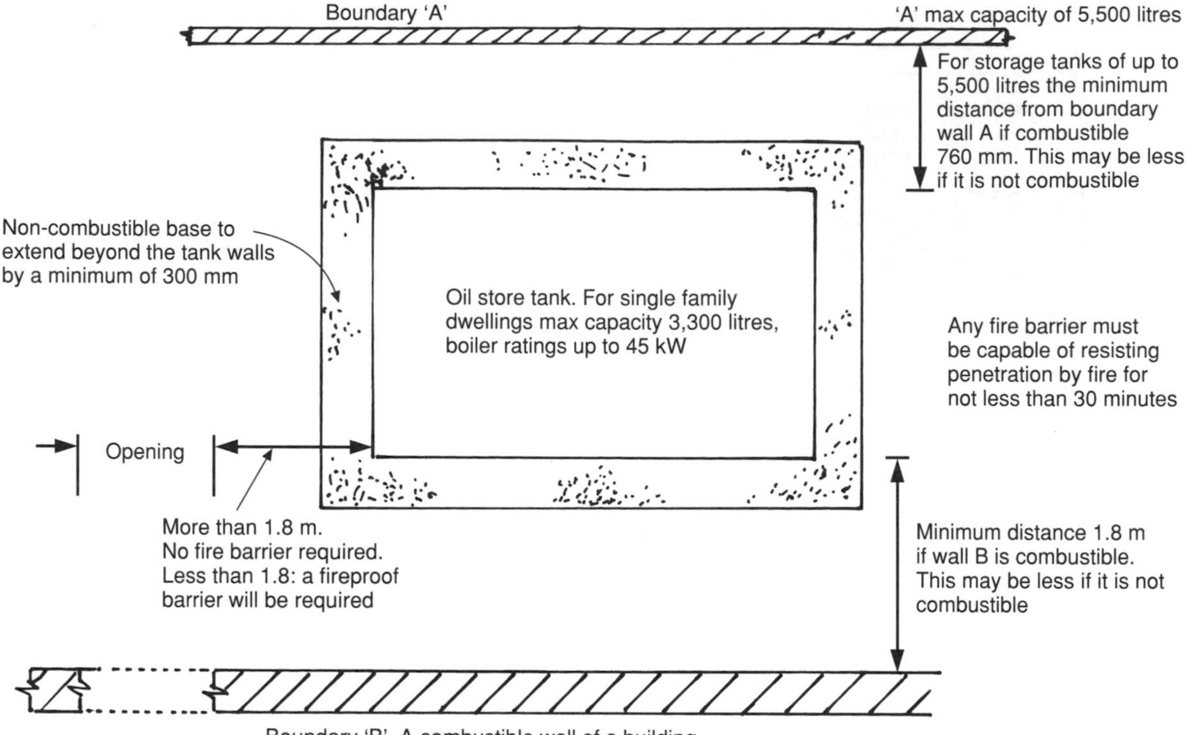
Boundary 'A'
'A' max capacity of 5,500 litres
For storage tanks of up to 5,500 litres the minimum distance from boundary wall A if combustible 760 mm. This may be less if it is not combustible
Non-combustible base to extend beyond the tank walls by a minimum of 300 mm
Oil store tank. For single family dwellings max capacity 3,300 litres, boiler ratings up to 45 kW
Any fire barrier must be capable of resisting penetration by fire for not less than 30 minutes
Opening
More than 1.8 m. No fire barrier required. Less than 1.8: a fireproof barrier will be required
Minimum distance 1.8 m if wall B is combustible. This may be less if it is not combustible
Boundary 'B'. A combustible wall of a building

(b) Plan illustrations of Building Regulations requirements for oil storage tanks adjacent to combustible walls fences, etc.

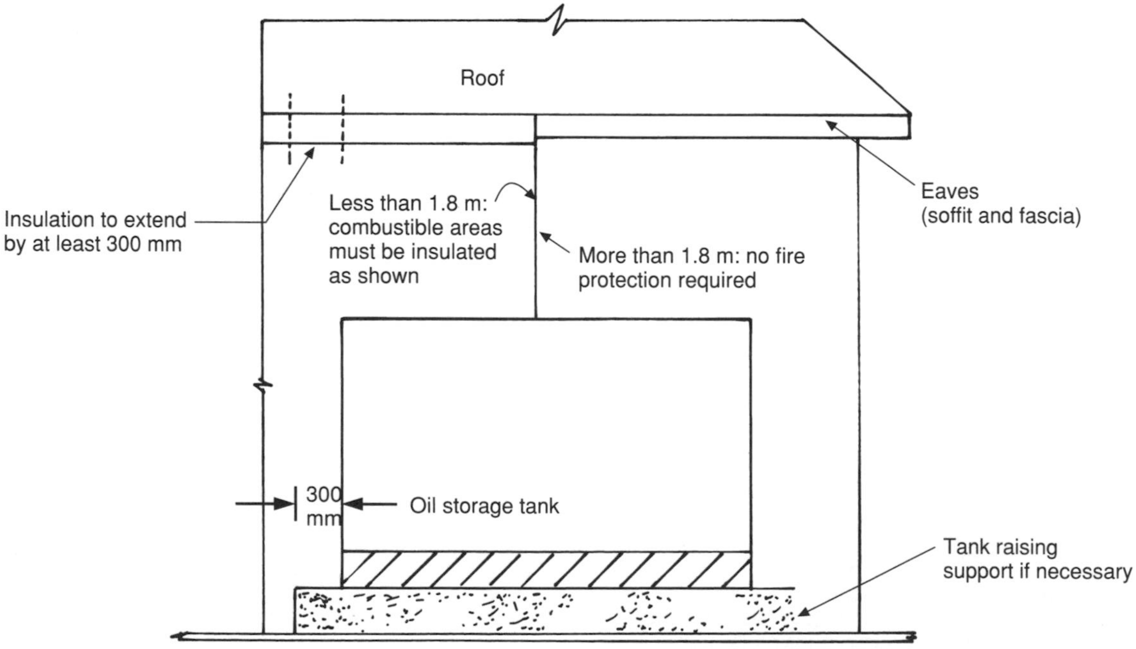
Roof
Eaves (soffit and fascia)
Insulation to extend by at least 300 mm
Less than 1.8 m: combustible areas must be insulated as shown
More than 1.8 m: no fire protection required
300 mm
Oil storage tank
Tank raising support if necessary

(c) Elevation showing the requirements of fire safety relating to tanks adjacent to roofs

**Fig. 6.4** (*cont'd*)

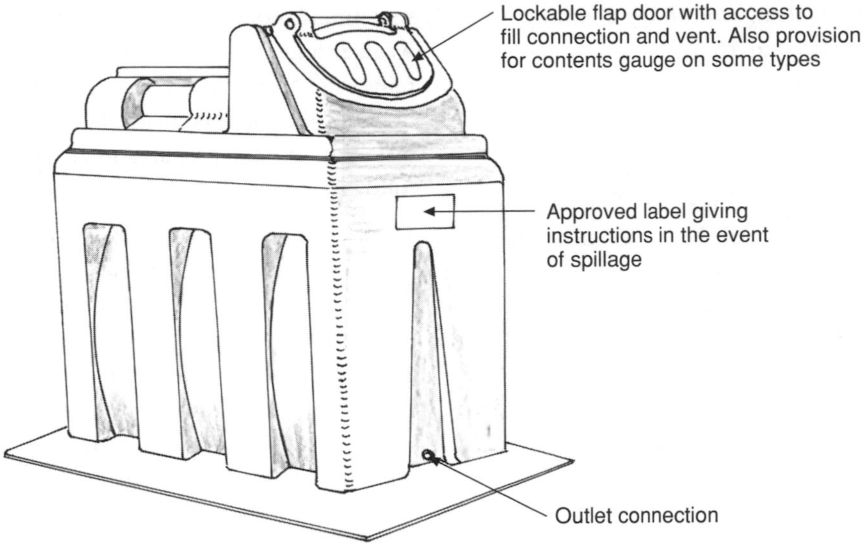
Lockable flap door with access to fill connection and vent. Also provision for contents gauge on some types
Approved label giving instructions in the event of spillage
Outlet connection

(a) Double skin integrally bunded plastic tank

These tanks are manufactured from a special grade of polythene which is unaffected by oil fuels. Capacities vary from approximately 1,200 litres (260 gals) to 5,000 litres (1,100 gals) depending on the manufacturer. Provision is made for all the ancillary components in the moulding. As with water cisterns made of plastic materials the base must be fully supported. A smooth finished concrete base or carefully levelled paving slabs meet the necessary requirements.

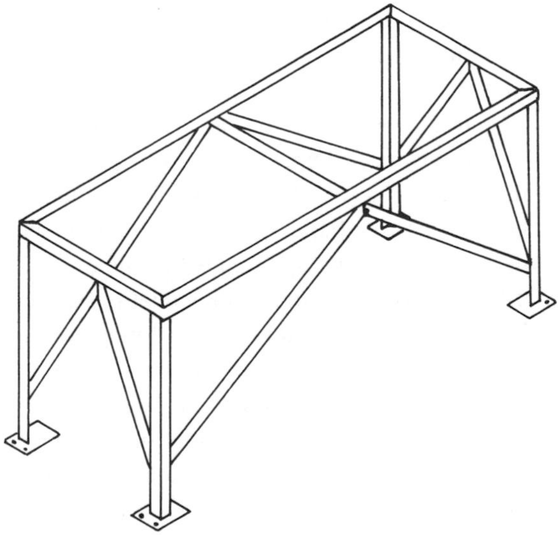

(b) Angle iron frame for supporting small oil tanks

Suitable for tanks made of both polythene and steel, bearing in mind a steel plate must be provided to give full support to tanks made of polythene. It is important that this frame is well painted with a suitable corrosion-resistant paint before the tank is fitted. The feet must be supported on a level concrete base.

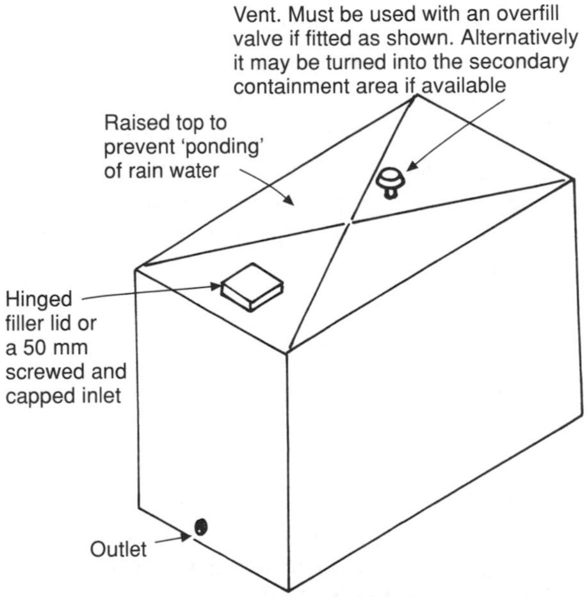
Vent. Must be used with an overfill valve if fitted as shown. Alternatively it may be turned into the secondary containment area if available
Raised top to prevent 'ponding' of rain water
Hinged filler lid or a 50 mm screwed and capped inlet
Outlet

(c) Steel tank fabricated with steel plates with welded joints. These tanks have been largely superseded by those made of plastic and except for specialist applications are rarely used

**Fig. 6.5** Fuel storage tanks and their support.

Reference to this is dealt with in the Building Regulations Parts 15 and 16.

*Fire risks*

Unlike petrol the oil used for heating will only ignite if it is vaporized or broken up into small droplets and mixed with air. However, it is a major risk in hot weather if it begins to vaporise. For this reason stringent safety regulations have to be observed when dealing with oil storage both inside and outside buildings. Figs. 6.5 (a) (b) and (c) illustrate the main requirements for fuel storage that comply with the Building Regulations Parts J 15-6.

*Fuel storage*

Bulk storage of fuel oil is not considered a fire risk itself, but it is necessary to protect contents of the storage vessel from a fire which may originate externally to it. If the vessel is housed in a room or chamber it should have at least a one-hour fire resistance and be provided with adequate high- and low-level ventilation. The access door must also have a one-hour fire resistance rating and should operate outwards from inside the chamber without a key. When oil is stored in a garage the foregoing applies, the chamber being inside the garage. Any lighting should be of the bulk head or wall pattern with switches located externally of the chamber.

*Fuel tanks*

Figure 6.5 illustrates typical types of fuel storage tanks and their method of support. The fact that polythene tanks are permissible for oil storage may be surprising as plumbers are usually taught that oil and its derivatives are harmful to plastic materials. Certain grades of polythene are, however, not adversely affected by fuel oil and can be safely used for this purpose. Fig. 6.6 illustrates a typical commercial oil storage tank fitted externally of a building showing all the main safety requirements and ancillary equipment. Tanks fitted like this are sometimes roofed in to protect them from the weather and avoids the necessity of pumping out rain water from the secondary containment area. Some of the main points to bear in mind when installing oil storage tanks in any situation are as follows:

1. They must have a slight slope towards the drain or sludge cock to enable any water or debris to be drawn off — the general recommendation is 6 mm in every 300 mm.
2. The capacity of the storage vessel should be as large as possible as most suppliers reduce the cost for large deliveries. For domestic properties capacities of not less than 1,250 litres are recommended. To take advantage of cheaper rates for larger oil deliveries, a tank capacity of not less than 2,700 litres is normally specified. Most oil suppliers these days provide a top-up service, and even if only a small quantity of oil is required its cost usually depends on the capacity of the tank.
3. Some consideration must be given to the distance of the tank from the road or hard standing where the delivery vehicle can park. Most tankers carry a length of flexible hose 32 mm diameter × 36 m long so that any tank within this measurement can be filled without the necessity of fixed extension pipes to the boundary of the building. The foregoing relates to domestic premises. Extended fill pipes are often necessary in large commercial or public buildings where the storage vessel may not be accessible to the tanker. The termination point depends on such variables as the type of oil used and the length and diameter of the pipe. The advice of the oil supplier should be sought in such cases.
4. Adequate space must be provided for the maintainance of storage vessels, especially those made of steel, which require periodic painting. Any loose scale or rust must be removed and treated with an inhibitor prior to repainting. Red oxide or bituminous paints are recommended.

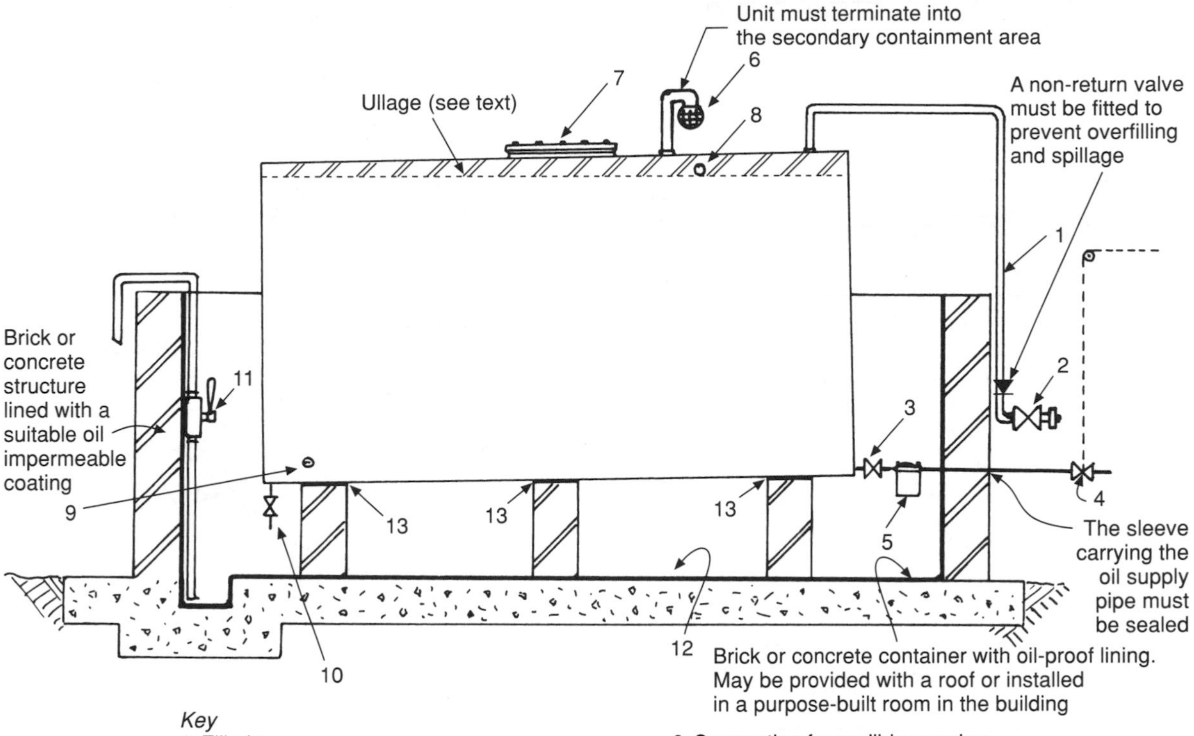
Unit must terminate into
the secondary containment area
6
7
8
Ullage (see text)
A non-return valve
must be fitted to
prevent overfilling
and spillage
1
2
3
4
5
Brick or
concrete
structure
lined with a
suitable oil
impermeable
coating
11
9
13
13
13
10
12
The sleeve
carrying the
oil supply
pipe must
be sealed
Brick or concrete container with oil-proof lining.
May be provided with a roof or installed
in a purpose-built room in the building

*Key*

1 Fill pipe
2 Gate valve NRV and cap assembly
3 Oil draw-off valve
4 Remote operated fire valve
5 Oil filter
6 Vent pipe
7 Manhole, normally only required for vessels of more than 4,500 litres capacity
8 Connection for audible warning
9 Remote contents gauge sensor
10 Sludge valve
11 Sump pump
12 Oil-proof lining
13 Damp-proof membrane

**Fig. 6.6** Typical oil storage requirements for a public or commercial building.

*Low-temperature effects on fuel oil*

In very cold weather fuel oil may be less viscous and become very 'waxy', i.e. it begins to solidify. Normally 28-second oil is not affected but heavier oils are, and it is necessary to maintain their fluidity in very cold weather by raising their temperature. Figure 6.7 shows some of the methods used.

### Oil storage tank accessories

Figure 6.6 illustrates the components associated with oil storage vessels. A main control valve, usually a gate valve, must be fitted as close as possible to the tank outlet to isolate the supply when necessary. A sludge or drain cock must also be provided to allow debris, rust particles and water to be periodically drained off, the main source of water being due to the formation of condensation.

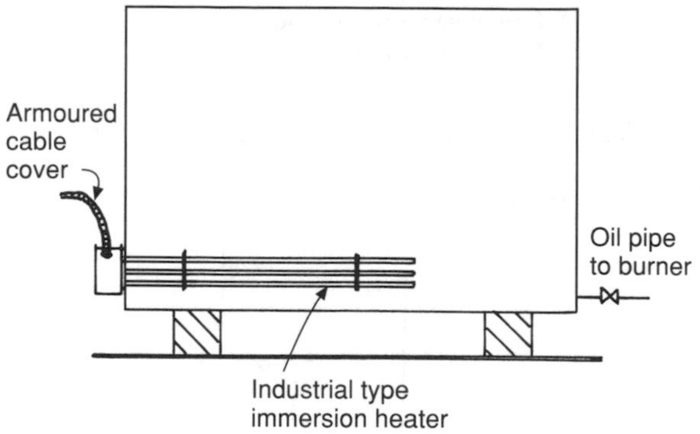


(a) During periods of low temperature

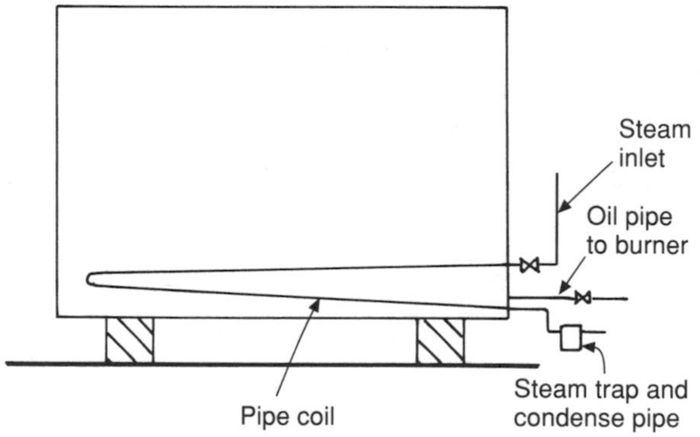


(b) Oils with a high viscosity require constant heating

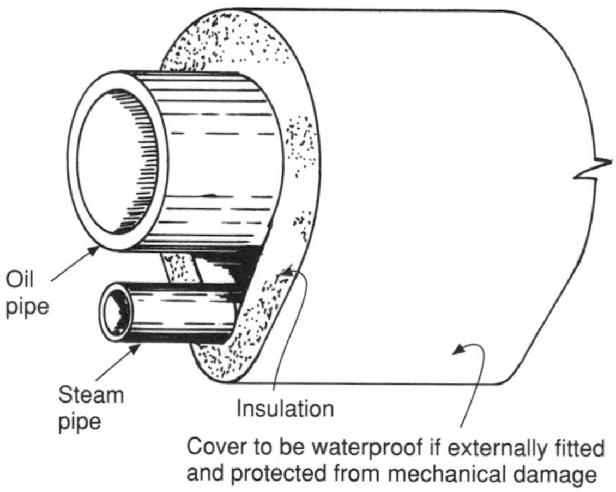


(c) Tracing an oil supply pipe where a steam supply is available. An alternative to steam is the use of an electrical tracing tape

**Fig. 6.7** Methods of keeping oil fluid. Thermostatic control of temperatures will depend on the viscosity of the oil.

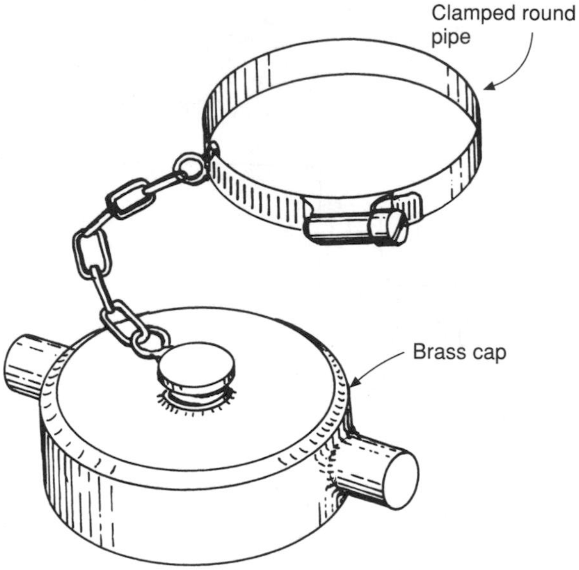


**Fig. 6.8** Cap for sealing the end of an oil fill pipe.

It is suggested that the outlet of this valve is plugged to avoid oil loss due to unauthorised operation of the valve. An open vent is necessary to maintain atmospheric conditions in the tank.

Easy access to the top of the storage vessel is provided, for example by suitable ladders or steps. Filling is achieved through a circular inlet 50 mm in diameter, oil being delivered through a trigger-operated nozzle similar to those used on petrol pumps. If it is not possible to fill the tank by direct means, use a fill pipe complete with valve and plug to avoid oil spillage when the filler hose is disconnected (see Fig. 6.8). Some types which may be used are provided with a lock to prevent tampering by unauthorised persons.

*Oil filters*

The amount of solids in the oil causing sediment is negligible, but after a period of time deposits on the base of the tank may be disturbed e.g. particles of rust in the case of steel tanks, possibly after an oil delivery. To avoid such deposits causing trouble in the burner, a main filter must be provided adjacent to the outlet valve. Figure 6.9 illustrates some of the types used. In the case of atomising

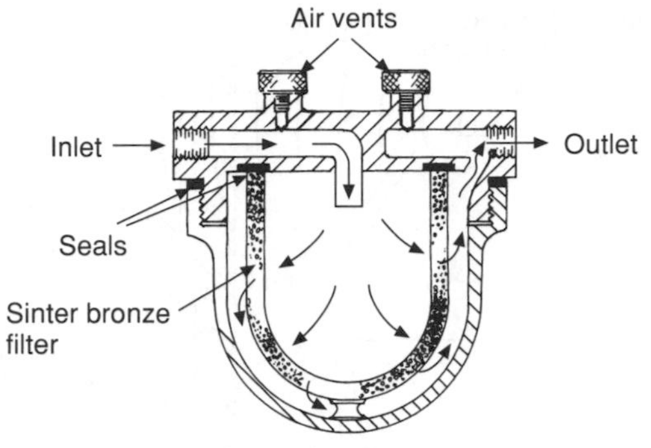


(a) Sinter bronze type

All oil filters are clearly marked inlet and outlet. They must be fitted correctly to avoid air locking.

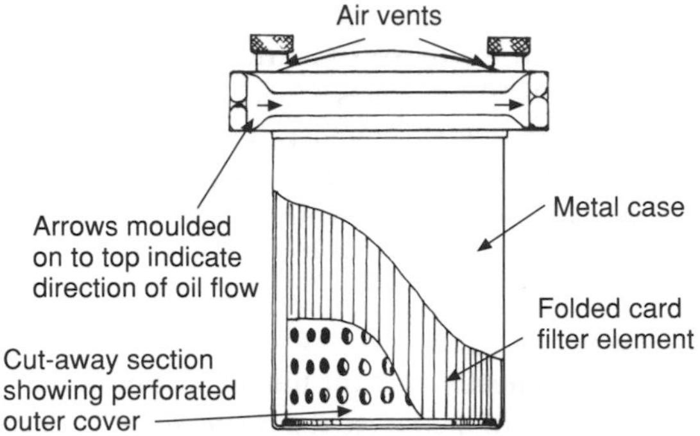


(b) Cardboard element type. Some types may be reused after cleansing; others must be renewed

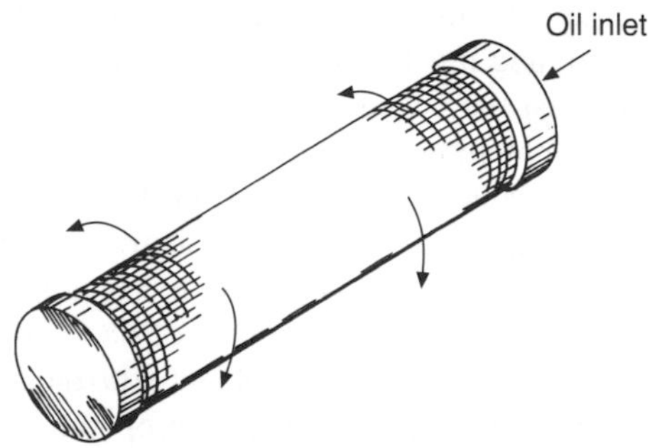


(c) Fine wire mesh strainer usually used with vaporising burners. The strainer only is shown as it is normally housed in the burner control box

**Fig. 6.9** Oil filters.

burners, additional filters are also component parts of the oil pump and nozzles.

Although oil fuels are 99.9 per cent free of solids when delivered, a filter must be used on the fuel pipe outlet to prevent any debris or sludge deposited in the base of the tank contaminating the fuel pipe to the burner. Filters must be cleaned at regular intervals, preferably with clean oil of the type in use. Under no circumstances should petrol be used.

### Tank overfill prevention

*Audible alarms*

An audible filler alarm is often necessary if the contents gauge is out of sight when filling the tank. An automatic switch permanently fitted in the tank is temporarily connected by a cable to the delivery vehicle. This triggers an audible warning when the tank is full, allowing the driver to stop the flow of oil immediately. These are basically a float- or lever-operated valve fitted in the top of the tank. Manufacturers of tanks approved by OFTEC (Oil Firing Technical Association) must supply one of these valves with the tank. They are often fitted with an audible warning alarm as well.

*Oil contents gauges*

The two most simple of these are the direct-reading float-operated gauge and sight gauges, the latter usually being employed for domestic work. Both these are illustrated in Fig. 6.10. A simple indirect system, primarily for domestic use and operated by radio signals, is shown in Fig. 6.11. The read-out does not show the contents in litres but simply displays the oil level on a scale of 1–10. A reading of 5 indicates the vessel is half full. The transmitter, via a probe fitted in the vessel, sends radio signals which are picked up by the receiver, which is fitted into a 13 amp socket outlet. In large commercial installations it is necessary to be able to read the tank contents in the control or boiler room itself, and direct-reading gauges would be unsuitable for this purpose. There are many types of indirect gauges: some operate electrically by means of sensors fitted in the tank, others are self-powered

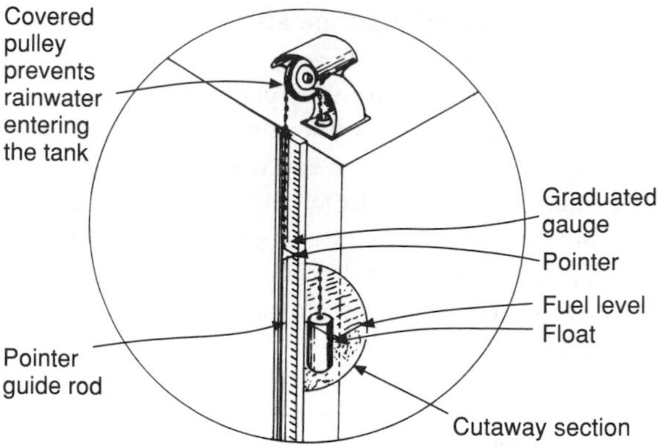


A float-operated circular gauge is an alternative but the depth of the tank must be specified when ordering.

(a) Direct-reading float gauge

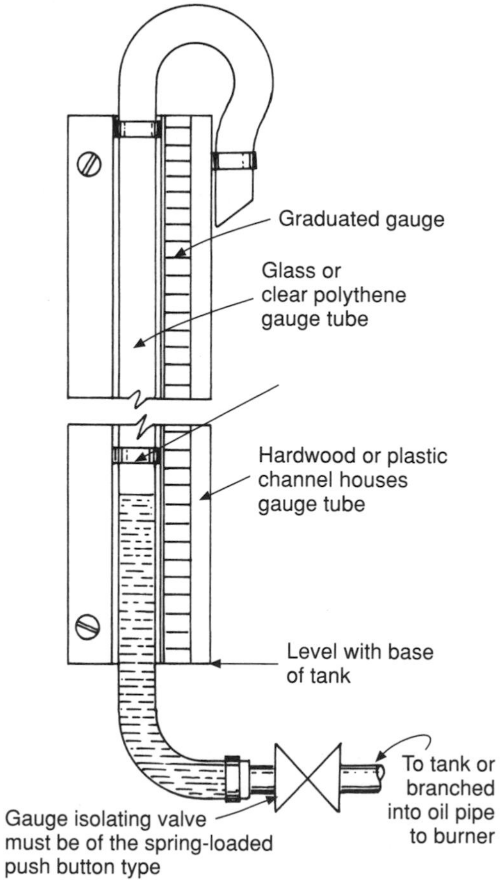


(b) Sight glass

**Fig. 6.10** Direct-acting fuel gauges.

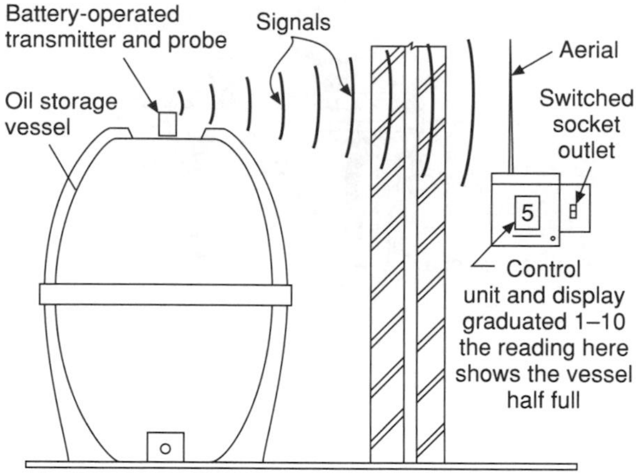


**Fig. 6.11** Radio-operated fuel contents gauge.

and operate on the pressure exerted by the oil in the tank. Figure 6.12 illustrates a typical example of this type. Due to the differences in tank dimensions and capacities, the manufacturer should be consulted as to the suitability of the gauge used for a specific purpose.

*Fire valves*

These automatically shut off the oil supply in the event of a fire in the boiler room and should comply with BS 5839 Part 1. Some of the main points relating to the foregoing are as follows (others are illustrated). Shut-off valves not being automatic can only be used in addition to, and not in place of, fire valves. Fire valves must be capable of withstanding the same pressure tests as the pipes to which they are fitted, and must be manufactured in such a manner that they cannot be rendered inoperative by over-tightening a sealing device or gland. They must be capable of a positive shut-down, even if debris is deposited on the valve seating. Heat-sensitive devices such as fusible links should normally operate at temperatures of 68 °C to 74 °C. In cases where the ambient (surrounding) air temperature exceeds 49 °C, the operative temperature of the heat sensor, e.g. fusible link, may be increased to 93 °C. A fire valve must be fitted preferably as near as possible to the entry of the building. If this is not practical it should be fitted at the point where the fuel pipe enters the boiler room. All fire valves must be capable of being manually

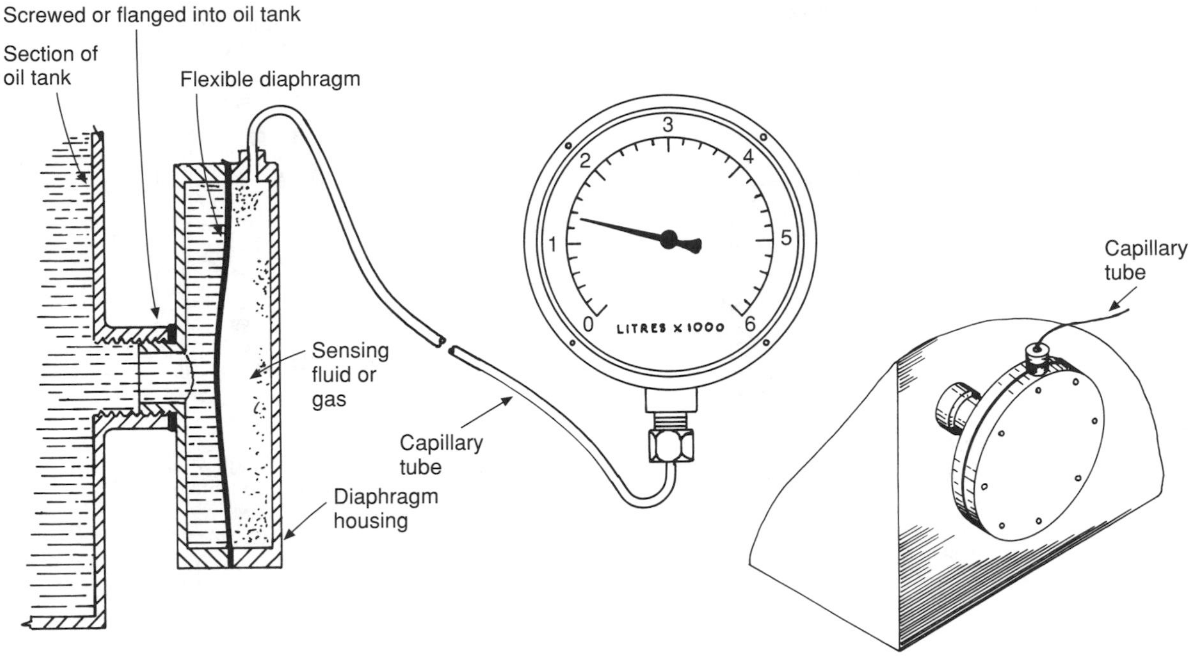


**Fig. 6.12** Pressure-operated remote reading fuel contents gauge. Pictorial illustration of the diaphragm housing fitted to the tank. Pressure of the oil in the tank is exerted on the flexible diaphragm pressurising a sensing fluid or gas which is converted into a dial gauge reading. The diaphragm housing is shown here in the base of the tank, but by using a purpose-made extension it can be completely immersed inside the tank via a manhole cover.

reset in the event of accidental operation. Details of these valves and their associated components are shown in Figs 6.13, 6.14 and 6.15. Figure 6.16 shows details of some components used with free fall fire valves.

*Oil supply pipes*

The most common and convenient material for this purpose is soft copper tube — 8 or 10 mm diameter is suitable for most domestic installations where comparatively short runs are necessary. The main points to remember when using soft copper tube are as follows: (a) Make sure it is laid in such a way that there are no high points resulting in air locks giving persistent trouble. (b) If it is buried it must be deep enough to prevent it becoming damaged by gardening activities — or alternatively suitably ducted. (c) Manipulative compression joints or brazed capillary fittings are the approved method of jointing. Soft soldered joints are no longer permissible. (d) Joints underground should be avoided. Where this is not possible, they should be tested prior to back-filling and, if possible, be accessible.

Pipes made with oil-resistant plastic are also available but they have their limitations. Because they are affected by sunlight they cannot be used above ground and must be terminated by copper or steel pipes. Twin-wall plastic pipes are made in larger sizes giving greater security to the environment. It is recommended that all underground fuel pipes are surrounded by pea shingle or sharp sand.

### Pipework systems for oil supply

The one-pipe system is usually fitted in domestic properties where the boiler is on the ground floor, and oil can be delivered to the burner by gravity. The base of the tank should be approximately 600–700 mm above the burner, as this normally provides sufficient pressure to overcome the frictional resistance of the pipeline. In exceptional circumstances where the oil store is a long way

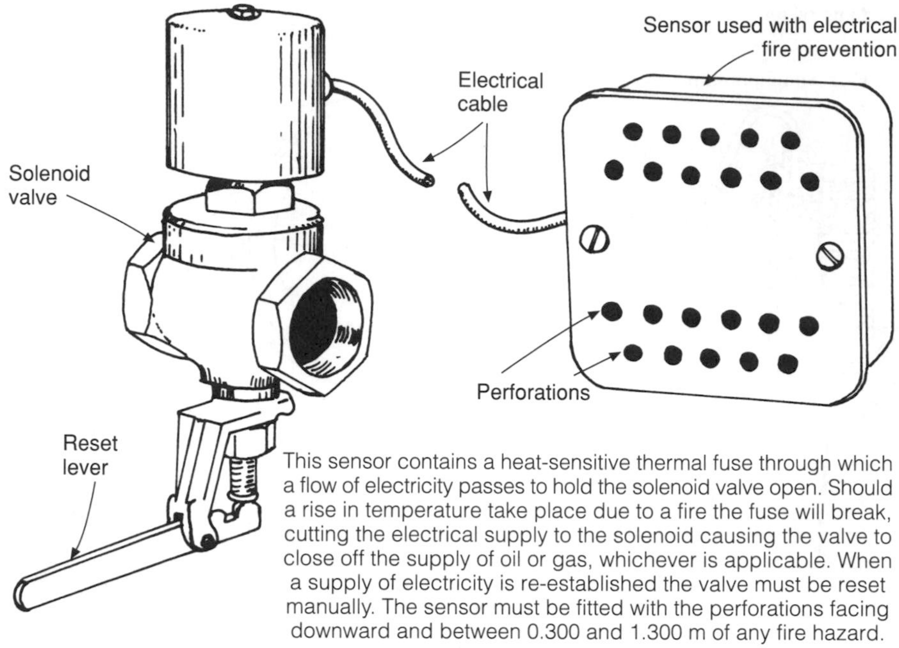

Sensor used with electrical fire prevention
Electrical cable
Solenoid valve
Perforations
Reset lever
This sensor contains a heat-sensitive thermal fuse through which a flow of electricity passes to hold the solenoid valve open. Should a rise in temperature take place due to a fire the fuse will break, cutting the electrical supply to the solenoid causing the valve to close off the supply of oil or gas, whichever is applicable. When a supply of electricity is re-established the valve must be reset manually. The sensor must be fitted with the perforations facing downward and between 0.300 and 1.300 m of any fire hazard.


**Fig. 6.13** Electrical system of fire safety control for oil or gas.

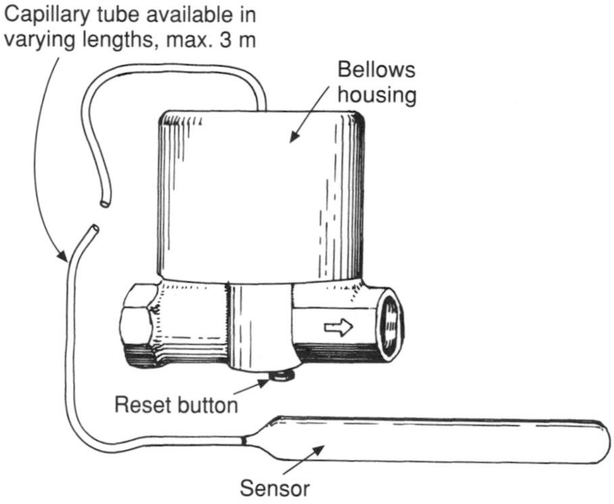


**Fig. 6.14** Heat-sensitive fluid fire valve. These are suitable for small, mainly domestic installations and work on the expansion of a heat-sensitive fluid which closes the valve at a predetermined temperature. If the valve has closed it may be reset manually by pressing the reset button. The sensor must be situated inside the boiler casing at least 300 mm from hot surfaces. The valve should preferably be fitted externally and suitably protected from mechanical damage.

from the burner, it may be necessary to raise the level of the tank or enlarge the diameter of the supply pipe. It should be noted that the oil supply to vaporising burners, having no integral oil pump, must always be gravity fed, see Fig. 6.17(a). In situations where the burner is at a higher level than the storage vessel, there are several alternatives. The following do not apply to vaporising burners.

(a) A two-pipe system may be employed as shown in Fig. 6.17(b).
(b) Some manufacturers produce specially adapted pumps which make a two-pipe system unnecessary.
(c) The use of a component marketed as the 'Tiger Loop System' shown in Fig. 6.18. Its use overcomes the problem previously mentioned, but also enables the release of air bubbles and minute particles of debris, adding to the pump's efficiency. As small quantities of flammable vapour may be released from the equipment when it is in use, it must always be fitted externally of the boiler room.

The main reason for these three alternatives is to avoid a vacuum in the oil line.

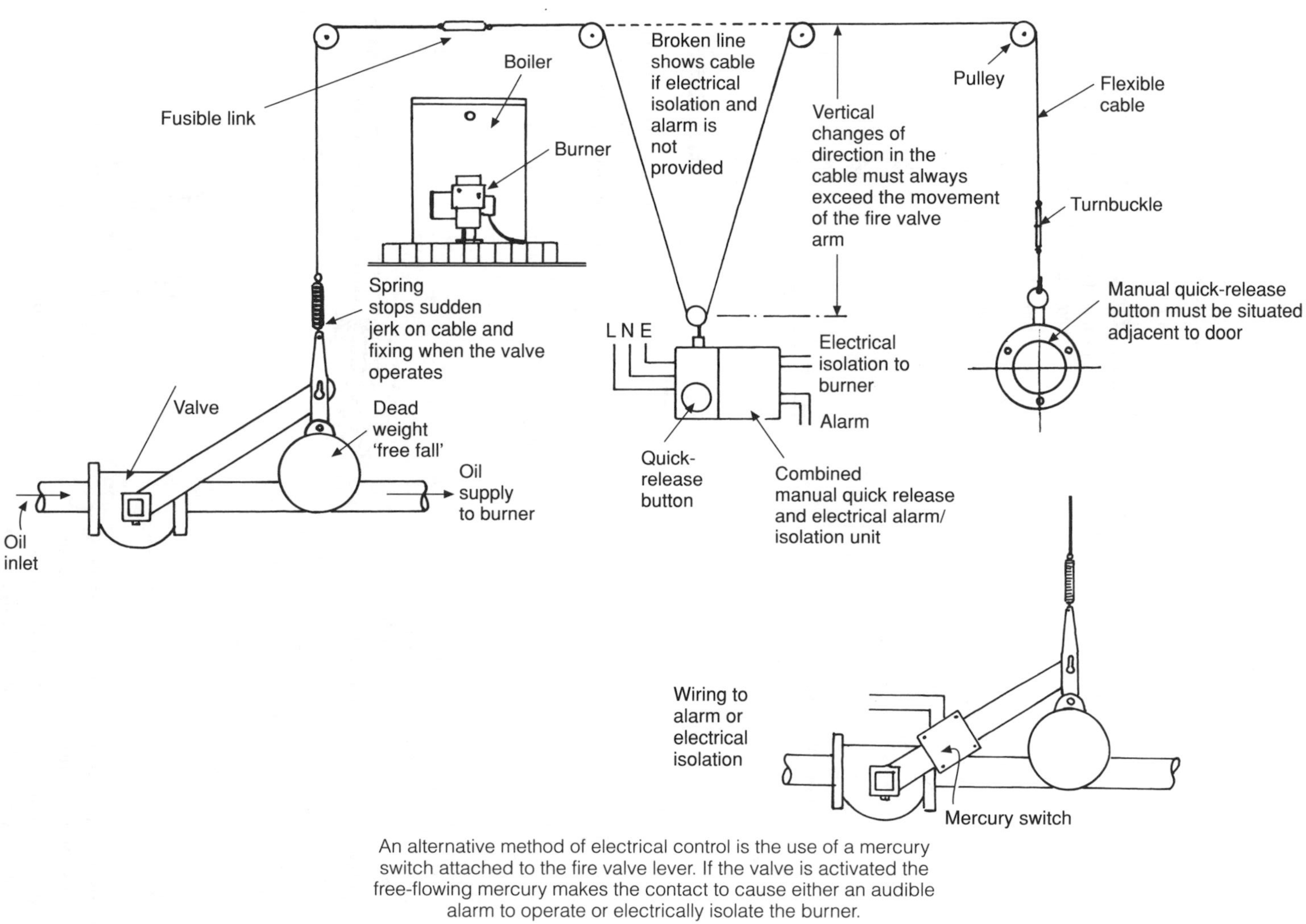
Fusible link
Boiler
Burner
Broken line shows cable if electrical isolation and alarm is not provided
Vertical changes of direction in the cable must always exceed the movement of the fire valve arm
Pulley
Flexible cable
Turnbuckle
Manual quick-release button must be situated adjacent to door
Spring stops sudden jerk on cable and fixing when the valve operates
L N E
Electrical isolation to burner
Alarm
Valve
Dead weight 'free fall'
Oil supply to burner
Oil inlet
Quick-release button
Combined manual quick release and electrical alarm/ isolation unit
Wiring to alarm or electrical isolation
Mercury switch
An alternative method of electrical control is the use of a mercury switch attached to the fire valve lever. If the valve is activated the free-flowing mercury makes the contact to cause either an audible alarm to operate or electrically isolate the burner.

**Fig. 6.15** Free-fall fire valve.

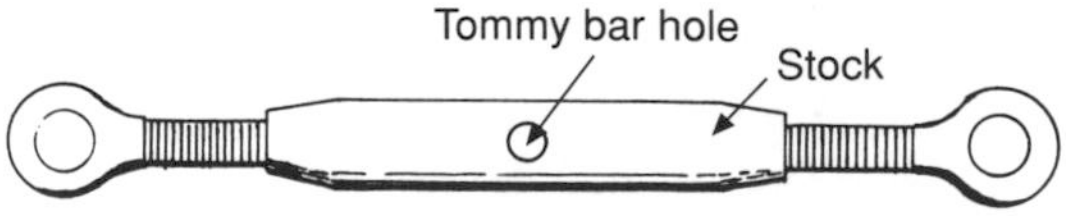


(a) Cable tensioner

By rotating the stock the right- and left-handed threaded ends permit the cable to be correctly tensioned.

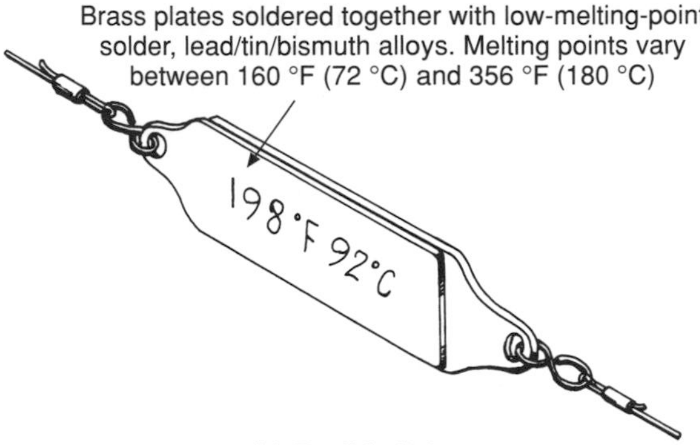


(b) Fusible link

The fire valve will close if the solder melts due to an excessive rise in temperature adjacent to the boiler.

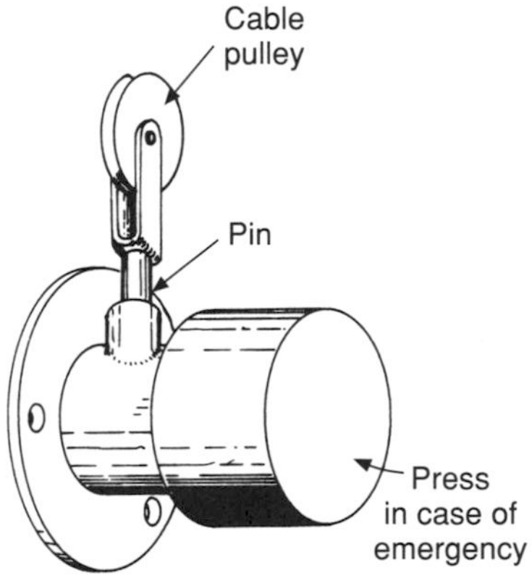


(c) Manual quick-release button

Should be fitted in such a way that it is not necessary to actually enter the boiler room to operate. Combination quick-release and electrical isolation types are available. To test the fire valve to ensure it closes this component may be activated and the pin will be released. It can be re-engaged by pressing the button and simultaneously replacing the pin in its housing.

**Fig. 6.16** Accessories used with free-fall fire valves.

Oil pumps are made to pump more oil than is necessary to supply the nozzle, the excess being recirculated in the pump itself. When the oil is supplied by gravity feed the pump is designed to undertake this work, but if it has to draw oil from

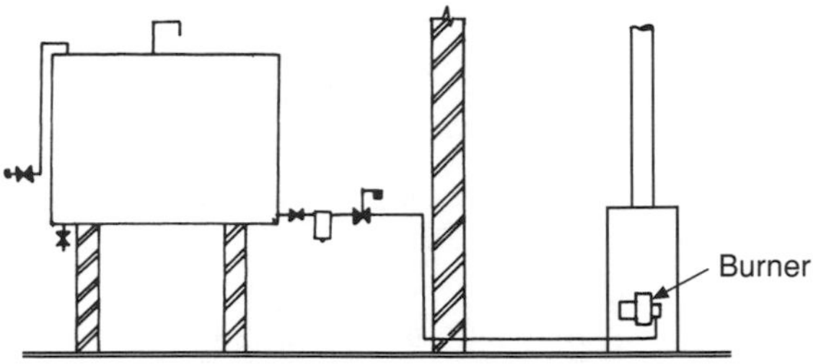


(a) Normal installation of oil tank where the oil flows to the burner by gravity

Both the height of the tank above the burner and the length of the supply pipe influences its diameter. The manufacturer's recommendations must always be complied with on this point.

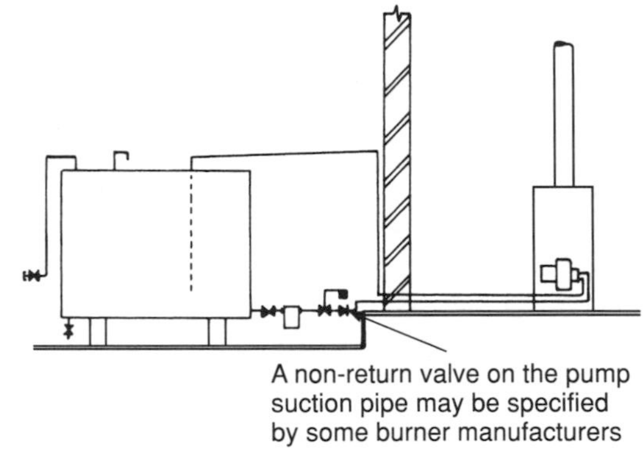

A non-return valve on the pump suction pipe may be specified by some burner manufacturers


(b) Two-pipe system of oil supply

**Fig. 6.17** Oil pipe supplies to burners.

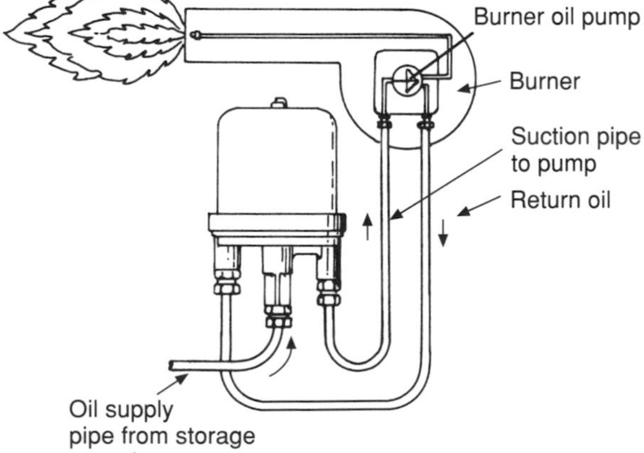


**Fig. 6.18** 'Tiger loop' valve for oil supply to burners.

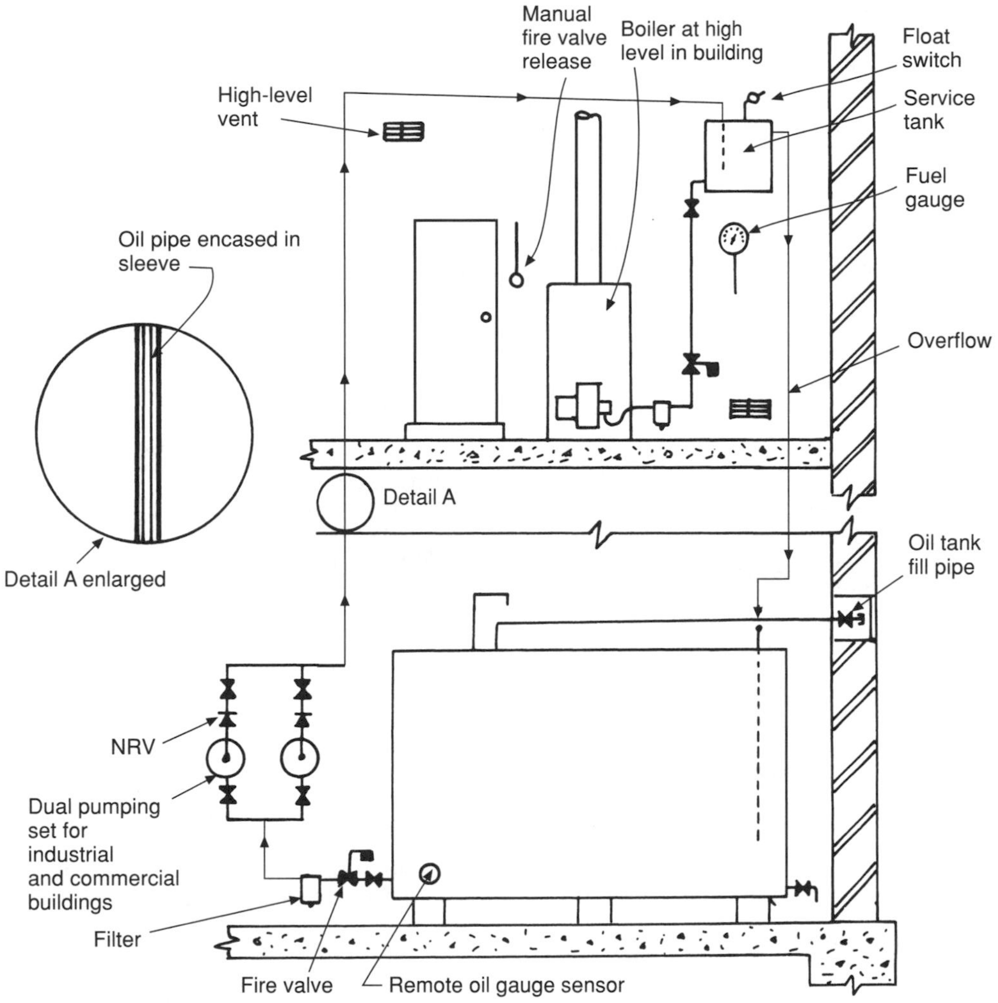


Note that no audible warning or tank overfill prevention valves are shown in this illustration.

**Fig. 6.19** Service tank system.

a tank at a lower level, the extra work will cause it abnormal wear. Figure 6.19 illustrates an oil supply system in a large commercial building where the boiler room is at high level, possibly on the roof of the structure. The burners are gravity fed from a service tank to which oil is independently pumped from the main storage tank.

### *Draught stabilisers*

Many types of burner are provided with a draught stabiliser, which, to obtain maximum efficiency, must be carefully regulated to ensure a constant draught in the flue (see Figs 6.20(a), (b) and (c)). It should be noted that stabilisers are also used with solid fuel boilers where there is excessive updraught on the flue. The type shown in Fig. 6.20(c) is designed to act as both a stabiliser and explosion door. This relieves pressure in the flue should any unburned gases ignite.

## Types of oil burner

### *Vaporising burners*

These boilers were once very popular for domestic work, but apart from a few exceptions they are no longer in use, mainly due to their limited heating

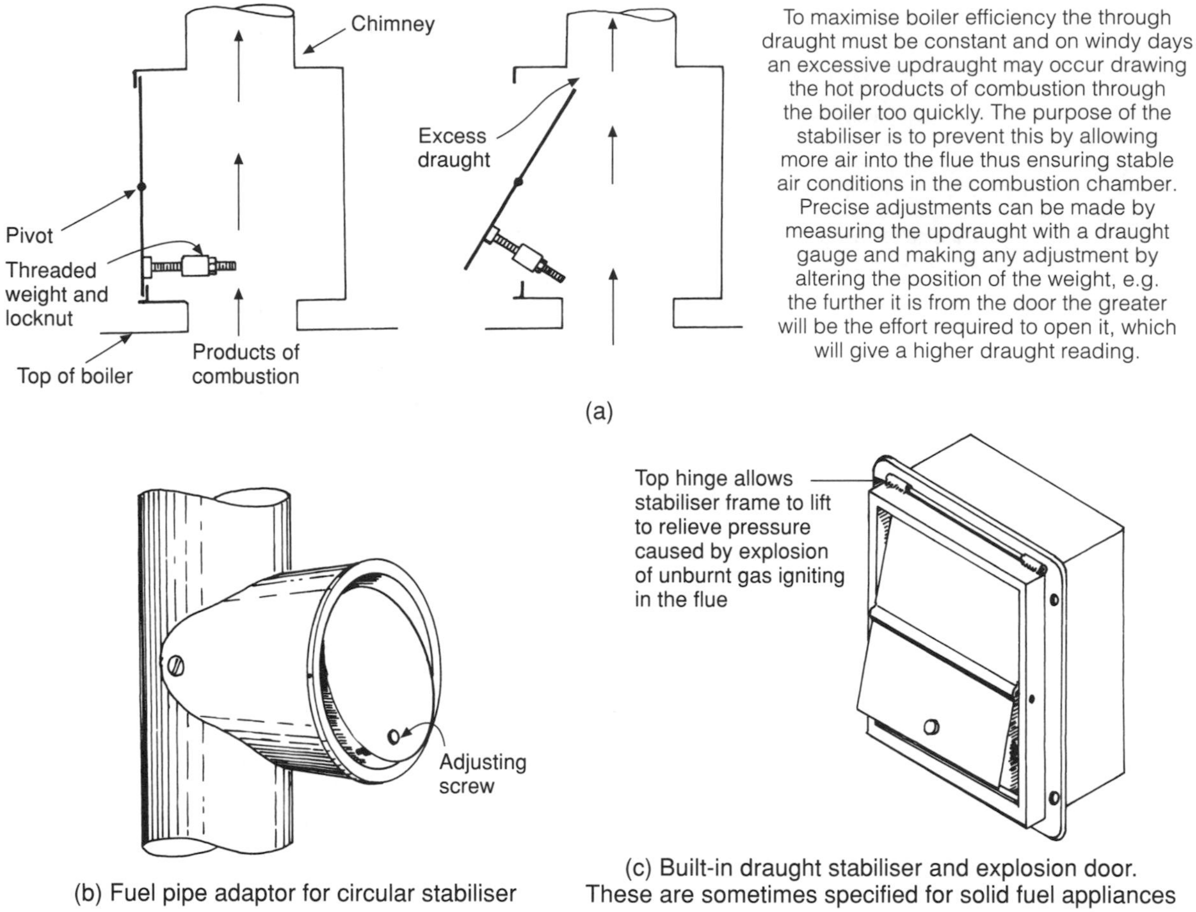

Chimney
Excess draught
Pivot
Threaded weight and locknut
Top of boiler
Products of combustion
To maximise boiler efficiency the through draught must be constant and on windy days an excessive updraught may occur drawing the hot products of combustion through the boiler too quickly. The purpose of the stabiliser is to prevent this by allowing more air into the flue thus ensuring stable air conditions in the combustion chamber. Precise adjustments can be made by measuring the updraught with a draught gauge and making any adjustment by altering the position of the weight, e.g. the further it is from the door the greater will be the effort required to open it, which will give a higher draught reading.
(a)
Top hinge allows stabiliser frame to lift to relieve pressure caused by explosion of unburnt gas igniting in the flue
Adjusting screw
(b) Fuel pipe adaptor for circular stabiliser
(c) Built-in draught stabiliser and explosion door. These are sometimes specified for solid fuel appliances


**Fig. 6.20** Principles of draught sterilizers.

output and the fact that modern atomising burners, which can be controlled electrically, are much smaller and quieter than older models. Because of this they can now be accommodated in the boiler casing and fitted into modern kitchens. Some types are made that enable servicing to be carried out from outside the building. Of the main types of vaporising burner once commonly in use, only the sleeve type has any application in building services, where it is used in some types of cooking stove; because no electrical supply is necessary this does have some advantage in rural areas. For further information reference should be made to the companies listed at the end of the chapter. A small market exists for pot vaporising burners, although they are not used now for water heating. Although they are very efficient, wall flame burners are no longer made for use in the UK due to their limited output.

*Atomising burners*

Sometimes called pressure jet burners, these are currently the most popular method of utilising oil for heating and hot water supply. They are available in both floor-standing or wall-hung mode with conventional or balanced flues. The burner is a self-contained unit, being bolted on to the front plate of the boiler by a flange. Some of the larger types used in industry are also provided with a leg or stand for additional support. A typical burner is shown in Fig. 6.21; whatever its heat output the same basic components are used. The rear end of the burner houses a fan which rotates on a spindle driven by an electric motor; the same spindle also drives the

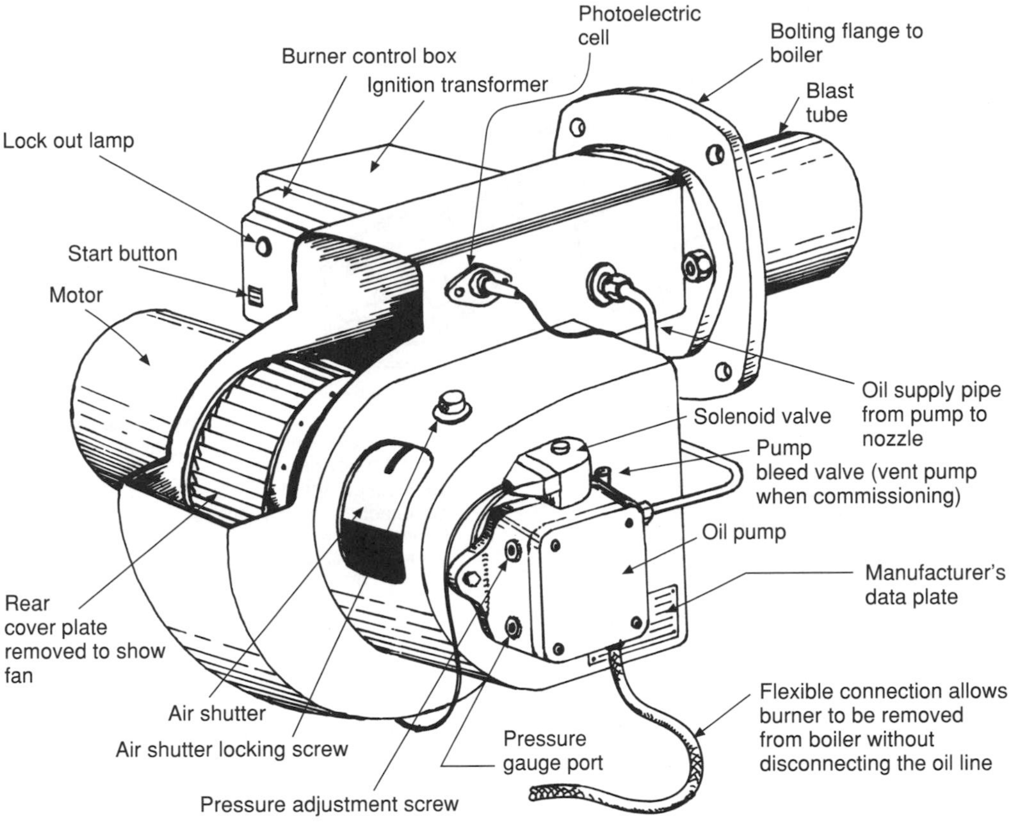


**Fig. 6.21** Typical atomising oil burner.

oil pump. Some manufacturers make the spindle in such a way that the oil pump is connected to it by a hard rubber sleeve. In the event of oil-pump seizure the sleeve will break, preventing damage to the motor. The delivery side of the pump is connected to a solenoid valve, the purpose of which is to positively close the supply of oil on burner shut-down. From the solenoid valve the oil is delivered at high pressure to the nozzle, which breaks it down into a fine oil mist so that it can mix easily with the air from the fan enabling combustion to take place. Figure 6.22 shows the end of the blast pipe which is fitted with a swirl or diffuser plate. As the air passes through it, this plate imparts a rotary motion to the air; this, and the fact that the nozzle is designed to give the oil mist a contrarotating motion, ensures thorough mixing of the oil and air to form a highly combustible compound. Ignition is achieved by a spark produced

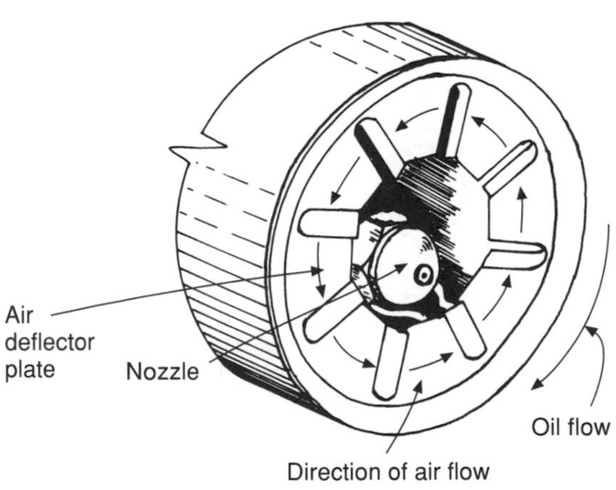


**Fig. 6.22** Air deflector plate and its relationship with the nozzle. The rotative motion of the air flow through the deflector plate and the contrarotation of the oil through the nozzle produce a flammable oil mist which is ignited by a spark across the electrodes.

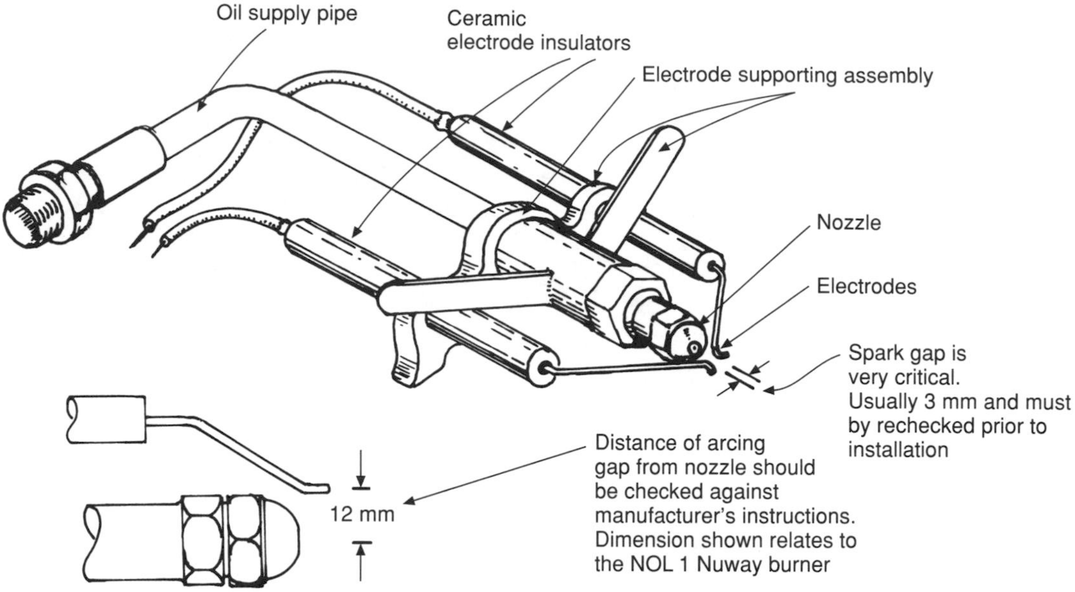


**Fig. 6.23** Electrode assembly housed in blast tube.

by a pair of electrodes shown in the nozzle assembly in Fig. 6.23. The position of the electrodes in relation to the nozzle is fairly critical and must be positioned as recommended by the manufacturer's data when the burner is commissioned.

All oil burners must be provided with a fail-safe device which prevents the combustible oil/air mixture entering the combustion chamber and flue without igniting. If this happened and the mixture accidentally ignited, the result could be a very serious explosion. Originally protection against this is provided by a flue thermostat, but a much more effective device called a photoelectric cell or resistor, which actually sees the flame, is now employed. Figure 6.24 shows this component. Reference should also be made to Fig. 2.46.

*The control box*

The control box is to the oil burner what the brain is to humans, its purpose being to coordinate

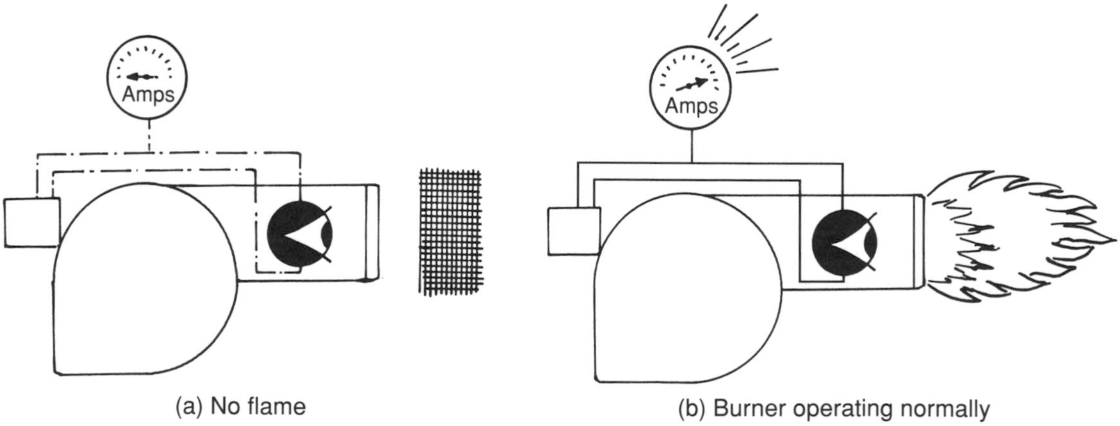


**Fig. 6.24** Fail-safe equipment. The photoelectric resistor or cell is illustrated as an eye. If the burner does not fire it will not see a flame and will send a signal to the control box to go to lockout.

the signals relayed by the remote controls to ensure that the burner fires and shuts down when necessary. In the event of failure to fire, it will automatically stop the fan and oil pump momentarily until the unburnt oil/air mixture has vacated the combustion chamber; this is called the 'purge' period. The control box will then again attempt to start the burner. If this meets with failure, the control box is able to recognise there is a fault in the system, causing its integral fail-safe mechanism to operate. This is called 'lockout' and a red light showing on the control box will indicate this. When the fault is located and rectified the burner can only be restarted manually by pressing the appropriate button; a typical control box is seen in Fig. 6.25.

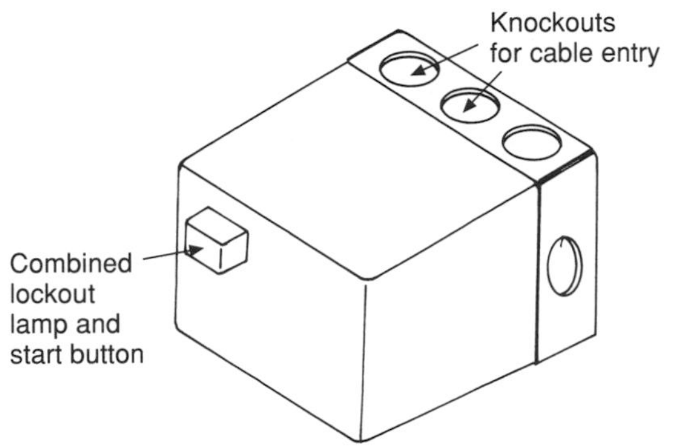


**Fig. 6.25** Oil burner control box.

*Oil pumps*

The reader may be aware that pumps vary considerably depending on the work they are required to do. There are three basic pump types:

(a) The reciprocating pump, which has many applications, typical examples being raising water from a well and pressure testing water installations.
(b) Centrifugal pumps used in heating systems, cold water and fire services in tall buildings.
(c) Gear pumps, shown diagrammatically in Fig. 6.26 which are widely used in situations where good suction and pressure characteristics are required; for example, to force a flow of oil through the burner nozzle.

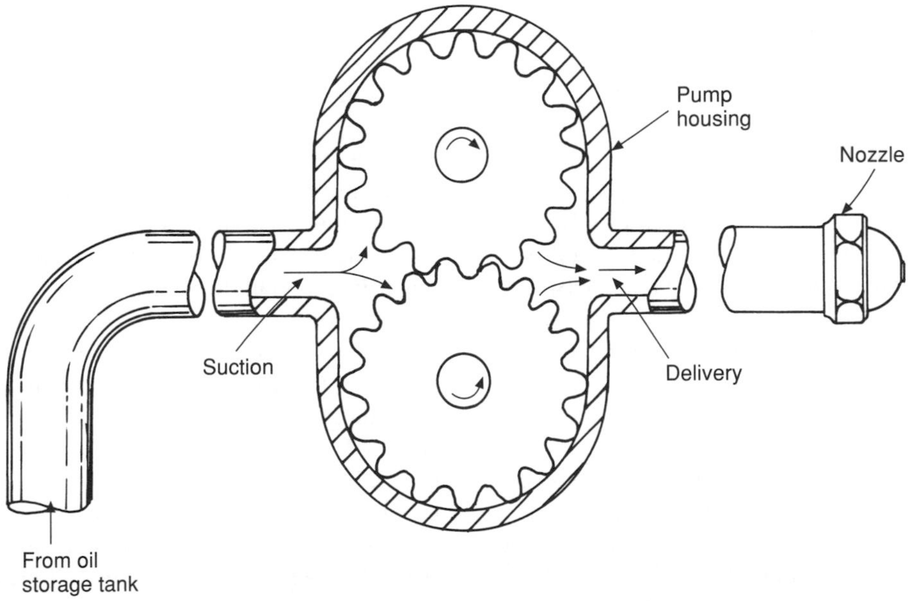


**Fig. 6.26** Oil burner pumps. Basic principle of an oil pump used on atomising oil burners. Gear pumps are used to develop the high oil pressure necessary to force the oil through the nozzle orifice. As the gear wheels rotate the disengagement of their teeth creates a negative pressure and oil is sucked into the pump housing, where it is carried round by the gear teeth to the delivery side of the pump and the nozzle.

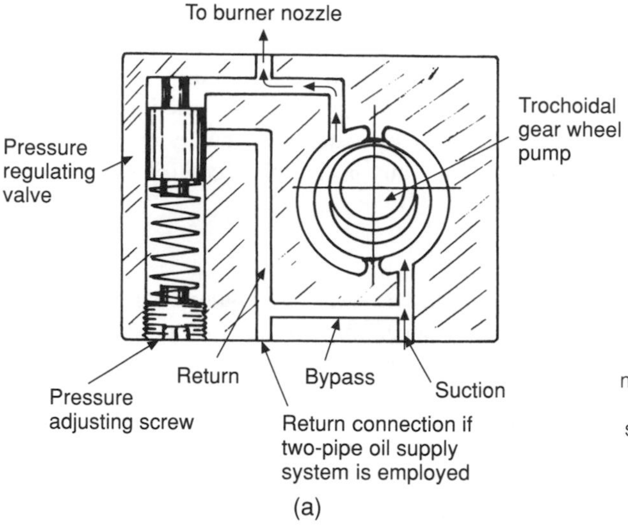


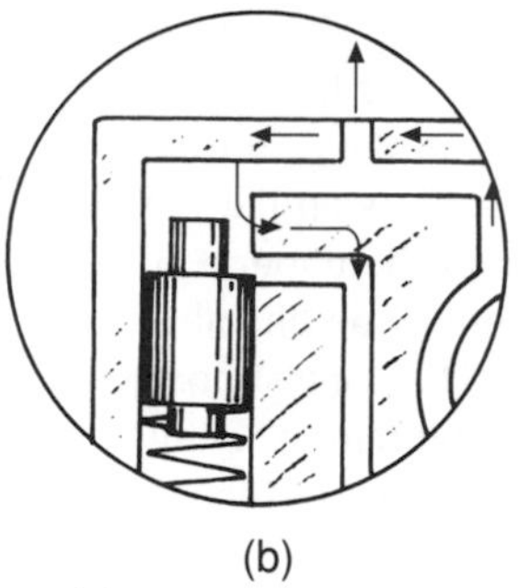
(b)
The purpose of the regulating valve is to maintain a constant oil pressure at the nozzle. Most pumps deliver more oil than required and the excess is either recirculated in the pump or returned to the oil tank when a two-pipe system is employed. This inset shows the regulating valve open to the return and bypass.

**Fig. 6.27** Pressure regulation of oil pump.

Burner oil pumps also incorporate a filter and a regulating valve which both maintains a stable oil pressure and directs the volume of oil not used at the nozzle back to the suction side of the pump, via a bypass, or if a two-pipe system is used, back to the storage tank. Figure 6.27 illustrates the working principles of a typical oil pump.

Modern oil pumps are also provided with an integrally operated cut-off, or solenoid valve. This permits oil to pass to the nozzle on start-up only when the fan has developed sufficient air pressure to ensure the oil/air mixture is correct. Conversely, when the oil burner stops, the motor speed falls, and both oil and air pressure will fall evenly as the motor slows down. Unless the flow of oil is positively stopped during this period, a sooty, pulsating flame will be momentarily produced and nozzle dribble may also occur, both of which are very undesirable.

*Oil burner nozzles*

A nozzle plays a very important part in the process of oil firing, and unless it is functioning correctly it will be impossible to achieve a clean, stable flame burning with maximum efficiency. Figure 6.28 shows all the components which make up the nozzle unit. The sintered bronze filter is provided to prevent any solid particles that may remain in the oil after passing through the main filter adjacent to the storage tank and pump filter. From the nozzle filter the oil runs along the outside of the cone and through the slots which give it a rotative motion, ensuring complete mixing with the combustion air provided by the fan. In practice the oil is forced through the nozzle at such a high velocity that a 'tube' of oil is formed in the nozzle orifice. This tube expands on leaving the nozzle and breaks up into very fine droplets which easily mix with air. Nozzles are marketed to give a variety of spray angle patterns and their hourly flow rate is usually quoted in US gallons or litres at a specified pressure. The spray angles and patterns are designed to match the boiler combustion chamber, the burner manufacturers taking this into account when supplying a burner for a specific boiler. In view of this it is important to replace worn-out nozzles with another exactly the same, therefore they are stamped with the relevant data for easy identification. It is important to ensure that nozzles are treated with care and kept in their protective casing until they are fitted. Avoid touching the tip of the nozzle — the orifice is very easily blocked; they are best handled by holding them across the spanner flats.

Scratches on the face must be avoided at all costs. Nozzles do wear; the high oil pressure at which the oil passes through them can cause erosion in the orifice resulting in it becoming enlarged. This will upset the oil/air ratio causing poor combustion. It is generally accepted that a nozzle will remain serviceable provided good burning and flame

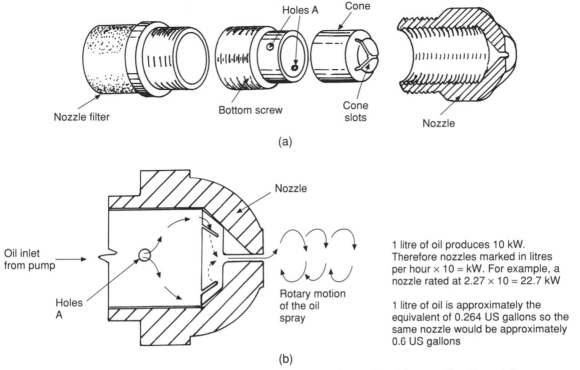

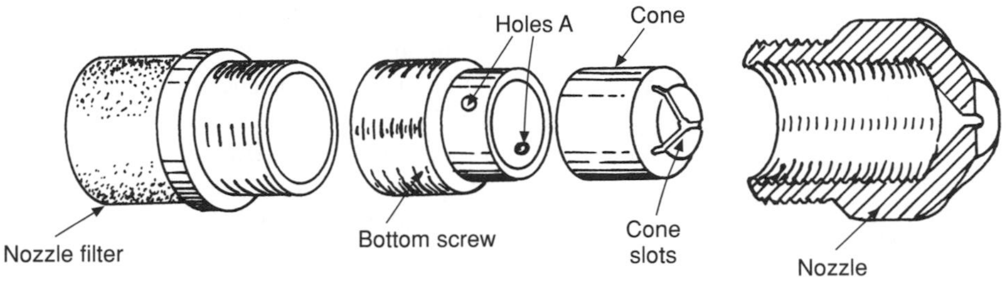


(a)

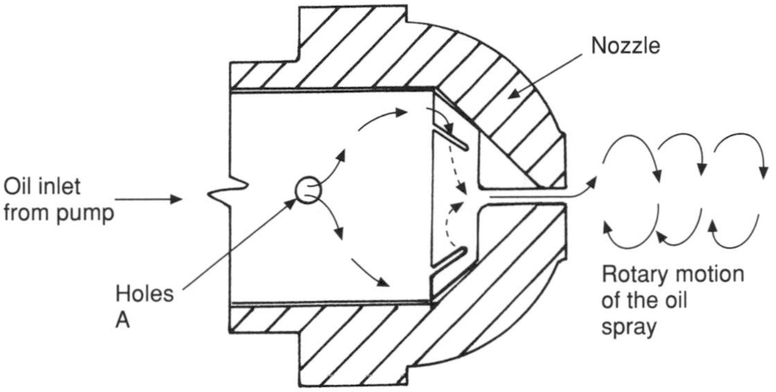


1 litre of oil produces 10 kW. Therefore nozzles marked in litres per hour × 10 = kW. For example, a nozzle rated at 2.27 × 10 = 22.7 kW

1 litre of oil is approximately the equivalent of 0.264 US gallons so the same nozzle would be approximately 0.6 US gallons

(b)

Oil is pumped through the holes in the bottom screw, passes over the outside of the cone then through the cone slots which are made in such a way that they impart a rotary movement to the oil as it passes through the nozzle. The contrarotative flow of air from the fan mixes with the oil to produce a highly flammable oil mist.

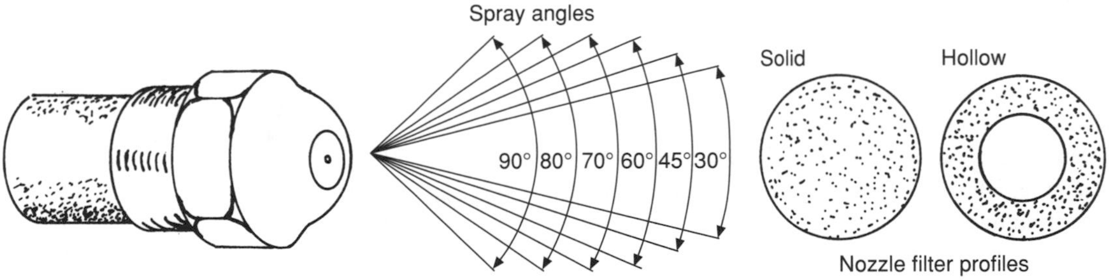


(c) Typical spray pattern

Nozzles are specified by the volume of oil they pass per hour, their spray angles and flame pattern. They are supplied with the burner and are matched with the boiler combustion chamber to give maximum firing efficiency.

**Fig. 6.28** Burner nozzles.

characteristics can be maintained. This can be determined by using the standard tests described later in this chapter.

*Preheaters*

Maximum burner efficiency can only be achieved if the oil is of the correct viscosity at the nozzle. It has been stated that at low temperatures 35-second oil starts to solidify and become waxy, and even 28-second oil tends to thicken in cold weather. This means that burner efficiency will vary in differing weather conditions. If the viscosity increases due to low temperatures, the result will be larger oil drops which make its combustion more sluggish, and a higher oil flow through the nozzle, which, because there is no corresponding increase in the air for

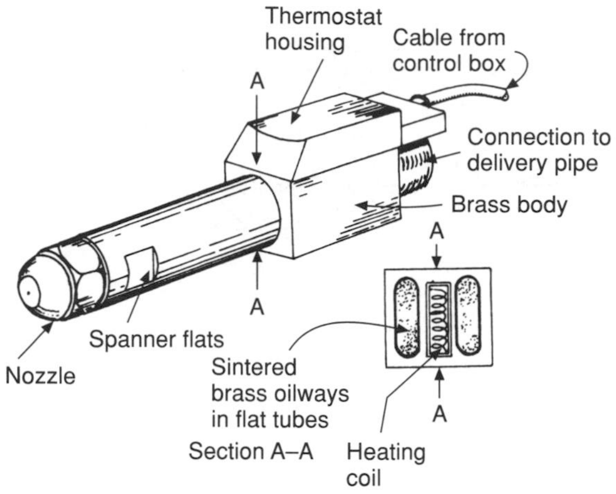


**Fig. 6.29** Oil preheater.

combustion, will result in a sooty flame. The presence of soot in any boiler, irrespective of the type of fuel used, means that combustion is incomplete, resulting in a loss of efficiency. To overcome this problem the oil must be preheated to approximately 65 °C prior to entering the nozzle assembly. This will result in precise atomisation and clean, stable, soot-free combustion. A typical preheater is shown in Fig. 6.29 and will normally improve efficiency of the burner. Preheaters incorporate a heater element controlled by a thermostat which maintains the supply of oil at the nozzle at a temperature of 65–70 °C. They are an optional extra on new oil burners and may also be fitted to existing burners subject to consultation with the manufacturer.

*Transformers*
Step-up transformers must be used to increase the normal 204 V supply to approximately 10,000 V — there are some variations, depending on the manufacturer. This voltage will 'jump' the gap between the electrodes, causing a spark which will ignite the oil/air mixture.

## Installation of oil-fired boilers

Floor-standing boilers should be sited on a solid hearth similar to those required for solid fuel appliances. Provision must be made for water connections, maintenance and cleaning. A suitable flue must also be available, except for boilers having a balanced flue, in which case the terminal position will be similar to those required for gas appliances. Boilers situated in outbuildings or specially constructed compartments must be provided with adequate frost protection, e.g. frost thermostats and effective pipework insulation. As with all heating appliances, it is essential to ensure that any mechanical ventilation in the building does not adversely affect the operation of the flue.

*Air requirements for combustion and ventilation*
This is illustrated in Fig. 6.30. The ventilation requirements shown apply if the boiler is fitted in a small compartment. In larger rooms the ventilation requirements may be less. Manufacturer's data will give air supply requirements precisely, but as a guide an average domestic burner needs about 60 $m^3$ of air per hour for good combustion. To this must be added sufficient air for ventilation of the compartment itself.

*Commissioning and testing*
A short length of flexible armoured pipe supplied with the boiler is used to connect the oil pump to the supply pipe so that the burner unit can be

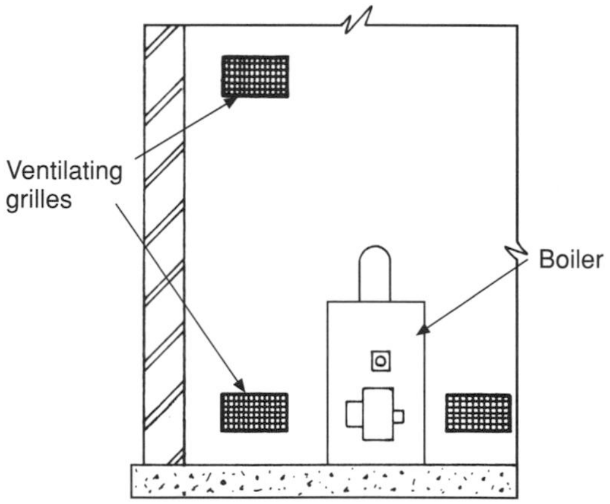


**Fig. 6.30** Ventilation requirements. For open-flued appliances a permanent supply of air for combustion at the rate of 550 $mm^2$ per kW input minus the first 5 kW must be provided. Air for ventilation, if taken from inside the building, must have a free area of grille 11.00 $mm^2$ per kW. When taken from outside the free area of each grille must be 550 $mm^2$ per kW.

easily removed for servicing without disconnecting the oil pipe. A stop valve should be fitted at this point — especially in commercial installations. The oil pipework and the pump must be completely purged of air, the first step being to disconnect the delivery pipe at the pump. All valves on the fuel pipe should then be opened allowing oil to discharge into a container until a good flow, free of any air bubbles, is observed. The delivery pipe can now be reconnected. Next the oil pump must be purged by starting the motor and opening the bleed screw until, as with the suction pipe, an air-free flow of oil is discharged. During the two foregoing operations have a supply of rag or cotton waste at hand to mop up any oil spillage to reduce fire risks.

The burner is next removed from the boiler and placed in a position so that the delivery pipe is vertical. The nozzle, if fitted, must be carefully removed and the leads from the transformer must also be temporarily disconnected at this stage. On some burners it may be necessary to remove the blast tube to achieve this. The burner should then be restarted so that oil is pumped into the delivery pipe. When it reaches the top, switch off the power supply and carefully refit the transformer leads and nozzle. A small quantity of oil should be seen to discharge through the nozzle orifice, proving that the discharge pipe and nozzle are completely purged of air. Some of these operations are illustrated in Fig. 6.31. Finally, before refitting the burner to the boiler, a final check must be made on the position of the nozzle in relation to the swirl plate, and that of the electrodes in relation to the nozzle. Figure 6.32 illustrates the effects of not carrying out this operation correctly.

*Testing combustion and burner efficiency*

Tests must now be conducted to ensure the burner is functioning correctly and efficiently. The oil pressure developed by the pump must be checked as shown in Fig. 6.33. Pressures usually vary between 100 psi (7 bar) and 140 psi (10 bar) depending on the pump used. The correct operating pressure for any unit will be found in the manufacturer's instructions.

Prior to conducting combustion and efficiency tests the burner should be allowed to run for 5–10

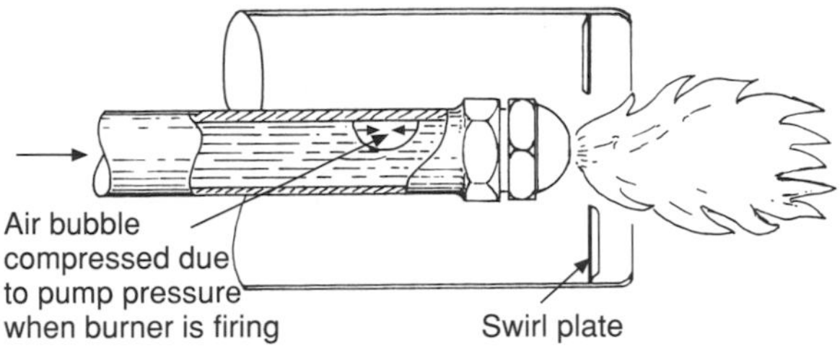
Air bubble
compressed due
to pump pressure
when burner is firing
Swirl plate

(a)

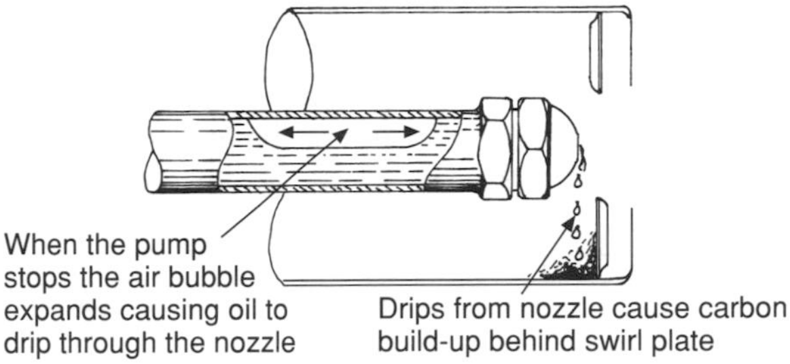
When the pump
stops the air bubble
expands causing oil to
drip through the nozzle
Drips from nozzle cause carbon
build-up behind swirl plate

(b)

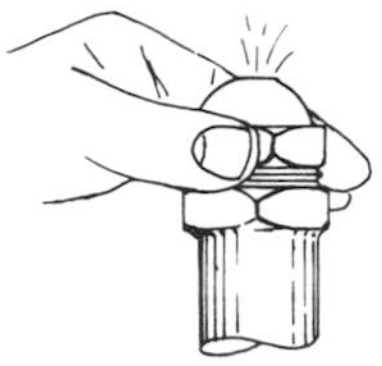

(c)

To ensure any air is purged in the delivery pipe first remove the blast pipe and turn the burner so that the oil delivery pipe is vertical and full of oil. Screw in the nozzle by hand so that oil is seen to pass through the orifice prior to tightening with spanners.

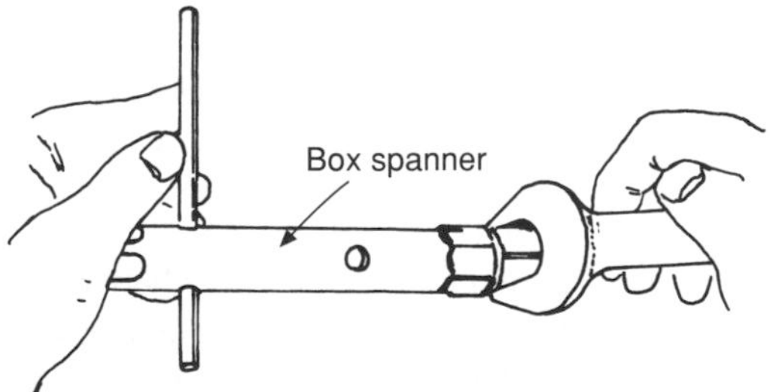
Box spanner

(d)

Finally tighten the nozzle on to the delivery pipe. Do not overtighten and use the correct size spanners.

**Fig. 6.31** Effect of entrapping air in the nozzle delivery pipe.

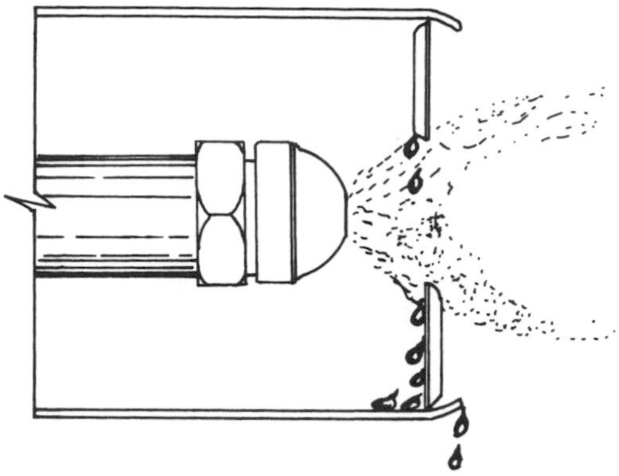

(a) In this case the nozzle assembly is set too far back into the blast pipe so that the atomised oil impinges on the diffuser plate

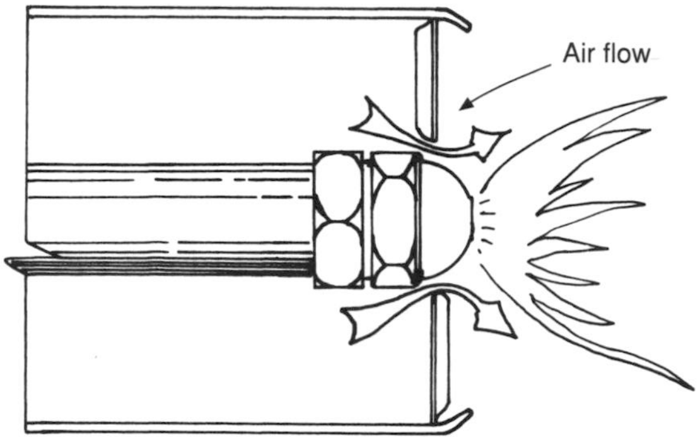


(b) If a nozzle is too far forward the air velocity will be too great. The burner may not fire under these circumstances, but if it does the flame will be very ragged as shown

**Fig. 6.32** Effects of incorrect positioning of nozzle assembly.

minutes to warm the flue. To avoid soot deposits in the boiler during this period, open the air shutter until the flame pulsates — then close it slightly. The flame can usually be seen through either a small glass window in the front of the boiler, or a small hole in the top. When the flame is burning properly it should look similar to a blowlamp flame with yellowish tinges. No smoke should ever be seen from the chimney. All the following tests are made through a small hole below the stabiliser if one is fitted. For boilers not having a stabiliser a small hole must be drilled in the flue pipe about

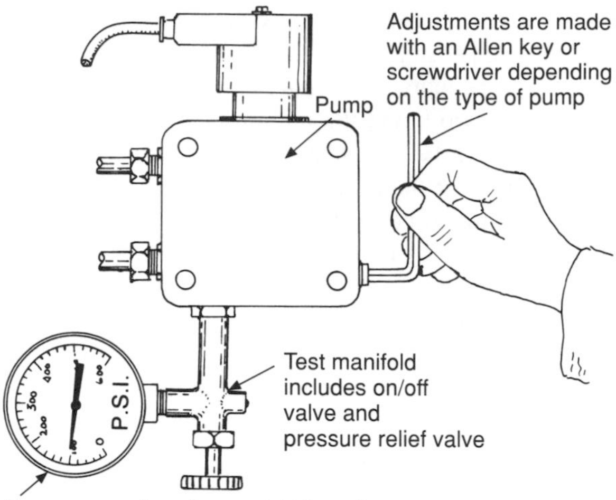


**Fig. 6.33** Checking and adjusting the oil pressure at the pump.

150 mm from the top of the boiler. This hole must be sealed when testing has been completed.

All the following tests must be conducted both when commissioning and at least once a year or when the boiler is serviced.

*Chimney updraught* This test is made using a draught gauge shown in Fig. 6.34 to indicate the updraught or 'pull' on the flue. They are graduated in inches or millibars water gauge (WG). There are slight variations for different burners, but the usual requirements are 0.04 inches WG. If a draught stabiliser is provided with the boiler, it can be adjusted to maintain a steady updraught — if not it may be necessary to fit one where excessive updraught readings are indicated.

*Smoke testing* (see Fig. 6.35) This is done with a pump which extracts a sample of flue gas and draws it through a filter paper. The standard test needs 10 strokes on the pump — more or less will give a false reading. The filter paper is then removed from the pump, and a spot, varying in shade between almost white to black, will be seen. This is compared with the shades on a standard card, known as the Bacharach scale, which is numbered from 1 to 10. If the spot on the filter paper is too dark, it is a sure indication that there is excessive

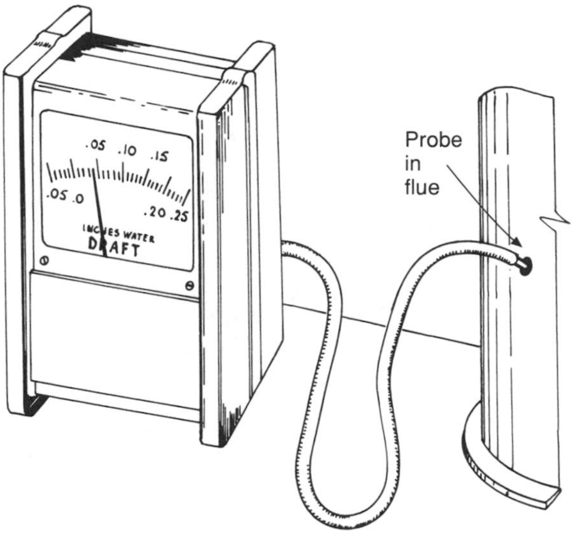


**Fig. 6.34** Draught gauge shown reading .04 in WG which is acceptable for most burners. Note that a reading to the left of .0 indicates downdraught. Gauges reading in mm are available, but most manufacturers currently quote draft requirements in inches WG.

soot in the flue gases and more air must be admitted to the burner by opening the air shutter. If the filter paper is unshaded it is an indication that too much air is being passed and the air shutter requires closing down. When it is properly adjusted the shade on the filter paper should match no. 1 or 2 on the scale. If adjusting the air shutter fails to give a good reading, the position of the nozzle assembly in the blast pipe should be rechecked, and in the case of an existing boiler the nozzle may need replacing.

If the previous two tests give good readings then the burner should be operating efficiently, but this must be checked by comparing the flue gas temperature against the $CO_2$ content. Figure 6.36 shows the type of dial thermometer used for flue gas temperature measurements. Net stack temperatures of approximately 230–240 °C are normally specified by manufacturers. The net temperature is that minus the ambient air temperature — thus a flue temperature reading of 260 °C when the air temperature is 20 °C will have a net stack temperature of 240 °C. Excessive flue temperatures may be the result of:

**Table 6.2** Effect of smoke on burner performance (refer to Fig. 6.35(b)).

| *Bacharach smoke scale no.* | *Burner rating* | *Sooting produced* |
|---|---|---|
| 1 | Excellent | Extremely light if at all |
| 2 | Good | Slight sooting which will not increase stack temperature appreciably |
| 3 | Fair | May be some sooting but will rarely require cleaning more than once a year |
| 4 | Poor | Borderline condition, some units will require cleaning more than once a year |
| 5+ | Very poor | Soot rapidly and heavily |

(a) Excessive draught through the boiler, in which case the draught stabiliser setting should be rechecked.
(b) Dirty, carboned or sooty surfaces, remedy — clean.
(c) Incorrect setting of air shutter.
(d) Overfiring — possibly due to incorrect or worn nozzle.

Figure 6.37 illustrates apparatus for testing the $CO_2$ content of the flue gas. The fluid contained in the gauge absorbs $CO_2$, and, prior to use, the adjustable scale should be set at zero against the fluid level. A hand pump is used to sample the flue gases and the standard test requires 18 pumps, no more, no less. On removing the probe, the gauge is turned twice through 180° to ensure thorough mixing of the $CO_2$ prior to taking a reading. Shaking the gauge is not recommended.

It should be noted that the fluid contained in the gauge should be changed at regular intervals, as its ability to absorb $CO_2$ diminishes, depending on the incidence of use.

A reading of between 8 and 11 per cent, the highest number applying to modern boilers, should be the aim; a reading of less than 8 per cent will indicate an excessive air flow through the flue. Readings in excess of 11 per cent are likely to result in soot formation within the combustion chamber, and possible condensation in the flue. The actual efficiency of the boiler can be read directly

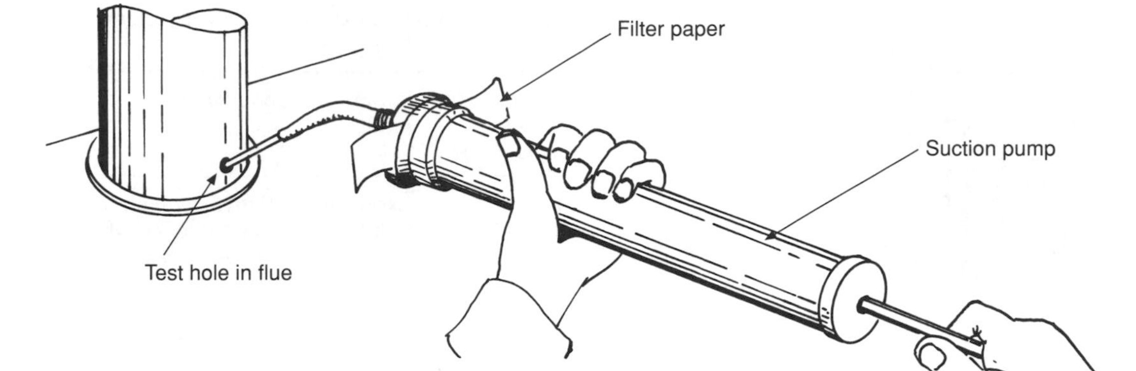

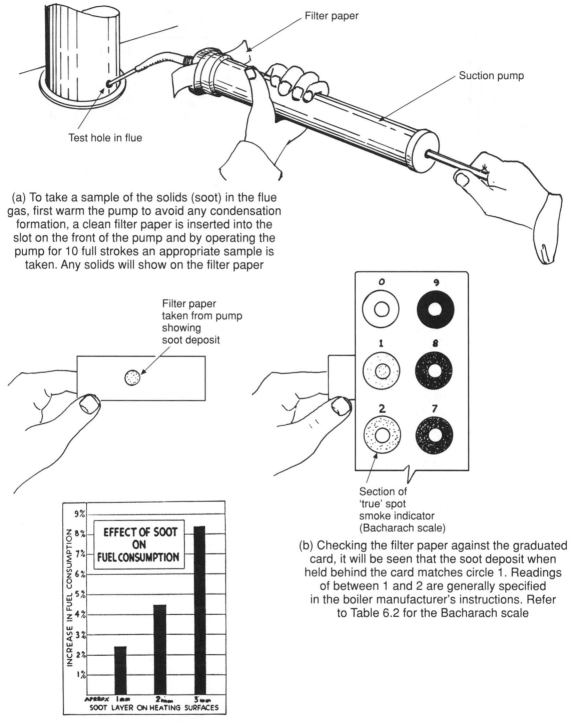

(a) To take a sample of the solids (soot) in the flue gas, first warm the pump to avoid any condensation formation, a clean filter paper is inserted into the slot on the front of the pump and by operating the pump for 10 full strokes an appropriate sample is taken. Any solids will show on the filter paper

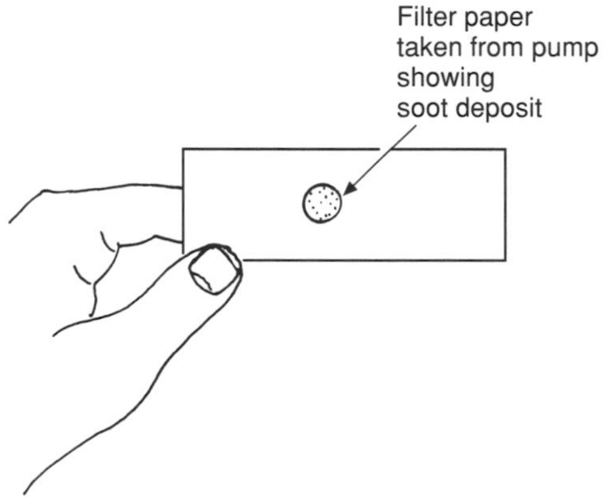


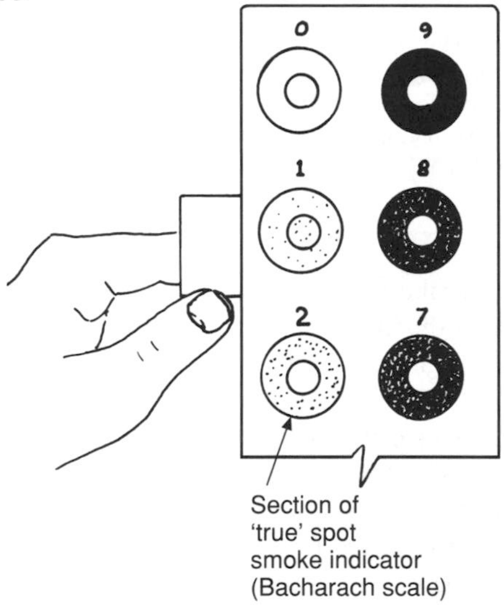


(b) Checking the filter paper against the graduated card, it will be seen that the soot deposit when held behind the card matches circle 1. Readings of between 1 and 2 are generally specified in the boiler manufacturer's instructions. Refer to Table 6.2 for the Bacharach scale

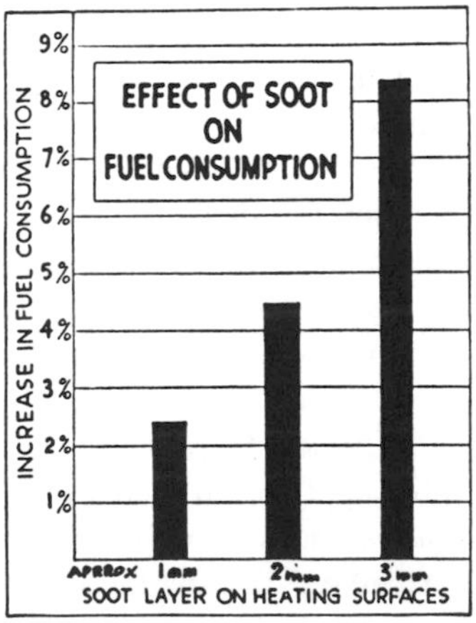


**Fig. 6.35** Smoke testing oil-fired boilers.

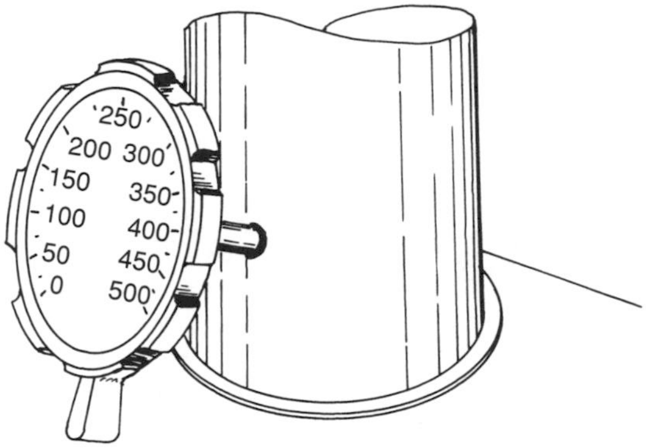


**Fig. 6.36** Checking the flue gas temperature with a rotary thermometer. The aim should be between 200 and 250 °C with a $CO_2$ reading of 11 per cent.

from a standard scale which cross-references flue gas temperatures with the $CO_2$ percentage, see Table 6.3, or it can be calculated using the following formula:

$$\text{Flue loss (per cent)} = \frac{0.477 + 0.072}{\text{per cent } CO_2} \times \left(\begin{matrix}\text{flue gas} \\ \text{temp.}\end{matrix} - \begin{matrix}\text{ambient} \\ \text{air temp.}\end{matrix}\right)°\text{C} + 6.2$$

(It should be noted that the figures 0.477, 0.072 and 6.2 are constants and relate directly to the type of fuel burned — in this case class C or D oil.)

Example: Calculate the efficiency of a boiler where the flue gas reading is 260 °C, with a $CO_2$

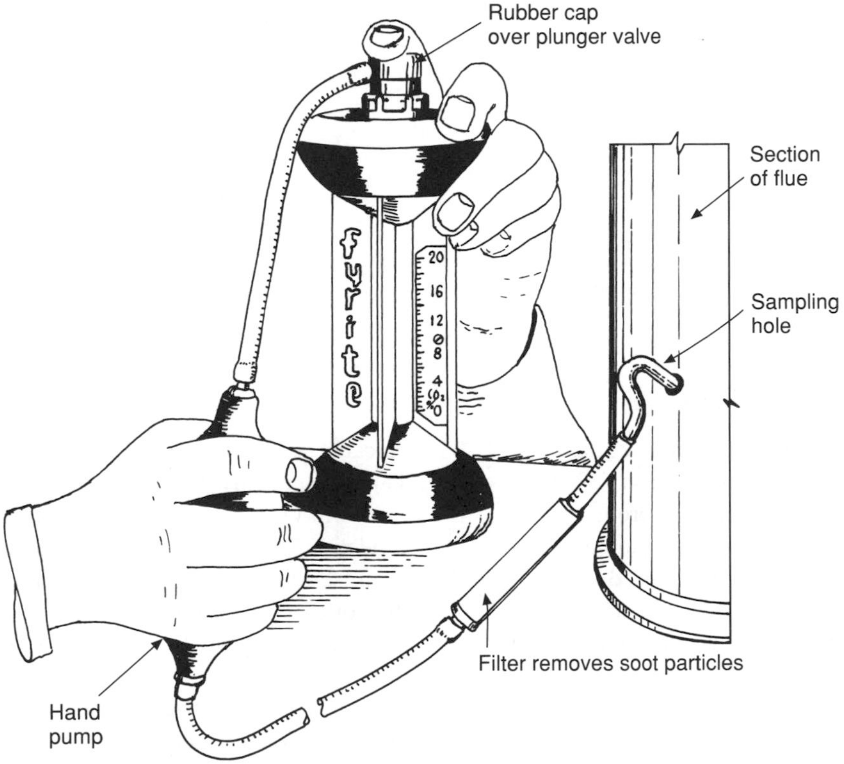


To ensure an accurate $CO_2$ test:

1. Zero the scale to the top level of the fluid.
2. Operate the burner for 5–10 minutes.
3. Insert sampling tube into test hole.
4. Place rubber cap over valve and depress.
5. The bulb of the aspirator pump is squeezed 18 times in succession.
6. The rubber cap is removed and the valve will automatically close.
7. The indicator is turned over twice to ensure thorough mixing of the fluid and $CO_2$.
8. Place the indicator on a level surface and read off percentage of the $CO_2$ on the scale.

**Fig. 6.37** Sampling flue gases with a $CO_2$ indicator.

**Table 6.3** Flue gas temperatures and percentage of $CO_2$.

| *$CO_2$ content (%)* | *Efficiency (%) at flue gas temperature of: (°C)* | | | | |
|---|---|---|---|---|---|
| | *114* | *204* | *260* | *316* | *371* |
| 14 | 88 | 86 | 83 | 81 | 79 |
| 13 | 88 | 85 | 83 | 80 | 78 |
| 12 | 87 | 85 | 82 | 80 | 77 |
| 11 | 87 | 84 | 81 | 79 | 76 |
| 10 | 86 | 83 | 80 | 77 | 74 |
| 9 | 85 | 82 | 79 | 76 | 73 |
| 8 | 85 | 81 | 77 | 74 | 70 |
| 7 | 84 | 80 | 75 | 70 | 67 |
| 6 | 82 | 78 | 73 | 68 | 63 |
| 5 | 81 | 75 | 69 | 66 | 58 |
| 4 | 77 | 70 | 63 | 57 | 50 |

percentage of 11 and an ambient air temperature of 15 °C:

$$0.477 + 0.072 = 0.549$$
$$0.549/11 = 0.05$$
$$0.05 \times (260 - 15) = 12.3$$
$$12.3 + 6.2 = 18.5$$

This would give a flue loss approximately 18.5% and 81.5% efficiency.

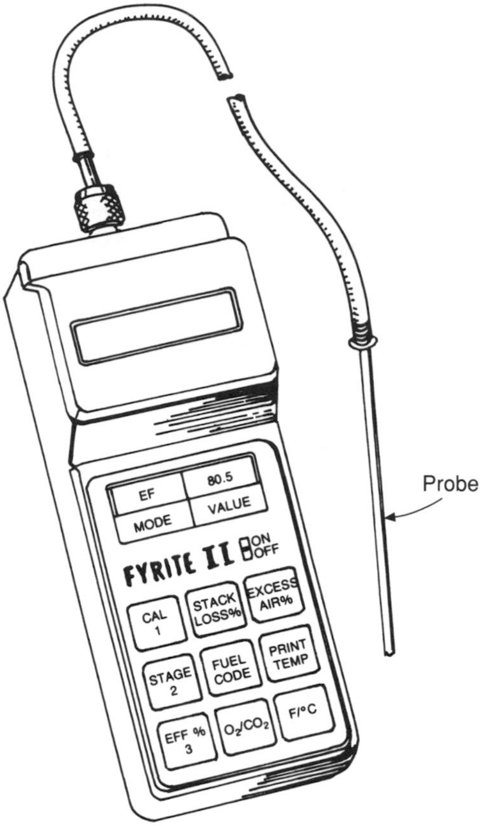


**Fig. 6.38** Digital combustion analyser.

### Digital combustion analysers

The method described for testing combustion efficiency using flue thermostats and $CO_2$ analysers has been used for many years. If properly maintained, their accuracy is acceptable for most installations and they are widely used in the plumbing and heating industry. Where greater degrees of accuracy are required, sophisticated computerised analysers are obtainable, some directly showing the test results on a small screen — others actually print out the relevant details. When this equipment is used the batteries must be changed on a regular basis to ensure accurate readings. Figure 6.38 illustrates a typical example of this type of combustion testing equipment.

### Flues for oil boilers

As with gas, the temperature of the products of combustion of oil is comparatively low and similar precautions must be taken to prevent condensation. This is especially so when burning class 'D' oil, as due to its higher sulphur content mixing with condensate, a dilute solution of sulphuric acid will form, seriously corroding the internal surfaces of the boiler. If a draught stabiliser is fitted, it must be carefully adjusted so that an excessive amount of cold air is not drawn into the flue, thus cooling the products of combustion below their dew point. Flues constructed to conform with Part J of the Building Regulations are satisfactory for oil burning, but in pre-1965 buildings, a suitable lining will be necessary. Flue terminals are not normally required, although it is recommended that some means of preventing rainwater entering the flue is employed. A terminal known as a 'Chinaman's hat' shown in Fig. 6.39, is suitable for this purpose.

In situations where downdraught is a serious problem, it can often be solved by fitting an OH

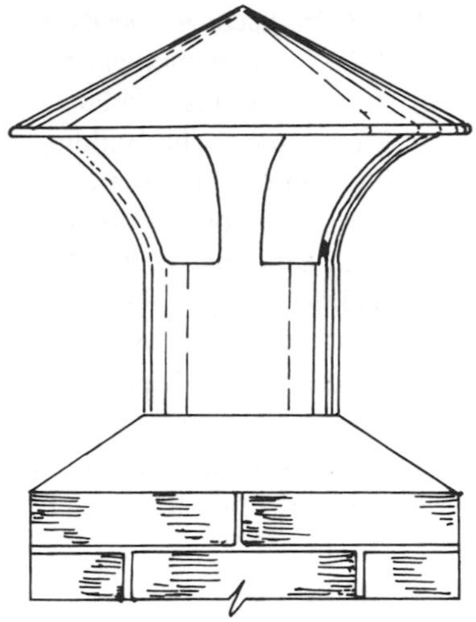

**Fig. 6.39** Chimney terminal. Sometimes called a Chinaman's hat due to its shape. Its purpose is to prevent rainwater entering the flue. It is not suitable as a terminal for gas flues.

cowl shown in Fig. 6.40. Generally they not only solve downdraught problems but also improve the flue updraught.

### Servicing oil-fired boilers

Servicing schedules are always supplied by burner or boiler manufacturers, usually as an appendix to the installation instructions. They may differ slightly depending on the type of boiler and burner with which it is supplied. The following, however, lists the main servicing points relating to boilers fitted with atomising burners:

1. Inspect the general condition of the flue and its cleanliness. Any signs of excessive sooting will give some indication as to the efficiency of the installation. The owner of the premises should be notified if the flue requires repairing or sweeping, as this is not usually the responsibility of the maintenance engineer.
2. The boiler flueways and combustion chamber must be cleaned. Special wire brushes are necessary for this purpose, as some boilers, having small flueways, are difficult to clean. If the combustion chamber is lined with firebricks they must be replaced if badly cracked or damaged.

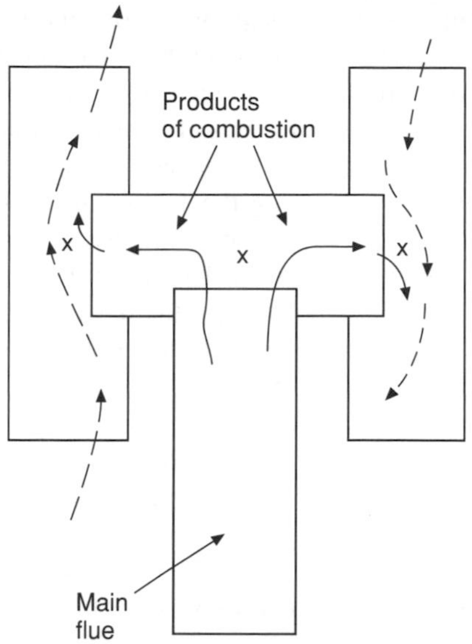



(a) This cowl is designed to eliminate downdraught and will increase the updraught in a flue. It will be seen that wind pressure shown by the arrows with dotted tails causes three points of negative pressure shown at x. This is due to the extensions of the main flue and horizontal section into the outlets which in effect form a type of Venturi effect. Suitable for all types of fuel, these cowls are made in galvanised and stainless steel, also terracotta. The last are very heavy and require a proper scaffold for their installation. When used for gas flues the outlets must be provided with mesh guards to prevent the entry of birds

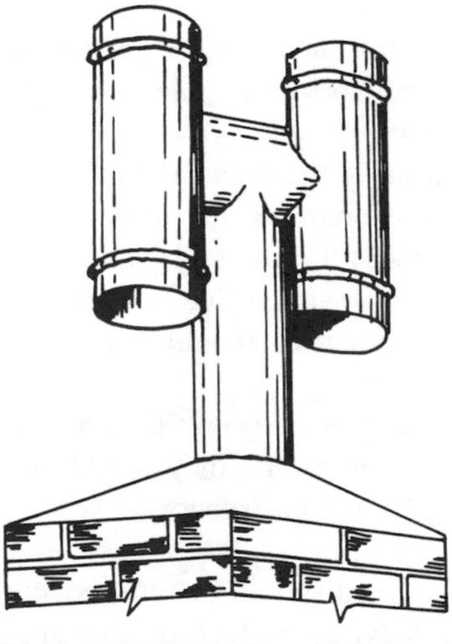

(b) Pictorial illustration of OH cowl

**Fig. 6.40** OH cowl.

3. Isolate the electric supply, remove the burner and clean the nozzle assembly. Nozzles may be stripped down and cleaned, but great care is necessary to ensure the very small components are not scratched or damaged — some recommend changing the nozzle annually. The electrode insulation must also be inspected for damage such as cracks. When the assembly is replaced, its position in the blast tube and that of the electrodes in relation to the nozzle must be carefully measured against the manufacturer's recommendations.
4. Clean the window of the photoelectric cell or resistor and ensure it is correctly located in its housing.
5. Apply some light oil to the motor bearings if necessary — some motors are lubricated for life.
6. All filters should be cleaned in paraffin.
7. Check all oil pipelines for leakage and rectify as necessary. This is important as any slight leak will vaporise, and apart from an unpleasant smell, could constitute a serious fire risk.
8. The electrical components and wiring of the installation should be checked for security, damage and correct operation.
9. The fire valve should be operated to ensure it functions correctly, and in the case of the drop-weight type, the cable and pulley fixings must be secure, free from corrosion and have freedom of movement.
10. The fuel tank should be visually examined for leakage and corrosion and any sludge should be drained off.
11. After ensuring the boiler is full of water (in unoccupied premises the system may have been drained) fire the boiler and after about 10 minutes when the flue is warm, combustion and efficiency tests should be conducted and recorded.
12. Any fail-safe devices on the burner must be operated to ensure they would function correctly in any emergency.

A service record should be retained showing the date of each service, the $CO_2$ reading and stack temperature.

*OFTEC (Oil Firing Technical Association)*

As from 1 April 2002 an amendment to the Building Regulations covering combustion appliances came into force. The Building Act of 1984 requires a person carrying out certain types of building work to give notice or deposit full plans with the local Building Control Officer. Appropriately qualified people registered with OFTEC are exempt from these requirements, providing a suitable record is kept, as they are classified as competent persons. This applies to any work carried out under the Building Regulations Part J in relation to all heat-producing appliances.

### Further reading

'Guidance note for the Control of Pollution (Oil Storage) (England) Regulations 2001' from DEFRA, Nobel House, 17 Smith Square, London, SW1P 5DU.

BS 5410 Part 1 Installation up to 44 kW for hot water and space heating installations.

BS 5410 Part 2 Installations of 45 kW and above output capacity for space heating, hot water and steam supply service.

BS 799 Oil-burning equipment.

BS 799 Part 3 Automatic and semi-automatic atomising burners.

BS 799 Part 5 Oil storage tanks.

BS 4543 Part 3 Chimneys for oil-fired appliances.

BBT Thermotechnology UK Ltd, Cotswold Way, Warndon, Worcester WR9 9SW. Tel. 01905 754624. www.worcester-bosch.co.uk

*Burners and control equipment*

NU-WAY Ltd, PO Box 1, Vines Lane, Droitwich, Worcester WR9 8NA Tel. 01905 794331. www.nu-way.co.uk

Danfoss Randall Ltd, Ampthill Road, Bedford MK42 9ER Tel. 01234 364621. www.danfoss-randall.co.uk

Riello Ltd, Unit 6, The Ermine Centre, Ermine Business Park, Huntingdon, Cambridgeshire PE29 6WX Tel. 01480 432144. www.rielloburners.co.uk

*Vaporising burners* Don Heating Products (Division of Gazco), Osprey Road, Sowton Industrial Estate, Exeter EX2 7GJ Tel. 01392 444070. www.gazco.com

*Testing equipment*

Shawcity Ltd, Pioneer Road, Faringdon, Oxon SN7 7BU Tel. 01367 241675. www.shawcity.co.uk

*Fire protection*

Falcon Landon Kingsway, Unit 5, Bold Business Centre, St Helens, Merseyside WA9 4TX Tel. 01925 290 660.

Teddington Controls Ltd, Daniels Lane, St Austell, Cornwall PL25 3HG Tel. 01726 74400. www.tedcon.com

*Oil storage vessels*

Balmoral Tanks, Balmoral Park, Loirston, Aberdeen, Scotland AB12 3GY Tel. 01224 859000. www.balmoral-group.com

*Technical information and training*

Oil Firing Technical Association, Foxwood House, Dobbs Lane, Kesgrove, Ipswich IP5 2QQ Tel. 0845 6585080. www.oftec.co.uk

**Self-testing questions**

1. Identify the two main classifications and viscosity of oil used for domestic, commercial and light industrial use.
2. State the effects of low temperature on the viscosity of oil.
3. List two materials suitable for oil pipelines.
4. State why the oil storage vessel must fall towards the sludge cock.
5. Explain how a combustible oil mist is produced in an atomising burner.
6. List the procedure that must be followed to ensure the oil supply is purged of air from the storage vessel through to the nozzle when the burner is commissioned.
7. Explain the term ‘purge period’ in connection with atomising boilers.
8. From Table 6.3 determine the combustion efficiency of a burner having a flue temperature of 204 °C and a $CO_2$ reading of 11 per cent.
9. State the effect on boiler efficiency if a smoke test gives a reading of nil on the smoke scale and the net flue temperature is found to be 420 °C.
10. List the items that must be checked when carrying out an annual service on an atomising oil boiler installation.
11. Describe the terms (a) bunding and (b) ‘ullage’ in connection with oil storage vessels.
12. To protect the environment list the situations where secondary containment of oil storage tanks is required to comply with the Building Regulations.

# 7 Sanitary appliances

After completing this chapter the reader should be able to:

1. Describe the installation and working principles of WC macerator units.
2. List and describe the functions of the appliances necessary in commercial and industrial buildings for sanitary and ablutionary purposes.
3. Explain the need for thermostatic mixing valves and describe their working principles.
4. Understand the methods of fitting all types of shower units.
5. Identify the main types of urinal and their suitability for various situations.
6. State the basic principles of automatic flushing cisterns.
7. Understand the need for conservation of water and describe the methods used to avoid wastage with automatic flushing cisterns and ablutionary fittings.
8. Understand the requirements of ablutionary and WC facilities for the disabled.

## Introduction

This chapter deals mainly with sanitary appliances more commonly found in industrial and commercial premises than in domestic use, the one exception being WC macerating units.

## WC macerators

These units are designed for where it is not possible to make a normal connection from the outlet of a WC to the main discharge stack. The function of the macerator is to break up the solids which may be discharged when a WC is flushed, simultaneously pumping them and the flushing water to the main stack or drain via a 20 mm or 32 mm pipe, depending on the type of unit used.

Pipe sizes and their length depend upon the pumping capacity of the unit. Some macerators are suitable for WCs only, others can pump the discharge from a wash basin and WC or a complete bathroom. Figure 7.1 shows the general arrangement

**Fig. 7.1** Pictorial illustration of WC suite fitted with macerator unit.

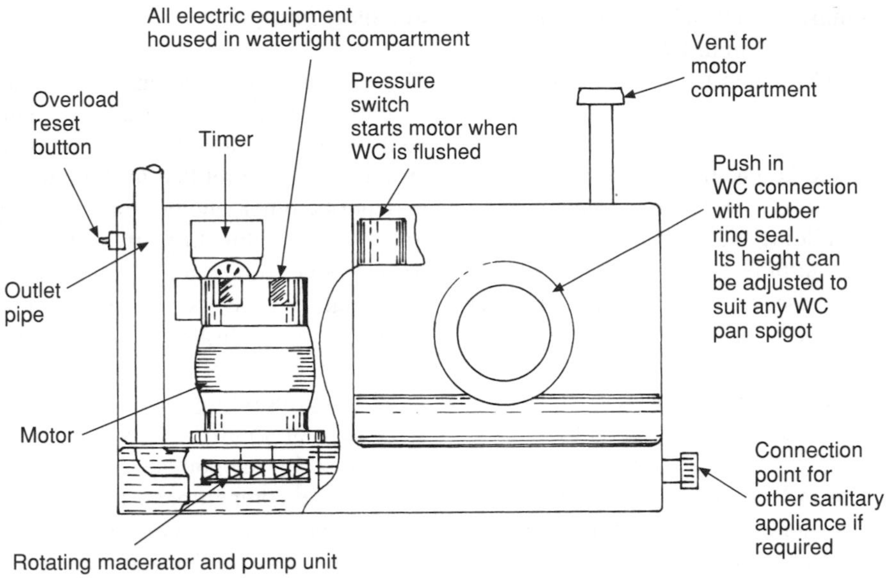


**Fig. 7.2** Cutaway section of macerator unit.

of the unit in relation to the WC. Figure 7.2 illustrates the main components of the macerator and pumping unit.

The following observations relate to all types of macerator irrespective of whether they are used with WCs, ablutionary appliances only or a combination of both. Prior to selecting a macerator, it is important to establish the types of appliances for which it will be used, and the lengths, both horizontal and vertical, of the main discharge pipes. It is important to refer to the manufacturer's specifications on these points.

*Discharge pipes*

As there are various types of macerator available the following should be used for general guidance only. With most types an increase in vertical height will reduce the length of horizontal run that is possible. One manufacturer states that for every 1 m of vertical pipework, 10 m of the maximum horizontal length must be deducted. This takes into account the extra work the macerator pump has to do. For example, if the maximum horizontal length pumped by a macerator is 50 m and the job requires a vertical lift of 2 m, the maximum horizontal length will be 30 m. Typical discharge pipe arrangements are shown in Fig. 7.3. In most

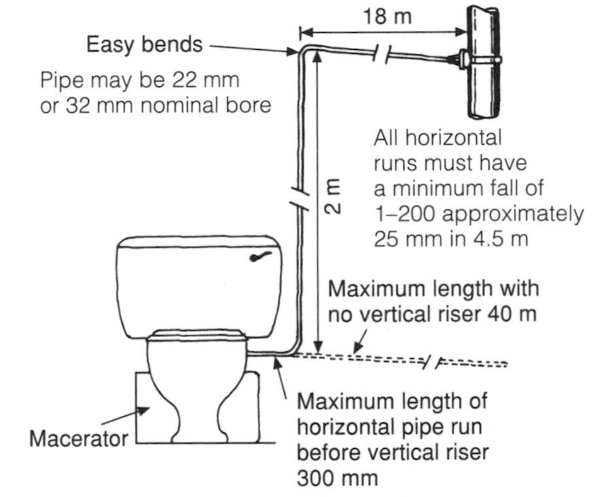


For discharge pipes having a vertical run, start with a base of 40 m maximum length and deduct from this 10 m for every 1 m vertical riser. In the example shown, the riser is 2 m high, 1 m must also be deducted for each bend.

**Fig. 7.3**

cases the diameter of the discharge pipe is 22 mm, and it must be copper or chlorinated PVC to BS 7291 Part 4. The use of these materials will prevent sagging and the build-up of residual water in the pipe leading to blockages. Connections to the main discharge stack may be achieved by a suitable branch or a 100 mm × 32 mm clamp-on connector and reducing fitting to 22 mm. Horizontal runs should have a 1:200 fall and if they are longer than 12 m the diameter must be increased to 32 mm to avoid self-siphonage. Only one vertical lift is permitted and should be taken off not more than 300 mm from the macerator. If the macerator is at high level in the building, necessitating a vertical drop on the discharge pipe, self-siphonage could take place. In such cases an approved air admittance valve must be fitted at the highest point in the pipe run. The falls and pipe diameters from all appliances discharging into the macerator must comply with BS EN 12056 Part 2, e.g. a basin discharge pipe will be 32 mm in diameter.

The operation of the motor is fully automatic, by means of a pressure switch and timing device. Units designed to deal with discharge of WCs and ablutionary appliances are provided with the necessary connections, which also incorporate non-return valves to prevent the back-up of foul water into the appliance discharge pipes. To install the macerator unit it is placed in position behind the WC and the pan spigot is pushed into the outgo socket on the macerator which is sealed by a synthetic rubber joint. Brass screws must be used to secure the WC to the floor, as although the motor can normally be serviced with the WC in position, should it be necessary to remove the unit as a whole, the WC must be capable of being withdrawn easily. The electrical connections must conform to the Institution of Electrical Engineers (IEE) standards, the usual arrangement being to provide a fused unswitched socket outlet having a 5 A fuse rating. The discharge pipe must be fitted in such a way that any vertical pipework is taken directly off the macerator or within a maximum of 300 mm. Changes of direction must be made using large-radius bends. For those units with a 20 mm outlet it is recommended that 22 mm copper pipe with machine-made bends is used. It must be clearly understood that these macerators are entirely mechanical in operation and being liable to occasional failure they should not be installed in buildings where no alternative WC accommodation is available. The advice of the local authority's Building Control Officer should be sought in cases where any doubt exists concerning their installation.

### Flushing troughs

These are used in buildings such as factories or schools where at peak periods the appliances may be in almost continuous use. A normal flushing cistern should be refilled in 2 minutes, but this may not be quick enough for peak usage. Flushing troughs enable this difficulty to be overcome because, due to the large volume of water held in the trough and the speed with which it is replaced, the WCs can be flushed almost continuously. The tanks themselves are made of galvanised steel or a suitable thermosetting plastic material with a base and height of approximately 255 mm. Lengths are determined by the number of WCs served and the width of the compartment. It is normal to restrict the number of WCs served from one trough to six. Due to the possible demand made on the water in the trough, it should be supplied via a float-operated valve with a minimum nominal diameter of 25 mm. It may be necessary to provide two valves if the supply is insufficient to meet the demand.

When filled with water these tanks are heavy and must therefore be well supported. If the WC partitions are constructed of brickwork, little support will be necessary, but if only lightweight partitions are provided then very strong additional supports will be needed. If there is any doubt about the strength of fixings obtained with brackets and screws, short lengths of galvanised angle iron should be built into the wall.

The actual flushing mechanism is very similar to that of an ordinary cistern, the only difference being that a measuring chamber is necessary to limit the volume of flush. Figure 7.4 shows a typical siphon for use with a flushing trough. When the cistern is flushed, water is drawn from the trough and the measuring chamber, in which is drilled a small hole. Water is siphoned out of the measuring chamber more quickly than it can enter, and when air is admitted to the siphon via the dip pipe, it breaks the siphon thus stopping the flush. If the siphon is made of metal, the small hole sometimes becomes corroded and fails to allow water to enter the measuring chamber. This will result in failure of the flushing arrangement and is the first thing to check if the siphon fails to operate.

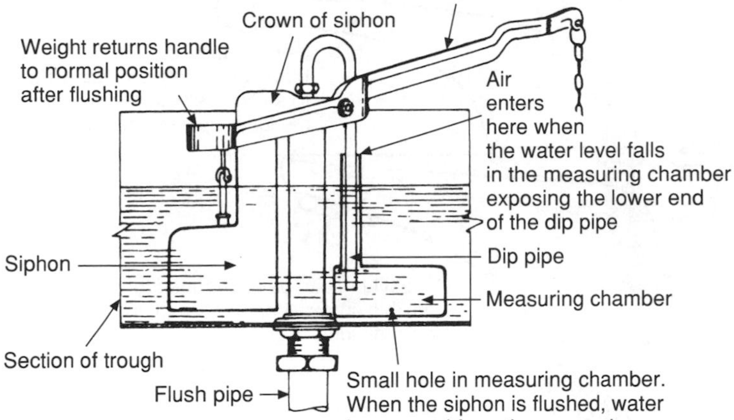
Flushing lever arm
Crown of siphon
Weight returns handle to normal position after flushing
Air enters here when the water level falls in the measuring chamber exposing the lower end of the dip pipe
Siphon
Dip pipe
Measuring chamber
Section of trough
Flush pipe
Small hole in measuring chamber. When the siphon is flushed, water is removed from the measuring chamber faster than it can enter this hole. Siphonage ceases when air enters the crown of the siphon via the dip pipe

**Fig. 7.4** Siphon for flushing trough. Do not attempt to reduce this volume of water to 6 litres by shortening the dip pipe, as this may be insufficient to clear the contents of the trap in existing WCs. Those produced now have been modified to function with a smaller flush volume and any new installations will have to be designed to comply with the Water Regulations.

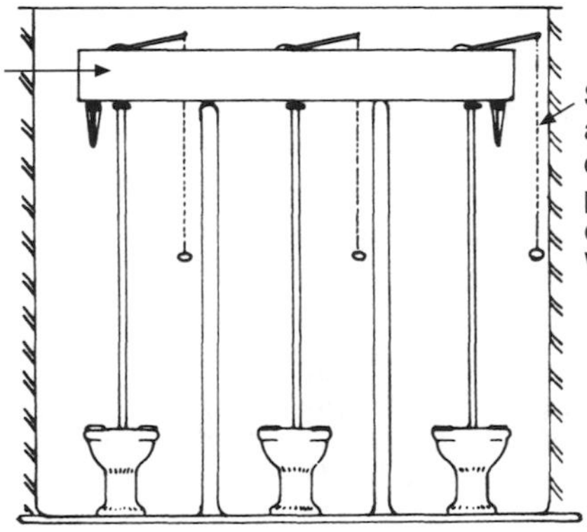
Trough supported on both brackets and division walls.
Sep-arate chain pull for each WC

**Fig. 7.5** Application of flushing trough.

Figure 7.5 illustrates a typical flushing trough serving three WCs.

## Ablutionary fittings

### *Taps*

Figure 7.6 illustrates a modern monobloc tap. They are designed for use in domestic and commercial premises, e.g. hotels. The mixing arrangement is similar to that of the non-thermostatic shower shown in Fig. 7.13. It is important that a careful check on the water temperature is made on the completed installation to avoid possible scalding Details of these taps must be retained by the customer, so that if maintenance becomes necessary the taps can be identified. Most of them embody a cartridge containing on/off and mixing components, also a non-thermostatic temperature control which is set by the plumber during installation. A temperature check must be made when the taps are commissioned.

There are several alternatives from which to select a suitable arrangement for hand washing in commercial and industrial premises. Figure 7.7 shows a pipework arrangement employing a secondary circulation so that hot water is delivered to the draw-off point almost instantaneously and at a safe temperature. To avoid waste of water some form of automatic shut-off must be provided. One method is the use of an electronic sensor which will automatically open or close the tap when it detects no movement by the user. These taps usually incorporate a timing device which it is set by the plumber when the work is commissioned. Basins

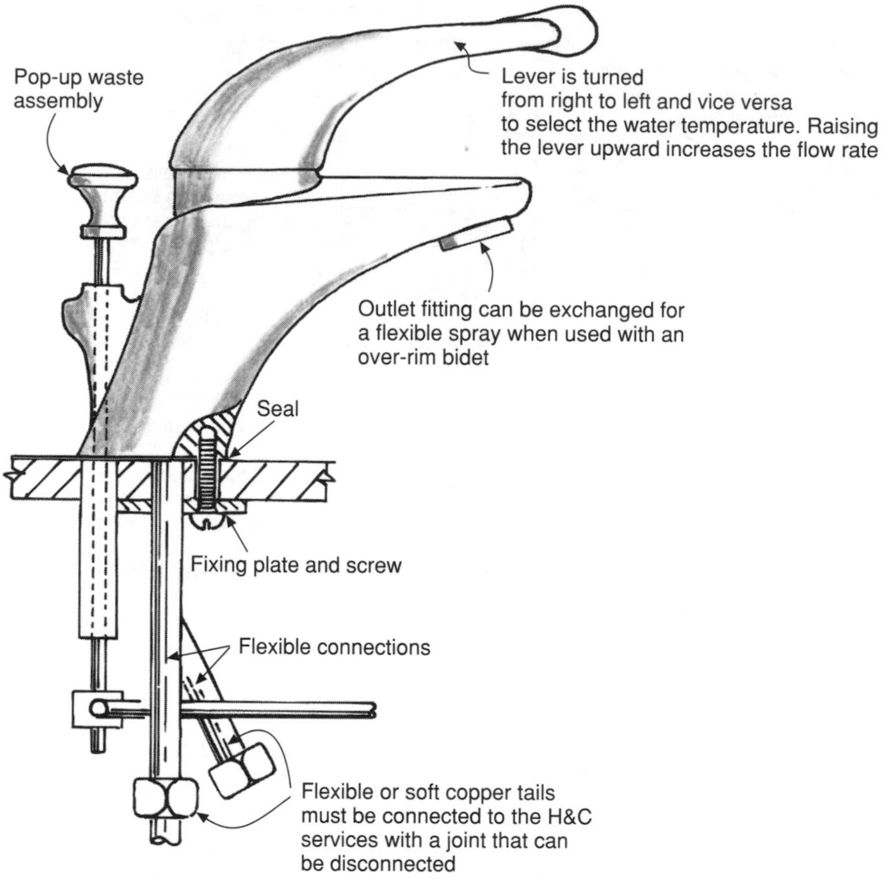

Pop-up waste assembly
Lever is turned from right to left and vice versa to select the water temperature. Raising the lever upward increases the flow rate
Outlet fitting can be exchanged for a flexible spray when used with an over-rim bidet
Seal
Fixing plate and screw
Flexible connections
Flexible or soft copper tails must be connected to the H&C services with a joint that can be disconnected

**Fig. 7.6** Modern monobloc basin/bidet mixing valve. Equal pressures are recommended.

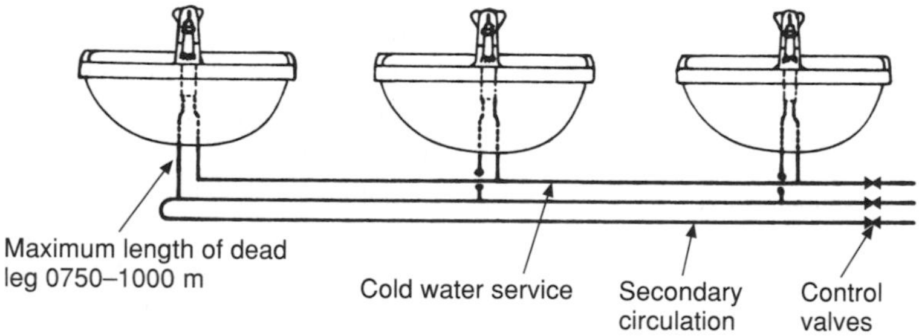

Maximum length of dead leg 0750–1000 m
Cold water service
Secondary circulation
Control valves

**Fig. 7.7** Pipework details to monobloc laps on a range of basins. The hot water supply to each draw-off point must be provided with some form of temperature control unless the whole installation is controlled by a master valve supplying mixed hot and cold water at handwashing temperature.

used for hand washing only are sometimes smaller than those for domestic properties, and are usually supported on a purpose-made frame or built in brackets. Fixings for any appliance in public buildings must be robust enough to resist carelessness and deliberate vandalism. The use of screw-on brackets is seldom suitable in these conditions. Where it is necessary to seal the gap

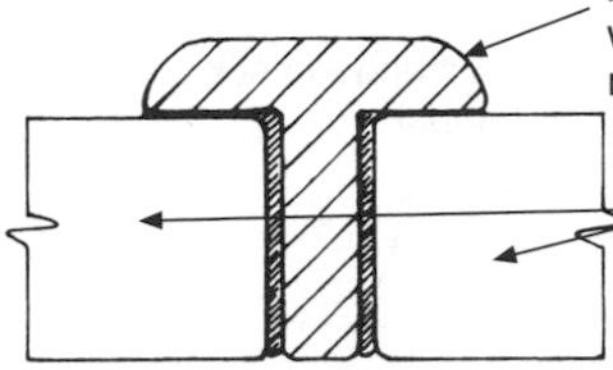
Clayware cover strip bedded on to basin edges with non-hardening mastic
Edges of wash-basins. Note that similar arrangements are made to seal the edges of stall urinals

**Fig. 7.8** Sealing the edges of basins in a range.

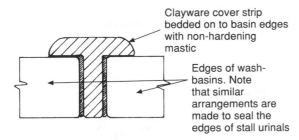

between adjoining basins, a glazed clayware cover strip is used as shown in Fig. 7.8. This prevents the ingress of filth to an area difficult or impossible to keep clean. These cover strips should be made good on the edges of the basins, using a material that allows for their easy removal should a basin have to be replaced. A suitable non-hardening mastic should be used for this purpose as it is not easy to match these strips should they be broken.

*Industrial hand-washing facilities*
Washing facilities for industrial usage may have to withstand rough treatment, and for this reason ablution troughs made of heavy glazed clayware or stainless steel are commonly used. Figure 7.9 shows a typical example. These troughs are suitable for wall fixing or, to save wall space, they may be fitted back to back and are then known as an island arrangement. An alternative to ablution troughs is the circular ablution fountain illustrated in Fig. 7.10. Due to its shape it allows the maximum number of people to wash at the same time. The water supply, which must be thermostatically controlled at a suitable temperature for hand washing, is discharged in the centre, forming an unbrella-like spray. The spray is operated by depressing the foot ring which is connected to a valve inside the column. Maintenance of the valve or discharge pipe is effected via the access panels in the column.

*Non-concussive valves*
This is a cheaper, tried and tested method of avoiding wasting water. They are designed in such a way that the head must be depressed to permit the valve to open. On releasing the head, the water will flow for a short period only prior to automatically closing thus avoiding the possibility of a tap being left running. The term *non-concussive* is derived from the fact that the original self-closing valves were spring loaded, causing sudden closure of the supply often resulting in water hammer. Modern taps of this type are made in such a way that they close gradually. It is essential that when they are fitted or repaired they are carefully regulated, to the manufacturer's instructions, to close slowly. Plugs must be provided for basins fitted with non-concussive taps. Unfortunately, plugs often disappear from public toilets, and the type which are locked into the waste fitting and therefore cannot be removed should be used.

## Anti-scald valves

These valves are essential to prevent scalding in sheltered accommodation and public buildings. Although non-thermostatic controls are currently permissible, it is likely that future regulations will require the use of thermostatic control for all hot water installations, as it offers greater protection against scalding. The usual form of temperature control used in these valves is a thermostatic bellows similar to that shown in Fig. 7.18. They can be used within certain limitations on supplies of differing pressures, usually up to 2 bar, although it is always preferable that the pressure on both hot and cold supplies is the same. The manufacturer's specifications should always be checked before fixing these valves. It should be noted that prior to the 1986 Water By-laws it was mandatory that the hot and cold supplies in any mixer valve, with the exception of the 'biflo' type, should have equal pressure. The main reason for this was (a) to reduce the risk of pollution and (b) to prevent water wastage arising from the possibility of a main pressure cold water supply backing up into the low-pressure hot water system and causing the feed cistern to overflow. The reasons for prohibiting mixed pressures no longer apply due to the installation of check valves. It is, however, much easier to balance a mixed supply of H&C water if both pressures are equal.

The valve shown in Fig. 7.11 is designed for controlling the supply of hot water to wash basins, baths and bidets. One valve can be used to control

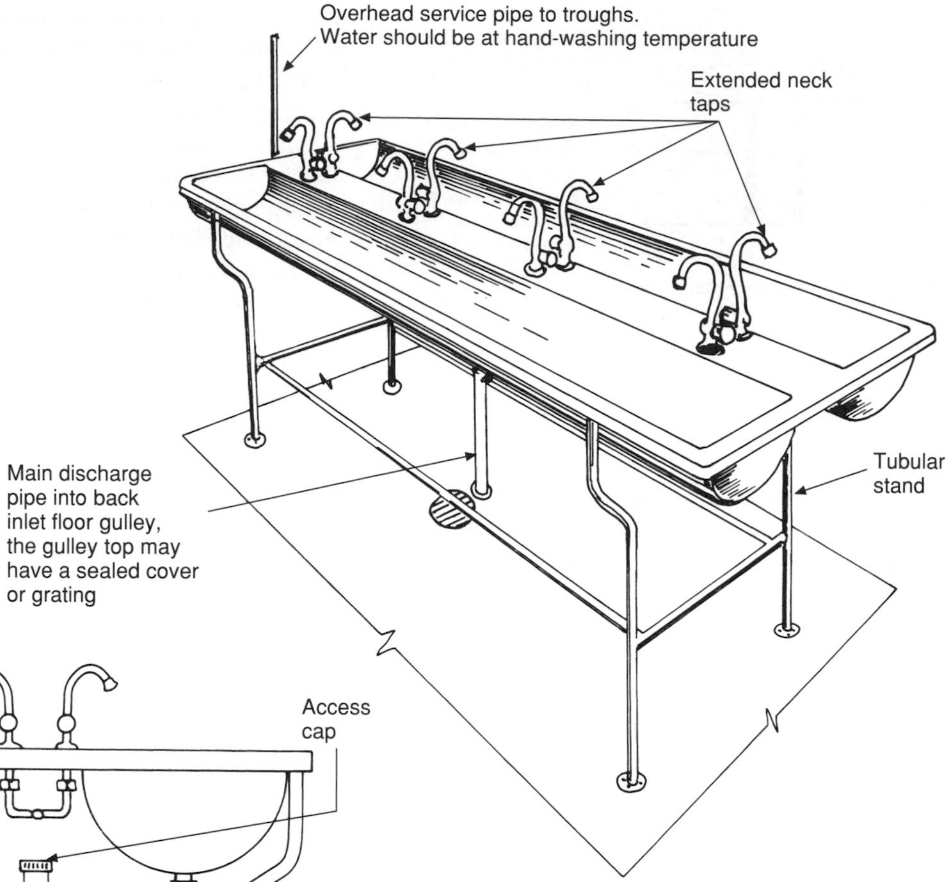


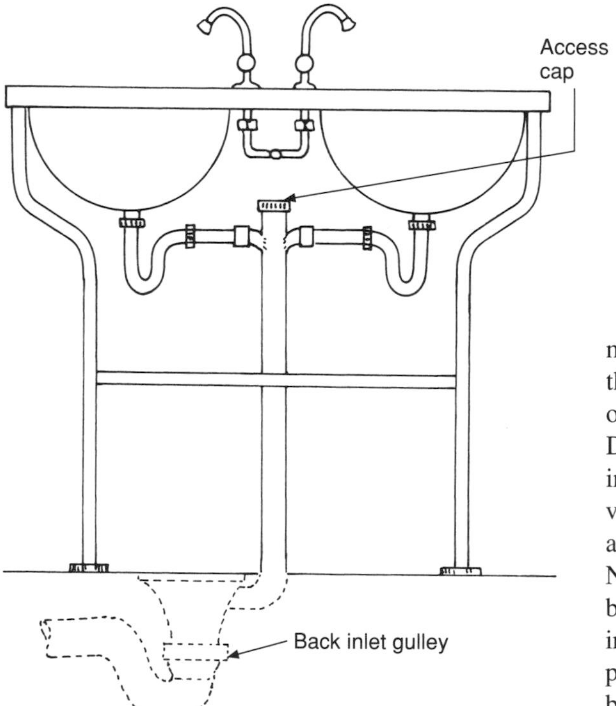

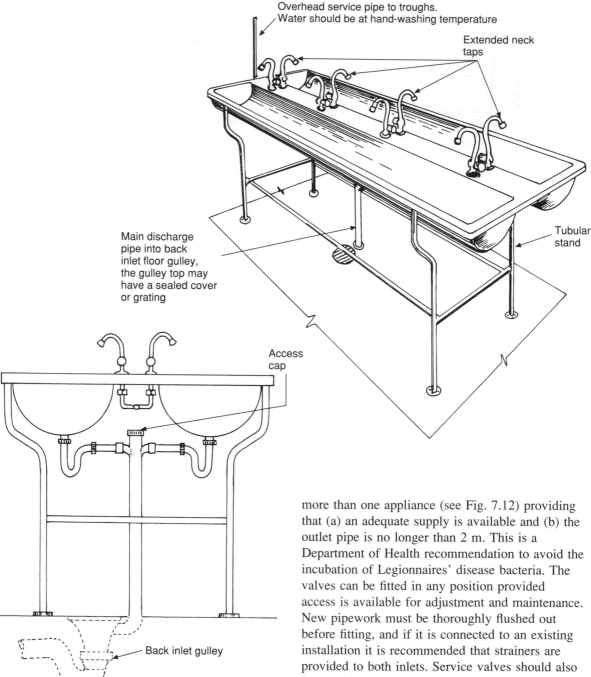

End view of island showing discharge pipe and water supply arrangement. The back inlet gulley is not essential, but would be useful in cases where it is necessary to wash down the floor area.

**Fig. 7.9** Typical industrial island-type washing trough.

more than one appliance (see Fig. 7.12) providing that (a) an adequate supply is available and (b) the outlet pipe is no longer than 2 m. This is a Department of Health recommendation to avoid the incubation of Legionnaires' disease bacteria. The valves can be fitted in any position provided access is available for adjustment and maintenance. New pipework must be thoroughly flushed out before fitting, and if it is connected to an existing installation it is recommended that strainers are provided to both inlets. Service valves should also be fitted in all cases to avoid major shut-downs when maintenance is necessary. Always comply with the manufacturer's instructions relating to installation, commissioning and testing. The recommended temperature settings are 40–41 °C for wash basins and bidets and approximately 43–44 °C

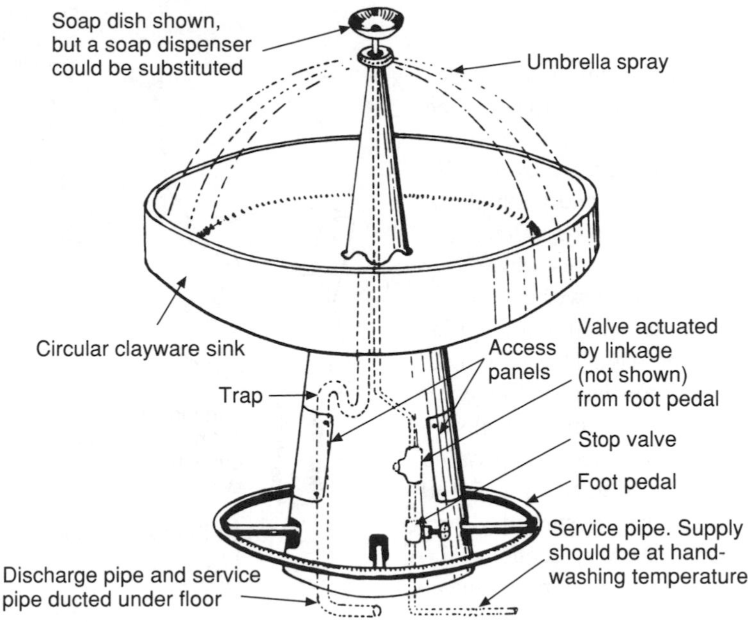


**Fig. 7.10** Washing fountain.

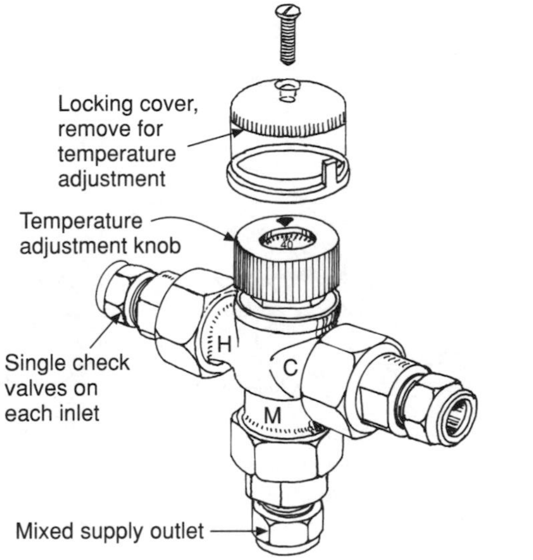


**Fig. 7.11** Thermostatic mixer valve for wash basins, bidets and baths. Some types include ball servicing valves; they are essential in commercial and public buildings to avoid a major shut-down when maintenance/servicing becomes necessary.

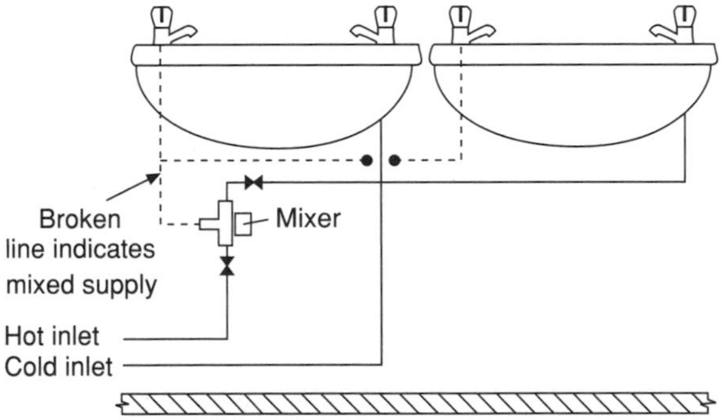


**Fig. 7.12** Application of the thermostatic mixer valve. Note the maximum length of the mixed supply is 2 m.

for baths. When the recommended temperature has been set the locking cap provided must be fitted to prevent any unauthorised tampering with the setting. After installation a check should be made within the period of time recommended by the manufacturer to ensure the mixed water temperature is within the prescribed limits. Temperature checks must also be carried out after any servicing operation. Always use reliable instruments for this purpose.

Before dismantling a valve suspected of malfunctioning, first check the strainers are clean, the check valves are in working order and any isolating valves are fully open.

## Showers

A wide variety of shower equipment is available and it is important to discuss with customers their requirements and expectations before a choice is made. Showers have many advantages over baths, e.g. they occupy less floor space, and washing the body with running water is said to be more hygienic. In small domestic properties where there is insufficient space for a shower cubicle, a shower and mixer tap combination as shown in Fig. 7.13 can be fitted to the bath. These are not thermostatically controlled, however, and unless carefully regulated can cause scalding. It is now possible to purchase a similar mixer incorporating thermostatic control to the shower. See Fig. 7.13(b).

Showers operating under the pressure of a storage cistern in domestic housing are very economical. A shower lasting 4–5 minutes will use approximately 28–36 litres, whereas an average bath will use approximately 110–140 litres. However, some pumped power showers are capable of producing flow rates of 20 litres per minute. Many operate at a minimum pressure of 1 bar and are often used with multiple shower heads which use more water than a bath and are not water conservation appliances. While this does not preclude the use of all power showers, they should be selected very carefully after consultation with the customer.

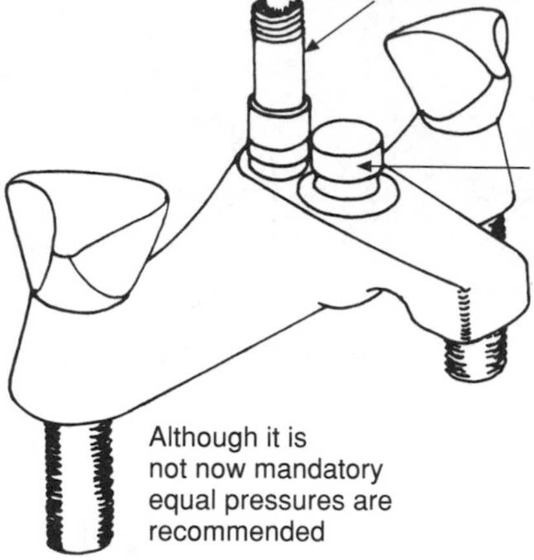
Flexible hose connection to shower rose
Manual selector for bath or shower outlet. Also acts as single check valve. Where fitted directly to main services, as with an unvented hot water supply, a further check valve must be fitted in the shower outlet connection
Although it is not now mandatory equal pressures are recommended

(a) Bath mixer and shower combination

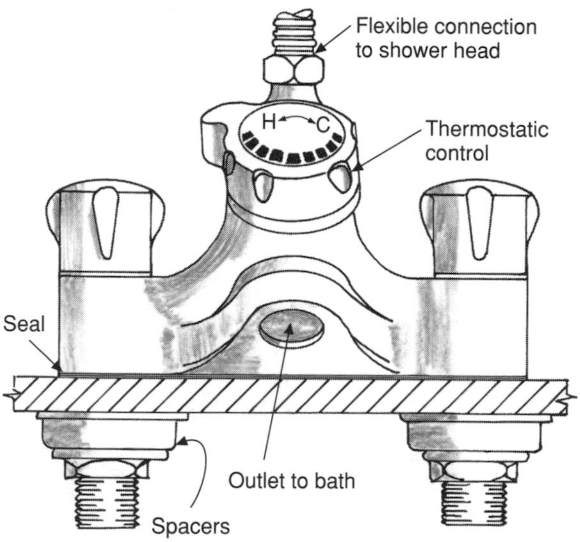
Flexible connection to shower head
H
C
Thermostatic control
Seal
Outlet to bath
Spacers

Note that showers of this type may require a pressure-reducing valve where pressures exceed 5 bar.

(b) Bath shower mixer with thermostatic shower

**Fig. 7.13**

*Shower trays*

Details of the types available and the methods employed to ensure water tightness between the wall surface and the top end of the tray are essential. Some types of shower tray are produced with an upstand as shown in Fig. 7.14. Properly installed they also solve seepage problems. Acrylic shower trays are normally provided with adjustable feet, which allows the top edges to be carefully levelled; this is very important if purpose-made enclosures are to be fitted. When showers are installed on suspended wooden floors, shrinkage of the timbers, especially when new, can cause many problems and for this reason one-piece shower units are recommended. These incorporate the tray and sides of the shower in one complete prefabricated unit, sometimes called a 'pod', which has no joints so cannot possibly leak. An alternative to this is to fit the shower tray in a prefabricated metal safe as shown in Fig. 7.15, which properly installed is very resistant to leakage. After fitting the shower tray make sure it is protected from damage by other trades. Cardboard packing secured with masking tape and covered with a dust sheet is usually satisfactory.

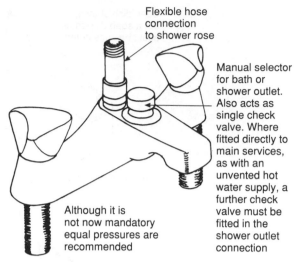

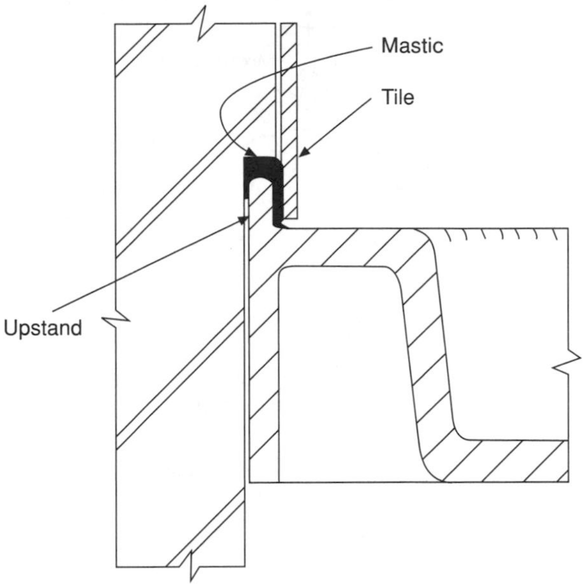


**Fig. 7.14** This shower tray is manufactured with an upstand, which, properly installed, solves seepage problems.

*Pollution risks*

Showers are classified as a category 3 risk for back-siphonage and require a backflow device complying with type AU2; this can be an air gap or check valves. Those with fixed shower heads have an air gap well in excess of the minimum requirements. Those with flexible hoses require double check valves. Provided the shower mixer inlets and the mixed outlet to the head are fitted with approved single check valves this will meet the requirements of the Water Regulations. An additional safeguard is the use of a hose restraining ring shown in Fig. 7.16. Fittings with flexible hoses that can fall into a WC or bidet pose a category 5 risk of pollution, and reference should be made to Fig. 3.34.

*Shower discharge pipes*

The connection of shower traps to the discharge pipes when it was mandatory to use traps having a 75 mm seal was always difficult due to lack of space. The Building Regulations now permit traps having a 50 mm seal on flat-bottom appliances such as shower trays, baths and sinks. Shower traps are very prone to blockage with hair and soap residues and the very low flow rate in the discharge pipe allows the build-up of soap deposits, especially in the trap. The use of bottle traps, which can be dismantled for cleaning by removing the grating and dip pipe, makes maintenance a lot easier providing the trap has been correctly installed and the discharge pipework checked for leakage prior to tiling or panelling the shower cubicle. Bottle traps, by virtue of their design, are not self-cleansing.

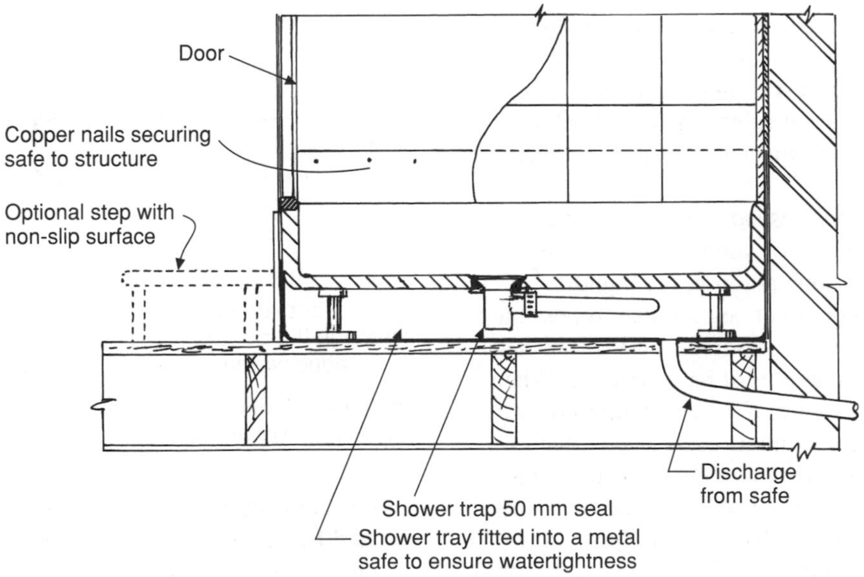


**Fig. 7.15** Shower trays installed on suspended floors.

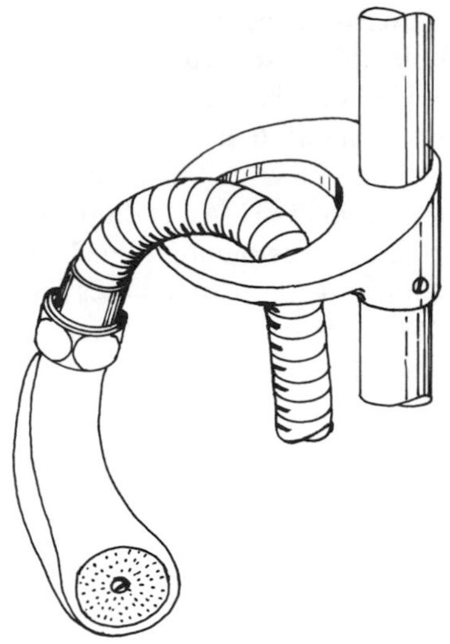

**Fig. 7.16** Hose restraining ring. This must be fitted in such a way as to prevent the shower rose falling below the flood level of the appliance or into an adjacent WC or bidet.

*Shower mixing valves*

Those of the non-thermostatic type are still available and many are in current use. Figure 7.17 illustrates their working principles. Thermostatic mixer showers work on the movement of a heat-sensitive bellows or bimetallic coil, both of which automatically open or close the inlets to maintain a predetermined temperature. The temperature of water for showering is normally between 38 and 40 °C and thermostatic valves will maintain this temperature range plus or minus one or two degrees, thus ensuring safety. A typical bellows-operated thermostat mixer is shown in Fig. 7.18.

*Commissioning and testing of thermostatic controls*

Digital thermometers of known accuracy are recommended for this purpose, but liquid thermometers should not be used due to possible breakage. Temperature readings must be taken at the normal flow rate after it has stabilised. When testing the temperature of a spray fitting take a sample drawn from, and as close as possible to, the outlet, taking a reading only after the container and its contents are at a stable temperature. Any thermostatic valve that has been readjusted must be recommissioned and retested in compliance with the manufacturer's instructions. It is important that records are kept of the results, one copy being retained by the fitter or plumber or their employer, the other given to the client or customer for safe keeping. The position of the mixer unit in the

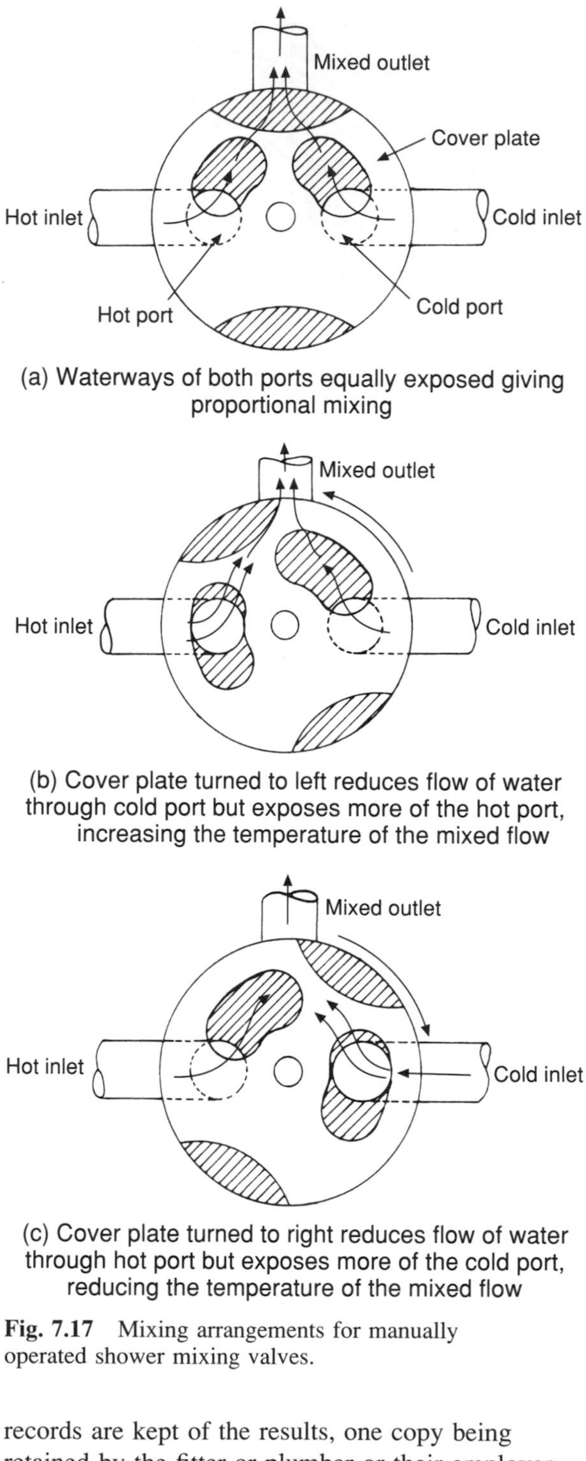


(a) Waterways of both ports equally exposed giving proportional mixing

(b) Cover plate turned to left reduces flow of water through cold port but exposes more of the hot port, increasing the temperature of the mixed flow

(c) Cover plate turned to right reduces flow of water through hot port but exposes more of the cold port, reducing the temperature of the mixed flow

**Fig. 7.17** Mixing arrangements for manually operated shower mixing valves.

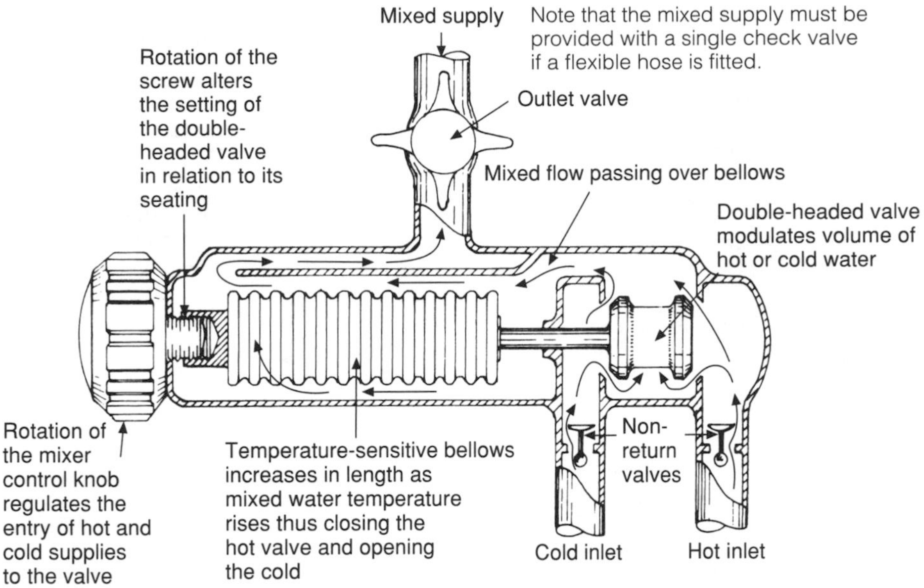


**Fig. 7.18** Working principles of thermostatic mixing valves.

shower cubicle is important. Its height from the base of the shower tray is usually 1.45 m. It should be positioned in the shower cubicle on the right or the left in such a way that the user's hand does not have to pass through the shower spray to adjust the temperature. The mixer valve and its associated pipework may be built into the wall providing the pipes are suitably protected against the corrosive effects of cement mortar or plaster. This avoids having the rather bulky mixer protruding into the cubicle. It is essential, of course, to check first if the wall is thick enough to accommodate the mixer unit. Most manufacturers supply a suitable casing with units designed for built-in fixing, which ensures that any servicing can be carried out with a minimum of inconvenience. Where it is not possible or advisable to fit concealed units, surface-mounting types are available. While such an installation does not present such an attractive appearance, both the pipework and the mixer unit are available for easy maintenance and this is the recommended practice in public or industrial premises. Polished stainless steel or chromium-plated copper pipe should be used where pipework is exposed on the surface of a shower cubicle.

The arrangement of the shower spray head has two main alternatives, both of which are illustrated in Fig. 7.19. The type with a flexible pipe (Fig. 7.19(a)) enables the height of the shower rose to be adjusted and also permits its removal for hair washing. The fixed type (Fig. 7.19(b)) should be fitted, unless otherwise specified, so that the rose is a minimum height of 2 m from the floor of the shower tray. The minimum recommended head of water for the shower to be effective is 1 m from the shower rose to the base of the cistern (see Fig. 7.21 below).

*Pumped showers*

When the installation of a shower is considered in an existing building, this is one of the points that must be ascertained before work is commenced. If the minimum head is not available, several alternatives are possible. One is to raise the position of the feed cistern so that sufficient pressure is obtained; another is to fit a pump on the outlet of the mixer. The pump can be installed at floor level, in a false ceiling or even under the bath, the only prerequisite being that it is accessible for maintenance.

*Cavitation*

This term is used to describe where a mixture of water and air finds its way into a pump. It is a very

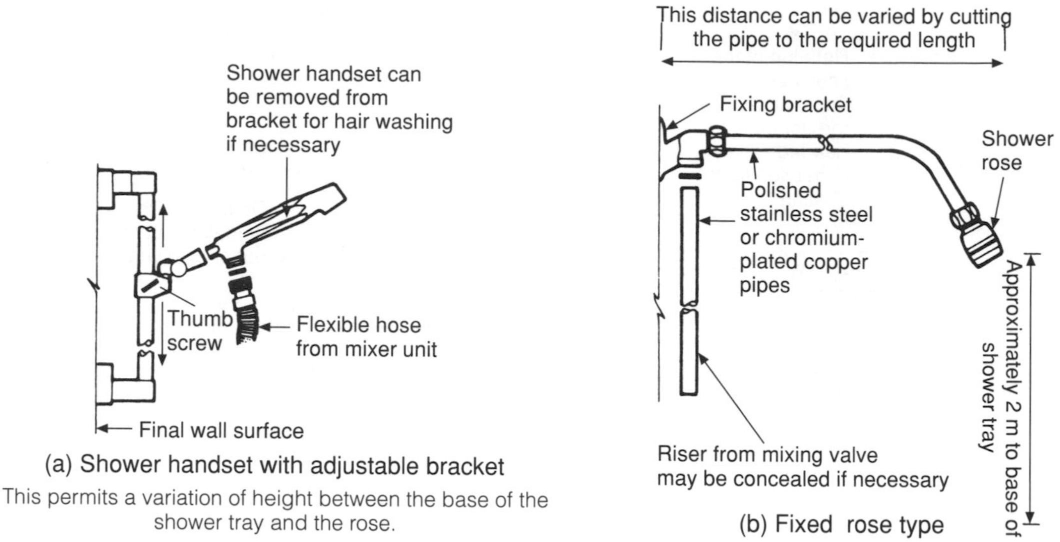
Shower handset can be removed from bracket for hair washing if necessary
Thumb screw
Flexible hose from mixer unit
Final wall surface
(a) Shower handset with adjustable bracket
This permits a variation of height between the base of the shower tray and the rose.
This distance can be varied by cutting the pipe to the required length
Fixing bracket
Shower rose
Polished stainless steel or chromium-plated copper pipes
Approximately 2 m to base of shower tray
Riser from mixing valve may be concealed if necessary
(b) Fixed rose type

**Fig. 7.19** Shower head arrangements.

undesirable situation as it leads to loss of pump efficiency and possible damage. Water contains air which it picks up as it falls through the atmosphere as rain, and to a lesser degree when it is stored. When it is under a positive pressure the air remains in the water, but when it is heated and subjected to a negative pressure, as it is on the suction or inlet side of a pump, the air and any other gases the water contains can no longer be held in solution and escape via the vent pipe as shown in Fig. 7.20. This

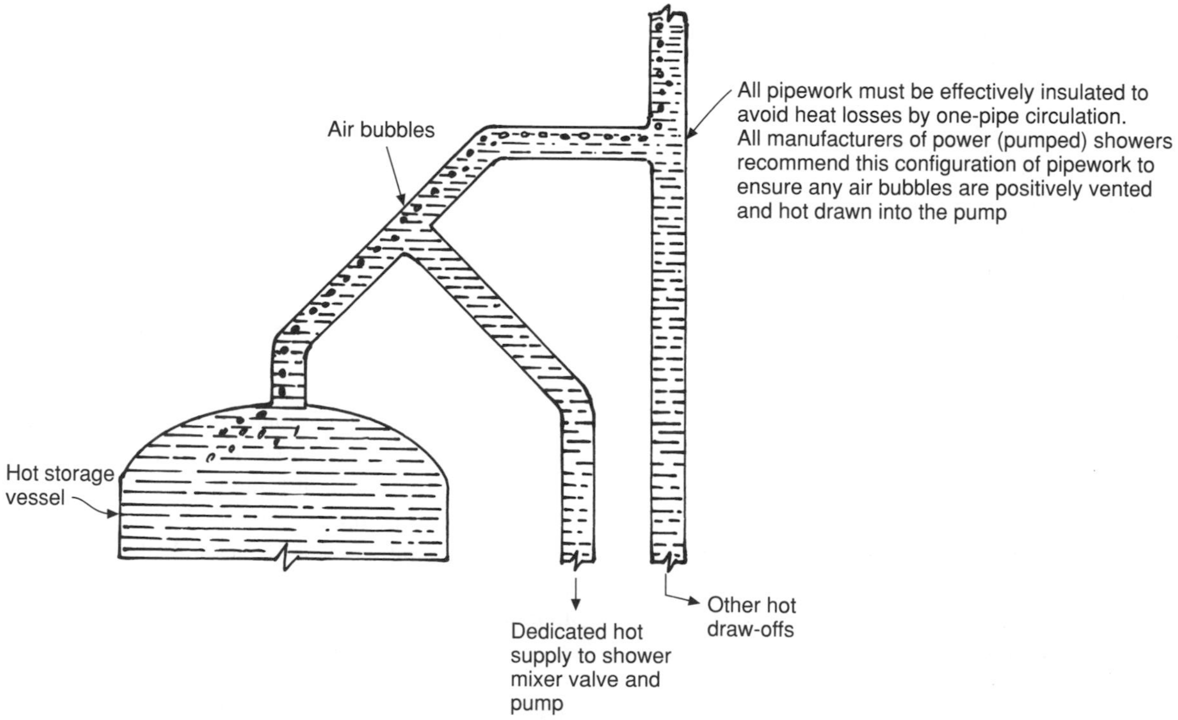
All pipework must be effectively insulated to avoid heat losses by one-pipe circulation.
All manufacturers of power (pumped) showers recommend this configuration of pipework to ensure any air bubbles are positively vented and hot drawn into the pump
Air bubbles
Hot storage vessel
Dedicated hot supply to shower mixer valve and pump
Other hot draw-offs

**Fig. 7.20** Connection of hot water services to pumped showers.

ensures positive venting and it will be seen that the possibility of air being drawn into the pipe is unlikely. One pipe circulation is likely to take place but this effect will be minimised if the pipes are adequately insulated. Yet another fact that must be taken into account is that the boiling point of water is lowered when the water in the pump suction pipe is at a lower pressure than the atmosphere. This will further permit the water to release yet more air and other gases into the atmosphere. Apart from the foregoing reduction of heat losses and furring, there are yet other reasons for maintaining the temperature of stored water at not more than 65 °C. A shower mixer may also be connected directly to a main water supply providing there are adequate safeguards against pollution risk.

### *Pipework schemes for showers*

The main types of pipework schemes recommended by Mira Showers are shown in Figs 7.21(a)–(h). These relate mainly to domestic work. When the hot storage vessel is a long distance away from the shower, it may be necessary to install a secondary circulation to avoid a long wait for hot water to reach the mixer valve. This will certainly be required in buildings where several showers are needed, e.g. changing rooms in a sports pavilion, or an industrial establishment where, due to the nature of the work, the employer provides bathing facilities. Such an installation is illustrated in Fig. 7.22. Although it is now permissible to provide mixer valves with varying H&C pressures, it is accepted that equal pressures are more user-friendly; however, always refer to the manufacturer's instructions on this point. Although non-thermostatic controls are currently permissible it is likely that future regulations will require the use of thermostatic control for all installations to protect users against scalding.

### *Electric showers*

These showers have become very popular for domestic use in recent years, mainly due to their relatively low installation costs. They are also useful where an isolated shower unit is to be installed necessitating a long run of hot water pipe. Their rating in kW varies from 7.5 to 10.0 at normal mains voltage, the larger ratings giving quite a hot shower. Most electric showers will raise the temperature of the incoming water through

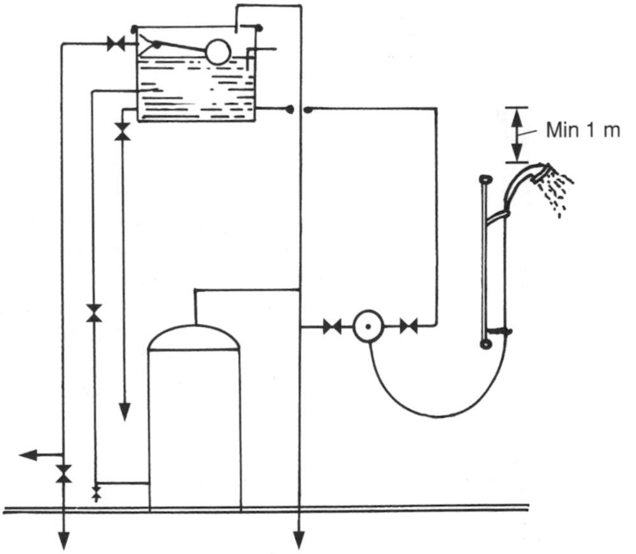

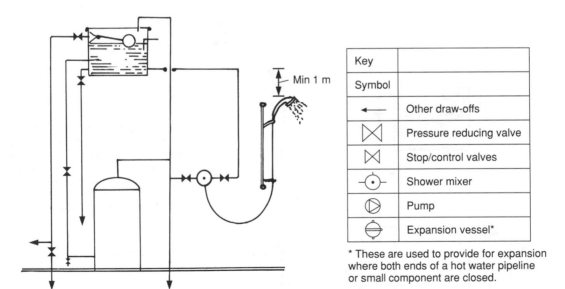

| Key | |
|---|---|
| Symbol | |
| ← | Other draw-offs |
| | Pressure reducing valve |
| | Stop/control valves |
| | Shower mixer |
| | Pump |
| | Expansion vessel* |

* These are used to provide for expansion where both ends of a hot water pipeline or small component are closed.

(a) A traditional gravity head system. Both the hot and cold services must be gravity fed from the feed cistern. This arrangement is very satisfactory and economic in terms of water conservation

**Fig. 7.21** Illustrations based on the recommendations of 'Mira Showers'.

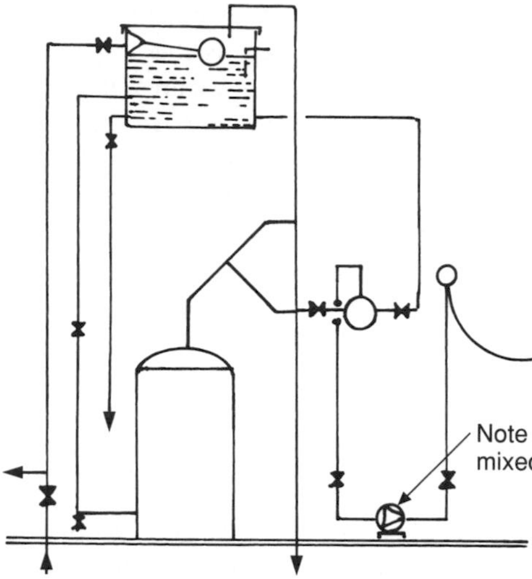


(b) Single impeller pumped showers. Used where a shower is required but the available head of water is insufficient

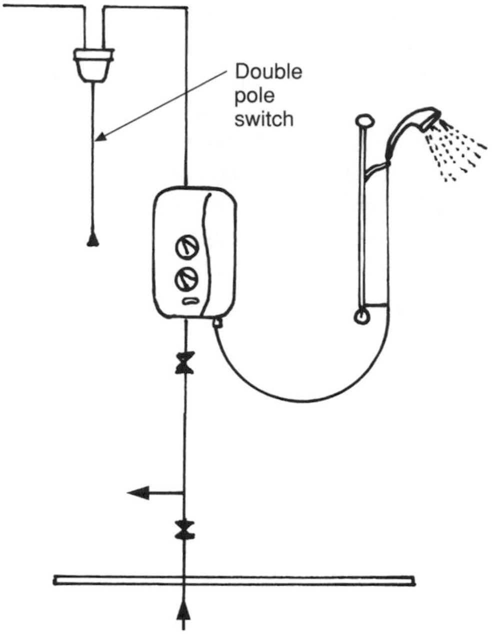


(c) Instantaneous electric shower. These are supplied direct from the main and are useful in situations where there is no water storage. The main disadvantage is the fluctuating water temperature which sometimes occurs *when* other draw-offs are opened. To avoid this a pumped version similar to the all-in-one type will avoid this problem assumming a cold water storage vessel is available or can be fitted

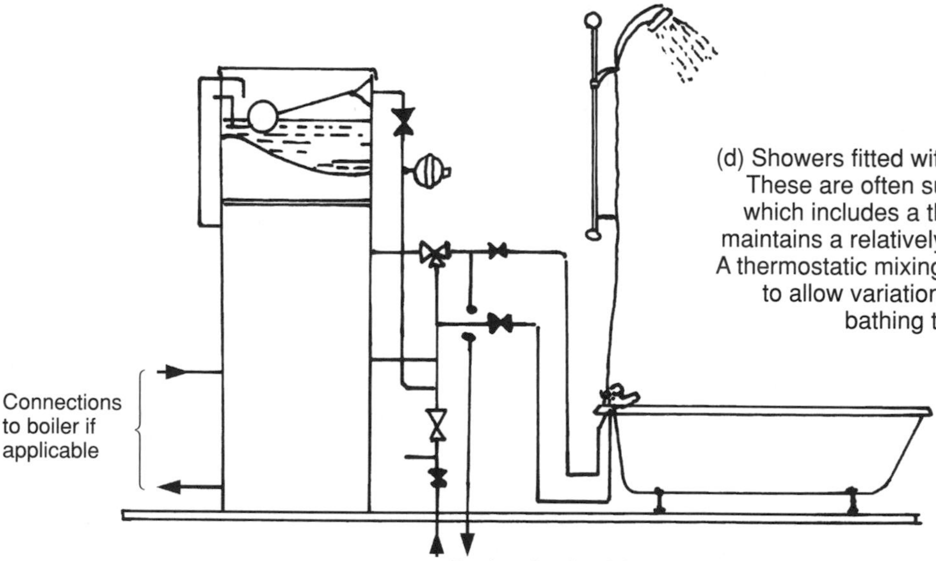


(d) Showers fitted with thermal storage units. These are often supplied as a package which includes a three-way valve which maintains a relatively constant temperature. A thermostatic mixing valve is recommended to allow variations of showering and bathing temperature

Further details of these units are shown in Fig. 4.17

**Fig. 7.21** *(cont'd)*

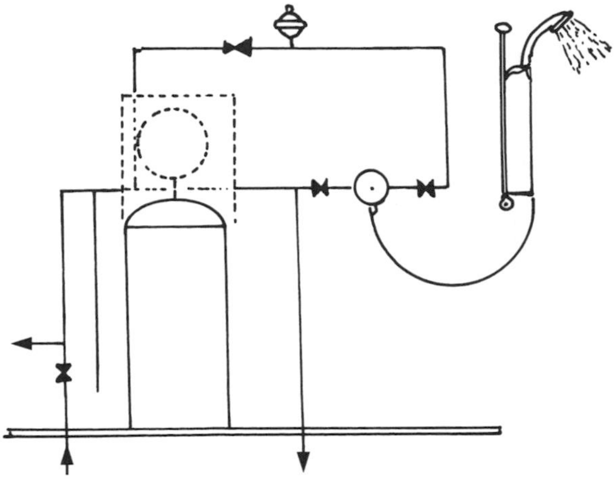

(e) Unvented main pressure showers. Only a competent person as defined by the Building Regulations Part G schedule is allowed to work on unvented systems. It may be necessary to fit an extra pressure-reducing valve if the cold supply bypasses that fitted to the unit. In cases of doubt the manufacturer should be contacted

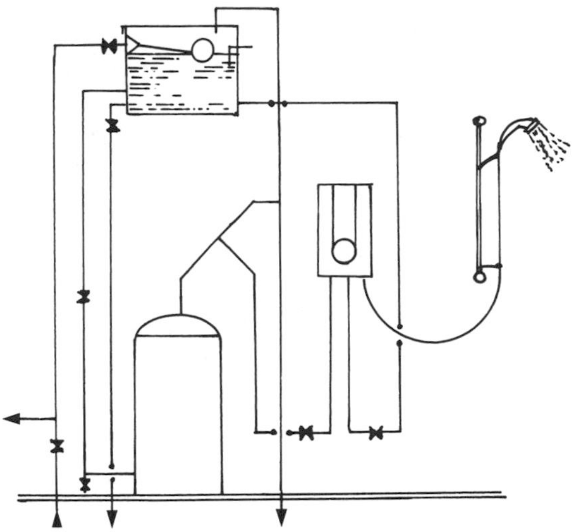

(f) All-in-one power showers. These are often a convenient alternative to (b). Both the pump and mixer controls are housed in a sealed unit

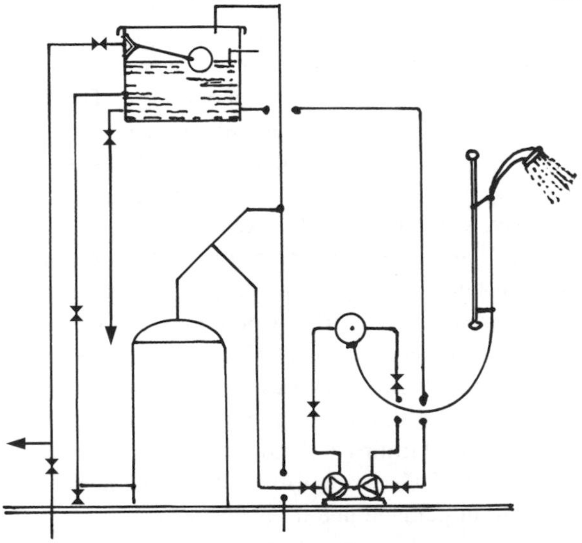

(g) Dual impeller pump scene. This arrangement can be used as an alternative to (a) or where it is necessary to convert an existing shower to a power shower where it is not possible to fit a pump on the outlet from the mixing valve

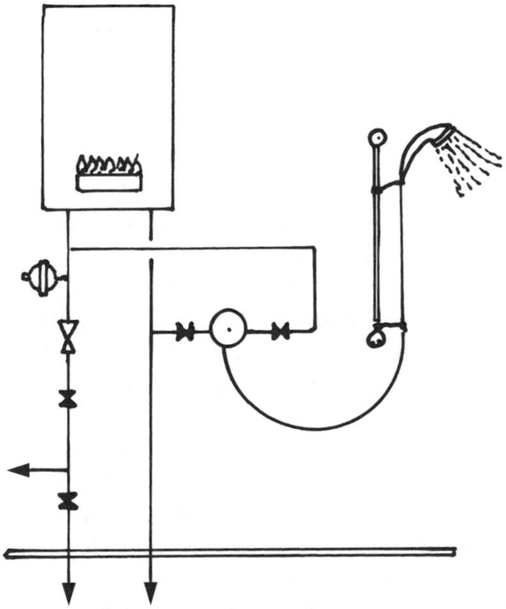

(h) A suitable thermostatic shower mixer must be provided where a multi-point gas heater or combination boiler are fitted. The appliance must be fully modulating and provide equal pressures to both the hot and cold services. A pressure-reducing valve is fitted to ensure pressures do not exceed 5 bar

**Fig. 7.21** (*cont'd*)

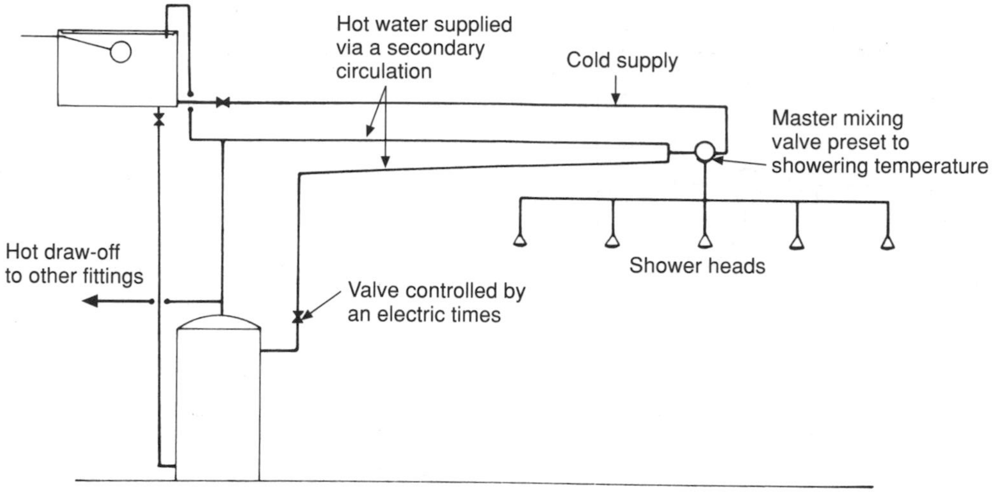


**Fig. 7.22** Layout for a range of showers.

32–35 °C. The temperature of the main water supply is variable and is usually between 10 °C and 15 °C, but it can drop in very cold weather to 5 °C. This means that a shower with a low electrical rating might have difficulty in heating the water to showering temperature at high flow rates. A typical electric shower heater is shown in Fig. 7.23.

The electrical part of the installation must be carried out by a competent person and comply with the IEE Regulations. This is important because cable sizing, voltage drop and earthing have to be considered prior to installation. Electric showers are, like electric immersion heaters, high rating appliances and must be wired directly to the main consumer unit, with a separate fuse or miniature circuit breaker. An adequate system of earthing should be in place (see Chapter 11). An isolating double pole switch must be fitted to isolate the unit for repair or maintenance and a double check valve must be provided on the *outlet and not the inlet.* This allows the water in the heater to expand without causing a pressure build-up. Installation of a water-treatment device or water softener is recommended in very hard water areas. Prior to installation the flow rate and static pressure of the water supply should be ascertained to ensure the compatibility of the unit. This is easily carried out using a pressure gauge and flow meter. Metered supplies do incorporate check valves and in such cases the manufacturers recommend the use of a mini expansion vessel on the water inlet to the shower unit. Most instant electric heaters incorporate a means of compensation for variations in supply pressures to maintain an even flow and stable temperature. This can be a simple device such as (a) a flow regulator, (b) a pressure-reducing valve or (c) a modulating solenoid valve. The last is the most effective in maintaining both a constant outlet supply and temperature. In situations where there is insufficient main pressure to operate the shower, a pumped supply taken from a storage cistern will increase the pressure, but it is essential that sufficient storage is available to supply any other existing draw-offs. A minimum storage of 120 litres is recommended by most electric shower manufacturers in such cases. If the appliance is to be used for the disabled, old or infirm, a temperature-limiting thermostat must be employed. Two other devices are also built into the appliance to prevent damage due to overheating:

(a) The thermal cut-off which automatically switches off the power supply if the water reaches a temperature of approximately 50 °C.
(b) The pressure switch which ensures that an adequate supply of water is available before the heater element is energised.

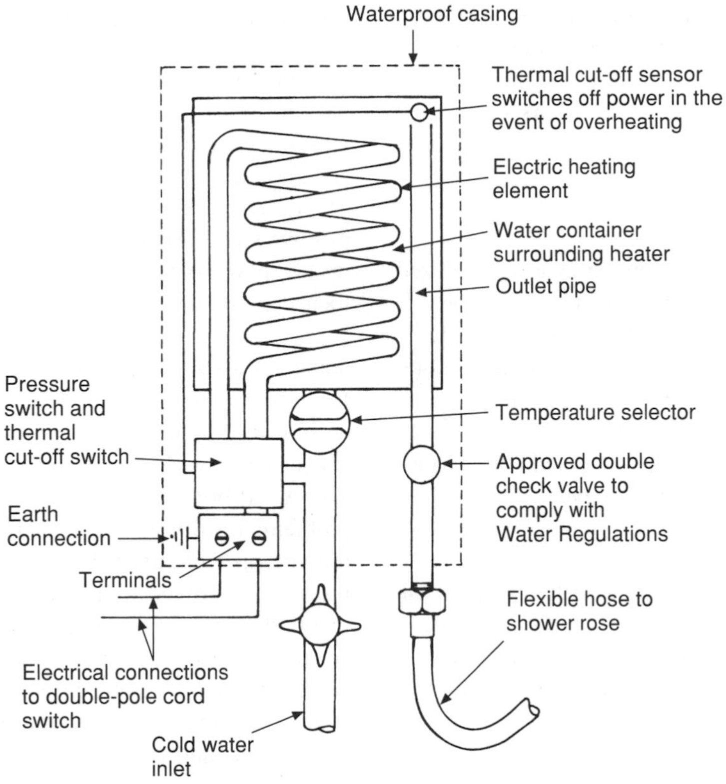


**Fig. 7.23** Diagrammatic section of electric instantaneous heater suitable for a shower.

*Electrical switching*

A fixed power supply with a double-pole cord-operated pull switch is essential. An ordinary pull switch simply breaks one conductor or wire, whereas the double-pole type breaks both the conductors, isolating the heater completely, and provides the high degree of safety needed where electrical appliances are installed in surroundings subject to moisture and dampness. Overcurrent devices may be cartridge or enclosed rewirable fuses; an alternative is a miniature circuit breaker. Always check the fuse rating against the manufacturer's recommendations. As an example, 7 kW units require a fuse rating of 30 A; 8.3 kW outputs require a fuse rating of 45–50 A.

*Maintenance of shower units*

Reputable manufacturers of shower equipment provide adequate information on maintenance, either with the unit or on request. The part of the installation which usually requires maintenance is the mixer where the O-ring seals and washers become worn after a period of use. Replacements are available from the manufacturer or a good plumbers' merchant. It will, of course, be necessary to quote the type and model number of the mixer. Before any maintenance work is undertaken the floor of the shower tray must be protected with a suitable covering from damage by shoes or boots. The outlet grating should be sealed to prevent any screws or small parts of the mixer falling into the discharge pipe while work is in progress as these will be difficult if not impossible to retrieve. Any scale formation on the working parts of the unit should be carefully removed using a suitable descaler. The type used for descaling kettles is suitable for this purpose. Abrasive materials or acid solutions may do irreparable damage and should not be used. Where loss of temperature control occurs in thermostatic showers it is usual to renew the temperature-sensing cartridge, as it is usually this that is at fault. The outlet temperature must always be verified using a suitable thermometer on completion of any maintenance work on the mixer unit.

### Urinals

Urinals are made of the following materials: glazed clayware, stainless steel or, in the case of some bowl types, vitrified china. Fibreglass urinals are made, but are considered by some to be unsuitable as they lack the strength and durability of the materials previously mentioned.

There are four main groups of urinals, known as the stall, slab, bowl and trough types.

*Stall urinals*

These were the original form of earthenware urinal and are still in use in old public buildings. While they have many advantages they are very heavy and expensive. Few, if any, manufacturers now list them.

*Slab urinals*

As the name implies, this type of urinal is built up of rectangular slabs of glazed clayware which are bedded onto a separate channel. The usual flushing arrangement is a sparge pipe running the whole length of the urinal in which is drilled a series of holes. When the cistern is flushed, water passes through these holes to cleanse the slab. It is important that the sparge pipe is fitted in such a way that the water impinges on the slab to ensure that the whole surface is cleansed. Slab urinals are illustrated in Fig. 7.24; they may be supplied with or without division pieces. From a sanitary standpoint, they are better without them as they are not cleansed by the sparge pipe, but they do provide some privacy for the user. Their main disadvantage is the number of joints required in their construction.

*Joints for slab and stall urinals*

It is usually the job of a bricklayer to fix and bed these urinals, and while they are concerned with giving a neat appearance to the finished joint, even good bricklayers do not always realise the problems resulting from defective jointing. For this reason a plumber should be aware of the processes involved. Cement mortar should be of one part fine washed sand to one part cement to ensure that it is waterproof. Sometimes white cement is used to point the joints to improve their appearance. It is essential that all joints are properly bedded as in Fig. 7.25. Simply pointing the joints when the slabs are in position is not satisfactory as the pointing medium is soon washed away leaving an unsealed area open to the entry of foul water. Some channels are made with a stepped rebate which reduces seepage problems. Channels for slab urinals are made having an integral fall and are numbered to indicate the correct fixing sequence.

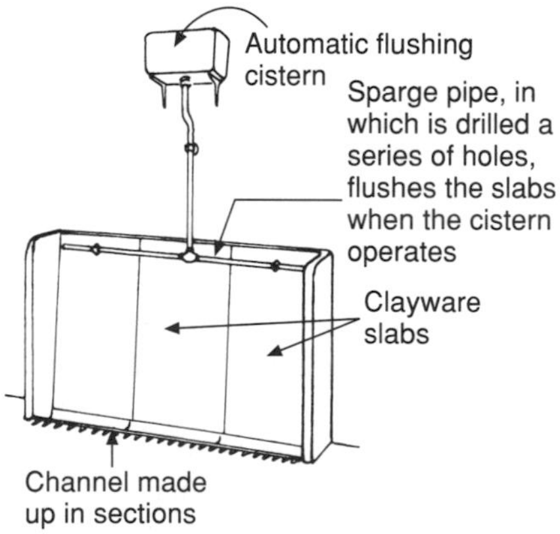


(a) Arrangement of slab urinals

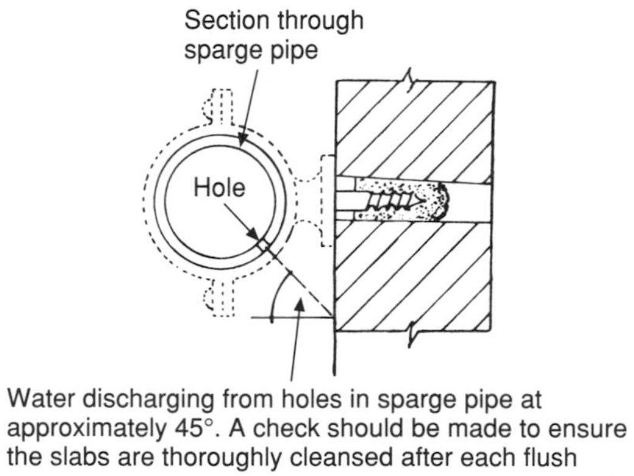


(b) Fixings for sparge pipes

**Fig. 7.24** Slab urinals.

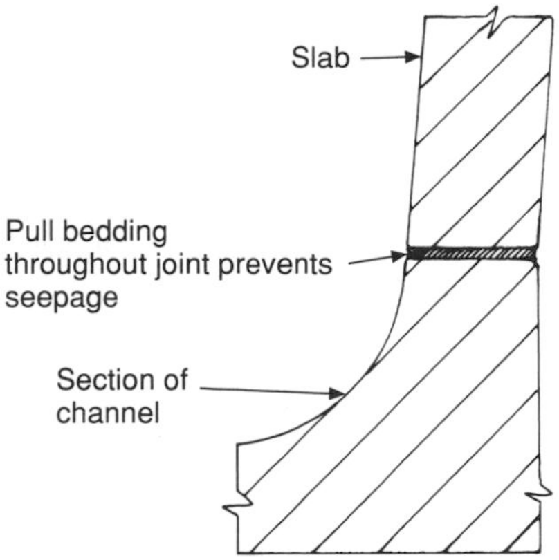


**Fig. 7.25** Correct method of bedding urinal slabs.

*Bowl urinals*

No matter how carefully the joints on slab urinals are made, any movement of the structure in which they are installed may cause them to crack. While this is serious in any case, if urinals are fitted on the upper floors of buildings the result of leakage will be especially unpleasant. For this reason they are sometimes built into a lead safe in a similar way to shower trays. This is expensive and for this reason bowl or trough urinals are often specified because they are made as a single unit (see Fig. 7.26). This overcomes the jointing problem presented by slab urinals. The main objection to bowl urinals is the possibility of fouling the floor, but this can be overcome by fitting a floor gulley which enables the whole area to be periodically washed down. As with wall-hung urinals the branch discharge pipes can be fitted above the floor level, and if mechanical joints are used the discharge pipe system can easily be dismantled for cleansing. The discharge pipes should be large enough to accommodate the deposits of scale built up by uric acid and, in hard water districts, calcium carbonate. The number of urinals in the range also has an influence on the discharge pipe diameter.

*Stainless steel trough urinals*

Stainless steel has much to commend it for the construction of urinals as it is less prone to damage by vandalism. It is easily cleaned and complete units can be fabricated by welding under factory conditions making site jointing of sections unnecessary — except for fitting the outlet. Urinals made of stainless steel may be of the wall-hung trough type shown in Fig. 7.27, which gives access above floor level to the discharge pipework making maintenance easier. They are also constructed with an integral channel for floor

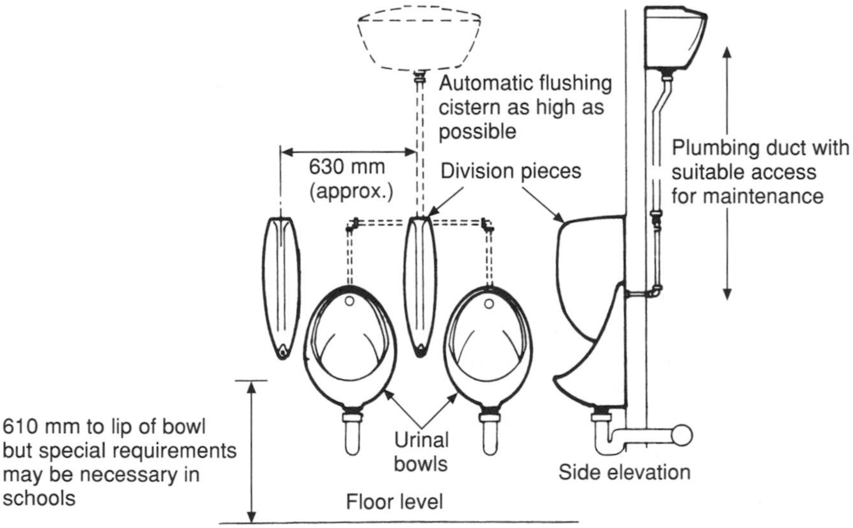


**Fig. 7.26** Bowl urinals. Those parts of this illustration shown as broken lines may be concealed in a plumbing duct. Use of such a duct provides a neat installation, cleaning is easier and there is less opportunity for vandalism.

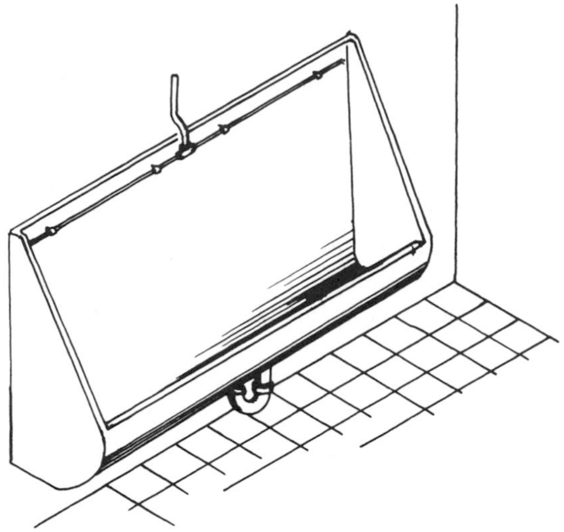

**Fig. 7.27** Stainless steel trough urinal.

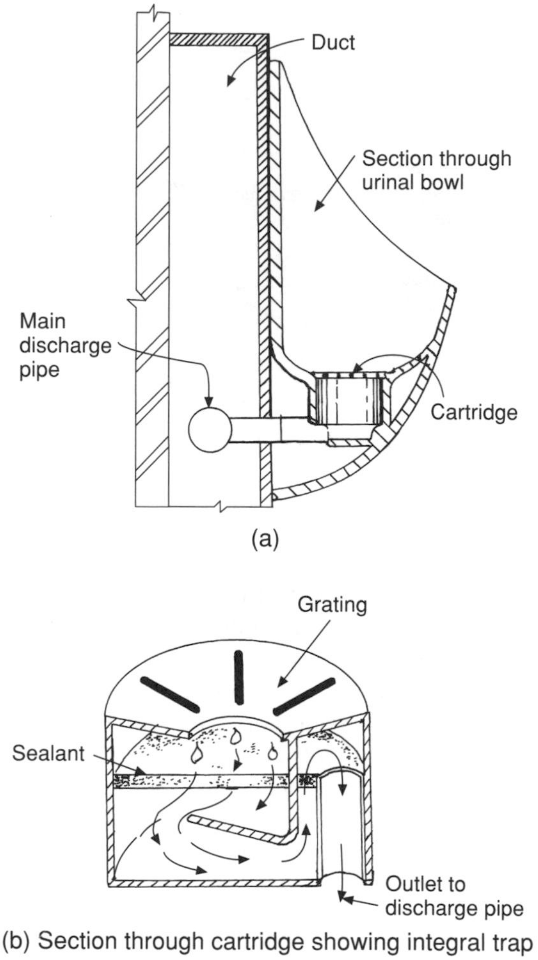


**Fig. 7.28** Bowl type waterless urinals.

fixing, and in this mode are not unlike slab urinals in appearance.

Bowl urinals can also be fabricated of stainless steel and, although expensive, they are less prone to vandalism than those made of clayware.

*Waterless urinals*

This is a comparatively new type of sanitary appliance which obviates the need for water for flushing urinals, thus saving water and reducing installation costs. A special renewable cartridge is necessary which contains an integral trap, a cross-section of which is shown in Fig. 7.28 together with its application in a bowl urinal. They are also suitable for installation in specially adapted trough urinals. Extensive tests have been successfully carried out in Germany and the USA to ensure they comply to the high standard of sanitation required. The absence of water prevents chemical reactions with urine, which avoids the objectionable odour associated with urinals. The cartridge contains a special biodegradable scaler which allows the urine to percolate into the cartridge and then floats on top of it. This provides a barrier preventing odours from the drain and the contents of the cartridge entering the washroom. A decrease in the speed of draining the urinal indicates the cartridge should be changed; frequency of changing will depend on the frequency of use. Heavy usage, such as in public toilets, may necessitate a change every 3–6 weeks. In areas of medium use such as offices and industrial premises, the cartridge may only need changing every 17–16 weeks.

*Urinal waste fittings*

The sizes of waste outlets recommended in BS EN 12056 (Part 2) are indicated in Table 8.1 (page 266). Ranges of more than six urinals must be provided with two outlets. The joint between the channel and the discharge pipe is made using a fitting similar to those shown in Fig. 7.29. They are designed to prevent debris that collects in the channel gaining access to the discharge pipe causing a blockage. If the outlet was flat it could become covered and

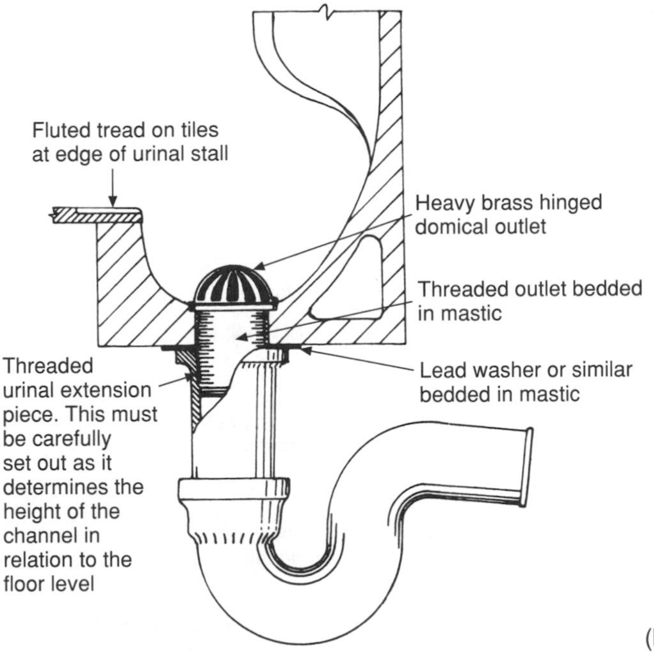


(a) Joint between urinal channel and cast iron drain

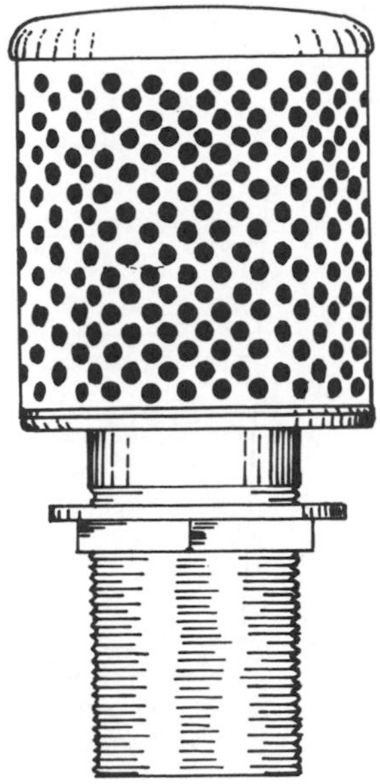

(b) Perforated stainless steel urinal waste fitting. These are more effective than traditional domical gratings in preventing blocked discharge pipes

**Fig. 7.29** Urinal wastes.

cause flooding. The outlets shown may be made of good-quality brass, stainless steel or high-density polythene. The last is not recommended as its service life, due to its inherent weakness, is likely to be shorter than metal. Domical gratings are hinged in such a way that they can be opened to expose the trap for rodding should this become necessary.

Due to the general lack of accessibility to discharge pipes from slab urinals, and the stresses to which they are subjected when cleaning is carried out, they should be made of a strong material such as copper or cast iron. With bowl or wall-hung urinals, the discharge pipes are usually exposed and sometimes PVC is used to reduce costs. Urinals can be a source of trouble, from both blockage in the discharge system and objectionable smells. Every care must be taken to ensure that the system is correctly installed with plenty of access to the discharge pipework and an efficient flushing cistern. A lockshield hose union tap should also be provided in urinal closets to enable the floor to be periodically washed down. The floor should have a slight fall towards the urinal channel, unless bowl urinals are installed, in which case, as stated previously, a suitable floor gulley must be provided.

*Cleaning urinal discharge pipes*

Due to the nature of urine, which can cause a build-up of lime deposit, and the ingress of debris, urinal traps and discharge pipes frequently become blocked. Reference should be made to Chapter 8 (pages 282–3) for details on maintenance requirements.

*Urinal flushing cisterns*

Until the 1986 Water By-laws took effect, urinal flushing cisterns were regulated to flush three times per hour whether or not the building was occupied. This obviously led to a great deal of wasted water, and since 1986 full automatic control has been mandatory to ensure urinals are flushed only when the building is occupied. The Water Regulations specify the volume of water permissible for urinal flushing. For a single urinal bowl automatically flushed, up to 10 litres of water per hour is permissible. With single appliances, however, it is more usual to use manual flushing apparatus such

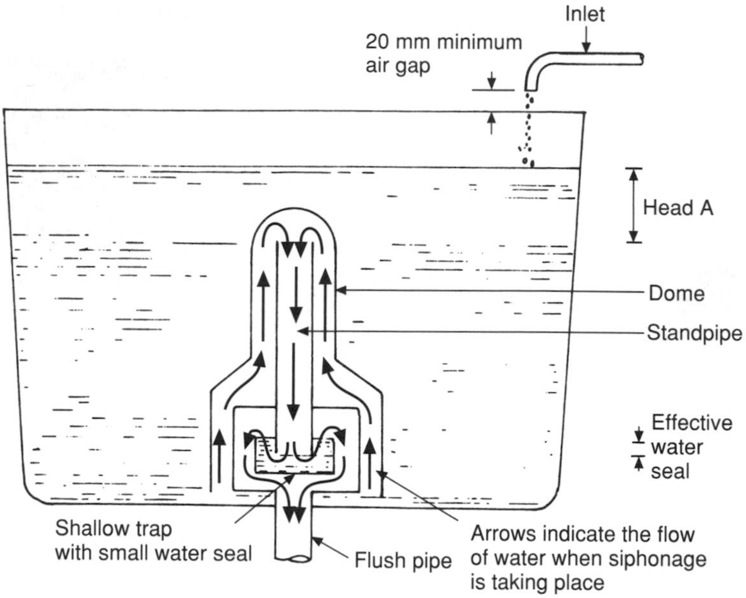


**Fig. 7.30** Automatic flushing cistern. The action of an automatic flushing cistern is quite simple. As the water level rises pressure shown as head A increases until it overcomes the water seal in the shallow trap, causing it to overflow into the flush pipe. This causes a lowering of the air pressure in the standpipe and the dome and the water contained in the cistern is forced under the dome and into the flush pipe by atmospheric pressure. The action continues until the cistern is emptied and air enters the base of the dome. The shallow trap is resealed during the flush.

as a flushing cistern or valve. Where slab urinals or ranges of bowls are used the allowance is 7.5 litres of water per hour or for each 700 mm of urinal slab width. Where urinals are flushed by means of a valve it should not deliver more than 1.5 litres of water per bowl or slab position each time it is flushed.

In most cases where urinals are ordered, the supplier provides a suitable flushing cistern and all the associated pipework. The capacity of the cistern is normally 4.5 litres per bowl or 700 mm width of stall. The action of a typical automatic flushing cistern is illustrated in Fig. 7.30. The associated caption explains its working principles.

*Automatic flushing cistern control*

These cisterns will flush only when the water level reaches a point where its pressure 'blows' the shallow trap. By limiting the water inlet into the cistern to periods when the building is occupied, the flushing requirements can be controlled.

*Pressure control valves (hydraulic valve)*

An effective device for saving water in urinal flushing cisterns is a valve patented under the name Cistermiser (see Fig. 7.31). This valve allows water to enter a urinal flushing cistern only when the building is occupied. It is automatic, passing water only when a pressure drop occurs in the pipeline to which it is fitted due to one or more taps being opened. The valve operates in the following manner. When pressure is applied at the inlet port it will immediately press against the valve seat side of the diaphragm, thus holding the valve firmly shut. Water will then flow slowly through the adjustable restrictor into the bellows chamber, compressing the spring and air contained in the collapsible bellows until the pressure on both sides of the diaphragm is equal. The valve is now held shut only by the return spring and water pressure on the valve seat. If the pressure drops on the inlet side of the valve this change will immediately be transmitted to the valve seat side of the diaphragm, and the valve will open causing a further pressure drop on the valve seat of

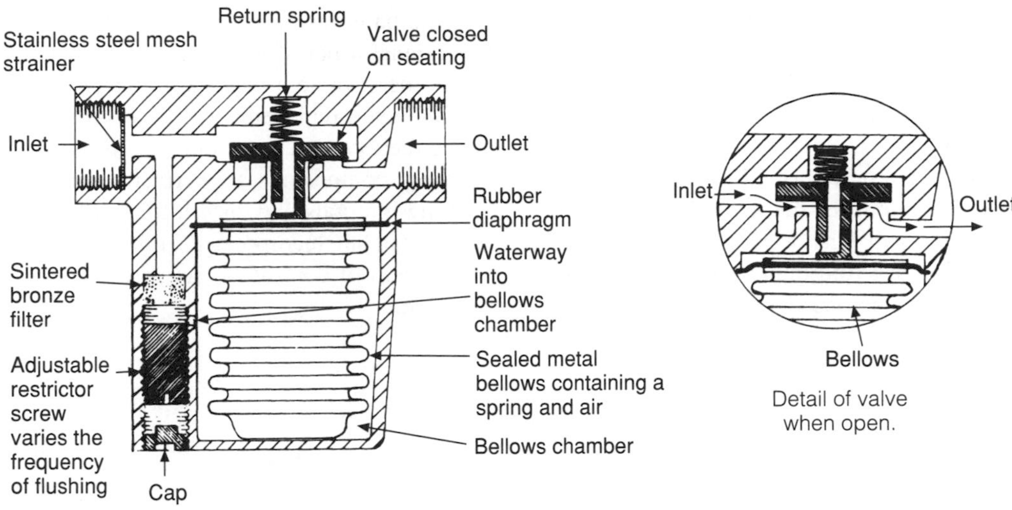
Return spring
Stainless steel mesh strainer
Valve closed on seating
Inlet
Outlet
Rubber diaphragm
Waterway into bellows chamber
Sintered bronze filter
Adjustable restrictor screw varies the frequency of flushing
Sealed metal bellows containing a spring and air
Bellows chamber
Cap
Inlet
Outlet
Bellows
Detail of valve when open.

**Fig. 7.31**

the diaphragm. The valve will stay open until water flows back through the restrictor and pressures on both sides of the diaphragm again equalise and the return spring closes the valve. The length of time that the valve remains open, and consequently the quantity of water that flows into the cistern at each operation, is controlled by the setting of the adjustable restrictor which is accessible from the outside of the valve. If a valve of this type is fitted to an existing installation having a disc valve or pet cock, they should be removed to avoid interference with its operation.

*Electrical control*

These work on the infrared ray principle shown in Fig. 7.32 which, in very simple terms, detects

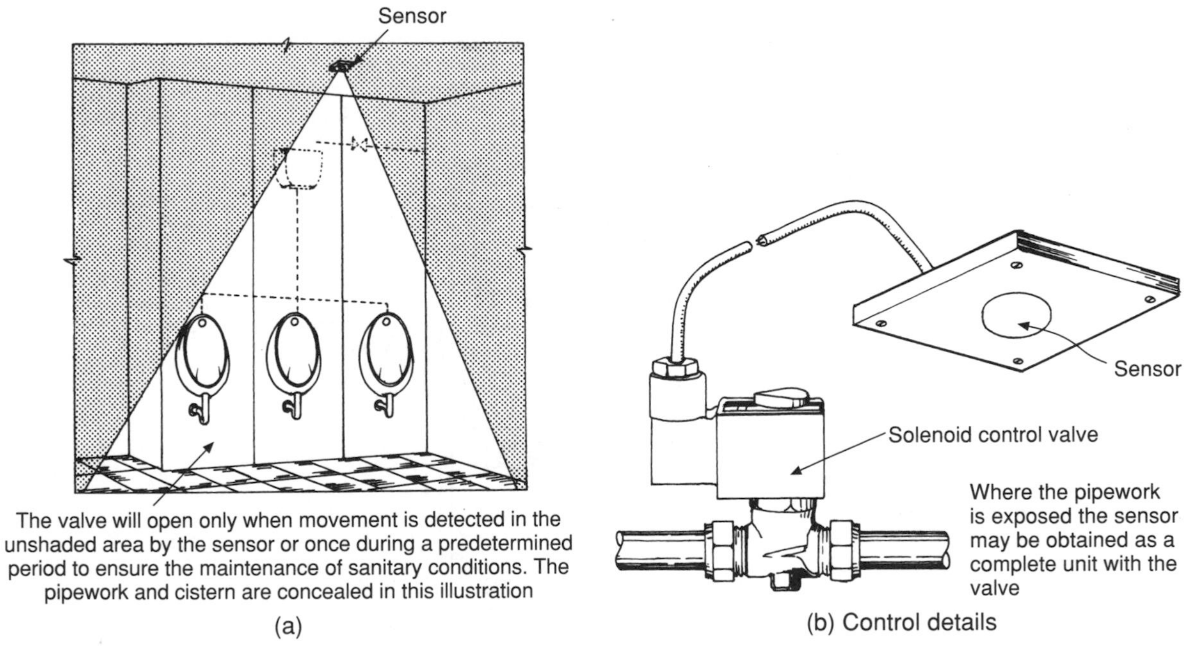
Sensor
The valve will open only when movement is detected in the unshaded area by the sensor or once during a predetermined period to ensure the maintenance of sanitary conditions. The pipework and cistern are concealed in this illustration
(a)
Sensor
Solenoid control valve
Where the pipework is exposed the sensor may be obtained as a complete unit with the valve
(b) Control details

**Fig. 7.32** Electrical control of urinal flushing.

movement, causing a solenoid valve to open allowing water to flow into the flushing cistern for a predetermined period. Most systems using this type of control can detect, by means of an inbuilt microprocessor, whether or not the cistern has flushed during a period of 12 or 24 hours. This ensures that premises not normally used, e.g. at weekends, will flush at least once during the prescribed period to avoid unpleasant odours. This form of control is electrically operated, either by long-life batteries or mains supply using a step-down transformer. Those operated by batteries should incorporate a visual warning system to indicate when replacement is necessary. To ensure their correct functioning, it is essential that these components are correctly installed. Flushing intervals and cistern fill can be manually set by switches incorporated in the control unit. This system of control is very effective and trouble free.

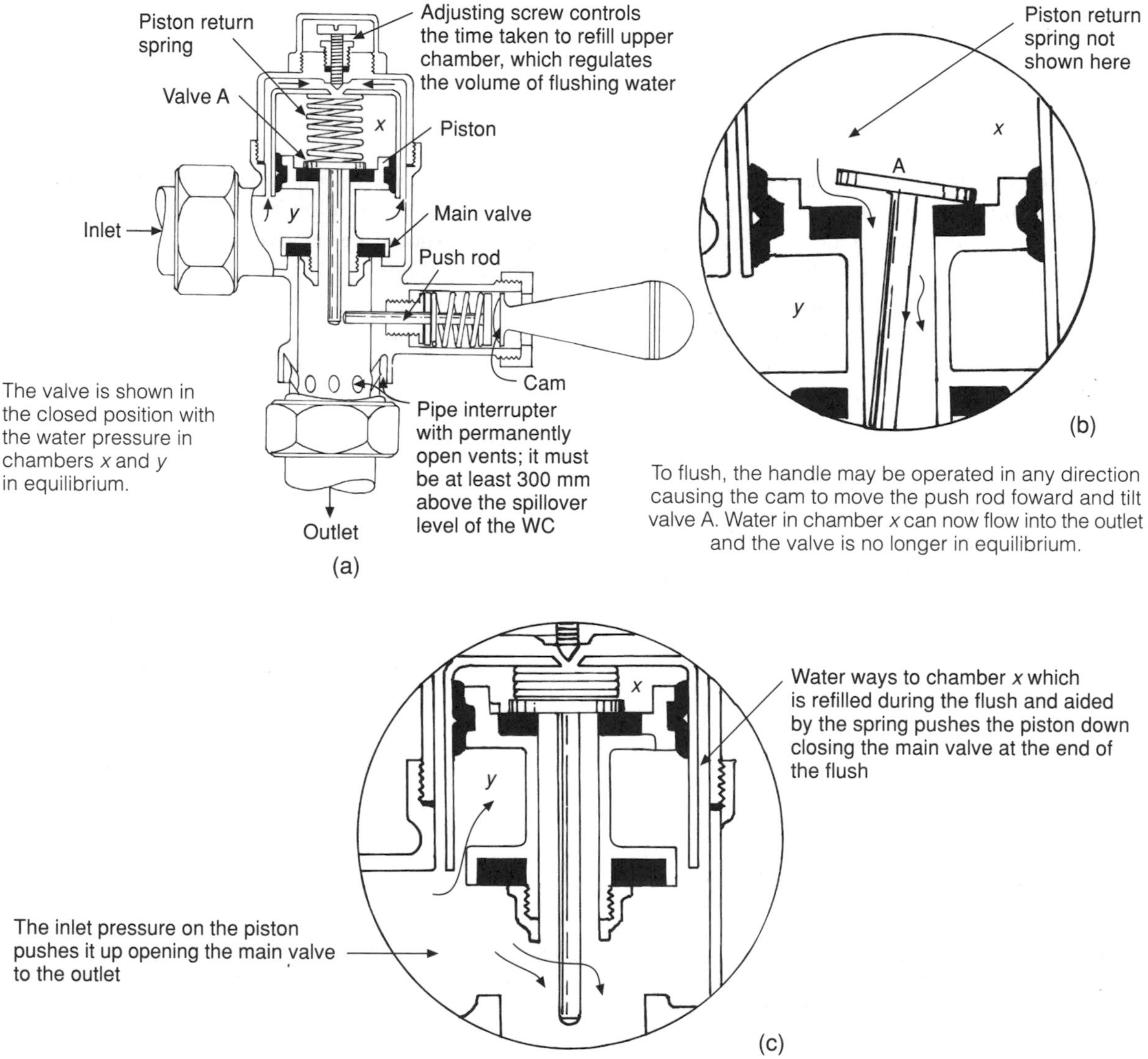


**Fig. 7.33** Flushing valve.

*Flushing valves*

These are a type of equilibrium valve, and unlike a flushing cistern which requires time to fill, they can be used continuously. Until the 1999 Water Regulations took effect these valves were not permitted in the UK, mainly due to possible wastage of water and category 5 pollution risks. As the regulations now permit overflowing water from flushing cisterns to disharge into the WC, any malfunction of a flushing valve will only do the same. These valves are not permissible in domestic properties or any other building where a minimum flow rate of 1.2 litres per second cannot be achieved; this is the minimum required for these valves to function effectively. Any valve supplied by a pressure-flushing cistern which is connected directly to the main supply must be made or provided with a pipe interrupter permanently vented to the atmosphere. There are slight variations in the design of these valves, but the one shown in Fig. 7.33 is fairly typical and embodies their main characteristics. The inlet may be supplied by a separate cistern designated only for use with these valves and such cisterns must be supplied through a type AG air gap. Figure 7.34 shows a suitable layout. They may also be used with an individual pressure-flushing cistern connected directly to the main supply. To avoid serious pollution risks only flushing valves provided with a pipe interrupter are permissible where this system of supply is employed. Although the use of these cisterns is a new concept in the UK, it is claimed they are very effective using the smaller volume of flushing water now mandatory.

*Pressure-flushing cisterns*

These are designed for use with flushing valves where the supply is taken from the main. They are constructed and operate in a similar way to an expansion vessel in an unvented hot water system. When empty the vessel contains air at atmospheric pressure, but as it fills the air is compressed until it is at the same pressure as the water supply. At this point the automatic inlet valve will close. When the flushing valve is operated, water is released under the pressure of the compressed air into the WC.

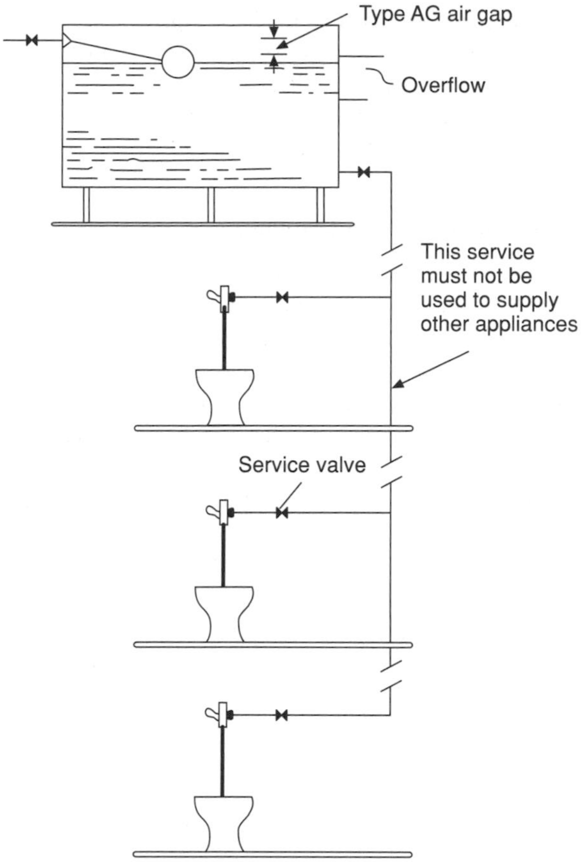


**Fig. 7.34** Installation of flushing valves fed from a dedicated storage cistern.

## Disposal of chemical wastes

Industrial wastes containing chemicals need special attention and the plumber is quite often involved with the discharge from laboratory sinks in schools and hospitals and should be aware of the methods used. Due to the action of concentrated acids and alkalis, it is important that these are diluted as much as possible before being discharged into the drain. Bottle traps having a large water capacity are sometimes used, which effectively dilute any corrosive chemicals. Another method is to use an acid receiver which is simply a large water container made of high-density polypropylene, glass or glazed earthenware. The discharge from the container is turned into an open channel leading to a suitable gulley. Pipes used for the discharge of chemicals are usually made of high-density polythene with fusion-welded joints or chemical-resisting glass pipes. Fittings are used having push-fit joints which permit the removal of the pipe periodically for cleansing.

### Hospital appliances (general)

All such appliances must be made to be robust, easily cleaned and as trouble free as possible. They are traditionally made of fireclay but many are now made of good-quality stainless steel. The basic design of the bedpan washer/hopper, sometimes called a sluice, shown in Fig. 7.35 has not changed for many years. The contents of the bed pans are emptied into the slop hopper (note its similarity to a WC) and flushed away. Both urine bottles and bed pans are washed up in the sink forming part of the unit and are usually dried in a steam-heated steriliser.

*Plaster sinks*

These are used to prepare plaster casts for the support of broken bones. Figure 7.36 illustrates the components of a typical clayware appliance. These too can be made of stainless steel, although to a different design.

### Sanitary accommodation

This subject is dealt with in the Building Regulations 2000 in Part G of Schedule I which relates to hygiene. Sanitary accommodation may be defined as a room or space in a building which contains a water closet or urinal, whether or not it also contains ablutionary fittings such as a wash basin. No sanitary accommodation may open directly into a habitable room unless it is used solely for sleeping or dressing purposes. If the situation relates to a private dwelling and the WC is the only one available to its occupants, the

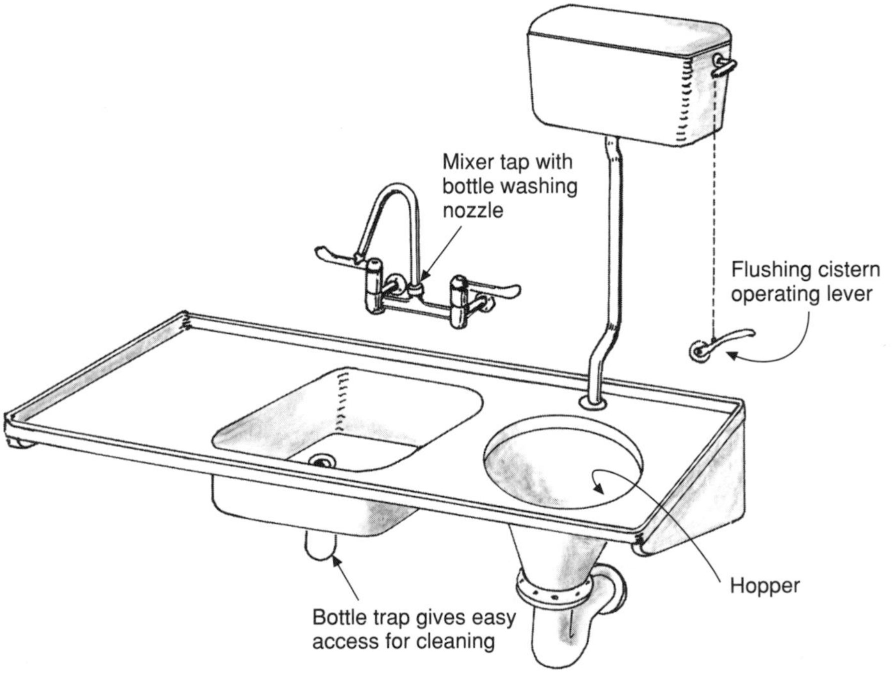


**Fig. 7.35** Hospital bedpan sluice. These appliances are used in hospitals to cleanse bedpans after use. It is usual in most cases to sterilise and dry them for reuse in a steam steriliser. The appliance shown is constructed of stainless steel, but those made of clayware are still available. The front edge can be supported on legs if necessary.

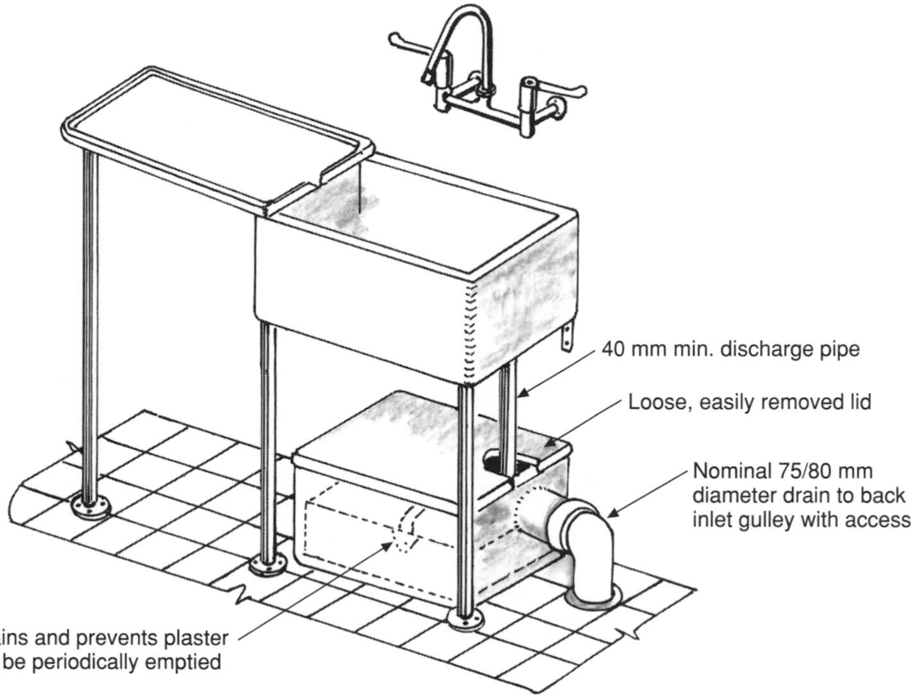


**Fig. 7.36** Plaster sink.

accommodation must be constructed so that it is possible to enter it without passing through the bedroom or dressing room (see Fig. 7.37). The foregoing does not apply if there is other sanitary accommodation on the premises that can be used by the occupants. In simple terms, this means that, should there be another WC on the premises, the one in the bedroom or dressing room can only be used by the persons using this room. No sanitary accommodation may open directly into a kitchen or room in which food is prepared, or a room used for trade or business. The usual arrangements made to meet this requirement are:

(a) To site the accommodation so that it can be entered from the open air — not a very convenient method in poor weather conditions.
(b) To construct it so that it can only be entered through a ventilated lobby as shown in Fig. 7.38.

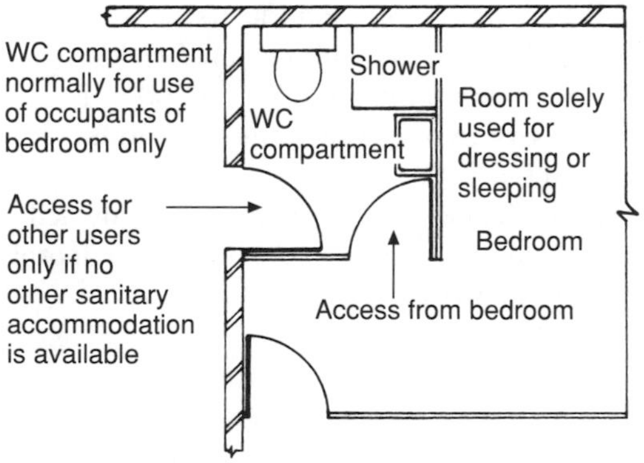


**Fig. 7.37** Regulations relating to ventilation and access to WC compartments.

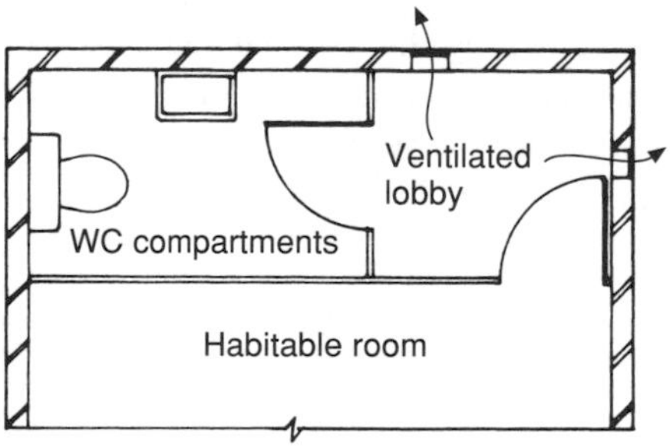


**Fig. 7.38** Access to WC compartment only via a ventilated lobby.

*Ventilation of sanitary accommodation*

All sanitary accommodation must have a window, skylight or similar means of ventilation opening directly to the open air, where an equivalent of one-twentieth of the floor area must be capable of being opened. In buildings where the sanitary accommodation cannot be ventilated by natural means, mechanical ventilation will be necessary, giving at least three air changes per hour and discharging into the open air. In large buildings containing sanitary accommodation requiring ventilation, a ducted system is employed. It is essential with such an installation that two extractor units are provided so that if one breaks down the other can be put into service, providing continuity of ventilation. Whether or not dual extractors are required for one area of sanitary accommodation in private dwellings depends on the local authority. The foregoing relates mainly to public and multi-storey housing, but it is not uncommon to encounter similar situations where mechanical ventilation is necessary in smaller properties, a typical example being where a WC apartment is constructed under a flight of stairs. There are many small extractor fans suitable for wall or ceiling fitting which are available for this purpose. The electrical supply can be arranged in such a way that when the light is switched on the extractor operates. These extractors are provided with an adjustable timing device which enables them to continue running after the light has been switched off. Adjustment of the timing device depends on the length of time the extractor should continue to operate to give the necessary air change. This can be calculated by dividing the volume of air removed per second into the volume of air contained in the apartment. When these units are situated in a ceiling the extracted air must be ducted into the atmosphere. It is not permissible to allow it to discharge into the roof space.

### Washing and WC facilities for the disabled

This is an important subject and one in which plumbers should have a basic knowledge. Disabled people are increasingly mobile and special facilities are necessary in the area of sanitation. The Building Regulations, Approved Document Part M, and BS 8300 deal with this subject in a very comprehensive way and provide all the necessary information for those carrying out this type of work. The type of facilities necessary will depend on the degree of disability. Figure 7.39 shows a standard layout for a disabled persons' toilet in public and commercial buildings. Some toilets have wheelchair access, which is not shown here. A suitable bathroom arrangement in private dwellings or care homes for the severely disabled is shown in Fig. 7.40. Steps should be avoided to enable easy access for wheelchairs, especially in shower compartments, and it is possible to provide a watertight floor having a built-in outlet. The trap and the discharge pipe will be built into the floor and it is important that this section of pipework is air tested prior to the final wall and floor finish. Hinged arm and hand rails are also fitted to assist movement for the disabled and the elderly in sanitary compartments. The following gives general guidance for this type of work, but reference must be made to the documents previously mentioned.

(a) Lever-operated mixing valves, taps and flushing cisterns are preferred for invalids, the elderly and sufferers of arthritis.
(b) Approved thermostatic valves must be provided to prevent scalding from ablutionary appliances.
(c) Coloured grab rails and toilet seats provide contrast to people with impaired vision.
(d) A minimum turning circle of 1,500 m is necessary to turn wheelchairs in a toilet or bathroom.
(e) Grab rails (see Fig. 7.41) must be provided as required. They must be supplied with approved (very robust) fixings.

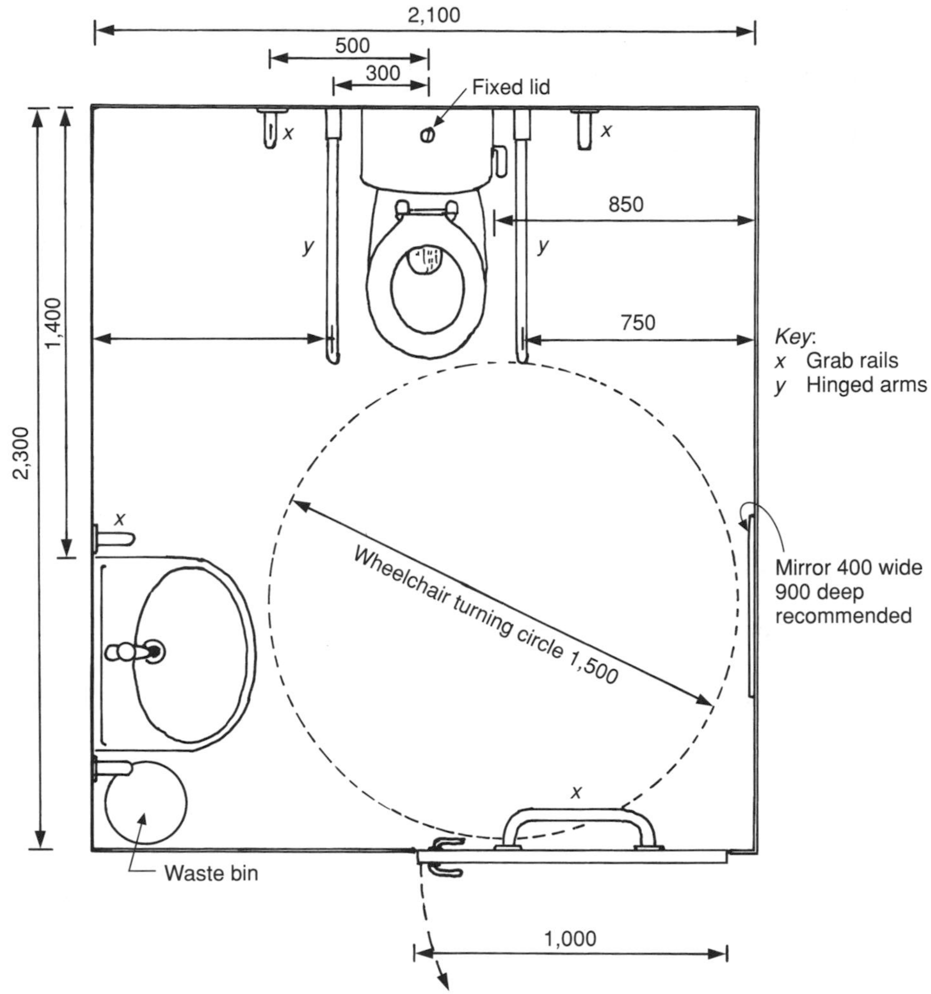


**Fig. 7.39** Toilet suitable for disabled persons (mm, not to scale).

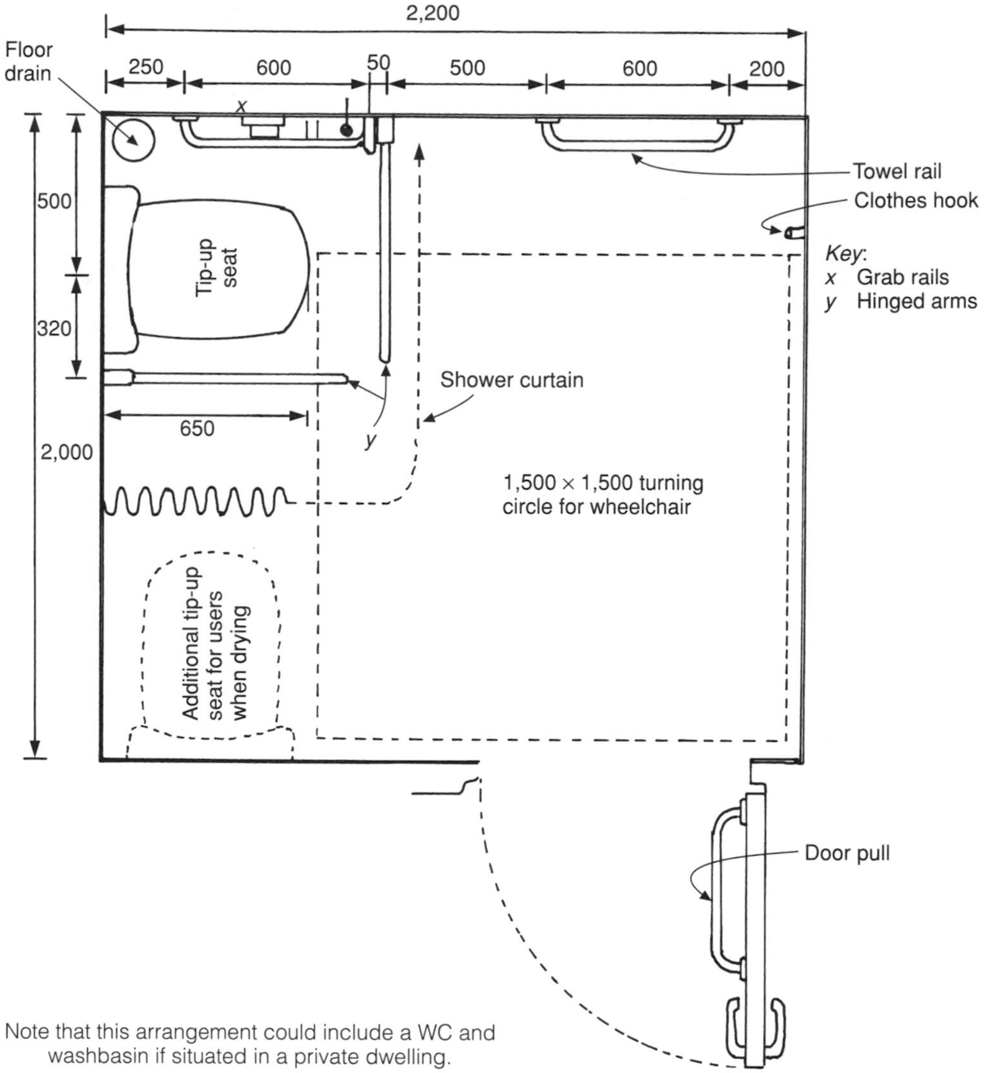


**Fig. 7.40** Shower cubicle for disabled persons (mm, not to scale).

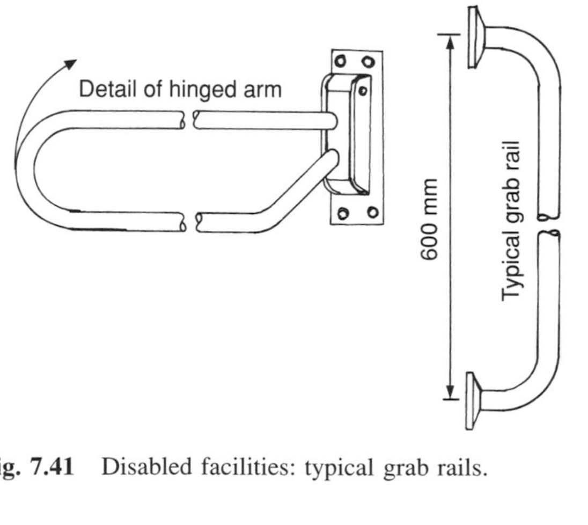


**Fig. 7.41** Disabled facilities: typical grab rails.

## Further reading

*Showers*

Aqualisa Products Ltd, The Flyers Way, Westerham, Kent TN16 1DE Tel. 01959 560020. www.aqualisa.co.uk

Mira Showers, Kohler Mira Ltd, Cromwell Road, Cheltenham GL52 5EP Tel. 01242 221221. 'A guide to domestic pumped shower systems'/ 'Installing mixer showers'/'Electric showers' (three booklets). www.mirashowers.com

*Sanitary fittings and appliances*

BS 6465: Part 1. Sanitary Fittings.

BS EN 1111 and BS EN 1287 British and Euopean Standards for thermostatic mixing valves.

Armitage Shanks Ltd, Rugeley, Staffordshire WS15 4BT Tel. 01543 490253. www.armitage-shanks.co.uk

Ideal Standard (UK) Ltd, The Bathroom Works, National Avenue, Kingston-Upon-Hull HU5 4HS Tel. 01482 346461. www.ideal-standard.co.uk

*WC macerators*

Saniflow Ltd, Howard House, The Runway, South Ruislip, Middlesex HA4 6SE Tel. 020 8842 0033. www.saniflow.co.uk

Technical Advice on Sanitary Appliances from the Building Centre, 26 Store Street, London WC1E 7BT Tel. 020 7692 4000. www.buildingcentre.co.uk

Thermostatic Mixing Valve Manufacturers Association (TMVA), Westminster Tower, 3 Albert Embankment, London SE1 7SL Tel. 020 7793 3008. www.tmva.org.uk

## Self-testing questions

1. (a) State the recommended fall for horizontal discharge pipes for a WC macerator.
   (b) Assuming a macerator to be capable of pumping a horizontal distance of 50 m, state the actual length of discharge pipe if it includes a vertical run of 3 m and three bends.
2. Specify the type of flushing arrangements recommended for ranges of WCs with a high incidence of use.
3. List the advantages of spray taps for hand washing.
4. State the maximum length of a dead leg for spray tap installations and give the reason for imposing this limit.
5. (a) State the recommended temperature of water for showering.
   (b) List the advantages of a shower bath compared with a bath.
6. Specify the type of shower mixer you would recommend for use in an old people's home and state the reasons for your choice.
7. Identify the reasons for providing an independent cold water supply to a shower.
8. (a) State the type of urinal with which a sparge pipe is usually fitted.
   (b) Outline two ways of effecting savings on the consumption of water used in urinal flushing cisterns.
9. Explain the working principles of an automatic flushing cistern.
10. In what circumstances in WC compartments is mechanical ventilation mandatory?
11. Describe the requirements of the Water Regulations in relation to appliances with outlets fitted with flexible hoses.
12. Describe how backflow is prevented with flushing valves operated by a pressure cistern connected directly to the main supply pipe.
13. (a) Describe the maintenance checks necessary in the event of failure of a mixing valve.
    (b) State the final operation that must be carried out on completion of any maintenance.
14. Explain why double check valves are fitted on the outlet, not the inlet, side of electric showers.

# 8 Sanitary Pipework

After completing this chapter the reader should be able to:

1. State the basic requirements of the Building Regulations relating to sanitary pipework.
2. Explain the basic principles of discharge pipe systems.
3. Identify the causes of trap seal loss due to defective design.
4. Recognise the main features of above-ground sanitary pipework systems.
5. Sketch and describe simple details showing discharge pipe arrangements for single and ranges of sanitary appliances.
6. Recognise the importance of ventilating sanitary pipework where necessary.
7. Describe the correct procedures for testing and commissioning sanitary pipework installations.

## Building Regulations relevant to sanitary pipework

Regulations are necessary in the construction industry to ensure that a building and its components are safe and suitable for the purpose for which they are designed and do not cause offence or nuisance in the environment. Part H of Schedule I to the Building Regulations 2002 and the associated Approved Document H cover the main legislation governing building drainage, and it is important for plumbers to be aware of the principal regulations related to their work. The following is a summary of the above-mentioned documents relating to discharge and ventilating pipes.

(a) Provision must be made in the drainage system (this includes above- and below-ground drainage) to prevent the destruction of trap seals, which would result in the admittance of foul air to the building.

(b) All discharge pipes must be of adequate size for their purpose and *must not* be smaller in diameter than the outlet of the fitting discharging into it.

(c) The internal diameter of a pipe carrying the discharge from a urinal must not be less than 50 mm excluding bowl types. Further information on urinal discharge pipes is included in Table 8.1. In all other cases of pipes carrying excremental matter the minimum internal diameter is 75 mm.

(d) All pipes and fittings used for the discharge of soil or waste and the ventilation of above-ground systems must be made of suitable materials having the required strength and durability for this purpose.

**Table 8.1** Internal diameter and depth of seal of traps serving sanitary appliances, as recommended in BS EN 12056 Part 2.

| *Type of appliance* | *Minimum nominal diameter (mm)* | *Seal depth (mm)* |
|---|---|---|
| Wash basin | 32 | 75 |
| Bidet | 32 | 75 |
| Sink | 40 | 75 |
| Bath | 40 | 50 |
| Shower tray | 40 | 50 |
| Urinal bowl | 32 | 75 |
| Urinal stalls, 1 or 2 in range | 50 | 75 |
| Urinal stalls, 3 or 4 in range | 65 | 50 |
| Urinal stalls, 5 or 6 in range | 75 | 50 |
| WC | — | 50 |

(e) All joints must be made in such a way as to avoid obstructions, leaks and corrosion.
(f) Bends must have an easy radius and should not have any change of cross-sectional area throughout their length.
(g) Pipes must be adequately secured to the building fabric without restricting their movement due to thermal expansion.
(h) The system must be capable of withstanding an air test when subjected to a minimum pressure equivalent to 38 mm head of water. (Details of testing procedures are illustrated later in this chapter.)
(i) Pipework and fittings must be accessible for repair and maintenance, and means of access must be provided for clearing blockages in the system.
(j) Every sanitary appliance must be fitted to its outlet, having an adequate water seal and access for cleaning. Where the appliance is made with an integral trap, e.g. a WC, it must be capable of being removed to provide access to the discharge pipework, unless other suitable cleansing provision is available. This precludes the use of cement mortar for bedding WCs on solid floors which was never good practice due to possible damage to the pan caused by expansion of the cement while hardening.
(k) No discharge pipes from sanitary appliances may be fitted on the exterior of the building except:
    (i) Where the building was erected before the Building Regulations came into force and is altered or extended.
    (ii) On low-rise buildings of up to three floors.
(l) No discharge pipes on the exterior of the building may discharge into a hopper head or above the grating of an open gulley.
    The regulations are in effect insisting that:
    (a) Only gulleys of the back inlet type are acceptable.
    (b) Discharge pipes must be connected directly to the underground drain or main discharge stack.

## Standards relating to sanitary pipework

Apart from the Building Regulations, to which it is obligatory to comply, the other main document relating to this subject was BS 5572. It became the basis for the design of modern sanitary pipework systems, but due to the harmonisation of European standards, it has been superseded by BS EN 12056 Part 2. All reference in this chapter relates to Part 2. Those who are familiar with BS 5572 will notice few changes in the illustration of the basic pipework systems, although there have been changes in name. To avoid confusion, where relevant both names are shown, except in the fully ventilated system, where due to what appears to be an error in BS EN 12056, only the BS 5572 name is shown.

### *Requirements of sanitary pipework systems*

The basic principles of overground sanitation and its associated pipework for lower-rise dwellings have been covered in Book 1 of this series. It is proposed in this volume to extend these principles to enable them to be applied, not only to multi-storey dwellings, i.e. blocks of flats, but also to industrial, commercial and public buildings of up to five floors. In some cases the text and illustrations go a little beyond this requirement so that the subject can be treated in a comprehensive manner.

The object of any waste disposal system is to remove waste water from the building as quickly and quietly as possible and in such a way that no nuisance or danger to health is caused. Objectionable bacteria multiply very quickly in waste matter and sewage, hence the need for quick and efficient disposal. One of the most important factors relating to the removal of waste water is the discharge pipe diameter, and this has a direct effect, coupled with the fall of the pipe, on the retention of the trap seals. Any loss of the seals will admit foul air into the building, and the underlying principle of all sanitary pipework systems is to prevent this from happening. Pipes having a small diameter are the most likely to run full bore and cause self-siphonage.

Another factor that is often overlooked is the noise caused by the operation of sanitary appliances and their discharge pipes. It has been discovered that this can give rise to and aggravate nervous tension in some people, therefore systems should be designed having regard to this problem. Inefficient ventilation and anti-siphon traps are the principal causes of excessive noise.

## Sanitary pipe sizes

The minimum diameter of traps for sanitary appliances is also shown in Table 8.1. In most cases the diameter of the waste is the same, although there are some exceptions, depending on circumstances, which are dealt with later.

It will be noticed that no mention has been made of WC outlet diameters. The reason is that, depending on the WC design and type, the outlet diameter may vary. A standard wash-down WC has an outlet of approximately 90 mm, and although good practice requires a discharge pipe to have the same diameter as the trap to which it is connected, it is not essential in the case of WCs. This is because of the relatively short period for which a WC discharges, and the fact that the pipe is unlikely to run full bore.

It is often considered good practice to oversize urinal discharge pipes so that they are of a larger diameter than the actual trap, as the ammonia in urine, especially in hard water areas, causes scale to build up very quickly inside the pipe, and by increasing its diameter the frequency of descaling is reduced.

## Trap seal loss

Pressure fluctuations causing seal loss are much more likely to occur in multi-storey developments and public and industrial buildings than in small dwellings as the discharge pipes are longer, larger numbers of sanitary appliances are employed and the incidence of usage is greater. It is therefore necessary to design sanitary pipework to suit the type and purpose of the building in which it is to be installed.

Unstable air conditions in the discharge pipe system are the principal cause of trap seal loss, which causes self-siphonage, induced siphonage and compression. These three causes of seal loss are described and illustrated in detail in Book 1, but as they are necessary for a full understanding of the following text, they are repeated here very briefly. Self-siphonage is said to happen when the discharge from an appliance causes loss of seal in its own trap, while induced siphonage relates to the loss of seal in a trap caused by the discharge of another appliance connected to the same discharge pipe. Compression takes place when a falling body of water in a vertical pipe compresses the air below it, forcing out the trap seals of the lower appliances. Factors other than unstable air conditions in the pipework which affect trap performance can be summarised as follows:

(a) The shape of the appliance has a considerable bearing on its trap performance; the U shape of a wash basin makes it prone to seal loss as there is no 'rill' or tail off to reseal the trap.
(b) Where an appliance is fitted with an integral overflow, air can be drawn into the waste via the overflow and in some cases helps to maintain equilibrium of the air in the discharge pipework.
(c) The type of waste fitting used can influence the trap's performance: if its grating offers resistance to the flow of water, it will reduce the volume of discharge and there will be less likelihood of the pipe running full (see Fig. 8.1). A special waste fitting with a

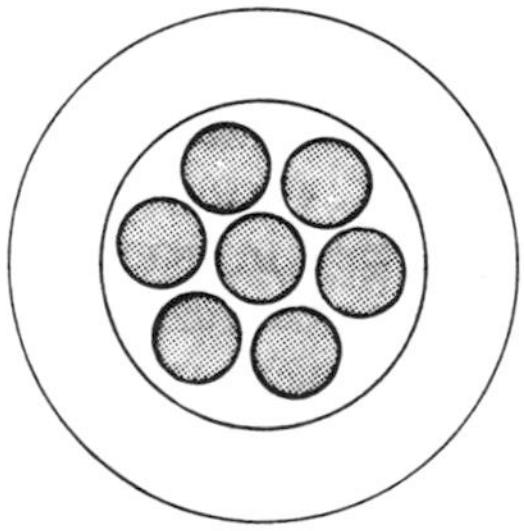

(a) Grating offers more resistance

(b) Grating offers less resistance

Note that the total area of the outlet holes in the waste fitting (a) is less than for this waste fitting. This has the effect of restricting the volume of water discharged and reduces the possibility of self-siphonage.

**Fig. 8.1** Basin waste grids.

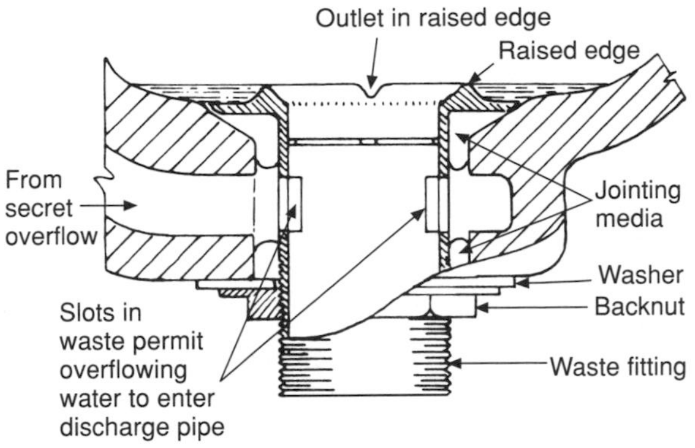


**Fig. 8.2** Patent basin waste. The raised edge causes a small quantity of water to be retained momentarily when the basin is discharged. This water enters the waste after the discharge and reseals the trap should some seal have been lost due to self-siphonage.

raised edge is shown in Fig. 8.2; this holds back a small quantity of water from the final discharge which serves to top up the trap seal. Even if an appliance has an integral overflow, this should not be used as a substitute for a ventilating pipe.

(d) Bends should be reduced to a minimum as these slow up the flow, causing a solid plug of water to form more quickly than in a straight length of pipe.

### Ventilation of sanitary pipework

A ventilating pipe may be defined as a pipe connected to a discharge pipe at one end, the other being open to the atmosphere which prevents positive or negative pressures being set up in the system, thus ensuring the retention of the trap seals. Providing branch wastes, especially those serving wash basins, are kept within the limits of length, number of bends, diameter and falls recommended later in this chapter, no ventilation of individual appliances will be necessary. Ventilation may be necessary, however, due to the number of appliances and their incidence of use, their distance from the main stack and the number of bends employed on the branch discharge pipe. The designation of modern sanitary pipework is in fact determined by the type of ventilation employed.

### Compression and its prevention

Compression has been defined earlier in this chapter. Special precautions are essential to avoid a build-up of positive air pressure in the pipes, especially those serving high-rise developments. The ground floor appliances in tall buildings will be especially exposed to compression, and for this reason ground floor ablutionary appliances should be discharged into back inlet gulleys. The use of either a long-radius bend, or two 135° bends where the discharge stack joins the underground drain, is essential, permitting the air to flow easily into the drain in front of the incoming water. For the same reason, offsets should not be used in the wet part of the stack unless they are ventilated. Generally, the effects of compression in very tall buildings can be avoided in one of two ways. One is to increase the diameter of the main discharge pipe to 150 mm nominal bore, which has the effect of reducing the possibility of the formation of a solid plug of water. There is, however, a disadvantage in using this method, because, if the discharge pipe diameter is increased, the diameter of the underground drain must also be increased to the same size, which might have the effect of reducing the velocity of flow in the drain to such a degree that it would not be self-cleansing. The other and more common alternative is to use a nominal 100 mm pipe with selective venting. By siting these vents at strategic points in the system, the use of a 100 mm discharge stack can be extended to most applications in modern buildings. The single stack system and its vented modifications are described in the following text. Note that where the term 'group of appliances' is used, it should be interpreted as comprising one bath, one sink, one or two wash basins and one or two WCs with a flushing capacity of 7.5 or 6 litres.

### Modern pipework systems

Prior to the adoption of the 1965 Building Regulations, one was permitted to fix discharge pipes to the exterior face of buildings. This was not only unsightly but it increased maintenance costs if the pipework required painting to protect it from corrosion. It was also often found that a dripping tap could, on a frosty night, cause an ice plug to

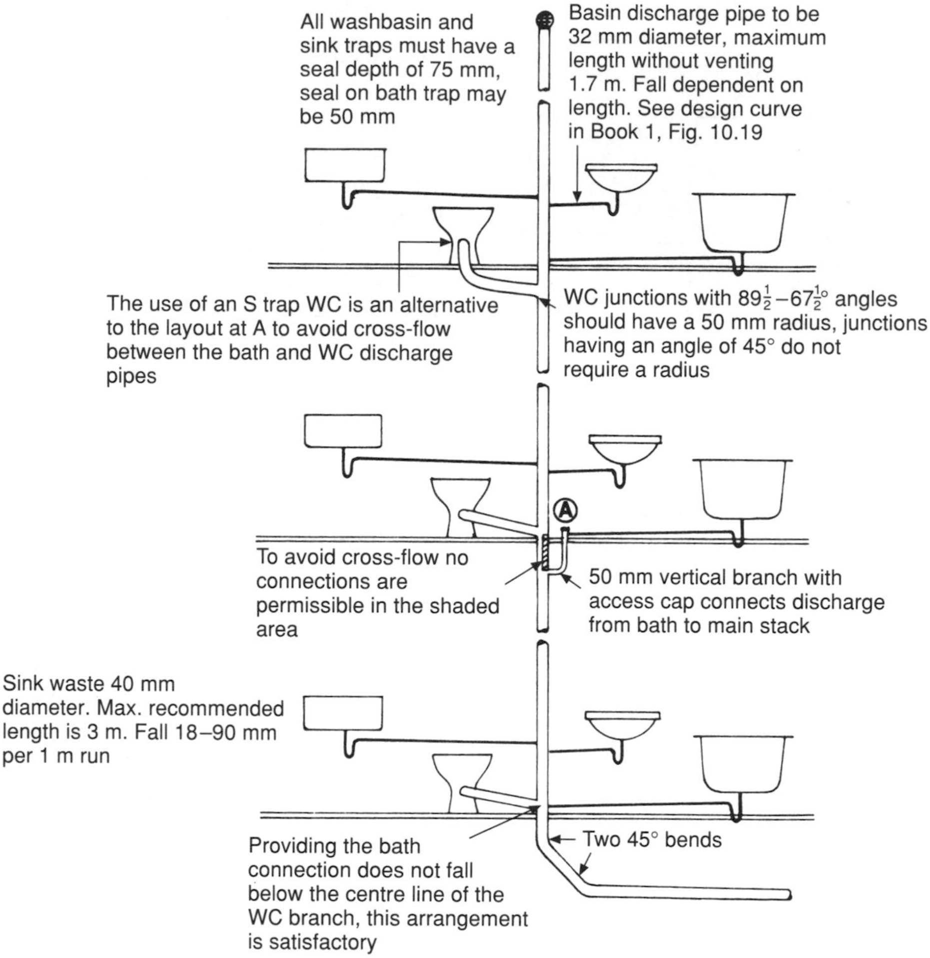
All washbasin and sink traps must have a seal depth of 75 mm, seal on bath trap may be 50 mm
Basin discharge pipe to be 32 mm diameter, maximum length without venting 1.7 m. Fall dependent on length. See design curve in Book 1, Fig. 10.19
The use of an S trap WC is an alternative to the layout at A to avoid cross-flow between the bath and WC discharge pipes
WC junctions with $89\frac{1}{2}$–$67\frac{1}{2}$° angles should have a 50 mm radius, junctions having an angle of 45° do not require a radius
A
To avoid cross-flow no connections are permissible in the shaded area
50 mm vertical branch with access cap connects discharge from bath to main stack
Sink waste 40 mm diameter. Max. recommended length is 3 m. Fall 18–90 mm per 1 m run
Two 45° bends
Providing the bath connection does not fall below the centre line of the WC branch, this arrangement is satisfactory

**Fig. 8.3** Primary ventilated stack system (single stack system) in a three-storey building. The recommendations for sink discharge pipes also apply to bath wastes.

form in the main discharge pipe causing it to split. The 1965 Building Regulations insisted that all discharge pipes should be fitted inside the building, with the exception of work on existing buildings or extensions where it might be difficult or costly to comply with this requirement. Subsequent regulations permit the exterior fixing of discharge pipes on buildings of up to three floors. Although this simplifies installations and releases floor space which would otherwise be occupied by a duct, it may involve the householder in a considerable amount of inconvenience, especially if the system is subjected to frost damage.

*Primary ventilated stack system*

Previously known as the single stack system, this system is used wherever possible, as the complete absence of trap ventilating pipes makes it less costly to install and enables it to be accommodated easily into internal pipe ducts. Due to the low flow rates in domestic dwellings, it can be used successfully in such buildings of up to five storeys, using a standard 100 mm main discharge stack, having one or two groups of appliances on each floor. Figure 8.3 illustrates this system in low-rise dwellings. Figure 8.4 shows a much larger scheme suitable for a block of flats. In buildings having

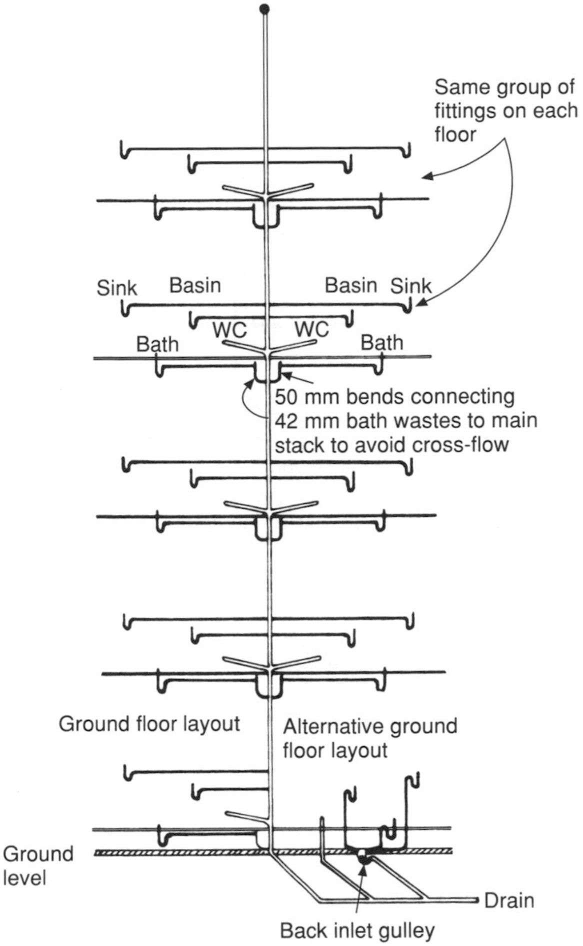


**Fig. 8.4** Primary ventilated system in a five-storey block of dwellings. Note that two groups of fittings are permissible on each floor without stack ventilation. If the building has more than five floors or there are more than two groups of fitting per floor, stack ventilation will be necessary; see Fig. 8.6 if a 100 mm diameter main discharge pipe is used. An alternative is to increase its size to 125 or 150 mm.

more than five floors, the basic principles are the same, but when siphonage to an appliance or a group of appliances is likely, ventilation of the branch discharge pipe may be necessary. If compression is possible, the main discharge stack must be ventilated.

*Stub stack system*

The use of this system is confined to single-storey buildings or to a group of appliances on the ground floor. Traditionally, in such cases, the WC is connected directly to the drain, while the ablutionary fittings discharge into a back inlet gulley, and while this arrangement is acceptable, it does have some disadvantages. The possible exposure of discharge pipes is unsightly on the exterior of a building and can lead to blockage of the external pipes by formation of ice in frosty weather due to dripping taps. Such pipes are often run at a steep angle or vertically, resulting in noisy discharges. Where an appliance is situated some distance from the gulley, due to the length of the discharge pipe, self-siphonage of the trap seal may take place. The use of the stub stack system is illustrated in Fig. 8.5 and solves most of the foregoing problems, and in some cases can be more economical in cost. The only limitations imposed are:

(a) The floor level of the WC apartment should not be in excess of 1.3 m.
(b) The distance between the highest connection on the stack and the invert must not exceed more than 2 m (Part H of the Building Regulations).
(c) Arrangements must be made to ventilate the underground drain elsewhere in the system.

It is important to remember that the underground drain is satisfactorily vented.

*Secondary ventilated stack system*

Shown in BS 5572 as the ventilated stack system and illustrated in Fig. 8.6, the main feature of this system is the cross-venting of groups of appliances, either from the WC branch or from the main discharge pipe adjacent to the appliances. This type of system is normally employed for high-rise housing of more than five floors. In buildings of this type, compression is the main cause of seal loss, especially when appliances on the upper floors are discharged. The use of the ventilating pipe permits the relief of any pressure build-up in the stack caused by a falling body of water. There is little danger of trap seal loss in each appliance if basic principles are embodied in the system, e.g. the careful grouping of appliances, limitation of the

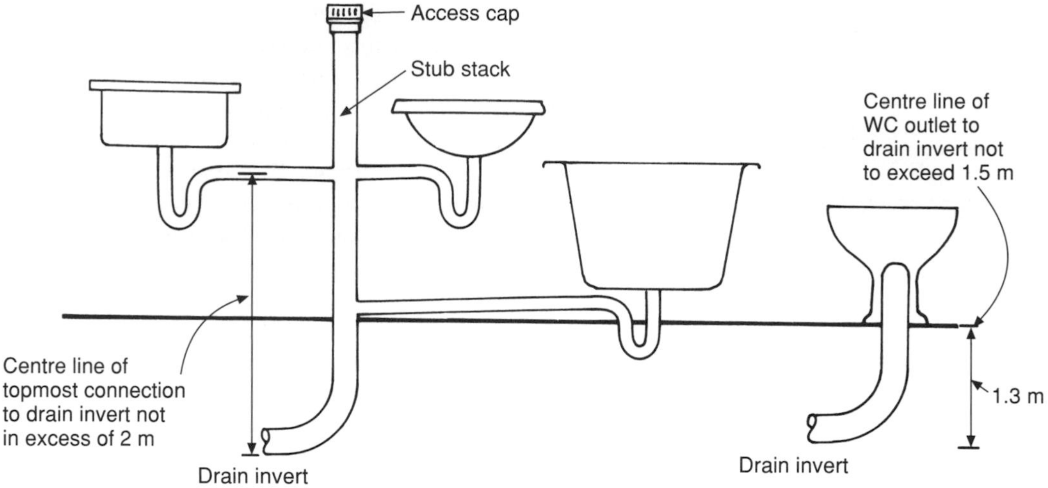


**Fig. 8.5** Stub stack system. Note that an open vent must be provided at the highest point of the drain. The measurements given here comply with Part H of the Building Regulations.

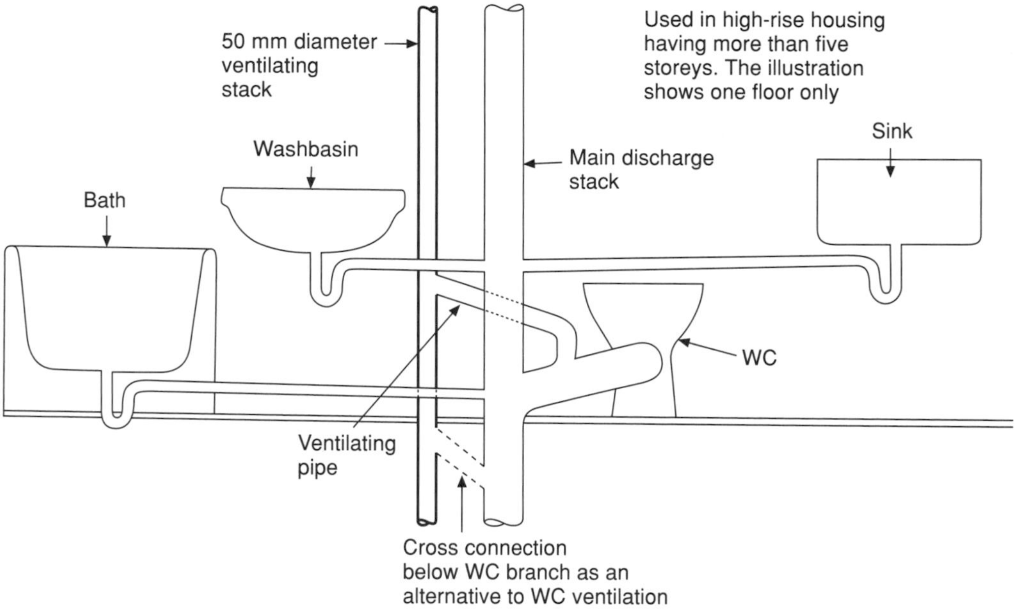


**Fig. 8.6** Secondary ventilated stack system.

lengths and falls of branch discharge pipes and the use of the recommended branch waste diameters.

*Modified primary ventilated stack system*

Formerly known as the modified single stack system, this system can assume many forms, depending on the number of floors it serves, the number of appliances connected to it, the length of the branch discharge pipes, the number of bends on each and the incidence of appliance usage. In general, however, it may be defined as a primary ventilated stack system with branch wastes of

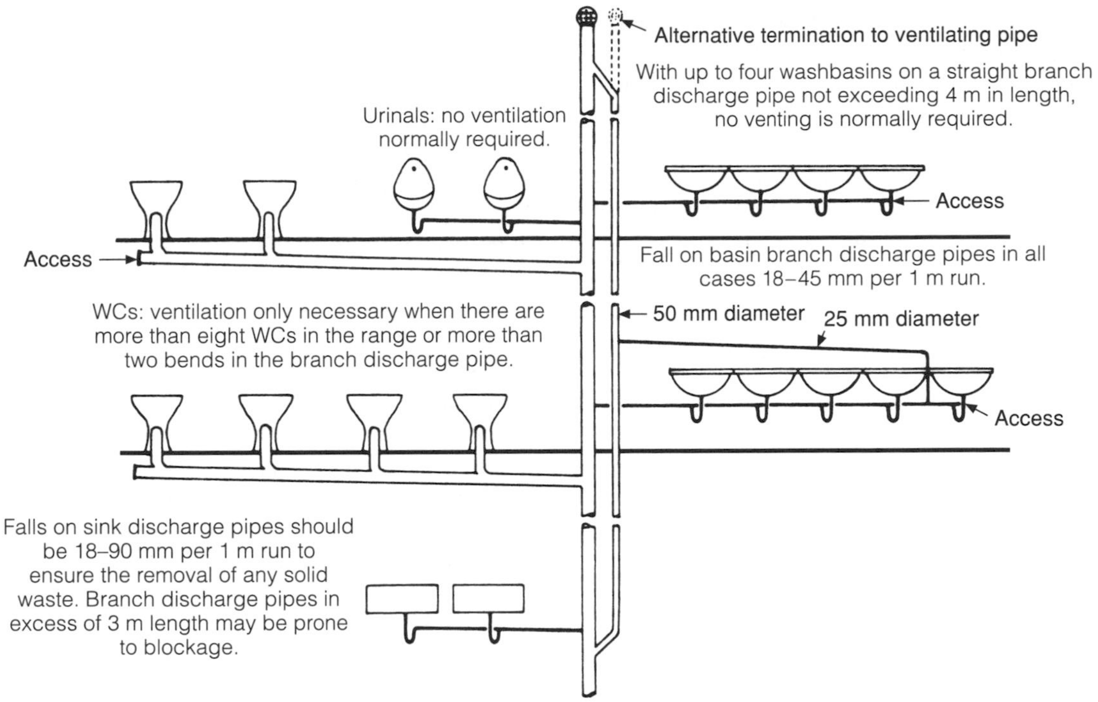
Alternative termination to ventilating pipe
With up to four washbasins on a straight branch discharge pipe not exceeding 4 m in length, no venting is normally required.
Urinals: no ventilation normally required.
Access
Access
Fall on basin branch discharge pipes in all cases 18–45 mm per 1 m run.
WCs: ventilation only necessary when there are more than eight WCs in the range or more than two bends in the branch discharge pipe.
50 mm diameter
25 mm diameter
Access
Falls on sink discharge pipes should be 18–90 mm per 1 m run to ensure the removal of any solid waste. Branch discharge pipes in excess of 3 m length may be prone to blockage.

**Fig. 8.7** Modified primary ventilated stack system.

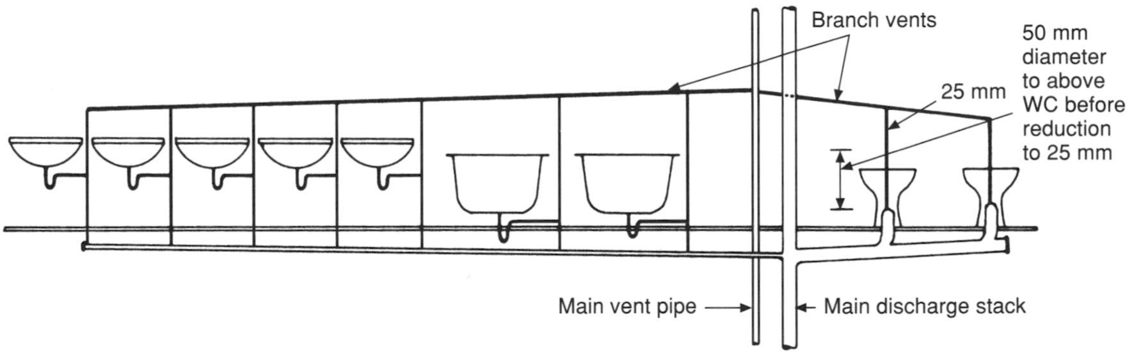
Branch vents
50 mm diameter to above WC before reduction to 25 mm
25 mm
Main vent pipe
Main discharge stack

**Fig. 8.8** Ventilated system. The general layout of pipework on one floor only of a hospital ablutionary annexe. If the main stacks serve a similar arrangement of appliances on three or more floors such a system of pipework may become surcharged, necessitating the ventilation of every trap.

greater length than those recommended, or with branch wastes serving ranges of fittings where some danger of loss of trap seal exists due to siphonage. To overcome this, ventilation of certain appliances or groups of appliances is necessary as illustrated in Fig. 8.7. Note that with this type of system, the ventilating pipe is quite independent of the main discharge stack unless its upper end is turned into the main stack above the highest appliance. In many cases there is obviously no need to carry two pipes above the eaves.

*Ventilated system (Ref. BS 5572)*

The ventilated system is only used in the form shown in Fig. 8.8 when the incidence of usage is said to be 'congested'. In simple terms, this means that the appliances, being in almost continuous use, may cause surcharging of the pipework and subsequent seal loss unless each trap is ventilated. Apart from some modifications the system is very similar to the pre-1965, one-pipe, fully ventilated system. At the time of writing, there appears to be no clear listing in BS EN 12056 for this system, and until verification should be referred to as shown.

### Branch positions in discharge stacks

The branch discharge vent connection for a range of WCs is shown in Fig. 8.9. The possibility of cross-flow is greater in large multi-storey buildings than in low-rise single domestic dwellings. To avoid this, all opposing branches must be carefully positioned in the main discharge stack as illustrated in Fig. 8.10. Note that the distance between opposed branches is greater when one of the opposing branches is a WC. This is because, unlike other sanitary fittings, a WC discharges water very quickly and increases the area in which cross-flow can take place.

*Multi-discharge pipe adaptor*

These are usually listed in manufacturers' catalogues as 'collar bosses'. They can be used in multi-storey dwellings where branch discharge pipes from a group of fittings such as a bath, basin, sink and WC, all on the same floor, are connected to the main discharge stack. A typical fitting of this type is shown in Fig. 8.11 and it will be seen that its use avoids the necessity of making several separate branches into the stack. What is possibly more important is the fact that branch discharge pipes can enter the stack in close proximity to the WC branch without the possibility of cross-flow. Being a cross-sectional illustration only two connections are shown; there are in fact four at 90° to each other, which allows for considerable flexibility in use. As is common with PVC fittings for sanitary pipework, these connections are blanked off, the blanks on the connections selected for a specific installation being cut out with a hole or pad saw. Possibly the only disadvantage with fittings of this type is the fact that they are not self-cleansing and it is possible for soap deposits to build up in the annular chamber. To enable an obstruction to be removed all discharge pipe connections to the adaptor should be capable of being disconnected easily. Push-fit rubber ring joints or mechanical joints of the compression type are most suitable. Some thought should also be given to the siting of these fittings, as in most modern bathrooms they are housed behind panels, in which case means of access

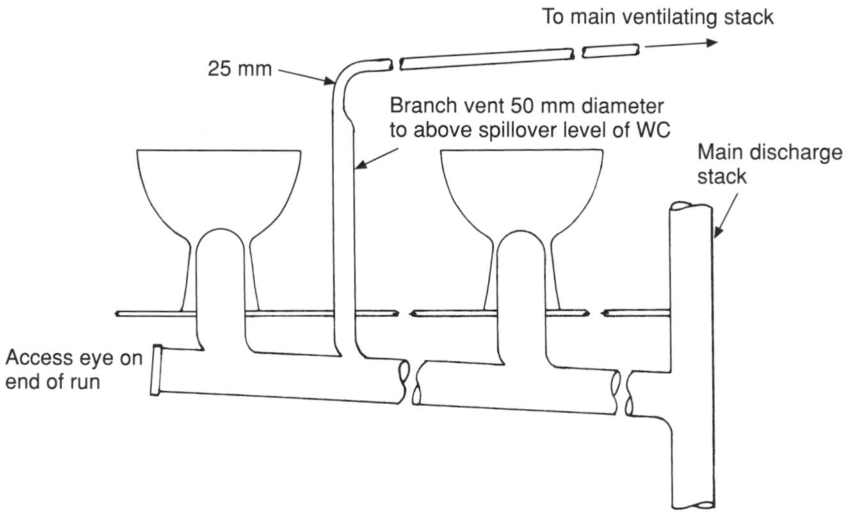


**Fig. 8.9** Range of WCs showing branch discharge vent connection. Ventilation of WC branch discharge pipes is not normally necessary unless there are more than eight WCs or the pipe contains more than two bends.

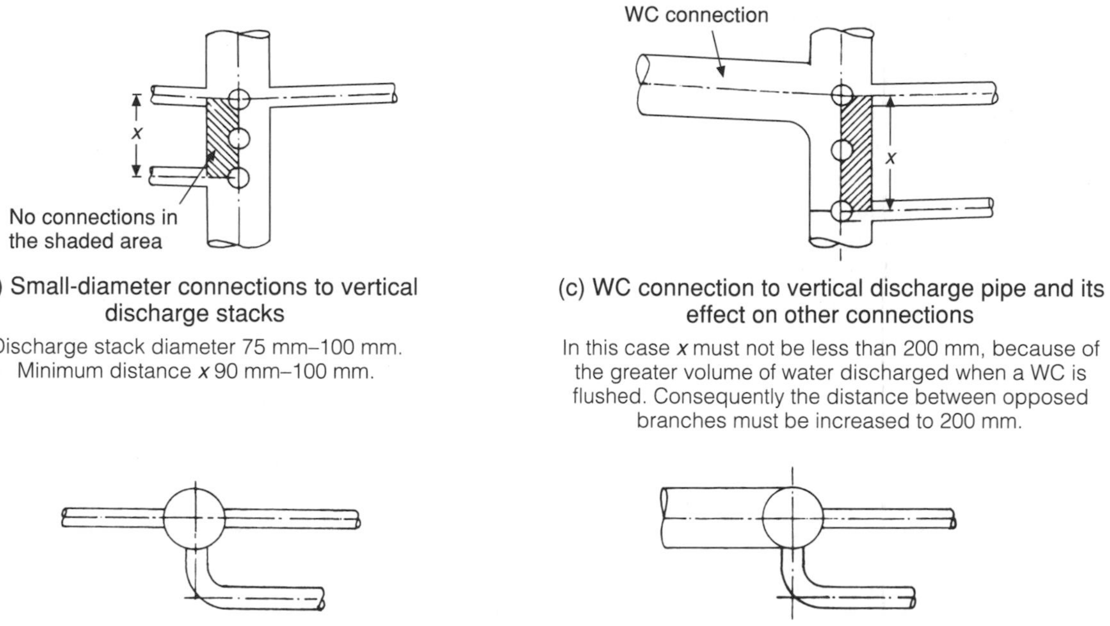


(a) Small-diameter connections to vertical discharge stacks

Discharge stack diameter 75 mm–100 mm. Minimum distance *x* 90 mm–100 mm.

(c) WC connection to vertical discharge pipe and its effect on other connections

In this case *x* must not be less than 200 mm, because of the greater volume of water discharged when a WC is flushed. Consequently the distance between opposed branches must be increased to 200 mm.

(b) Plan view of (a)

(d) Plan view of (c)

**Fig. 8.10** Permissible connections of branch discharge pipes to main discharge stack to avoid cross-flow.

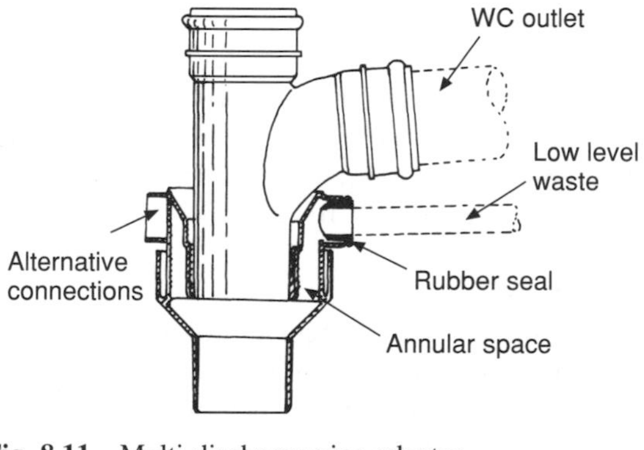


**Fig. 8.11** Multi-discharge pipe adaptor.

should be provided. For details of other applications of these fittings reference should be made to the manufacturers listed at the end of this chapter.

### Ventilation of branch discharge pipes to single appliances

It has already been stated that if the principles of the primary ventilated stack system are adhered to, no ventilation of branch discharge pipes is necessary. The following applies where single appliances are situated some distance from the main stack, requiring longer branch pipes than recommended.

*WCs and urinals*

These appliances do not normally require ventilation when fitted singly as the discharge pipes are unlikely to become surcharged. The recommended limit on the length of a discharge pipe for a WC is 6 m, and there will be few instances in practice where it is longer than this. Urinal wastes should not exceed 3 m in length, due not so much to the possibility of trap seal loss, as to the build-up of scale on the inside of the pipe. Even pipes serving bowl urinals having such small diameters as 32 mm or 40 mm are unlikely to run full, but should siphonage take place the rill or tail-off of each flush is sufficient to reseal the trap.

*Baths and showers*

Siphonage does not normally cause trap seal loss with these fittings as the discharge pipes are adequately sized to ensure they do not run full. As baths and showers are flat-bottomed appliances, any seal loss, should it take place, would be replaced by the tail-off of the flow of water which occurs at the end of each discharge. Because of this, and the difficulty of installing a trap having a 75 mm seal under these appliances, it is now permissible to use one having a 50 mm seal. The gradient or fall of the discharge pipe is not critical, being between 1 and 5° (18–90 mm per 1 m run). The main danger with these appliances is not so much loss of trap seal as the build-up of soap deposits causing a blockage in long discharge pipes. For this reason they should not exceed 3 m in length.

*Sinks*

Because particles of solid waste are often discharged from a sink, common examples being tea leaves and vegetable peelings, a tubular trap (not a bottle type) should always be used as it is considered to be more self-cleansing. Normally, no special precautions are necessary to preserve the seal in a sink trap as sinks, like baths and showers, are flat-bottomed appliances, and the tail-off or rill reseals the trap should it be siphoned. The maximum recommended length of a sink waste is 3 m, this limitation being imposed not to avoid the formation of a solid plug of water but to minimise the possibility of blockage and reduce noise. Where food macerators are installed an increase in the fall of the discharge pipe is recommended. The absolute minimum is normally 6° (1 in 10), as shown in Book 1, but many manufacturers recommend 15° as this increases the velocity of flow, ensuring any solid particles of food waste are removed. These appliances reduce solid waste to a mass of particles creating ideal conditions for blockage. The reader is referred to Book 1, Chapter 10, where this topic is dealt with.

*Wash basins*

Detailed information relating to discharge pipes for wash basins is given in Book 1, page 265. Generally, the maximum length of waste permissible is 1.7 m, having a fall of not less than 1° (18 mm per 1 m run). If it is impossible to keep within these limits due to the distance of the basin from the main stack, or if an excessive number of bends are used, several alternatives are available to overcome the problem. A resealing trap can be used, although they tend to be noisy and for this reason are not recommended. Other alternatives are illustrated in Figs 8.12(a), (b), (c). They all relate to single basins having a discharge pipe of not more than 3 m in length, diameter of 32–40 mm and a fall of not less than 18 mm per 1 m run. The one exception is that in Fig. 8.12(a) which may have a fall of up to $2\frac{1}{2}$° or 45 mm per 1 m run. This is because it contains no bends. The introduction of bends in a discharge pipe having higher velocities than those recommended can cause the pipe to run full with subsequent siphonage of the trap. In all cases where the diameter of the discharge pipe exceeds that of the trap, as in Fig. 8.12(a), adequate access must be made to remove any sludge deposited due to the low-velocity flow of water.

## Ranges of appliances

A series of similar appliances connected to a common waste is called a *range*, the most common examples being WCs, urinals or wash basins. Ranges occur only in commercial or public buildings where sanitary accommodation must be provided for large numbers of people. As the appliances have a higher incidence of usage than those in private dwellings there is a greater danger of trap seal loss, and some ventilation of sanitary pipework, especially that serving wash basins, will be required.

*Ranges of basins*

In factories and offices adequate provision for washing must be provided by law, and as many people are often employed in such premises washing facilities must be plentiful. In workplaces of an industrial nature wash fountains or washing troughs are quite common as they can be used by a greater number of people at the same time and occupy less space than a range of wash basins. In general, wash basins are fitted in sanitary accommodation for offices because more space is usually available for ablutionary facilities. The best and most sanitary method of connecting a basin range to a main discharge stack is to use a common waste pipe like that shown in Fig. 8.13. Providing that not more than four basins are fitted to the common waste, which has a diameter of 50 mm and

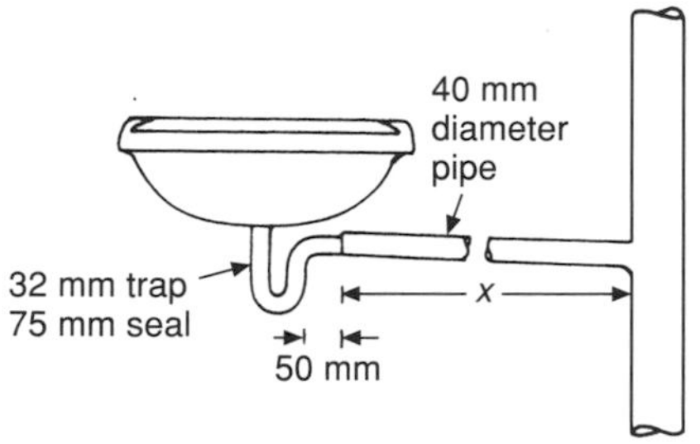


(a) Increase in discharge pipe diameter

All the arrangements shown here are suitable for discharge pipes exceeding 1.7 m in length, but less than 3.0 m (shown as *x* in the illustration).

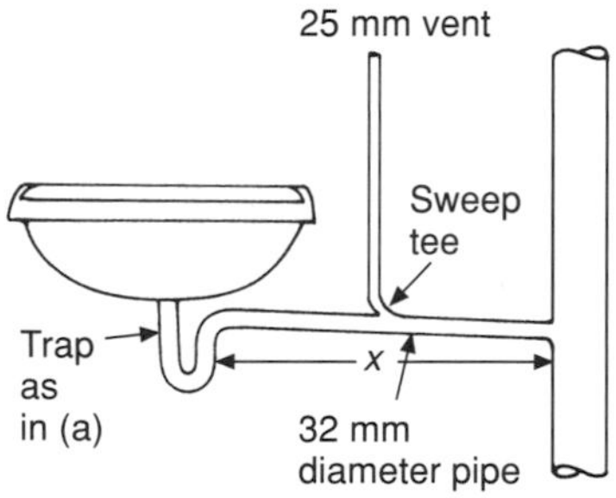


(b) Alternative to (a) using a 32 mm branch discharge pipe

Trap vent to be turned into main ventilating stack or vent portion of main discharge stack above the highest appliance.

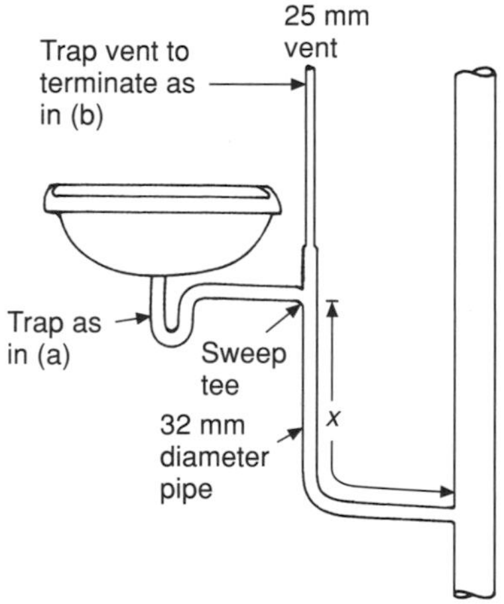


(c) As (b) but with a low-level connection to the main stack pipe

**Fig. 8.12** Single basin connections.

does not exeed 4 m in length, no vent is necessary. The low angle and the relatively large common waste ensure that it does not flow full bore. If a range of more than four basins is contemplated, or the main pipe exceeds 4 m in length, a vent must be fitted as shown in Fig. 8.13. The number of changes of direction in the branch discharge pipe must be kept to a minimum, and if any bends are necessary they must have a large radius. The reason for this is that they can reduce the flow rate to such an extent that a solid plug of water may form, leading to induced siphonage.

Spray taps are often used for hand washing and, as they have such a low flow rate, the discharge may not remove the grease and soap sediment in the discharge pipe. When these taps are specified BS EN 12056 recommends the use of a common discharge pipe having a diameter of not more than 32 mm maximum nominal bore as this has the effect of increasing the velocity of the discharge, washing out any soap deposits. The maximum length of the discharge pipe is unlimited, but for practical purposes it should be as short as possible to reduce problems with soap deposits. As the discharge pipes are unlikely to run full, traps of 50 mm seal depth are permissible. Basins fitted with spray taps should not be provided with plugs when small waste diameters are used as this could lead to surcharging if the basins were accidentally filled. It should be noted that many manufacturers have stopped producing spray taps in favour of modern non-concussive or infrared sensing taps. In such cases the diameter of the discharge pipe must be increased accordingly.

Although modern design has largely eliminated fully ventilated systems, such a system will be necessary where the incidence of usage is high and the branch discharge pipes run full. An experienced designer would anticipate these conditions and specify the type of ventilation required. Reference should be made to Fig. 8.8, part of which shows a fully ventilated range of basins.

### *Ranges of WCs*

The normal size of a branch pipe for a WC range is 100 mm. Ventilation of these branches is seldom necessary except where more than eight WCs are fitted on one branch or where the branch has a

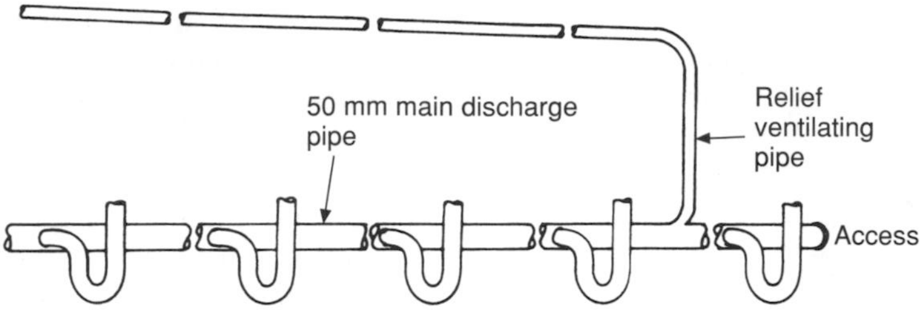


**Fig. 8.13** Ventilation of a range of appliances using the modified single stack system.

significant number of bends. Even then, unless the usage of the appliances is 'congested' (in which case a ventilated system should be fitted), it is sufficient to ventilate the branch discharge pipe only, in a similar way to that of a range of wash basins where the main discharge pipe only is ventilated.

*Ranges of urinals*

As with WCs, it is seldom necessary to ventilate urinal wastes due to the relatively low flow rate. It is recommended that where bowl urinals are fitted, the individual pipe from each bowl to the main discharge pipe should be as short as possible with a fall of 1–2$\frac{1}{2}$° (18–45 mm per 1 m run). Falls on the main discharge pipe may be 1–5° (18–90 mm per 1 m run).

### Special notes relating to ventilating pipes

Ventilating pipes do not normally carry waste water, the one exception to this being a wet vent (see Fig. 8.14). In systems constructed of cast iron, condensation in a long vertical section of the pipe may cause corrosion, resulting in particles of rust falling and blocking the lower bend in the stack. By connecting a branch discharge as shown, any debris will periodically be washed away. The position of the connection between a branch ventilating pipe and a main ventilating pipe is important and must always be made above the flood level of the appliance, or appliances if a range is under consideration. Figure 8.15 shows the correct method of connection so as to avoid the discharge of waste water through the branch ventilating pipe should an obstruction form in the appliance discharge pipe. The lower end of the main vent stack can be

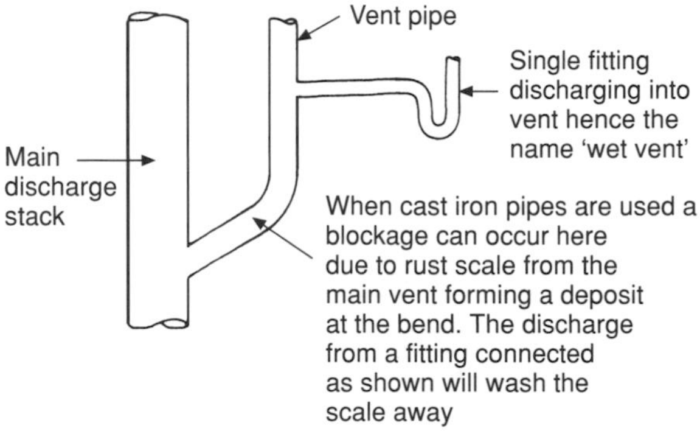


**Fig. 8.14** Wet vents.

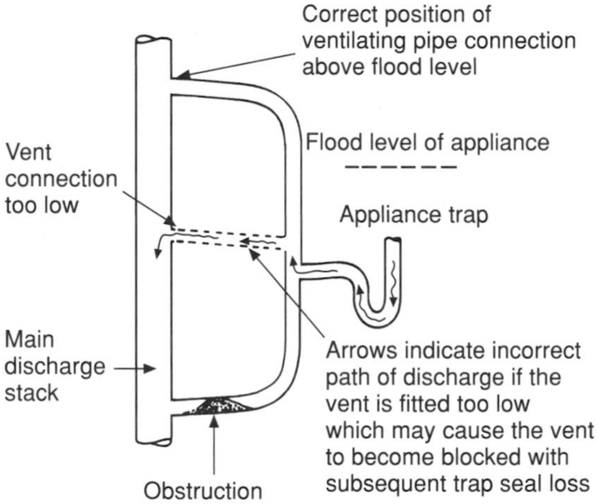


**Fig. 8.15** The effect of incorrect vent pipe connections.

connected to the drain in a convenient inspection chamber or, alternatively, above ground level in the main discharge stack as shown in Fig. 8.16. In buildings of more than five storeys, ground floor

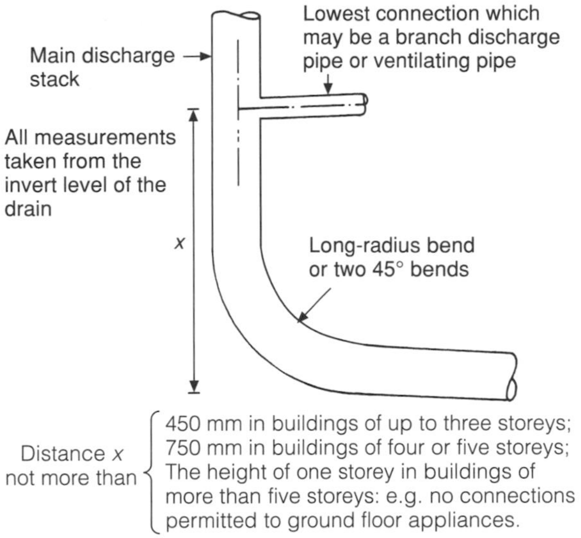
Main discharge stack
Lowest connection which may be a branch discharge pipe or ventilating pipe
All measurements taken from the invert level of the drain
x
Long-radius bend or two 45° bends
Distance x not more than
450 mm in buildings of up to three storeys;
750 mm in buildings of four or five storeys;
The height of one storey in buildings of more than five storeys: e.g. no connections permitted to ground floor appliances.

**Fig. 8.16** Low connections on discharge stacks.

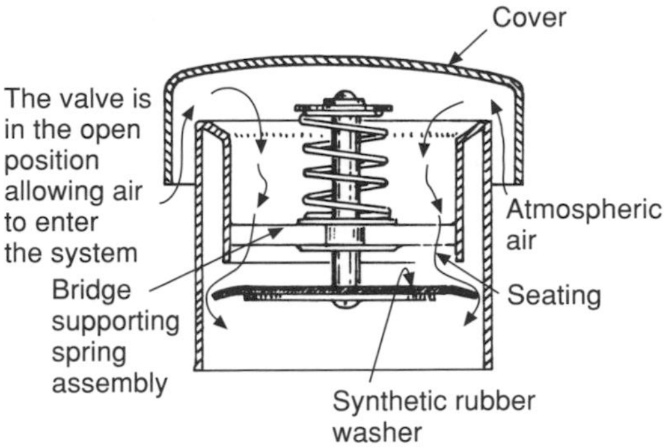
Cover
The valve is in the open position allowing air to enter the system
Atmospheric air
Bridge supporting spring assembly
Seating
Synthetic rubber washer

**Fig. 8.17** Automatic air-admittance valve. These valves only admit air to a discharge pipe when it is subjected to a negative pressure. Atmospheric pressure acting on the top surface of the valve pushes it down, compressing the spring. When atmospheric pressure is restored the spring closes the valve to prevent the escape of drain air.

appliances should discharge into the underground drain. The object of these limitations is to avoid the base of the vent becoming surcharged, thus limiting its effectiveness in combating compression and also avoiding the entry of foaming detergent into the ground floor appliances. Any offsets or bends in the wet section of the main discharge stack should be avoided. If this is not possible they must be ventilated (see page 269). Ventilating pipes should terminate in the open air in such a way that nuisance and health hazards are avoided. This is usually achieved by carrying them 900 mm above and not within 3 m of any opening, window or vent. For further details of ventilating offsets and ventilating pipe terminals, reference should be made to Book 1, Chapter 10.

### *Air-admittance valves*

These are designed to automatically maintain equilibrium in an unventilated part of a system and must comply with BS EN 12380. The Building Regulations permit only valves complying to this standard or those approved by the British Board of Agrément. It is essential that installation complies with the terms of the certificate. It should be clearly understood that these valves do not take the place of a vent pipe, which not only maintains stable atmospheric conditions in the drainage system, but also ventilates the drain. These valves are quite simple in operation and a typical example is shown in Fig. 8.17. The spring is adjusted during manufacture so that it just overcomes the weight of the valve and holds it in the closed position. The slightest negative pressure in the discharge pipe causes the valve to open, thus maintaining atmospheric pressure in the discharge system. A typical situation where these valves may be an advantage is in a building having two main discharge stacks. It would only be necessary to terminate one externally of the building allowing for ventilation of the drain, the other could be accommodated in a suitable duct or the roof space using an air-admittance valve. As they are mechanical devices it is important that they are fitted in an accessible position. A similar type of device is also available for branch discharge pipes where for various reasons the requirements of BS EN 12056 relating to trap ventilation and discharge pipe maximum lengths cannot be met. It should be noted that not all authorities accept such mechanical devices in connection with discharge pipe systems and their use should be treated with care.

### *Ventilating pipe sizing*

This work is normally performed by a qualified plumbing designer or sanitary engineer and should not be carried out using rule-of-thumb methods. Circumstances alter cases and ventilating pipe sized

**Table 8.2** General guide to sizes of ventilating pipework.

| *Discharge stack or branch discharge pipe diameter* | *Size of main vent stack or branch ventilating pipe* |
|---|---|
| Less than 75 mm | $\frac{2}{3}$ diameter (not less than 25 mm) |
| More than 75 mm | $\frac{1}{2}$ diameter |

for one job may not necessarily be satisfactory for another, even if the situation appears to be the same. Some of the factors that must be considered are:

(a) The incidence of usage of the appliance.
(b) The number of appliances.
(c) The height of the building.
(d) Whether a group of single appliances or a range is being considered.
(e) Whether the building is domestic, public or commercial.

Bearing in mind the foregoing, it must be stated that in many cases the plumber carries out work in smaller premises where no proper design of the sanitary pipework is available. For this reason the following information is given based on the recommendations of BS EN 12056. In dwellings of up to five storeys where the appliances are grouped together, sufficient ventilation is provided via the main ventilation pipe. In buildings of greater height where the ventilated stack system is used, the main ventilating stack should be 50 mm. A 25 mm pipe is normally considered to be of sufficient size to ventilate single fittings and branch discharge pipes, but if it is 15 m or more in length, or contains more than five bends, then it should be increased to 32 mm in diameter. In the absence of specific information relating to the diameter of ventilation pipes, reference should be made to BS EN 12056. Table 8.2 gives a general guide which meets the requirements of the standard.

### Access for cleaning and maintenance

It is essential that adequate provision is made to enable obstructions in the pipework to be cleared with the minimum of inconvenience to the client. One of the main problems with modern waste water disposal systems having low flow velocities in the branch discharge pipes is that they allow soap and other debris to build up on the invert of the pipe, which over a period of time frequently causes a blockage. The incidence of blockage is higher in public or commercial buildings than in domestic premises and for this reason fewer access points are required in the latter type of building. It is not always necessary to provide access to every branch, and one cleaning eye in a suitable position may allow access to a group of appliances. Special consideration should be given to discharge pipes from appliances that are likely to cause trouble, i.e. urinals, public WCs, showers and sinks, especially those fitted with macerators, or those in canteen or restaurant kitchens into which large quantities of grease may be discharged. Good access is also necessary in multi-storey buildings so that sections of the sanitary pipework system can be tested as they are completed.

### Pipe ducts

In modern buildings having internally fitted pipework systems, some form of cover or ducting is often necessary to conceal the pipes. It is a fact that while many plumbers can appreciate a well-designed, neatly fitted pipe system, few members of the public can, and for this reason much plumbing pipework has to be concealed. Since ducts are used not only for plumbing systems but also for gas and heating pipework, they should be large enough to accommodate all the services, and provide sufficient space for maintenance work, especially the cleansing of discharge pipework, to be effected. Particular attention should be given to those items likely to require most maintenance, i.e. discharge pipes. Ducts should be constructed in such a way as to prevent the spread of fire and transmission of sound. This can be achieved by placing suitable packing material around the pipes where they pass through the floor of the duct. The material or method of sealing should not restrict thermal movement of the pipes. Water-mixed pastes and fire-resistant silicones are useful when irregular holes have to be filled. Special sleeves are also available containing flexible materials which swell, sealing any voids when heated.

In low-rise housing and conversions some builders give little thought to subsequent access to pipework, and plywood panels are often fixed over the pipes with nails and adhesives. However, a sufficiently conscientious and competent plumber can be involved at the planning stage to advise on the proper construction of small ducts, i.e. suitable panels *screwed* in position over access points. Otherwise, the householder may be faced with inconvenience and considerable expense in the future.

### Inspection and testing

While work is progressing, care must be taken to prevent debris of any kind from entering the system. Any open ends of pipework should be fitted with temporary plugs such as a well-packed wad of rag until the final connection is made. The installation must, on completion, be checked to make sure it has not been damaged in any way and has been securely fixed to the building fabric.

*Testing for soundness*

Sanitary pipework overground is tested with air. Testing should be carried out as one operation where possible, but on large installations it may have to be done in sections as the work proceeds. An air test is very searching and will detect any leak in the system. It will be found that it pays to take a little care when making the joints during installation in order to save a lot of trouble and expense searching for leaks on completion.

The procedure for testing is as follows. Plug any open ends with suitable plugs or stoppers, as shown in Fig. 8.18, and seal the plugs with water as air, being a gas, can pass through very small apertures. The water seals of any traps should be filled and a U gauge and test pump connected at a convenient point as shown, air being pumped into the system in a similar way to that used when testing a gas installation. The Building Regulations recommend a test pressure of 38 mm which must hold for a minimum period of 3 minutes. Should a leak be indicated by a drop in pressure, the joints should be tested with soapy water, again using the same technique as with gas installations. Leakage from cast iron stacks is usually due to defective joints, although blow holes in castings are not unknown and can be the cause of a drop in air pressure. If PVC pipe is used, any leakage is usually caused by the rubber seal having been

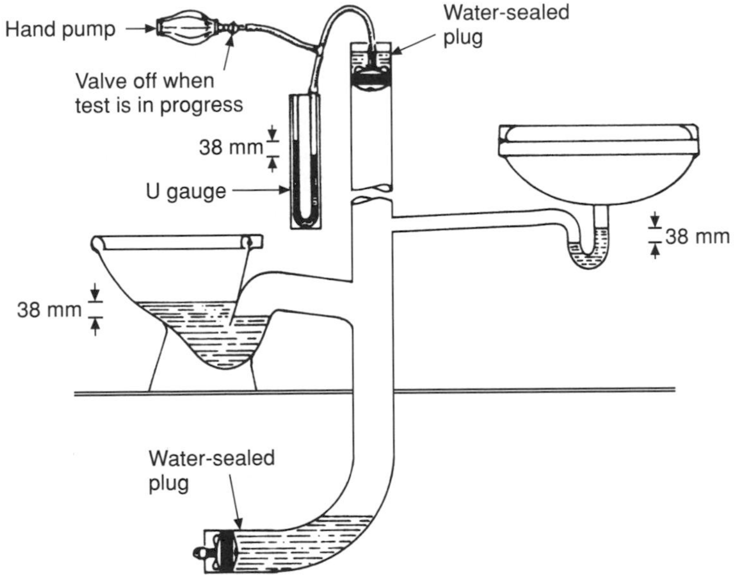


**Fig. 8.18** Air testing discharge pipework above ground. Test should be maintained at 38 mm for a period of 3 minutes.

displaced from its housing in the socket or damaged during the jointing process. Solvent-welded joints, if found to be leaking, would indicate very poor workmanship indeed.

*Water testing*

Some authorities may request that a system be water tested up to the flood level of the lowest fitting. This simulates the effect of a blocked drain where, in practice, the lower part of the system could be filled with foul water. Although this test is still relevant, most authorities accept the air test described.

*Performance tests*

The retention of trap seals should be tested under the worst possible working conditions. The appliances used for the test must be filled to their overflowing level and discharged simultaneously by pulling the plugs or flushing the cisterns. The seal losses due to positive or negative pressures in the stack should be noted. Each test must be repeated three times and a minimum seal depth of not less than 25 mm should be retained after each test. This can be checked by using a short length of clear plastic or glass tube and immersing it in the trap after each test (see Fig. 8.19) or a suitable dipstick can be made by coating a small piece of wood or metal with a dark matt paint. Gloss paint may give a false reading. If plastic tube is used, make sure it is touching the bottom of the trap when the test is made, then place the thumb over the top end. This will retain the water in the tube, and when it is withdrawn a true reading will be obtained. Table 8.3 is based on the recommendations of BS EN 12056 and shows the number of appliances that should be discharged simultaneously during tests in different types of buildings under various conditions, i.e. domestic, public and congested. The selected appliances should normally be close to the top of the stack and on adjacent floors, as this gives the worst pressure conditions.

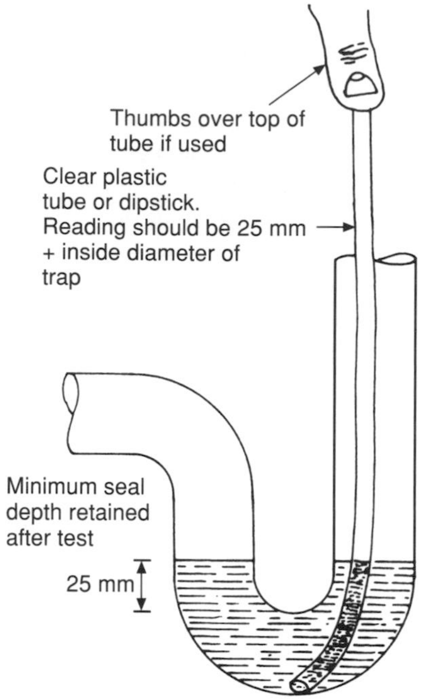


**Fig. 8.19** Method of checking trap seal depth after performance test.

## Maintenance

(a) Visually inspect all visible pipework, fixings and appliances for damage and security.

(b) Replace any access cover seals which may be damaged or decayed when the covers are removed for maintenance.

(c) Any hand-operated rods or cables used for removing blockages in discharge pipes should not damage the internal surfaces in any way. Powered equipment and kinetic rams should be used with extreme care. The rams are suitable for soft blockage only, as a stubborn blockage can cause a blow-back and injure the operator. Before using powered equipment or kinetic rams, any automatic air valves should be removed as waste matter may be forced out of the openings causing damage to decorations. Their use can also blow out a rubber ring-sealed joint.

(d) Soft blockages such as soap and grease deposits can build up in a long run of pipe where the flow rate fails to maintain a self-cleansing gradient. Such obstructions can often be cleared using a plunger, but in such cases a more effective method would be to flush out the system using a solution of hot water and soda.

**Table 8.3** Number of sanitary appliances to be discharged for performance testing.

| *Type of use* | *Number of appliances of each kind on the stack* | *Number of appliances to be discharged simultaneously* | | |
|---|---|---|---|---|
| | | *6/7.5/9 litres WC* | *Wash basin* | *Kitchen sink* |
| Domestic | 1 to 9 | 1 | 1 | 1 |
| | 10 to 24 | 1 | 1 | 2 |
| | 25 to 35 | 1 | 2 | 3 |
| | 36 to 50 | 2 | 2 | 3 |
| | 51 to 65 | 2 | 2 | 4 |
| Commercial or public | 1 to 9 | 1 | 1 | |
| | 10 to 18 | 1 | 2 | |
| | 19 to 26 | 2 | 2 | |
| | 27 to 52 | 2 | 3 | |
| | 53 to 78 | 3 | 4 | |
| | 79 to 100 | 3 | 5 | |
| Congested | 1 to 4 | 1 | 1 | |
| | 5 to 9 | 1 | 2 | |
| | 10 to 13 | 2 | 2 | |
| | 14 to 26 | 2 | 3 | |
| | 27 to 39 | 3 | 4 | |
| | 40 to 50 | 3 | 5 | |
| | 51 to 55 | 4 | 5 | |
| | 56 to 70 | 4 | 6 | |
| | 71 to 78 | 4 | 7 | |
| | 79 to 90 | 5 | 7 | |
| | 91 to 100 | 5 | 8 | |

*Note*: These figures are based on a criterion of satisfactory service of 99 per cent. In practice, for systems serving mixed appliances, this slightly overestimates the probable hydraulic loading. Flow load from urinals, spray tap basins and showers is usually small in most mixed systems, hence these appliances need not normally be discharged.

(e) Lime scale deposits form not only on the surface of appliances, but also on the inside of pipes. This is especially so in the case of urinals, and in extreme cases can completely block the discharge pipe. The usual method of removal is by using chemical descaling agents. These are usually corrosive in nature and a check should be made to ensure they will not damage the material on which they are used. **It is essential to carefully read any instructions shown on the container before use**. Acid-based descaling agents in contact with cleansing agents, such as bleach, containing chlorine will produce chlorine gas. All pipework systems must therefore be flushed out prior to the application of the descaling fluid, and adequate ventilation must be maintained during its application. Operatives must always use protective clothing, e.g. gloves, overalls and eye protection when working with these materials.

## Further reading

The Building Regulations 2002, Part H.

BS EN 12056 Part 2 Sanitary pipework.

BS 8313 Code for accommodation of building services in ducts.

BRS Digest 248 *Sanitary Pipework; design basis.*

BRS Digest 249 *Sanitary Pipework; design of pipework.*

BRS Digest 81 *Hospital Sanitary Services; some design and maintenance problems.*

*Soil and Waste Pipe Systems for Large Buildings*, Seminar Notes. Available from BRE Bookshop,

151 Rosebery Avenue, Farringdon, London EC1R 4GB Tel. 020 7505 6622. www.brebookshop.com

Information on fire-resistant sleeves: Palm Fire Products Ltd, 1 Station Road, Romsey, Hants SO51 8DP Tel. 0238 0227337.

## Self-testing questions

1. State the two main causes of noise in sanitary pipework systems.
2. List four factors that influence trap seal loss in modern pipework systems.
3. State the recommended diameters of branch discharge pipes serving baths, wash basins and sinks in a multi-storey building.
4. List and identify four pipework systems used for waste water disposal.
5. State which of these systems is the most economical and effective when the fittings are grouped round the main stack.
6. Describe and sketch three methods of preventing seal loss in a trap on a basin waste which is longer than the recommended design length.
7. Describe the type of trap recommended for use with a kitchen sink. Give your reasons.
8. Explain the term 'wet vent' and describe the circumstances in which it would be used.
9. Describe the procedure for air testing a system of discharge pipework.
10. (a) Describe the procedure for checking the depth of trap seal retained after a discharge has taken place.
    (b) State the minimum recommended depth of seal that must be retained.
11. Explain why a smaller discharge pipe is permissible for a range of wash basins fitted with spray taps.
12. Explain the term 'wet ventilating pipe'.

# 9 Underground drainage materials, fittings and systems

After completing this chapter the reader should be able to:

1. Select suitable fittings for various applications.
2. Describe and sketch methods of support and protection for drainage pipelines.
3. State why ventilation of underground drains is necessary and how it is achieved.
4. Identify methods of providing access to drains including those under buildings.
5. Recognise special drainage fittings and understand their use.
6. Differentiate between types of drainage schemes.
7. Explain and illustrate the basic principles of setting out drains and the necessity of providing self-cleansing gradients.
8. Describe the methods and equipment used for testing underground drainage systems.

### Regulations relating to building drainage

All work on drainage, whether new or alterations to existing schemes, is subject to the Building Regulations, Approved Document H and all relevant British Standards. Any work connected with underground drainage comes under the jurisdiction of the local authority planning department. Advice should be sought from and notice in writing given to this department before any drainage work is undertaken. In most cases drawings of the proposed work will be required and, to avoid delay, they should be deposited well before the job is started. Although they are not at present mandatory, the relevant British Standards should be studied to ensure that the installation meets the minimum requirements. BS 8301 is the standard for underground drainage and although it has been superseded by BS EN 752 Parts 1–4, its content is still valid and is currently accepted by Building Control Officers.

### Functions of drainage systems

The choice of materials now on the market for underground drainage is so vast that it is impossible to cover fully in this volume all the various joints, fittings and techniques used. Manufacturers of these materials, however, provide ample literature on the subject and the student who wants to obtain a greater depth of knowledge is recommended to contact them. Some of the companies prepared to send detailed information are listed at the end of the chapter.

The function of underground drainage systems is to convey waste water, which comes from the sanitary fittings via the overground discharge pipe system, to the sewer or, if applicable, other sources of disposal. It is essential that the pipes, fittings and methods of jointing should be suitable for the purpose for which they are used.

A good drainage system must be both self-cleansing and watertight. If it is not self-cleansing the system could become a breeding ground for harmful bacteria and would certainly give rise to unpleasant smells. If it is not watertight, the ground surrounding the drain would become polluted, again giving rise to unpleasant and unhealthy conditions.

The reader should understand that there are two main systems of drainage, the *conservancy system* and the *water carriage system*. The former is very rare in housing, the waste matter being retained in a suitable container and disposed of when necessary. As most permanent premises have some form of piped disposal system, the conservancy system is used only with temporary quarters such as building sites, touring caravans, etc., where no drainage is available. The water carriage system employs pipes to remove waste matter to the local authority's treatment works or, in an isolated area, a septic tank.

Although the topic of sewage treatment is beyond the scope of this volume it should be explained that a septic tank provides sewage treatment for single or small groups of premises. Its working principles are very similar to those of a large treatment plant and, properly maintained, it can be very effective in rural areas where no main drainage scheme exists.

### Health and safety

The hazards related to working in trenches are dealt with fully in Book 1 of this series, but the main points are as follows:

(a) Shuttering of the trench sides will be necessary to prevent their collapse. Whether it is open or closed will depend on the composition of the earth strata and the depth of the trench.
(b) Suitable guard rails must be erected.
(c) Adequate warning notices must be displayed.
(d) Trenches should be inspected for safety on a daily basis by a responsible person. This is very important, especially where the soil is sandy or very wet.

### Materials

Three main materials are used for underground drainage: clayware, cast iron and PVC. Drains made of clayware are seldom fitted by plumbers, therefore this material will not be dealt with to any great depth. Cast iron drainage has always been, and still is, mainly installed by plumbers. They understand the basic principles of drainage and from a practical point of view they have the necessary skills for its installation. PVC is sometimes installed by plumbers for both overground and underground drainage. It has many advantages over traditional materials, being much lighter in weight and very easy to install. The materials used for underground drainage and methods of jointing are dealt with in Book 1 of the series.

### Drainage fittings

*General*

The variety of drainage fittings produced is considerable, some only being encountered for special types of work. Those illustrated and described here are representative of the more common types the plumber is likely to come across during normal work. Producers of clayware, cast iron and PVC systems of drainage manufacture a complete range of fittings in each material together with adaptors to enable connections to be made to other materials. Relevant catalogues should be referred to in order to ensure effective connections.

The adaptor shown in Fig. 9.1 is useful when adapting existing clayware drains having differing

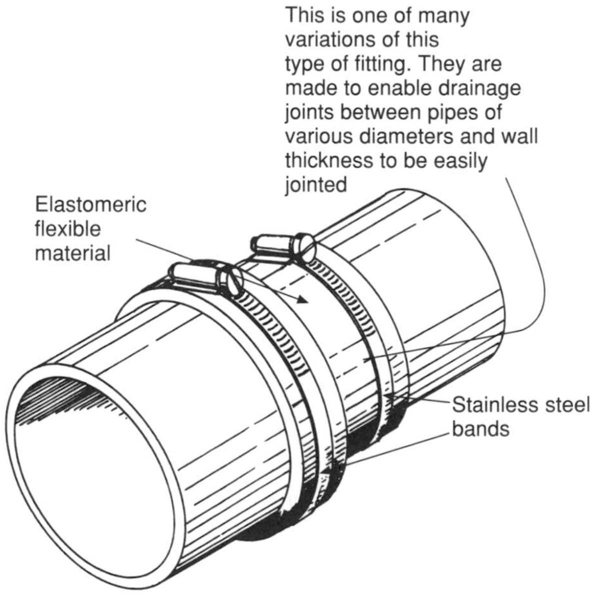
This is one of many variations of this type of fitting. They are made to enable drainage joints between pipes of various diameters and wall thickness to be easily jointed
Elastomeric flexible material
Stainless steel bands

**Fig. 9.1** Drainpipe adaptor.

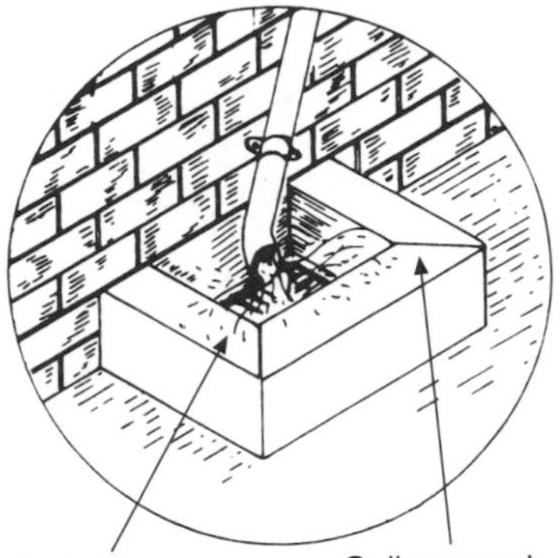


**Fig. 9.2** Obsolescent method of waste discharge into gulley. Many older buildings have this arrangement. But splashing caused by the discharge pipe emptying over an open gulley grating causes fouling on the interior and top of the gulley guard leading to insanitary conditions.

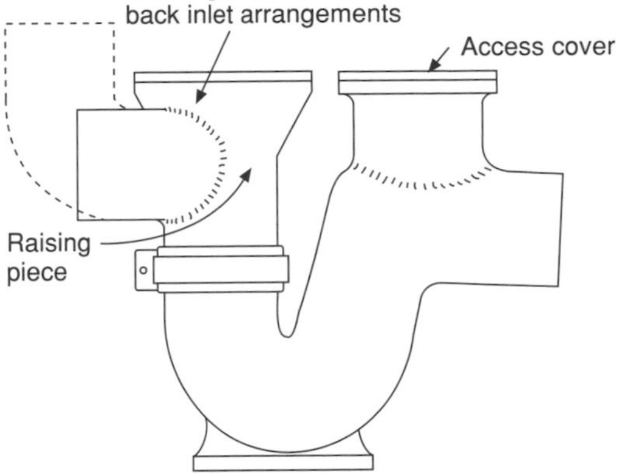


**Fig. 9.3** Typical cast iron back inlet gulley trap obtainable with or without access.

external diameters. These old drains are usually laid with cement mortar joints and it is near impossible to cut out the joint without breaking the socket. Adaptors such as this can be used for jointing socketless pipes of various materials and external diameters. Whenever connections have to be made between various materials, never attempt to 'make do' with non-standard adaptors, as this often leads to trouble in the future.

*Gulleys*

The purpose of a gulley is to provide a trap to prevent odours entering the atmosphere in cases where it is not possible, or desirable, to terminate a discharge pipe directly into the drain. Prior to *c.* 1950 the method shown in Fig. 9.2 was commonly used in domestic dwellings. This method of discharge is no longer permissible and was always undesirable. Modern buildings employ the use of back inlet gulleys where the discharge pipe terminates under the grating but over the water seal (see Fig. 9.3). A typical raising piece designed to fit a standard trap is also illustrated and shows how a gulley trap can be adapted to suit a variety of applications. Like other traps used on underground drainage systems, only a 50 mm depth of seal is necessary. It will be noted that this is less than that required for traps on small-diameter discharge pipes and is due to the fact that the risks of siphonage are almost negligible, as any discharge seldom permits the pipe to run full. Figure 9.4 shows two gulleys that are suitable for draining

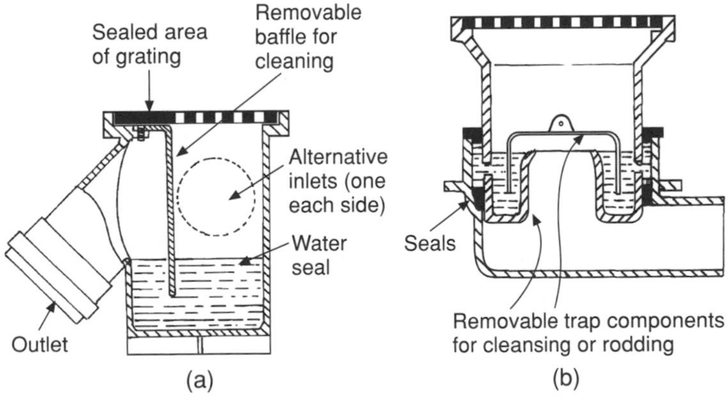


**Fig. 9.4** PVC bottle-type floor gulleys.

**Fig. 9.5**  Trapless gulley. Used for surface water only as an alternative to a rainwater shoe. Silt collecting in the base can easily be removed.

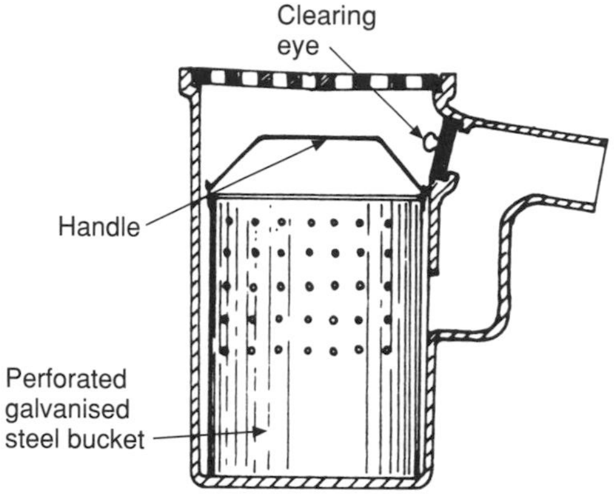

**Fig. 9.6**  Yard gulley. Used to collect the run-off from paved surfaces. The removable bucket prevents any silt being washed into the drain. It can be periodically removed for cleansing.

floors, such as commercial kitchens and sanitary annexes. Their short overall height makes them very suitable for this purpose. They are not self-cleansing but are provided with suitable access for routine maintenance.

Trapless gulleys (Fig. 9.5) and yard gulleys (Fig. 9.6) are often used for draining paved areas, as they provide for the easy removal of any silt that is washed off these surfaces. If silt is allowed to pass into the drain it will eventually build up and cause a blockage. Trapless gulleys must not be connected to foul water systems. They are usually used in conjunction with a trapped master gulley in the same way as rainwater shoes.

*Rainwater shoes*
These are trapless fittings used mainly to connect rainwater pipes to a surface water drain with access for cleansing both the drain and the rainwater pipe. Two or more such fittings are often used in conjunction with a master gulley (see Fig. 9.7).

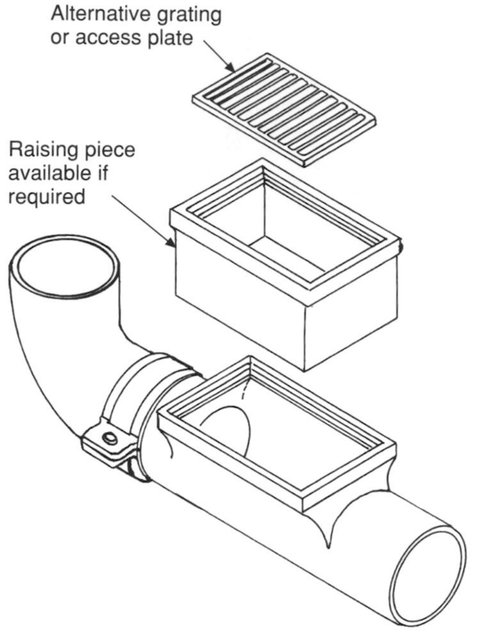

(a) Drain shoe. An alternative to a trapless gulley for surface water use

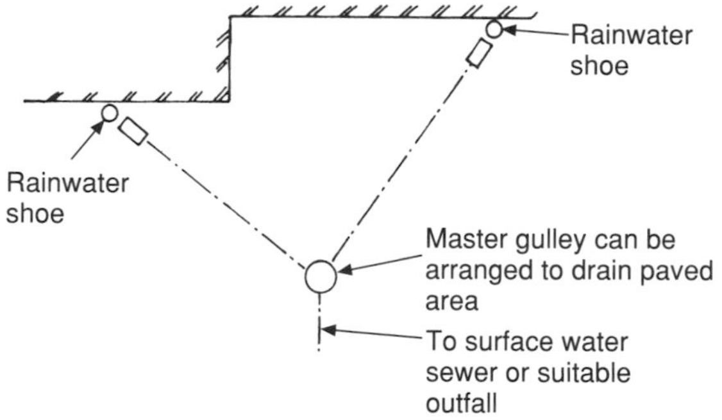

(b) Several of these shoes may be connected to a master gulley as shown

**Fig. 9.7**  Rainwater shoes.

*Intercepting traps*
An intercepting trap is illustrated in Fig. 9.8(a), but they are seldom necessary in modern properties as they were originally intended to provide a trap to prevent sewer gas from entering the house drain. It should be explained that the house drain is that part of the system which is the responsibility of the householder. The sewer may be broadly defined as a main drain carrying the discharge from house drains

to a sewage treatment plant; sewers are generally the responsibility of the local authority. In bygone days, the ventilation of sewers was not as thorough as it is today and, because sewage begins to decompose very quickly due to the action of bacteria, sewer gas was produced which is both flammable and, in heavy concentrations, toxic. In these circumstances it was often necessary to prevent access of sewer gas to house drains. Due to better systems of drain ventilation, sewer gas nowadays seldom constitutes a danger. The main disadvantages of intercepting traps are that:

(a) They are the most common cause of obstruction due to the collection of debris in the trap.
(b) They interrupt the smooth flow of water through the drain.

They are normally situated in an inspection chamber so that they are accessible when necessary. Figure 9.8(b) shows an improved interception stopper. As a result of back pressure in the sewer or incorrect fitting, the normal type provided often falls into the trap causing it to become blocked. This is impossible with the lever locking type.

*Bends*

A wide variety of bends is available in all drainage materials. The bends are specified by their radius and angles. They can also be obtained with access doors on either side or on the back. For full details of ranges of bends available, reference should be made to manufacturers' catalogues. Some typical cast iron bends are shown in Fig. 9.9.

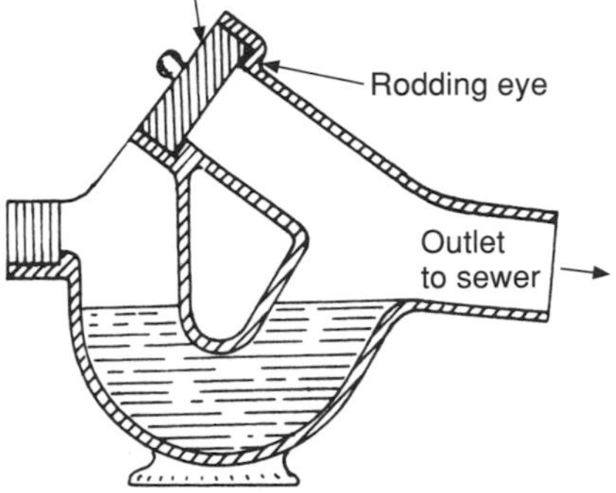


(a) Clayware intercepting trap. Made in clayware or cast iron. Sometimes called a disconnecting trap as its original function was to provide a trap separating the house drain from the sewer

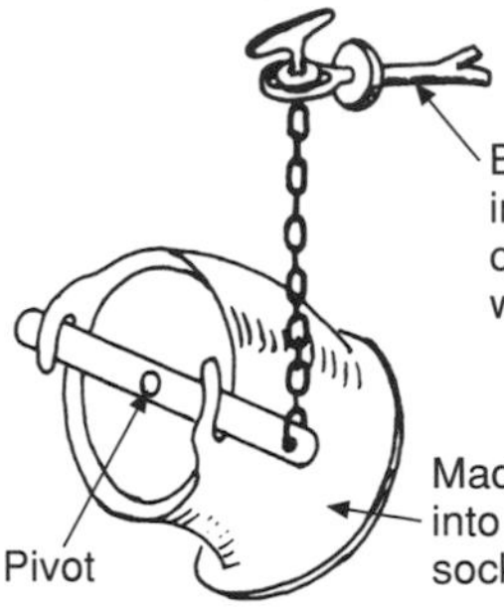


(b) Lever locking stopper for clayware interceptor

**Fig. 9.8** Intercepting trap and lever locking stopper.

*Duck foot or rest bends*

These bends, made of clayware or cast iron, are provided with a foot or rest and are designed to give extra support to offset the weight of a cast iron discharge or vent pipe (see Fig. 9.10).

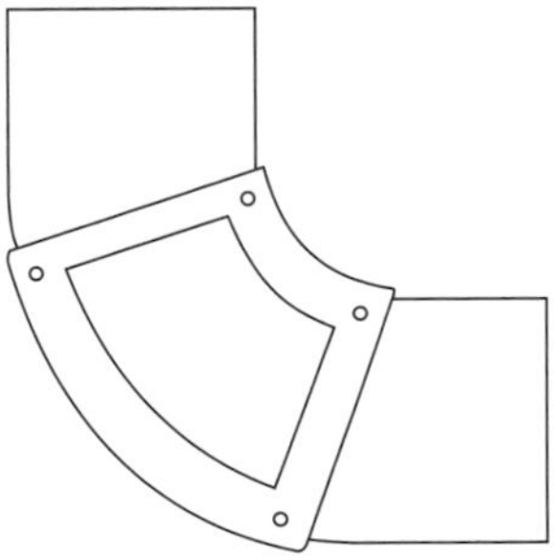

(a) Bend with access door

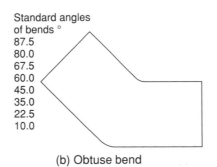

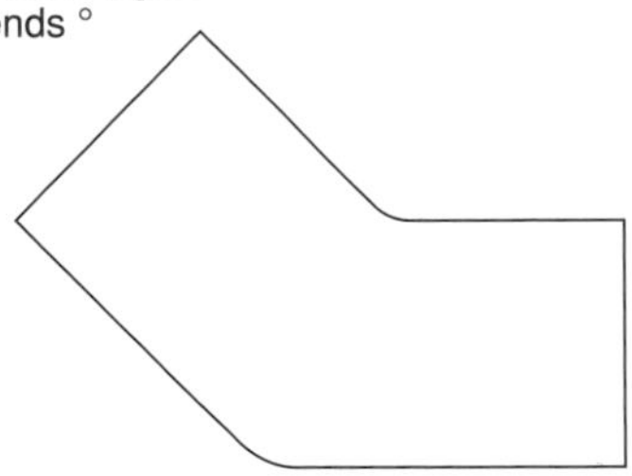


(b) Obtuse bend

**Fig. 9.9** Examples of typical cast iron bends.

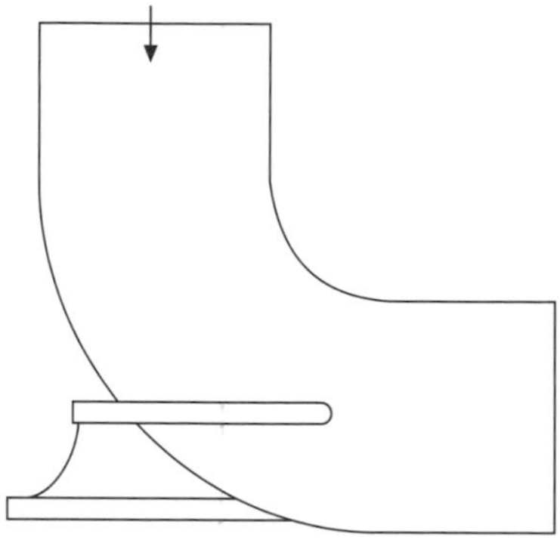

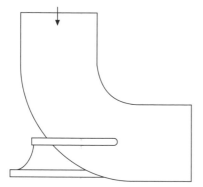

**Fig. 9.10** Rest bend. Used where the bend takes the downward thrust of a stack.

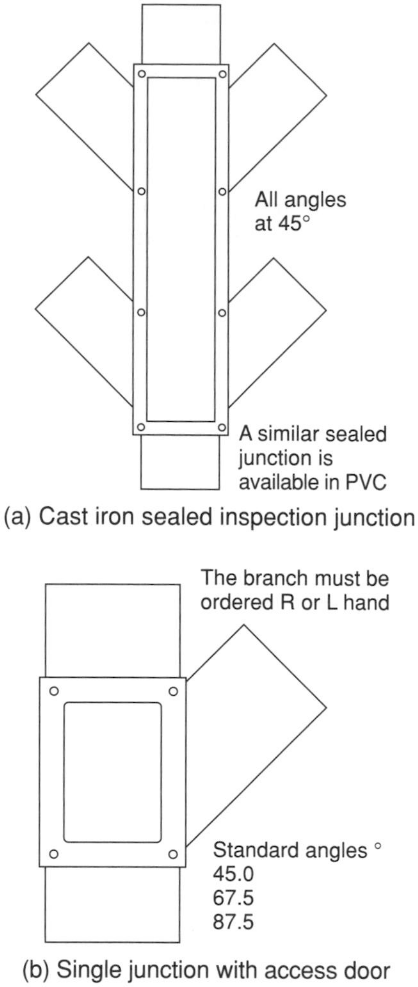


**Fig. 9.11** Sealed junctions for access to cast iron drains.

*Bends and junctions used in chambers*
Cast iron drainage systems were rarely used with open channels in inspection chambers and few, if any, manufacturers now make the necessary fittings. Sealed junctions of the type shown in Fig. 9.11, housed in brick or concrete chambers, are the normal method of providing access to underground and suspended cast iron drains. Producers of PVC systems also make sealed junctions and bends which are similar to those of cast iron and are mainly used for suspended drainage. For underground use, the open channel system, very similar to that of clayware, is generally used.

*Level invert open channel junctions*
A typical example of a clayware junction is shown in Fig. 9.12; those made of PVC are very similar. On all such junctions the branch joins the main channel at 45°. A wide selection of channel bends are available to enable a branch drain to enter the main channel at the same invert level.

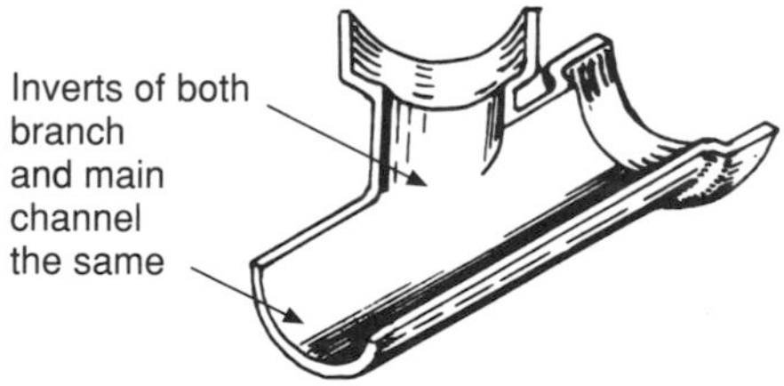


**Fig. 9.12** Channel junction.

*Stepped channel bends*
These are used in conjunction with the PVC channel shown in Fig. 9.13(a), or to make additional connections to an existing inspection chamber. The bend shown in Fig. 9.13(b) is used with this type of main channel and Fig. 9.13(c) shows their application.

*Drain chutes*
The purpose of these fittings (see Fig. 9.14) is to provide better access for rodding drains in deep manholes. They are normally only used with clayware drainage systems which employ open channels. Manufacturers of PVC and cast iron drainage do not list these fittings.

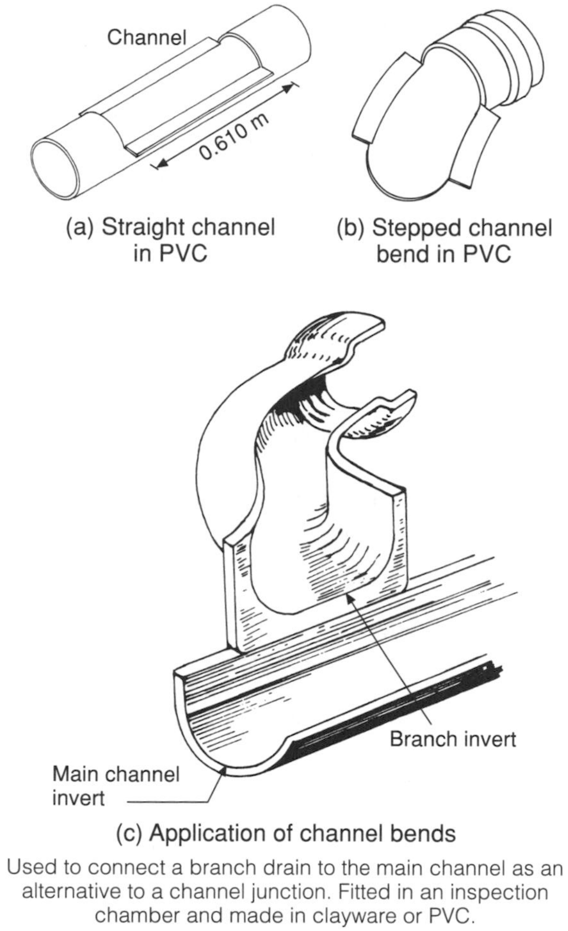


**Fig. 9.13** Straight channel, stepped channel and channel bends.

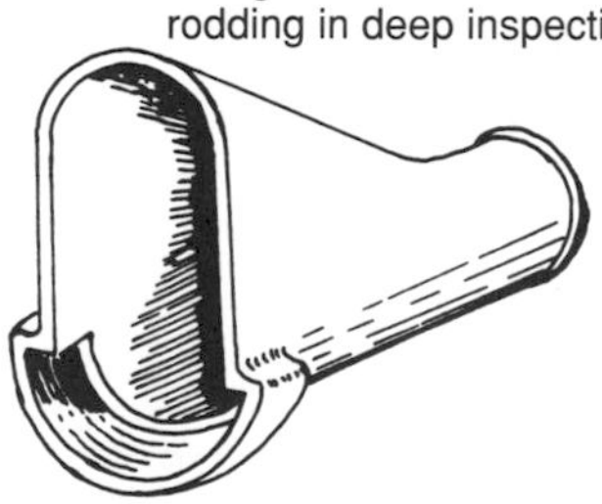


**Fig. 9.14** Drain chute.

*Rust pockets*
Rust pockets (see Fig. 9.15) are seldom necessary in modern drainage systems and few manufacturers list them. They are designed to retain the scale and rust deposits which collect at the base of cast iron vent pipes and which can cause the lower end of the pipe to become completely obstructed. The debris must be removed when periodic maintenance is carried out. In cases where the vent has a branch discharge pipe fitted to it, a rust pocket is unnecessary as the flow of water will wash away any debris. These fittings are normally only found in older public and commercial buildings.

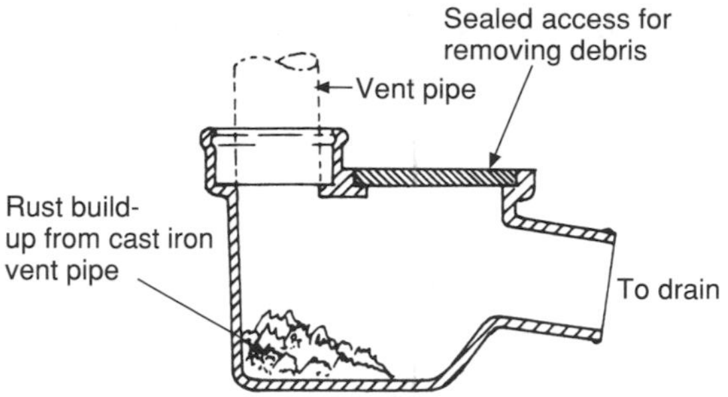


**Fig. 9.15** Rust pocket.

*Level invert taper pipe*
This is used when a change of diameter is made in a drain to avoid any disturbance of the flow in the drain (see Fig. 9.16).

*Anti-flood fittings*
These are used only on surface water drains which discharge into a watercourse or tidal river. During periods of heavy rainfall or high tides, it is possible that water may back up the drain and cause flooding in or around the building. The gulley illustrated in

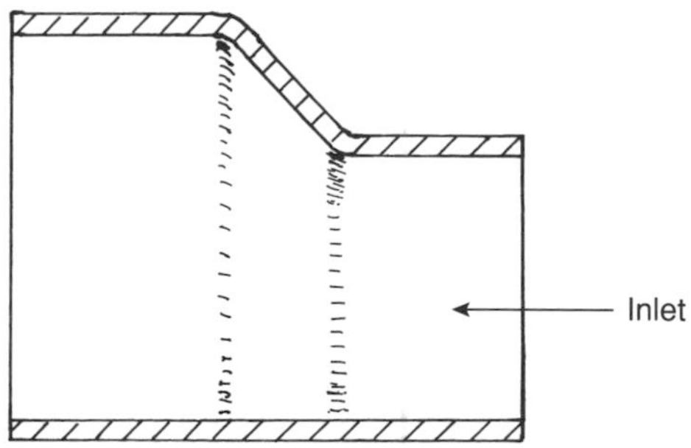


**Fig. 9.16** Level invert taper pipe used to connect a smaller diameter drain to a larger one.

Fig. 9.17(a) is made of cast iron with a rubber seating; they are also obtainable in clayware. If the water in the gulley rises the float engages on the seating making a watertight seal. The float seating and cage are usually made as a unit which sits on lugs cast into the body of the gulley enabling it to be withdrawn for inspection. An anti-flood interceptor (see Fig. 9.17(b)) is a similar device which is housed in an inspection chamber and serves to prevent a complete system of drains from flooding.

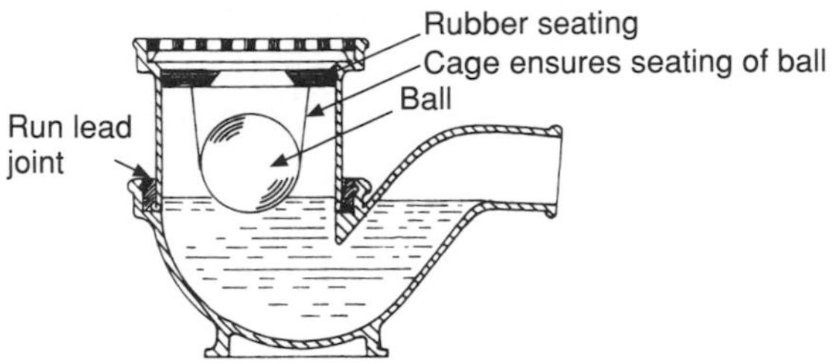


(a) Two-piece cast iron anti-flood gulley. Prevents flooding of surface-water drains when discharging into watercourses or tidal rivers

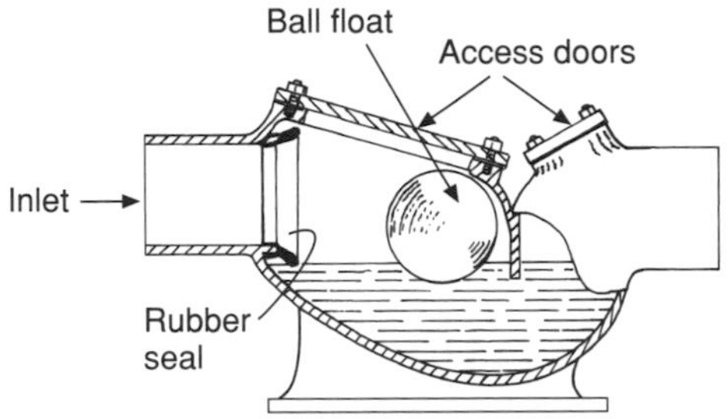


(b) Anti-flood intercepter

For use with surface water only. In the event of backflow the ball rises with the water until it seats on the rubber seal, thus preventing any further flow of water into the inlet.

**Fig. 9.17** Anti-flood fittings.

*Grease disposal*

Grease and cooking oils have long been recognised as being a major cause of blocked drains, especially in catering establishments and commercial buildings such as hotels where large volumes of grease and cooking fats are discharged into the underground drainage system. As the waste water cools, the grease it contains solidifies and becomes deposited on the pipe walls, where it builds up until a solid plug is formed, blocking the drain.

*Grease trap converters* Until comparatively recently the only method possible to separate the grease from the waste water was to install a grease trap. A typical example of a traditional grease trap is shown in Fig. 9.18. Providing these traps are correctly sized and regularly cleaned they are suitable for small installations. To deal with grease disposal in larger kitchens, the congealed grease is reduced by the addition of a biological agent which 'digests' some of the grease, making it more

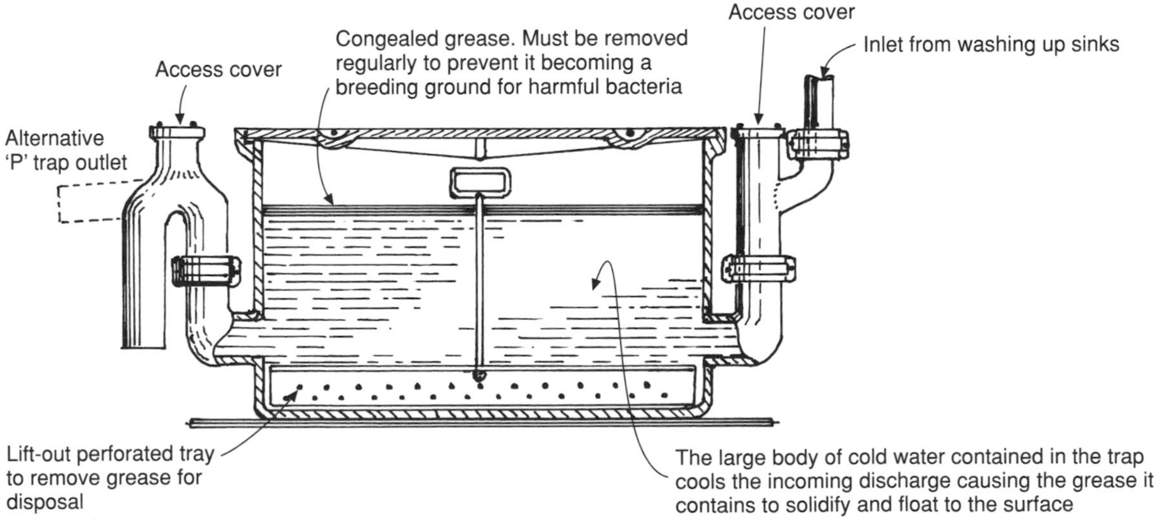


**Fig. 9.18** Cast iron grease trap.

soluble. Any grease remaining can be removed with a specially adapted vacuum cleaner. The success of this arrangement is dependent on correctly sizing the interceptor and an air-admittance valve which allows air to mix with the grease, making it softer and more easily removed. An influent control valve is also necessary with this system. It is basically a spring-loaded valve which monitors the speed of flow through the intercepting chamber allowing grease to separate from the water. The general arrangement of this scheme is shown in Figs 9.19(a) and (b) which show details of the interceptor. It is important that modern grease interceptors are accurately sized. Table 9.1 shows the procedures that must be carried out to ensure their maximum efficiency.

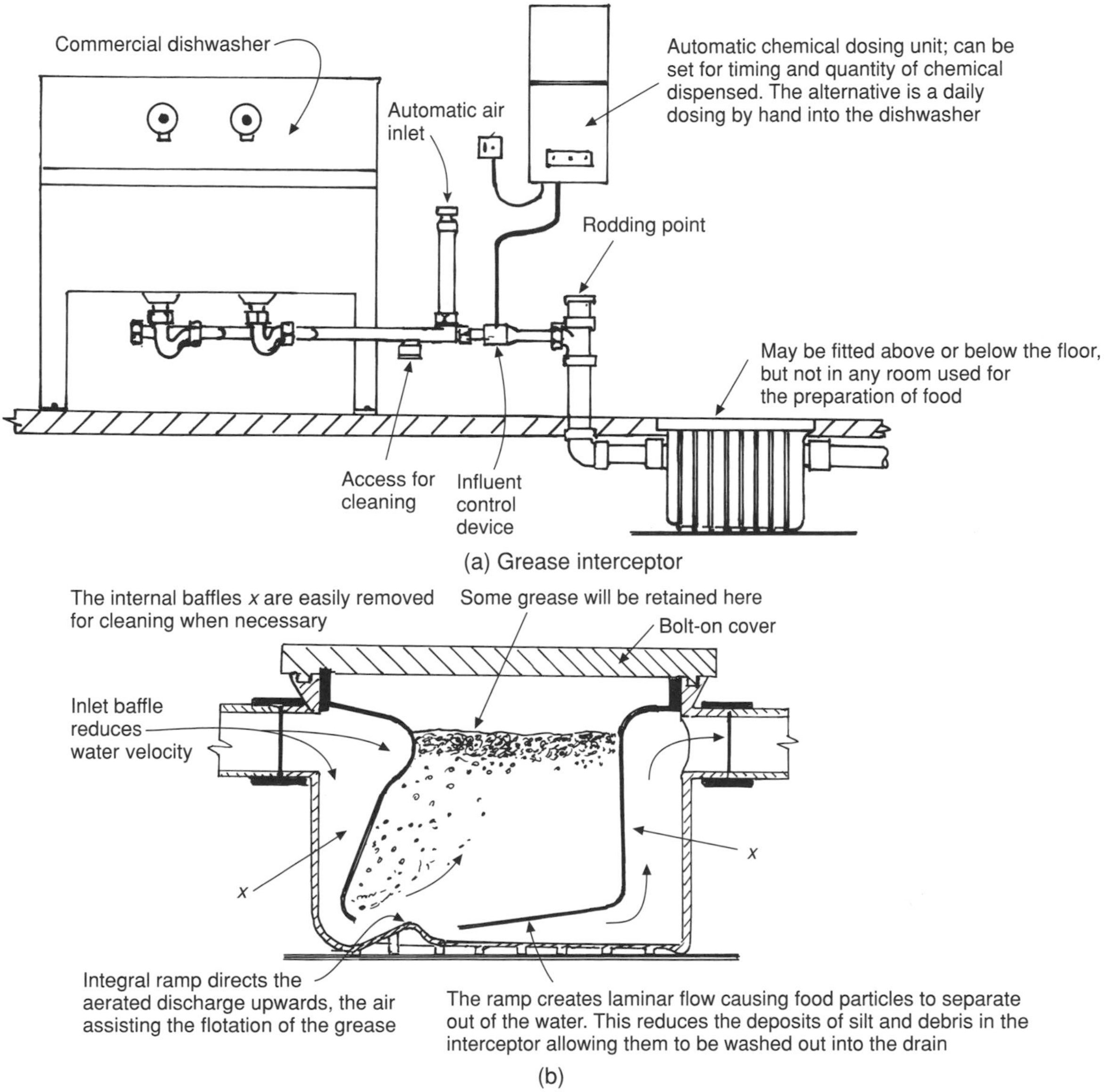
Commercial dishwasher
Automatic air inlet
Automatic chemical dosing unit; can be set for timing and quantity of chemical dispensed. The alternative is a daily dosing by hand into the dishwasher
Rodding point
May be fitted above or below the floor, but not in any room used for the preparation of food
Access for cleaning
Influent control device
(a) Grease interceptor
The internal baffles x are easily removed for cleaning when necessary
Some grease will be retained here
Bolt-on cover
Inlet baffle reduces water velocity
x
x
Integral ramp directs the aerated discharge upwards, the air assisting the flotation of the grease
The ramp creates laminar flow causing food particles to separate out of the water. This reduces the deposits of silt and debris in the interceptor allowing them to be washed out into the drain
(b)

**Fig. 9.19**

**Table 9.1** Interceptor sizing. (Reproduced courtesy or Hunter Plastics)

Interceptor sizing is based on two key factors. The first is the time taken for the appliance(s) that are connected to the interceptor to empty from when discharge begins. The second is the volume of waste water being discharged by the appliance(s).

These two factors combined provide a rate of flow through the interceptor and allow specification of the correct unit(s) for a particular application. The link between flow rate and efficient interceptor operation is the unique influent control device (ICD) which is essential on all Endura installations.

| *Step* | *Formula* | *Example* |
|---|---|---|
| 1 | **Determine cubic capacity in metres of the appliance(s) by multiplying length × width × depth** | **A sink 0.6 m long, 0.46 m wide and 0.2 m deep**<br>Cubic capacity: $0.6 \times 0.46 \times 0.2 = 0.055\ m^3$ |
| 2 | **Convert cubic metres into cubic litres — multiply by 1,000** | **Capacity in litres:**<br>$0.055 \times 1{,}000 = 55$ litres |
| 3 | **Determine the actual drainage load**<br>It is considered that a fixture is normally filled to approximately 75% of its full capacity with water as the items being washed displace about 25% of that content<br>Actual drainage load = 75% of appliance capacity | **Actual drainage load (ADL):**<br>$0.75 \times 55 = 41.25$ l |
| 4 | **Determine flow rate and drainage period**<br>In general, good practice dictates a drainage period of between 1 and 2 minutes maximum. Drainage period is defined as the actual time required to completely drain an appliance<br>$\text{Flow Rate} = \dfrac{\text{Actual Drainage Load}}{\text{Drainage Period}}$ | **For a 1 minute drainage period:**<br>$\dfrac{41.25}{60\,(s)} = 0.69$ l/s flow rate<br>**For a 2 minute drainage period:**<br>$\dfrac{41.25}{120\,(s)} = 0.344$ l/s flow rate |
| 5 | **Selecting the correctly sized interceptor**<br>An interceptor should be selected which has a flow rate capacity at least equal to the calculated flow rate. Where the calculated flow rate exceeds the Endura DSG16 capacity of 1.6 litres per second, a larger interceptor should be selected. Alternatively, more than one DSG16 can be used. In this case ensure that appliances are piped separately to each grease interceptor, so that the total capacity from each of the appliances does not exceed 1.6 litres per second | Two appliances of this size plus an additional 0.2 l/s could be specified with an Endura DSG16<br>**For a 2 minute drainage period:**<br>Up to four appliances plus 0.25 l as per this example |

Endura Grease Interceptor

| *Description* | *Dimension* |
|---|---|
| | DSG16 |
| Flow rate (l/s) | 1.6 |
| Grease capacity (kg) | 23 |
| Connection size | 110 mm |
| Dimension A | 600 mm |
| Dimension B | 445 mm |
| Dimension C | 107 mm |
| Dimension D | 307 mm |
| Dimension E | 414 mm |
| Meals per day (est.) | 50–75 |

The correct specification of the Endura grease interceptor can be determined by using the step-by-step method shown in the adjacent table.

The rule of thumb for any installation is that the interceptor being considered has a flow rate capacity at least equal to the calculated flow rate with a drainage period of no longer than two minutes.

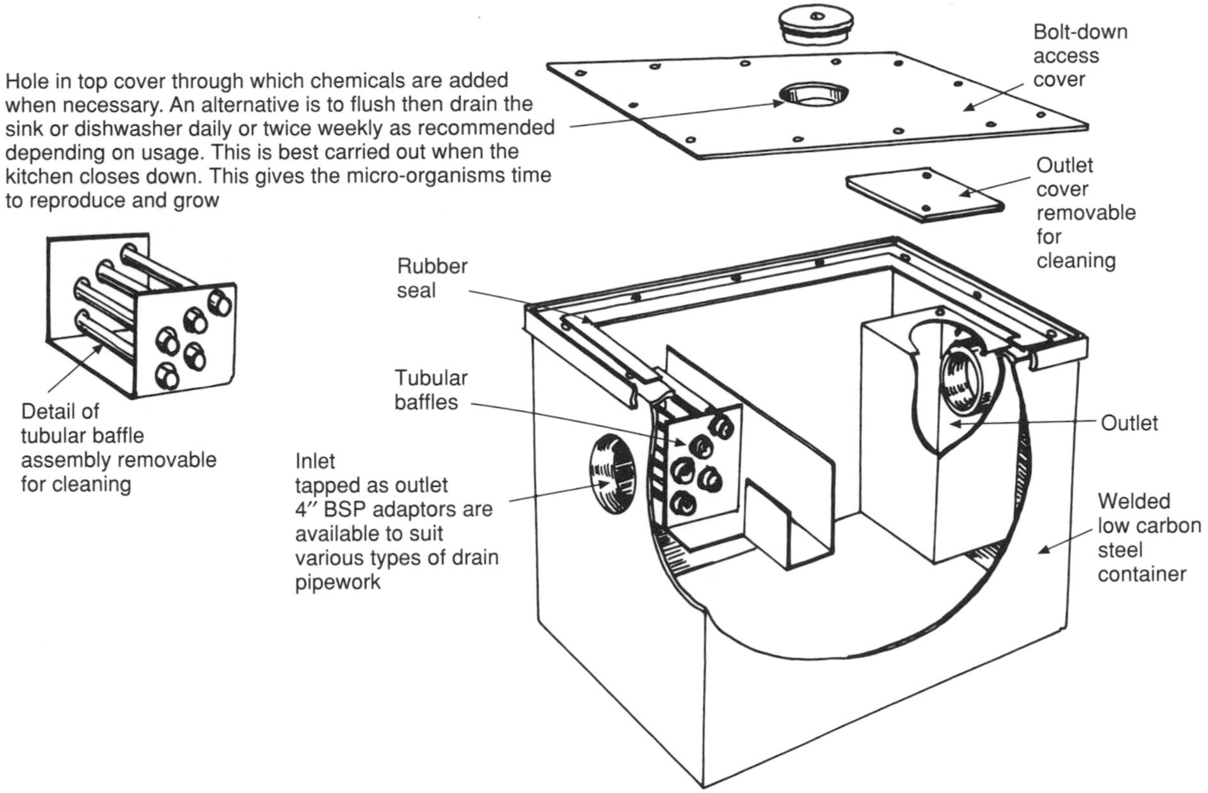
Hole in top cover through which chemicals are added when necessary. An alternative is to flush then drain the sink or dishwasher daily or twice weekly as recommended depending on usage. This is best carried out when the kitchen closes down. This gives the micro-organisms time to reproduce and grow
Bolt-down access cover
Outlet cover removable for cleaning
Rubber seal
Tubular baffles
Outlet
Detail of tubular baffle assembly removable for cleaning
Inlet tapped as outlet 4″ BSP adaptors are available to suit various types of drain pipework
Welded low carbon steel container

**Fig. 9.20** Patent grease converter constructed of welded low-carbon-steel plate and tube.

A similar grease converter is shown in Fig. 9.20 where it will be seen that the effluent passes over the baffles which cause the formation of globules of grease. As the grease builds up it falls from the baffles and floats to the surface of the water in the container. Chemicals are added to the water producing active micro-organisms, which degrade and break down the grease causing it to become soluble in the water.

It will be necessary to periodically remove the top cover and clean out any solids deposited in the base of the container — ideally every three months. Both when commissioning and after cleaning, a culture of micro-organisms must be established before breakdown of the grease is effective, and the manufacturer's instructions must be complied with in this respect.

No type of grease disposal unit will function efficiently if it is incorrectly sized. The most reliable method of achieving this is to determine the requirements of the kitchen, e.g. number of sinks and type of usage, and to seek the manufacturer's instructions. If the grease converter is working effectively, there will be little or no odour, and its content will have a consistency similar to a thick soup, with no surface grease or caking on the inside surfaces of the container.

No grease disposal appliance should be situated in areas where food is prepared or served, and adequate access must be made for maintenance and cleaning purposes.

### Support for above-ground drains

While drainpipes are usually laid underground, they sometimes pass through the basement of a structure; they are then referred to as suspended drainage. Only cast iron or PVC is used in these circumstances as other materials would not be strong enough. Brackets for suspended drainage

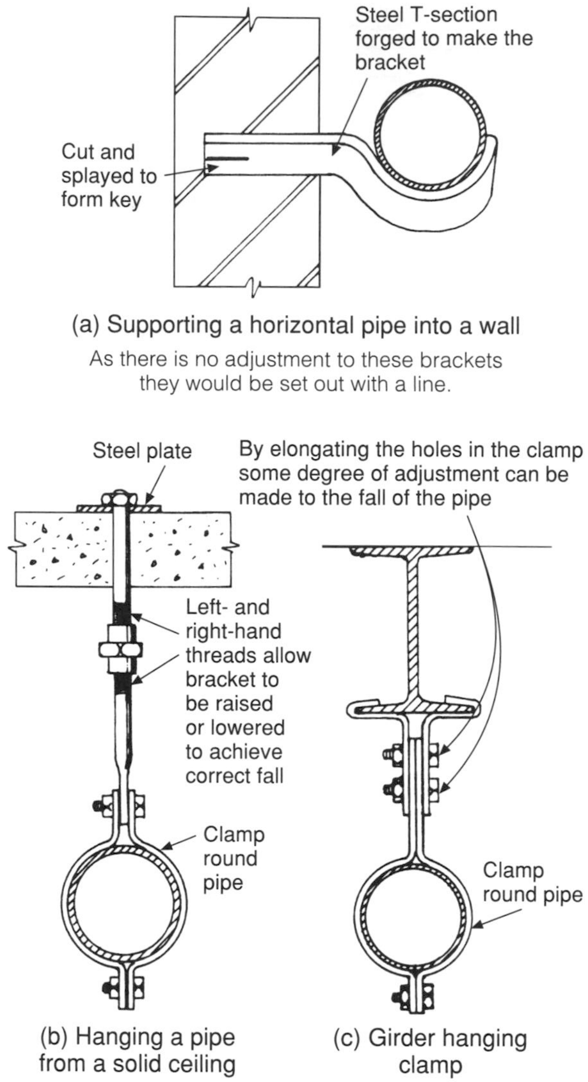
Steel T-section forged to make the bracket
Cut and splayed to form key
(a) Supporting a horizontal pipe into a wall
As there is no adjustment to these brackets they would be set out with a line.
Steel plate
By elongating the holes in the clamp some degree of adjustment can be made to the fall of the pipe
Left- and right-hand threads allow bracket to be raised or lowered to achieve correct fall
Clamp round pipe
Clamp round pipe
(b) Hanging a pipe from a solid ceiling
(c) Girder hanging clamp

**Fig. 9.21** Fixings for suspended cast iron drainage systems.

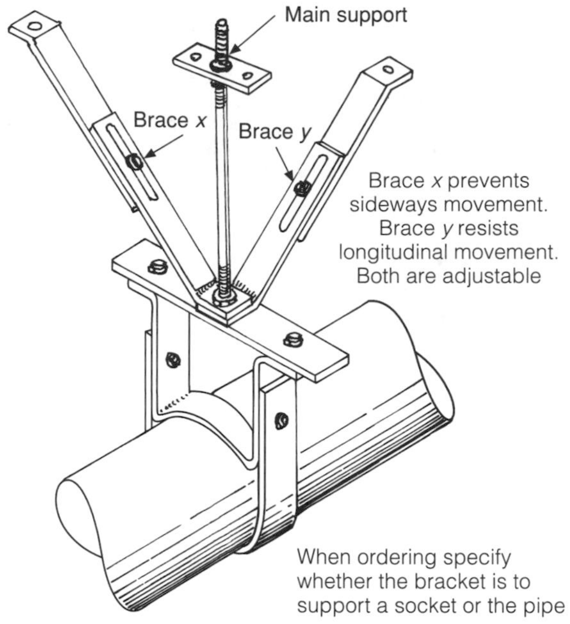
Main support
Brace x
Brace y
Brace x prevents sideways movement. Brace y resists longitudinal movement. Both are adjustable
When ordering specify whether the bracket is to support a socket or the pipe

**Fig. 9.22** Purpose-made support bracket for PVC suspended drains. Similar but stronger brackets are available for cast iron.

are usually purpose-made with provision for adjusting the falls on the pipeline to create a self-cleansing velocity. More will be said about this later. Figure 9.21 shows brackets, often made on site, for supporting drainage in a building. They are made of black low-carbon-steel bar or strip, angle iron and bright steel rod. Some method of heating will be required to forge the steel, oxy-acetylene equipment being ideal for this purpose. Suitable dies will also be necessary to cut the threads on the steel rod. An alternative is to use *studding* which is simply threaded rod which can be cut to the length required. This material is available in various lengths, diameters and types of thread from good engineering stockholders, tool dealers and some builders' merchants.

Cast iron drains are very heavy and any brackets fabricated for their support should be stout enough to take their weight. If PVC is used, more brackets will be required due to its lack of rigidity, but they can be made of thinner material as PVC is not as heavy as cast iron. A typical bracket for PVC is shown in Fig. 9.22. The spacing of brackets for both cast iron and PVC suspended drainage is shown in Fig. 9.23. Apart from any anchor brackets, intermediate fixings should allow the pipeline to move when it expands or contracts. A method of supporting cast iron drainage just above ground is shown in Fig. 9.24. The pipe is simply bedded on brick piers which must be carefully set out so that the sockets are clear of the piers, otherwise it will be difficult to make the joints.

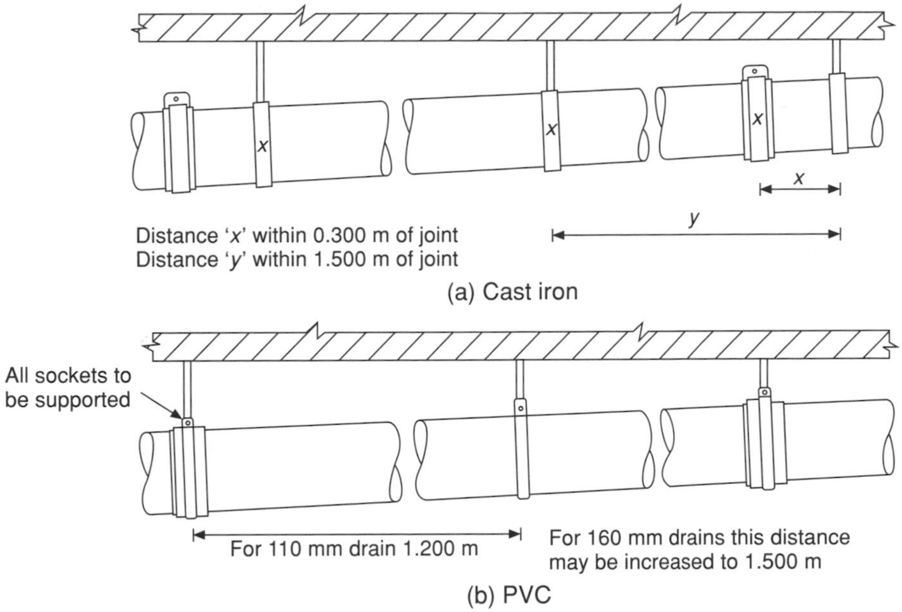


**Fig. 9.23** Recommendations for the support of suspended drains.

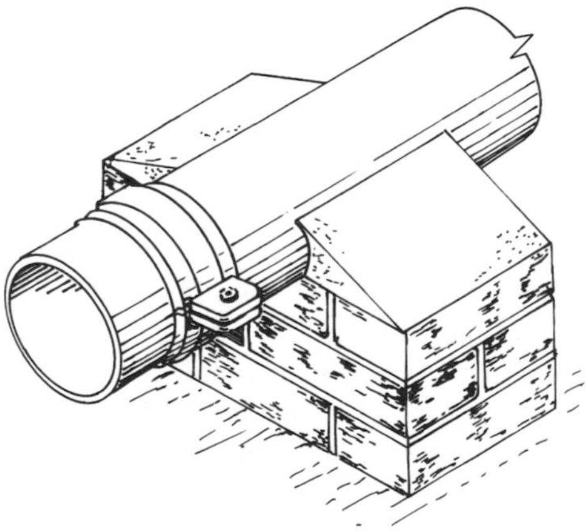

**Fig. 9.24** Support for drains above ground level on brick piers.

## Flexibility

Due to the damage to rigid drainage systems caused by the settlement of building and subsidence of the subsoil, all underground systems of drainage must be flexible. Provision must be made for both angular and telescopic withdrawal movement: Figure 9.25 shows the standard to which all underground joints must comply. Prior to the 1965

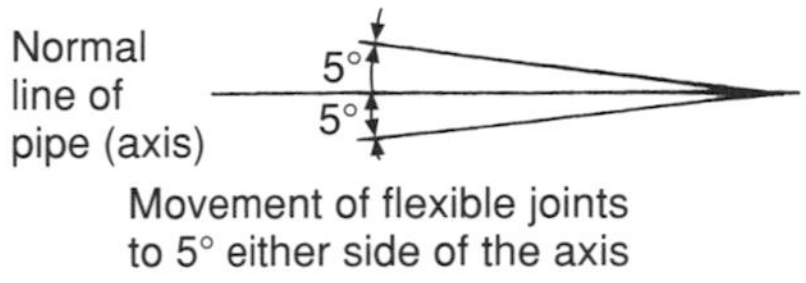


**Fig. 9.25** Minimum movement for flexible joints.

Building Regulations, cast iron and clayware drains were laid on a concrete bed, but due to the success of the methods employed for bedding PVC drains, similar methods are now employed with all materials.

## Bedding for underground drains

The word 'bedding' relates to the support of underground drainage and the material used for this purpose. It must be carefully laid in order to give a drain the necessary fall. For many years, when clayware and cast iron drains having rigid joints were mainly used, concrete was almost invariably employed as a bedding material. This 'traditional' method of bedding is shown in Fig. 9.26, but due to the types of joints and bedding materials currently in use and the emphasis on flexibility, rigid concrete beds are not normally recommended. The main

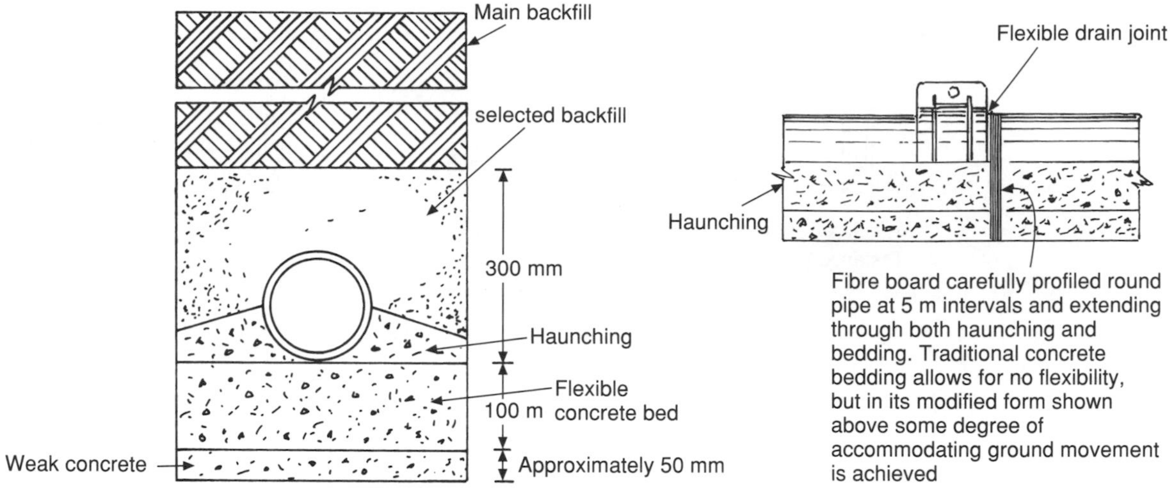
Main backfill
selected backfill
300 mm
Haunching
Flexible
100 m concrete bed
Weak concrete
Approximately 50 mm
Flexible drain joint
Haunching
Fibre board carefully profiled round pipe at 5 m intervals and extending through both haunching and bedding. Traditional concrete bedding allows for no flexibility, but in its modified form shown above some degree of accommodating ground movement is achieved

(a) Details for rigid pipes

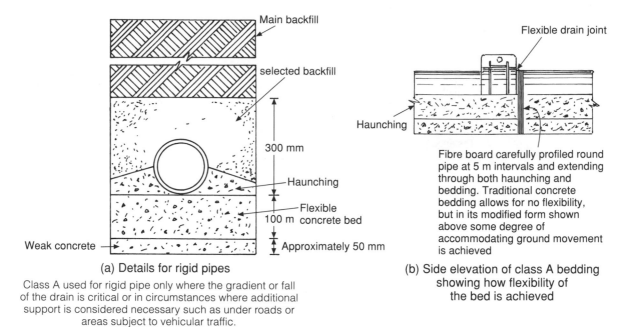

Class A used for rigid pipe only where the gradient or fall of the drain is critical or in circumstances where additional support is considered necessary such as under roads or areas subject to vehicular traffic.

(b) Side elevation of class A bedding showing how flexibility of the bed is achieved

**Fig. 9.26** Bedding for rigid drain.

disadvantage of a rigid bed is that any movement of the soil due to shrinkage or subsidence in made-up ground may cause the concrete to crack. This would result in cracking of the drainpipe and subsequent leakage. Cast iron, due to its strength, is far more resistant to damage from this cause than clayware, and for this reason, despite its high cost, it is commonly used in public and industrial buildings. After a considerable amount of research a series of bedding procedures were devised to take into account the many differing factors involved. These include the flexibility or rigidity of the pipes used, the effect of imposed loads, e.g. under roads where applicable, and the nature of the ground in which the drain is to be laid. References to specific methods of bedding and backfilling of underground pipelines as shown in Figs 9.26 and 9.27 are best obtained from the manufacturers of the pipes and the development associations listed in the further reading section of this chapter. All details of work to be carried out on underground drains must be submitted to the local planning authority and comply with Part H of the Building Regulations.

Any wooden pegs used for determining the gradient or fall of the trench must be removed prior to laying the drain as they would cause tilting of rigid pipes and distort the circular profile of PVC pipes. The main feature of current bedding systems is the use of granular infill around the pipes such as sharp sand or pea gravel. Sharp sand is easily recognisable as it is much more gritty to the touch than that used for bricklaying and contains a lot of very small sharp stones. Do not, however, confuse it with the aggregate used for concrete, as this contains larger stones which could damage the drain, especially if it is made of PVC. The object of the granular infill is to distribute any loading more evenly around the surface of the pipe thus avoiding its distortion, or, in the case of clayware, cracked pipes due to heavy loads. All drain trenches must be carefully backfilled and consolidated with selected fill to avoid damage to or distortion of the pipes. It is generally recommended that this should be done using a hand punner, similar to that shown in Fig. 9.28, for a distance of at least 300 mm above the top of the pipe.

### Protection of drains near or above ground level or subject to imposed loads

It is no longer mandatory to encase pipes in concrete if they are near the surface of the ground

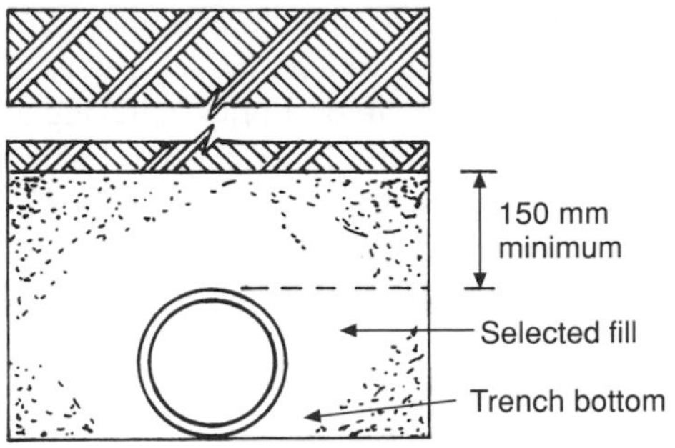


(a)

Class D used where the soil is free of stones and reasonably dry and firm where the trench bottom can be accurately trimmed by hand.

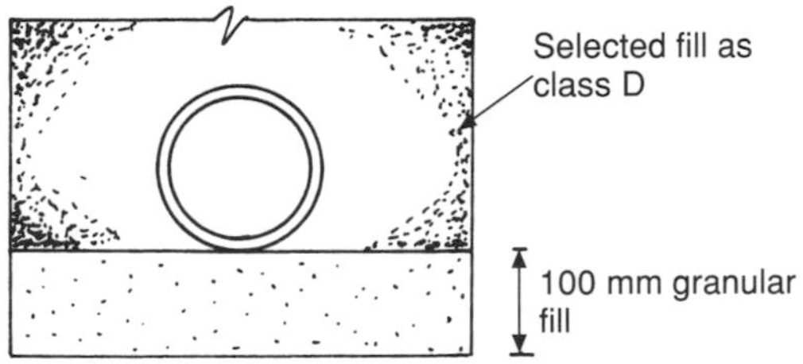


(b)

Class N, identical to class D except that the pipes are laid on a granular bed. Used in similar conditions to those of class D except where the trench bottom cannot be trimmed accurately.

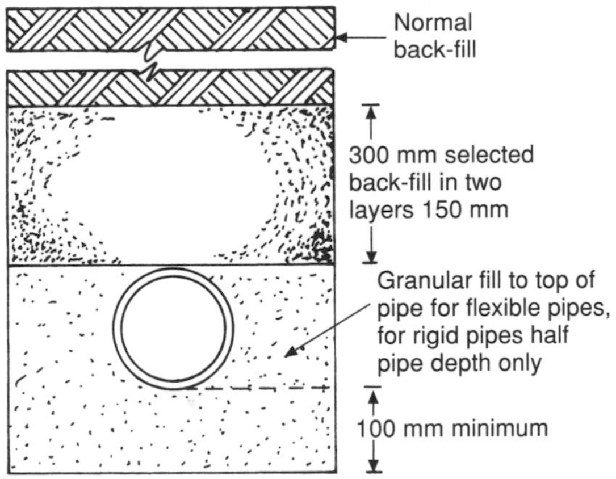


(c)

Bedding for flexible pipes. Class B granular fill to crown of pipe gives sufficient support to walls of pipe to prevent flattening and in the case of PVC pipes permits movement due to expansion and contraction.

Note that in all cases selected fill must be free of large stones, lumps of clay, pieces of timber, vegetable matter or frozen material of any sort.

**Fig. 9.27** Basic bedding systems for all types of drainage materials.

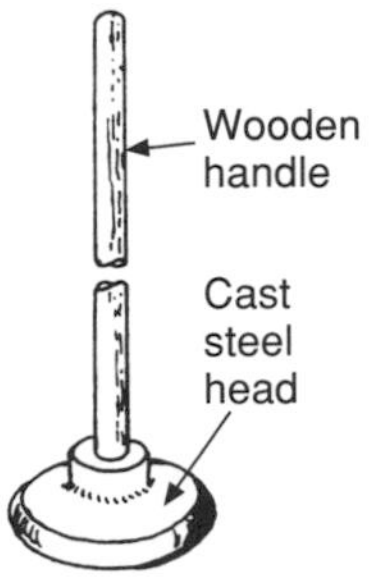


Punner used for light hand tamping (consolidating) backfill.

**Fig. 9.28**

unless it is considered necessary for their protection. If, however, this method is used, flexibility of the encasement must be maintained as illustrated in Fig. 9.29(a). Where the soffit of the pipe is between 600 and 300 mm of the surface, an alternative to complete encasement is to lay paving slabs over the granular fill (see Fig. 9.29(b)). This method has the advantage of allowing the pipe much more freedom of movement. Pipes laid under roads at depths of less than 900 mm from the soffit of the pipe to the surface may require a reinforced concrete slab extending beyond the sides of the trench to ensure the drain is not damaged by heavy traffic.

Drains laid below the bottom of a building foundation and within 1 m of it may be subjected to excessive stress. Reference should be made to the Building Regulations, Approved Document H, which specifies the methods that must be employed to prevent damage under such circumstances. Figures 9.30(a) and (b) illustrate typical cases.

*Drains under buildings*

In the past it was common practice to encase drainpipes under buildings in concrete, but since all drains must now be flexible, such encasement is no longer recommended. Much better provision is now made to prevent damage to the drain due to settlement of the structure, as shown in Figs 9.31(a) and (b). In the case of Fig. 9.31(a) any movement of the structure is accommodated by the rocker pipes, while in Fig. 9.31(b) the 50 mm gap around the pipe permits movement of the structure independently of the pipe. This latter method has the advantage that it is less likely to alter the gradient of the drain. Special care is needed to

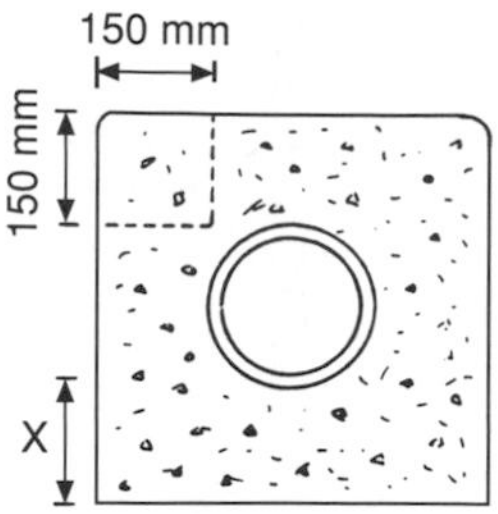


(a)

Recommended for drains above ground level, or just below it when it may be subject to vehicular traffic. The pipe is encased by a 150 mm surround of concrete.

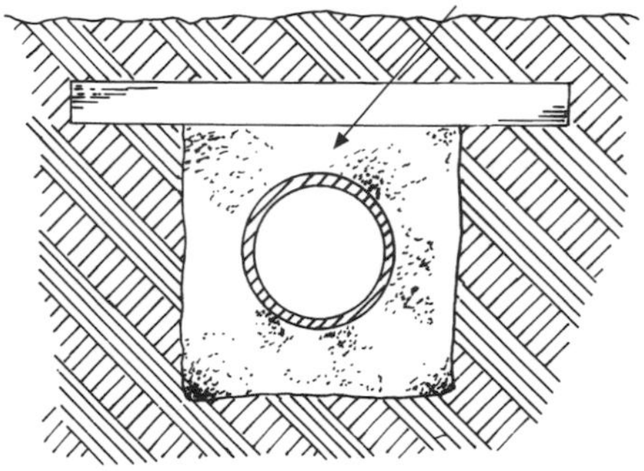


(b)

As an alternative shallow drains may be protected by means of a paving slab laid across the trench. This method is not suitable where heavy vehicular traffic is likely.

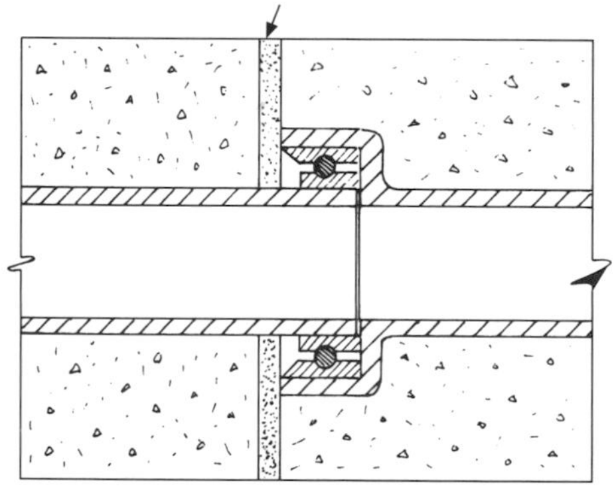


(c)

This illustration shows a flexible socketed clayware joint, but the same techniques apply to all drainage pipework materials.

**Fig. 9.29** Concrete encasement of drains. This is only necessary in the case of very shallow drains or those laid above ground in locations where they may be subject to damage.

relieve the mass of the structure over the opening in the brickwork through which the pipe passes. A concrete lintel as shown, or a brick relieving arch, satisfies this requirement. Figures 9.31(c) and (d) show the approved methods of bedding for pipes within the building structure. Cast iron is normally recommended for this purpose but other materials may be used.

### Types of drainage system

Although many plumbers do not actually install drainage systems, some knowledge of surface and foul water drainage is essential to understand fully the principles of sanitary plumbing.

There are three types of drainage system in common usage, known as combined, separate and partially separate. Each has its merits and demerits, but the local authority determines which type is used, basing its choice on its overall knowledge of local conditions. The reader should be aware that not only has foul water drainage to be considered but also that of surface water, e.g. water falling on roofs and paved areas.

*Combined system*

In this case, as its name implies, both foul and surface water are discharged into the same sewer (see Fig. 9.32). This system has the cheapest layout as it requires only one set of pipes, and during heavy rainfall both house drains and sewers are thoroughly flushed out. (Sewers are public authority drains to which private house drains are connected.) This system has often been used in the past where raw sewage was disposed of without treatment, e.g. discharged into the sea or a watercourse. Because of public alarm at this unhealthy state of affairs, it is now considered to be unacceptable and all foul water must be treated before the effluent is discharged in such a way. In many cases this, and the rapid growth of some urban areas, has put a serious strain on the capacity of existing sewage disposal plant, and most local authorities in urban areas now insist on the installation of separate systems. There is a further disadvantage: in storms and periods of very heavy rainfall, flooding and subsequent surcharging of the drains has been known to occur. Then again, in areas of undulating

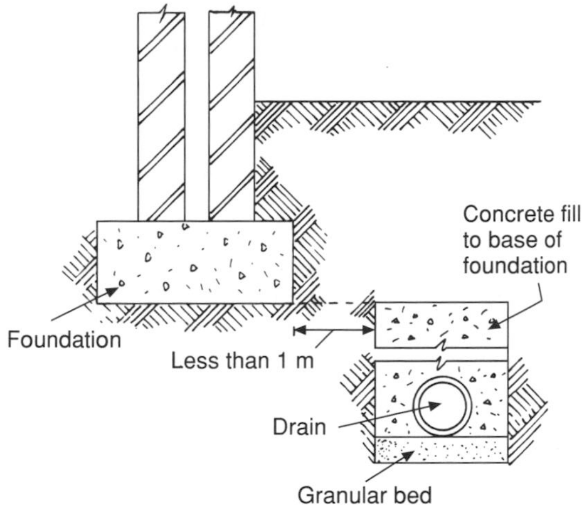


(a) Preventing instability of foundations less than 1 m from a drain trench

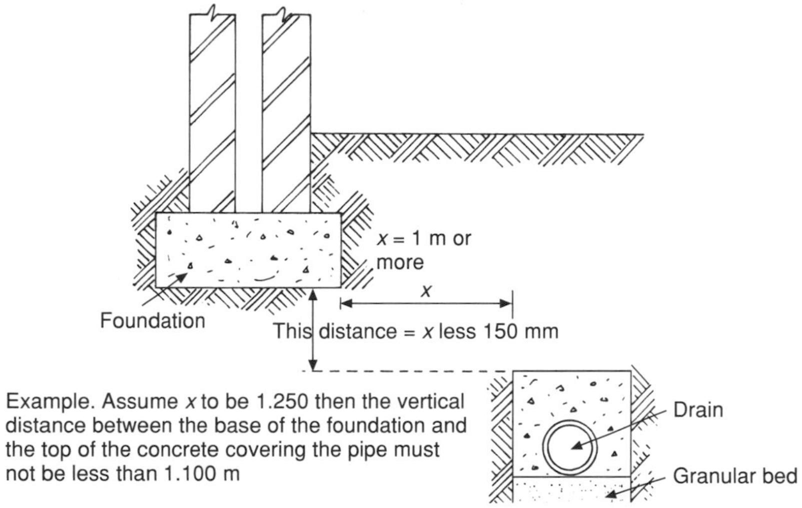


(b) Trenches more than 1 m from a foundation

**Fig. 9.30** Drains adjacent to building foundations.

country it is often necessary to pump sewage from one level to another so that it can be discharged into the sewage treatment plant. In the combined system it would be necessary to pump both foul and surface water, resulting in greater capital and maintenance costs for the equipment used.

*Separate system*

This system requires the use of two sewers (see Fig. 9.33), one carrying foul water to the treatment works, the other carrying surface water (which requires no treatment) to the nearest watercourse or river. It is expensive to install, but from the local authorities' point of view it is the most economical to operate because the volume of sewage to be treated is far smaller than the discharge from a combined system. The biggest danger is that cross-connections may be made, i.e. foul water may be connected to a surface water drain. There is little chance of flooding at times of heavy rainfall, but the foul water sewers are not flushed periodically with relatively pure water as in combined systems.

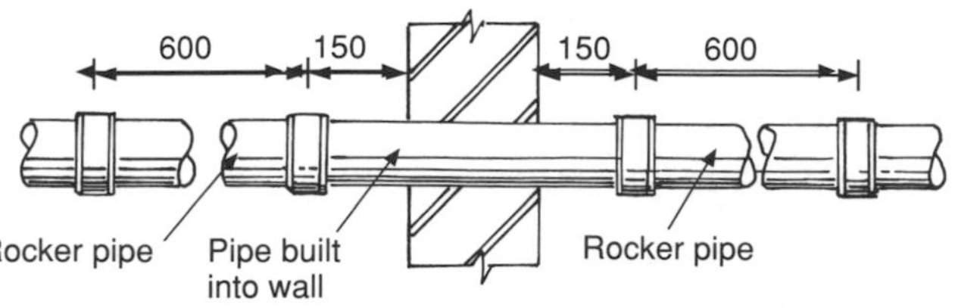


(a) In this case the pipe is built into the wall and any downward movement of the structure likely to cause damage on the drain is prevented by the 'rocker' pipes. All joints must be of the flexible type

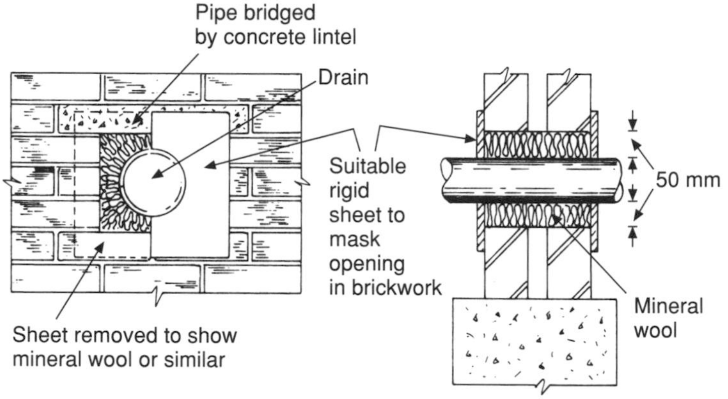


(b) Any movement of the structure will be absorbed by the mineral wool without damage to the pipe. This method may be used with no rocker pipes as *any* subsidence of the structure will be absorbed by the mineral wool

**Fig. 9.31** Protection of drains passing through foundations or under buildings.

It is the most commonly employed method of waste water disposal in new towns and urban areas, especially where large housing estates have been built, as pumping arrangements and sewage plants, which may already be overloaded, have only to cope with foul water. Local authorities now insist on the use of this system on new build work.

*Partially separate system*

This system probably originated when towns began to grow in size and local authorities found it necessary to try to reduce the loading on the combined system, which in most cases had hitherto been employed. This is something of a compromise between the previously mentioned systems. It requires two sewers, one carrying water from paved areas and part of the roof, the other carrying foul water and water from the remainder of the roof as shown in Fig. 9.34. Some authorities permit the water falling on the front part of the premises to be discharged into the surface water sewer, water from other parts of the roof and paved yards at the rear of the premises being discharged into the foul water sewer. The disadvantages of this system are similar to those of the combined system, but to a lesser degree.

*Soakaways* Soakaways are often used with partially separate systems to deal with water from paved surfaces or rainwater pipes having no connection to a surface water drain. As most urban areas and towns have surface water sewers, their use is limited to isolated areas out of reach of the main drainage system. In such cases foul water may be treated in a small individual sewage disposal plant known as a septic tank and the surface water

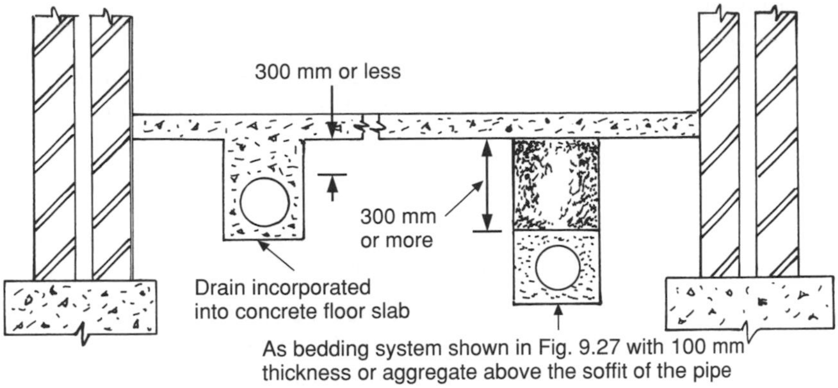


(c) Drains under floors in building

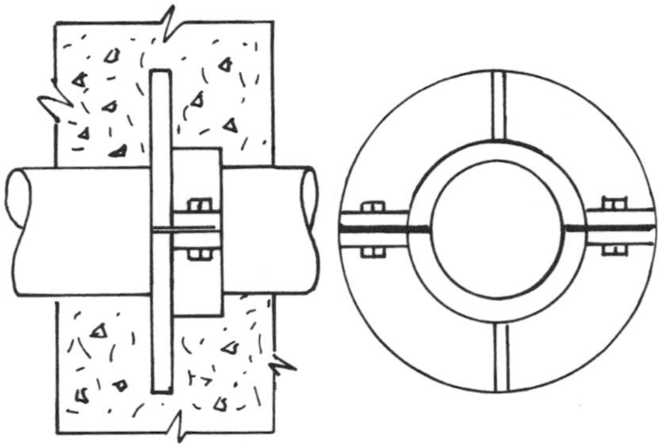

(d) Puddle flange, front and side elevations shown. The flange prevents entry of water into a building where a pipe penetrates the footings. The bolted sections of the flange must be sealed round the pipe with a waterproof mastic

**Fig. 9.31** (*cont'd*)

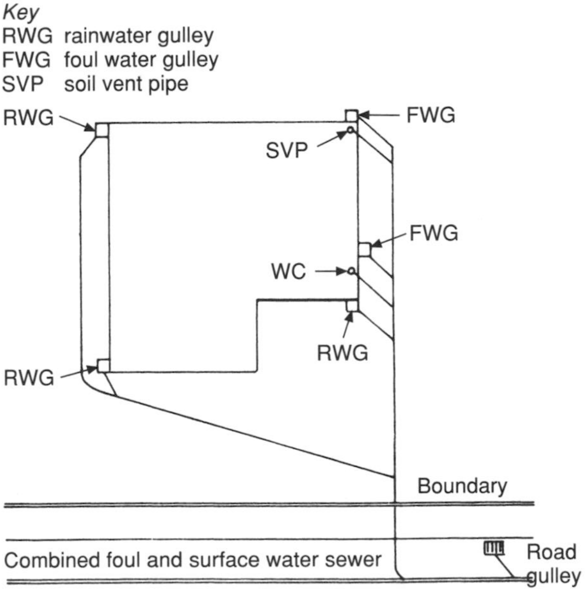


**Fig. 9.32** Combined system of drainage.

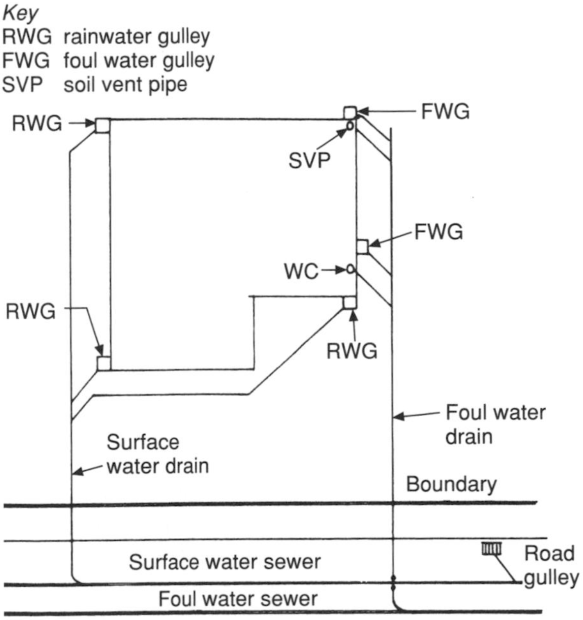


**Fig. 9.33** Separate system of drainage.

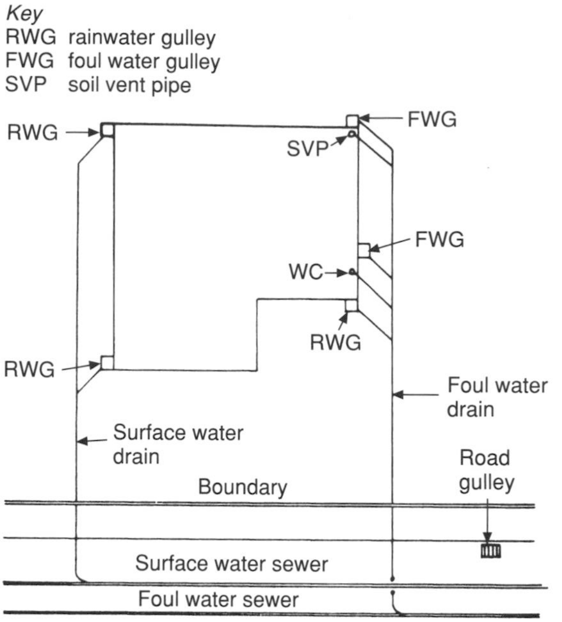


**Fig. 9.34** Partially separate system of drainage.

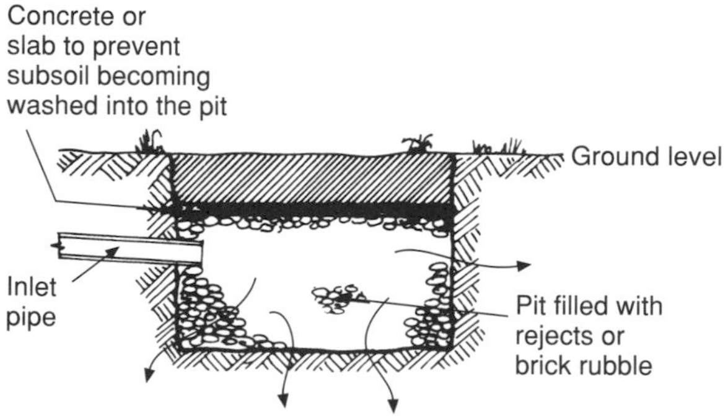


**Fig. 9.35** Soakaway.

drained into a convenient watercourse or a soakaway. See Fig. 9.35. Soakaways usually consist of a pit dug well away from the building, filled with stones or brick rubble to prevent the sides caving in, into which surface water, usually from the roof, is discharged. Soakaways are only effective in ground that is sufficiently permeable to allow water to soak into the surrounding subsoil. They should be sited on sloping ground so that, in the event of flooding, water will flow away from the building. Soakaways have to be approved by the local authority planning officer who often specifies their dimensions, the distance they must be from a dwelling and their construction. Reference should be made to BS EN 752-4 for further information.

### Ventilation of drains

Underground systems of drainage must be well ventilated for two main reasons.

(a) To maintain an equilibrium of pressure between the air inside the drain and that of the atmosphere. If, for instance, the air pressure in the drain was lower than that of the atmosphere, the trap seals of gulleys and WCs would very quickly be destroyed and foul air from the sewer would be admitted to the building and its precincts.
(b) To prevent the build-up of foul air and possibly dangerous gases. Sewage is very quickly acted upon by bacteria which thrive and multiply in dark airless environments. Their action on the sewage produces gases which not only have an objectionable smell but are also potentially dangerous due to their flammability. The action of the bacteria in breaking down the sewage is perfectly normal, but it must be confined to sewage disposal works and not allowed to happen in a house drainage system.

The system of ventilation used will depend upon whether or not an intercepting trap is used. Where it is, the inspection chamber in which it is housed must be ventilated to the atmosphere via either a fresh air inlet (see Fig. 9.36) or, in some cases

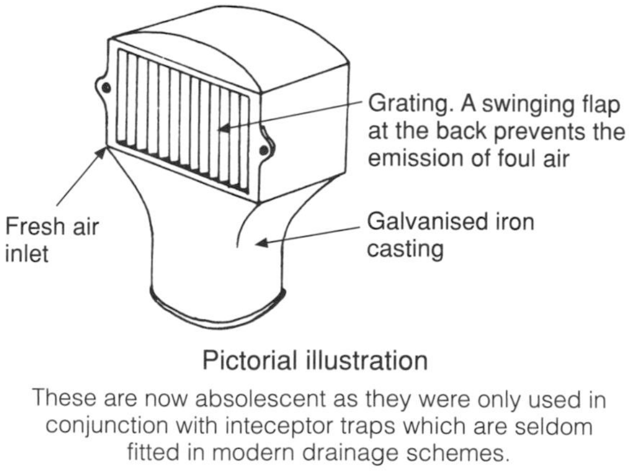


**Fig. 9.36** Fresh air inlet.

where a building has no frontage, a second vent pipe. The upper end or highest point of the drain should also terminate as a vent pipe which in effect acts as a flue through which drain air can be discharged to the atmosphere. Drain air is warmer than that of the atmosphere, and air movement through the system occurs due to convection. Where no interceptor is fitted, both the house drains and the public sewers are ventilated as shown in Fig. 9.37(b).

Figure 9.37 illustrates the two methods of drain ventilation. It must be emphasised that drain ventilation pipes should be fitted at the highest point of the drain, thus avoiding long unventilated branches. Both the vent pipe and fresh air inlet (if fitted) should terminate in such a way as not to allow foul air to enter the building via windows or ventilators.

The diameter of the main ventilating pipe is not specified in the Building Regulations, which

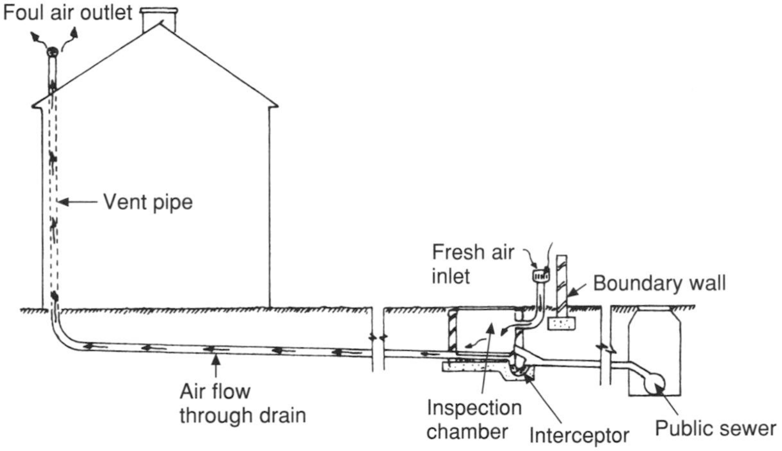


(a) The sewer is not ventilated using this method and was not generally used for housing stock after *c.* 1946

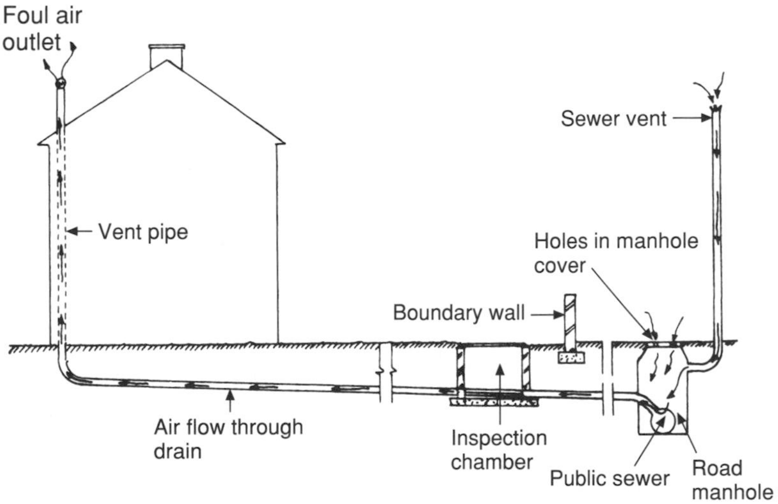


(b) Air admitted to sewer via sewer vents and holes in manhole covers allows both ventilation of the sewer and the house drains through the individual vent pipes

**Fig. 9.37** Ventilation of drains.

simply state that it should be of adequate size for its purpose. It is normal in modern practice, however, to fit a pipe of the same nominal diameter as the main discharge stack, although older buildings may be found with pipe diameters of only 75 or 90 mm.

### Access to drains

Although a good drainage system (whether carrying surface water or foul water) should be designed to avoid the possibility of a blockage, circumstances arise, often due to misuse, where this happens. It is therefore very important that adequate provision is made so that obstructions can be cleared with the minimum of trouble. Special attention should be paid to the areas where blockages are most likely, such as bends and junctions. The methods of obtaining access for rodding drains vary. Some drainage fittings are made to incorporate a rodding eye, or rodding eyes may be installed at suitable points in the drains. In some circumstances inspection chambers must be constructed. Where possible, rodding eyes should be used in preference to inspection chambers, as they are cheaper and cause less interference to the drainage flow. The following list is a good general guide to the requirements of the Building Regulations for means of access, but whether a rodding eye or inspection chamber is used will be at the discretion of the local authority:

(a) Changes of direction of the drain requiring a 135° bend or less.
(b) Junctions where the branch joins the main pipe at an angle of less than 135° or bends occur with a similar angle or less.
(c) The point of connection between a drain and a sewer or within 12 m of such a connection.
(d) When a drain exceeds 90 m in length.
(e) Where a change of gradient occurs in the drain, e.g. with ramps or backdrops.
(f) The highest point or head of the drain.

Rodding eyes may be used with advantage where the drain is comparatively shallow as, for example, at its head. Figure 9.38 illustrates such an arrangement.

Inspection chambers are traditionally constructed of brick, either engineering bricks or good-quality

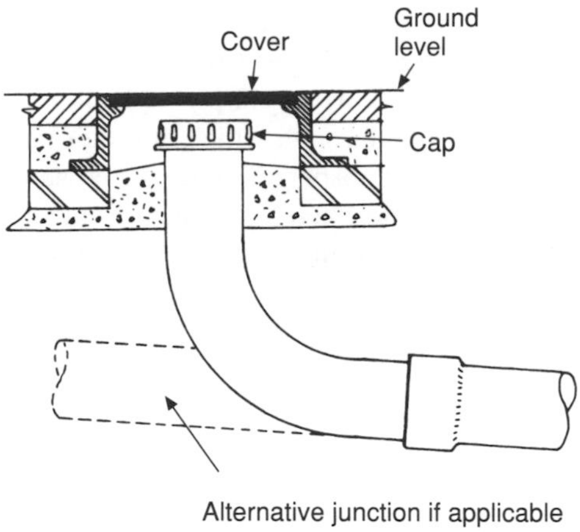


**Fig. 9.38** Rodding eye bend.

stock bricks rendered with an impervious layer of cement mortar. It is essential that these chambers are watertight, and in cases where a drain is at great depth there must be sufficient room to work in comfort. They are still often built using traditional methods, but for house drains and small schemes, due to the standard size of the chambers, precast concrete sections are used instead of bricks, which saves time and often costs less. Figure 9.39 shows a section through a typical brick-built or precast chamber showing details of the benching. Benching must be high enough to prevent solid matter being washed up on it and putrifying. Where the depth

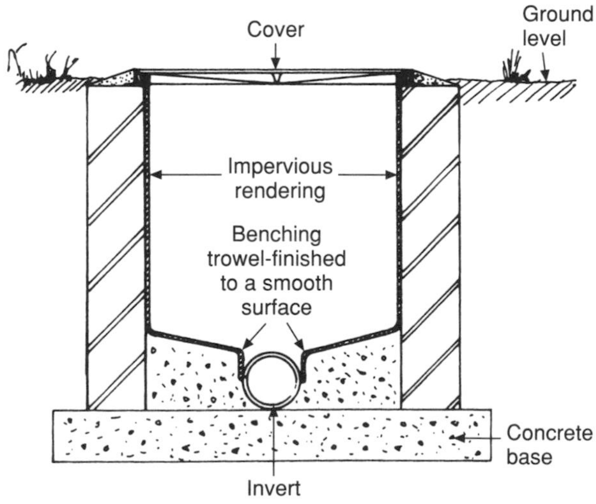


**Fig. 9.39** Section through typical inspection chamber constructed of brick or precast concrete.

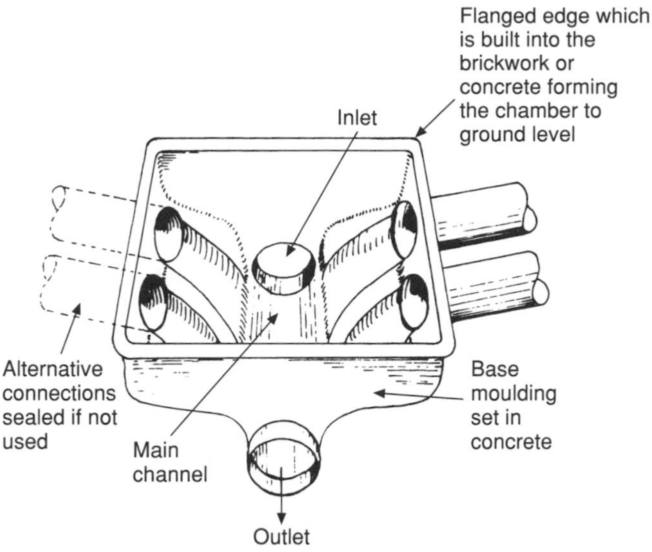


(a) Typical inspection chamber moulding made of PVC for plastic drainage systems

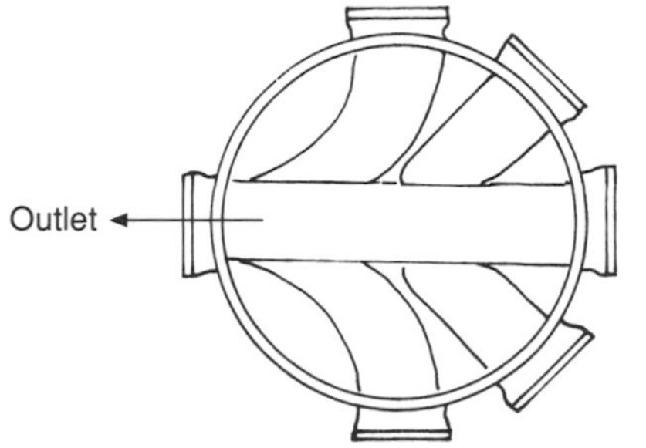


(b) Plan view of moulded base with five inlets and one outlet. Suitable plugs are provided for inlets not used.

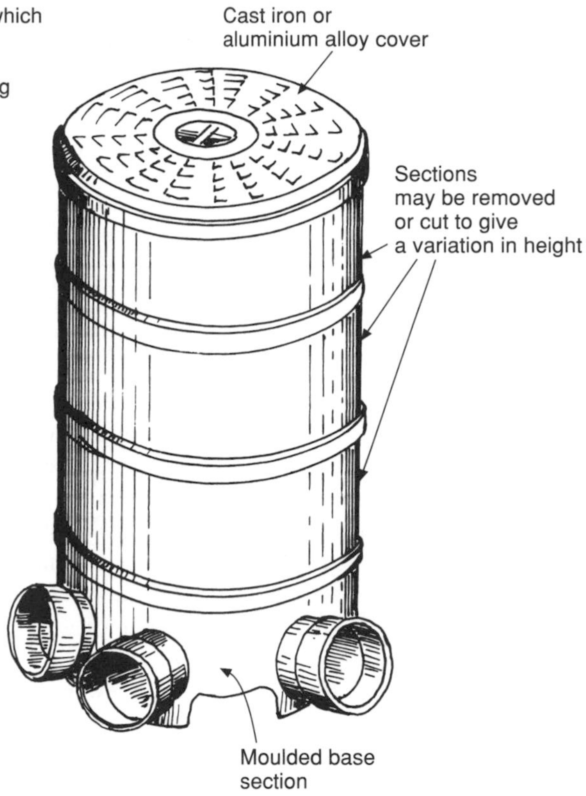


(c)

These inspection chambers are obtainable made of PVC, polypropylene or clayware. They are suitable for shallow drains only as if the invert is too deep difficulty will be experienced when it is necessary to clear blockages with drain rods. The initial cost of this type of chamber is relatively high, but their use reduces labour costs due to the comparative costs of traditional brick built chambers.

**Fig. 9.40** Inspection chambers.

of a drain is such that it is necessary to enter a manhole for maintenance purposes, a steel ladder or step irons conforming to BS EN 13101-2002 must be provided. When plastic systems of drainage are used, a chamber can be obtained as a complete unit having a number of blank sockets moulded into it which can be cut as required on site. The design of those illustrated in Fig. 9.40 is based on a traditional layout. As with PVC systems of discharge pipework, all manufacturers of PVC drainage produce their own system, the parts of which are not always interchangeable with those of others. Before altering or extending an existing system, a check should be made that the appropriate components are available.

### Connections to existing drains

It is sometimes necessary to make extra connections to existing drains; for example, a client may be having an extension built containing extra sanitary fittings. If an inspection chamber is already situated nearby, the easiest method of connection is to use a channel bend, shown in Fig. 9.13(b). If no such chamber is available it will almost certainly be necessary to construct a new one. This is a

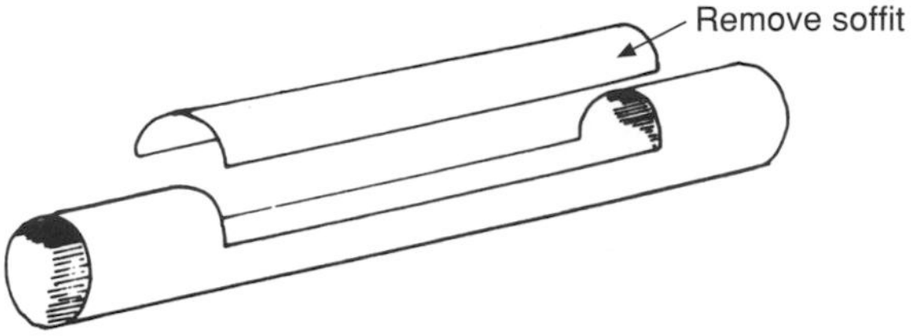


This leaves a channel around which an inspection chamber can be constructed, the new branch being connected by a channel bend. Note this method is not practicable for cast iron drains.

**Fig. 9.41** Additional branches in existing drains.

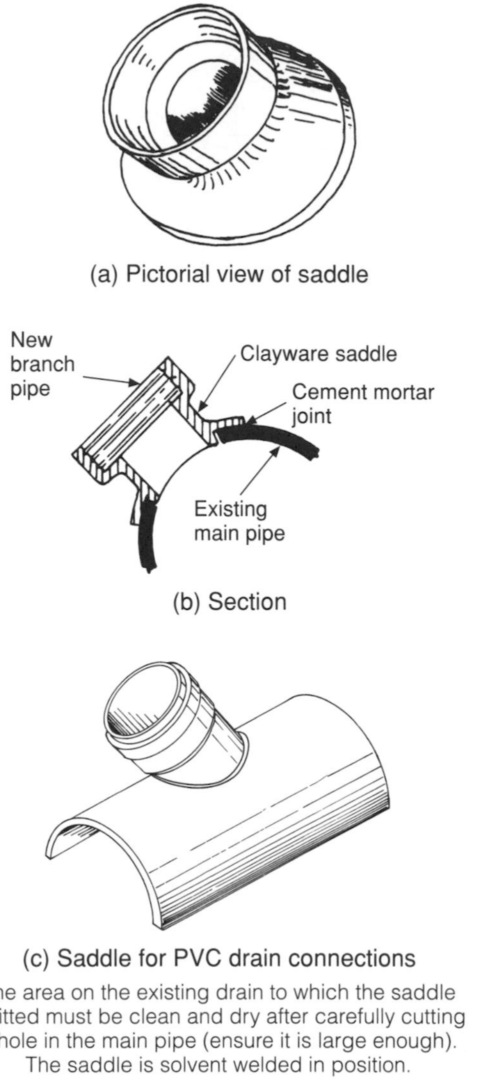
(a) Pictorial view of saddle
New branch pipe
Clayware saddle
Cement mortar joint
Existing main pipe
(b) Section
(c) Saddle for PVC drain connections
The area on the existing drain to which the saddle is fitted must be clean and dry after carefully cutting a hole in the main pipe (ensure it is large enough). The saddle is solvent welded in position.

**Fig. 9.42** Saddles used for making a new connection to an existing pipe. Note that when cast iron is used an appropriate adaptor will be necessary to connect to the saddle.

comparatively simple matter in the case of clayware or PVC systems, as once the existing pipe has been exposed, the soffit can be carefully removed as shown in Fig. 9.41 using a disc cutter. Care must be taken not to let any spoil enter the drain, a channel bend being used to make the new connection.

There are times when a direct connection has to be made to an existing pipe, a typical example being where a house drain joins a public sewer. A fitting called a *saddle* is the usual means of making such a connection if the main pipe is big enough (see Fig. 9.42). If it has a diameter of 225 mm or less this method should not be used, because a hole cut in a small pipe would weaken it to an unacceptable degree. In such cases a section of the pipe must be cut and a junction fitted using a slip socket. It should be noted that cutting a new junction into an existing pipe, especially if it is cast iron or clayware, is not an easy task and requires great care. However, the use of power cutting tools and the variety of fittings available have made it less formidable than previously. Remember that any connection using a saddle or junction, where there is no means of access, must enter the main drain or sewer at an angle of not less than 45°.

Local authorities usually insist that where a connection is made between a house drain and a public sewer, one of their own employees who specialises in this work carries it out.

#### Working principles of a good drainage scheme

Any drain or private sewer must be designed and constructed of a suitable size and gradient to ensure that it is self-cleansing and efficiently carries away the maximum volume of matter which may be discharged into it. With this requirement in mind the plumber should be aware of two important facts which affect the efficiency of the drain. These are the gradient, inclination or fall of the drain, and the quantity of water flowing through it. (The latter has an important bearing on the fall as will be shown later.) If the water is flowing very slowly in the

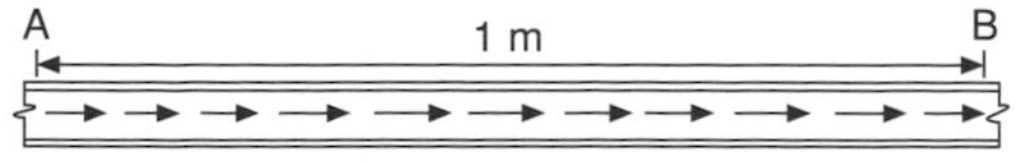


**Fig. 9.43** Explaining velocity. If the movement of a particle of water from A to B takes 2 seconds the flow rate may be expressed as 0.5 m per second.

**Table 9.2** Maguire's rule.

| *Diameter of pipe (mm)* | *Recommended fall* |
|---|---|
| 100 | 1 in 40 |
| 150 | 1 in 60 |
| 225 | 1 in 90 |

pipe, its velocity (speed of flow) is insufficient to carry with it any solid matter. If, on the other hand, the velocity is too great, the water leaves the solids behind. In both cases, solid matter can build up and subsequently cause a blockage. The quantity of water it carries and whether it is a constant or intermittent flow also influence the gradient of the drain. If, for example, the drain is laid to a very shallow fall, it may be self-cleansing if it is running half full, but if the rate of flow falls below this its velocity may be insufficient to keep the solids moving. The important point to understand from this is that the velocity is influenced by the quantity of water carried by the drain. To enlarge a little more on the term *velocity*, consider a length of drain 1 m long as shown in Fig. 9.43. Imagine a particle of water entering it at point A and the time it takes to reach point B. If it takes 2 seconds the water is said to have a velocity of 0.5 m per second; in other words, it takes 1 second to travel a distance of half a metre. The minimum velocity for a self-cleansing gradient is approximately 0.75 m per second when the drain is flowing quarter full at a reasonably constant rate.

*Sizing drainage pipelines*

Calculating the diameter for underground drainage pipes is seldom within the remit of the plumber; most house drains are 100 mm and this pipe diameter is normally large enough to drain a large number of domestic properties. Sizing pipes for main foul and surface water sewers is a specialist job normally carried out by a qualified service engineer. The main factors involved with all pipe sizing calculations are the hydraulic mean depth (HMD) which relates to the volume of water to be drained, the frictional resistance of the pipe walls and the inclination or fall of the pipe. These factors are embodied in all drainage pipe sizing formulae. Reference should be made to the information available from the drainage development associations listed.

There are two practical points to bear in mind that relate to the sizing of drains. It is often asked, 'Why not have drainpipes of smaller diameter?' (100 mm nominal is the smallest permitted for foul water). If this were so, the incidence of blockage would be increased, but, even more important, the drains might, on occasion, run full bore causing the traps in gulleys and WCs to be siphoned out.

The plumber is normally only concerned with short branch drains, and as these are almost invariably only 100 mm in diameter the question of selecting a drain size is largely eliminated. What is important here is the fall of such drains. As they normally carry only intermittent flows, the incline must be greater than those of drains with a steady flow to ensure that they are self-cleansing. The recommended falls for such drains are shown in Table 9.2 which illustrates what is known as Maguire's rule. These figures indicate that a 100 mm drain must have a fall of 1 m in every 40 m, and so on.

If the falls in Table 9.2 are adhered to, a self-cleansing gradient is assured with very low flow rates.

### Invert and soffit

A clear understanding of these two terms is essential. Most drainage measurements are taken from the lowest point of the invert and such a measurement is referred to as the 'invert level'. If one can imagine the cross-section of a circular pipe with a horizontal line drawn across its centre, the curve or arch above the line is called the *soffit*. The half-section below the line will appear as an arch in the upside-down position and is the *invert*. Figure 9.44 illustrates these two terms.

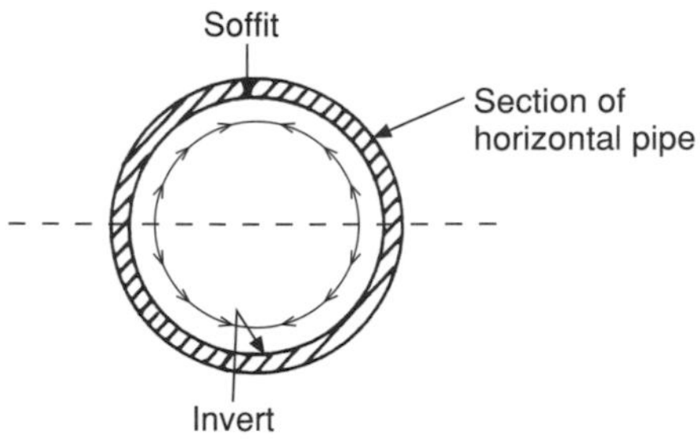


**Fig. 9.44** Invert and soffit of drain section.

### Determining drain levels

All levels, not just those relating to drains, relating to a building are measured from the site 'datum'. It is sometimes called a temporary bench mark, and consists of a peg firmly driven into the ground, usually protected with concrete and carefully levelled from the nearest Ordnance Survey bench mark. It is from the temporary bench mark (TBM) that the invert levels of drains are taken. As an example, assume that the invert of a drain run is to be a given distance below TBM; a peg will be driven into the bottom of the drain trench and levelled by the supervisor or site engineer using a surveyor's staff and theodolite or a laser level. If, however, such equipment is not available, the job can be done quite as effectively with a water level. This is a simple device using the principle that fluids find their own levels and consists of two graduated glass tubes connected by a suitable hose. Its use in ascertaining levels is shown in Fig. 9.45.

### Setting out the incline of a drain

For short runs the method involving the use of an incidence board is usually found satisfactory. To cut the board to the correct angle the required incline of the drain must first be found by the following simple formula:

$$\text{Incline} = \frac{\text{Fall (in m)}}{\text{Length (in m)}}$$

where the fall is the difference between the highest and lowest points of the drain.

For example, if a drain has a fall of 1 m in a 40 m length, the fall per metre run will be calculated as follows:

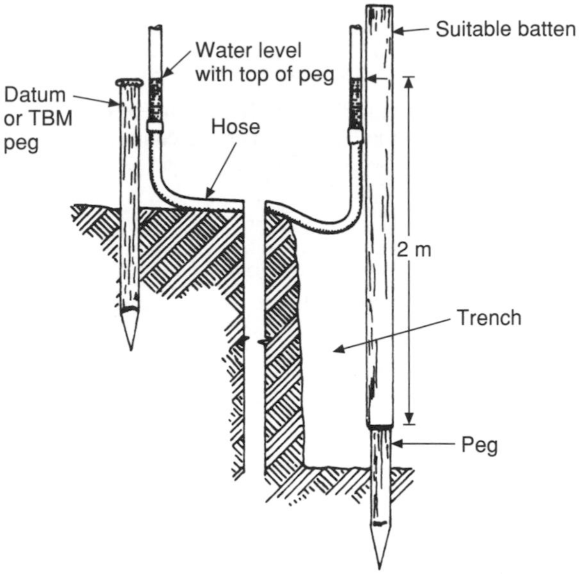


**Fig. 9.45** Transferring invert level from a temporary bench-mark (TBM). Note that TBM has a value of 100,000. Levels above this will therefore read 100,000+, levels below 100,000–, e.g. assume the drain invert level would be indicated in this case as 98,000–.

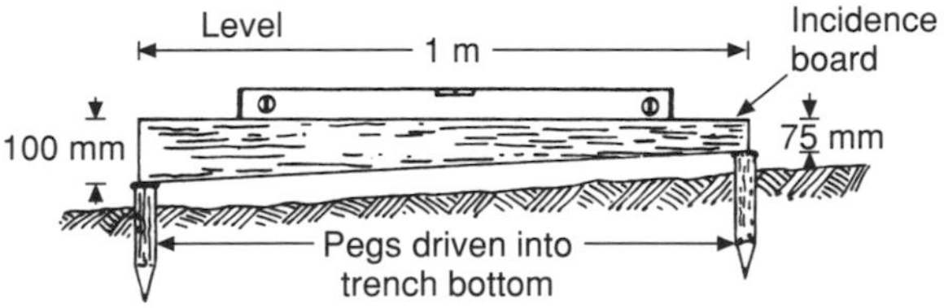


**Fig. 9.46** Levelling in short branch drains using an incidence board. If the top of the board is kept level its bottom edge will give the fall found in the calculation.

$$\begin{aligned}\text{Fall} &= 1 \div 40\\ &= 0.025 \text{ m (or 25 mm)}\end{aligned}$$

The drain will fall 25 mm in every metre. An appropriate gradient incidence board with measurements is shown in Fig. 9.46.

For setting out longer runs, sight rails and boning rods are used. Figure 9.47 shows a typical sight rail set up in drainpipes. Two will be required, one at the highest point of the drain and one at the lowest, and the difference in height between the two will be the required fall of the drain. The boning rod looks rather like a large T-square and is illustrated in

**Fig. 9.47** Sight rail set up in pipes.

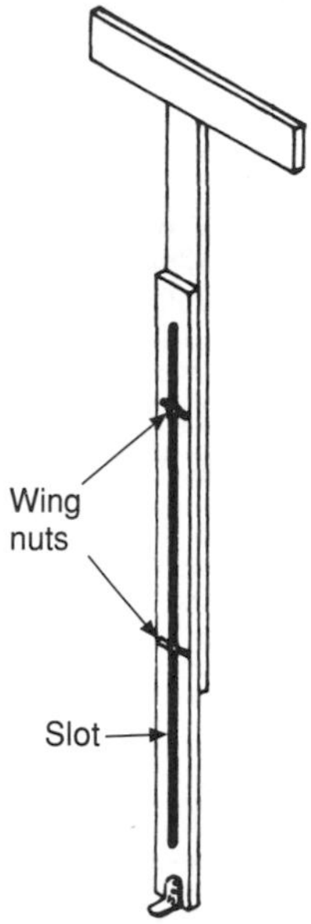


**Fig. 9.48** Purpose-made adjustable boning rod.

Fig. 9.48. The procedure is as follows: a trench is excavated, the bottom of which follows the gradient as closely as possible; pegs are then driven into the trench bottom at intervals of about 1 m. The required gradient is achieved by placing the boning rod on top of each peg and checking that the top is lined up with the sight boards. If the top of the boning rod shows above the boards then the peg must be driven down, if it is lower than the boards then the peg must be replaced by a longer one. Figure 9.49 illustrates the procedure. If necessary the trench itself can be excavated to the required gradient by boning in each section as it is dug. Where a granular bed is used the pegs must be removed as the drain is laid because they can damage pipes made of PVC if any movement of the ground takes place.

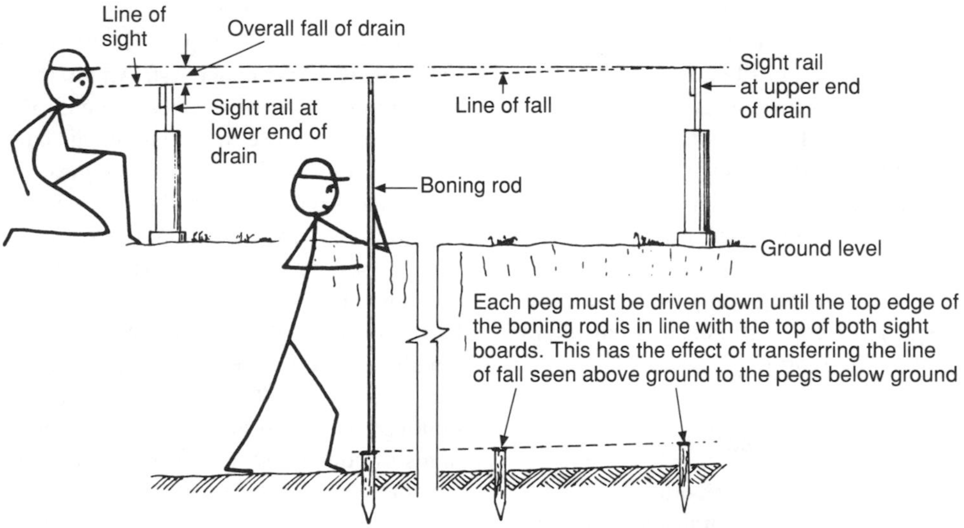


**Fig. 9.49** Boning in pegs for drain levels.

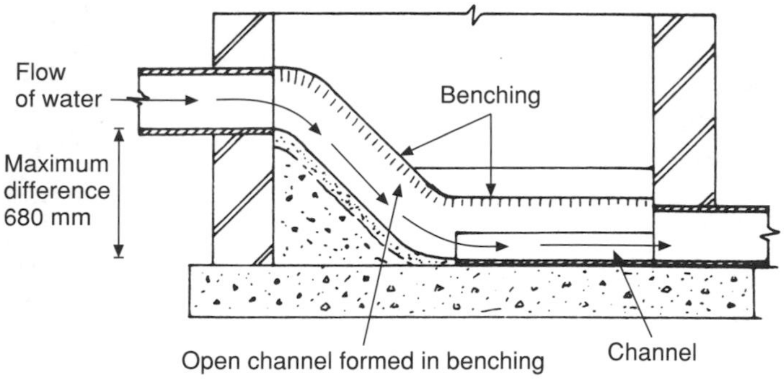


**Fig. 9.50** Drain ramp, used only for differences of up to 680 mm in invert levels. Not all authorities will accept this arrangement due to possible fouling of the benching.

## Ramps and backdrops

The gradient must be correct if a drain is to be self-cleansing. There are occasions where it is difficult to achieve a constant gradient, for instance where the drain passes through steeply sloping ground or when a drain discharges into a very deep sewer. Depending on the vertical distance between the inverts of the drains or sewers to be connected a ramp or a backdrop (sometimes referred to as a *tumbling bay*) can be considered. A ramp like that shown in Fig. 9.50 may be used if the difference in the invert levels is not in excess of 680 mm. If the difference is greater, a backdrop must be constructed (see Fig. 9.51).

In cases where the invert of the drain is lower than that of the sewer — for example, where the drainage of sanitary fittings situated in basements has to be considered — the use of a pump or sewage lift is essential. Due to both initial and possible maintenance costs of such equipment, drainage schemes should be designed in such a way that their use is unnecessary, certainly in all but very large buildings.

Mechanical sewage lifting equipment is usually installed by specialist contractors and is outside the scope of normal plumbing work.

## Testing drains for soundness

For obvious reasons it is very important that, on completion of the work, a new drainage system is thoroughly tested. The two main methods of testing underground drains for soundness are water tests and air tests.

The building inspector will require the drain to be tested for soundness after backfilling of the trench is completed. This is to ensure that any defects or damage sustained by careless backfilling will show up during the test. For this reason the contractors test the drain to their own satisfaction both before and after backfilling. Backfilling should be done very carefully, avoiding large heavy objects such as pieces of brick or concrete, especially in the first layers covering the drain as it can easily be damaged at this stage.

### *Drain-testing equipment*

Expanding plugs (see Fig. 9.52) are used to seal off the ends of drains under test. An alternative method of plugging, used when only one end of the drain is accessible, is to use the inflatable air bag or stopper shown in Fig. 9.53. This is floated down the drain and inflated with a pump similar to a bicycle pump when in the correct position. Great care should be taken to secure these plugs when water testing, as on completion of the test they can be swept away by the velocity of the water. When water testing lengths of drain sealed at the upper end of the trap, e.g. a gulley or WC, the air in the pipe must be removed as shown in Fig. 9.54. The only special items of equipment needed for air testing other than suitable plugs are a U-gauge (manometer) and hand bellows. This equipment has already been described in Chapter 8.

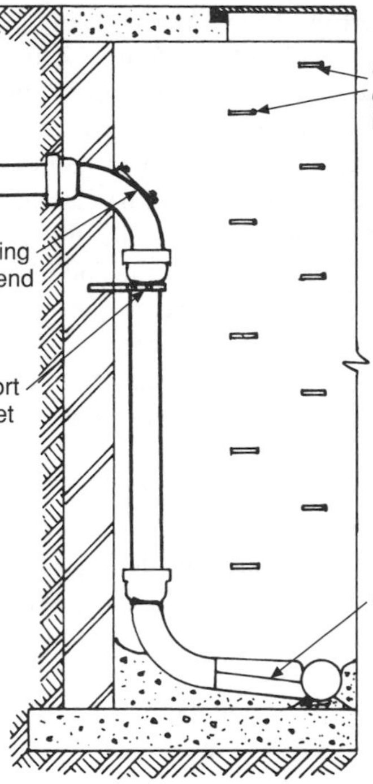


(a) Backdrop

Where cast iron or PVC drainage is used the vertical pipe may be fitted in the inspection chamber. In the case of clayware the pipe is fitted outside the chamber and is usually encased in concrete.

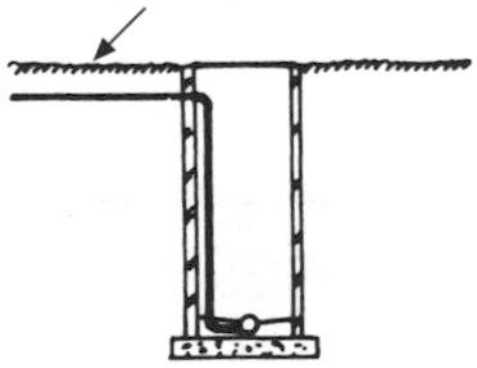


(b) High-level drain discharging into a low-level sewer

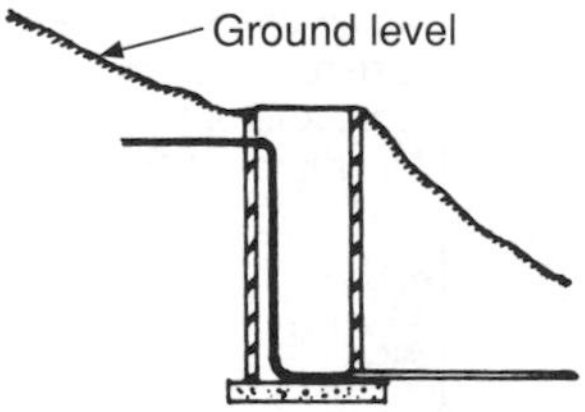


(c) Use of backdrop in steeply sloping ground

**Fig. 9.51** Backdrop inspection chambers.

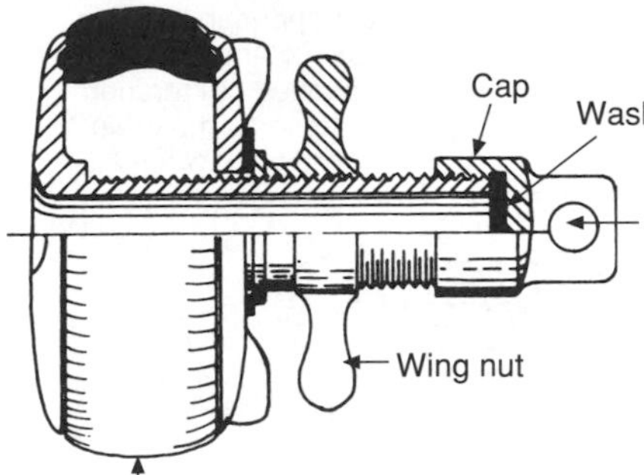


**Fig. 9.52** Drain plug.

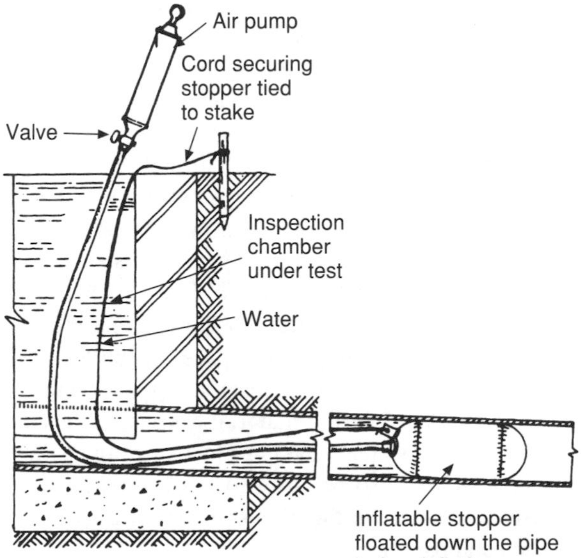


**Fig. 9.53** Testing a length of drain with an inflatable stopper. This method is suitable when access can only be made to one end of the drain, e.g. last inspection chamber before sewer connection.

*Testing procedures*

The Building Regulations 1985, Schedule 1, Part H specify the requirements for testing underground drainage systems. Drains may be water or air tested, but water is usually preferred as it relates more closely to the actual working conditions. It is not so severe as an air test, since air, being a gas, can penetrate smaller apertures or cracks than a fluid. Should an air test reveal a leak,

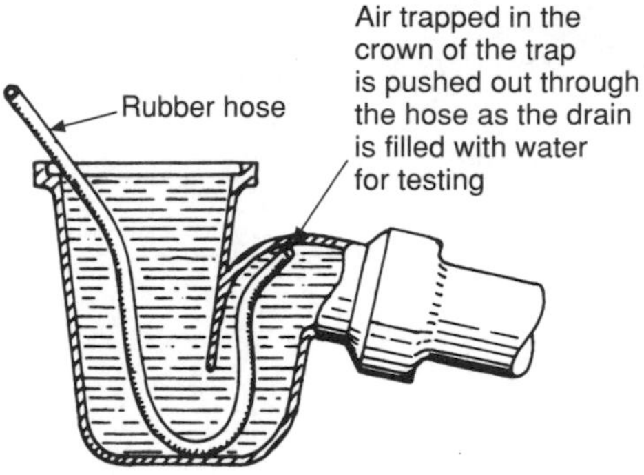


**Fig 9.54** Removing the air from the crown of a gulley trap when water testing drains.

it is not easy to find, especially if it is at some point underneath the drain. It will be noted in the following text that some latitude is allowed when testing with either air or water. If a small loss of water or air pressure is shown up during the test, it does not necessarily mean the drain is leaking. Several factors may account for small losses, such as temperature changes resulting in expansion or contraction of the water, air or pipes. This especially applies to PVC, which, having a high expansion rate, has the effect of lengthening the pipe run, thus providing more space for the testing media. In the case of water tests, trapped air and seepage into the jointing material can sometimes show a loss of water. Leaking plugs or drain stoppers should also be checked if a loss of air pressure, or water in the case of a water test, becomes apparent.

*Water testing* (see Fig. 9.55) When testing drains with water a minimum head of 1.5 m is required, measured from the invert of the drain at its highest point. The drain is then filled with water. Any air in the crown of trapped gulleys should be removed after which a gulley stopper should be clamped in position. The head of water to which the drain is subjected should not exceed 4 m including the 1.5 m above the invert of the pipe, as pressure in excess of this may result in split pipes. In some cases, to avoid subjecting a long drain to excessive pressure, it may be tested in sections. A period of approximately 2 hours is allowed for the water to settle and for any absorption by the pipe or jointing materials to take place. The drain should then be topped up and left for 30 minutes during which time any loss of water will be observed. The permissible loss of water during this test period is shown in Table 9.3. To give an example, if a drain has a nominal diameter of 100 mm and has a length of 20 m, the permissible loss will be

$$0.050 \times 20 = 1 \text{ litre}$$

**Table 9.3** Permissible loss of water.

| *Nominal diameter of pipe (mm)* | *Permissible loss of water in litres per metre run* |
|---|---|
| 100 | 0.050 |
| 150 | 0.080 |
| 225 | 0.120 |
| 300 | 0.150 |

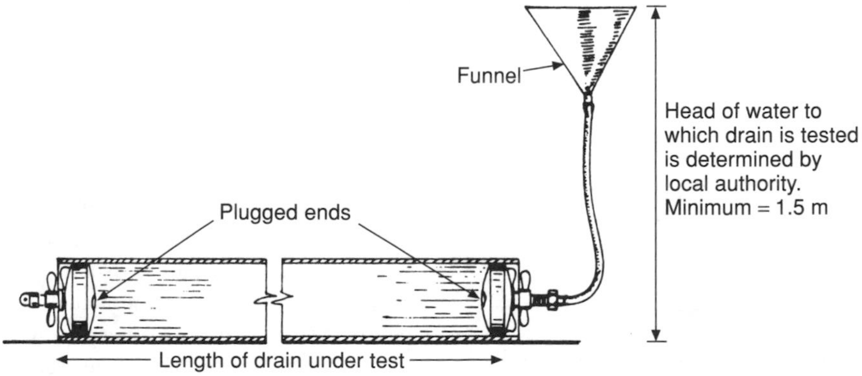


**Fig. 9.55** Water testing of drains.

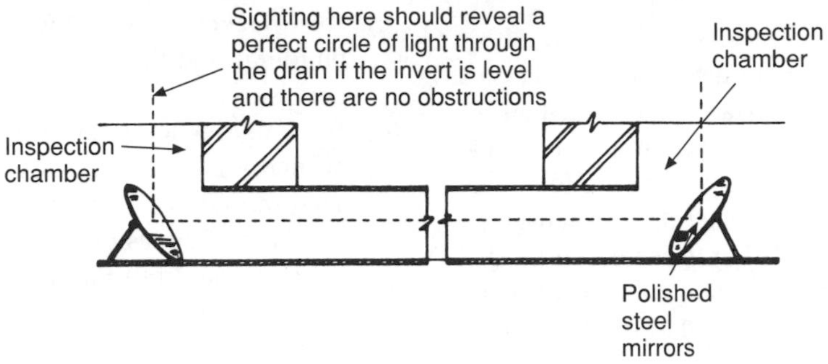


**Fig. 9.56** Mirror test.

If the water level has fallen during the 30 minute test period, and in the case of this example not more than 1 litre of water is required to bring it up to its original level, the test result is acceptable.

*Air tests* Air tests are conducted in a similar way to those on discharge pipework with all open ends on the system suitably plugged and air blown or pumped in until a pressure slightly in excess of 100 mm is recorded. Allow 5 to 10 minutes for the air to stabilise, after which adjust the air pressure as necessary. During a period of 5 minutes the pressure should not fall more than 25 mm for the test to be acceptable. If any gulleys or WCs are fitted prior to testing, a test pressure of 50 mm is the maximum obtainable without blowing the water seal in the trap. The test period is the same, but in this case a permissible pressure loss of only 12 mm is allowed. The reader should note that the test pressure and period of time are slightly different from the tests conducted on discharge pipework.

*Mirror and ball tests* Tests to check the alignment and general condition of the inside of the drain are sometimes conducted. They are seldom required by the local authority inspectorate but may be necessary to satisfy the clients or their representative, the clerk of works, on the standard of workmanship. The mirror test illustrated in Fig. 9.56 permits the drain to be sighted through by eye to check the invert level. A clear circle of light should be seen if the drain is absolutely straight and true. This test will also show up any defect such as material protruding from a joint or deposited on the invert. A similar test may be conducted with a ball slightly smaller in diameter than the drain. When it is rolled down the drain, any obstruction will, of course, arrest its progress. Both of these tests can only be applied if access is available at both ends of the drain.

*Testing existing drains* Chemical or scent tests are sometimes used to detect leaks in existing drains which should not be subjected to pressure testing. These smell or olfactory tests, as they are called, are often applied to determine whether a leaking drain is responsible for causing unpleasant odours inside a building. Chemical substances such as calcium carbide or oil of peppermint, when in contact with water, give off an unmistakable smell. Both these chemicals can be washed down the vent pipe with water. Special containers are also available for introducing the chemical to the drain. When charged they are washed into the drain via a WC or gulley. Figure 9.57 illustrates this equipment and how it operates.

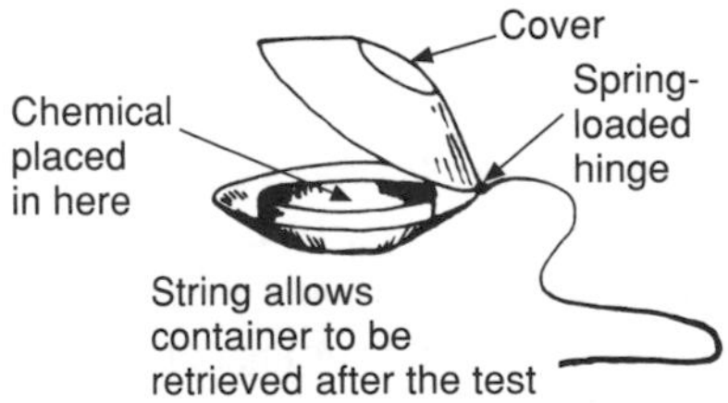


**Fig. 9.57** Chemical test. The chemical in liquid or soluble tablet is placed in the container, the two halves of which are then sealed with gummed paper. The container is then flushed down the drain and will open due to the action of moisture on the gummed paper, allowing the chemical to enter the drain.

### Tracing drains

It is sometimes necessary to determine whether a drain is carrying foul or surface water. Such an occasion could occur if a connection is to be made to an existing drain. Obviously, allowing foul water to discharge into a surface water drain could give rise to untreated sewage being deposited in local rivers or streams, thereby causing a nuisance and the possibility of danger to health. For this purpose a solution of fluorescein is poured into the drain via a gulley, WC or convenient inspection chamber and, when added to water, it imparts to it a bright green colour. Checks can subsequently be made at other points of inspection to determine the passage of water through the system. Any method of colouring the water may be used in the absence of fluorescein, a suitable substitute being whitening or a little coloured emulsion paint. Another chemical is potassium permanganate, which when added to water colours it purple enabling it to be easily identified.

### Maintenance of underground drainage systems

In a well-installed self-cleansing system of drains there is little that can go seriously wrong — when it does it is usually due to misuse.

In the case of public buildings such as hospitals and large commercial and industrial buildings where the possibility of a breakdown could be serious, consideration should be given to a planned maintenance scheme to ensure regular inspection of the drainage components. Each building complex will require its own individual schedule which is usually drawn up by the person in charge of maintenance. Some of the more general items are listed as follows:

(a) All gulleys should be checked to ensure the base of the trap is clear of debris prior to washing out with a suitable detergent. Where applicable, ensure all gratings are clean — remove and clean buckets in yard gulleys.
(b) Where an interceptor is fitted, check the trap is not blocked with debris and make sure the stopper in the rodding eye is sealed. Check that the flap in the fresh air inlet is unbroken and in working order.
(c) Rust pockets may be found in old public buildings and these must be cleaned if necessary.
(d) Brick and concrete inspection chambers are normally parged internally with cement and sand mortar. This should be checked for cracks and lack of adhesion, as broken pieces of parging could fall into the channel causing a blockage. In deep chambers inspect the step irons for deterioration and replace if necessary. Any double-sealed chamber covers inside a building should be cleaned and repacked with grease.
(e) To inspect the internal surfaces of drain pipes, the mirror test shown in Fig. 9.56 is a simple method but it can only be used on straight lengths of pipe with adequate access. A more effective method is the use of closed circuit television where a very small camera is drawn through the pipe recording pictorially its condition. Companies with the necessary expertise and equipment are normally employed to carry out work of this nature.

### Decommissioning

Where possible, any drains no longer in use should be removed. This may not be feasible for those underground, in which case the ends of the pipes must be securely stopped and filled as far as possible with lime slurry, which provides some degree of disinfection.

### Further reading

Building Regulations, Part H
BS EN 752 Drain & Sewer Systems outside Buildings Part 1 generalities & definitions; Part 2 performance requirements; Part 3 planning; Part 4 hydraulic design environmental considerations.
BS EN 877 Ductile iron pipes and fittings
BS 4660 Specifications for UPVC. Drain pipes and fittings for gravity drains and sewers.
BS 6087 Specifications for flexible joints for cast iron pipes and fittings.
BS 8000 Part 14 Workmanship on building sites, code of practice for below ground drains.

BS EN 1610 Construction and Testing of Drains and Sewers.

*Drainage systems and materials*

*Cast iron systems*

Saint Gobain Drainage Systems, PO Box 3, Sinclair Works, Ketley, Telford, Shropshire TF1 5AD Tel. 01952 262500. www.saint-gobain-pipelines.co.uk

*Plastic systems*

Wavin Plastics Ltd, Parsonage Way, Chippenham, Wiltshire SN15 5PN Tel. 01249 766600. www.wavin.co.uk

Marley Extrusions, Lenham, Maidstone, Kent ME17 2DE Tel. 01622 858888. www.marleyplumbinganddrainage.com

Hunter Plastics Ltd, Nathan Way, London SE28 0AE (drainage and grease disposal). Tel. 020 8855 9851. www.hunterplastics.co.uk

*Grease disposal*

Wade International UK Ltd, Third Avenue, Halstead, Essex CO9 2SX Tel. 01787 475151. www.wadedrainage.co.uk

The Cast Iron Drainage Development Association, c/o Wyatt International Ltd, Wyatt Hse, 72 Francis Rd, Edgebaston, Birmingham B16 8SP Tel. 0121 4548181. www.cidda.com

The Clay Pipe Development Association, Copsham Hse, 53 Broad St, Chesham Bucks, HP5 3EA Tel. 01494 791456. www.cpda.co.uk

BRS Digest 292 Access to domestic underground drainage systems.

BRS Digest 365 Soakaways.

BRE Bookshop, 151 Rosebery Avenue, Farringdon, London EC1R 4GB Tel. 020 7505 6622.

### Self-testing questions

1. State the procedures that must be complied with before any new drains are laid, and before extensions are made to those in existence.
2. Make a sketch showing the invert and soffit of an underground drain pipe.
3. List the principal materials from which underground drain pipes and fittings are made.
4. State the maximum angular movement either side of the horizontal position for flexible joints.
5. Explain the terms 'invert' and 'soffit' in relation to circular pipes.
6. (a) State the minimum depth of seal in a gulley or intercepting trap and describe the purpose of each.
   (b) In what circumstances would trapless gulleys be used?
7. State the two important details that should be included on any order for cast iron drain bends.
8. State the purpose of flexible bedding and outline its main features.
9. (a) Identify the difference between combined and separate systems of drainage.
   (b) Explain why the separate system is now more common in urban areas.
10. (a) State the circumstances in which a soakaway might be employed.
    (b) Describe a suitable site for a soakaway attached to a private house.
11. Give the reasons for ventilating drainage systems, and explain how this ventilation takes place.
12. List the requirements of the Building Regulations for means of access to drains.
13. State the minimum permissible diameter for a foul water drain.
14. (a) Define the term 'self-cleansing velocity' and identify its relationship with the fall of a drain and the volume of water it carries.
    (b) State the recommended fall for a branch drain carrying intermittent flows.
    (c) What is likely to be the effect of an excessive fall on the contents of a foul water drain?
15. (a) Make a sketch showing how air is removed from the crown of a gulley trap when testing drains.
    (b) Explain why it is sometimes necessary to trace drains, and state the methods used.
16. State the approved procedures for water testing an underground drain.

# 10 Sheet lead roof weatherings

After completing this chapter the reader should be able to:

1. Describe and illustrate with sketches how lead sheet gutters are fabricated.
2. Describe the various methods of discharging water from lead gutters into rainwater pipes.
3. Describe the techniques of forming solid and hollow rolls and joints for flat and pitched roofs covered with sheet lead.
4. Select and illustrate with sketches the methods of weathering dormer windows.
5. Mark out sheet lead for forming all types of welded joint.
6. Select suitable welding techniques for fabricating sheet lead weatherings.

## Introduction

Metal roof weatherings have been used throughout the ages both for covering whole roofs and for forming the components necessary to make a watertight joint between building materials such as brick, stone, timber, tiles or slates. This subject is very extensive due not only to the number of materials that can be used, but also to the varying techniques employed with each one. For many years the methods of weathering with sheet lead remained unchanged, but relatively recently many improvements have been made enabling thinner sheet to be used than hitherto. A greater emphasis is now given to lead welding, which when employed by skilled craftsworkers can result in considerable saving of time and materials. The basic joints, however, remain the same and it is perhaps a tribute to the plumbers of long ago that the skills evolved by them are still in use. Much of this chapter deals with traditional methods that plumbers may encounter in their everyday work.

The skills and knowledge in working with lead sheet are still necessary, especially for those engaged in jobbing or renovation work. This is recognised by the lead body in plumbing, the British Plumbing Employers Confederation, and an NVQ unit dealing with lead flashings is required for the completion of level 2. For those requiring a more comprehensive range of skills an NVQ level 3 module is available at some colleges or training centres.

The recommendations given in this chapter are based on those quoted in the *Sheet Lead Manual* volumes 1–3 published by the Lead Sheet Association (LSA). This body is the leading authority on good practice using lead sheet in buildings.

The text here relates to larger areas of lead sheet than those used for flashings. It is therefore important that the following facts are understood relating to the free movement of lead when it expands and contracts due to temperature variation. Figure 10.1 shows the basic principles that apply

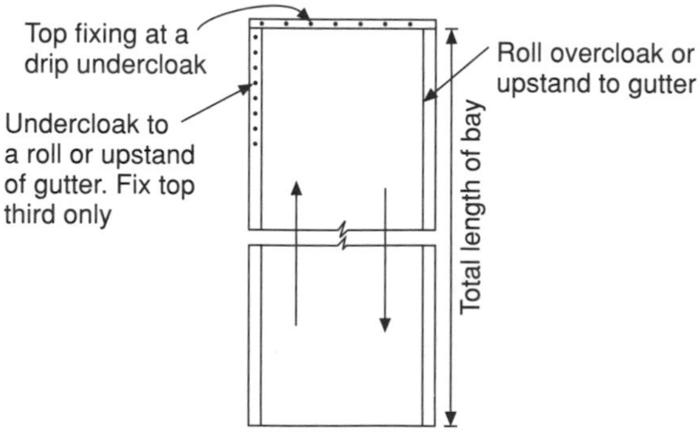


**Fig. 10.1** Basic principles for fixing large areas of lead, showing how it can expand and contract freely.

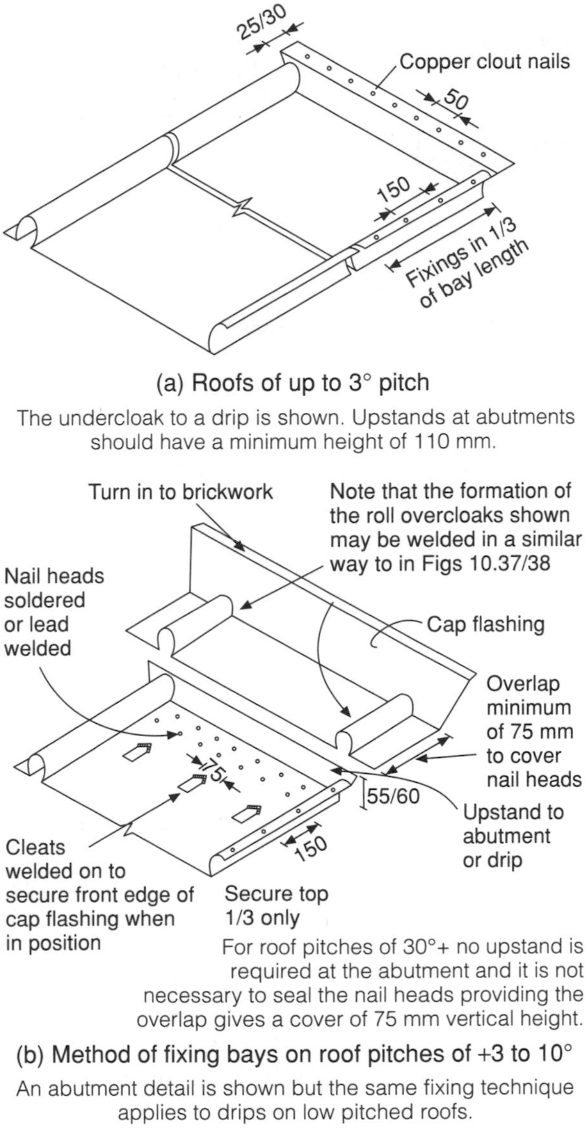

25/30
Copper clout nails
50
150
Fixings in 1/3 of bay length
(a) Roofs of up to 3° pitch
The undercloak to a drip is shown. Upstands at abutments should have a minimum height of 110 mm.
Turn in to brickwork
Note that the formation of the roll overcloaks shown may be welded in a similar way to in Figs 10.37/38
Nail heads soldered or lead welded
Cap flashing
Overlap minimum of 75 mm to cover nail heads
75
55/60
Upstand to abutment or drip
150
Cleats welded on to secure front edge of cap flashing when in position
Secure top 1/3 only
For roof pitches of 30°+ no upstand is required at the abutment and it is not necessary to seal the nail heads providing the overlap gives a cover of 75 mm vertical height.
(b) Method of fixing bays on roof pitches of +3 to 10°
An abutment detail is shown but the same fixing technique applies to drips on low pitched roofs.

**Fig. 10.2** Fixing bays.

when fixing large areas of sheet such as flat or pitched roofs, gutters and dormer tops. Figures 10.2(a) and (b) illustrate some of the practical applications of this principle. Note also that the roof pitch has an influence on the methods of fixing employed and the thickness of lead used. This is especially important when weathering vertical or steeply pitched areas, and smaller pieces of lead than those used on low pitches should be considered to avoid creep. Tables 10.1 and 10.2 show that generally the thicker the lead used the greater will be the area that can be covered in one piece. When costing a job the extra cost of thicker lead must be weighed against the greater number of joints required when using thinner sheet. This is especially important when considering valley gutters, as the number of drips used increases the width of the sole of the gutter at the upper end. This will require a larger area of lead and the possibility of centre roll, thus increasing labour costs. See Fig. 10.5.

**Table 10.1** Maximum superficial areas recommended for bays in lead gutters.

| *BS EN 12588 code no.* | *Length between drips (mm)* | *Overall widths (mm)* | *Drip heights (mm)* |
|---|---|---|---|
| 4 | 1,500 | 750 | 55 |
| 5 | 2,000 | 800 | 55 |
| 6 | 2,250 | 850 | 55 |
| 7 | 2,500 | 900 | 60 |
| 8 | 3,000 | 1,000 | 60 |

*Note*: 1. The minimum fall in all gutters is 1:80.
2. Increase in the drip height for code 7/8 accommodates the increased lead thickness.

**Table 10.2** LSA recommendations for bay sizes on roofs pitch at 10° or less.

| *BS EN 12588 code no.* | *Spacing of joints with the fall (rolls) (mm)* | *Spacing of joints across the fall (drips) (mm)* |
|---|---|---|
| 4 | 500 | 1,500 |
| 5 | 600 | 2,000 |
| 6 | 675 | 2,250 |
| 7 | 675 | 2,500 |
| 8 | 750 | 3,000 |

*Note*: The spacing for drips is the same as for gutters.

### Contact with other materials

Lead can be used with copper, stainless steel, aluminium and zinc or galvanised products without any significant electrolytic corrosion, with one exception. Where lead and aluminium are used together in a marine environment, the oxide on the surface of the lead reacts with the sodium chloride in a salt-laden atmosphere. This creates a caustic run-off which attacks the aluminium, especially where any water is trapped in crevices. For this reason aluminium or its alloys should not be used in a marine environment unless protected by a suitable paint.

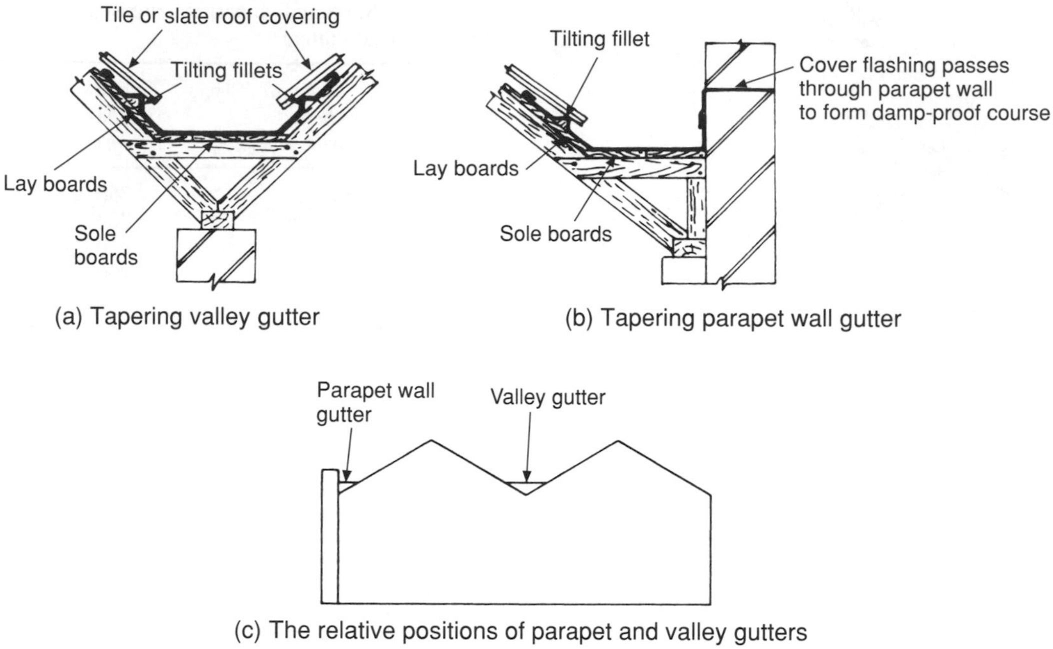


**Fig. 10.3** Sections through tapering valley and parapet wall gutters.

*Contact with timber*

The dilute solutions of organic acids found in hardwood, especially oak, can cause very slow corrosion of the lead. The use of underlays such as inodorous felt or building paper largely overcomes this problem. Any preservative or fire retardant solution with which the timber is treated is unlikely to cause corrosion, providing it is dry before the lead is laid. In wet weather the run-off from new cedar or oak shingles forms a dilute acid solution which will cause corrosion. The lead should be painted with a bituminous solution for a few years until the free acid has leached out of the shingles.

*Contact with mortars and concrete*

Where lead is in contact with new cement or concrete it should be treated with bituminous paint, typical examples being where lead is used as a damp-proof course or for cladding concrete walls. Where lead is turned into brickwork joints, e.g. for flashings, no protection is necessary as the free lime in the cement, which causes the corrosion, is carbonised very quickly.

**Metal lining of gutters constructed of timber**

Gutters formed with timber and lined with metal are usually referred to as *tapering valley* or *parapet wall gutters*, sections of which and the positions they occupy on a roof are shown in Fig. 10.3. The term *tapering* is used to denote the difference between these and the valley gutters dealt with in Book 1, Figs 11.50, 11.52 and 11.53. Both tapering valley and parapet wall gutters are usually found in old buildings constructed before the introduction of purpose-made profiled gutters. A variation may occur where two flat roofs drain into what is called a *box gutter* (see Fig. 10.4(a)).

The plumber must always advise the carpenter responsible for the wooden structure about the positioning of any drips and falls required. Assuming the plumber is working with lead sheet which is 2.4 m wide, the most economic way to cut it will be across its width, so the distance between the drips must be less than this to allow for the upstand and turndown. If the drips are too far apart the cutting of the lead will be uneconomical and its movement may be restricted, possibly causing it to crack at a later date. It cannot be stressed enough that most of the failures experienced with sheet metal weathering of gutters or flat roofs are due to insufficient freedom of movement caused by fixing the sheets too rigidly or covering excessively large areas with one piece of material. The fall of the gutter is also important as it is essential to remove the water as quickly as possible to avoid overflowing.

Falls in gutters and flat roofs should never be less than 1 in 80, although a pitch of $2\frac{1}{2}°–3°$ is recommended for roofs to avoid ponding. The depth of drips in gutters with a capillary groove is 40 mm, but those are not recommended, as instances have been found where the groove has become choked with debris causing the ingress of water into the building. The minimum recommended depth of drips in gutters is 50 mm. Figure 10.5(a) illustrates

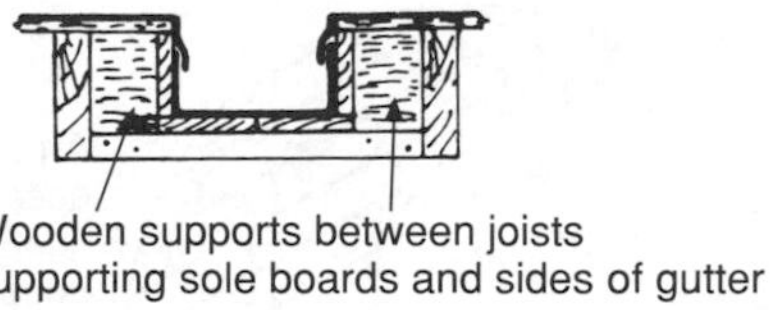


(a) Section of a box gutter between two flat roofs

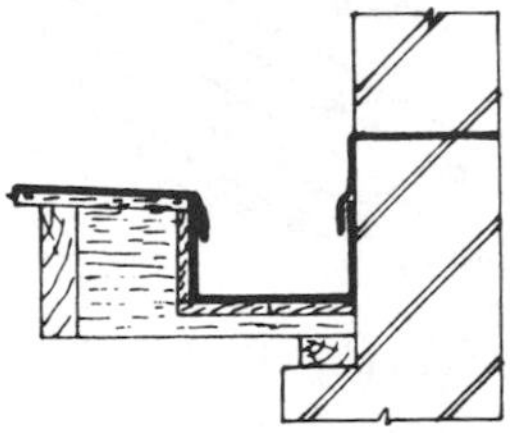

(b) Section of a box gutter between a flat roof and a parapet wall

**Fig. 10.4** Box gutters.

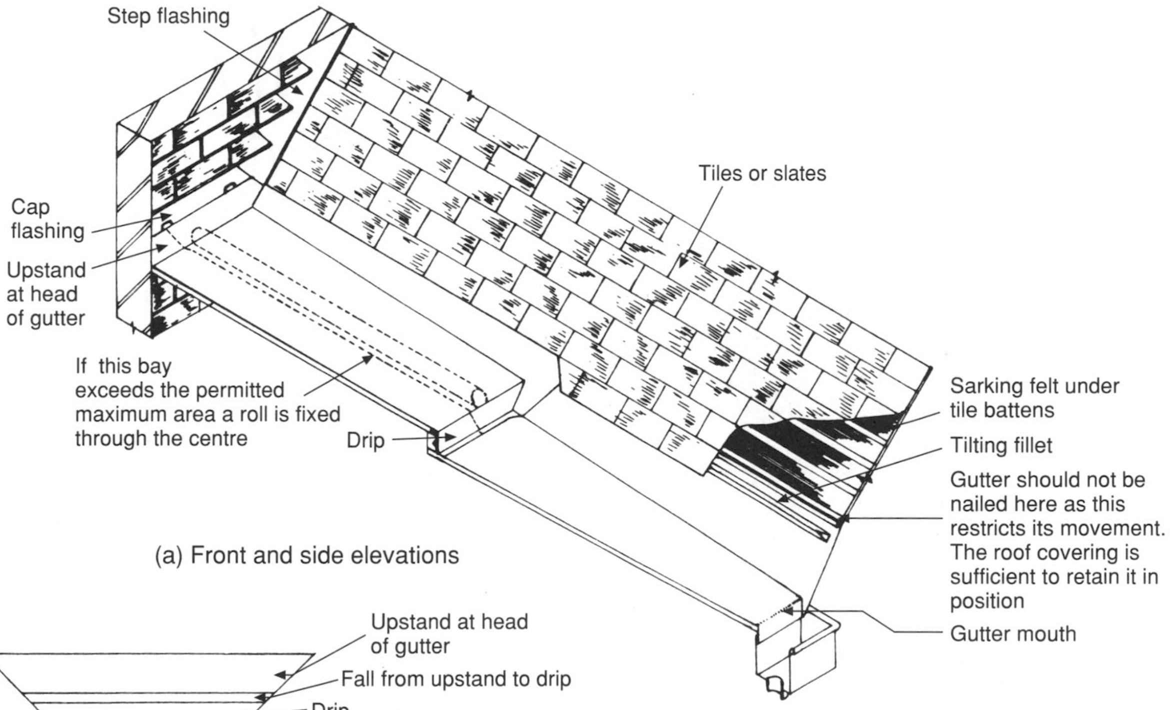


(a) Front and side elevations

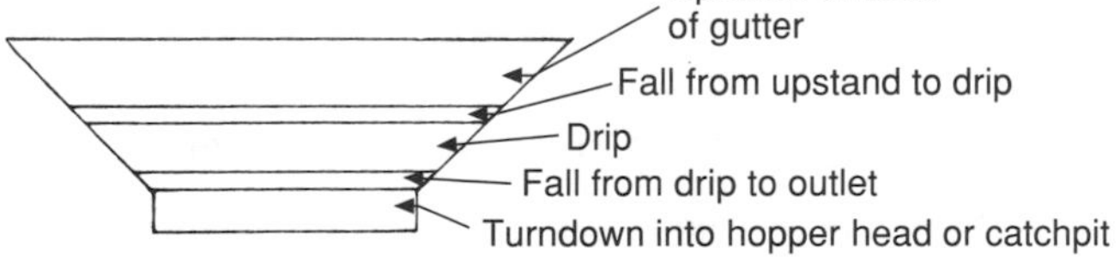


(b) Section of gutter from outlet end showing falls and front of drips

It becomes narrower towards the lower end hence the name 'tapering'. A box gutter retains the same width throughout its length.

**Fig. 10.5** Section through a tapering valley gutter.

a typical tapering valley lead-lined gutter laid between two sloping roofs. In the case of parapet wall gutters only one side will taper. A study of the front elevation in Figs 10.5(a) and (b) will indicate that the taper of such a gutter is due to its fall and any drips in its length. In long gutters it is sometimes necessary to split the upper bays by means of a roll to avoid exceeding the maximum recommended total area of the material. This does not apply to box gutters as they do not increase in width due to the fact that the upstands on both sides are at 90° to base. In all gutters most expansion will take place longitudinally and it is important that the upstands are not rigidly nailed. Any cleats used must allow freedom of movement throughout their length. When the tiles or slates are laid their weight is normally quite sufficient to keep the edge of the gutter in position.

### Gutter outlets

Two types of outlet are used to allow water from the gutter to be discharged into the drainage system. The chute method is used when the gutter discharges through a parapet wall into the hopper head as shown in Fig. 10.6. The other method is known as a catchpit, shown in Fig. 10.7, and is generally used where the rainwater pipe is fitted inside the building. The exception to this is where the outlet from the catchpit is offset through the wall directly into the rainwater pipe. The advantage

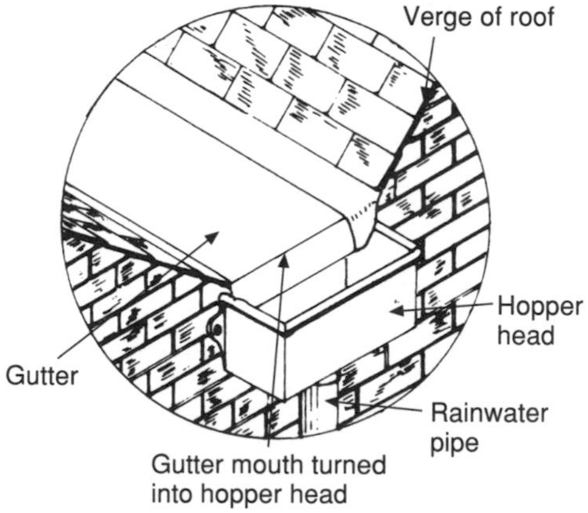


**Fig. 10.6** Rainwater chute. Used to discharge rainwater from a gutter directly into a hopper head.

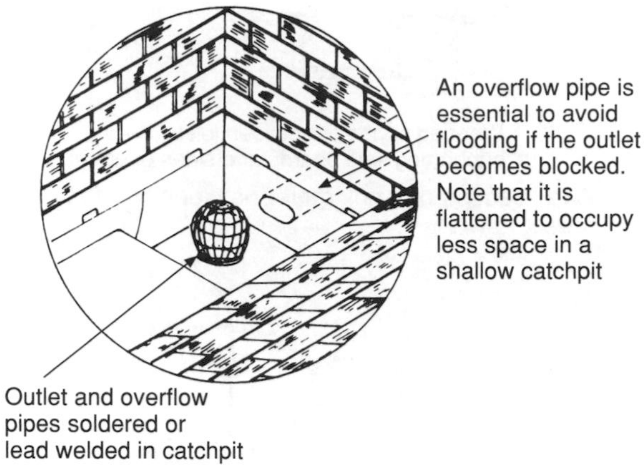


**Fig. 10.7** Catchpit. Generally used to remove water from roofs via an internal rainwater pipe.

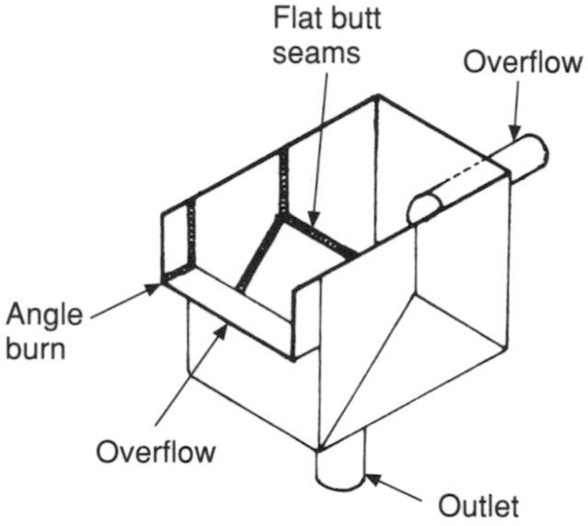


**Fig. 10.8** Lead-welded catchpit.

of using a catchpit is that as the water falls into it from the gutter its velocity is increased enabling it to be discharged more rapidly. A catchpit made of lead sheet is shown in Fig. 10.8 and lead-welded joints have been used in its construction. They may be bossed, but this is time consuming and the application of lead-welding techniques would be a more practical and economic proposition.

### Covering large flat areas with lead sheet

There are two main systems used for weathering large areas. Wood-cored rolls and drips are used on pitches of up to 10°–15°. On steeper pitches lap joints and hollow rolls are usually employed as this method is less labour intensive. The lower limit of fall recommended for flat roofs is $2\frac{1}{2}$°. Although

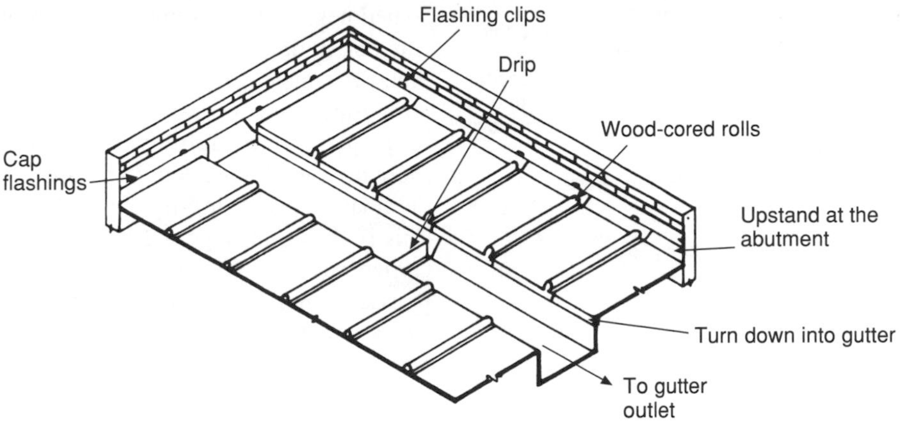


**Fig. 10.9** Typical layout of a simple flat roof discharging into a box gutter.

low-pitched roofs are often called flats, a fall is essential to enable the water falling on them to run off as quickly as possible; insufficient fall will result in 'ponding'. The amount of fall on a roof also influences the type of joint that may be made laterally across the flow of water. For those with a very low pitch drips will be necessary, but as the roof angle increases a lap joint may be used. The amount of fall on a flat roof is determined by what are called *furring pieces* which are cut and fixed by the carpenters across the joists before the decking is laid. The specifications recommended in Book 1 should be adopted, and the preparation of the surface, including laying the felt, should be carried out as described there prior to fixing any metal coverings.

As with gutters, the thickness of the lead used for covering large flat areas will influence the maximum superficial area that should be laid in one piece. The type of joint used with the fall on roof pitches of up to 15° is almost invariably the wood-cored roll. Figure 10.9 illustrates a part of a lead-covered flat to enable the reader to visualise the various details to which reference is made and their relative positions on the finished job. Figure 10.10 shows how to set out one 'bay' of sheet lead with approximate allowances for the joints. Table 10.2 gives the recommended distances between the joints on the roof. Do not forget the allowances for the joints which have to be added to the measurements given in the table.

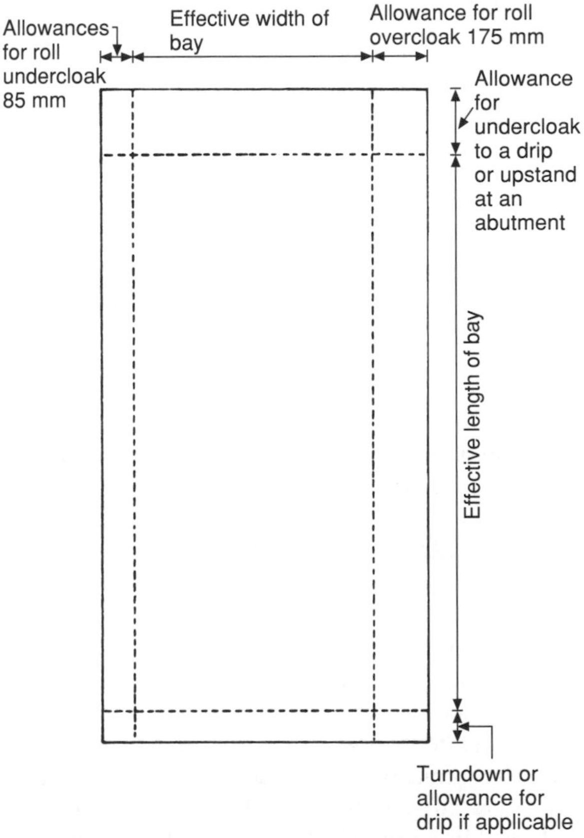


**Fig. 10.10** Setting out a bay for a lead flat showing the necessary allowances. Note that the allowances for the undercloak and overcloak will vary slightly according to the size of the wood roll. Refer to Table 10.2 for overall width and length.

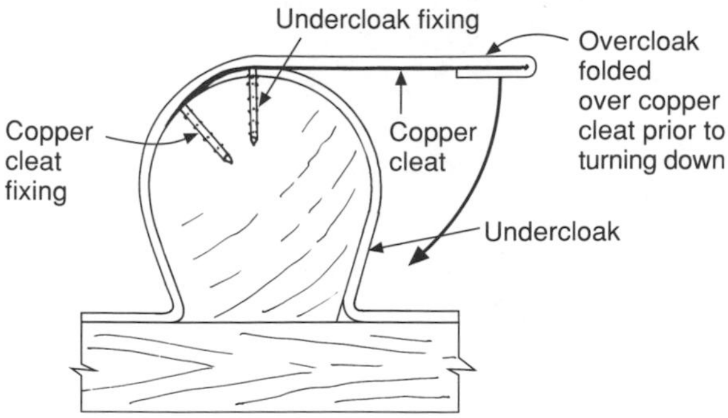


**Fig. 10.11** Alternative wood roll finish without splash lap.

## Wood rolls

The traditional method of forming lead sheet over wood rolls with splash laps is described in Book 1, Chapter 11. The purpose is to give the edge of an overcloak a greater degree of rigidity, preventing high winds lifting the lead. Rolls and splash laps are still widely used but the roll shown in Fig. 10.11 is an alternative when the appearance of the splash lap is undesirable, e.g. on or near vertical cladding. When using lead codes 4,5,6 the lead is folded over copper cleats fixed at intervals of 450 mm which resist the tendency of the edge to lift in high winds. When using codes 7,8 in moderately sheltered areas the copper cleats are unnecessary, as the thickness of the lead is considered sufficiently stiff to resist lifting. It is important that the fixings for the copper cleats do not penetrate the undercloak, as this will restrict its free movement.

### *Forming wood-cored rolls at abutments*

Rolls can be worked by the bossing process or by lead welding which is dealt with later in this chapter. Figure 10.12 shows the stages of bossing roll overcloaks at an abutment. This is a more difficult procedure than working undercloaks as the lead has to be worked all the way round the wooden roll. The formation of an undercloak is not a great deal more difficult than that of an external corner, and for this reason details are not included. During the final stages of working the overcloak the lead must be firmly held down with a wooden block as shown or it will lift and a hollow will form under the lead. This will cause the lead to be stretched, sometimes resulting in the formation of a hole when it is finally dressed to shape around the roll. The plumber's mate usually stands on the timber block during the whole operation, holding down the top of the partly formed roll with a piece of batten. This 'holding down' is essential and must be done during the final stages of driving the lead into the corner. This is shown in stage 4 and it must be carefully carried out to avoid splitting the lead. A lot of practice is required to produce an acceptable job. Many plumbers have specially shaped chase wedges for 'driving in', as this is difficult to accomplish with the ones normally supplied due to their rounded edge. The use of a well-greased drip plate is also essential during this operation to enable the lead to 'slide' easily over the undercloak when it is finally driven home.

The finished job should appear as shown in stage 5 with the lead driven home into the corner and trimmed prior to dressing out all the tool marks.

### *Roll ends at the eaves*

The stages of bossing the roll end are shown in Fig. 10.13. Some plumbers use a short-handled dummy over which the lead is bossed out. During this operation it is essential that the lead bay is not drawn forward leaving a gap at the abutment and this must be watched continually. As with the formation of the overcloak at the abutment, it is helpful to have a mate standing on a suitable timber in the bay to prevent this happening. Figure 10.14 illustrates the detail of a bossed roll intersection with a drip on a low-pitched roof. The rolls may be 'inline', or 'staggered' as shown in the inset. This is one of the most difficult bossing operations to do as lead has to be gained to successfully work in the corners shown at 'B'. For this reason the roll inline method is preferred as the lead bossed out of the roll end can be worked into areas 'B'.

## Hollow roll work

Hollow rolls are used on roofs covered with sheet lead having pitches upward of approximately 20°. These joints are easily damaged and must not be used in situations where foot traffic is anticipated. Joints across the flow of water are lapped, drips or welts being unnecessary on pitched roofs of this

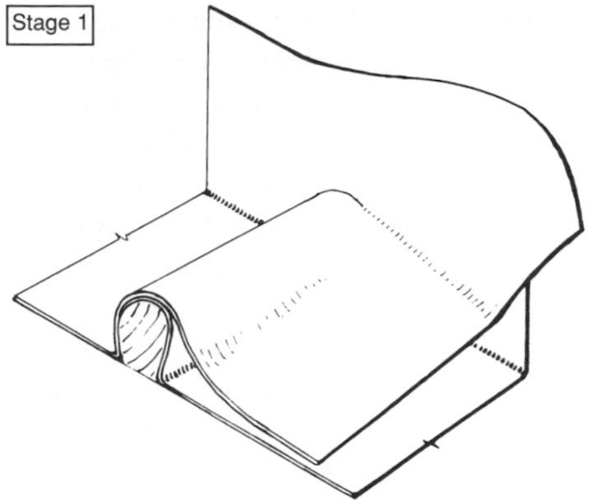


The lead is bossed to the height of the roll using the same technique as for an external corner. It must be worked to a little less than 90° to allow for the taper in the wood core.

Surplus lead from corner worked over the top of the roll. This will be needed when the roll is finally driven home.

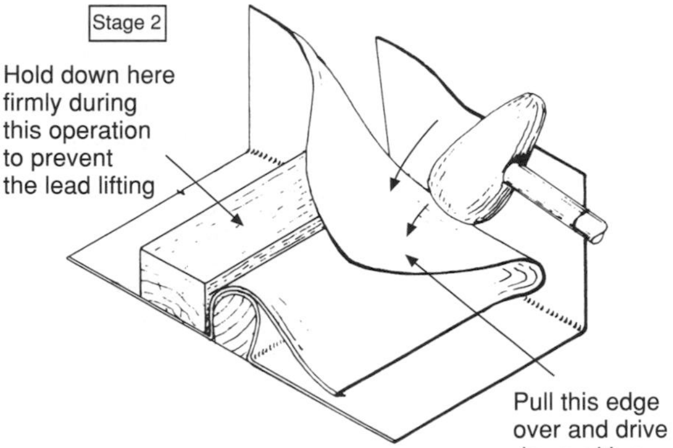


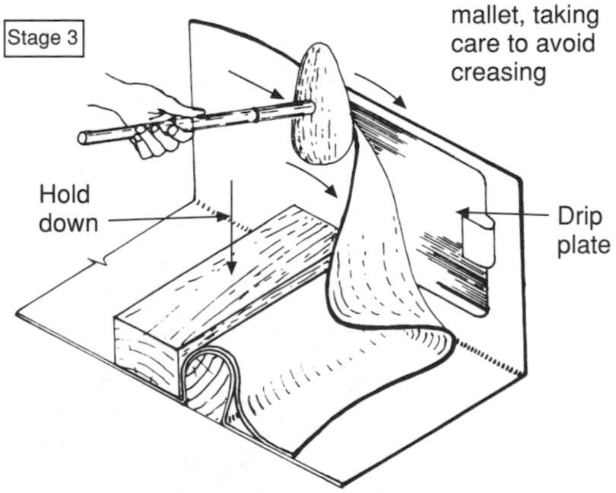


Insert a drip plate between the undercloak and overcloak and drive the lead along with a mallet as shown. It may be necessary to repeat this, and the operation shown in stage 2, two or three times prior to driving in as shown in stage 4.

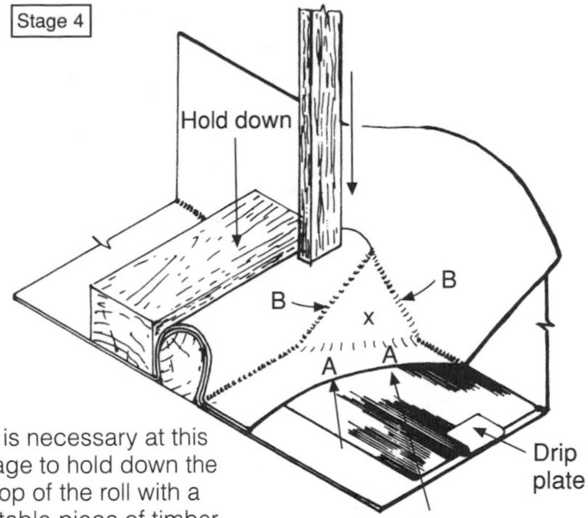


It is necessary at this stage to hold down the top of the roll with a suitable piece of timber.

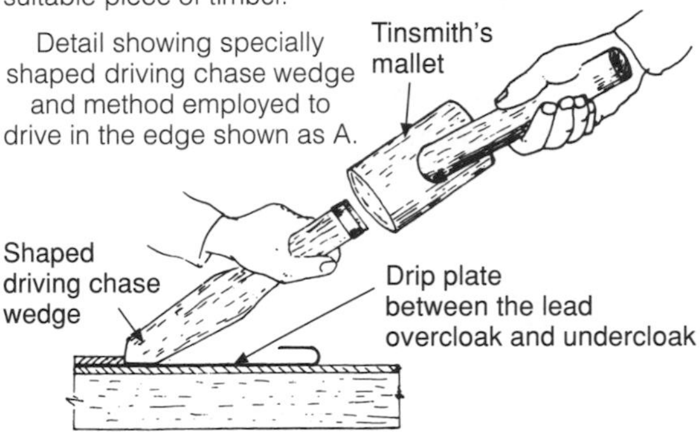


Detail showing specially shaped driving chase wedge and method employed to drive in the edge shown as A.

The final operation requires the triangle x to be worked into the corner with great care, by driving in the edge shown as A, triangle x is gradually diminished, chasing in at arrows B simultaneously. It is essential that the drip plate is well oiled or greased to enable the lead to slide over it. The lead must be worked all the way and any attempt to stretch it in the final stages will cause it to split.

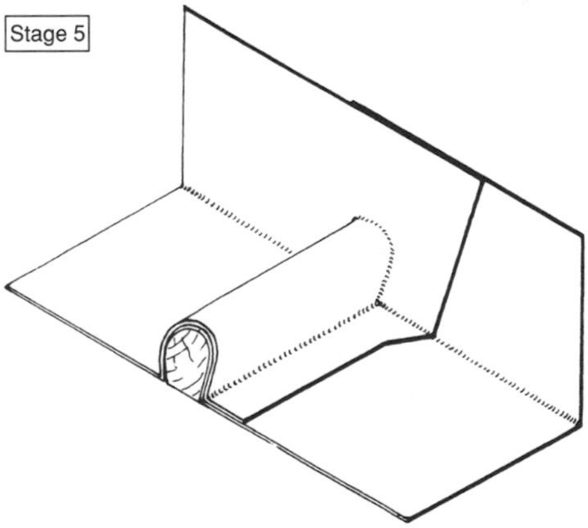


Showing the finished appearance with the lead worked neatly in position.

**Fig. 10.12** Bossed roll ends at abutments.

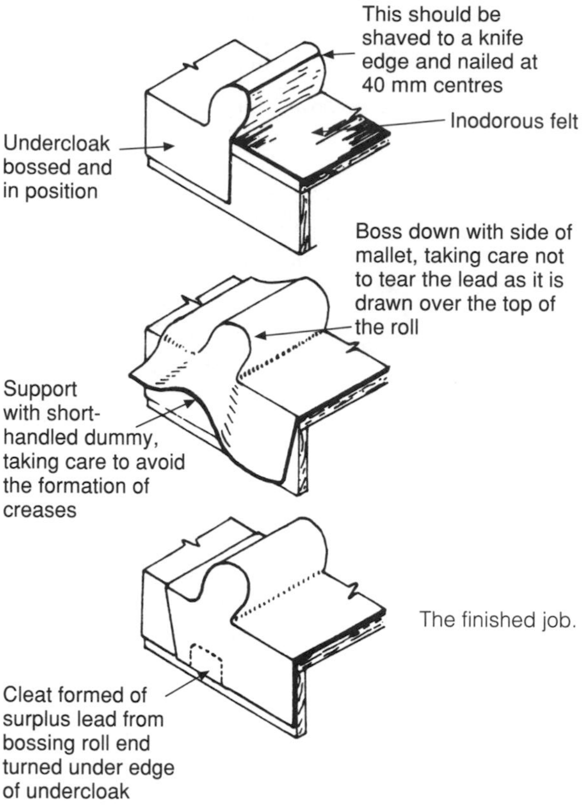
This should be shaved to a knife edge and nailed at 40 mm centres
Inodorous felt
Undercloak bossed and in position
Boss down with side of mallet, taking care not to tear the lead as it is drawn over the top of the roll
Support with short-handled dummy, taking care to avoid the formation of creases
The finished job.
Cleat formed of surplus lead from bossing roll end turned under edge of undercloak

**Fig. 10.13** Forming bossed roll ends.

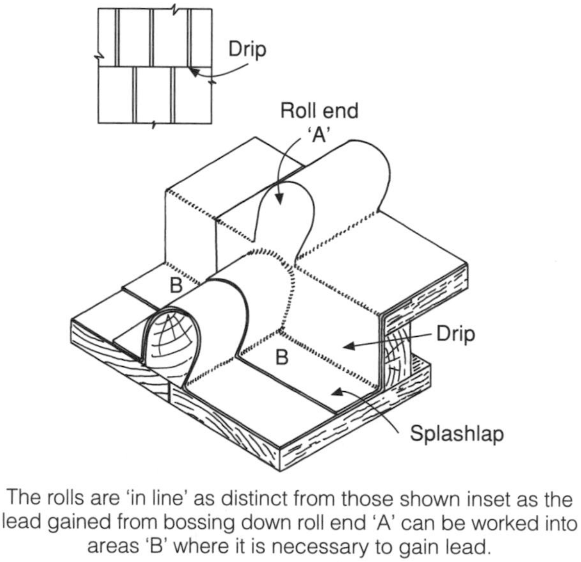
Drip
Roll end 'A'
B
B
Drip
Splashlap
The rolls are 'in line' as distinct from those shown inset as the lead gained from bossing down roll end 'A' can be worked into areas 'B' where it is necessary to gain lead.

**Fig. 10.14** Detail of bossed roll intersection on a low-pitched roof.

angle. Figure 10.15 illustrates the general method of weathering using this technique. The top edges of each sheet must be well secured by a double row of nailing to prevent it slipping or pulling away from its fixings. As this method of weathering is almost invariably confined to double-pitched sloping roofs, it is seldom necessary to form a hollow roll to an abutment. Where necessary, however, they can be bossed over a short length of wooden roll, using it as a former, or they can be lead welded using similar techniques to those used for wood-cored rolls. Details of these operations are not shown, but can be obtained from the LSA.

*Hollow roll finish at eaves and ridge*

One method employed to turn the roll over the eaves is to use a 32 mm bending spring which is

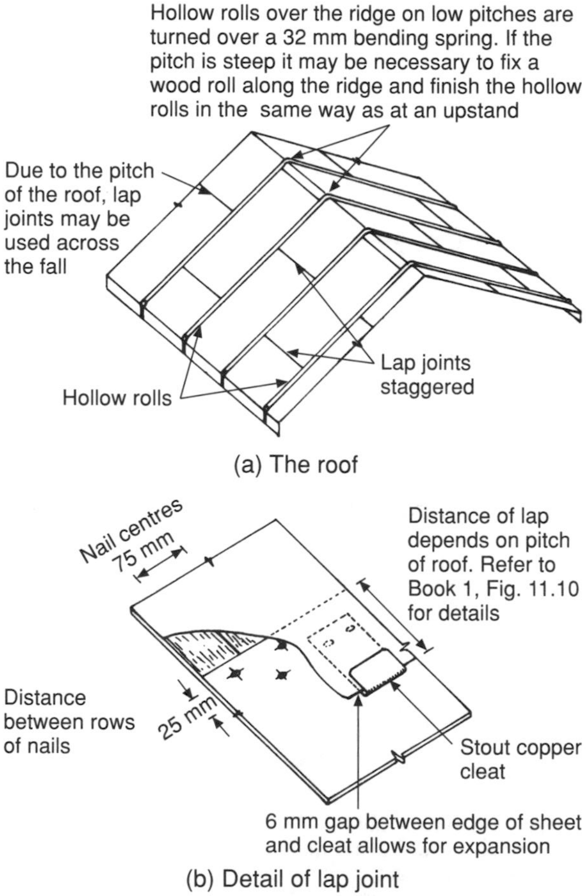
Hollow rolls over the ridge on low pitches are turned over a 32 mm bending spring. If the pitch is steep it may be necessary to fix a wood roll along the ridge and finish the hollow rolls in the same way as at an upstand
Due to the pitch of the roof, lap joints may be used across the fall
Hollow rolls
Lap joints staggered
(a) The roof
Nail centres 75 mm
Distance of lap depends on pitch of roof. Refer to Book 1, Fig. 11.10 for details
Distance between rows of nails
25 mm
Stout copper cleat
6 mm gap between edge of sheet and cleat allows for expansion
(b) Detail of lap joint

**Fig. 10.15** Double-pitched lead-covered roof using hollow roll jointing.

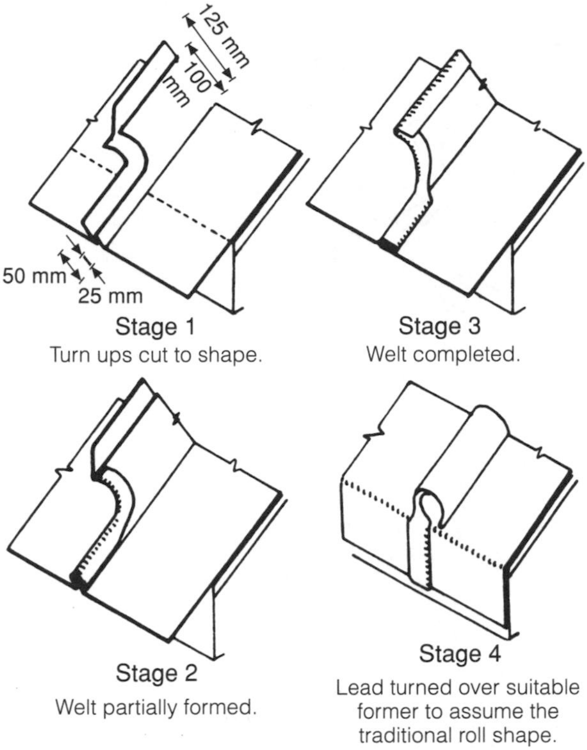


**Fig. 10.16** Stages in forming hollow roll ends using the welted method.

inserted into the roll and supports it as it is turned. The same method is used to turn it over the ridge where applicable, but if an angle of more than 30° is required in either case it will be difficult to remove the spring. An alternative method of finishing the roll at the eaves, by means of a welt, is illustrated in Fig. 10.16 and is both simple and effective. Another alternative is to weld 'in' a cover to the end of the roll.

### Dormer windows

These are used to give natural light to rooms built into a roofspace as in chalets or bunaglows. There are three main types of dormer: external, internal and partially internal. The external type is the most common as it generally affords more light and often allows for the enlargement of the room in which it is situated. The reasons for using the other types of dormers are usually aesthetic as they may enhance the appearance of the building. For example, partially internal dormers are often used with mansard roofs and can give a very pleasing effect. The three types of dormer are shown in Fig. 10.17.

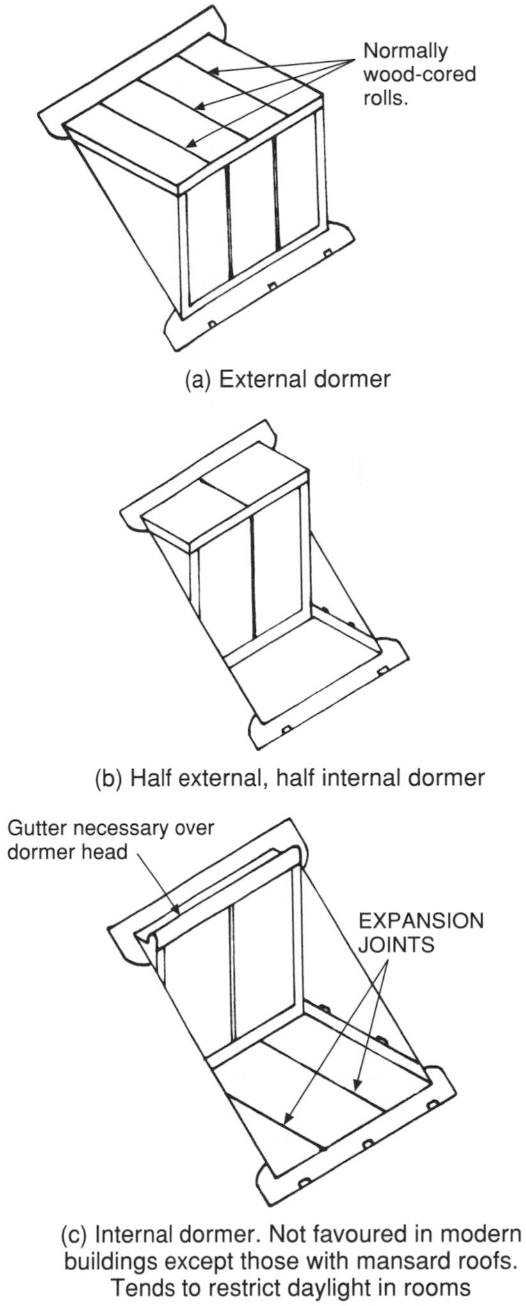


**Fig. 10.17** Types of dormer window.

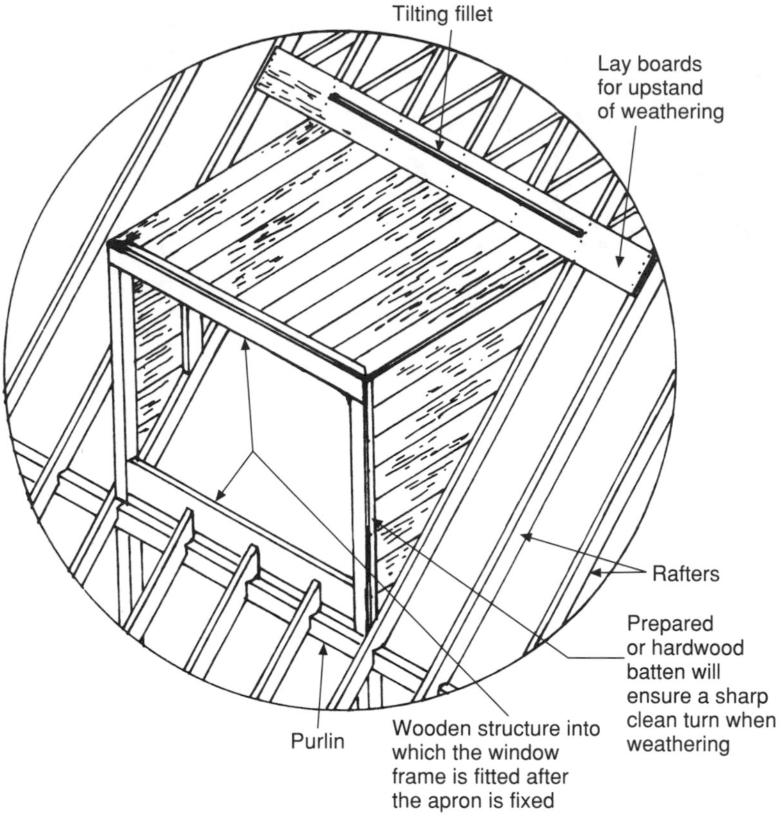


**Fig. 10.18** Wooden frame for a dormer prior to fixing apron and window frame.

As external dormers are the most commonly encountered, it is this type that will be discussed in detail, although the basic principles of weathering all types of dormers are very similar. Tile hanging (roof tiles fixed vertically) or shiplap boarding is sometimes used to weather the cheeks (the triangular sides of the dormer) and in such cases the roof tiles are weathered to the dormer cheek by means of soakers or cover flashing.

*Dormer weatherings*

It is an unfortunate fact that some builders, to save expense, often use the window frame as a structural part of the dormer. This usually involves nailing the frame through the apron, allowing water to penetrate the roof. The implications of this are not always immediately apparent, but in the long term it can result in serious timber defects which are expensive to correct. It is essential that a timber frame forming the dormer is first constructed by the carpenter, into which the window frame is fitted after the apron has been fixed. Figure 10.18 illustrates the wooden structure forming the dormer prior to fixing any weathering.

The bottom of the frame must be weathered first by means of the dormer apron. Figure 10.19 illustrates both the timber framework of a dormer and a section through the apron showing how the sill of the window frame should be fixed to avoid nailing through the apron. An alternative method that can be used with a sill having no groove is to use metal cleats screwed to the window frame and the dormer trimmer. They should be made of galvanised steel, but short lengths of 25 mm copper tube flattened to form a cleat provide a useful substitute.

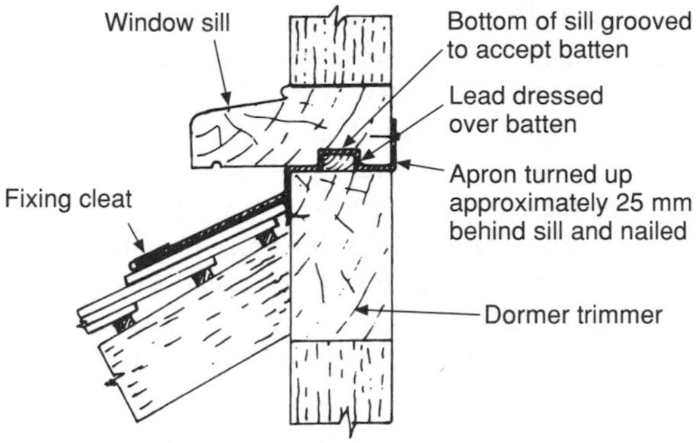


(a) Section through dormer apron

Shows how the window sill should be secured to the dormer frame without resorting to nailing through the weathering.

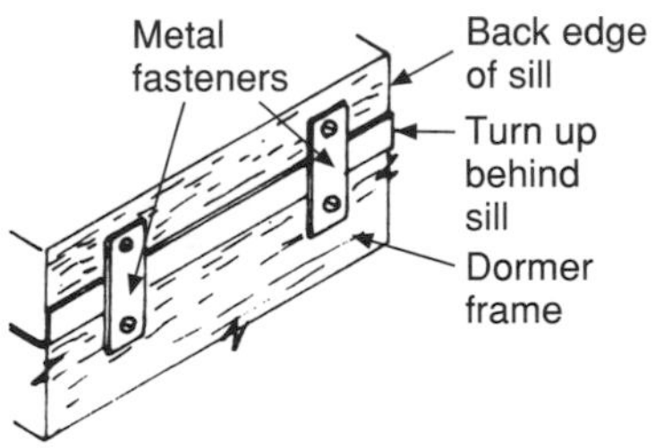


(b) An alternative method of securing the window frame to the dormer frame avoiding the necessity of grooving the underside of the sill

**Fig. 10.19** Dormer sill details.

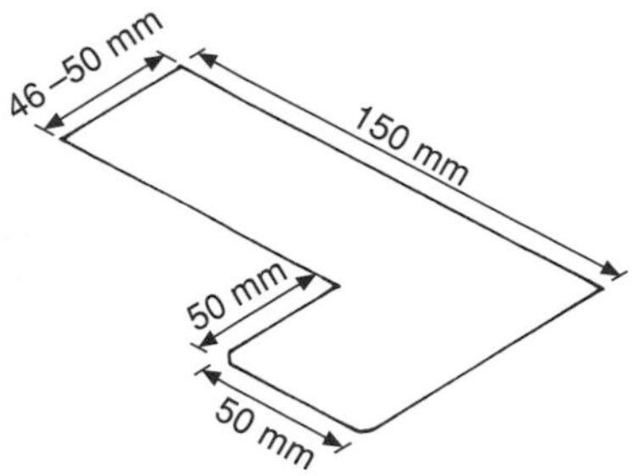


Setting out (approximate measurements only)

(a) Fixing the front edge of dormer apron: setting out

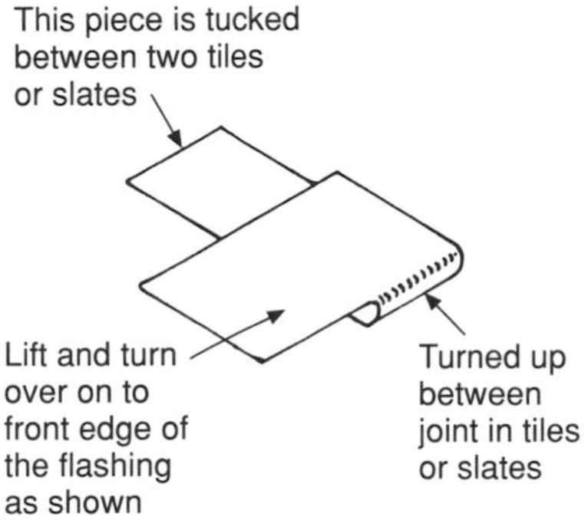


(b) Cleat turned prior to fixing

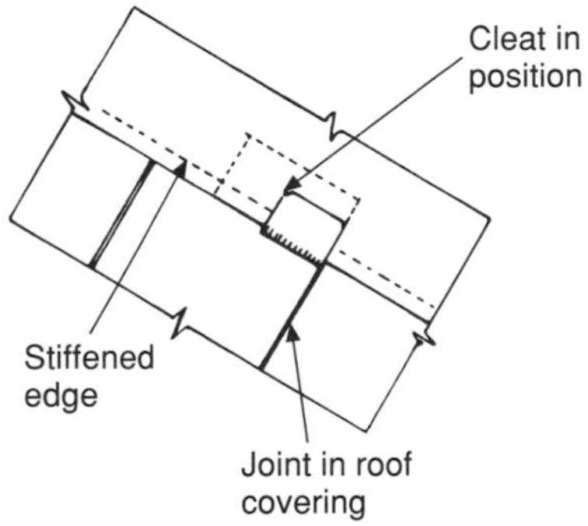


(c)

**Fig. 10.20** L-shaped cleat for securing the free edges of aprons.

It is important to secure aprons for two reasons: (a) to prevent creep and (b) to prevent high winds lifting the free edge. The fixing cleat nailed to the wooden frame or turned over a tile batten will prevent creep, and the L-shaped type shown in Fig. 10.20 secures the front edge. One end of the dormer apron is shown in Fig. 10.21 and it will be seen that it weathers not only the underside of the wooden sill but also the corner post. This is important as any seepage of water into the building at this point, although it may not show, will cause rapid deterioration of the wooden structure.

*Dormer cheeks*

The cheeks on small dormers may be fixed in one piece, providing the maximum superficial area does not exceed approximately 0.5 m$^2$. Where larger dormers are required the cheeks should be fixed within the limitations of those shown in Fig. 10.22. For larger areas a greater number of welts and laps will obviously be necessary. The method of fixing the top edge of the cheek is shown in Fig. 10.23 where copper or stainless steel nails are used as

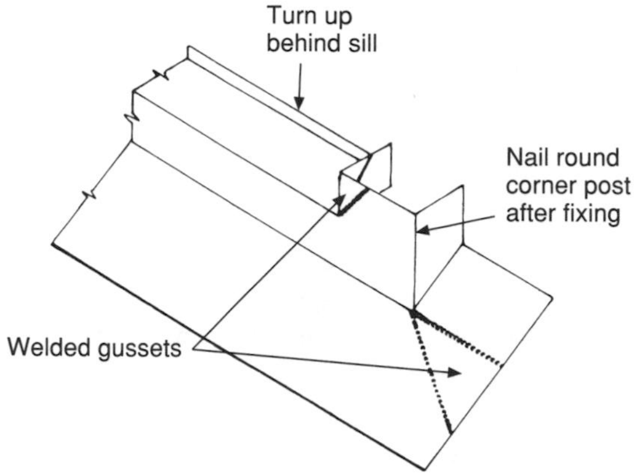


**Fig. 10.21** Apron detail round corner post.

prescribed. Three methods of securing the lower edges of any form of lead cladding including dormer cheeks are shown in Fig. 10.24. However, the most common method employed for dormers is to form an 'S' cleat, tucking one end behind the soakers, the other turned to secure the free edge of the cheek. Figure 10.25 illustrates a method of securing large vertical (or near vertical) sheets, including dormer cheeks, by means of a concealed copper cleat worked into the welt. They should be spaced at 450 mm intervals. Rigid fixings in the centre of the sheet, often called intermediate fixings, are avoided where possible due to the restriction on the movement of the lead. The

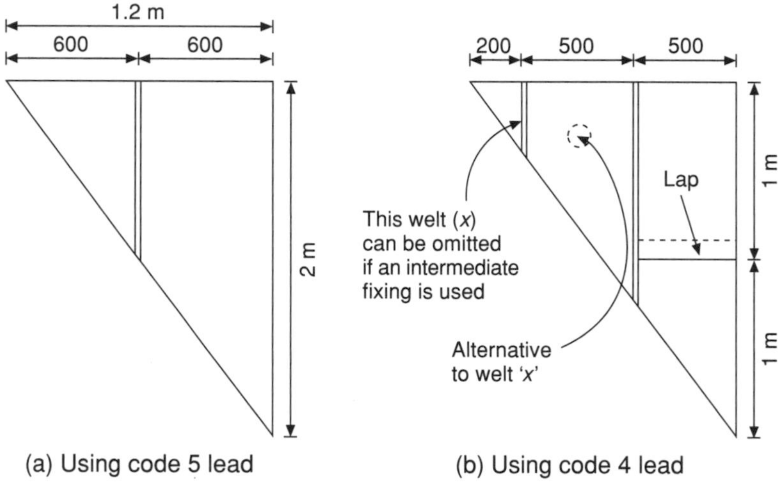


**Fig. 10.22** Dormer cheek cladding: recommended sizes.

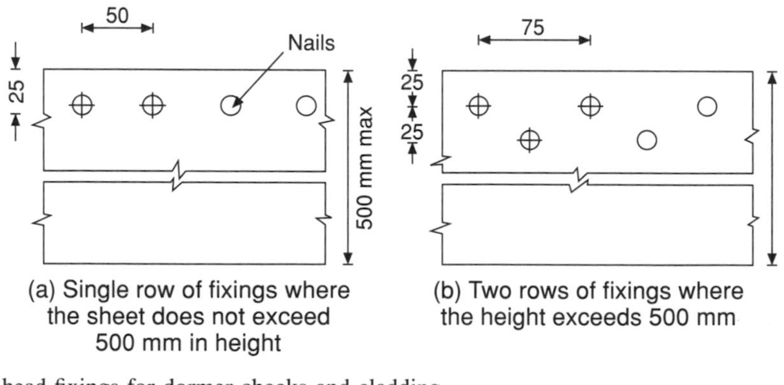


**Fig. 10.23** Top or head fixings for dormer cheeks and cladding.

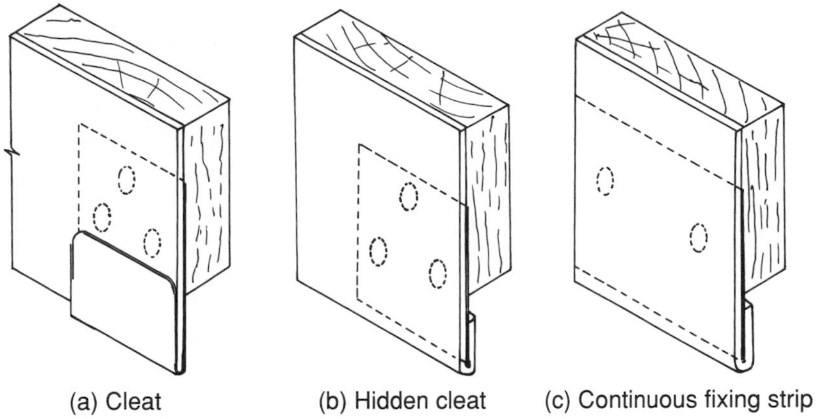


**Fig. 10.24** Methods of clipping the bottom edge of lead weatherings.

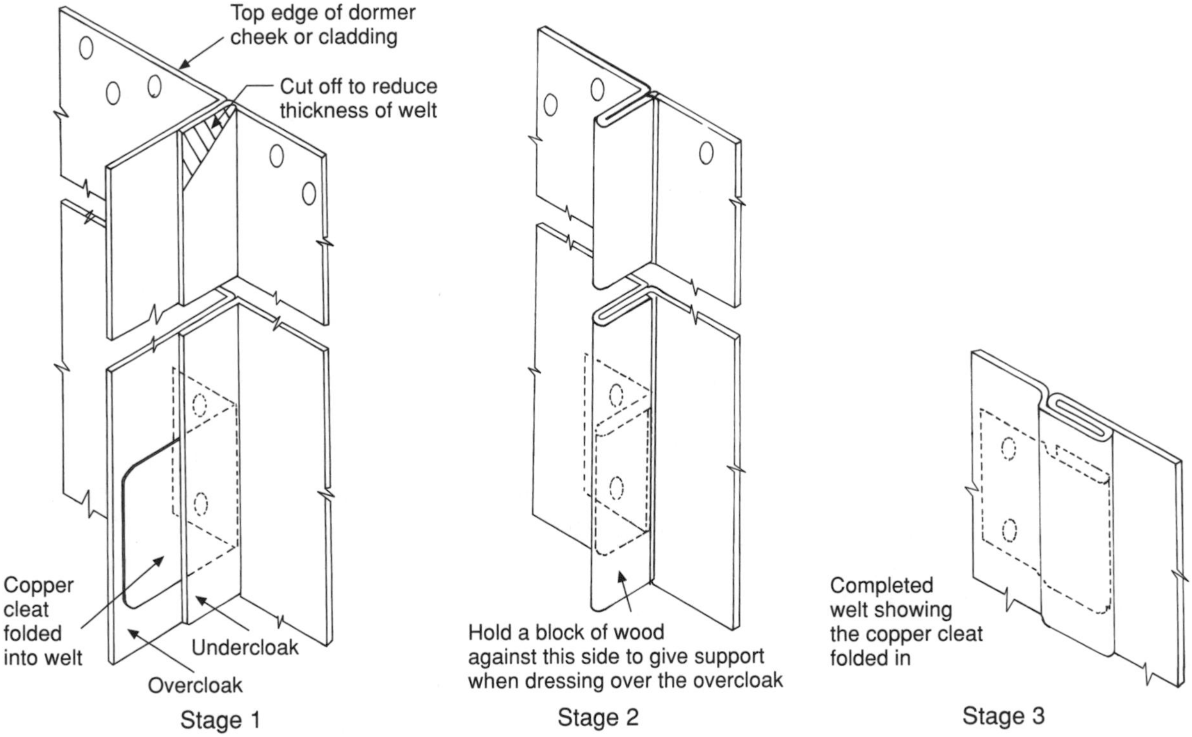


**Fig. 10.25** Details of vertical welted joints.

traditional solder dot is typical of this undesirable method of fixing. The method shown in Fig. 10.26 is far more satisfactory and providing the screw is not too tight, the elongated hole in the lead under the washer permits sufficient freedom of movement.

The sides of the cheeks are turned round the dormer frame so that the joint between this and the window frame can be weathered. The lead can then be nailed to the frame and finished by covering with timber moulding or allowing enough material to turn back onto itself, concealing the nails as shown

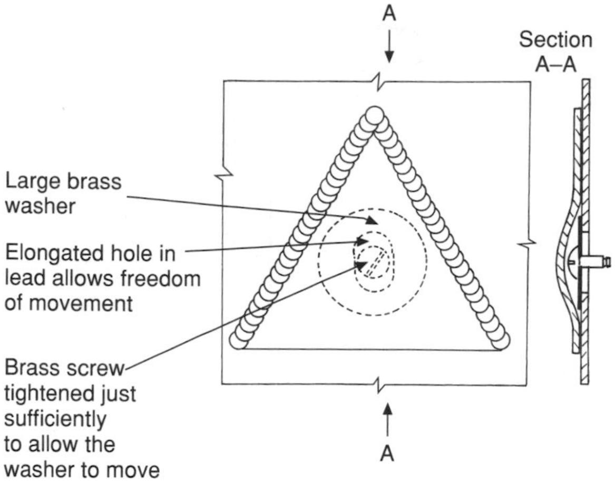


**Fig. 10.26** Fixing a lead sheet using a brass screw.

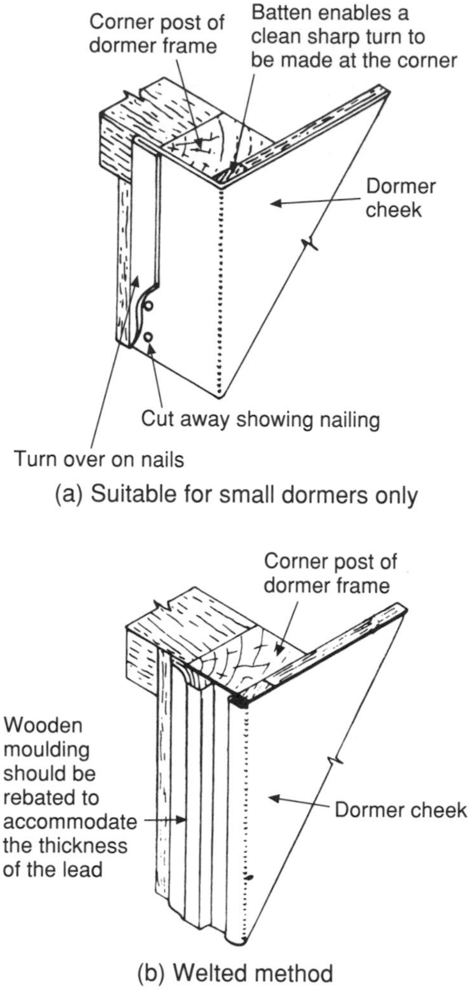


**Fig. 10.27** Methods of finishing the cheek to the corner post.

in Fig. 10.27. An alternative method using a welt to secure the facing lead to the cheek is preferable (Fig. 10.27(b)). Although the welt is shown on the front it is often more convenient to turn it back onto the cheek. The wooden moulding shown is not used where the lead is turned completely round the post.

*The right-angled triangle* The reader will see that the cheeks of a dormer have the form of a right-angled triangle, and it is useful to be able to calculate the length of the slope C (called the *hypotenuse*) between the two known lengths of the right angle A–B (see Fig. 10.28).

The following will show how measurement C can be determined. If a number is multiplied by itself it is said to be squared, e.g. $4 \times 4 = 16$. Another way of writing this is $4^2 = 16$. The number 4 is said to be the square root of 16. The square root of number 36 is 6, and in this case the square root of the number is said to have been extracted. It can be indicated thus:

$$6 = \sqrt{36}$$

This symbol over a number indicates that its square root must be found. If the foregoing is understood it is not a big step to determine the length of C in the right-angled triangle using the following formula:

$$C = \sqrt{A^2 + B^2}$$

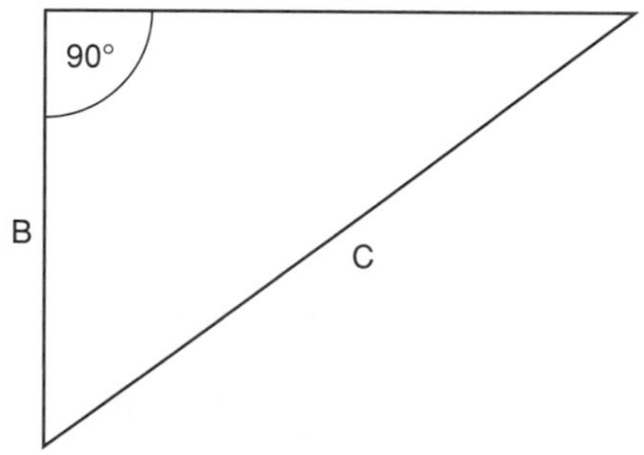


**Fig. 10.28** The right-angled triangle.

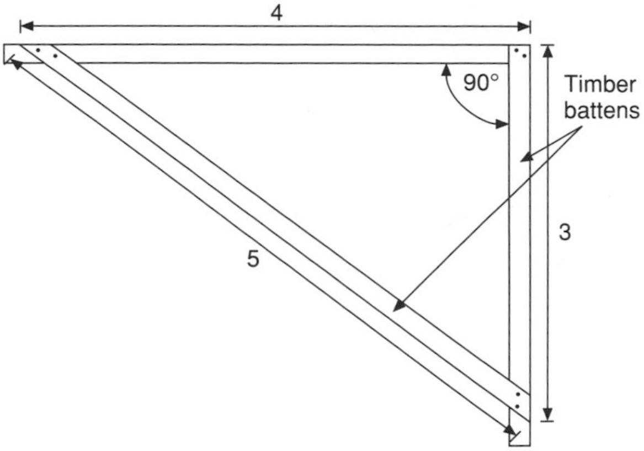


**Fig. 10.29** Construction of a site set square.

If the alphabetical symbols are given a numerical value of 3 and 4 respectively, the equation will become:

$$C = \sqrt{3^2 + 4^2}$$

The numerical value of C can now be ascertained by first squaring 3 and 4 thus:

$$C = \sqrt{9 + 16}$$
$$= \sqrt{25}$$

All that remains is to extract the square root of 25, which can be seen at a glance to be 5; thus C, the unknown side of the original triangle, will be 5. It is a fact that when the ratio of the measurements of a right-angled triangle are 3, 4, 5 they form a 90° angle, often called a *perfect square*. This knowledge is very useful as it enables one to remember how to determine the hypotenuse of a right-angled triangle, and also how to set out 90° angles. If a large square is required on site it can be made with three pieces of batten as shown in Fig. 10.29. This principle is often called the *3, 4, 5 rule* in the building industry and is known to have been used by builders in ancient civilisations. It must be said that in practice the ratios of a right-angled triangle are seldom 3, 4, 5. These numbers have been used because their square roots can be seen at a glance; more difficult examples can easily be ascertained using a calculator.

### *Dormer tops*

Figure 10.30 shows one side of a small dormer with all the components so far discussed in position.

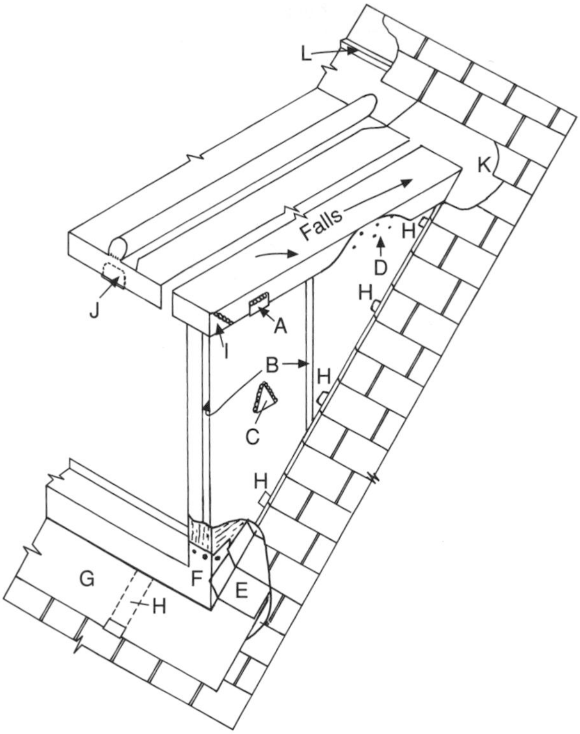


Key to details
A Cleat welded to turn down; the methods shown in Fig. 10.24 are an alternative
B Welts
C Intermediate fixing
D Head fixings
E Soakers
F Apron fixing detail
G Apron
H Cleats, cheeks and apron
I Corner welded or bossed
J Surplus lead from bossed overcloak turned under the undercloak
K Roof coverings cut away allowing water to clear
L Tilting fillet

**Fig. 10.30** Small dormer window showing the main details. The falls shown are suitable only if a single roll is provided.

It will be seen that the top of the dormer is similar to a small flat roof or canopy and the procedure for weathering is exactly the same. Whether the dormer top falls to the front or back depends on several circumstances. If the dormer cheeks are finished with tile hanging or shiplap boards, the usual procedure is to provide a fascia board and gutter as for the eaves of a building, as shown in Fig. 10.31. In such cases it is usual to arrange for the dormer top to fall towards the front so that water can be collected in a normal eaves gutter. A roll or batten is often fixed to the edge of the

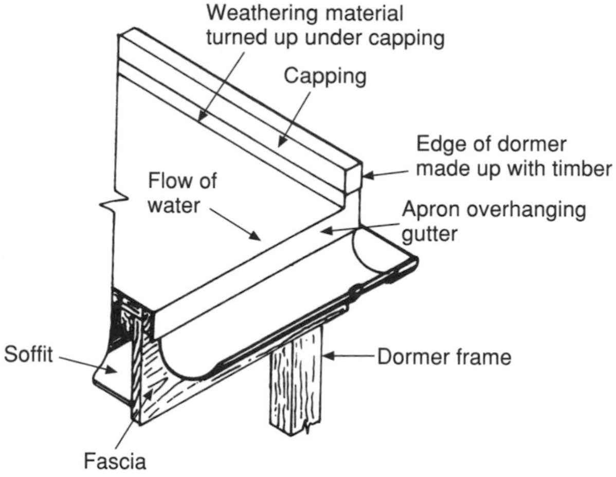


**Fig. 10.31** Finishing the verge of a dormer with tile-hung or shiplap boarded cheeks.

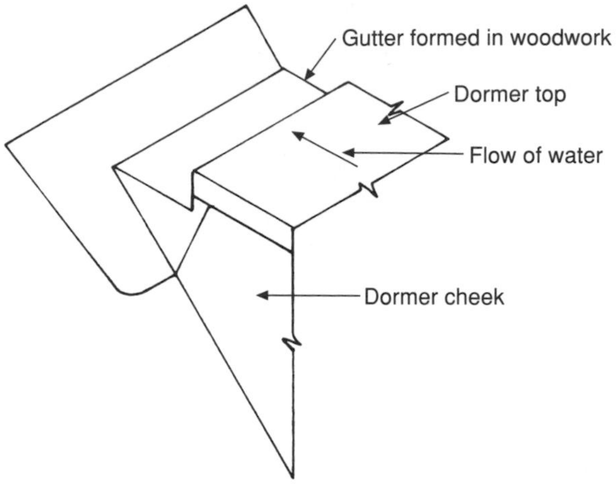


**Fig. 10.32** Dormer gutters. Where it is undesirable to discharge the flow of water into an eaves gutter on the front of the dormer, a proper gutter must be constructed and weathered in a similar way to a chimney back gutter.

dormer top to avoid the necessity of fixing gutters all round. In some buildings the use of eaves gutters is avoided as they require a certain amount of maintenance, and in cases where the metal roofing and cladding are features, their appearance would be objectionable. In these circumstances it is necessary to arrange for the dormer top to fall to the back to avoid water from the roof discharging over the windows. If the dormer is a small one with not more than one longitudinal joint in the roof covering, the fall can be arranged so that the water runs off the edges, but if there is more than one longitudinal joint, a gutter will be necessary at the intersection of the dormer top and the roof (see Fig. 10.32).

### Mansard roofs

The mansard roof is sometimes called a double-pitched roof, and it is thought that this type of structure originated in Holland. Mansard roofs can also be constructed with a flat top (see Fig. 10.33(a)), although these are uncommon in the United Kingdom. The double-pitched type is illustrated in Fig. 10.33(b) from which it will be seen that it provides more living space than one having a single pitch. When hand-made clay tiles were common it was sometimes possible to pick out those having more curve than others so that the angle between the two pitches could be weathered completely without flashing. With mass-produced tiles the curves are all identical and the point where the two pitches intersect must be weathered with a flashing as shown in Fig. 10.34. The flat-topped mansard roof (see Fig. 10.33(a)) must be weathered

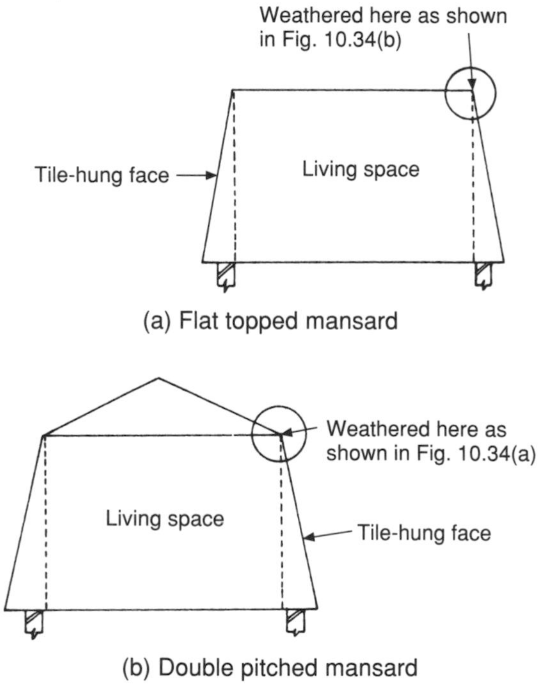


**Fig. 10.33** Mansard roofs.

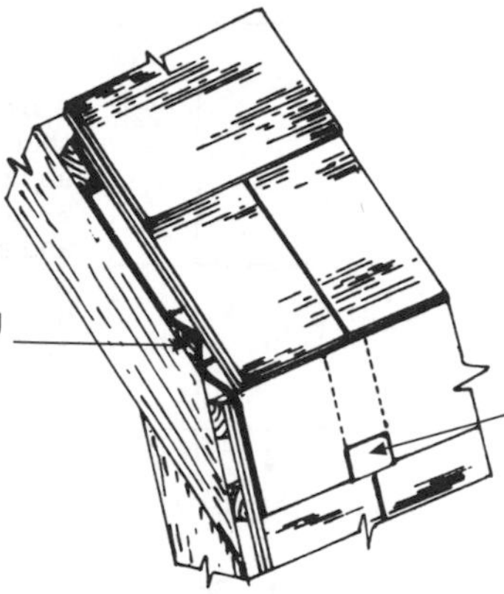


(a) Weathering the roof covering of a double-pitched mansard roof

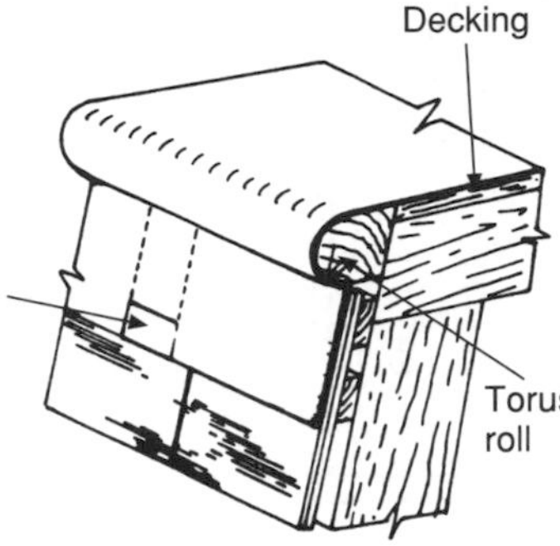


(b) Flat-topped mansard roof weatherings

The torus roll shown here is a common method of finishing the edge when lead sheet is used.

**Fig. 10.34**

like a flat roof, using traditional materials or asphalt. If the latter is chosen, it will still be necessary to use a metal edging strip, similar to that shown and having a doubled back edge left open to provide a key for the asphalt.

### Lead-welded details

Due to its very low melting point, lead can be easily welded to fabricate roofing components, both on and off the site, the latter enabling much of the work to be prepared in favourable conditions. A good example of this is where a number of chimney or pipe flashings are to be made. Most of the work can be done out of position in one operation which saves a considerable amount of time. Lead components do not need to be tailor-made quite so precisely for a specific piece of work and can be eased one way or the other to suit; for example, a number of chimneys which may have slight variations in size.

Chapter 1 refers to the two main techniques of lead welding; here it is shown how they are applied. These simplify some of the traditional methods, and the increased use of welding techniques reduces the necessity for many labour-intensive bossing operations. It should be noted, however, that not all bossed details take long and many experienced lead workers prefer bossing to lead welding for such details as undercloaks to rolls and drips and working down a roll and into a gutter or onto a fascia board. Conversely, most would agree that some details can be achieved more economically using welding. It must be stressed that the use of welding requires absolute competence to avoid the early failure of any joints.

Where possible welded joints should be made off the job where the most convenient position can be used and to avoid fire risks. If the joint is to be made in position use a lap joint, as the welding flame does not penetrate the lead. If this is not possible soak any woodwork with water or a flame retardant solution prior to welding. Always carry out any positional welding early in the working day so that plenty of time is left on completion to carry out a thorough inspection. The following illustrations show some of the techniques employed in the fabrication of lead gutters and flats using wood-covered rolls.

#### *Setting out for welded joints*

It will be seen that some careful setting out is necessary when forming the details for welding. It is recommended that the reader practise on a piece of card, cutting with scissors and folding to make the details shown. It is a practical operation and worth knowing that card can be usefully employed for making templates for repetitive work. The following illustrations, which supplement those in Book 1, are not exhaustive, but they show a typical selection of components that can easily be fabricated using the lead welding process.

Figure 10.35 illustrates how to set out and form a drip in a box gutter so that all the welding can be

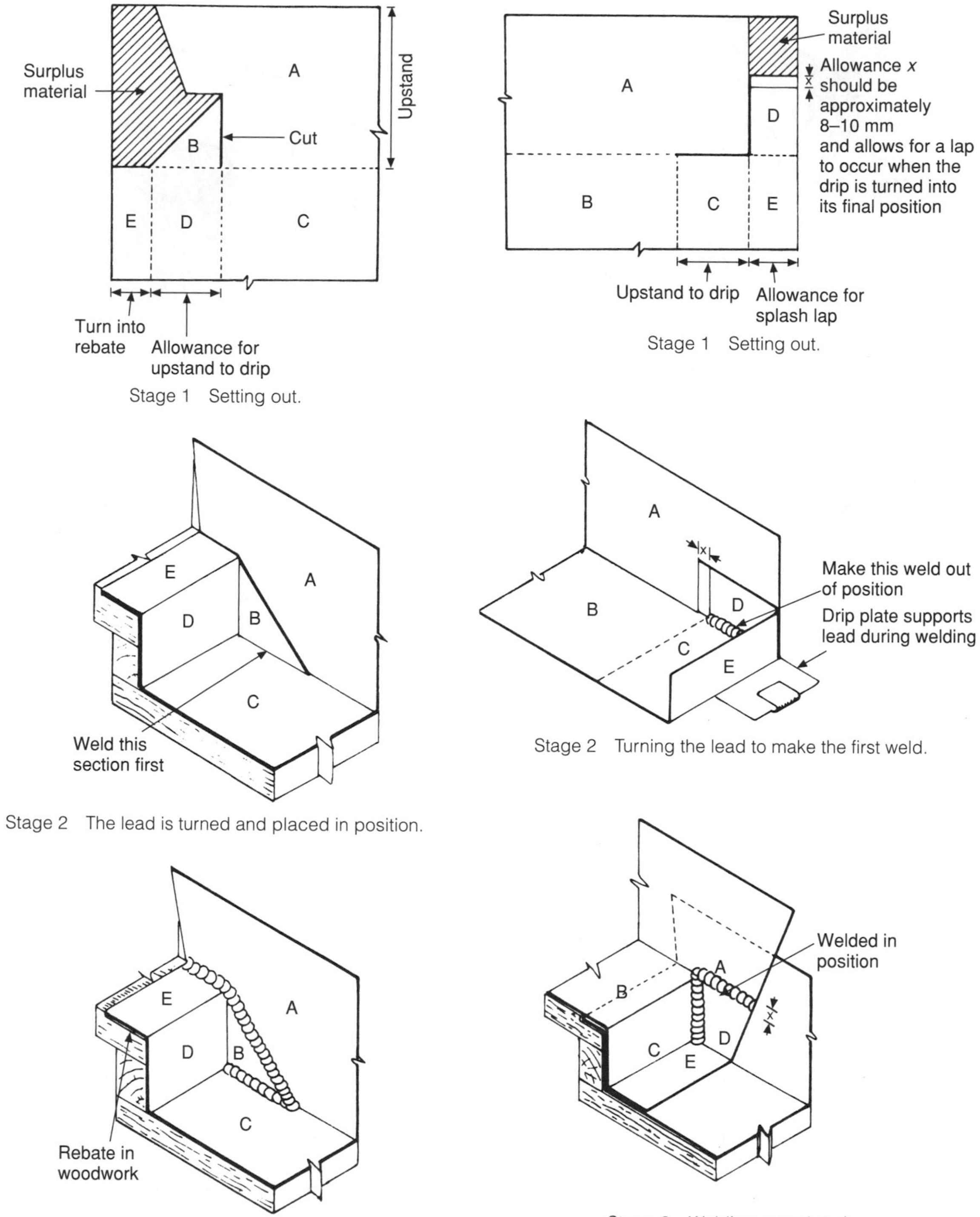


Stage 1 Setting out.

Stage 2 The lead is turned and placed in position.

Stage 3 All joints lead welded.

(a) Welded drip undercloak

Stage 1 Setting out.

Stage 2 Turning the lead to make the first weld.

Stage 3 Welding completed.

(b) Welded drip overcloak

**Fig. 10.35** Lead-welded drips in box gutters.

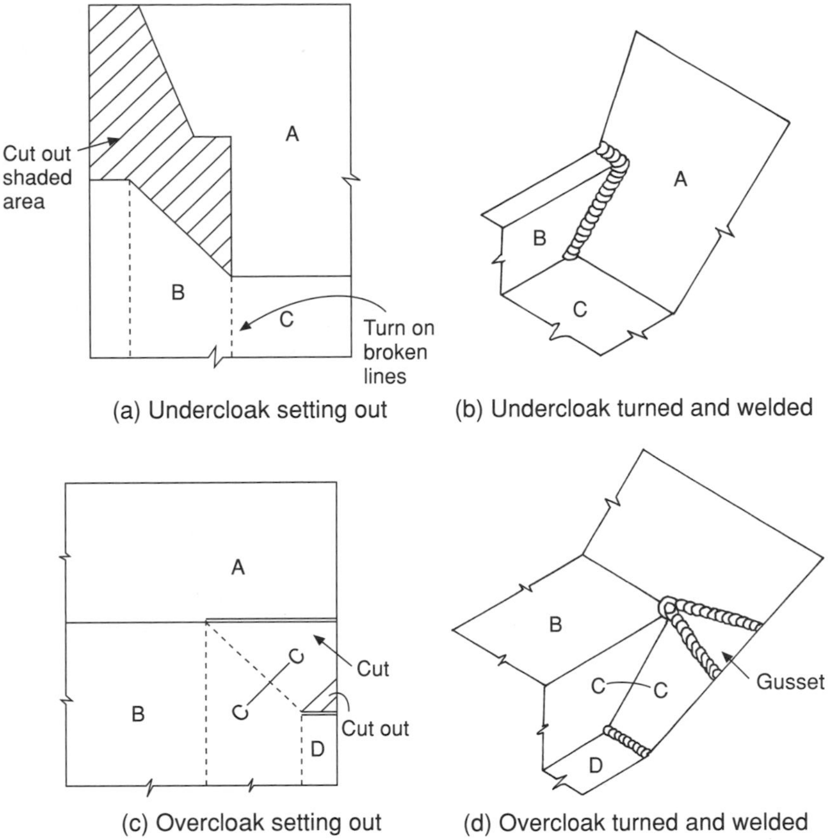


**Fig. 10.36** Welded drips in tapering gutters.

done with the lead in position. Figure 10.36 shows a similar arrangement for a tapering gutter. In this case, due to possible fire risks, the undercloak would be better welded out of position. The method of fabricating lead rolls at an abutment is shown in Fig. 10.37; both the undercloak and overcloak can be welded out of position. As with drips, the undercloaks can often be bossed up as quickly as they can be welded. A disadvantage with welded undercloaks is that they tend to be bulky and can make it difficult to achieve a neat finish on the overcloak.

A method of weathering rolls using lead welding, where they intersect with a drip, is shown in Fig. 10.38. The sequence of laying the lead is numbered 1–4. Both the undercloak (1) and the overcloak (2) can be bossed or welded using the technique shown in Fig. 10.37. Undercloak (3) is then laid and finally the overcloak (4). The detail for fabricating by welding (‘*x*’) is shown in Fig. 10.37(b); this is best done in position. The curved position can be welded by supporting it on a drip plate prior to turning it down so the two edges (‘*y*’) can be welded. As these are lap joints any fire risks are minimised.

The roll end at the eaves is not difficult to boss and, again, many plumbers prefer this method to welding. Figures 10.39(a) and (b) illustrate two methods commonly used where lead welding techniques are employed. Note the drip plate used to support the lead during the welding operation. As both of these methods employ the butt welding process there is a fire risk if they are welded in position. One method of overcoming this is to form the roll end, then draw the bay forward and support the area of welding with an offcut of the wood roll. When the welding has been completed, the bay can be pushed back and dressed down.

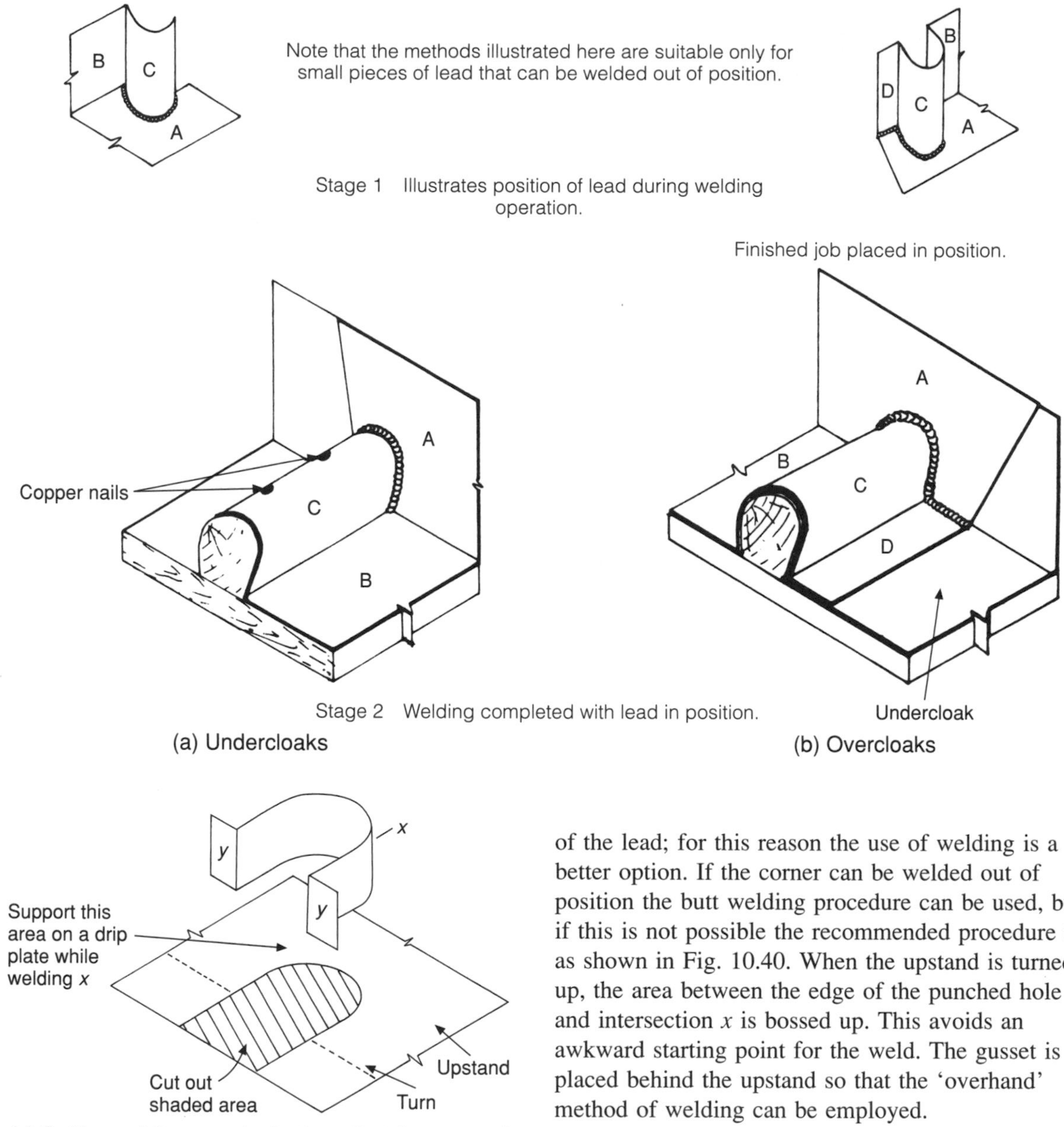


(c) Setting out the overcloak of a roll at the upstand to a drip or abutment using welding techniques
See text for details of procedures.

**Fig. 10.37** Lead welding wood-cored rolls.

*Internal corners*

These are time-consuming if bossed, and unless carried out with some understanding of the principles involved bossing can result in thinning of the lead; for this reason the use of welding is a better option. If the corner can be welded out of position the butt welding procedure can be used, but if this is not possible the recommended procedure is as shown in Fig. 10.40. When the upstand is turned up, the area between the edge of the punched hole and intersection *x* is bossed up. This avoids an awkward starting point for the weld. The gusset is placed behind the upstand so that the 'overhand' method of welding can be employed.

*Welded cleats*

Where copper cleats or nails are welded to a section of lead, they must be coated with solder — the plumbers term for this is 'tinning'. Bear in mind that copper has a higher conductivity than lead and should be heated to near welding temperature before the weld is attempted. Some practice will be necessary before this operation can be carried out successfully.

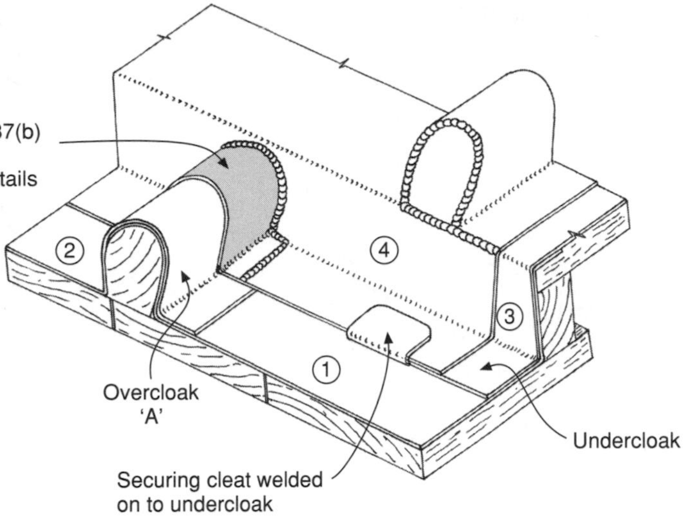


**Fig. 10.38** Sequence of laying lead-welded overcloaks at a drip or abutment and the roll end at the eaves or drip edge.

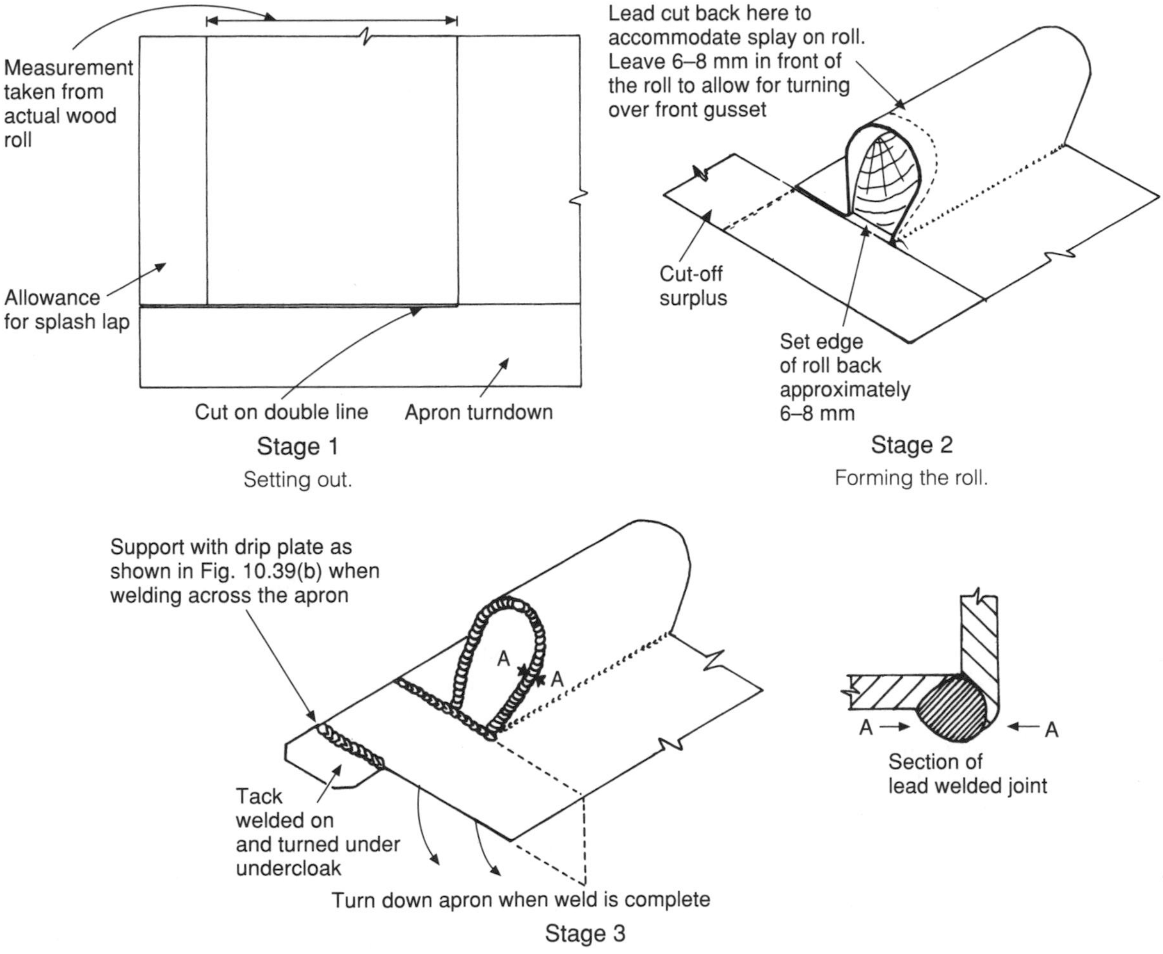


**Fig. 10.39(a)** Lead welding at roll ends at eaves.

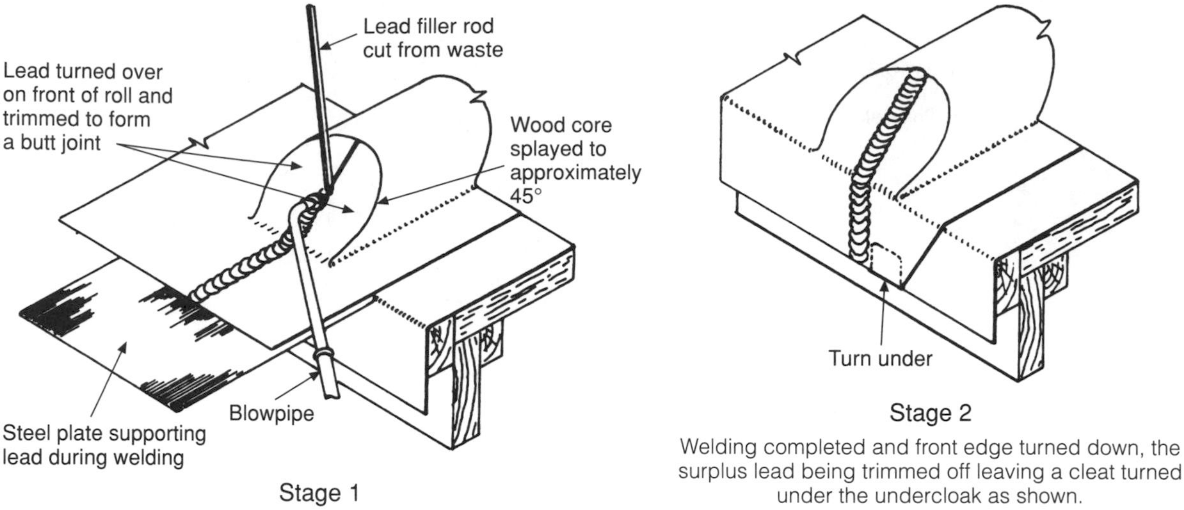


**Fig. 10.39(b)** Lead welding roll ends at eaves.

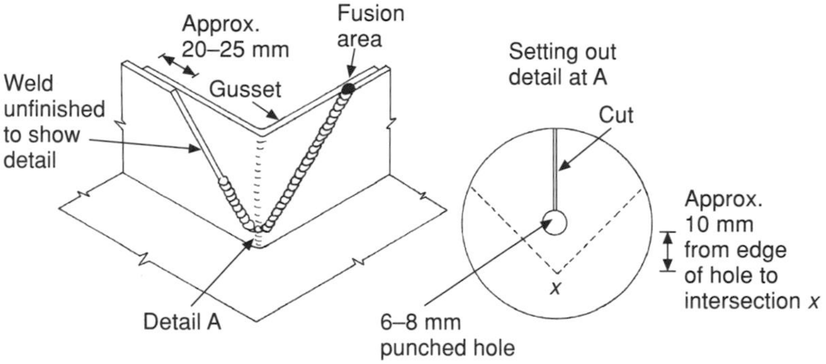


**Fig. 10.40** Fabricating an internal corner.

### Metal cladding of buildings with preformed panels

The term is generally understood to mean the covering of vertical surfaces of buildings as distinct from the weathering of sloping or flat surfaces. The technique has been used in Scandinavian countries for many years where the buildings are constructed of wood with a cladding of copper or aluminium for protection. Similar techniques have become popular in this country in recent years, not so much due to necessity, but to provide architectural features.

All the materials used for weathering buildings can be used for cladding, but one of the most common is lead, this being due to its good weathering qualities and to the ease with which it can be bossed or welded. Both copper and aluminium sheet are also used extensively for this purpose, and are often supplied in preformed panels which are usually fixed by specialists employed by the manufacturer. Lead cladding is more flexible because it can be fabricated, if necessary, on site by the plumber using traditional methods. Information about cladding with sheet lead is obtainable from the Lead Sheet Association who will advise on

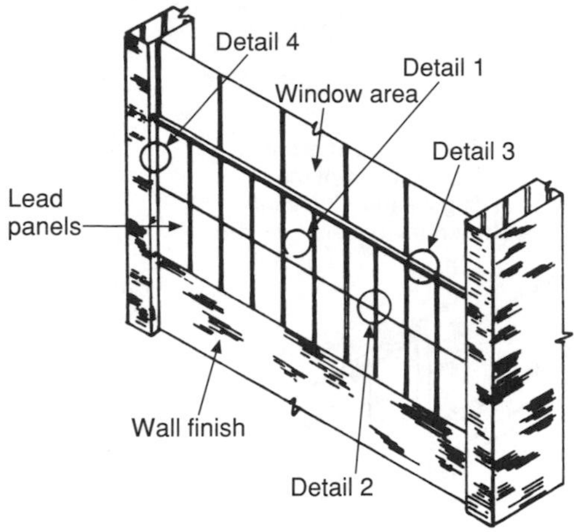


This illustrates part of a typical modern building and shows how lead-weathered panels are used to cover areas traditionally constructed with other materials.

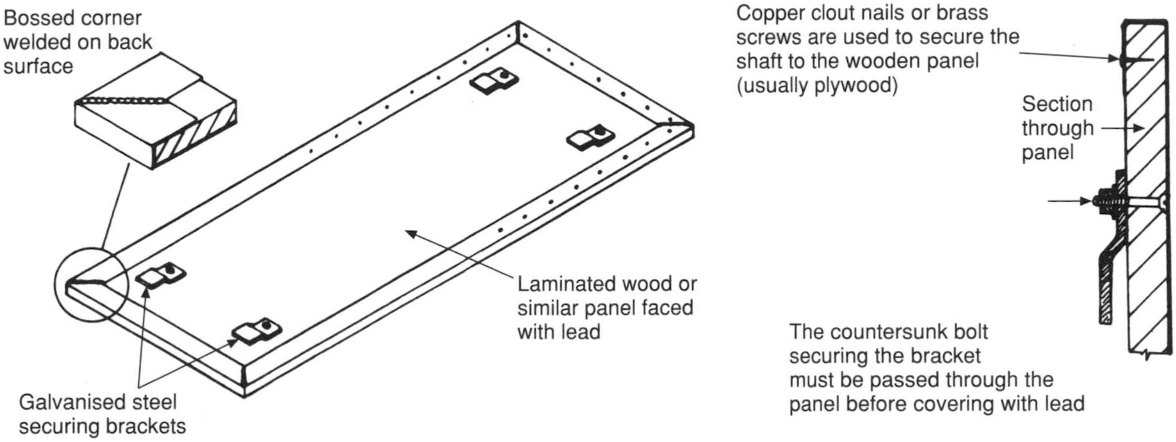


Detail 1 The fabrication of the panels

Note that provision is made only to secure the top half of the lead covering to the panel, which allows for thermal movement.

During fabrication the panel should be laid on a canvas sheet to avoid scratching or indentation of the face side.

**Fig. 10.41** Lead cladding for buildings.

specific techniques. Figure 10.41 shows some of the details encountered in cladding with preformed panels.

The panels are covered off site with lead sheet as shown in the illustration. The lead is nailed only on the top half of the panel, which allows the lead to expand freely; the maximum areas of sheet are similar to those relating to dormer cheeks.

The panels are hung on steel bars bolted to timbers which are securely fixed to the fabric of the building. These timbers are weathered with a strip of lead having turned edges to prevent the

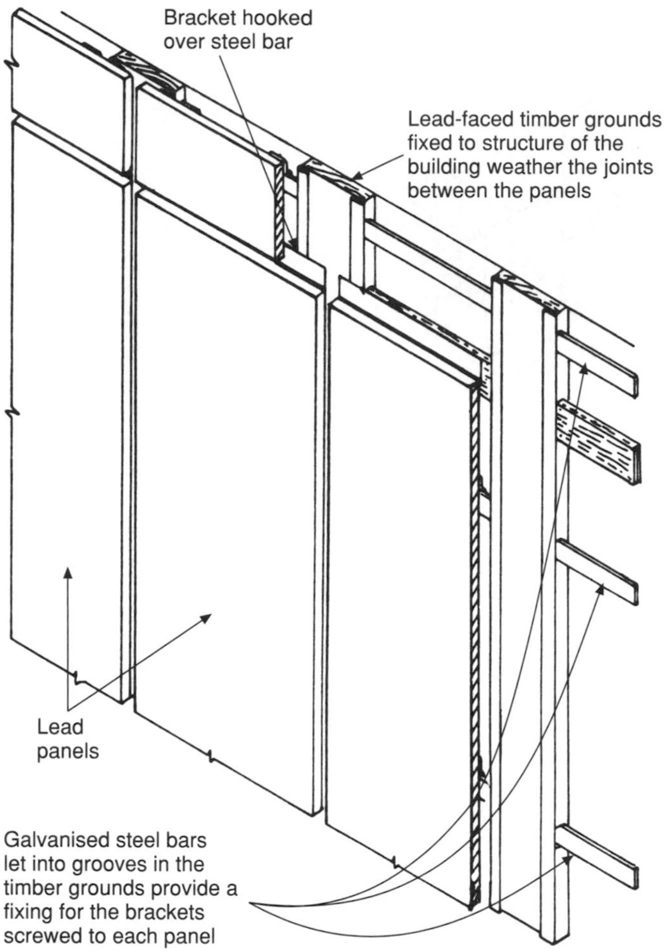


Detail 2

The lower edge of each panel is weathered with a strip of lead welded on to the top of the panel immediately below.

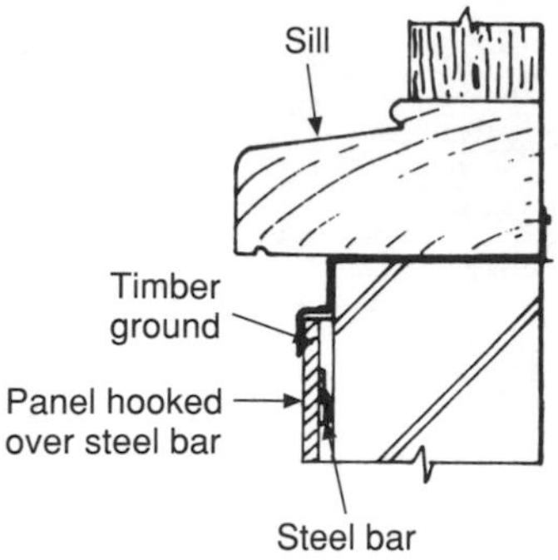


Detail 3 One method used to finish the top of panels under a sill

An apron, fitted in a similar way as a dormer apron, is turned over the top edge of the panels.

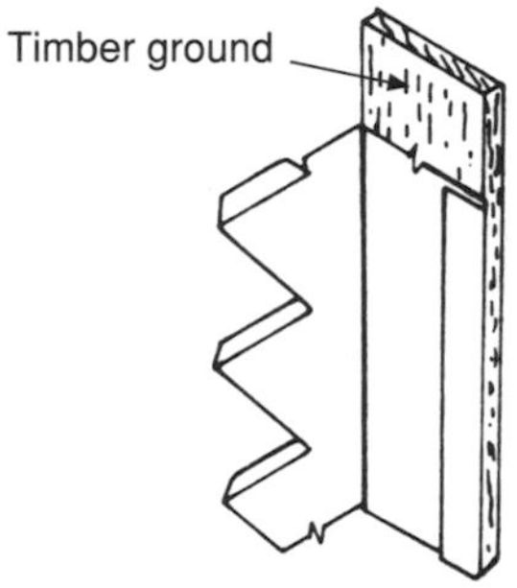


Detail 4 The way edges of the panels are finished where they abut the brickwork

Angled flashing strips are used, one side being turned into the brickwork joints, the other on to a timber fixed to the face of the building.

**Fig. 10.41** *(cont'd)*

ingress of water behind the panels. The top of the panels can be weathered by means of a cover ashing or, as in the example shown, the window apron.

### Patenation oil

Lead sheet forms its own protective patina when exposed to the atmosphere, but in damp conditions the formation of an uneven white carbonate will appear on newly fixed lead. As the protective coating forms the carbonate film disappears. Apart from not being pleasing to the eye, especially on external cladding, the white carbonate will be washed off by rain causing unsightly stains on the fabric of the building. To prevent this a coat of patenation oil can be applied, preferably at the end of each day's work. It should be applied with a soft cloth working from top to bottom of the work. One coat is usually sufficient but if the surface of the lead is marked by foot traffic during subsequent builders' work, a second coat may be applied after all other operations have been completed.

### Inspection and maintenance of sheet weatherings

A well-designed and correctly fitted sheet metal roofing installation should have a long working life, and except in the event of a serious breakdown in the metal itself, should be as effective and long lasting as tiles or slates.

In cases where maintenance is necessary, in almost every case it is due to: faulty workmanship, incorrect thickness of metals used and poor design, a typical example of the last being insufficient provision made for expansion occurring in hot weather, thereby causing fatigue cracking. Lead sheet is possibly more prone to 'creep' than other sheet materials due to its mass, but this can be minimised if it is well fixed and the maximum superficial areas in relation to its thickness have not been exceeded. Both creep and fatigue cracking are dealt with in Book 1, Chapter 10.

Although regular inspection of roofs covered with any material is rare, it is recommended, as a fault detected in its early stages is more easily corrected and damage to the substructure minimised.

Possibly the most common fault found is where a turn into the brick- or stonework of the structure has pulled out of the joint. This is easily remedied by refixing, but the cause of the defect should be investigated and is usually due to one of the following:

(a) Insufficient turn in.
(b) In the case of cap flushing the maximum recommended length has been exceeded.
(c) The bottom edge of the flashing is not properly supported.

Where lead weatherings are fixed in exposed positions, e.g. ridges and verges, the effect of high winds can cause the lead to be torn from the decking. It should be noted, however, that this would certainly be due to insufficient or incorrect fixings.

During an inspection of lead any signs of cracking should be investigated. Such defects are almost always due to poor design and improper fixings. A repair may be effected by patching with wiping solder, or preferably by welding. It must be stressed, however, that in all such cases no amount of patching will permanently solve a problem caused by poor design. If it is possible to gain access to the area immediately below the roof, the underside of the decking and its supports can be examined for dampness or staining which will indicate any ingress of water. Any sign of corrosion should also be carefully investigated, paying special attention to the eaves of moss-covered tiled roofs where the acidic run-off may cause local pitting corrosion.

The only maintenance a well-designed roof requires is periodic cleansing, especially if there are deciduous trees nearby where the falling leaves could possibly lead to blocked gutters. If necessary any catchpits in gutters and hopper heads should be checked for cleanliness; also ensure the wire guards preventing debris entering the rainwater pipes are in good condition.

### Further reading

BS 6915 Specification for design and construction of fully supported lead sheet roof and wall coverings.
Lead sheet manuals & data sheets The Lead Sheet Association, Hawkwell Business Centre, Maidstone Rd, Pembury, Tunbridge Wells, Kent TN2 4AH Tel. 01892 822773. www.leadsheetassociation.org.uk

### Self-testing questions

1. From Table 10.2 calculate the total superficial area of one bay of a flat roof covered with code 6 lead sheet, allowing 200 mm for the under- and overcloak for the drips, and 225 mm for the under- and overcloak to the rolls.
2. Describe the underlying principles of fixing lead sheet to avoid fatigue cracking.
3. Sketch and describe the type of transverse expansion joint that must be used across a low-pitched lead covered roof with a fall of $2\frac{1}{2}°$.
4. (a) State the advantages of hollow roll work on a lead-covered pitched roof.
   (b) Describe the procedure for laying such a roof and how the roll ends are finished at the eaves.

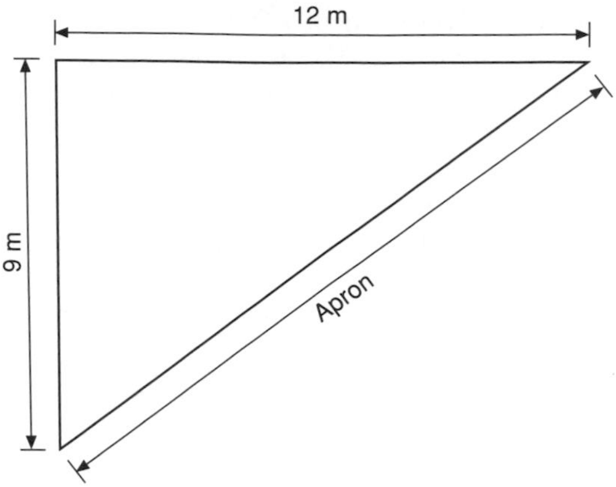


**Fig. 10.42**

5. Name the type of gutter that would be used in the centre of a large, low-pitched lead roof.
6. (a) Describe the 3,4,5 rule and its use in building.
   (b) Calculate the length of metal apron flashing for a triangular asphalt roof shown as a plan drawing in Fig. 10.42.
7. Describe one situation where an L-shaped cleat would be employed.
8. (a) Make a sketch showing a torus roll.
   (b) Suggest a position where this detail can be used with advantage.
9. (a) List three types of dormer in common use.
   (b) Describe two methods of fixing and weathering the front edge of a dormer cheek on to the window frame.
10. (a) List the advantages of using lead-welding techniques for sheet roofing.
    (b) Set out on paper the folding and cutting lines for a lead-welded drip overcloak.
11. (a) State the type of lead-welded joint recommended for positional work on lead-covered roofs and indicate the reason for your choice.
    (b) Describe an alternative to a 90° vertical joint when welding lead, and explain its advantages.
12. List the precautions and methods you would use when welding positional details on a lead-covered roof with a timber substructure.

# 11 Electrical systems

After completing this chapter the reader should be able to:

1. Recognise electrical supply systems in domestic properties.
2. Describe the components used to automatically cut off the supply in the event of failure of the conductors or appliances due to an electrical fault.
3. State the purpose of earthing and bonding.
4. Recognise potential dangers due to incorrect installation procedures in bathrooms, kitchens, and plumbing or heating appliances.
5. Understand the basic principle of power circuits.
6. Recognise the need for correct power ratings, fuses and cable sizes for plumbing appliances using mains electricity.
7. Describe the correct procedures for testing and isolating electrical supply systems.

The purpose of this chapter is to enable the student plumber to acquire a basic knowledge of electricity and its supply systems in order that electrically controlled and operated plumbing appliances may be installed correctly and safely. Even when electrical work is undertaken by a qualified electrician the plumber must be able to recognise the need for bonding appliances which can conduct electricity even when a supply is not connected directly to it.

Always remember that electricity is dangerous and is a potential killer; unlike water, gas or fire it cannot be seen or smelt, and the first indication of poor installation practice may be a severe shock which at best is unpleasant and painful and at worst fatal.

## Regulations governing the installation of electrical supplies

In the interest of safety it is essential that electrical installations of any kind must be carried out by a competent person and that any appropriate regulations or recommendations are strictly adhered to.

On 1 January 2005, Building Regulations Approved Document P became mandatory and applies to all trades carrying out domestic electrical installation work. It has been introduced to reduce the number of deaths, injuries and fires caused by unsafe and faulty electrical installations in the home. It also enforces BS 7671 (Wiring Regulations) which requires all installations to be tested and the issue by the installer of an appropriate completion certificate. Part P applies to all permanent installations from the consumer unit in:

(a) Dwellings.
(b) Combined dwellings and business premises, e.g. a shop having a common metered supply.
(c) Shared amenities and common access areas in flats.
(d) It also cover parts of electrical installation external of the main structure including garages, sheds, pond pumps and fixed garden lighting.

All proposed electrical work in a dwelling must be notified to a building control body prior to starting any work except when (a) it is carried out by a registered competent person who is qualified to self-certify the work and (b) the work is of a minor nature. It is not necessary to notify work such as, for example, the replacement or addition to socket outlets or lighting points. Work such as wiring from the consumer unit to supply, for example, a power shower or an electric immersion heater would be notifiable. Approved Document P lists the main areas which concern plumbers/heating installers. Copies of this document are available on loan from public libraries.

*IEE Regulations*
These are now incorporated in a British Standard which covers both the Electrical Wiring Regulations and the Health and Safety Act, parts of which relate to the safety of electrical installation work. The wiring regulations are very comprehensive, state the recommended practice and are continually updated. Although they are not statutory, no electricity supplier would make a connection to any installation not meeting their requirements.

*The Electricity Supply Regulations 1988*
These are statutory, but relate mainly to the installation of supplies rather than work carried out in a building and the installation of appliances.

*The Electrical Equipment Regulations*
These are also statutory and specify the type and safety standards of equipment and appliances used by the consumer. If these items meet the requirements of British Standards it can be assumed the manufacturer has complied with these regulations.

*The Electricity at Work Regulations 1989*
These relate to the safe use of electrical installations and power tools in the workplace. Many of the requirements of this act are embodied in the Health and Safety Regulations and apply to both employers and employees. Other safety publications relating to specific items of equipment also quote these regulations where they apply.

## Distribution of electricity

Electricity is best defined as an invisible source of energy which is conveyed from point to point via conductors. Apart from the various components of an appliance, these conductors are usually insulated copper wires, copper being used because it is a good conductor and its malleability enables the cables to be manipulated very easily. Insulation of the conductors is necessary as electricity will always take the shortest possible route. If the 'phase' or live conductor comes into direct contact with the neutral or earth wires, it will result in a short circuit causing overloading, which in a correctly designed installation will result in a safety device, usually a fuse or miniature circuit breaker (m.c.b.), operating to cut off or isolate the supply.

*Single-phase supplies*
The term 'phase' relates to the current-carrying cable sometimes called 'live'. It should be noted that the wiring regulations also refer to the neutral as live, so one must be clear as to the term 'phase'. The supply of electricity to domestic properties normally has only one phase wire, hence the term 'single phase'. In industrial premises three-phase supplies are often used which produce higher voltages, so that more power is available for large machines. The normal voltage for single-phase supplies is 240 V, and for three-phase 415 V. It must be stressed that any work on three-phase supplies should only be carried out by a qualified electrician.

*Types of supply*

*Direct current* In this case a flow of electricity takes place in one direction only. It should be noted that the flow is negative to positive, which is technically correct. However, for practical purposes, it is normally assumed that the flow takes place in the opposite direction, i.e. from positive to negative.

Direct current is not usually supplied for household use, unless it is provided by a private generating plant in areas where no main supply is available. Other examples include the use of

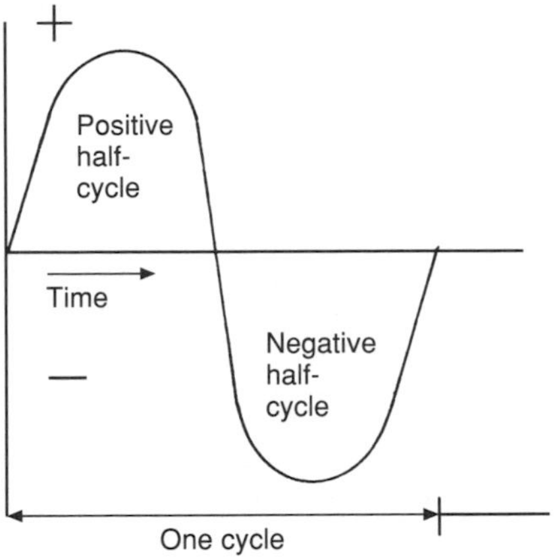


**Fig. 11.1** Alternating current. Alternating current does not flow from the positive terminal to the negative terminal consistently, but alternates for half the cycle. Note that the term *hertz* is commonly used for *cycle*.

dry batteries, e.g. for torches. It has very few applications in the type of electrical work undertaken by plumbers, apart from its use in some control systems. As a point of interest it was used in electric bell systems *c*. 1900 and it was usual for plumbers to carry out this work.

*Alternating current* This type of supply is normally provided by the electricity generating companies and, as its name implies, its direction of flow alternates continually in what are called cycles, see Fig. 11.1. The frequency, or number of times it changes direction, has been standardised in this country at 50 Hz (hertz) or cycles per second. Again, for all practical purposes, it should be assumed that the flow of current takes place from the live terminal through an appliance and returns via the neutral terminal.

### Electrical units

*Voltage (symbol 'V')*

The volt is the unit of electromotive force (emf) and just as in plumbing systems we know that the higher the head of water the greater will be the pressure, so with electricity the higher the voltage the greater will be its force to cause electrons to flow or drift along the conductor. An electric torch battery has a positive and negative terminal, and a potential difference is said to exist between them causing the electrons to flow from one to the other in a closed electric circuit. It should be noted that there is a drop in pressure called voltage drop as electricity flows around the circuit, and in long circuits this may be significant and require the use of a cable having a greater cross-sectional area. Although this is unlikely to have much effect on the type of work carried out by a plumber, it is important to be aware of this factor and pay attention to it.

*Current (symbol 'I')*

The ampere is the unit of current and is normally abbreviated to 'amp' and may be defined as the electrical unit of quantity and the amperage or volume of current passing through a conductor may be equated with the volume of water in litres passing through a pipe. Do not confuse the amp with the watt or kilowatt which are units of power. It is very important that a conductor or cable is of sufficient size to permit the current to pass through it without offering too much resistance, as this will lead to overheating and possible fire risk.

*Resistance (symbol 'R')*

The ohm (Ω) may be defined as the unit of resistance which opposes the flow of current. Resistance to current flow depends on the type of conductor used. Materials offering the least resistance to the current flow are termed good *conductors*; insulating materials are those that have great *resistance* to current flow. Most metals are good conductors. Good insulators include wood, glass, ceramics, paper, plastic and rubber. Providing the temperature does not change, the rate of drift of the electrons (the current) flowing through a wire is directly proportional to the potential difference (the voltage difference which makes the electrons move) between the two ends of a wire. This relationship is known as *Ohm's law*. A typical example of this law is to assume a current of 4 A is flowing through an electric element which is connected to a 240 V supply. When the supply voltage is halved, i.e. reduced to 120 V, then the current is halved and becomes 2 A. In each case the ratio of the voltage to the current is the same, namely:

$$\frac{240}{4} = 60 \quad \text{and} \quad \frac{120}{2} = 60$$

and thus Ohm's law can be stated in the form:

$$\frac{\text{Voltage}}{\text{Current}} = \text{A constant, i.e. } \frac{V}{I} = R$$

When the potential difference is measured in volts and the current is measured in amps, the constant, $R$, will be the resistance in ohms of the conductor. Thus, to find the resistance of a wire subjected to a mains supply of 240 V and a current of 6 A, since:

$$\frac{V}{I} = R, \ R = \frac{V}{I} = \frac{240}{6} = 40$$

Thus the resistance of the wire is 40 Ω. Another example of the use of Ohm's law is to calculate a fuse rating for an appliance. In this case the formula must be rearranged thus, since:

$$\frac{V}{I} = R, \text{ then } I = \frac{V}{R}$$

Assuming a voltage of 240 V with an appliance having a resistance of 80 Ω, then:

$$I = \frac{240}{80}, \quad I = 3\text{ A}$$

that is, the fuse rating will be 3 A. An easy way to remember Ohm's law, and to transpose the equation so that any one value can be found if the other two are known, is shown in Fig. 11.2.

*Power*

Electrical energy is measured in kilowatts (1,000 W) and for metering and billing purposes a kilowatt-hour is referred to as one unit. Power in watts is the product of voltage × current. The cost of operating electrical appliances can be calculated very simply and is shown as follows.

An electric water heater having a resistance of 40 Ω is supplied by a 240 V ring main. Determine its current rating and calculate its cost per hour, assuming the cost per unit is 7.5p. Since the amperage of the appliance must first be ascertained, Ohm's law equation is used in the following form:

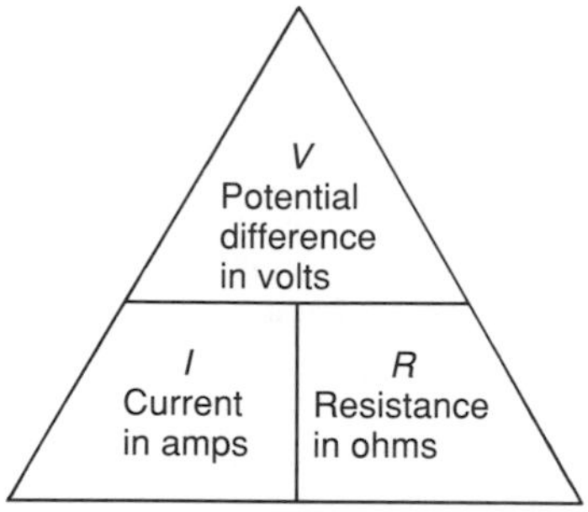


Cover the quantity to be found and the remaining two symbols will give the required formula

Ex 1 to find $V$ multiply $I$ by $R$
Ex 2 to find $R$ divide $V$ by $I$
Ex 3 to find $I$ divide $V$ by $R$

**Fig. 11.2** Ohm's law equation.

$$I = \frac{V}{R}$$

$$\therefore I = \frac{240}{40} = 6\text{ A}$$

As power in watts is the product of volts × amps, thus:

$$240 \times 6 = 1{,}440$$

to convert this to kilowatts:

$$\frac{1{,}440}{1{,}000} = 1.44\text{ kW}$$

This is approximately 1.5 kW and at 7.5p per unit the approximate cost of running this heater will be 11.25p per hour. It is appreciated that the electrical units and the calculation involved are unfamiliar to many plumbers and it is suggested that a little practice in them may be helpful. This can be done by using the information given by suppliers of electrical appliances of various types and also local electricity company shops and showrooms.

*Simple electrical circuits*

Before a flow of electricity can take place a complete circuit must be established from the positive terminal to the negative terminal as shown in Fig. 11.3. The consuming component, which may be a lamp, motor, heater, etc., is usually referred to as the 'load'. Any interruption by a break in the circuit, usually by operating a switch, and the

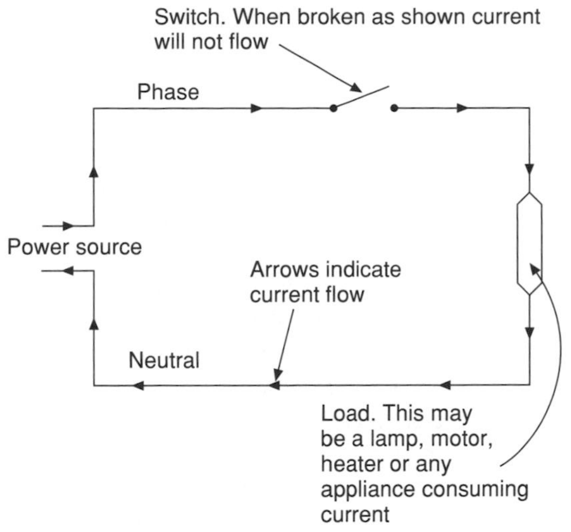


**Fig. 11.3** Simple electric circuit.

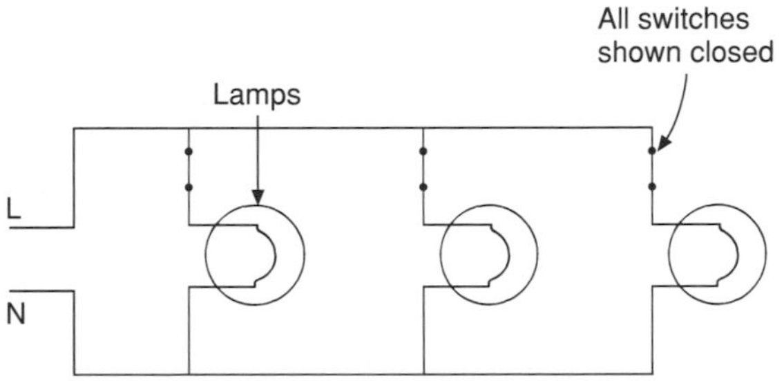


**Fig. 11.4** Lamps wired into a parallel circuit. This system allows the loads, in this case lamps, to be supplied by the full voltage and each can be independently switched.

current will cease to flow. From this it will be seen that a current will only flow through an unbroken circuit from positive to neutral. Figure 11.4 shows a parallel circuit with three lamps controlled by one switch. With this type of circuit each lamp takes the full voltage and power from the supply. A series circuit is shown in Fig. 11.5. If three lamps are used again it will be seen that they give less illumination, due to the fact that, in effect, the current has to be shared between them. It is therefore unlikely that such a system would be used for lighting, but the term must be understood, as many control systems, especially those in boilers, employ components wired in series.

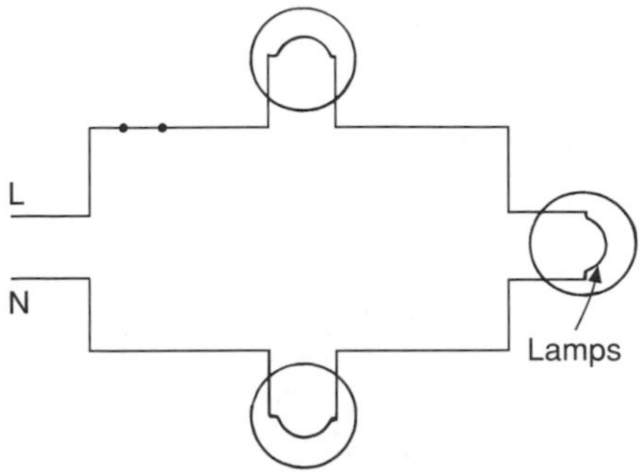


**Fig. 11.5** Lamps wired in series. In this case the available voltage must be shared so that if 24 V are supplied, each lamp will only take 8 V. This system would not be suitable for lighting circuits as the illumination would be dimmed and independent switching would not be possible.

## Electrical safety

*Overcurrent*

Overcurrent or overloading a conductor takes place where the current of electricity exceeds that which the conductor is safely capable of carrying. Just as a water pipe will only deliver a given quantity of water in litres for a given pipe size, so an electrical conductor of a given capacity can only carry a given current. If this is exceeded the conductor will become hot due to its resistance to the current flow. This could destroy the insulation, leading to the possibility of a fire which may be well established before the conductors make contact, causing the circuit protective device to operate. It must be quite clear that any cable selected for a specific task must be of adequate size for the current it is to carry.

*Short circuit*

This is said to take place when, for various reasons, the neutral or earth cables are in direct contact with the phase. One reason for this could be careless connection of a phase to the neutral or earth conductor or breakdown in the cable insulation due to ageing or damage. It can also happen accidentally, as shown in Fig. 11.6, during the course of one's work. It is worth mentioning that moderately priced instruments are available enabling concealed pipes and cables to be located before work is commenced.

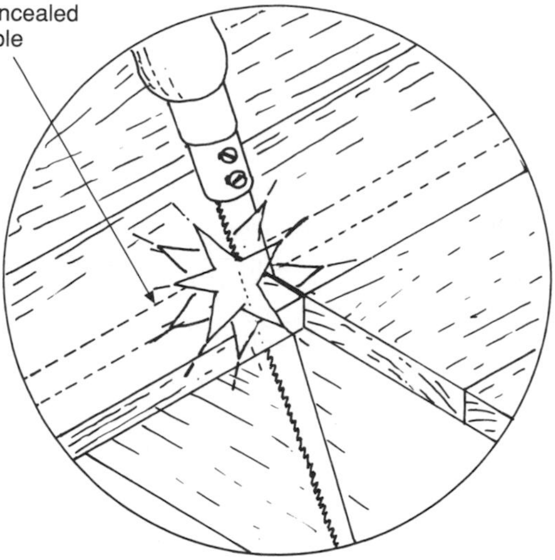


(a) Removing flooring using a padsaw. The use of a circular saw set for the correct depth would avoid most accidents of this type

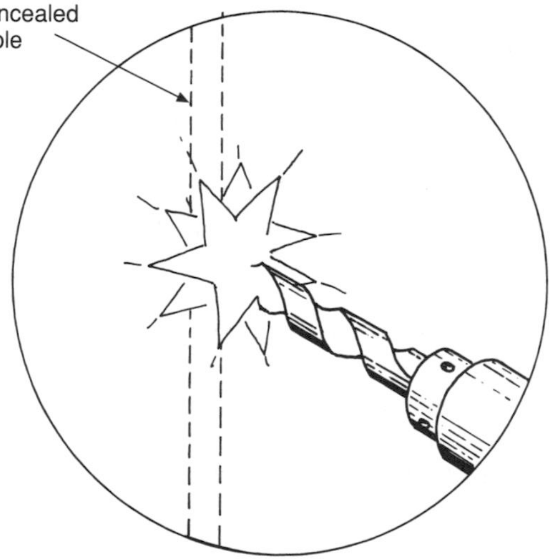


(b) Drilling walls can often result in the accidental damage to electrical cables. Avoid drilling near power sockets or switches and on a line above or below them

**Fig. 11.6** Accidental damage to concealed cables. This usually results in short-circuiting the cables.

*Circuit protection devices*

Isolation of a circuit in the event of overcurrent or a dead short is provided by a fuse or miniature circuit breaker (m.c.b.), both of which are designed to automatically isolate or cut off the supply of electricity in the event of a fault.

*Fuses* A fuse consists of a single strand of wire which is designed to overheat and melt under overload conditions. There are two main types, rewirable and cartridge fuses. The rewirable type is simple and cheap. Should a fuse of this type 'blow' it can be replaced with wire obtained on a card obtainable from most ironmongers or electrical suppliers. The card will usually have fuse wire for 5, 10 and 15 A circuit control, the 5 A being the smallest in cross-sectional area and mostly used for lighting circuits. It is important when replacing fuse wires that the correct type is selected. An indication of the circuit controlled by the fuse should be provided on the cover of the consumer unit where the fuses are located. The main disadvantage with rewirable fuses is that they take at least three times their rated value to operate. The alloy from which the wire is made tends to deteriorate with age, and if it melts due to overloading, white heat is momentarily achieved, constituting a fire risk. Cartridge fuses consist of a single wire completely enclosed in a glass or ceramic tube, both ends of which are sealed with a brass cap. The fuse rating is usually marked quite clearly on the tube. In comparison to fuse wire they are expensive to replace, but they operate more quickly and there is little fire risk.

*Miniature circuit breakers* These are basically an automatic switch used instead of fuses in the consumer unit to control various ring and lighting circuits. While they are more expensive than fuses they are considered safer — they do not present a fire risk and can easily be reset when a fault is corrected. A much heavier type of circuit breaker, called a residual current circuit breaker (r.c.c.b.), is used in conjunction with earth electrode earthing systems to isolate, not individual circuits, but the whole installation. More details are given on these components in the section 'Earthing and bonding'. Figure 11.7 illustrates the main circuit protective devices used in housing. Automatic isolating switches are often incorporated in plumbing equipment such as sink macerators to prevent overloading the motor in the event of jamming.

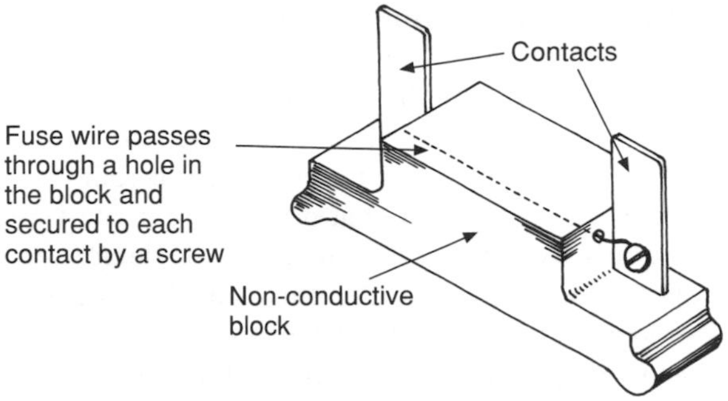


Fuses must always be correctly rated for the circuit or appliance they protect.

(a) Renewable wire fuse. In the event of overload current on the circuit the fuse wire will melt automatically breaking the circuit

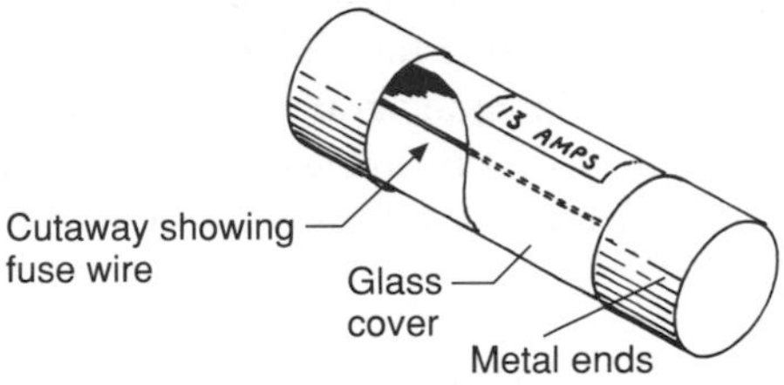


(b) Cartridge fuse. These work on the same principle as a wire fuse but in this case it is enclosed in a glass tube. They are designed to protect against faults in individual appliances and are used in fused plugs and switches

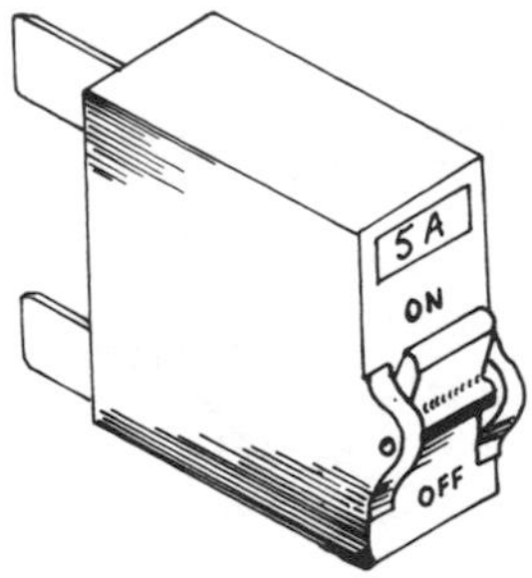


(c) Miniature circuit breaker. These are used in the distribution panel instead of fuses. They are more expensive, but much quicker acting. In the event of circuit overload, a bimetallic strip operates a solenoid which trips a switch to isolate the circuit. They can then only be reset manually

**Fig. 11.7** Circuit protective components.

They are used for the same reasons in many types of washing and dishwashing machines. They work on a different principle to m.c.b.s and should not be confused with them.

### Earthing and bonding

The surface of our planet is referred to as the 'earth' and we use this as our reference, calling the earth 'zero potential'. As humans usually stand on the earth they are 'at earth' or 'zero potential'. If, however, they touch a metal object which has become electrically charged, they may receive an electric shock, as the current flows to earth through them.

In order to reduce the possibility of metal cases of electrical equipment becoming charged, due maybe to an internal fault, e.g. the breakdown of insulation on a line conductor, we 'earth' the metal case via a circuit protective conductor (c.p.c.) (the earth wire in the three-core cable feeding the equipment). An excessive current flowing to earth under fault conditions will cause the fuse feeding the equipment to blow. Providing the c.p.c. is of adequate size and has sound connections, there will be no substantial rise in voltage of the metal case and the fuse will blow in 5 seconds.

The c.p.c. of every final circuit in an installation is taken back to the earth terminal at the fuse board, and then to the main earth terminal at the incoming supply. Under fault conditions this current will flow back to the supply transformer, either through the ground or via the sheath of the supply cable. The neutral is connected to earth at the supply transformer — this provides a fault path back to the supply. Any metalwork which is earthed and associated with electrical equipment is referred to as an 'exposed conductive part', examples being the metal case of a boiler, sink, water heater, etc.

Other items of metalwork could possibly become live due to a fault elsewhere and be connected by plumbing pipes, metal frame of the building, metal ducting of the air-conditioning system, etc. Now if under these conditions someone touches this metalwork, referred to as 'extraneous conductive parts', and are themselves earthed, they could get a fatal shock.

To prevent this difference of potential existing between the adjacent metalwork, we bond the items together — known as supplementary bonding in a bathroom, for example, and main bonding at the main service position. These bonding conductors must be of adequate size to allow the current to flow to earth without causing the metalwork to rise in potential above 50 V.

*Earthing systems*

It was common practice in days gone by to connect the main earth conductor to the water service, as being metal it provided a good conductor to earth. This is no longer permissible due to the fact that both water services and the main are, in many cases, made of polythene, which is not a conductor of electricity.

*Solid earthing* This is the TNS system (see Fig. 11.8). In most modern buildings where the electrical supply is underground the consumer's earth wire is connected to the armoured incoming cable. This armour not only protects the cable from accidental damage but also acts as an effective earth.

*TT system* In some rural areas where the supply is taken from overhead cables an earthing rod is used. This is a copper or steel rod of approximately 12 mm in diameter which is driven into the ground so that any current leakage is dissipated and eventually finds its way back to the supplier's neutral point. Bearing in mind that the path of a current to earth must have negligible resistance and that certain types of soil and the amount of moisture it contains will affect its conductivity, an automatic device for isolating the supply must be provided. This is called a residual current circuit breaker (r.c.c.b.) which is a heavy-duty, very sensitive circuit breaker that will automatically shut off the supply in the event of any earth leakage. These devices work on the principle of a coil which operates in a magnetic circuit: while the flow through the phase and neutral conductor is balanced, as under normal conditions, the coil will remain inactive. In the event of earth leakage this balance is lost causing the coil to react and automatically trip an integral switch to the off position which must be manually reset to restore the supply. If the r.c.c.b trips continuously it is

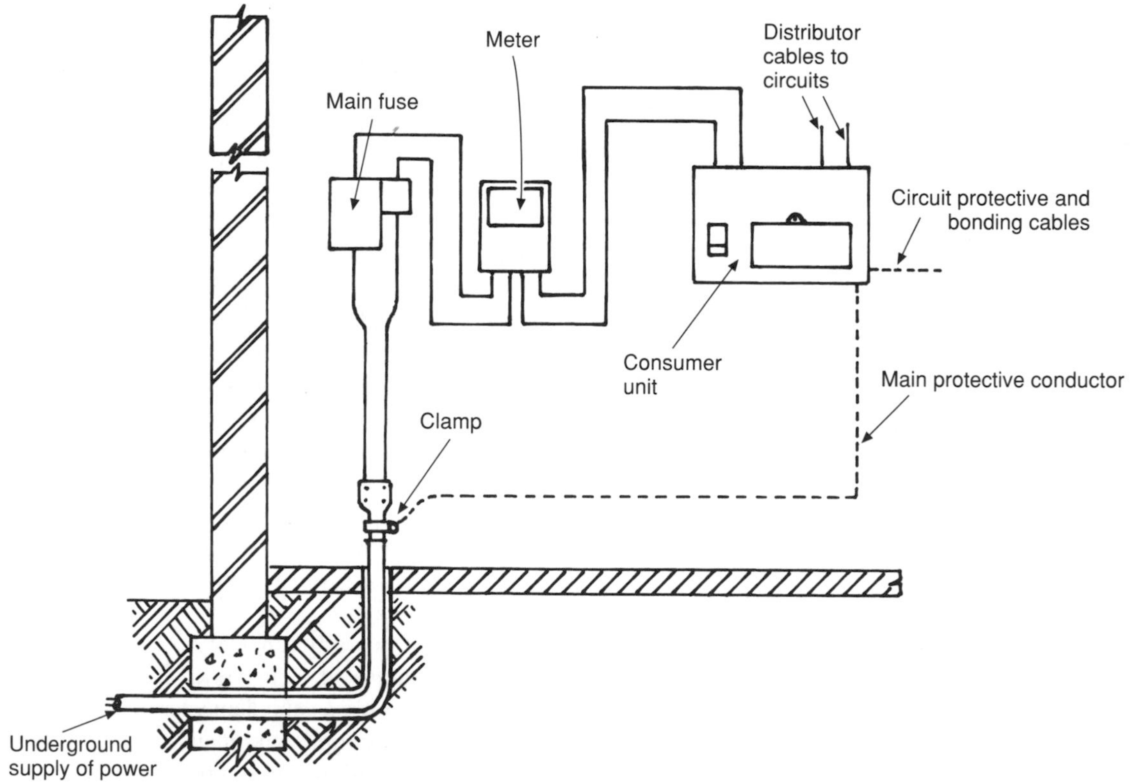


**Fig. 11.8** TNS system of earthing. With this type of earthing arrangement the main earth cable is connected by means of a clamp to the metallic armoured sheath protecting the current-carrying cables.

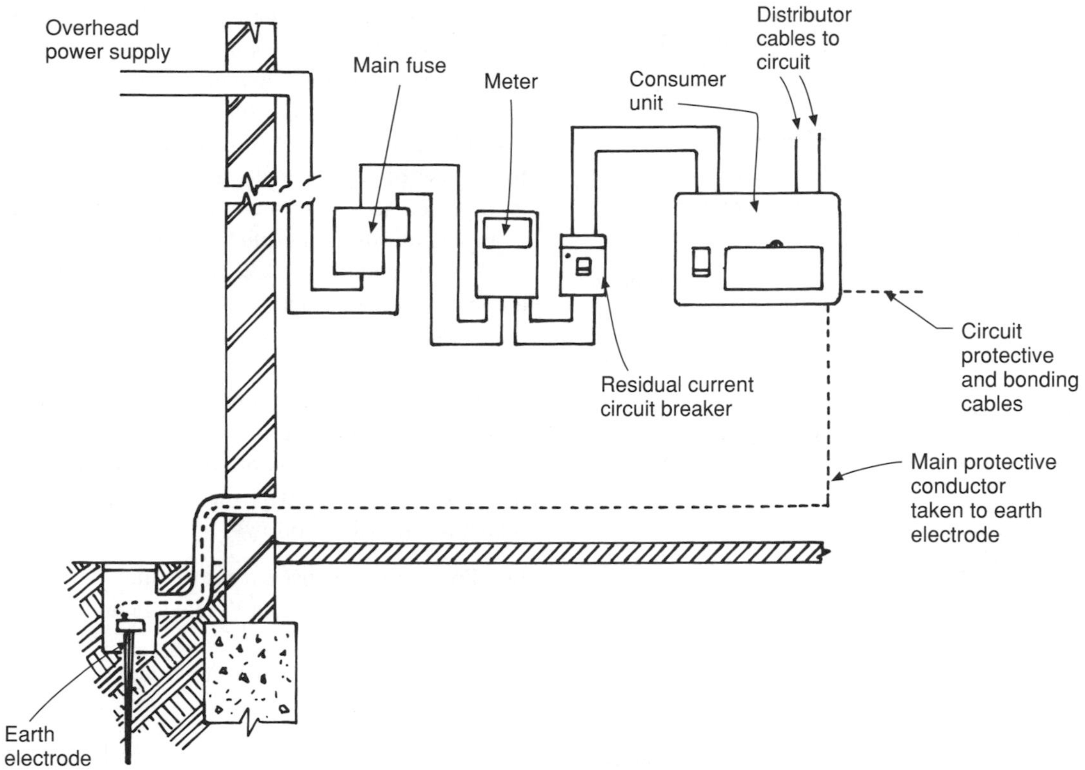


**Fig. 11.9** TT system of earthing. This is used where the authorities' supply is taken from overhead cables which provide a phase and neutral supply only. Earthing in this case is by a metallic rod driven into the ground.

indicative of a fault or short circuit in the system. This earthing system is illustrated in Fig. 11.9.

*Protective multiple earthing* This method of earthing (the TNCS system) shown in Fig. 11.10 uses the electrical supplier's neutral conductor as an earth, which in effect means that any current leakage goes straight back to the supplier's neutral point. One of the advantages of this system is that there is little resistance to the current leakage, which means the safety cut-out devices will be rapidly activated.

*Equipotential of phase and earth conductors* The word 'equipotential' in this context means that any bonding and earthing conductors must be of sufficient cross-sectional area in the event of a fault to carry away any flow of current with little or no resistance. If, for example, a short circuit took place between a phase conductor fused for 13 A and an earthing conductor capable of carrying only 10 A, it would be of little use, as the fuse would not melt and cut off the current.

*Equipotential bonding* Any conductive material which is not a part of an electrical installation but which, due to a fault, can become energised is termed an *extraneous conductive part*. An example of this could occur where a fault has developed in an electric heater installed in a hot storage vessel causing all the metallic parts of the hot water system, and possibly the heating system as well, to become live. A similar situation could arise in connection with a metal boiler casing. From this it will be seen that although things like metal baths, sinks, taps and metal pipes have no direct contact with electrical supplies, due to a fault they may become energised. Under these circumstances anyone touching them as shown in Fig. 11.11 could suffer an electric shock, possibly fatal.

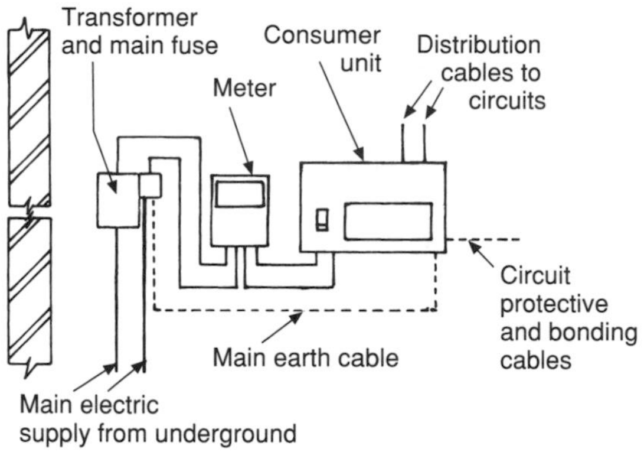


**Fig. 11.10** TNCS system of earthing. This system differs from TNS only in that the main earth cable is connected to the neutral cable at the supply authorities' transformer.

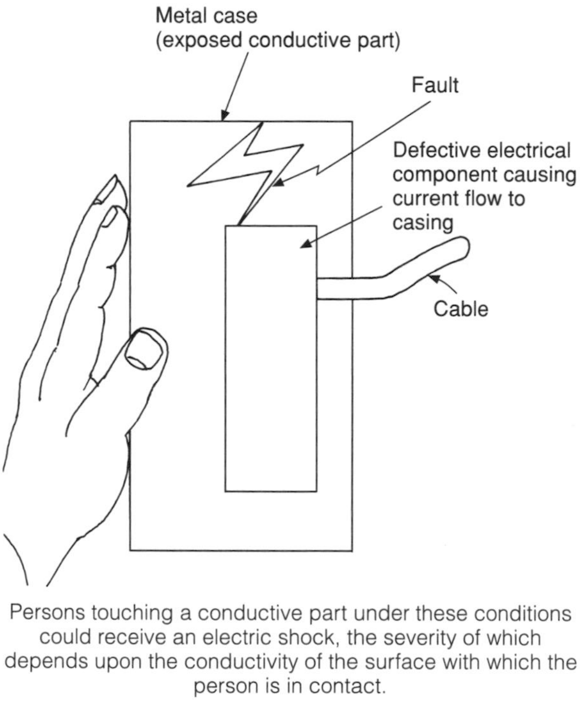


**Fig. 11.11** Earthing extraneous conductive components. This illustration shows the necessity of earthing or bonding exposed conductive parts.

It must also be remembered that water is a conductor of electricity due largely to impurities it may contain, so extra care needs to be taken when electrical connections are made to any equipment that is also in contact with water. See Fig. 11.12(a) which generally indicates the basic principles of equipotential bonding. A typical bonding clamp suitable for a bonding cable to a pipe is shown in Fig. 11.12(b). If any clamps are removed during the course of repair or alteration work, they must be replaced securely on completion of the work. If any additions or modifications to existing work are carried out, a check should be made with a competent electrician as to whether any bonding is necessary.

*Consumer unit* This component is installed on a panel adjacent to the incoming electrical main and provides the means of control and protection necessary to comply with the IEE Regulations which are listed as follows:

(a) Protection against excess loading.
(b) Protection against earth leakage and fault conditions.
(c) Provision for isolating individual circuits and a main switch.

Protection against excess loading, earth leakage and fault conditions is provided by suitably rated fuses or m.c.b.s. Isolation of the supply is provided by a double-pole main switch.

### Distribution of supplies in the property

The conductors to the various sub-circuits serving both power and lighting points are taken from the consumer unit. The term *power* here means the supplies to socket outlets and fixed appliances such as fires or hot water heaters. Each of these sub-circuits has its own fuse or circuit breaker. A 5 A fuse or equivalent m.c.b. is provided for lighting and 30 A for power circuits. It is not proposed to deal with lighting circuits here as this is normally the province of electricians, but it is important that the plumber is able to recognise power circuits and know the methods used to make additional connections to them.

*Ring circuits* These are the most common form of power distribution in modern premises and there are slight variations in some types, mainly depending on the floor area covered and the current-carrying

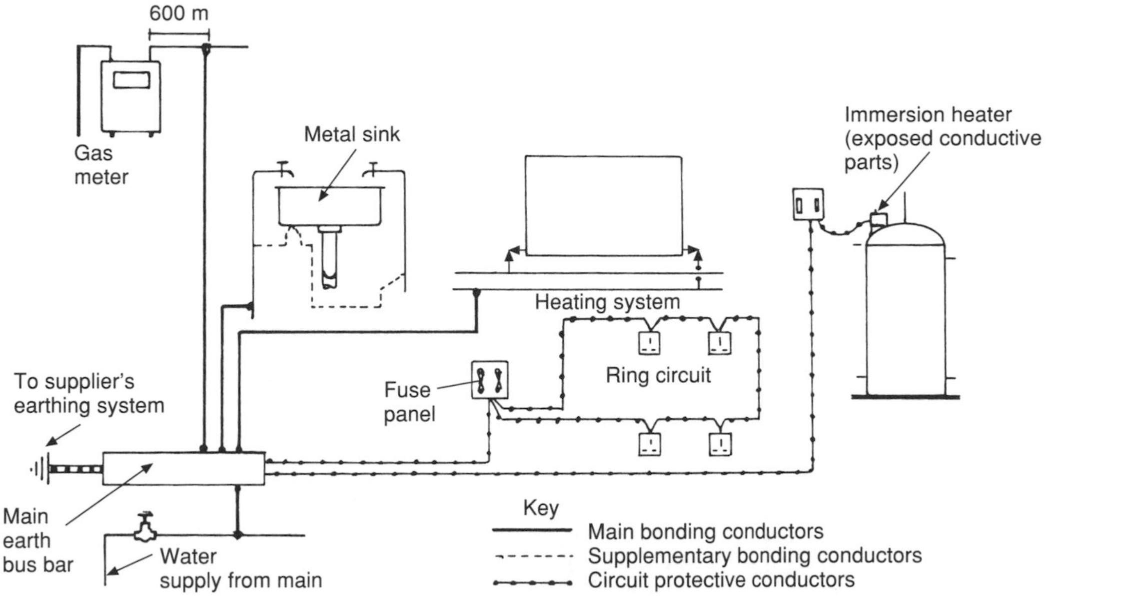


(a) Earthed equipotential bonding

The purpose of bonding metallic components is to ensure that if accidental contact is made between a part of the electrical installation through which an electrical leak is passing and other metallic or conductive components, they may become energised. Touching them could result in a severe electric shock. The illustration shows the bonding and earthing cables used to ensure any electrical leakage is safely conveyed to earth.

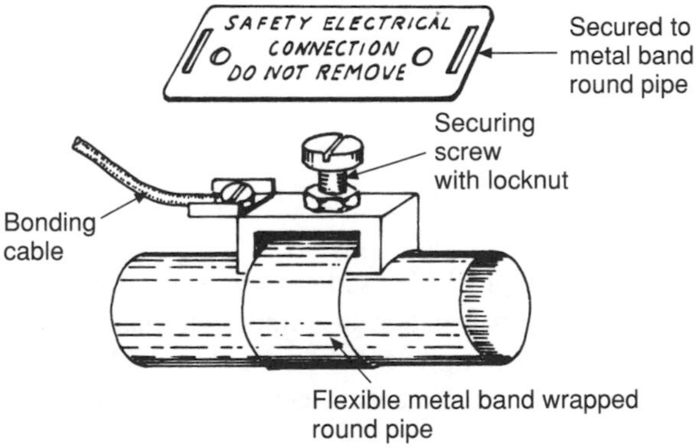


(b) Electrical bonding clamp

**Fig. 11.12** Earthing and bonding.

capacity of the main conductors. The one illustrated in Fig. 11.13 is possibly the most representative of those commonly used. The supply of power for domestic heating equipment is frequently taken from such a system. As its name implies it is simply a cable laid out in the form of a ring which carries current to each socket outlet. There is no limit to the number of socket outlets on any one ring, but the floor area served must not exceed 100 m$^2$. This limit is to prevent the connection of too many appliances which could result in overloading. It is permissible to connect a spur to the ring main as shown, but the number of non-fused spurs on any one ring must not exceed the number of socket outlets and stationary apparatus connected directly in the ring. Figure 11.14 illustrates the wiring of a fused plug used in connection with a three-pin socket outlet in the ring. Another form of outlet is a fuse switch shown in Fig. 11.15. Figure 11.16 shows two approved methods of wiring an additional spur to an existing ring circuit. Non-permanent connections such as electric irons, radios, hair dryers, etc. are provided with a flexible cable and fused plug which is inserted into a socket outlet. Appliances which are permanently connected such as fires, boilers, water heaters, etc. are supplied with power via a fused switch which incorporates a fuse under a sealed cover. This cover, usually secured with a screw, is removable to allow access to the fuse. Whenever electrically controlled or operated

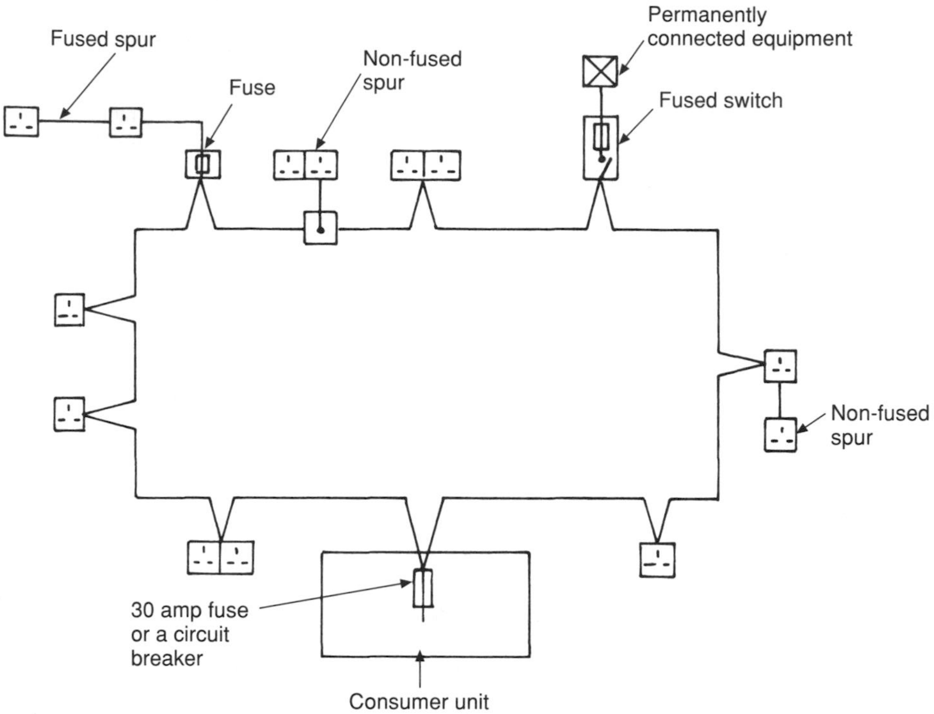


The main feature of this type of ring main is that there is no limit to the number of socket outlets. The floor area must not exceed 100 $m^2$. The number of non-fused spurs must not exceed the total number of socket outlets or permanently connected equipment connected to the ring. The total number of fused spurs is unlimited. Permanently connected equipment must be protected against overload by either a fuse having a maximum rating of 13 amps or a suitable circuit breaker.

**Fig. 11.13** Typical ring main circuit: note neutral conductors not shown.

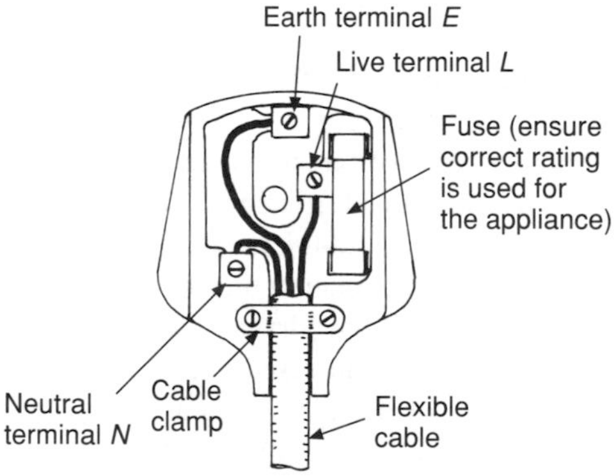


Colour code for *L* brown
Colour code for *N* blue
Colour code for *E* green with yellow stripe

**Fig. 11.14** Three-pin plug used for connecting movable electrical equipment to a socket outlet.

plumbing equipment is used, or any electrical appliance for that matter, it is important that a fuse of the correct rating is fitted to avoid possible damage to the equipment or danger under fault conditions.

The principle of electrical ring circuits is based on the 'diversity' factor: that is to say, an assumption is made as to the number of appliances in use and the power they consume at any one time. While it is very unlikely that the ring conductors will overload due to the use of electric kettles, fires, radios, etc. very high-rated appliances such as immersion heaters and cookers might overload the circuit, causing the safety cut-out devices to be activated. For this reason such appliances are normally provided with their own separate circuits.

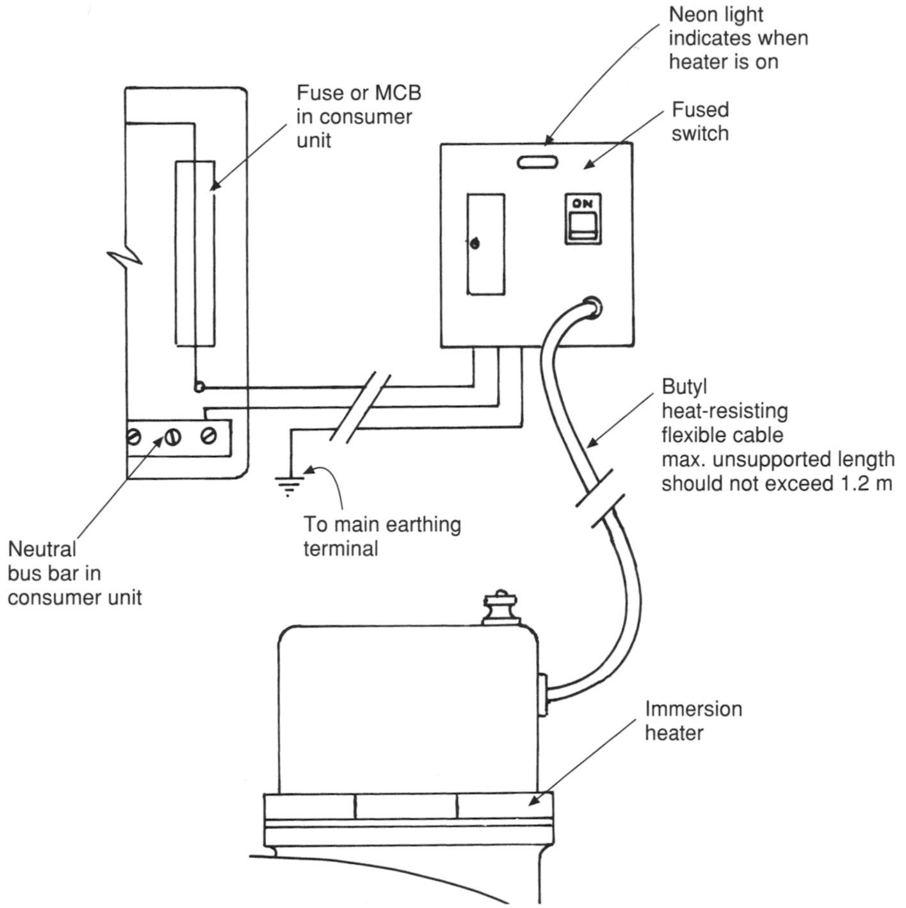


**Fig. 11.15** Electrical supplies to water heaters. Because heaters of this type require a high amperage to avoid overloading the ring main, they are supplied with power direct from the consumer unit having a separately fused circuit.

*Circuits supplying electric immersion heaters*
It is not recommended that these heaters are connected to a ring main, due to their very high electrical loading. The types of heaters available are dealt with in the section on hot water supply (pages 151–3). Figure 11.15 has illustrated the wiring from the consumer unit to the heater, and Fig. 11.17 shows how the element and thermostat are connected. It will be seen that the heater is supplied by a separately fused circuit taken from the consumer unit, and is usually controlled by a 20 A double-pole switch adjacent to the heater. If the immersion heater is situated in a bathroom the switch must be fitted externally.

All switches controlling electrical equipment in bathrooms must be fitted externally due to the damp, humid conditions normally found there constituting a possible danger from electric shock. One exception is for switches at ceiling level, provided they are operated by a cord pull. See Fig. 11.18(a).

A double-pole switch differs from a single-pole switch in that both the phase and neutral conductors are switched, see Figs 11.18(b) and (c). A single-pole switch only isolates the phase conductor. A typical example of single-pole switches is those used for lighting circuits.

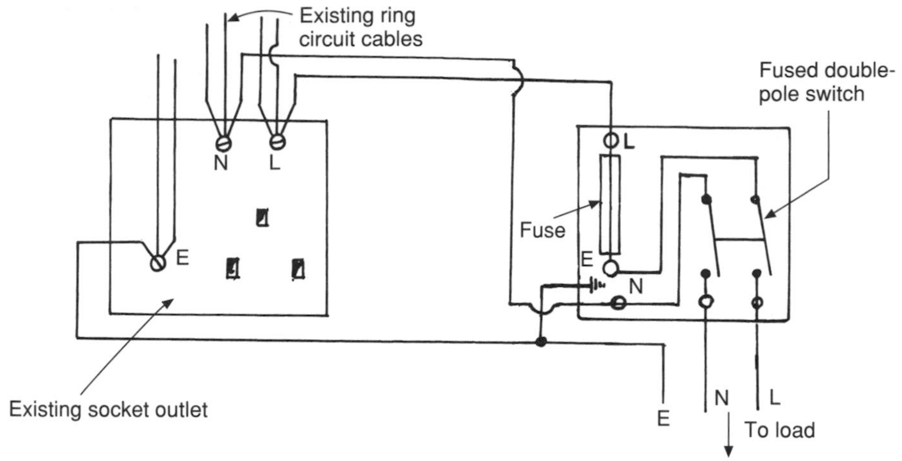


(a) Illustrates the cable connections from an existing socket outlet to a fused switch controlling an additional appliance. Additional socket outlets may be installed in a similar way.

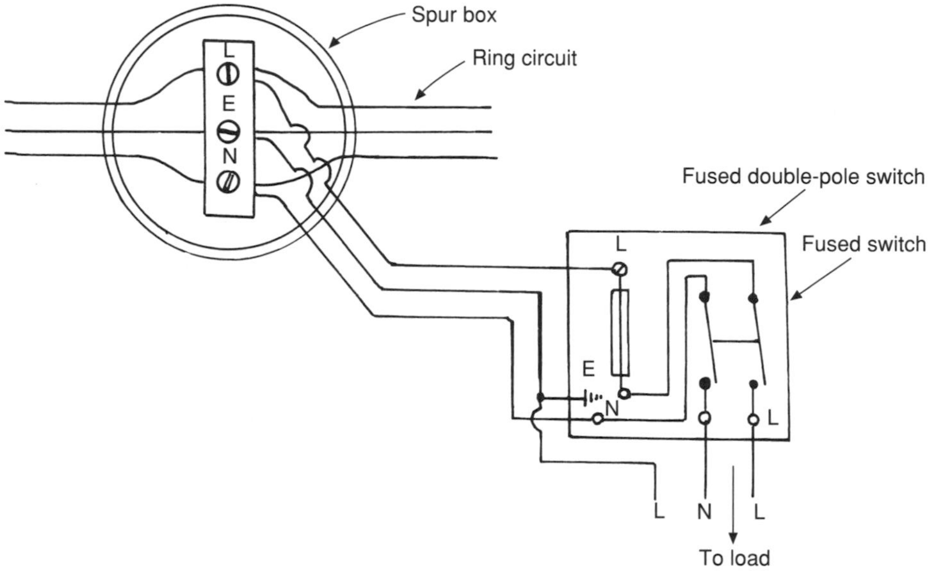


(b) Shows a method of making a direct connection to an existing ring main using a spur box

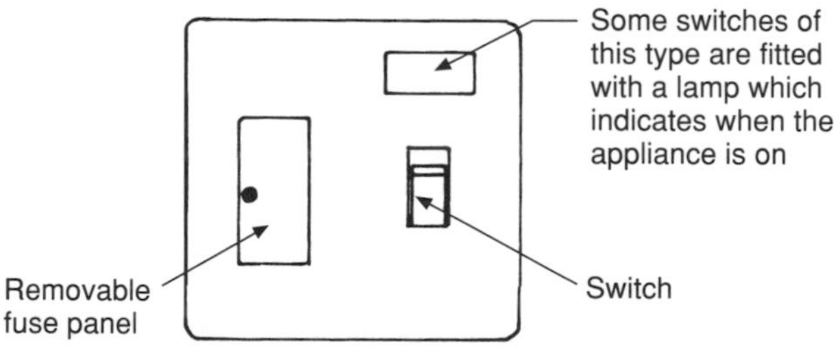


(c)

This switch may be used with a surface mounted box or where the cables are chased into the wall, a concealed box is used so the switch is flush with the surface of the wall.

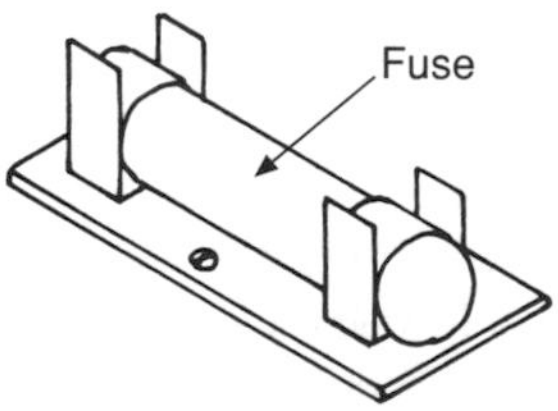


(d) Fuse panel removed showing fuse holding clamps which also act as contacts

**Fig. 11.16** Showing how additional connections may be made to existing ring main.

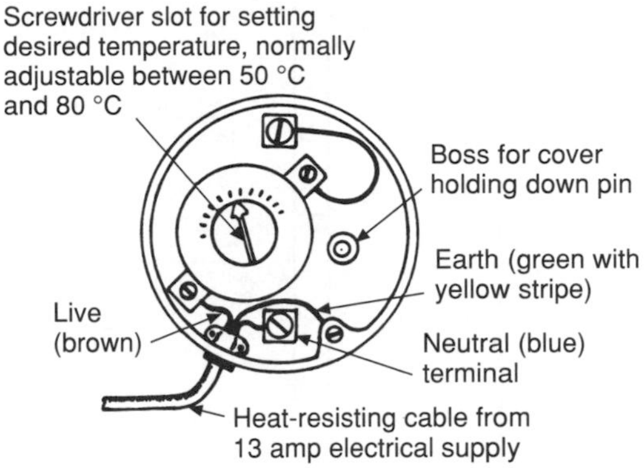


**Fig. 11.17** View of electric immersion heater with the cover removed.

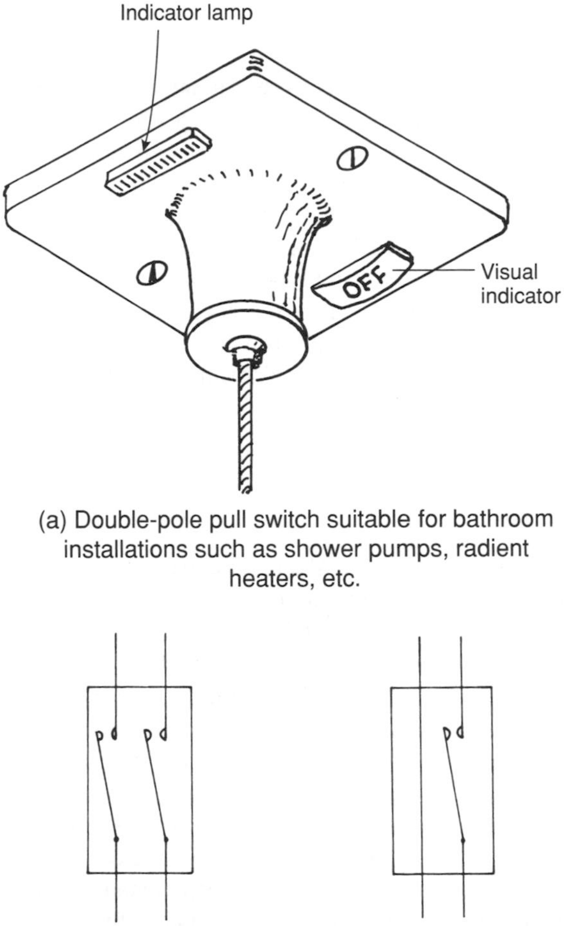


(a) Double-pole pull switch suitable for bathroom installations such as shower pumps, radient heaters, etc.

(b) Double pole
Both line and neutral to the appliance are switched.

(c) Single pole
Line only switched.

**Fig. 11.18** Types of switches.

**Current-carrying capacity of conductors (cables)**

It is not always appreciated that wiring conductors of a given size through which a current of electricity is flowing has limitations on the capacity that can be carried without an excessive amount of resistance. Failure to observe this fact will result in overheating, causing possible fire risk and destruction of the insulation.

Wiring conductors are measured by their cross-sectional area (c.s.a.) and typical examples are those used for lighting circuits which have a c.s.a. of 1 $mm^2$. Such circuits are normally protected by a 5 A safety device. Conductors used for ring circuits, because of the possibility of heavy loading, have a c.s.a. of 2.5 $mm^2$ and are protected by a fuse of 30 A. It should be noted the two previous examples given relate to copper PVC sheathed wires. The other main factor to be considered is what is called voltage drop. Assuming a cable has been selected which is capable of carrying the required amperage but the cable lengths are very long, some account must be taken of this factor. Just as the frictional resistance the pipe walls offer to a flow of water, so the voltage or pressure of electricity will be reduced progressively over long runs of wiring. This will have a similar effect as using wires of inadequate c.s.a., and while as a general rule cables of 1 $mm^2$ and 2.5 $mm^2$ c.s.a. are satisfactory for lighting and ring circuits, cables for cookers and showers are usually 6 or 10 $mm^2$. In cases of doubt the IEE Regulations should be consulted as they contain tables which show the voltage drop per amp per metre run. These regulations recommend that the voltage drop in any final circuit should not be in excess of 4 per cent. To illustrate this in practical terms it could be said that if the supply at the inlet to a cable is 240 V the voltage at the furthermost outlet should not be less than 232 V.

*Protection, insulation and fixing of cables*

The methods and types of cables used vary considerably and depend mainly on the type of building, the purpose of the cable and the service conditions to which it will be subjected.

It is not proposed to deal with steel conduits and mineral-insulated copper sheathed systems, as

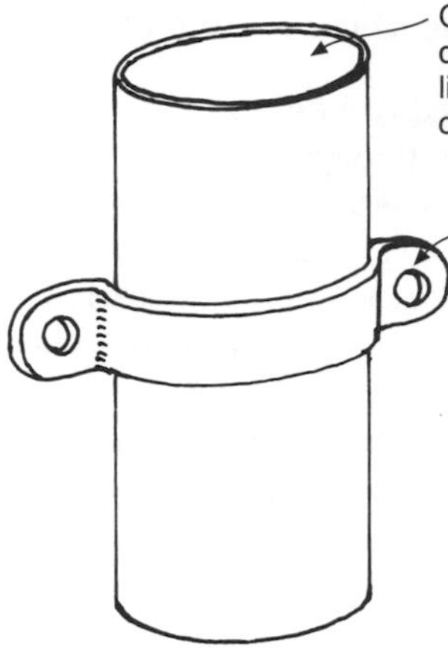
Oval section designed to limit depth of chase in wall
Proper clips must be used. Turned-over nails may damage the conduit and prevent withdrawal of the cables. Rubber grommets should be used at both the top and bottom of any conduits to avoid the cables chafing against any sharp edges

**Fig. 11.19** Light conduit.

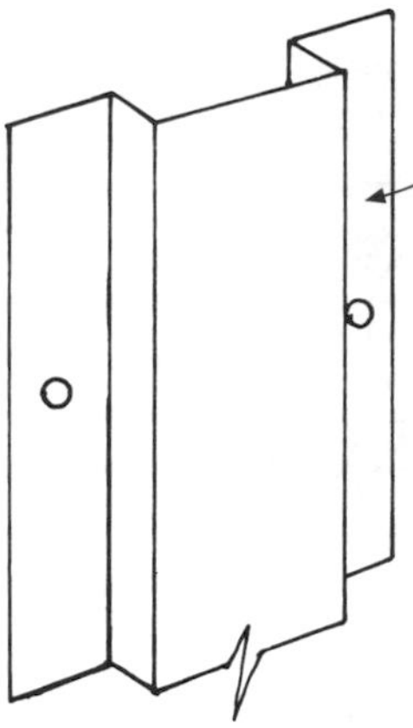
Usually nailed to the wall

**Fig. 11.20** Channel section cable cover.

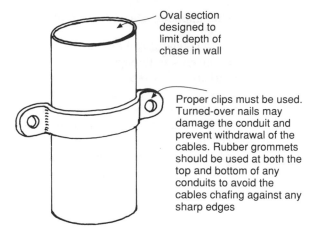

these are the subject of specialised work and do not normally fall into the province of work undertaken by the plumber in domestic premises.

*Conduit*

The main purpose of the conduit is to permit the cable to be withdrawn if necessary, and if made of metal it may also offer some protection to the cable from mechanical damage. Any mechanical conduit must be effectively earthed, and for this reason, PVC, being a non-conductor, is generally used except in industrial and commercial installations.

*Oval conduit*

Figure 11.19 illustrates this light conduit that is used for concealing cables in walls under the finished plaster. Its oval shape reduces the depth of chasing that is necessary, and if no power tools are available such as angle grinders, a bolster is probably the best hand tool to use to cut the chases, unless the wall is made of concrete or very hard brickwork, in which case a hard/cold chisel is recommended.

An alternative to oval conduit is the use of channel section PVC shown in Fig. 11.20. It normally makes cutting away unnecessary, as having little depth it can be covered by a normal thickness of plaster. Any concealed cable should always be run straight up or down when feeding any electrical components. Most building operatives are aware of this, therefore chances of piercing a cable with a nail or screw are minimised. Concealed cables running horizontally across a wall are not recommended. Where it is necessary to conceal cables run on the surface of a wall, a light type of PVC trunking is available, the details of which are shown in Fig. 11.21. It is not suitable for situations where it could be damaged, but for domestic and light industrial work it does afford some degree of protection for cables and provides a very neat finish to exposed cables. Another advantage is it can be cut and fitted with tools normally carried by a plumber.

## Cables

*Flat section insulated sheathed cables*

These cables are commonly used for domestic installations and are suitable for both surface and concealed work where they must be housed in the light types of conduit previously described. Figure 11.22 shows a typical section of this type of cable, the conductors of which are made of copper with the outer insulation being PVC. Due to the fact it is non-flammable and will resist attack by most acids and alkalis, and to some degree solvents, it is a very convenient material. It is manufactured with single, twin or twin and earth core for normal use. Unlike the phase and neutral cores which are insulated, the earth wire is bare except for the PVC outer sheath. For this reason, when it is connected to any exposed earth the wire must be protected

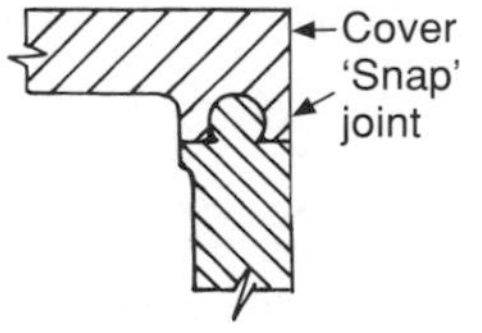


(a)

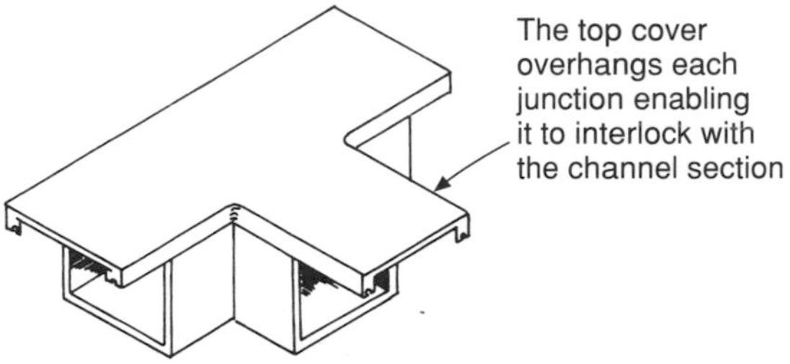


(b) This system produces a complete range of components such as junctions and bends

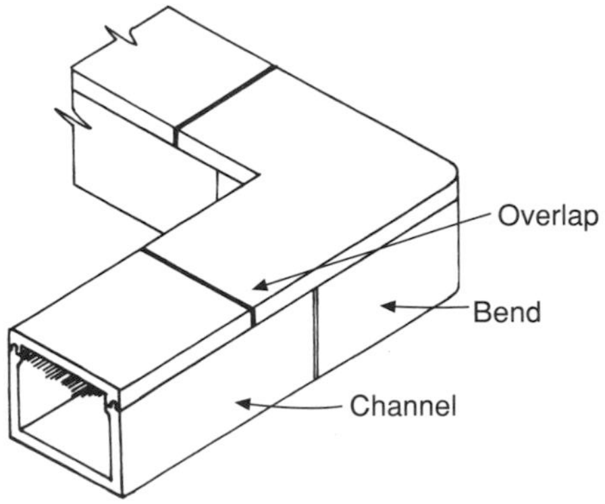


(c) Showing how the sections overlap to ensure a rigid joint

**Fig. 11.21** Light PVC trunking. The illustration shows one of many light trunking systems for the containment of cables on wall surfaces. It may be fixed on smooth surfaces by a self-adhesive backing strip or, alternatively, traditional fixings such as screws. It is basically a channel which accommodates the cables, the top cover having a 'snap' closing arrangement shown at (a).

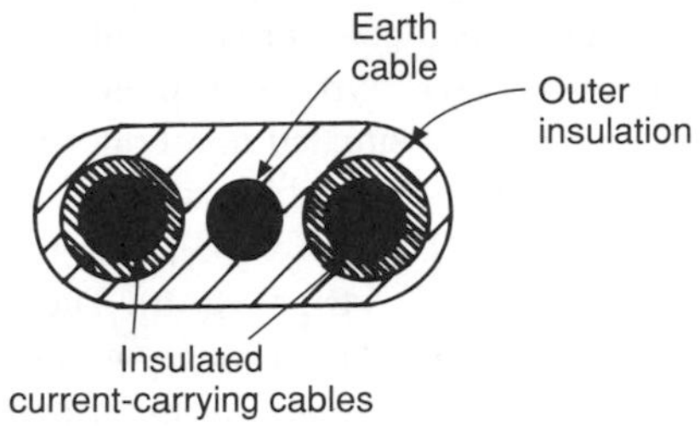


(a)

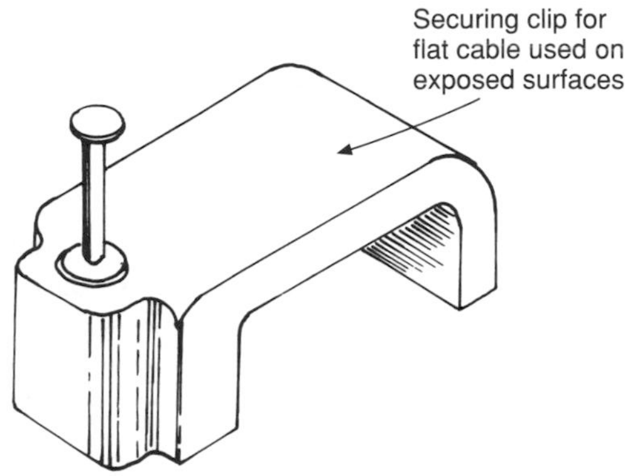


(b)

**Fig. 11.22** Section of flat section cable for general-purpose work. Cable-carrying capacities vary as to the length and temperatures to which it is subjected. Generally 2.5 c.s.a. cable is suitable for ring circuits including any spurs; 1.5 is generally used for lighting circuits.

with a corrosion-resistant and identifying sleeve which is available from electrical stores and is made with the correct earthing colours. The phase and neutral cables are normally red and black, red indicating phase. This type of cable can also be obtained with three insulated conductors and an earth cable. This is extremely useful for wiring components where two live cables are necessary, a typical example being room thermostats. It is not permissible to use the earth wire as a conductor at any time, even if the appliance is double insulated and no earth is necessary.

*Bell wire*

This term relates to cables used for electric bells which operate on very low voltages. It can in fact be used for any purpose where very low voltages are employed, but must not be used for 240 V, as it usually has only very light insulation.

*Sheathed flexible cables*

Sometimes called cords these are similar to the PVC sheathed cables previously mentioned, except for the fact they are circular in section and the conductors do not consist of one solid cable but are made up from many fine strands which make it very flexible and suitable for situations where trailing cables are used; typical examples are wiring to pumps and controls situated in a boiler casing. For normal use three-core cables are used with insulation colours of brown (phase), blue (neutral) and green with a yellow stripe (earth). Those having multiple cores employ a larger range of colours for identification purposes. The use of multi-core cable avoids the unnecessary doubling up of standard three-core cables in control systems.

Where any cables are used in situations where temperatures are higher than normal, i.e. cables connecting an electric immersion heater to a fused switch, butyl or ethylene–propylene insulation is used as these plastics are more high-temperature resistant than PVC. It is important, however, that no cable should be fixed to or adjacent to hot surfaces as no plastic material used for insulation is capable of withstanding prolonged intense heat without suffering damage. For this reason special care is necessary when installing cables in boiler casings. Figure 11.23 illustrates this type of cable and a typical type of fixing clip. Table 11.1 gives details of cable ratings and sizes.

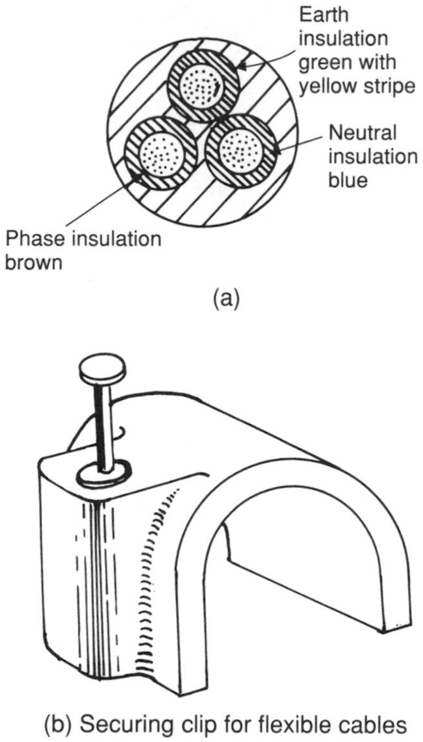


**Fig. 11.23** Flexible cables (cords).

*Cable fixings and installation techniques*

Because they are usually unsupported, it is very important that flexible cables are firmly secured

**Table 11.1** Cable and flex sizes and ratings. Conductors in cable and flexes are described by their cross-sectional area, which will give a rough guide to their current-carrying capacity.

| *Cross-section ($mm^2$)* | *Capacity* | *Domestic usages* | *Fuse* (amps)* |
|---|---|---|---|
| *Twin with c.p.c.* | | | |
| 1.0 | 11 A | Used for lighting circuits | 5/6 |
| 1.5 | 14 A | | 10 |
| 2.5 | 18 A | Socket outlets, fixed equipment | 15/20 |
| 4.0 | 25 A | Socket outlets, fixed equipment | 15/20 |
| 6.0 | 32 A | Showers and cookers | 30/32 |
| 10.0 | 43 A | Showers and cookers | 45 |
| *Flexes* | | | |
| 0.5 | 3 A | | 2/3 |
| 0.75 | 6 A | | 5 |
| 1.0 | 10 A | | — |
| 1.25 | 13 A | | 13 |
| 1.5 | 16 A | | — |

* The relative fuse sizes to the above cable ratings give a 'rule of thumb' method of cable sizing.

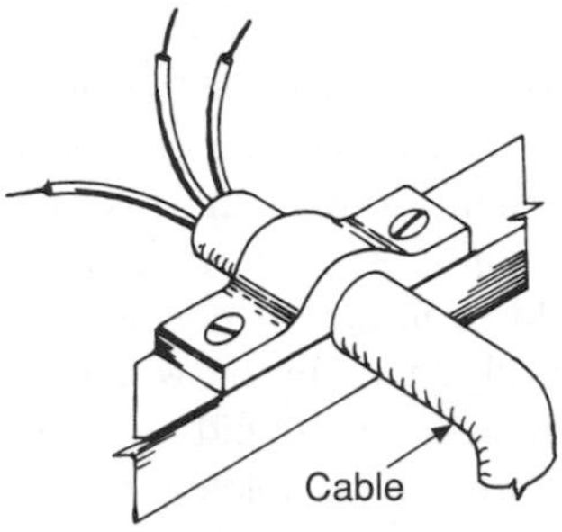


(a) Clamping arrangements used to secure cables in 13 amp plugs, immersion heaters, pumps, etc.

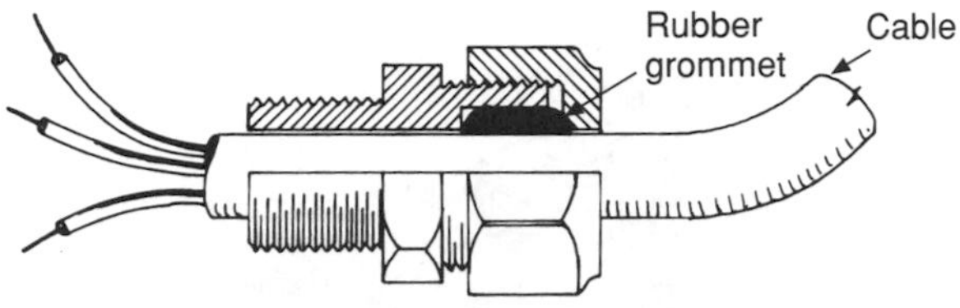


(b) This type of clamp is similar to a copper compression fitting. By tightening the nut the rubber grommet is squeezed on to the cable. It may be threaded directly into a component having a suitable thread or in some cases secured by a backnut (not shown)

**Fig. 11.24** Security of cables at terminations.

**Table 11.2** Details of cable supports (refer to Fig. 11.24(b)).

| | *Spacing of support for cables* | |
|---|---|---|
| *Overall cable (mm)* | *Horizontal (mm)* | *Vertical (mm)* |
| Up to 9 | 250 | 400 |
| 10–15 | 300 | 400 |
| 16–20 | 350 | 450 |

to the component without placing undue strain on the terminal connections. Figure 11.24 shows some of the methods of achieving this. Table 11.2 gives details of spacing of supports for cables.

To reduce the possibility of damage, runs of flexible cord should be as short as possible, and when unsupported should never exceed 1.2 m in length. Cables under suspended floors can be clipped to a joist where it runs in the same direction, but if not, a hole is bored through the joist at least 50 mm below its top edge through which the cable is threaded. This ensures they are unlikely to be penetrated by nails or screws when the finished flooring is fixed. In situations where it may be difficult to obtain good surface fixings, the following alternatives may be used:

(a) A light batten screwed to the wall to provide a ground for the fixing clips.
(b) Light plastic trunking previously described.
(c) The use of quick-drying purpose-made adhesives which provide a secure fixing to the cable throughout its length.

When stripping the insulation from cables prior to fixing to terminals always check that the insulation and the conductors are undamaged. Special care is necessary when stripping flexible cords as the fine wires forming the conductors are easily cut and damaged. It is recommended that correctly adjusted, cable-stripping pliers are used and the setting is checked on a scrap of wire before use. Figure 11.25 shows how side-cutting pliers are used to strip the outer cable insulation.

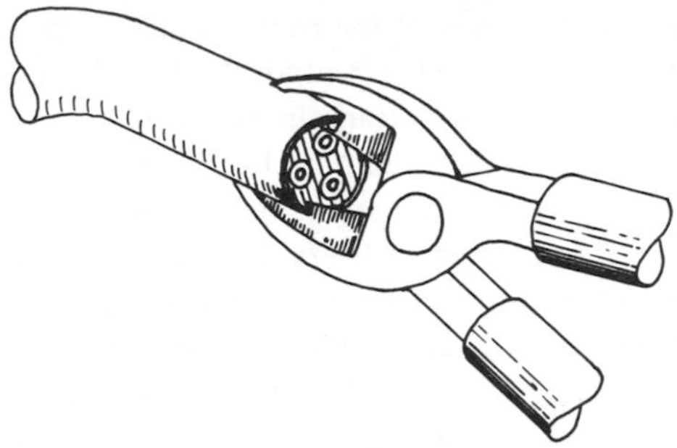

In all cases the following points must be observed:
(a) Care must be taken not to damage the conducting cables or the remaining insulation.
(b) Remove only sufficient insulation necessary to make a satisfactory connection to the terminal.

**Fig. 11.25** Cable stripping. Side-cutting pliers may be used to cut the outer insulating sleeve of flexible cord prior to peeling it back to expose the insulated cables. Stripping with a knife is possible, but the cable insulation may be damaged using this method unless extreme care is applied.

*Termination of conductors to components*

Cables are connected to components by means of terminals which vary depending on the type of component, the c.s.a. of the cables, and whether the component is used in an industrial or domestic environment. It is important, however, that any connections are secure, that no bare cable is exposed outside the connecting point and the cables are not subject to stress. Figure 11.26 shows a variety of terminals used for terminating cables to electrical components. Where flexible cables are used, the strands of wire should be twisted together to avoid loose strands protruding from the terminal.

*Safe isolation*

Prior to any work being carried out on existing installations, or when servicing electrically controlled plumbing equipment, it is essential to isolate the supply. In domestic properties any switches should be in the off position, with the fuses that control the circuit on which one is working removed. Fuses should be stored in a safe place until the work is completed. Special care must be employed if work is carried out in industrial

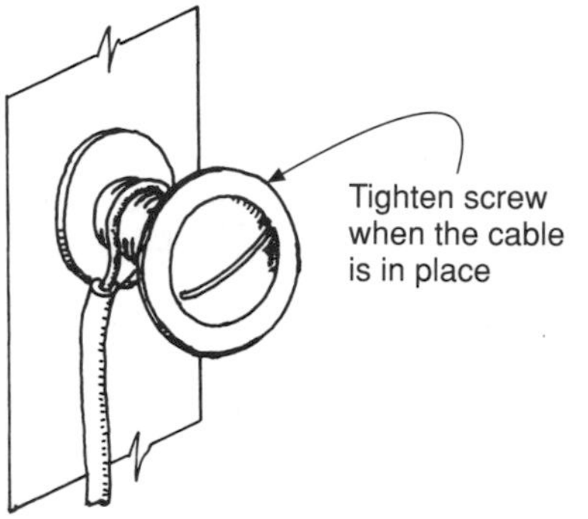


(a) Type of terminal common to domestic heating controls

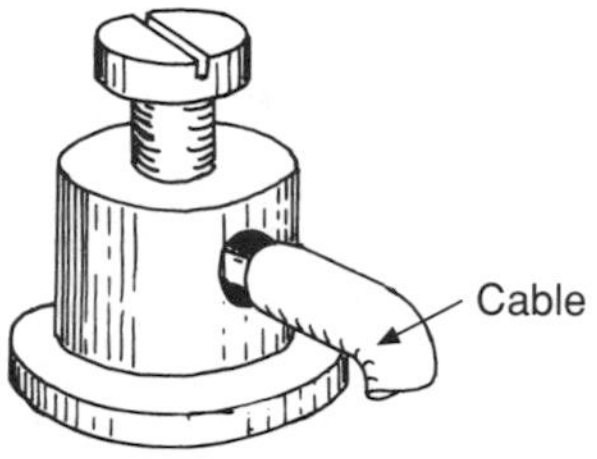


(c) Pillar-type terminals

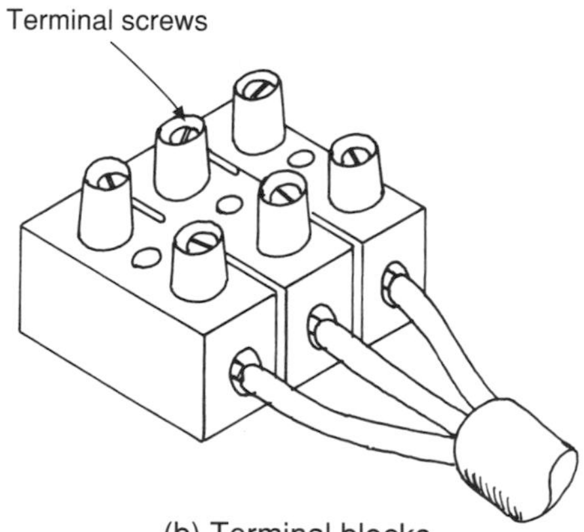


(b) Terminal blocks

These are basically lengths of brass tube encased in a plastic moulding. The cables are secured in the insulated tube by the terminal screws as shown. These blocks are obtained in lengths containing 7 or 10 connections which may be easily cut with a sharp knife depending on the number required. They are made in three sizes to accommodate a range of cables having differing cross-sectional areas. Terminal blocks should always be housed in a suitable box and not simply wrapped in insulating tape.

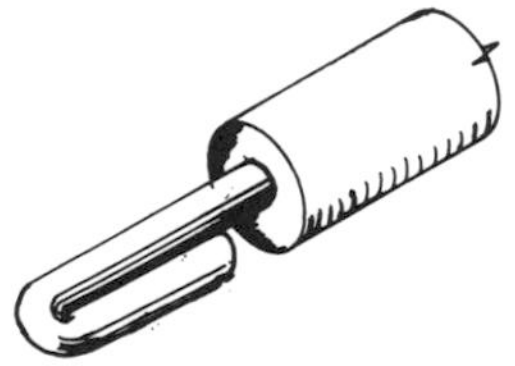

(d) Illustrates the method of doubling the conductor to ensure maximum security

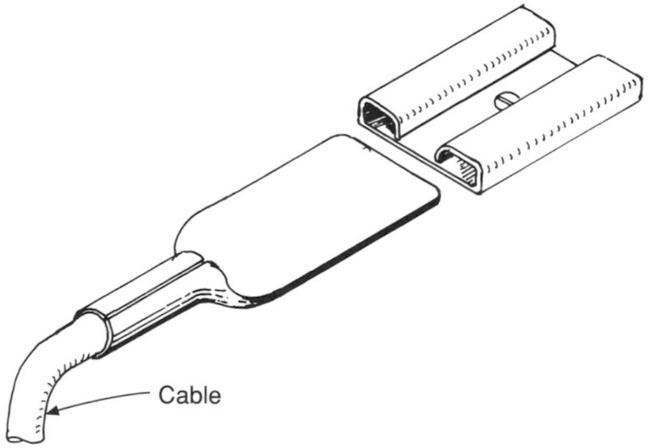


(e) Spade connection commonly used for extra low voltage work, e.g. 12–24 V

**Fig. 11.26** Terminations. This term is used to denote the connections of cables to components.

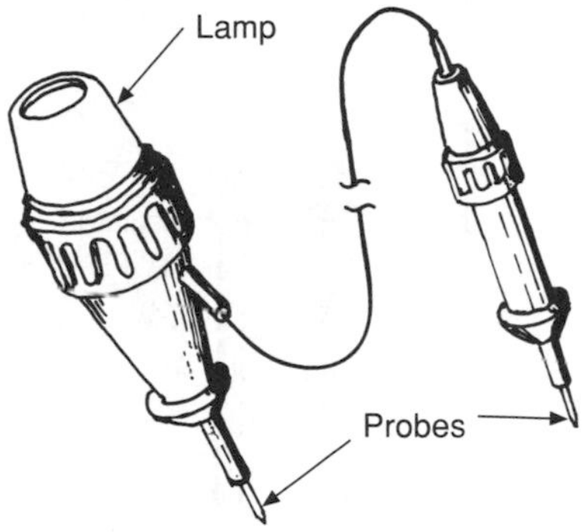


**Fig. 11.27** Approved test lamp. Principal use is to check whether the phase conductor of a circuit is energised. To use, the probes are held against the phase and neutral terminals of a switch or appliance, if the lamp lights the phase is obviously live.

premises. The first step is to inform a responsible person that the electrical supply will be shut off prior to isolating the supply. A prominent notice should be displayed on the switch or fuse box warning personnel that electrical work is in the process of being carried out and the power must not be switched on. If it is possible to lock the fuse box do so, and keep the key along with any fuses in your toolbox or overall pocket. Finally a test must be made to check the system is isolated by using an approved test lamp shown in Fig. 11.27 or a voltmeter. Circuit testing screwdrivers are not always reliable.

### Inspecting and testing electrical installations

Any work carried out on an electrical installation must be tested to comply with the recommendations of the current edition of the *Wiring Regulations*, which covers all the requirements for both visual inspections and tests carried out using special instruments. It is recommended that more detailed information is obtained on the subject from publications listed in the further reading section of this chapter. Only those tests applying to electrical work carried out by a plumber are dealt with here, and it is important they are performed in a logical sequence. Testing procedures shown do not include three-phase installation.

#### *Visual inspection*

The main points to consider here are:

(a) All connections to terminals are properly secured.
(b) Cables are correctly supported and fixed having, where necessary, suitable protection against heat and mechanical damage.
(c) Where applicable, electrical components, switches, etc. should be clearly marked for easy identification.
(d) All protective devices such as fuses must be verified as to their correct rating.
(e) Checks should be made to ensure that all earthing and bonding cables are fitted and checked for security.

#### *Installation tests*

Testing complete electrical installations is not usually carried out by a plumber. It is, however, necessary to be aware of the associated testing procedures. A multifunction test meter is the most suitable for the kind of work a plumber is likely to undertake, and tests voltage resistance and current over a wide range of applications. Such an instrument is shown in Fig. 11.28. They are very sensitive and care must be taken when they are handled; they must be carefully stored when not in use. They are definitely not the type of tool to be thrown into a tool bag. It should be noted that these instruments can be seriously damaged if used incorrectly, and it is important that instructions for their use are read and understood. The following tests must be carried out in the sequence shown. The reader should note that an instrument called a 'Mega' is necessary for 'insulation testing'.

#### *Continuity of conductors*

This test is carried out to ensure that all conductors or wires are not broken and any connection such as terminals are secure. Figure 11.29 illustrates the method of carrying out this test. A slight resistance may be noted during this test, depending on the length of cable, but if it does not exceed more than a few ohms, the test results are acceptable.

#### *Insulation resistance test*

This test is made to ensure the insulation of both phase and neutral conductors is intact, and capable

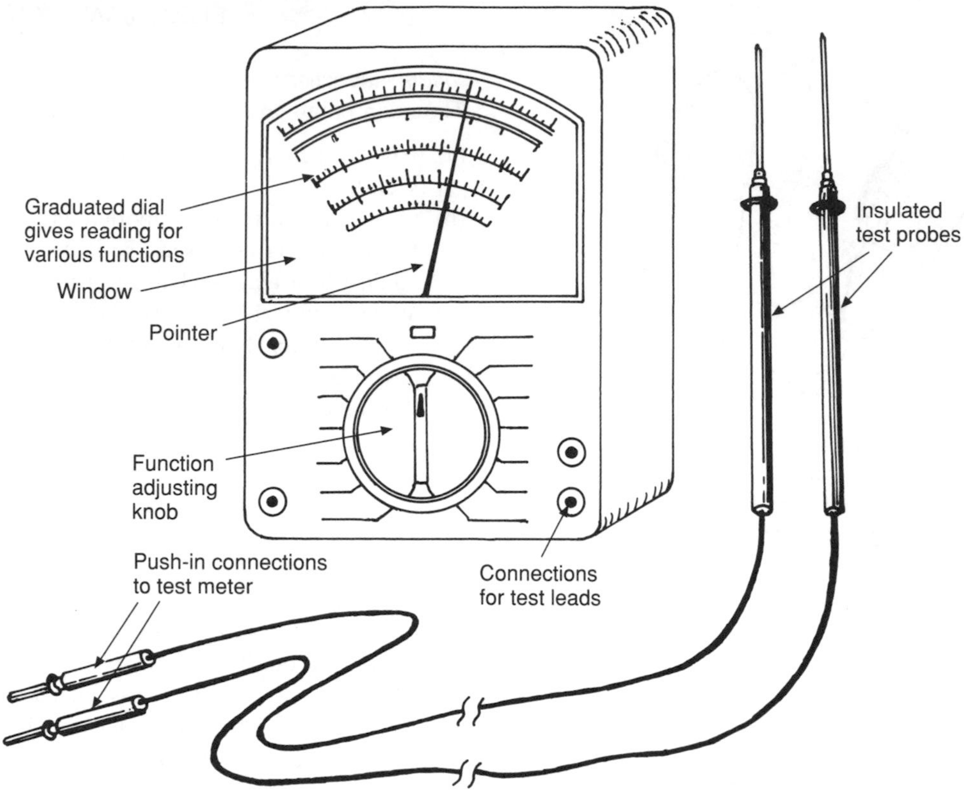


**Fig. 11.28** Multifunction test meter. This type of meter is suitable for most of the test functions required for the installation and testing of heating controls and equipment. It is not suitable for all the test procedures described, some of which require an instrument called a 'Mega'.

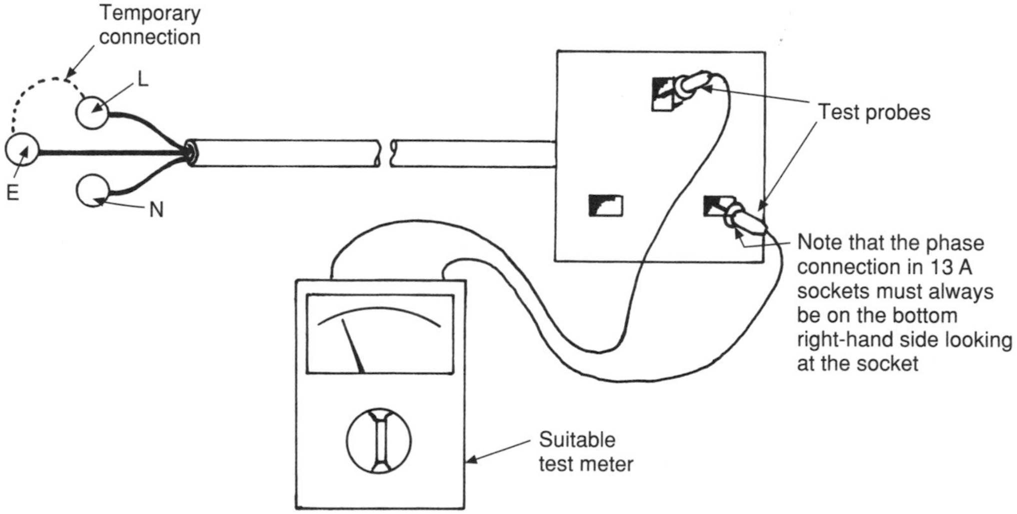


**Fig. 11.29** Testing the continuity of conductors. Note that the main supply is isolated while carrying out this test. This test is conducted to ensure that all the current conductors are sound, e.g. unbroken, and any connections are secure. The illustration shows the phase cable, connected to a socket outlet being tested. In this case the earth cable is used as part of the circuit and is temporarily connected to the phase terminal in the distribution panel. This test can also be carried out using a battery and bulb.

of preventing fault conditions (short-circuiting) between the current-carrying conductors and earth. If, for example, a cable has been subject to abrasion during installation, or inadvertently cut in such a way that the insulation is damaged, this test will detect it. Prior to conducting the test, any pilot or indicator lamps must be disconnected to avoid inaccurate readings being shown. Any components that may be damaged by the high test voltage employed should be temporarily disconnected. This includes most of the components used in heating control systems. The tests must be conducted with all fuses in place, with switches and circuit breakers closed. Where it is not practical to disconnect current-using equipment, the local switches controlling them must be in the open position. The tests are carried out using an insulation tester with a scale reading in ohms. The illustration in Fig. 11.30 shows how an insulation test is conducted on an electric immersion heater circuit.

*Polarity testing*

The test must be carried out to ensure that all circuit protection devices, e.g. fuses and switches, are connected to the phase conductor only. Figure 11.31 illustrates the possible result of reversed polarity, and this test is carried out to ensure a dangerous mistake such as this does not occur. Testing may be carried out as shown in Fig. 11.32 using a test meter, but it can be done using an electric bell or test lamp powered by a battery.

## Ancillary components

The following relates to some of the common components used with electrical equipment common to plumbing installations. The object is to enable the reader to identify them and understand their working principles.

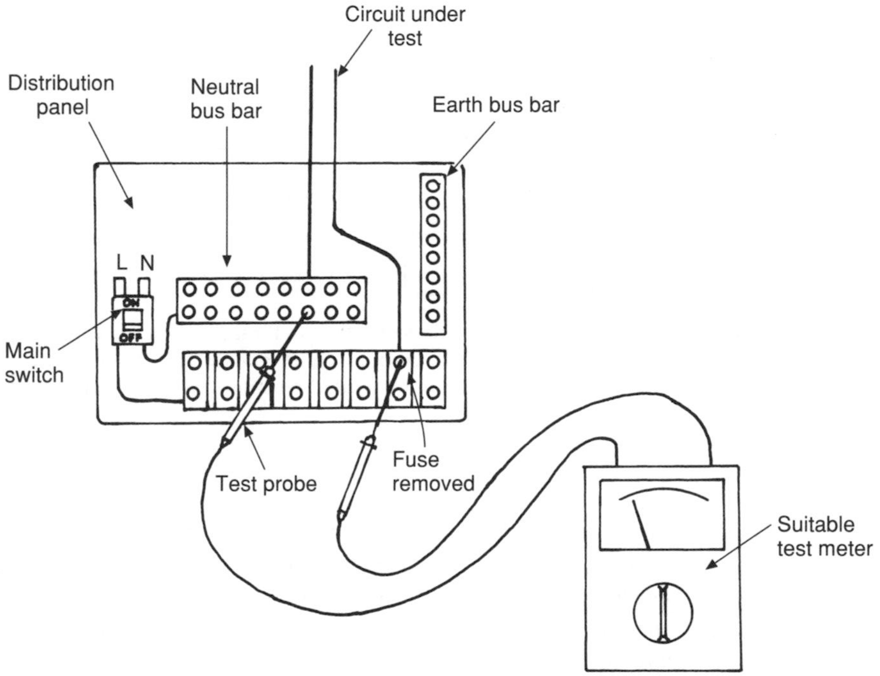


**Fig. 11.30** Insulation resistance test. The illustration shows this test between the phase and neutral cables. The circuit protective cable (earth) must also be checked in the same way. Between the phase and earth terminals for main supply installations the test voltage should not exceed 500 V. The minimum insulation resistance should not be less than 0.5 MΩ.

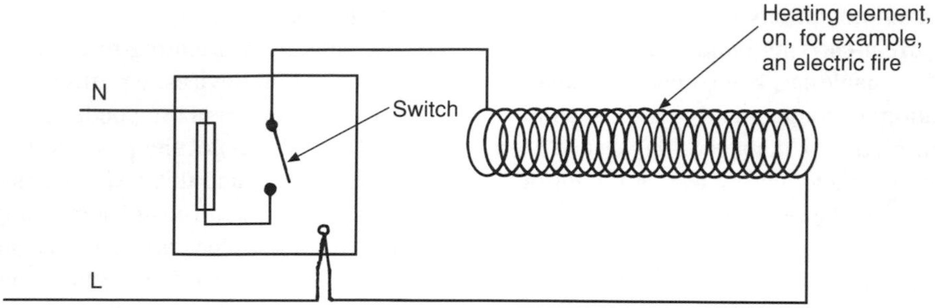


**Fig. 11.31** Illustrating the dangerous effects of incorrect polarity. This switch has been incorrectly wired so that although it is open and the circuit is switched off and the element does not glow, it is still live and anyone touching it could be subjected to a possibly fatal electric shock. The fuse is also unlikely to function should a fault occur.

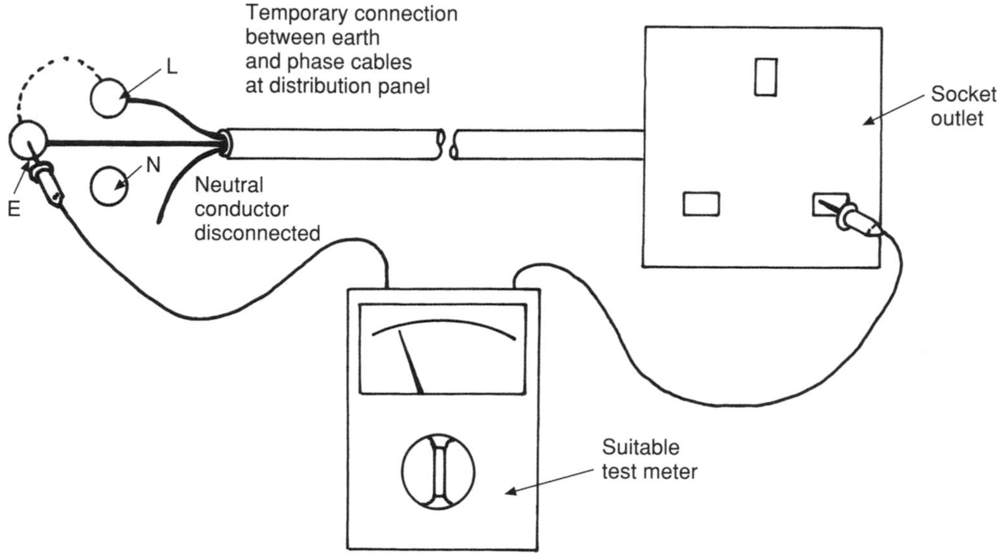


**Fig. 11.32** Polarity testing. This test ensures the correct polarity of the circuit throughout. It will be seen that this and the continuity test are very similar and in some cases are carried out simultaneously. It is essential that all fuses, circuit breakers and switches in the phase cable are in position and in the closed position prior to conducting this test.

*Resistors*

The resistance to the flow of current in a conductor produces heat, two good examples being the element in an electric immersion heater, fires or kettles and lighting where the element offers so much resistance it is heated to white heat. Those resistors used in circuit boards are of varying size and look very similar to a cartridge fuse. They are extensively used in the control systems of modern boilers.

*Capacitors*

The plumber will be familiar with this component where it is fitted to the motor of the pumps used for domestic heating. In this situation its purpose is to provide the starting torque for the rotor on the type of electric motor used. This electric motor will not start on the single-phase current, and to overcome this a means must be introduced to give the starting torque. This is effected by the capacitor in the way illustrated in Fig. 11.33. The solid line shows the

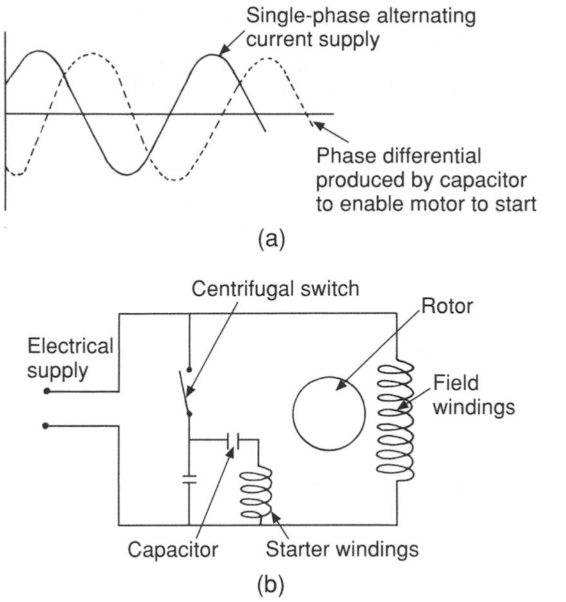


**Fig. 11.33** Capacitors — how a capacitor is necessary to start a single-phase electric motor. An electric motor will not start on single-phase current. A capacitor is normally used on small motors used for heating pumps, washing machines, etc., to provide a secondary phase to start the motor. When the rotor is turning the integral centrifugal switch contacts open, closing off the current to the starter windings.

alternating current (a.c.) flow from a single-phase supply. The broken line shows the effect of introducing a capacitor. Figure 11.33 shows a diagrammatic layout of a capacitor-start-type electric motor. When the motor is energised both the starting and running windings cause the rotor to revolve. At a predetermined speed the centrifugal switch will break the supply to the starter winding because, once rotating, its motor will enable the running winding to supply the necessary torque or momentum.

One of the most common causes of domestic heating pump failure is a faulty capacitor, usually indicated by the humming sound from the pump and no rotation of the motor. Care must be exercised when handling capacitors as they store an electric charge, and even when isolated they are capable of delivering an unpleasant electric shock. The terminals of a capacitor should be earthed to remove the charge when the capacitor is disconnected for maintenance purposes.

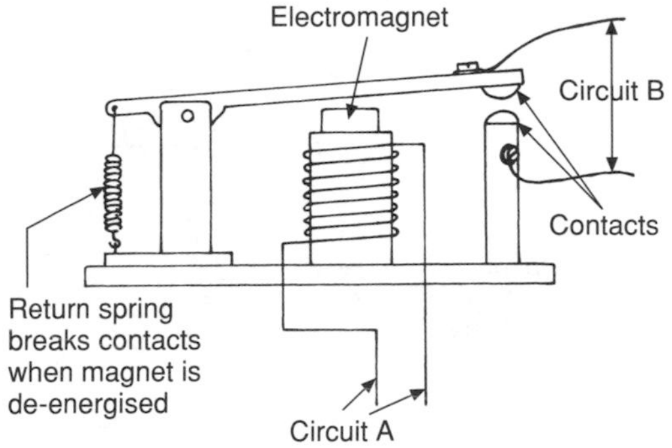


**Fig. 11.34** Application of electromagnet. Electromagnet acting as a relay; when circuit A is energised it makes the contacts of circuit B. This principle has many applications in the automatic functions of gas- and oil-fired equipment.

*Electromagnets*

These consist of a wire wound onto an iron core and when the coil is energised, the resulting magnetic field in the core is similar to that of the familiar permanent magnet. A simple experiment may be carried out by winding a length of fine insulated wire onto a wire nail about 75 mm long and connecting the two ends of the wire to a torch battery. The result will be a fairly powerful magnet. Electromagnets have many applications, one of the most common being used in an electric bell. Yet another is to operate a relay, which is basically an electric switch, when the coil of the relay is energised. This opens or closes another switch to control a separate circuit to that which operates the relay. In this way a low-voltage circuit can be used to control a higher voltage. Figure 11.34 shows a simple relay used in this way. Relays are widely used in automatic control systems and may be obtained with various contacts which can be either opened or closed by the single operation of a relay. Relays are also used in safety devices, when the magnetic force resulting from excessive current can be made to overcome the force of a retaining spring, resulting in the isolation of the voltage source. This forms the basis of many of the automatic cut-out devices on plumbing equipment.

Solenoids are another application of electromagnets. Figures 2.34, 2.38 and 2.45 show typical examples of their use.

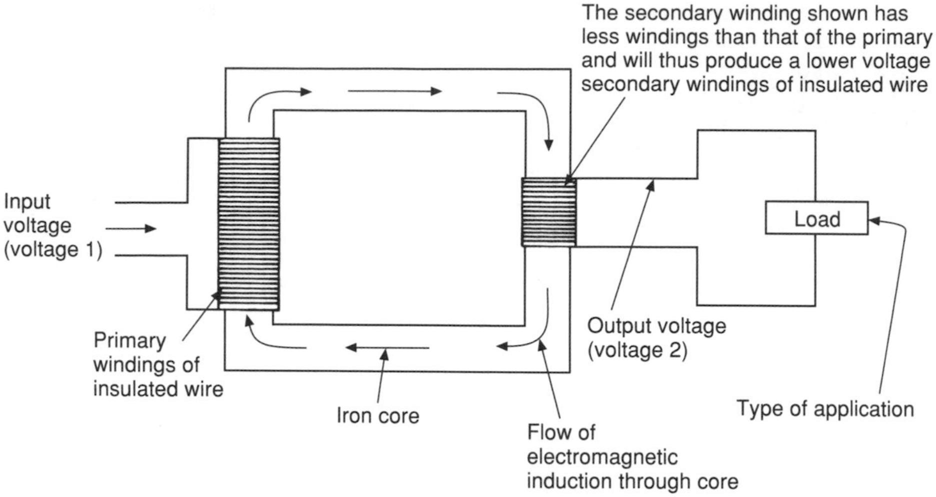


**Fig. 11.35** Basic transformer. A transformer provides an outlet voltage of a different value to that of the input. They can be used to step up or increase the voltage to provide the spark for ignition devices, or step down where the output or secondary voltage may be less than that of the primary.

*Transformers*

Electricity is produced in power stations and distributed by the national grid at high voltages, which must be reduced to 240 V for domestic supplies. It is quite often necessary to reduce this voltage still further in the interest of safety, the reduction to 110 V for power tools being a common example. Many components in heating equipment used by the plumber require an increase in the main voltage, a typical example being to produce a spark for ignition devices. The component used for reducing or increasing an a.c. voltage is known as a step-down or step-up transformer, which lowers or increases the voltage respectively. A transformer works on the principle of electromagnetic induction and consists in its basic form of two independent windings magnetically interlinked by an iron or iron alloy frame or core. When the current flows through one of the windings, called the *primary winding*, the magnetic field *induces* a voltage into the other winding, called the *secondary winding*. If there are more turns of wire on the secondary winding than on the primary, the voltage across the secondary will be proportionately greater. If the secondary has fewer turns of wire than the primary then the converse is true. The two coils are placed in close proximity. Figure 11.35 illustrates in diagrammatic form a simple transformer.

## Regulations relating to electrical appliances in bathrooms

Amendment No. 3 to BS 7671, Requirements for Electrical Installations, is now mandatory. The amendment covers section 601 on locations containing a bath or shower. Water, being a conductor of electricity, poses a real danger when it comes into contact with electrical appliances and switches, and previous legislation has been stringent. However, the latest regulations are very specific. Bath and shower rooms are divided into four zones: 0, 1, 2, and 3. Figure 11.36 shows these zones in connection with a shower. The lower the numeral, the greater will be the possibility of a dangerous situation. Thus zone '0' relates to a situation where the appliance is likely to be completely submerged, which is unusual with building services. The following is a guide to the types of controls and switches which may be used in the various zones. There are, however, certain exceptions where they are incorporated into the equipment designed for use in a specific zone. Normally the following applies:

Electrical equipment used in bathrooms is coded IPX or marked with a symbol as shown. The minimum degree of protection in the various zones is as follows:

Zone 0            IPX 7 or ₒₐ
Zone 1 and 2      IPX 4 or △. Where water jets may be used
                  for cleaning in public buildings, equipment
                  to IPX 5 or △ △ must be used.
Zone 3            As for Zone 1 and 2 but only if water jets are
                  used for cleaning.

(a) Plan view

Classified as
Zone 1 unless a tool,
e.g. a screwdriver, is
necessary to gain access;
it is then outside of zones

(b) Elevation

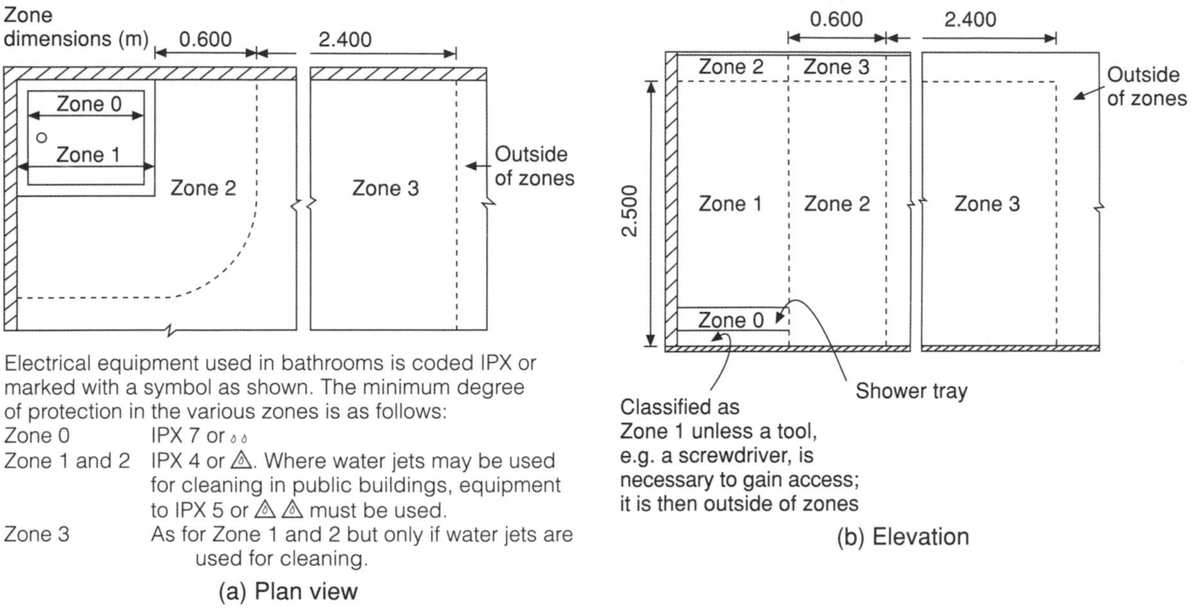

**Fig. 11.36** Electrical switches and controls in shower rooms.

Zone 0. No switch or controls normally permissible.

Zone 1. Only switches of up to 12 V d.c. or 30 V 'ripple-free' d.c. may be fitted, with all safety sources being installed outside of zones 0–1–2. The term ripple-free relates to certain types of rectifying equipment which are integral to the appliance.

Zone 2. The recommendations are the same as for zone 1 except for showering supply outlets complying with BS EN 60742 Parts 2 to 4.

Zone 3. As for zones 1 and 2 and additionally low-voltage socket outlets incorporating a safety isolating transformer.

The following fixed electrical current-using equipment may be used in zone 1, providing it meets the relevant standards and is suitable for the zone.

(a) A water heater, e.g. electric shower units.
(b) A shower pump.
(c) Other equipment which meets the requirements of IEE Regulations 412-06 (certain types of extractor fan may be suitable).
(d) Low-voltage current-using equipment.

All the foregoing are suitable for inclusion in zone 2, as are protection lighting, extractor fans, heaters, and whirlpool bath equipment. Insulated pull-cord switches complying with BS 3676 are suitable for use in zones 1, 2 and 3. See Fig. 11.18(a).

All the appliances listed in zones 1 and 2 may be used in zone 3, which also permits equipment other than fixed current devices which meet the requirements of IEE Regulations 412-06. They must be protected by a suitable residual current device affording automatic disconnection of the supply, and earth equipotential bonding. Where possible they should be used in a non-conducting location, e.g. out of reach of splashing or accidental contact with water.

**Further reading**

*IEI Regulations*
Details of publications relating to electrical work can be obtained from
The Institution of Engineering and Technology,
  Savoy Place, London WC2R 0BL
  Tel. 020 7240 7735. www.theiet.org
MK Electric, The Arnold Centre, Paycocke
  Road, Basildon, Essex SS14 3EA
  Tel. 01268 563000. www.mkelectric.co.uk

*Electrical testing equipment*
Megger Ltd, Archcliffe Rd, Dover, Kent CT17 9EN
Tel. 01304 502101. www.megger.com

**Self-testing questions**

1. Name the regulations to which all electrical installation work must comply.
2. Identify the difference between direct and alternating currents.
3. Identify the electrical terms used to define pressure of current flow, quantity of current and resistance to flow.
4. Assuming an appliance is supplied with a voltage of 240 and has a resistance of 60 Ω, state the amperage of a commercial cartridge fuse that can safely be used.
5. State the effect on a cable subject to overcurrent.
6. State the reasons for earthing electrical systems and explain the three main methods used.
7. Define the term 'equipotential bonding' and explain why it is necessary.
8. State the reason why high rating appliances such as electric water heaters and cookers are not recommended to be connected to a ring circuit.
9. Specify the type of insulation necessary for flexible cables where higher than normal temperatures may be encountered, e.g. adjacent to boilers or water heaters.
10. List the procedures taken to ensure safe electrical isolation when rewiring an electric immersion heater from an existing fused switch.
11. List the visual checks that must be made on an electrical installation prior to commissioning.
12. (a) State the purpose of conducting a polarity test after having installed a fused switch from an existing supply.
    (b) Describe the possible effect of incorrect polarity.
13. State the maximum floor area in a building that can be covered by one ring main.
14. (a) Explain the term 'spur' in connection to a ring main circuit.
    (b) Explain the regulations relating to the addition of 'spurs' and fixed appliance to a ring main.
15. State the maximum permissible voltage drop on a cable taken from the consumer unit to the furthermost power outlet on a circuit.

# Appendix A: Assignments

**Dormer window weathering assignment**

A dormer window is to be weathered in lead sheet, the relevant details of which are shown in Fig. A.1.

1. State the recommended codes of lead sheet for weathering:
   (a) The apron.
   (b) The cheeks.
   (c) The top.

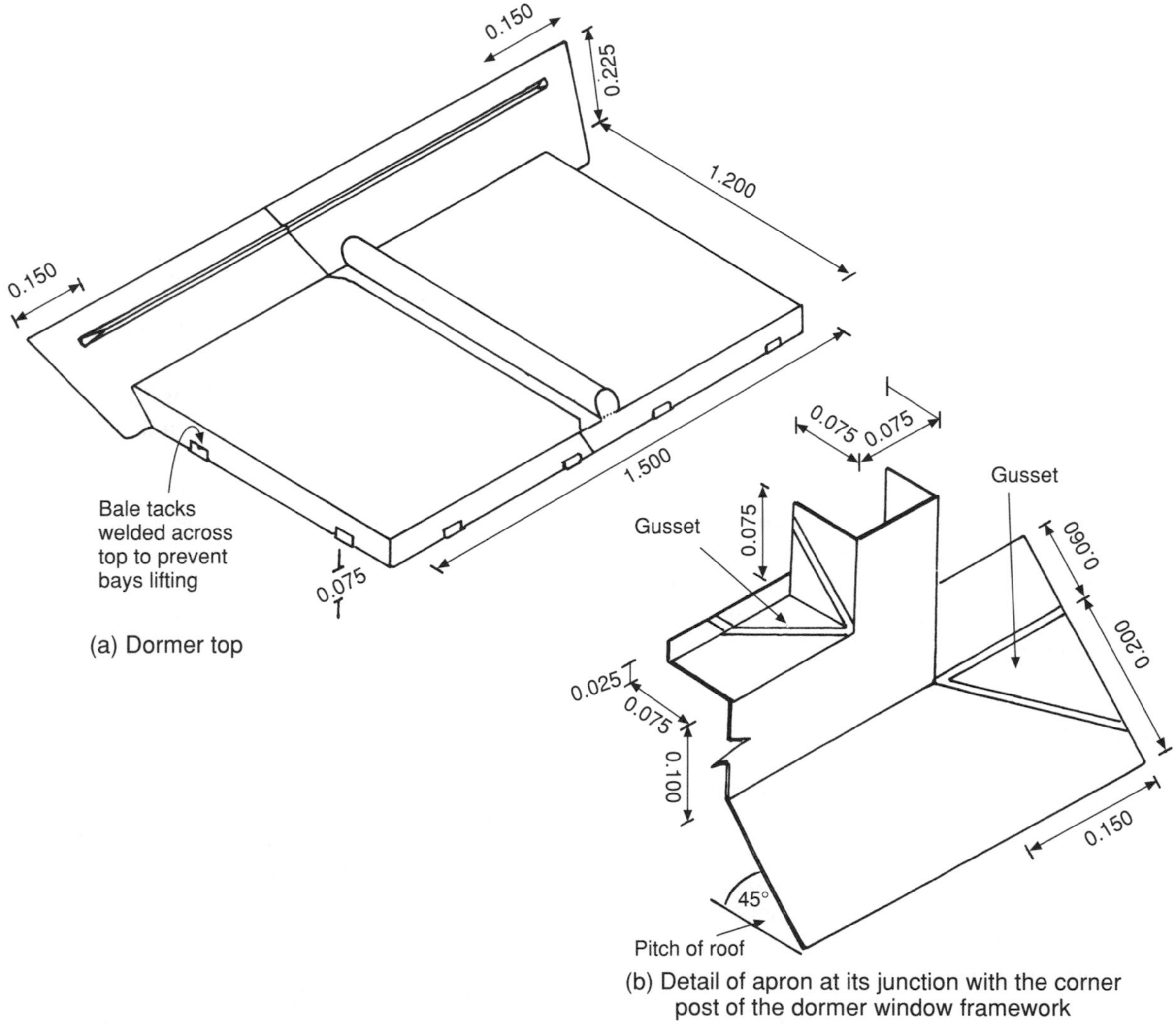


**Fig. A.1** Weathering dormer windows.

2. (a) From the measurement shown in Fig. A.1(b) specify the width of the lead sheet required to form the apron.
   (b) Set out half full size on paper the cuts and folds necessary to form the apron using lead welding techniques round the corner post and on to the roof.
   (c) Cut and fold the paper, make and insert the necessary gussets, securing them by masking or adhesive tape, and check the measurements and roof pitch against the details shown.
3. (a) Name three intermediate fixings that could be used to secure the dormer cheeks.
   (b) Make a sketch showing two methods of securing and finishing the front edge of the cheeks to the corner posts.
   (c) Specify the type of nails and cleats used for fixing lead sheet work.
4. To prevent high winds lifting the lead, the drip edges are shown secured by bale tacks welded to the edge. An alternative method of finishing this detail is the use of a torus roll. Make a sectional sketch through a torus roll including the lead weatherings.
5. Show by means of a sketch a suitable method for fabrication by lead, welding the roll overcloak at the junction with the roof. The weld is to be made out of position.
6. From the measurements shown in Fig. A.1 calculate the total area of lead required to weather the dormer top, making an allowance of 85 mm for the undercloak and 175 mm for the overcloak. Make an allowance of $12\frac{1}{2}$ per cent for waste.

### Hot and cold water supply assignment

The water service in a domestic property, which has been unoccupied during the winter months, is to be commissioned, and the estate agent responsible for the property has also asked for a general report on the hot and cold water services. Compile a report listing the remedial action necessary to correct the faults and defects you have found in the system.

1. Although the water was shut off and the pipework drained when the building became unoccupied, it is noted that some of the pipework is below the level of the drain cock and may have suffered frost damage. As much of the work is concealed below the floor level, describe how it could be tested for soundness prior to admitting water to the system.
2. The original cold water storage cistern has at some stage been replaced with two smaller ones. Figure A.2 illustrates these and the feed and expansion cistern. State the reasons why the cisterns and pipework as shown contravene the Water Regulations.
3. The cold water supply to a shower in the ground floor cloakroom is found to be connected directly to the cold water service pipe, and the water temperature at the shower rose is found to be very difficult to adjust. It is also noticed that there is a discharge from the overflow of the cisterns in the roof space when the shower is operating. State on your report the cause of these problems and how they should be rectified.
4. It is also noticed that when the cold tap on the bath is turned off suddenly, a loud noise is heard throughout the entire cold water system indicating it is subject to water hammer. Despite changing the washer and reducing the flow rate at the stopcock, the noise persists. List two possible components that could be used to solve this problem.
5. It is noted that a large quantity of water has to be drawn off from the hot supply to the wash basin in the downstairs toilet before water at a usable temperature is delivered. This is due to a very long draw-off. Describe a less wasteful method of providing hot water if an electrical power supply is available.

### Oil-fired heating assignment

A sealed system of heating is to be installed in the office suite of a medium-sized industrial building. The system is to be heated using an atomising oil fired boiler.

1. (a) State the recommended maximum working temperature of the heating system.
   (b) In the event of thermostat failure, what provision must be made to ensure the safety of the installation?

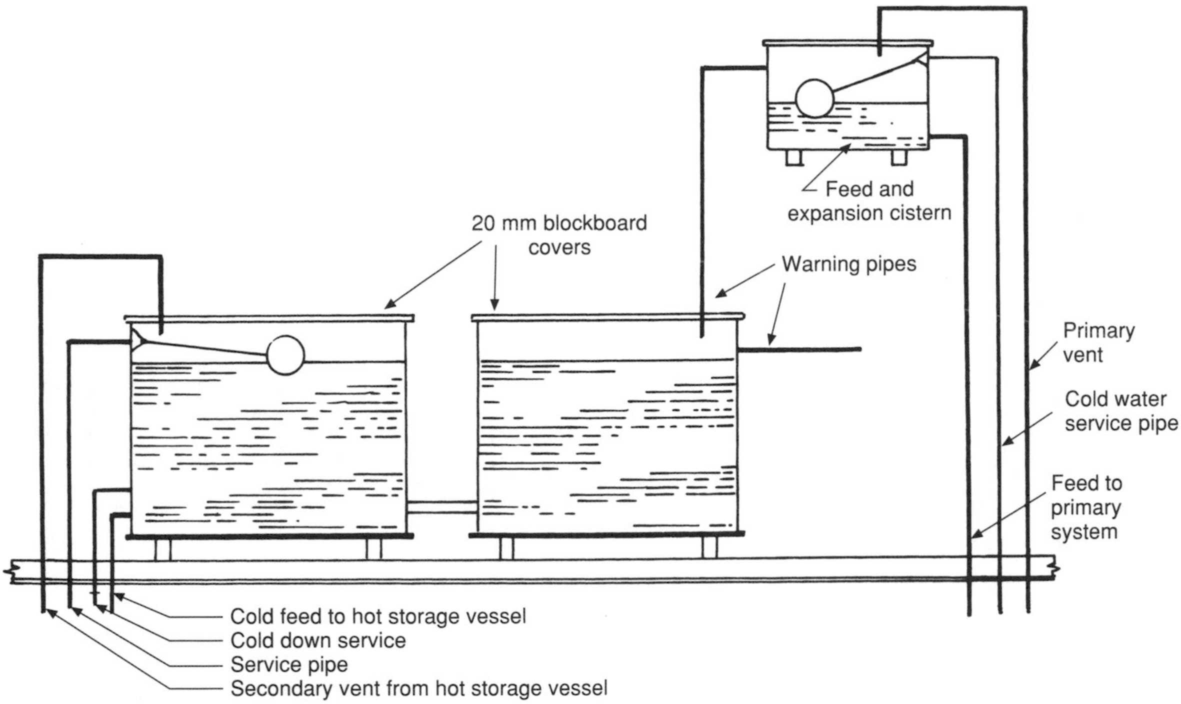


**Fig. A.2** Illustrating cold water storage cisterns and pipework in roof space. No insulation or valves shown.

2. (a) Specify two types of heat emitters that would be suitable for a sealed heating system and explain the reasons for your selection.
   (b) Specify two types of pipework systems that would be suitable for the heat emitters you have selected.
3. (a) Explain how the system should be filled from the main cold water service to comply with the Water Regulations.
   (b) Describe an alternative method of filling the system.
4. From the illustration shown in Fig. A.3, list the components A to J and describe their function.
5. Describe the procedures taken when commissioning the atomising burner, including all the tests necessary to ensure its efficient operation.
6. While conducting the tests it is found the chimney is subject to an excessive updraught. Specify a suitable component that must be fitted in the flue to overcome this problem and explain its working principles.
7. The owners of the premises have indicated that they would like to enter into a maintenance agreement. Prepare a schedule indicating all the items that must be checked during an annual service on the installation.

### Sanitation and drainage assignment

A three-storey office building is to be extended, necessitating an increase in the sanitary accommodation. Due to problems involved in connecting the new sanitary pipework to that in existence, it has been decided to install an additional discharge pipe serving a new sanitary annexe on the first floor. A 6 m length of drain will also be necessary from the stack to the existing PVC drain. The connection at the drain will be made by constructing an additional inspection chamber.

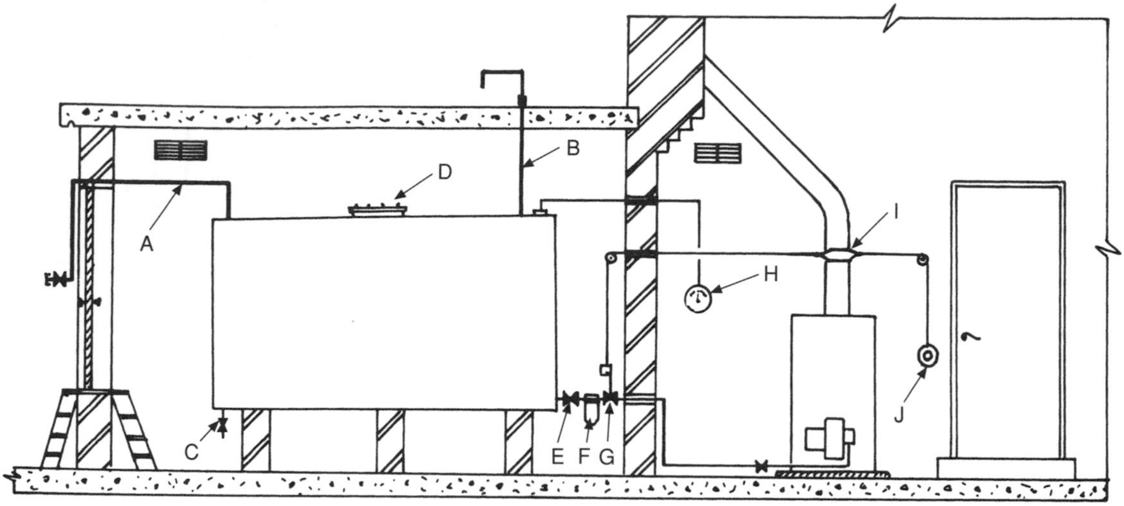


**Fig. A.3** Oil storage and boiler installation.

1. (a) If the invert of the existing drain is 1.2 m below ground level, and the gradient is 1 in 40, what will be the invert level at the upper end?
   (b) Describe a suitable method of checking the invert on a short length of drain.
   (c) Describe the type of bedding necessary for flexible underground drainpipes to comply with the Building Regulations.
2. (a) Explain how you would connect the new branch into the existing drain to cause as little inconvenience as possible to the occupants of the existing building.
   (b) Select a suitable component for making the actual connection of the new branch drain to the existing main drain.
   (c) Describe an approved method of water testing the underground drain on completion.
3. The first and second floors of the extension are constructed of reinforced concrete, with holes left at the points where the main discharge stack passes through them.
   (a) State how the gap where the discharge pipe passes through the floor is sealed to prevent the spread of fire in the event of an emergency.
   (b) State the recommendations relating to (i) the main discharge stack where it terminates as a vent to the atmosphere, and (ii) the type of bend used at the base of the stack.
4. (a) Explain why an excessive number of bends on a branch discharge pipe may cause siphonage of the appliance traps.
   (b) In a situation where the maximum recommended length of a basin waste is exceeded, describe three acceptable methods of overcoming the possible siphonage of the trap seals.
5. An automatic flushing cistern is fitted serving three bowl urinals in the male toilet.
   (a) State two methods of controlling the frequency of flushing which comply with the Water Regulations.
   (b) List the advantages and disadvantages of bowl urinals.
6. Describe the approved method of conducting soundness and performance tests on the completed sanitary pipework installation.

# Appendix B: NVQ in plumbing — Levels 2 and 3

### NVQs in Mechanical Engineering Services — Plumbing

National Vocational Qualifications (NVQs) are a system of certification which has been introduced in England and Wales. They have replaced the long-established method of assessing awards by external examinations, which were accepted by industry for many years, in recognition of the achievement of a craft qualification. NVQs are based on continuous assessment rather than examinations and, like their Scottish equivalent, Scottish Vocational Qualifications (SVQs), reflect standards that have been agreed by the plumbing industry. They have been developed jointly by the British Plumbing Employers Council, plumbing employers, organisations and the Joint Industry Board (JIB) for Plumbing.

In England and Wales there are two levels of NVQ in plumbing: level 2 and 3 in Mechanical Engineering Services — Plumbing. (The scheme described here is related to Vocational Scheme 6129 Basic Plumbing Studies, Technical Certificate.)

### Structure

Each level has a specified number of units of competence, which are divided into elements. Each element of competence has performance criteria and range statements. The element defines the standards which have to be met and the range defines the circumstances in which the standard must be applied.

### Subjects covered

*Units*

001 Maintain the safe working environment when undertaking plumbing work activities

002 Maintain effective plumbing working relationships

003 Contribute to the improvement of the plumbing work environment

004 Contribute to the improvement of plumbing business products and services

005 Install non-complex plumbing systems and components

006 Decommission non-complex plumbing systems

007 Maintain non-complex plumbing systems and components

008 Plan complex domestic plumbing work activities

009 Install complex domestic plumbing systems and components

010 Commission and decommission complex domestic plumbing systems

011 Service and maintain complex domestic plumbing systems and components

*Optional units — oil systems and appliances*

030 Plan domestic plumbing oil heating systems work activities

031 Install domestic plumbing oil heating systems and components

032 Commission and decommission domestic plumbing oil heating systems

033 Service and maintain domestic plumbing oil heating systems and components

*Optional units — gas systems and appliances*

034 Plan domestic plumbing gas systems work activities

035 Install domestic plumbing gas systems and components

036 Commission and decommission domestic plumbing gas systems

037 Service and maintain domestic plumbing gas systems and components

*Additional units*

040 Design domestic plumbing systems
041 Specify programmes for working on domestic plumbing systems

*Additional units — sheet weathering systems (lead)*

050 Plan plumbing sheet weathering work activities
051 Install plumbing sheet weathering components
052 Maintain plumbing sheet weathering components

*Additional units — fire control systems*

060 Plan domestic plumbing fire control systems work activities
061 Install domestic plumbing fire control systems and components
062 Commission and decommission domestic plumbing fire control systems
063 Service and maintain domestic plumbing fire control systems and components

## Assessment

The syllabus (which summarises what has to be assessed) is dictated by the element of competence, the performance criteria and the range statements. Generally, the assessment for each unit comprises:

- practical observations
- oral questions
- written questions

Written evidence is required for the practical observations.

There is no need to assess all elements of a unit at once. Each element can be assessed separately. When a candidate has satisfied all the performance criteria for each element in a unit the candidate will be credited with that unit. The candidate must meet all performance criteria successfully for each element and satisfy the necessary knowledge requirements.

There is no requirement for units to be completed in a particular order. Candidates can complete separate elements in several units before completing any one unit. Units and elements can also be assessed in any order. In addition, units do not have to be completed within a certain time.

*Work-based assignments*

An integral part of the course is the continuous work-based assessment of the candidate's work which can be carried out in the workplace or at an appropriate centre. Assessments are made on a range of practical activities and on documentary evidence provided by the candidates to prove their competence. The evidence may be in the form of photographs or written statements by employers or clients. Where assessment in the workplace is not possible, it can be carried out through simulated activities and tasks in an approved centre.

It is, however, mandatory that some work must be workplace orientated. A workplace recorder will certify competence of the candidate in the workplace subject to the agreement of the centre assessor. The recorder must be suitably qualified and experienced and able to meet the requirements of the plumbing industries qualifications and assessment strategy.

The requirements for the award of the Technical Certificate in Plumbing are the same as those for NVQ levels 2/3. They are available to people who are unable to meet the retraining and educational requirements of the modern apprenticeship scheme. All the practical work and assessments for this course are taken in a training/further educational establishment

*Activity record*

An activity record is a written description of the activities that have been undertaken by the candidate during the completion of a unit.
The record lists details of the following:

- Activities undertaken, including production techniques used accompanied by appropriate sketches.
- Any problems or difficulties encountered and how they were resolved.
- Knowledge of what was done and of any relevant legislation and company policies that applied to the activities.
- How the job was planned, including any liaison with other personnel, co-contractors and customers.
- All safety precautions taken.
- Materials and equipment used.

*Related knowledge and understanding*
A candidate's performance may be demonstrated directly. However, in many cases satisfactory performance alone is not enough and the candidate must show an understanding of the task being done. Standard assessments have been prepared by the awarding body to allow candidates to show that they have this knowledge and understanding.

The assessments comprise a series of questions which require a short written answer or a sketch. Sometimes this will need to be supplemented by oral questions, which will be asked by the assessor. Centres may sometimes provide their own assessments.

*Accreditation of prior learning (APL)*
Candidates can use evidence from work or other activities undertaken before starting the NVQ in the assessment of a unit. The evidence must satisfy the performance criteria and range statements in the same way as evidence gathered while working towards the NVQ.

*Portfolio of evidence*
The 'portfolio of evidence' is the documentation which the candidate submits to the assessor for assessment of an element or unit. This will usually be a file or folder in which the candidate keeps all their evidence. It should be emphasised that it is the candidate's responsibility to keep the portfolio and add the appropriate evidence.

The portfolio should include a completed record of assessment and the supporting evidence. The evidence can be anything that illustrates the candidate's competence. Forms of evidence can include authenticated photographs, job cards, time sheets, appraisals from line managers or supervisors, testimonials, video and audio tapes and computer disks.

The portfolio must be accessible to the reader. It should be divided into sections which relate to the units and elements of the NVQ. The information must be coherent and include a contents page, dates on all entries and titles and headings to describe the contents. All related entries should be cross-referenced, indicating the relevant units and elements of the standards.

## Sources of information

Summit Skills is the agency responsible to the government for building services training and works closely with other training organisations. Its main aims are to raise the standard of training through City & Guilds external verifiers who ensure national standards are maintained. It also identifies training provision and supports employers in recruitment. Centres delivering the industries' NVQs are responsible for arranging, monitoring, evaluating and maintaining the assessment verification and certification system.

Further information on NVQs and SVQs can be obtained from the following organisations.

City & Guilds London Institute,
1 Giltspur Street,
London EC1A 9DD
Tel. 0207 294 2468. www.cityandguilds.com

Scottish Qualifications Authority (SQA)
58 Robertson Street,
Glasgow G2 8DQ
Tel. 0845 279 1000. www.sqa.org.uk

Summit Skills,
Vega House,
Opal Drive,
Fox Milne,
Milton Keynes MK15 0DF
Tel: 01908 303960. www.summitskills.org.uk

# Appendix C: Central heating control systems

The illustrations shown in this appendix shows all the main central heating control systems recommended by the Energy Saving Trust (EST). They comply with the mandatory requirements of the Building Regulations Part L and in some instances the 'best practice' which exceeds those which are mandatory. To select a suitable control system follow the guidance notes shown.

The semi gravity systems shown here in the illustrations 13–16 are not suitable for solid fuel boilers an no heat leak is provided. Electrical control systems suitable for solid fuel appliances are dealt with in chapter 4.

TACMA
THE ASSOCIATION OF CONTROLS MANUFACTURERS

# THE ASSOCIATION OF CONTROLS MANUFACTURERS

## HOW TO COMPLY WITH THE REVISED BUILDING REGULATIONS 2001/2002

### THE GUIDE

This guide supports the official guidance covering the use of controls in boiler based, gas and oil fired, domestic central heating installations within the revised Building Regulations, Part L1 for England & Wales and Part J for Scotland, 2002.

If followed, the guidance and example systems shown will provide confidence that designs and installations fully comply with the relevant parts of the Building Regulations.

### CONTROLS FOR NEW DWELLINGS

**Part L1, England and Wales: Part J, Scotland**

**1)** Provide installations that are fully pumped.
**2)** Provide independent time & temperature control to both the heating and hot water circuits. *Part J, Scotland requires the installation of 7-day timing.*
**3)** Provide control systems with a boiler interlock. *See note 1.*
**4)** Install an automatic by-pass valve if a bypass is fitted. *See note 2.*
**5)** Split the heating circuit into zones using either:
a) room thermostats or programmable room thermostats in all zones.
b) a room thermostat or programmable room thermostat in the main zone and Thermostatic Radiator Valves (TRV's) on all radiators in the other zones.
**6)** Ensure installation of a cylinder thermostat & a zone valve to control stored hot water. *See note 3.*
**7)** Provide time control by the use of either:
a) a full programmer with separate timing to each circuit.
b) two or more separate timers providing timing control to each circuit
c) programmable room thermostat(s) to the heating circuit(s), with separate timing of the hot water.
**8)** For dwellings with a total usable floor space greater than 150m$^2$, then:
a) the heating circuit should be split into a minimum of two zones plus a hot water service zone and:
b) each zone should be separately timed by the use of a multi-channel programmer or multiple heating programmers or programmable room thermostats
**9)** Any Boiler Management Control System that meets the specified zoning, timing & temperature requirements is a wholly acceptable alternative.

### CONTROLS FOR EXISTING DWELLINGS

**Part L 1, England and Wales.**

When a new installation is fitted, the controls should be as for a new dwelling.

When replacing the boiler and/or the hot water vessel, the opportunity to improve the controls should be considered. To be confident that the requirements of the Building Regulations are met this would entail the following:

**1)** Provide fully pumped installations with a boiler interlock. *See note 1.*
**2)** Install an automatic by-pass valve, if a bypass is fitted. *See note 2.*
**3)** Provide time and temperature control to both the heating & hot water circuits, for fully pumped installations.
**4.1)** Separate the space heating system into zones and:
a) if a room thermostat is already fitted fit TRV's on at least all radiators in the sleeping areas and check if a new room thermostat or programmable room thermostat should be fitted.
b) if there is no room thermostat, install either a room thermostat or programmable room thermostat and fit TRV's on at least all radiators in the sleeping areas.
**4.2)** Provide time control by the use of either:
a) a full, standard or mini programmer, or
b) one or more separate time switches, or
c) programmable room thermostat(s).
**5)** For semi-gravity installations, the recommended option is to convert it to fully pumped. The controls should then be as detailed above.
**5.1)** If either the new boiler or hot water storage vessel can only be used on a pumped circuit, the installation must be converted to fully pumped. Install controls as detailed above.
**5.2)** If the new boiler or storage vessel is designed for gravity hot water, or it is impractical to convert to a fully pumped installation, provide the following controls:
a) a cylinder thermostat & zone valve to control the hot water and provide a boiler interlock. *See note 3.*
b) a room or programmable room thermostat.
c) a programmer or time switch.
d) TRV's on at least all radiators in the bedrooms.
**6)** Any Boiler Management Control System that meets the specified zoning, timing & temperature requirements, is a wholly acceptable alternative.

***Note 1:** Boiler interlock is achieved by the correct use of the room thermostat(s) or programmable room thermostat(s), the cylinder thermostat and zone valve(s) in conjunction with the timing device(s). These should be wired such that when there is no demand from the heating or hot water both the pump and boiler are switched off. The use of TRV's alone does not provide interlock.*

***Note 2:** Although the regulations may not always require a bypass to be fitted, the performance of any installation with multi-zoning or TRV's is improved by the use of an automatic by-pass valve. **(See Good Practice Guide 302)***

***Note 3:** The use of non-electric hot water contr ollers does not meet this requirement.*

| To select a control system | Temperature & time control | 3 port valve control set |
|---|---|---|
| a) Select time and temperature control requirements by reading down the adjacent column.<br><br>b) Select the valve control set by reading across the column headings. | Programmer and standard room and cylinder thermostats to the heating and hot water circuits. | 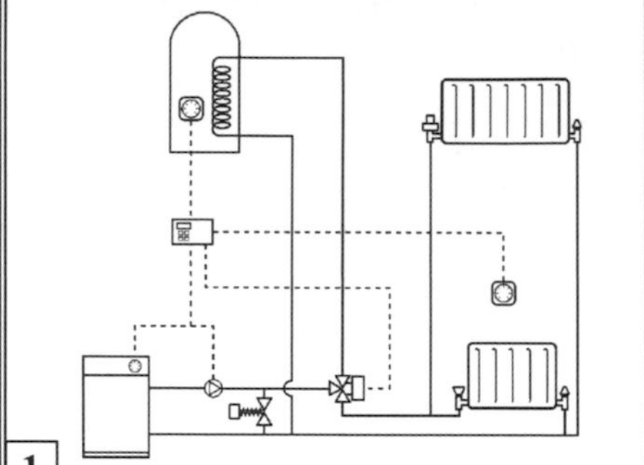<br>1 |
| c) For semi-gravity installations refer to the end column.<br><br>d) For combination boilers see overleaf. | Programmable room thermostat to the heating circuit(s), with timer and standard cylinder thermostat to the hot water circuit. | 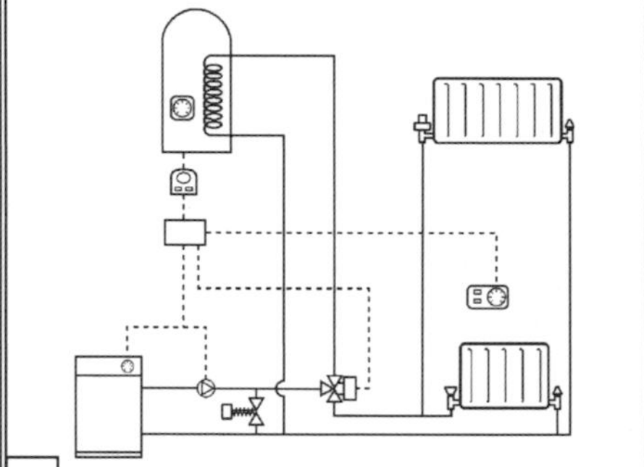<br>2 |
| **Control Symbols**<br>Room thermostat<br>Programmable room thermostat<br>Cylinder thermostat<br>Basic timer<br>Mini programmer<br>Standard programmer<br>Full Programmer<br>Multi-channel programmer<br>Boiler management controller<br>Temperature sensor<br>Two port valve<br>Three port valve<br>Auto by-pass valve<br>Junction box<br>Thermostatic radiator valve<br>Wheel-head valve<br>Lock shield valve<br>Pump<br>Boiler thermostat | Multi-channel programmer or multiple timers to the heating and hot water circuits, with standard room and cylinder thermostats. | 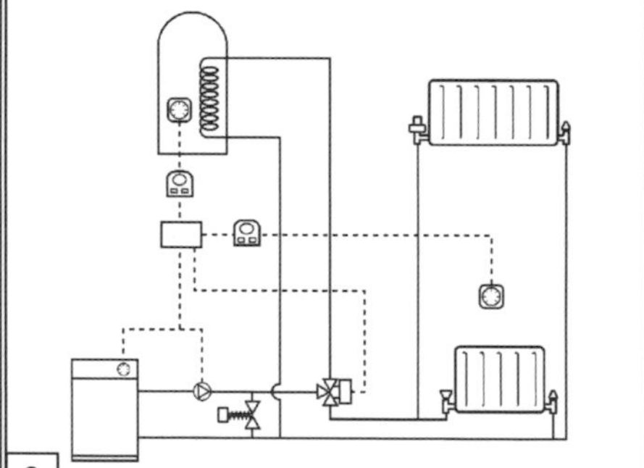<br>3 |
| | Boiler management control systems. | 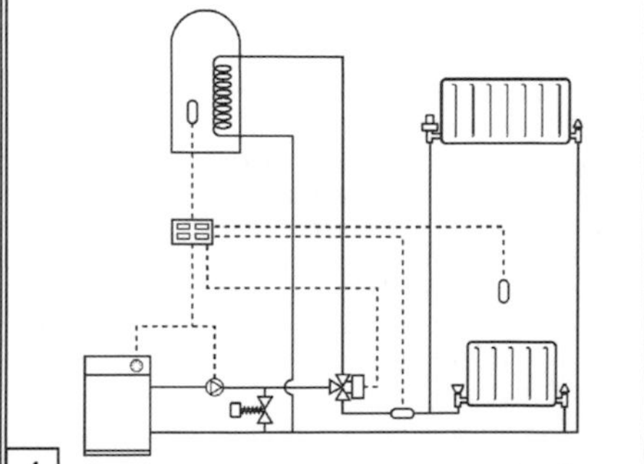<br>4 |

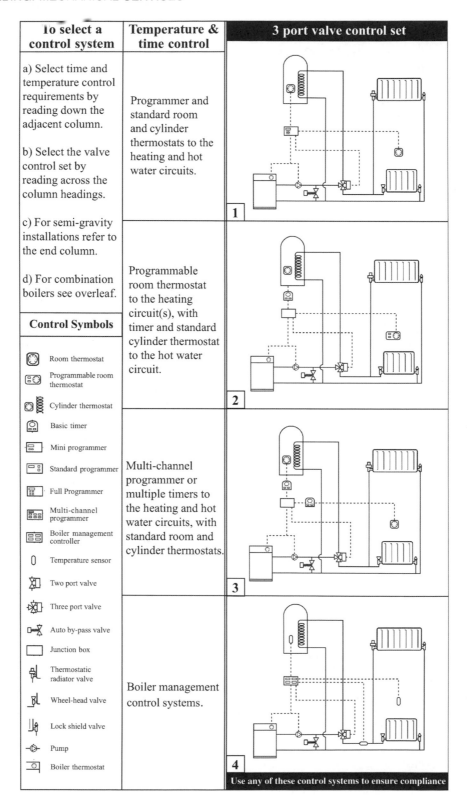

**Use any of these control systems to ensure compliance**

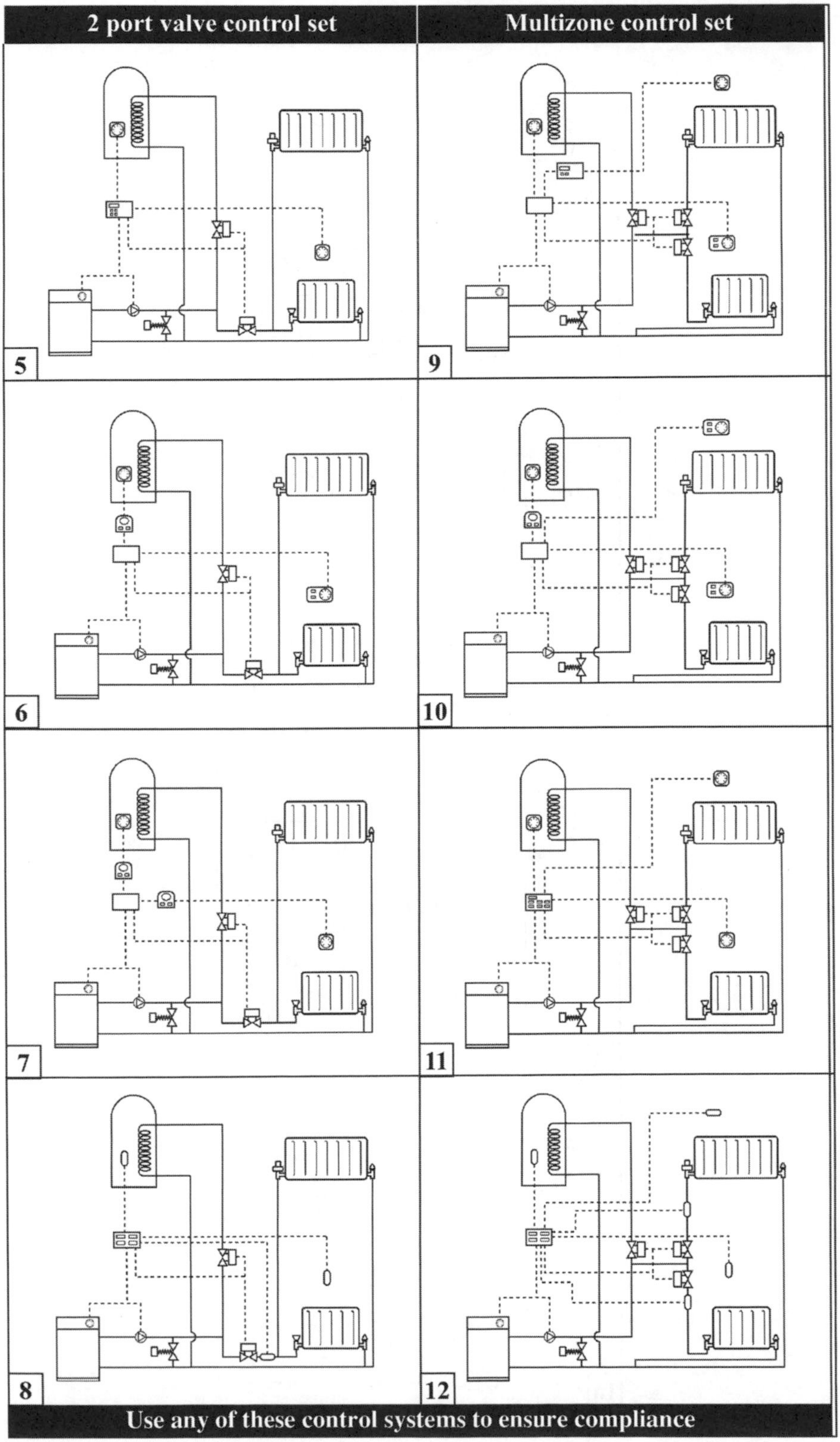

2 port valve control set
Multizone control set
5
9
6
10
7
11
8
12
Use any of these control systems to ensure compliance

**Semi-gravity control set**

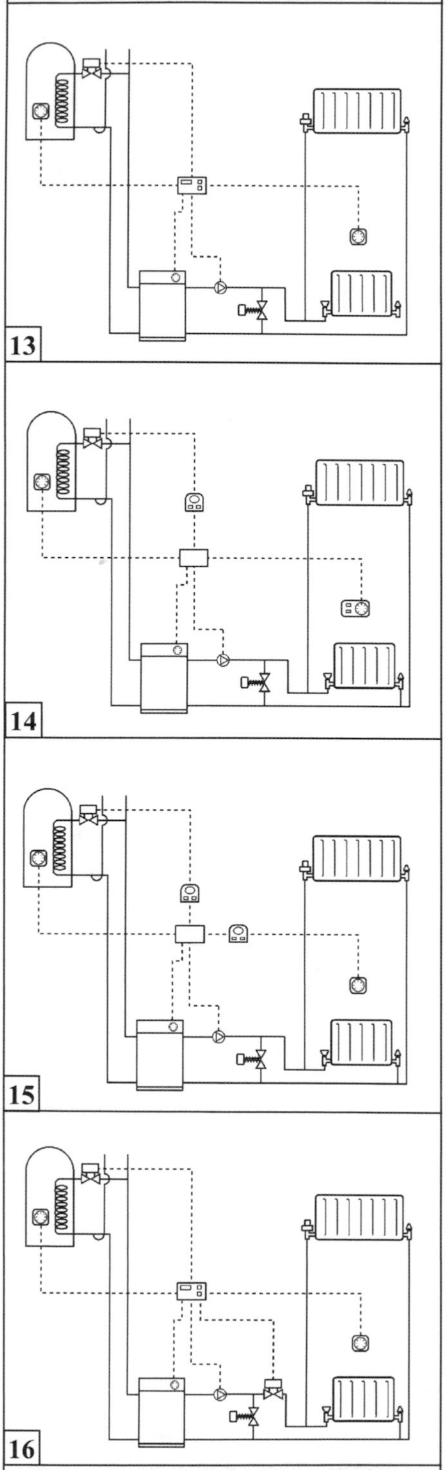


**Use only on existing installations**

## Control systems for installations incorporating a Combination boiler

Comprising a timer and room thermostat for time and temperature control of the heating circuit, interlocked to the boiler.

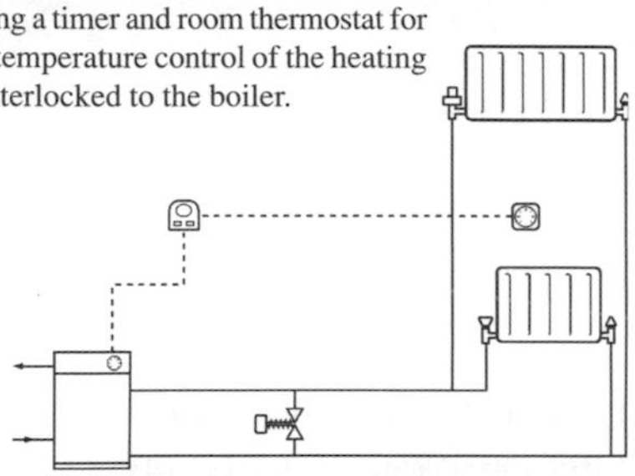

17

Comprising a programmable room thermostat for time and temperature control of the heating circuit, interlocked to the boiler.

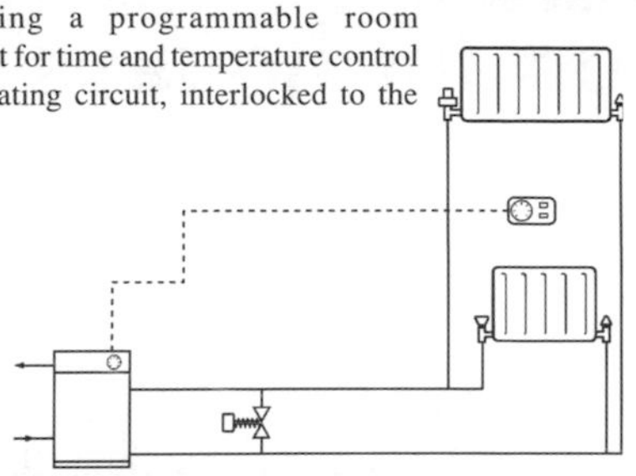

18

**A Combination boiler with TRV's only does not comply**

**For further information on the revised Building Regulations contact TACMA on www. heatingcontrols.org.uk or email: BRegs@beama.org.uk**

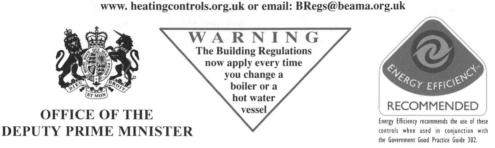

DIEU ET MON DROIT

**OFFICE OF THE DEPUTY PRIME MINISTER**

**WARNING**
**The Building Regulations now apply every time you change a boiler or a hot water vessel**

ENERGY EFFICIENCY™ RECOMMENDED

Energy Efficiency recommends the use of these controls when used in conjunction with the Government Good Practice Guide 302.

The control systems detailed in this guide meet the requirements of the Building Regulations in England, Wales and Scotland that came into effect in 2002 concerning the control of gas and oil fired hot water central heating installations. The guidance and example system designs provide the best advice on how to meet the Building Regulations for both new and existing dwellings with confidence. Other system designs could be used but designers and installers would need to show that these also meet the Building Regulations in full.

This guide has been prepared by the members of the Domestic Heating Controls Group of TACMA, that represents the interests of the manufacturers of timing controls, room & programmable room thermostats, thermostatic radiator valves and zone valves for the control of domestic central heating.

TACMA THE ASSOCIATION OF CONTROLS MANUFACTURERS

TACMA
Westminster Tower,
3, Albert Embankment,
LONDON. SE1 7SL
Tel. 0207 793 3008
Fax. 0207 793 3003

www.heatingcontrols.org.uk
email: BRegs@beama.org.uk

**Danfoss Randall Ltd.**
Tel. 01234 364621
www.danfoss-randall.co.uk

**Honeywell Control Systems Ltd.**
Tel. 01344 656000
www.honeywell.com/uk/homes.htm

**Horstmann Controls Ltd.**
Tel. 0117 9788700
www.horstmann.co.uk

**Invensys Climate Controls Ltd.**
Tel. 0845 130 5522
www.climate-eu.invensys.com

**Myson Heating Controls**
Tel. 01914 917530
www.myson.ie

**Pegler Ltd.**
Tel. 01302 560560
www.pegler.co.uk

**Reliance Water Controls.**
Tel. 01386 47148
www.rwc.co.uk

**Siemens HVAC Products**
Tel. 01784 461616
www.landisstaefa.co.uk

**Sunvic Controls Ltd.**
Tel. 01698 812944
www.sunvic.co.uk

BRPL1/cJc/07 2003.

# Appendix D: Whole house boiler-sizing method for houses

This is a simplified method of assessing the heating requirements in standard types of building. It enables the user to identify the correct rating of any replacement boiler, bearing in mind that in many cases boilers have been oversized in the past which has led to a greater initial cost and when used without thermostatic valves leads to waste of energy. The easy-to-follow instructions are fairly comprehensive and the U values shown are suitable for most post-war domestic properties. In the event of any doubt relating to buildings falling outside this parameter, reference should be made to BS 5449 and BS EN 12828. Worksheets are available from the Energy Savings Trust and can be downloaded free of charge.

## The aims of this method

Replacement boilers are rarely sized correctly. Oversized boilers cost more to purchase and generally operate less efficiently resulting in higher running costs and increased emissions to the atmosphere. This 'whole house' procedure provides the busy heating installer with a simple but reasonably accurate method of sizing which is both quick and easy to use.

This method is aimed at typical dwelling types found in the UK as indicated by the U-values and window areas in the tables opposite. Where the dwelling is untypical, then a more detailed procedure should be used. The method should only be used for gas, oil and LPG boilers up to 25 kW and should not be used for combination boilers or solid fuel heating systems. It is based on a number of assumptions:

- a design internal temperature of 21°C (included in the location factor)
- design external temperatures, dependent on the location of the property (included in the location factor)
- an allowance of 10% for intermittent heating (included in the location factor)
- an allowance of 5% for pipe losses (included in the location factor)
- a ventilation rate of 0.7 air changes per hour (included in the 0.25 ventilation factor)
- an allowance of 2 kW for heating hot water

### Instructions

Complete section 1 to 7 by filling in the shaded boxes from actual measurements, or from the tables. The unshaded boxes should be filled by calculation.

**Assess the shape of the house**

**A** Simple rectangular shaped dwellings can be covered by the worksheet opposite.

**B** Small extentions or loft conversions, with up to two radiators, simply note the radiator output and add them in at section 7 of the worksheet and use the worksheet for the remainder of the house.

**C** More complicated dwelling shapes should be divided into rectangular boxes. Use a separate worksheet for each main box (e.g. divide an L-shaped property into two boxes, and do two calculations).

### Section 1

**Take internal measurements (in metres) of the overall length, width and room height.** The width is generally taken across the front of the property and the length is a front-to-back measurement. Also note the type of dwelling, the number of floors (excluding any loft conversion) and the number of external walls along the length and width.

### Section 2

**Calculate the total external wall area (including windows) in square metres.** Count the number of 'length' external walls along the length measurement and the number of 'width' walls along the width measurement, e.g. a semi-detached will have only one external wall along its length but two along its width. The wall is not regarded as external where any extensions join the main property. Where there is a single storey extension on a two storey house then take half the wall area as external. The whole wall is still regarded as external when it is attached to an unheated garage.

### Section 3

**Using values from tables 1, 2, and 3, calculate the heat losses from the windows and masonry walls.**

If the property has an unusually large number of windows then use the actual window area taken from measurements.

### Section 4

**Calculate the roof and floor areas using the length and width.** If the dwelling is a mid or bottom flat then use a **roof** area of zero. If it is a mid or top flat then use a **floor** area of zero. Calculate the heat losses from the floor and roof using table 4.

### Section 5

**Add boxes A, B, C, and D and multiply by the location factor from table 5** to reach the overall heat losses from conduction through the fabric. This factor includes an allowance for design temperatures, intermittent heating and pipe losses.

### Section 6

**Using the floor plan area, calculate the volume of the dwelling.** Calculate the overall ventilation heat losses again using the same value from table 5.

### Section 7

**Calculate boiler output (in kW)**

- Add the fabric (E) and ventilation (F) losses and a further 2000 Watts for heating hot water.
- Add in the results of any separate worksheets for extensions or box-shapes identified in stage 1.
- Simply add in the radiator outputs (in Watts) for loft conversions or small extensions.
- This gives the required boiler output in Watts. To convert to kW divide by 1000. This calculation worksheet includes all the necessary factors and no further additions should be made to the end result. Choose a boiler that is closest, but above, the calculated output, eg for a calculated output of 8.7 kW install a 9 kW boiler.

# Boiler sizing worksheet

## Assess the dwelling shape

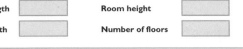

**A. Simple rectangular dwelling** Use worksheet alone.

**B. Extension and loft conversions** Use worksheet and add on radiators sizes in section 7.

**C. Non-rectangular dwelling** Divide into sections and repeat calculations.

## 1. Take three measurements (in metres)

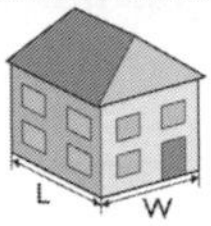

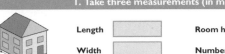

**Length** ☐ **Room height** ☐

**Width** ☐ **Number of floors** ☐

## 2. Calculate TOTAL external wall area

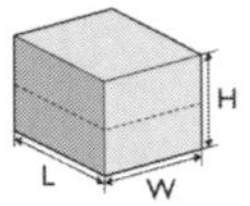

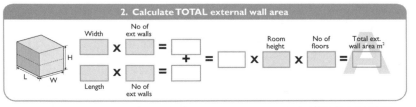

Width ☐ **X** No of ext walls ☐ **=** ☐

Length ☐ **X** No of ext walls ☐ **=** ☐

**+** **=** ☐ **X** Room height ☐ **X** No of floors ☐ **=** Total ext. wall area m² ☐ A

## 3. Calculate wall and window heat losses

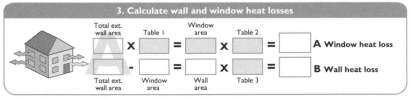

Total ext. wall area A **X** Table 1 ☐ **=** Window area ☐ **X** Table 2 ☐ **=** ☐ **A Window heat loss**

Total ext. wall area A **-** Window area ☐ **=** Wall area ☐ **X** Table 3 ☐ **=** ☐ **B Wall heat loss**

## 4. Calculate floor and roof heat losses

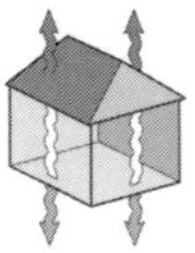

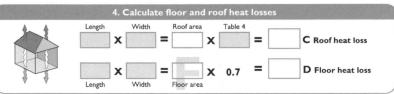

Length ☐ **X** Width ☐ **=** Roof area ☐ **X** Table 4 ☐ **=** ☐ **C Roof heat loss**

Length ☐ **X** Width ☐ **=** Floor area F **X 0.7** **=** ☐ **D Floor heat loss**

## 5. Add up fabric heat losses

**A + B + C + D =** ☐ **X** Table 5 ☐ **=** ☐ **E Total fabric heat loss (W)**

## 6. Calculate ventilation heat loss

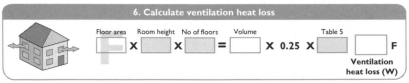

Floor area F **X** Room height ☐ **X** No of floors ☐ **=** Volume ☐ **X 0.25 X** Table 5 ☐ ☐ **F Ventilation heat loss (W)**

## 7. Calculate boiler output (in kW)

**E + F =** ☐ **+** Water heating (W) **2000** **=** ☐ **+** Add in any extension ☐ (From separate worksheet or radiators sizes) **=** Divide by 1000 to get kW ☐ **BOILER OUTPUT**

---

Worksheet ______

Date ______

Sheet ______ of ______

Name ______

Address ______

______

______

Type of dwelling ______

| Table 1 Window Factors | |
|---|---|
| Detached | 0.17 |
| Semi-detached | 0.2 |
| Mid terrace | 0.25 |
| Flat | 0.25 |

| Table 2 Window U-Values | |
|---|---|
| Double glazed wood/plastic | 3.0 |
| Double glazed metal frames | 4.2 |
| Single glazed wood/plastic | 4.7 |
| Single glazed metal frames | 5.8 |

| Table 3 Wall U-Values | |
|---|---|
| Filled cavity wall | 0.45 |
| Unfilled cavity wall | 1.6 |
| Solid wall 220 mm | 2.1 |

| Table 4 Roof U-Values | |
|---|---|
| Pitched < 50 mm insul. | 2.6 |
| Pitched 50-75 mm insul. | 0.99 |
| Pitched > 75 mm insul. | 0.44 |
| Flat uninsulated | 2.0 |
| Flat 50 mm insul. | 0.54 |

| Table 5 Location Factors | |
|---|---|
| North & Midlands | 29 |
| Northern Ireland | 26.5 |
| Scotland | 28.5 |
| South East & Wales | 27 |
| South West | 25 |

# Index

# The Dressmaker's Secret

Charlotte Betts

piatkus

PIATKUS

First published in Great Britain in 2017 by Piatkus

1 3 5 7 9 10 8 6 4 2



A CIP catalogue record for this book is available from the British Library.

TPB ISBN 978-0-349-41414-0

Typeset in Caslon by M Rules
Printed and bound in Great Britain by
Clays Ltd, St Ives plc

Papers used by Piatkus are from well-managed forests and other responsible sources.

MIX
Paper from responsible sources
FSC® C104740

Piatkus
An imprint of
Little, Brown Book Group
Carmelite House
50 Victoria Embankment
London EC4Y 0DZ

An Hachette UK Company
www.hachette.co.uk

www.littlebrown.co.uk

*For Oliver*

# Acknowledgements

It is often said that it takes a whole village to raise a child and it's the same with publishing a book. I may have given birth to *The Dressmaker's Secret* but I was assisted by many others.

The story was conceived whilst on a writing retreat with Carol McGrath, Deborah Swift and Jenny Barden, all historical novelist friends. I was searching for inspiration when I read about the extraordinary life of Princess Caroline of Brunswick. Full of excitement, we discussed my initial ideas, over a glass of wine or two, in the simmering heat of a Greek summer.

My agent, Heather Holden-Brown, was endlessly supportive as I wrestled to refine the story into something resembling an outline fit to present to my publishers.

Research into Caroline's story and the Villa Vittoria in Pesaro continued over many months. Local architect Roberta Martufi, who had some years before been involved in renovation works at the villa, patiently answered my email enquiries.

My wonderful writing group, WordWatchers, made time to read the whole manuscript of *The Dressmaker's Secret* and offered constructive criticism. As always, their suggestions were sweetened with copious supplies of tea and cake.

Once the manuscript was delivered to Piatkus, my lovely editor, Dominic Wakeford, made helpful suggestions for revisions and then Lynn Curtis copy-edited the revised manuscript with a light and careful hand.

My very grateful thanks to all of the above and, not least, to my husband and family who listened with only slightly glazed eyes when I banged on about Caroline of Brunswick, yet again.

# Chapter 1

## *January 1819*

**Pesaro, Italy**

As Ma and I approached Pesaro for the first time I was still simmering with resentment and the atmosphere between us was as frosty as the January afternoon. For days now the coach had lurched along the rutted road up one side of the snow-dusted mountains, only to plunge perilously down the other while we clung, white-knuckled, to the travelling straps. Despite the trials of the journey we'd barely exchanged a word since Arezzo.

Ma huddled into the seat opposite me with a blanket wrapped around her diminutive form. I stared out of the window, mulling over our quarrel as the mountains gave way to undulating hills and farmland. Finally, we reached the sea at the town of Fano, where the other passengers alighted. Ma and I remained shrouded in uneasy silence as the coach continued along the coast road. How many times had the wind blown us into a strange town to make a new start? Perhaps this time, *please* this time, we'd stay for good. My mother and I had always drifted like thistledown on the breeze, landing for a few precious months in a village or town until a panicky squall of her perpetual unease whisked us away again.

Despite my annoyance with Ma for insisting we move on yet again, my spirits lifted as I studied the coastline. A bracing wind buffeted the coach as we bowled along. I had never lived near the sea before and a prickle of unexpected excitement made me fidget. White horses danced along the tips of the waves rolling in from the sea and I itched to run down to the water's edge and feel the salty spray on my face. Dragging down the window, I leaned out.

'Emilia, it's too cold!' protested Ma.

An icy gust of wind snatched my breath away and I laughed and clutched at my hat as my hair whipped across my cheeks.

A warning shout, almost drowned by the pounding of the sea, made me glance at the road behind. Ears back and necks outstretched, two piebald ponies galloped towards us at breakneck speed. Behind the runaways was a high-perch phaeton, careering from side to side while its passengers shrieked in terror. Within seconds the ponies had bolted past and the little carriage crashed against our coach, scraping along one side with an ear-splitting screech. As the impact threw me backwards I caught a glimpse of the driver, his features set in a rictus of fear as he struggled to bring the ponies under control.

Ma screamed as we veered off the road and, with a bone-shaking crash, thumped into a tree. I was thrown violently to the floor and pain exploded in my shoulder. Outside, there was a yell and a clatter. A horse whinnied in terror. Then there was silence except for the creak of leather as our swaying coach settled.

I exhaled slowly.

Ma, her round face bleached with shock, looked at me mutely with a question in her grey eyes.

Rubbing my shoulder, I pulled myself to my feet and opened the door. My French and Italian were fluent since we'd spent my lifetime travelling but Ma still had a noticeable English accent. She always preferred me to communicate with others since she was frightened of drawing unwanted attention.

Our coachman was already calming his horses as they tossed their manes in fright.

I ran ahead to where the phaeton lay upside-down with its wheels still spinning. The piebald ponies lay tangled in the traces, pawing the grass, eyes rolling and flanks heaving.

The driver, a little younger than myself, lay sprawled on the ground fingering a trail of blood that trickled from beneath his sandy hair.

'Are you all right?' I called.

He nodded. 'Idiot ponies!' His eyes were pale blue and since his Italian was as heavily accented as Ma's, I surmised he was English.

I caught sight of what appeared to be a bundle of clothing tossed on the ground but then it moved. I hurried to investigate and discovered it was a little girl. Black ringlets lay across her face and I smoothed them away to reveal a lump the size of a pigeon's egg on her forehead. Her eyelashes fluttered as I gathered her up in my arms.

A faint cry came from behind the phaeton and then a foot encased in a scarlet boot waved in the air.

I sat the little girl on the ground and ran to help.

A stout middle-aged lady lay on her back, her skirts rucked up above her stockings to expose plump thighs. 'Victorine! Willy!' she called. She had a guttural accent, German perhaps.

I helped her into a sitting position and pulled down her skirts to restore her modesty. She appeared to be unharmed, though her outrageously high-crowned hat, adorned with enormous ostrich plumes, had slipped sideways over her eyes. Disconcertingly, the black wig she wore under the hat had slipped, too.

'Where is my son Willy?' she asked. Her lips were painted a garish vermilion, her eyebrows blackened and cheeks heavily rouged.

I brushed a clump of mud off her shoulder. 'The driver? He has a cut to the head but is otherwise unhurt.'

She looked wildly around. 'And Victorine ... Is she safe? Where is my little treasure?'

'Don't worry, I have her.' I returned to the child and lifted her up.

The woman held out plump, beringed hands. 'Come to Mamma, my darling!'

Surprised, I placed the whimpering child in her arms. No amount of white lead or rouge could make her look like the girl's mother rather than her grandmother.

Victorine howled, burying her head in the woman's pillowy bosom.

Our coachman ran his hands over the piebald ponies' fetlocks and then released them from their traces. 'You won't be driving that rig home,' he said to Willy. 'The wheel is buckled.'

The young man stuck out his bottom lip and kicked at the offending wheel. 'Useless thing!'

'Help me up!' called the woman.

I took the wailing child while Willy pulled his mother to her feet. She was short, even smaller than Ma, but rotund, with a bosom that formed a pronounced shelf under an embarrassment of chins.

Handing the reins to Willy, the coachman rocked the phaeton and heaved it the right way up. It was a most extraordinary vehicle, shaped like a conch shell, heavily gilded and decorated with mother-of-pearl. The inside was padded with blue velvet and embellished with silver fringing.

We stood watching in a shivering huddle, while Willy and the coachman dragged it back onto the road.

Victorine clung to me, hot tears soaking my shoulder as I attempted to soothe her. 'Would you all like to come in the coach with us to Pesaro?' I said over the child's noisy sobs. 'I expect you'll find a wheelwright there to repair the phaeton.'

'Someone will have to pay for the damage to my rig, too,' said the coachman, looking pointedly at Willy. 'The front's all stoved in and the paintwork's ruined. I shouldn't wonder if it's off the road for days.'

The woman waved her hand. 'It shall all be taken care of. Take us to Villa Vittoria and you will be reimbursed for your trouble.'

'I'll tie your horses to mine and they can trot along beside us.'

We trooped back to the coach and climbed inside. 'I am Emilia Barton,' I introduced myself, 'and this is my mother, Sarah. We're going to settle in Pesaro.'

'Come to dinner tomorrow,' said the woman, rocking the coach as she plumped her not inconsiderable weight down onto the seat beside Ma. 'I wish to make amends for inconveniencing you.'

Ma glanced at me in alarm but I refused to meet her gaze. I knew she wouldn't want to go; she preferred us to keep ourselves to ourselves. 'We'd like that,' I said.

'Villa Vittoria is at the foot of Monte San Bartolo above the town. Anyone will give you directions.'

I'd have a battle with Ma but I was determined to grasp this opportunity of making acquaintances in the town that was to become our new home. Victorine curled up on my lap; her sobs had subsided into the occasional hiccough by the time the coach rolled away. Willy sat beside me staring sulkily at his feet.

I looked out of the window, eager for my first glimpse of Pesaro. We passed a cluster of terracotta-tiled houses amongst vineyards and olive groves, and it wasn't long before we drove through the gateway of the walled town, clattered across a colonnaded piazza with a granite fountain set in its centre and then passed along a narrow street before turning into the courtyard of an inn. I sat Victorine beside her mother and Ma and I descended from the coach.

The woman leaned out of the window, her hat and wig still askew. 'Don't forget! Villa Vittoria tomorrow.'

Victorine waved at us as the coach trundled back out of the courtyard.

Ma glanced up at me, a muscle flickering at the corner of one eye, and a wave of pity washed over me for her constant state of anxiety. We had always been so very different, despite being travelling companions and so dependent upon each other.

The aroma of fried onions drifted from the inn and the exhaustion, unhappiness and frustration I'd felt over the previous days suddenly melted away. We had arrived and this was to be our new beginning.

I held out my hand to her. 'Shall we go in, Ma?' I said.

The next day we braved the bitter north wind to walk up the steep hill to Villa Vittoria. Vineyards, olive groves and mature oaks lined the country road. Our host at the inn had told us to look out for the other elegant villas, the *casini di delizia*, which had been built by the wealthy on the Colle San Bartolo in the seventeenth century. In those days the Della Rovere family had held a flourishing court in Pesaro.

Ma had put on her best dress of bronze wool, but her shoulders drooped and worry creased her brow. 'We should be out finding work,' she said, 'not paying social calls.'

As itinerant dressmakers we lived from hand to mouth and were permanently concerned about where we'd next find gainful employment.

'We'd be in a better financial position if you hadn't made us spend most of what we earned from the Conti bride's wedding dress on the coach fare from Florence,' I said. 'Still, the lady we helped after the carriage accident may be persuaded to give us introductions.'

'Perhaps. But ... ' Ma caught my sleeve. 'Don't walk so fast, Emilia!'

'But what?'

Frowning, she said, 'There was something odd about the lady in the phaeton, wasn't there? Her rings indicated she has a wealthy husband and her velvet pelisse was of excellent quality. But then there was that peculiar hat and those scarlet boots . . . Not at all suitable for a lady of her age.' Ma pursed her lips. 'And she wears paint.'

'Many women wear paint.'

'But it's sorely out of fashion now and women of consequence and breeding don't wear it applied in such a haphazard way. I can't think how her maid allowed her to go out looking like that.'

Ma, always discreetly dressed, never resorted to even a dusting of rice flour on her nose. She frequently said it was important to impress on our clients that we had impeccable taste. Once upon a time, before I was born, she'd been personal maid to a lady of the aristocracy and had learned how to conduct herself in a genteel fashion. Occasionally, she spoke wistfully of the natural elegance of her kind mistress and the grand houses they'd lived in.

Ma glanced up at the winter sky. 'We must leave in good time. It would be very disagreeable to have to find our way down this hill after dark.'

The wind was bitingly cold and I was relieved when we came to the avenue of cypress trees, which the innkeeper had informed us led to Villa Vittoria. We walked briskly through an orchard until we arrived at a substantial stone house whose new roof was embellished with a large cupola. The grounds were littered with piles of sand and stacks of timber, and workmen climbed the scaffolding heaving blocks of stone on their shoulders. A colonnaded gallery to one side of the building was in the process of being infilled, while a matching wing was being added to the other.

A pair of soldiers stood either side of the front door, dressed in the gaudily striped uniform of the Papal Guard complete with helmets and halberds. Since Napoleon's defeat, the Congress of Vienna had returned the Marche to Papal rule but I couldn't imagine why soldiers were on watch at Villa Vittoria. Our steps faltered but, since we were not challenged, I lifted the iron doorknocker and let it fall.

Two great dogs loped towards us, barking loud enough to wake the dead. Ma and I froze to the spot.

A man hurried after them. 'Titus! Bruna! Come here, you brutes!'

'We're here to visit your mistress,' I said, once he had the dogs under control.

A maid opened the door, took our pelisses and indicated we were to follow her. I heard the clatter of pans and of women's voices raised in argument from behind one of the doors leading off the hall. In the air hung the mouth-watering scents of fried garlic and baking bread.

A child's high-pitched laughter rang out as the maid opened the door to the *salone* and announced us. 'Ma'am, Signora and Signorina Barton.'

Ma and I paused on the threshold. The woman we'd rescued from the phaeton was on her hands and knees on the floor, skirts tucked up, pretending to be a horse. Victorine sat on her back, shrieking with delight as her mount bucked and whinnied. A bruise stained the little girl's forehead but otherwise she seemed none the worse for her tumble of yesterday.

Ma glanced at me, clearly shocked by such unladylike behaviour.

'Come in, come in!' Victorine's mother rolled over so that the little girl slid off her back and dropped onto the floor in a fit of giggles.

We stepped into the *salone*, a generously proportioned room with a high, beamed ceiling and arched windows. It was richly decorated in the Turkish style. A log fire blazed in the great stone fireplace and portraits in gilt frames adorned the walls. Our hostess could clearly afford to live in comfort.

'I hope you have recovered from the unfortunate accident, Signora?' I said.

She shrugged, her fat little legs encased in sagging silk stockings stretched out on the floor in front of her. 'It was only a tumble, though Willy was very naughty to give us such a fright. I had to scold him severely.' She held out her hands. 'Pull me up! We've been playing horses for the last half-hour,' she said, 'and I'm quite done in.'

'I'm not surprised,' I answered, grasping her hands. Victorine giggled and came to add her efforts to mine as I heaved her mother to her feet.

She sat down heavily on an overstuffed chair and waved us towards a sofa laden with tasselled pillows. There was a spider's web in her hair and her skirt was smudged with dust. 'Angelica, my dear, come and meet my saviours, Signora and Signorina Barton.'

A small cough came from behind me and I saw a dark-haired lady sitting in the corner of the room.

'This is the Countess Oldi, my lady-in-waiting,' said our hostess.

Ma glanced at me, her eyes wide.

Countess Oldi, a handsome woman of about thirty, came forward as we made our curtseys. 'Do you have family and friends in Pesaro?' she asked.

'None at all.' We had no friends or family anywhere. Just for a moment I pictured my friend Giulia's happy smile and anger with Ma for making us move on yet again rose up like bile.

'Come to Mamma,' our hostess said, pulling Victorine onto her lap. 'I have been here only two years but the climate is good and the scenery delightful. In the summer I bathe in the sea ... so good for the health.' She leaned forward and said in a theatrical whisper, 'And it is very cheap to live in Pesaro.'

The room was overheated and, as my fingers began to thaw, I wondered how to broach the subject of introductions. I glanced at Ma, who sat with her eyes lowered and hands folded. Clearly, she wasn't going to assist me so I took a deep breath, driven on by our urgent need. 'My mother and I are both talented dressmakers,' I said. This was no time for false modesty since we might never again find a wealthy lady so clearly under an obligation to us. 'We've recently completed an extensive trousseau for Signorina Lucrezia Conti in Florence. She's marrying the Duke of Mantova's brother and only the most skilled work was acceptable. We have references and will be happy to show these to any interested parties. Perhaps some of your friends ... '

The Signora clapped her hands together and her face lit up like a child's. 'I was forced to dismiss my lady's maid a while ago and my wardrobe is sadly out of order. I have a fancy for a new ballgown. Emerald, perhaps, in the Russian style with gold lacing. You might make it for me?'

Ma, her expression a model of controlled restraint at the very thought of such a garment, said, 'We'll call on you at your convenience with our pattern books.'

I let out my breath, relieved that I'd succeeded in my mission. I realised then I didn't know our hostess's name. Our unconventional reception had distracted me and it was awkward to ask her now.

Victorine began to fidget, winding her fingers through her mother's hair and pressing kisses on her cheeks to gain her full attention.

Ma whispered to me, 'You see, I knew she couldn't have a lady's maid to dress her properly.'

'I'm hungry!' said Victorine. 'When is Papà coming home?'

'Soon, my treasure. Why don't you go to the kitchen and wheedle a piece of bread?'

The little girl slid off the sofa and skipped from the room.

The savoury aroma of roasting meat wafted in from the kitchen. I was hungry, too, and wondered when we might eat. I didn't have long to wait.

The front door slammed and there was the sound of hearty male laughter and footsteps clattering across the hall.

'Ah, the Baron has returned from his hunting!' Our nameless hostess smoothed the wiry curls bunched over her ears and turned expectantly to the door.

It burst open and two men entered. The taller of the pair, aged about thirty-five with thick black hair and an extravagantly curled moustache, came to lift her hand to his full lips.

The pungent scent of horseflesh rising from the men's mud-splashed clothing made Ma's nose wrinkle.

Victorine ran back into the room and clasped the Baron around his knees. 'I was waiting for you, Papà!'

The Baron bent to kiss her. 'And who have we here?' He turned to look at Ma and me. At well over six feet tall, his splendid physique and confident presence seemed to fill the room.

'This is Signora and Signorina Barton,' said Countess Oldi. She turned to us. 'My brothers, Baron Bartolomeo Pergami and Luigi Pergami.'

'Ah, yes!' said the Baron. 'The two Good Samaritans. I believe we owe you our gratitude and apologies after Willy drove you off the road?'

'Not at all,' I mumbled, overawed to find him looming over us. I couldn't help thinking that he was the most unlikely husband for the little dumpling of a woman we had helped. He must have been at least fifteen years younger.

'Let's eat!' he said.

We trooped into the dining room where we found Willy, an elderly woman dressed in black, and a soberly dressed young man with a mop of unruly dark curls all awaiting our arrival. No one introduced us and Willy ignored us.

Menservants in gold waistcoats under embroidered black coats lined with scarlet silk pulled out our chairs. A variety of soups, pies, roasted game birds, fricassées and puddings were laid before us.

Covertly, I eyed the young man who sat beside me. A little older than I, he had an aquiline nose that would have graced a Roman soldier. Although not conventionally handsome, it was the hint of suppressed laughter in his eyes and his wide smile that caught my attention.

'I am Alessandro Fiorelli,' he said, 'Victorine's tutor.' His voice was warm and mellow.

For two heartbeats I gazed into his amber eyes. 'Emilia Barton,' I said, heat suffusing my cheeks.

The Baron and his brother began to relate the tale of their morning's hunting, verbally sparring with each other, and their laughter reverberated around the room. The old woman silently watched them as she chewed her dinner. Willy drank a great deal of wine.

'Signorina Barton, Victorine told me how you saved her,' said Signor Fiorelli. 'Where do you come from?' he asked. 'Your mother is English, I believe?'

'Where do I come from?' I said. 'Everywhere. And nowhere.' I couldn't begin to remember, never mind name, all the places where we'd lived since we rarely stayed anywhere for more than a few months. 'We're both English,' I said, 'though I've never been to England.'

He looked at me with a puzzled expression on his face but must have sensed I didn't care to discuss the matter for he changed the subject.

I drank sparingly of the thin red wine, anxious that I might say something foolish under its influence. I watched Signor Fiorelli's well-shaped mouth as he spoke and admired his dark hair and smooth olive skin. It was so very different from my pallid English complexion and strawberry-blonde tresses.

Time seemed to have no meaning while I was talking to Signor Fiorelli and when Ma attempted to catch my attention, I'm ashamed to say I ignored her.

Once we'd finished our dinner I glanced out of the window at the darkening sky. 'We must leave at once, before it's too dark to see our way back,' I said, alarmed.

Ma bit her lip with anxiety.

'Don't worry,' said Signor Fiorelli. He smiled at our hostess. 'With your permission, Ma'am, I shall escort these ladies back to the inn.'

A short while later we were outside in the cold. We walked quickly and when Ma stumbled in her haste to match our longer strides, Signor Fiorelli took her arm.

'I hope we're not taking you out of your way?' I said.

'Not at all. We pass my family house on the edge of the town.'

'Since we are strangers,' I said, 'I wonder if you know of anyone here who has a cottage available to rent?'

'I shall ask my mother,' he said. 'She always knows what is going on in the town.'

Signor Fiorelli pointed out his home as we passed by, a sizeable house with a neat garden.

Soon we arrived at the inn. 'I shall call on you in a few days,' he said, 'if I have any news.'

'There's one other thing,' I said. 'I'm embarrassed to admit this but I don't know the name of your employer. The maid didn't introduce us and then it was too late to ask.'

Signor Fiorelli laughed. 'I expect the maid assumed you knew of the Princess.'

'Princess?' Ma said.

'Indeed,' said Signor Fiorelli. 'Your hostess is Her Royal Highness, the Princess of Wales, daughter-in-law of your English King.'

Ma clutched his wrist. 'Not Princess Caroline of Brunswick?'

'The very same!'

'*Well*,' said Ma, 'that explains a very great deal!'

# Chapter 2

The following day we took our dinner in the restaurant in the Piazza del Popolo.

'My hands are frozen,' Ma complained as we settled ourselves at a table by the fire. I looked longingly at the next table where two men were enjoying roast veal. As usual, Ma and I ordered *ribollita*, bean soup, the least expensive item on the menu. I wondered how long it would be before we found ourselves a commission to replenish our dwindling resources.

'Tell me about the Princess of Wales,' I said as we waited for our soup. 'Why is she in Pesaro? I know little of what happens in England.'

Ma sighed. 'I haven't been home for so very long.'

How curious that although she rarely mentioned England, she still called it 'home'.

'It's a strange story,' she continued. 'The Prince Regent is often called "the First Gentleman of England", although, if you ask me, he's anything but!' She leaned towards me and spoke in an undertone. 'Still, having met his wife, perhaps that's easier to understand. Surprisingly, Princess Caroline was popular with the people.'

'What is easier to understand?'

Ma raised her eyebrows. 'Does she behave like a princess?'

I laughed. 'Not at all. She's ... ' I was lost for words. 'Unconventional,' I said at last.

'Quite. She has no sense of her proper position in life and is always happy to mix with the common people, despite being the Prince of Wales's cousin.'

'I thought she might be German.'

'It was a political marriage. She was past the first bloom of youth when she arrived in England and only met the Prince three days before they were married.'

'I shouldn't have cared for that!'

'As it turned out, neither did he. The gossip at the time was that when he first greeted her, he reeled back and called for brandy.'

I frowned. 'Why?'

'Apparently she was not fastidious about her person,' was my mother's prim response. 'Anyway,' she continued, with a gleam in her eyes, 'he drank for the whole three days before the wedding, had to be held up during the ceremony and then collapsed, dead drunk, into the grate on his wedding night.'

I tried not to laugh at the picture of events she presented. 'Hardly the behaviour of a gentleman! But why is the Princess in Pesaro?'

'The couple wrangled for years. They had a daughter, Princess Charlotte, nine months after the wedding.'

I smiled inwardly. The Prince of Wales wasn't totally incapable on his wedding night then.

'Within months they were living apart,' said Ma. 'The Princess of Wales behaved disgracefully, flirting with all and sundry. The Prince Regent wanted to divorce her and investigated her for alleged infidelity, despite certain ... ' Ma glanced at me from under her eyelashes '... irregularities in his own private life. Nothing was ever proved against her but she left the country. I didn't know she'd settled here.'

Our *ribollita* arrived and I ate mine slowly. If Princess Caroline was married to the Prince of Wales, what was Baron Bartolomeo Pergami to her, if not her husband? He was Victorine's father and the Princess was her mother so the child must have been born out of wedlock. And then there was the Princess's son ...

I put down my spoon. 'So, Willy is the future King of England?'

Ma spluttered into her soup. 'Oh dear me, no! That sullen boy isn't her son at all. The Princess adopted him when he was a baby. As for Victorine ... ' She broke off, her cheeks blazing. 'Oh dear, it's all most irregular. And now this woman of highly questionable morals is our best chance of immediate employment.'

'Ma,' I said, 'do you never want to go home to England?' She only ever spoke of the past reluctantly and if I pressed her.

She stared at her soup plate, suddenly very still. 'I can't,' she said, after a brief pause.

'Why not?'

'It holds nothing but unhappy memories for me.'

'But we must have family there ... '

'No!' She took a steadying breath. 'I grew up in an orphanage.'

'I thought you said your family lived in Essex?'

Her cheeks turned scarlet. 'Before I was orphaned.'

'What about my father's family?'

'I never met them. There was a quarrel and if they're anything like him, we're better off without them.'

I loved Italy and had no desire to leave. It was only that, sometimes, I felt there was a part of me I couldn't quite make out. My recollections of my father were vague since he'd abandoned us sixteen years ago when I was five. If I closed my eyes I could still hear echoes of Ma's cries when he hit her. I remembered the sour smell of his breath, acrid with tobacco and red wine. I used to hide on the floor between my bed and the wall but sometimes he'd drag me out by my hair and shout at me, his face scarlet with rage. I shuddered. It did no good to think about those terrible times. I did wonder, though, if the lack of closeness between Ma and myself was because she saw his likeness in me. I certainly didn't look like her so I suspected I must take after him.

Once we'd finished our *ribollita* we whiled away an hour visiting the cathedral, admired the exterior of the grand Ducal Palace and the red granite fountain in the cobbled piazza, and then there was still enough daylight for a wind-blown stroll down by the sea. It was dark when we returned to the inn and a serving maid handed Ma a note as we came in.

'It's from Signor Fiorelli,' she said. 'There's a cottage he'll take us to see at ten o'clock tomorrow morning.'

I held out my hands to the fire so she wouldn't see the spark of pleasure in my eyes. She didn't like me to have friends, but I wanted to know Signor Fiorelli better and was determined Ma wouldn't prevent it.

I dressed with care in my violet wool walking dress with the embroidered collar. The sun shone and I was full of anticipation at the thought of seeing Signor Fiorelli again. After smoothing my curls, I put on my bonnet, pinched colour into my cheeks and went down to the public room.

Ma waited for me there, her fingers plucking anxiously at her skirt. 'There you are, Emilia! I've been having second thoughts about staying in Pesaro,' she began.

My heart descended into my calfskin boots. Not again! 'Ma . . . '

'We cannot work for the Princess of Wales.' She spoke breathlessly, the words tumbling out of her.

'We need the money!'

'Listen to me, Emilia.' She gripped my hands. 'Think what it might do to our reputation if we become involved with that unsuitable household. No lady of quality would wish to use our services then.'

'The Princess is a member of the Royal Family. How could anyone find fault with that?' I sighed, exasperated.

Ma lowered her gaze. 'It may be indelicate to mention it but she and the Baron have a daughter and the Princess is married to someone else. How can we countenance such flagrant behaviour without being considered immoral ourselves? Besides, it's not just that.'

'What then?'

She glanced over her shoulder at the deserted room. 'We may have been followed here from Florence.'

'Ma, please don't make us move on again! You *promised* we'd settle this time.'

'But we had to make such a fuss to secure ourselves places on the first coach to leave . . . someone may remember us. And your red hair is so noticeable. If your father asks after us, he'll find out where we went. We must move on, more discreetly this time.'

'No! Why must you *always* do this, Ma?' I could hardly breathe for the fury that gripped me. 'I'm tired of roaming like a gypsy and I'm sick of being poor because we constantly spend our savings on travelling. And all because you have some misguided notion that my father wants to harm us.'

'He does, I'm sure of it!'

'He left us *sixteen* years ago, Ma! He can't hurt you anymore and your fears are utterly nonsensical.'

'You know we had to leave Florence because a man had been asking about us.'

'It was probably an enquiry for our dressmaking services.'

Her lips folded together in a stubborn line. 'I'm sure it was your father.'

I paced across the room. 'What possible reason could he have for wanting to harm us, after all this time?'

She opened her mouth to speak and then closed it again.

'You didn't let me say goodbye to Giulia in Florence.' Resentment coloured my voice. 'She was the best friend I've ever had and you insisted we left overnight without even giving me the chance to explain to her. I'd promised to make her wedding dress.' I fought back sudden tears. 'She'll be broken-hearted and think I didn't care about her.'

The door creaked open and I wiped my eyes as Signor Fiorelli entered.

'Good morning, ladies!' He came to greet us with outstretched hands and a wide smile on his handsome face. 'We have a cottage to see.' His voice faltered as he saw my tears and Ma's flushed cheeks.

'It's kind of you to arrange it,' I said, forcing a smile. Ma opened her mouth to protest and quickly I said, 'Shall we go?' I allowed Signor Fiorelli to escort me from the room, hoping Ma wouldn't balk.

A moment later her footsteps pattered after us and I let out a sigh of relief.

We crossed the piazza and reached the point where the River Foglia flowed into the Adriatic. I listened intently while Signor Fiorelli pointed out places of interest and all the while I prayed Ma wouldn't be difficult. Her groundless fears always came to nothing and, this time especially, I wanted to stay.

After a while we came to the harbour, bustling with boats and stalls selling mackerel and sardines. Fishermen called out to each other as they unloaded their catches and the air was full of the bracing scent of the sea.

'It's here,' said Signor Fiorelli, pointing to a higgledy-piggledy row of houses beside the harbour, all colour-washed in different shades of terracotta, ochre and cream. I was delighted when he stopped outside a primrose yellow cottage with a tiled roof.

'It belongs to my cousin's great-aunt, Prozia Polidori,' he said. 'She's too old now to manage on her own and she's gone to live with her daughter. The family want to keep the cottage and its furniture so are happy for it to be rented.'

The front door led into a sunny parlour furnished with a bookcase and rocking chairs to either side of the fireplace. A door opened into a dining room with an old-fashioned kitchen behind, leading to a walled yard with a pump and a privy. Up the winding staircase from the dining room were two bedrooms, not large but perfectly adequate.

'It's perfect!' I said as we descended into the dining room again. 'And this table is large enough to use for our sewing. The light in the parlour is excellent, isn't it, Ma?' I held my breath.

She looked hesitant. 'Is it available on a short lease?' she asked. 'If we don't find work here we shall have to move on.'

'I'm sure Prozia Polidori will find that acceptable,' said Signor Fiorelli. 'She lived in this house from the day she was married and frets about it remaining empty.'

'Then, if the rent is not too high . . . ' I said.

Signor Fiorelli mentioned a sum that seemed perfectly reasonable.

Impulsively, I hugged Ma. 'We'll take it!' My spirits soared as Signor Fiorelli shook my hand.

He beamed at me. 'I knew, if anyone could help, it would be my *mamma*.'

'When can we move in?'

'As soon as you like,' he said.

My new bedroom overlooked the harbour. I stood by the window looking out at the forest of masts and watching the fishermen mending their nets. Seagulls wheeled overhead. Humming, I opened one of my travelling bags. It never took long to unpack since we always travelled light.

Peggy, the calico rag doll I'd carried everywhere with me for as long as I could remember, smiled up at me. Her woollen hair, tied into two pigtails, was far redder than mine and she wore a spotted muslin dress, replaced from time to time according to what leftover scraps remained after completing a commission. Of course, I was far too old for dolls, but Peggy had always been there to comfort me when I was lonely and was the eternally faithful companion of my childhood. I lifted her out of the bag, kissed her freckled nose and set her on the pillow.

The remainder of my possessions was soon stowed away. I carried my sewing box downstairs and found Ma unsuccessfully trying to light the fire in the kitchen.

'The paper's damp,' she said. There was a smudge of soot on her cheek.

I kneeled in front of the grate, rearranged the kindling and blew on it until the paper caught.

'It's burning well now,' I said a while later as I held out my palms to the dancing flames. 'I'll light the fire in the parlour, too, and then everything will be comfortable.'

'I'll boil some water and start the cleaning.'

I didn't say that everything looked perfectly clean already; it would have been pointless. Ma always scrubbed everything from top to bottom when we moved into a new place.

'I'll buy something for supper.' I buttoned my pelisse and collected the shopping basket.

'Tuck your hair into your bonnet,' said Ma. 'Anyone looking for a redheaded girl will soon be on our trail if you go out like that.'

Sighing, I pushed a stray curl behind my ear. 'My hair is less likely to lead anyone to us than your English accent,' I said tartly.

'That's why I ask you to do the talking.' Ma turned away to lift a pan of water onto the fire.

Signor Fiorelli had told us that the market was held on Tuesdays and Saturdays but I was pleased to find shops nearby. I bought bread, olive oil, beans and polenta then counted out the remaining coins carefully. Just enough. I bought a small chicken. It was an extravagance but I wanted to celebrate our new home. And perhaps Signor Fiorelli might be persuaded to join us for dinner.

Ma scolded me when I returned. 'We can't afford meat until we have work.'

'Then we must hope the Princess sends for us soon.'

'I told you, we can't work for her.' She swept the floor vigorously, thumping the furniture with the broom as she muttered under her breath.

'The grocer was happy to display our card,' I said. We were often successful in finding clients that way when we moved to a new town.

'You didn't put our address on it?' Ma's expression was tense.

I sighed. 'You know I never do. I paid the grocer to make a note if anyone showed interest and said we'd call by to check. Then I introduced myself to the baker's wife. I'm to go back tomorrow because she might have some plain sewing for us, chemises and petticoats.'

'That won't bring in enough to pay the rent,' said Ma.

Somehow she always managed to spoil the small triumphs in life. It was too exhausting to keep wrangling with her.

Later that afternoon Ma was sitting in a rocking chair beside the kitchen fire, a pair of spectacles perched on the end of her nose while she darned her stockings. I was making polenta for dinner, wondering how to approach Signor Fiorelli with an invitation to share our chicken the following day.

There was a knock on the front door and Ma dropped her scissors in alarm. 'Who can that be?'

I wiped my hands and hurried to find out.

'Don't answer it, Emilia!' called Ma. 'We don't know anyone here.'

A lady stood on the doorstep with a covered basket in her hands. She wore plain but well-cut clothes and a smart bonnet over her greying hair. 'Signorina Barton?' She smiled at me with warm brown eyes.

I nodded.

'I am Signora Fiorelli. My son Alessandro told me that you were moving in today.'

I opened the door wide. 'Please, come in out of the cold! I understand we have you to thank for finding us this pretty cottage?'

Signora Fiorelli's eyes twinkled. 'And I see my son was correct when he said you have the face and hair of a Botticelli angel.'

'He said that?' I flushed with pleasure.

'Indeed he did. Several times, in fact, so I thought I'd better come and see for myself.'

I heard Ma's footsteps behind me and pulled her forward to greet our visitor. 'Signora Fiorelli, may I introduce my mother, Signora Barton?'

Ma bobbed a curtsey.

'I hope you are settling in?' said our visitor.

'Yes, thank you,' Ma replied.

Signora Fiorelli cast her gaze rapidly around the room and I was relieved we'd lit the parlour fire and made everything tidy.

'May we offer you some refreshment?' I asked.

'Another time, perhaps? The family is waiting for their dinner but I've brought you something for tonight in case you've been too busy to cook.' She uncovered her basket and held out a dish to Ma. 'Rabbit stew. It only needs heating.'

'How kind!' said Ma.

After Signora Fiorelli had gone I carried the stew into the kitchen. My pulse raced and I couldn't stop smiling. Alessandro Fiorelli thought I had the face of a Botticelli angel.

# Chapter 3

The following morning, I was returning to the cottage with a parcel of cotton lawn together with several items of the baker's wife's undergarments to use as a pattern when I saw Signor Fiorelli hurrying towards me. I couldn't prevent a smile from spreading across my face.

'A note for you from the Princess,' he said. 'I was on my way to deliver it.'

I unfolded it and sighed. 'She asks us to call tomorrow.'

Signor Fiorelli frowned. 'I thought you'd be pleased.'

'It would be an excellent commission but . . . '

'What is it?'

I decided to tell the truth. 'Ma is very cautious,' I said. I was too embarrassed to mention the extent of her irrational fears. 'She's worried it might affect our reputation if we work for the Princess since her household is somewhat . . . ' I hesitated, not sure how to put it. 'Irregular,' I said at last.

'The Princess is very hospitable,' said Signor Fiorelli. He shrugged. 'Perhaps some of her parties are a little high-spirited. She takes a lively interest in whoever she meets, whatever their station in life, and is generally well liked. What is it that Signora Barton objects to?'

'The Princess isn't married to the Baron, is she?'

He shook his head. 'Bartolomeo Pergami is her steward.'

'But Victorine is the Baron's daughter?'

'Yes, indeed.' He looked puzzled but then his expression cleared. 'Ah!' He grinned. 'I understand now! Victorine is not the Princess's child.'

'But . . . '

'Victorine has been encouraged to call the Princess "Mamma" because her real mother lives elsewhere. Since the tragic death of her own daughter eighteen months ago, the Princess has taken the little girl to her heart. She even named the Villa Vittoria after her.'

'I remember I read in the newspaper about the heir to the British throne dying in childbirth.'

'Princess Charlotte's death was a terrible shock,' he agreed. 'The Princess is only now beginning to recover her usual good spirits.'

'So Ma can have no justifiable objection to our working for her.' Relief made me bold. 'I wonder . . . '

'Yes?'

'Would you care to join us for dinner? We're having roast chicken.'

Signor Fiorelli beamed. 'I should be delighted! But now it's time for Victorine's geography lesson.' He pulled a serious expression but there was laughter in his eyes. 'Today I shall teach her about England and its strange inhabitants with their peculiar customs.'

'You had better teach me, too,' I said, 'since I am far more Italian than English.'

Signor Fiorelli laughed as he walked away.

I returned home in high spirits to break the good news to Ma.

'I'm still not sure,' she said, doubt written on her face.

I dropped my parcel on the dining-room table. 'Why not?' I asked, irritated.

'If you ask me, the Baron runs a pretty rackety household.'

'We need the work.'

'I know but . . . ' She gnawed at her fingernails.

'Ma, we can't stay here if we don't pay the rent next month and we certainly haven't enough saved for another coach fare. Besides,' I said, 'you promised me we'd settle this time.'

Ma sighed heavily and opened the baker's wife's parcel. She examined the chemises and petticoats we were to copy. 'Go and see the Princess then!' she said. 'You can take her measurements and advise her on styles . . . but for goodness' sake, make sure she chooses something tasteful or she'll frighten off any other potential clients. Meanwhile, I'll make a start on these.'

'I'll do that,' I said. 'By the way, I invited Signor Fiorelli to share our chicken later on.'

Ma looked at me through narrowed eyes. 'I hope you aren't growing too fond of that young man?'

'I hardly know him,' I said, turning my attention to the baker's wife's petticoat.

The sun shone as I knocked on Villa Vittoria's front door, sewing bag in my hand. I was hoping to see Signor Fiorelli again. We'd had a very pleasant dinner the previous evening and he'd even managed to make Ma smile with his light-hearted conversation. He'd told us his father was a doctor and that he had an elder sister and seven younger brothers and sisters.

The maid showed me into the salon, where Countess Oldi sat in her usual place.

The Princess lay on a day bed with a handkerchief clutched in her hand. She presented a sadly changed appearance from our previous meetings.

I curtseyed and she sat up, her eyes red and her rouged cheeks smudged with tears.

'My dear, I'm so pleased you've come,' she said. 'You're just in time to divert me from sinking into another depression of the spirits.'

'Your Royal Highness, I do hope you are quite well?' I said.

'We are not at court here in my little cottage in the country. You may call me Ma'am.' She sighed. 'My dearest child passed away a little while ago and not a day goes by without my shedding tears of sorrow.'

I was at a loss to know what to reply to this. I hesitated. 'Would it help to talk about her?'

The Princess swung her legs over the side of the day bed. 'Sit with me.'

I perched on the seat beside her and couldn't help noticing that her dress was stained. Ma would have been shocked.

'My Charlotte was the flower and the hope of the nation,' she said, 'and the only good thing to come out of the union with that wretched husband of mine.' She shuddered. 'If you knew the slights and insults that the Prince of Wales has thrown at both my person and my position, you would never believe it.'

'Although I'm English my knowledge of these matters is negligible,' I said. 'I've never visited the country.'

'Then you will not have been poisoned against me by wicked lies.'

'Indeed not!'

'Charlotte was only twenty-one when she died.' The Princess dabbed at her eyes. 'I waited and waited for news of my first grandchild's arrival but no letter came. Then a courier from England was passing through on his way to Rome to present a letter to the Pope. No one interesting travels through Pesaro without my being informed and the courier was brought to me in the hope that I might gain some information about my grandchild.' She twisted her handkerchief in her hands, unable to go on.

'If it's too painful to speak of . . . '

The Princess shook her head. 'My angel had died in childbirth and my grandson with her. And that venomous brute, the Prince of Wales, didn't even have the courtesy to write and tell me of our daughter's passing.' She bowed her head as a fresh paroxysm of sobs overwhelmed her.

I was deeply shocked. Almost without thinking, I put my arms about her and she rested her head on my shoulder. Her shoulders shook as I patted her back. What kind of a husband was the Prince to fail in such a duty? Surely the two parties should have been united in grief at the loss of their child, whatever else had happened between them?

'And now,' sobbed the Princess, 'not only have I lost my daughter but also any means of regaining my rightful position. Once she'd become Queen, my Charlotte would have welcomed me back to England and I would have taken my proper place at Court. If not for my dearest Baron and his family, I don't know what I should do now.'

A rustle of skirts came from the corner of the room. Countess Oldi was watching us intently and I became aware that it was impertinent of me to touch the Princess. Gently, I released her.

'You are a kind girl,' she said, lifting her sodden handkerchief to her eyes. 'You must be much the same age as my own sweet child was?'

I nodded and handed her my clean handkerchief.

'She was pretty, like you, with blue eyes and fair hair. Everybody loved her.' The Princess sighed heavily and made a visible effort to smile. 'But you have not come here to see me weep.'

'Would you prefer me to return another day?'

Shaking her head, she said, 'I must go on with my life and I still have much to be grateful for.'

I took my sketchbook from my sewing bag, hoping to distract her from her unhappiness. 'Shall we talk about your new ballgown?' I said. 'I've sketched some designs for you to look at.'

We spent the next hour discussing styles, looking at swatches of silks and samples of trimmings. I used all my skills in diplomacy to guide her away from unsuitably low necklines, transparent fabrics and garish gold braid and fringing, finally fixing on a style in amethyst and mustard silk that was elegant, if a trifle more flamboyant than I would have advised.

'We require an initial payment to cover the cost of materials,' I said as I took her measurements. 'Then, this afternoon, I'll write to our stockists in Florence for the silk and the trimmings. As soon as they arrive I'll prepare the gown for your first fitting.'

'Very good,' said the Princess. She turned to Countess Oldi. 'Angelica dear, will you find the Baron and tell him I need funds for the new gown to give to Signorina Barton?'

Countess Oldi nodded and left the room, leaving the door ajar.

'My wardrobe is in grave disorder,' said the Princess. 'I haven't been in any state of mind to care. I believe I mentioned that I had to dismiss my maid, Louise Demont?'

'You did, Ma'am.'

'Cruelly, it happened at the same time as Charlotte was taken from me and I was brought so low. I thought Louise, so elegant and respectful, was my friend. I haven't been able to bear the thought of taking on another maid since.' Tears glittered in her eyes again. 'I trusted her and she betrayed me.'

'How very dreadful that must have been for you.'

'One of my couriers, Giuseppe Sacchini, stole gold Napoleons from my box. I dismissed him, of course, but discovered that Louise was his lover and accomplice, so she had to go too. She retaliated by spreading untruths about me and Sacchini. I've never been so taken in by anybody.' The Princess shook her head dolefully.

'I'm sorry to hear that.'

'Louise wrote all my letters since I have such a poor hand. Now the Baron looks after my financial affairs and no one will cheat me again.' Her face brightened. 'He is a splendid figure of a man, isn't he? So tall and handsome!'

'Why, yes,' I said, hardly knowing how to answer such a question. This was one part of our discussion I decided not to relay to Ma.

High-pitched laughter came from the hall. The door was thrown back and Victorine came skipping in, her black curls flying. 'Mamma, Mamma!' she called. 'Look what I have made for you!' She climbed onto the Princess's knee and thrust a sheet of paper at her.

'Well, what a very fine dog you have drawn,' said the Princess.

'Silly Mamma! It's a cat, can't you see?'

I was studying the way the Princess's eyes had lit up and how tenderly she kissed the little girl's cheek so that I didn't notice, at first, Signor Fiorelli standing in the doorway.

'Good morning,' he said. 'I hope my charge does not disturb you, Ma'am?'

'How could she, the little darling?' The Princess smothered Victorine in noisy kisses until she was helpless with giggles.

'I am going to take her on her daily walk,' said Signor Fiorelli. 'Perhaps we may escort Signorina Barton back to the town when you have finished your consultation?'

'By all means,' said the Princess. 'We are waiting for the Baron first. Will you see what keeps him? Victorine may remain with me.'

Signor Fiorelli gave a small bow and returned to the hall.

'Such a charming young man, don't you think, Signorina Barton? If only I were ten years younger . . . ' The Princess sighed and gave me a sly glance.

Covered in confusion, I picked up the child's drawing from where it had fallen to the floor. 'Does your cat have a name, Victorine?' I asked.

'Beppo,' she said, twisting one of her curls around her finger. 'He lives in the kitchen because Mamma doesn't like cats.'

We chatted about her drawing for a while and it was easy to see why the Princess loved this little girl with her sparkling eyes and happy nature.

Signor Fiorelli returned with the Baron, who handed me a purse full of coins before I took my leave.

I curtseyed to the Princess. 'I'll send a note as soon as your ball-gown is ready for its first fitting,' I promised.

'Come and see me before then,' she said. 'We'll take a drive into the countryside or make toast by the fire.' She gave me a wavering smile.

As I walked out of Villa Vittoria I reflected that the life of a princess was not always an enviable one.

# *Chapter 4*

Two weeks later Ma had gone to the market and I was alone in the cottage. I sat on a rocking chair, sewing by the light of the parlour window and looking out for her while I listened to the keening cry of the seagulls.

Since we'd finished the chemises and petticoats for the baker's wife we'd received two more commissions. I had cut a pattern ready for the Princess's ballgown and was in daily expectation of receiving the silk from Florence. We were earning enough money for the occasional sweet confection from the *pasticceria* or some sausage for our dinner. Ma seemed less anxious than usual and I was happier than I'd been for weeks.

Signor Fiorelli regularly called on us. Frequently he brought Victorine with him and we'd walk along the beach searching for shells. I'd grown fond of the child and sometimes I let her play with Peggy. She laughed at the rag doll's carroty plaits and chattered to her, just as I had once upon a time.

A movement outside the window made me put down my needle. Ma was hurtling towards the cottage, her mouth ajar and gasping for air.

I hurried to open the door for her as she scrabbled at the lock from outside.

She stumbled into the room, banging the door shut behind her. Her basket was empty.

'Whatever happened?' I asked. 'Were you robbed?'

Supporting herself against the door, she looked up at me with terrified eyes. 'He was following me!'

'Who was?'

'I don't know.'

I felt as if a lead weight had suddenly settled on my chest. 'If you didn't recognise him, then it can't have been my father.'

She slammed the shutters over the parlour window and bolted them with trembling fingers. 'I called in to the grocer's shop,' she said, still panting, 'to see if there had been any interest in our card.'

'And was there?'

'The grocer said a gentleman, a *foreigner*, had asked about a dress for his wife. His name was John Smith and he was staying at the Albergo Duomo.'

'Did you go to the hotel?'

She nodded. 'But there's no John Smith staying there. And when I left the hotel there was a man watching me from the other side of the street. He was much too tall for an Italian. I set off for the market but then I heard footsteps behind me. I turned to look but there was no one there.'

'I expect he'd turned up a side street.'

She twisted her hands together. 'I started to run and I heard him coming after me.' Her breath caught on a sob. 'Emilia, I was so frightened.'

'Ma, you must be mistaken.' I made an effort to curb my impatience. 'If a man had wanted to catch you, he would have done so.'

'It wasn't like that,' she said, her mouth set in a mulish line. 'I hid behind a market stall.'

'There was no one following you when you ran down the street just now.'

'I told you, I lost him in the market. But we're wasting time.' She ran into the dining room and wrenched open the door to the staircase. 'We must pack immediately,' she said.

'No!' Furious, I caught hold of her shoulders. 'Ma, you have to stop this! You live in a perpetual state of fear. It's ruining our lives.'

'But I saw him!'

'You may have seen a man and thought he was following you but it wasn't my father. Why would anyone else have cause to follow us?'

'Emilia, come upstairs and help me pack.'

Something inside me turned to iron. 'I will not,' I said, my hands balled into fists.

She looked at me uncertainly, one foot on the stairs. 'You have to.'

'I'm tired of roaming from place to place because of your strange fancies,' I said. 'You go if you want but I'm staying here in Pesaro.' I watched her begin to tremble and sink down onto the stairs.

Tears welled up in her eyes. 'You know I can't speak Italian properly. I need you.' Her chin quivered and she began to weep.

'Ma,' I said, gently this time, 'you're perpetually caught up in a web of fear of your own making. Please, help me to understand.'

She sniffed, staring at her feet. The silence stretched almost to breaking point and then she lifted her tear-stained face. 'It isn't only your father who wants to find us.'

I stared at her.

She closed her eyes.

'Who else does? Tell me the truth, Ma. I'm not a child any longer.'

'No,' she said. 'Perhaps it is time. God knows, it's a hard burden to bear on my own.'

A frisson of excitement mixed with disquiet ran down my spine. I went to the cupboard and took out a bottle of wine and two glasses. 'Sit down,' I said. I poured the wine and pushed a glass towards her. 'Drink that to steady your nerves.'

She grasped the glass in trembling hands and drank the wine straight down.

'Tell me,' I said.

She heaved a deep sigh. 'You already know I was a lady's maid to a Lady Langdon in Grosvenor Street in London. One morning her baby son was found dead in his cradle. My mistress was distraught. Sir Frederick seemed to blame his wife for their son's death and their marriage became very troubled. Sometimes he hit her.'

I shuddered, remembering my father's violent nature.

'One day they had a terrible quarrel and Sir Frederick locked my mistress in her bedroom for weeks.'

'How dreadful!'

Ma nodded. 'Lady Langdon decided to run away to friends in Paris and asked me to help her escape. She promised me a year's salary if I'd accompany her and, fool that I was, I agreed. You see, the money meant that your father and I could afford to marry at last.'

'How did you help her to escape?'

'I took her bedroom door key from the cupboard in the kitchen and had a copy made. Sir Frederick was at his club that night. We tip-toed downstairs with our baggage. We were in the hall when we heard Sir Frederick coming up the front steps.'

'He'd returned earlier than expected?'

Ma shivered at the memory. 'I knew he'd turn me off without a reference. Lady Langdon told me to take her travelling bags and run out of the kitchen door while she returned to her room, in case he looked in on her. She said not to wait in case she couldn't leave until later but to take the coach to Dover. I was to stay at the inn there until she arrived.'

'What happened?' I asked.

'I waited five days,' said Ma. 'I was nearly mad with worry. I sent a note to Joe, my intended, asking him to make enquiries.'

'And did he?' I asked.

Ma burst into tears again. 'Sir Frederick had found his wife's clothes on the riverbank. My mistress drowned herself.'

'But she wanted to go to her friends,' I said.

'Sir Frederick had told everyone she'd drowned herself because she was grieving for her baby and he'd posted notices offering a reward for my capture, accusing me of absconding with my mistress's clothes and jewellery.'

'That wasn't true!' Ma was always honest.

'But it was!' Her face crumpled. 'Joe said the only thing to do was for us to marry and go to France as it wasn't safe to stay in England. Later on, we hadn't enough money so we had to sell Lady Langdon's possessions.'

'It was so very long ago,' I said, taking her in my arms. 'We've travelled extensively since then. I promise you, Sir Frederick couldn't find us even if he did come looking. Now dry your tears. This is a special day, the first day you stop being scared of your own shadow and begin to live without fear.'

Some of the lines on her forehead eased a little. 'I suppose you're right.'

I hugged her. 'Now give me a smile!'

It was a poor effort but it gave me a glimmer of hope for better understanding between us in the future.

I was hemming Widow Mancuso's mourning dress while daydreaming about Signor Fiorelli's mischievous smile when hooves clattered past. A shadow fell over the parlour window as a horse and its rider passed by and a moment later there was a pounding on the front door.

Ma looked up from her sewing, wide-eyed and motionless.

Even though we'd agreed her fears were groundless, old habits died hard. 'It's all right,' I said. 'I'll go.' Hastily, I plucked some loose threads off my skirt.

The Baron, dressed in a scarlet uniform and plumed hat, stood on the doorstep, holding the reins of his black stallion. He bowed. 'The

Princess's compliments. She requests that you accompany her for a ride in her carriage.'

'The Princess?' I said, astonished.

'She awaits you.' The Baron waved his hand up the street and I saw a barouche and four. Victorine was standing on the seat and waving at the boats in the harbour. The Princess, sitting opposite Countess Oldi, waved at me.

'How delightful,' I said. 'I'll fetch my pelisse.'

The Baron nodded. 'Victorine says will you bring Peggy?' He mounted his horse and set off towards the barouche.

I closed the door behind him. 'Well!' I said. 'When I last saw the Princess she suggested I might take a drive with her, though I didn't imagine it was really her intention. But I can hardly refuse, can I, Ma?'

'I suppose not,' she said.

I put on my bonnet and pelisse and tucked Peggy under my arm.

'Don't be long,' said Ma. 'You must finish Widow Mancuso's dress tonight.'

I hurried down the street to where the Baron waited beside the barouche, his stallion dancing from side to side.

The coachman closed the carriage door behind me and returned to his perch at the front.

'I received bad news this morning and needed fresh air and good company,' said the Princess as we rolled away. She spoke in English. 'Victorine told me where you live and I thought I should like to enjoy the sympathetic company of my little dressmaker friend.'

Victorine slid off the seat beside the Princess and climbed onto my knee.

'Signorina Barton,' she said, 'may I hold Peggy?'

The Princess smiled indulgently, watching the little girl as she whispered into Peggy's ear. 'It does my heart good to see her happy when the world is so full of troubles.'

'Did you say you had bad news this morning, Ma'am?' I ventured.

'The Prince of Wales continues to send his spies to Milan to conjure up evidence against me,' she said. Her hands were clasped together so tightly in her lap that her knuckles were white.

I glanced at Countess Oldi who stared ahead with her customary bland smile. 'I'm not sure I understand, Ma'am,' I said.

'Have you heard of the Milan Commission?'

I shook my head.

'The British government will not allow the Prince of Wales to divorce me,' said the Princess. The corners of her mouth twisted in a bitter smile. 'How could they, when there are no grounds for it except for his personal dislike of me? So my husband, conniving hypocrite that he is, has set out to *find* grounds for divorce. He has set up the "secret" Commission in Milan to interview anyone who might provide proof of my so-called adultery.'

I caught my breath, astonished that the Princess discussed such matters with me.

Her mouth trembled. 'Dismissed servants, like Louise Demont, who have a grudge against me, men who provided my household with candles, an innkeeper, a sea captain whose boat I sailed on ... all these people are being questioned.' Her voice rose in indignation. 'They are asked to suggest others who will malign me and some grow so fat on the Prince's bribes they will never need to work again.'

'That's outrageous!'

'My husband has always been outrageous,' she said. Her jaw clenched. 'He's profligate, squandering his fortune on that seaside pavilion in Brighton or the latest fashions. He rarely pays his debts and would prefer to spend a morning buying paintings rather than discussing affairs of state. And then there are his many mistresses, though what they see in him now he has grown so very fat, I cannot imagine!' She glanced at the elegant figure of the Baron, riding his stallion alongside us, and gave him a fond smile.

The way the Princess described the Prince of Wales led me to believe he would be the worst kind of king, when the time came. But then, perhaps the Princess was hardly an ideal queen, either.

We drove through the gate in the massive walls to the old town. I enjoyed being in such a smart barouche as we threaded our way through the cobbled streets. Pedestrians stood back and touched their foreheads when we passed and I wondered what it would be like to have people notice you everywhere you went.

'How long is it since you were in England?' I asked.

'Five years now,' she said. 'And since my Charlotte has been taken from me, I've no desire to return. I am happy here.' She sighed. 'Or I would be, if my husband and his spies would leave me alone. Now there is no heir, I suppose he's more anxious than ever to rid himself of me.'

We stopped in the Piazza del Popolo so Victorine could look at the marble horses prancing in the fountain.

The Baron was attentive to the Princess, handing her down from the carriage and kissing her fingers.

'Thank you, *amore*,' she said.

I tried not to smile but he was so tall and elegant in his scarlet uniform and the Princess so dumpy that they made an ill-assorted pair.

Victorine ran to the red granite pool with its splashing fountain, her footsteps echoing across the piazza. She laughed in delight at the stone seahorses, frozen in motion as they frolicked at the edge of the pool. Spouting dolphins and mermen supporting a giant shell blew jets of water from their pipes.

The Baron lifted Victorine onto the low wall surrounding the pool and held her hand as she walked.

The Princess shivered. 'It's pretty in the summer but the water looks cold today.' She glanced at three or four ragged street children crouched in the square playing five-stones, and beckoned to the Baron. 'Will you buy bread for those children?' she said. 'They look hungry.'

Five minutes later, the ragamuffins, clutching crusts of bread in their grimy fists, waved as we set off again in the barouche. We drove through the gate on the opposite side of the walled town and down to the sea.

I breathed in deeply the invigoratingly salty air and watched the waves as they frothed onto the golden sand.

The Baron thundered off along the beach, his horse kicking up clumps of wet sand behind them.

'Catch him!' shouted the Princess to the coachman.

'Hold tight, Victorine!' I said, as the coachman cracked his whip. The barouche lurched forward and I clung to the side as we rattled along.

Victorine squealed, her eyes sparkling. 'Look at Papà! Go faster or we won't catch him!'

The Princess laughed, snatched off her hat and waved it in the air, hooting with delight.

The Baron was only a tiny figure in the distance now.

Countess Oldi, a terrified smile frozen on her face, gripped the door.

I was filled with exhilaration as we bounced over the sand at breakneck speed. The wind was in my hair and my pulse raced. I'd never travelled so fast in my life. A dog ran towards us and snapped at the horses, making them canter even faster and the Princess laugh louder.

At last, the Baron galloped back towards us.

The coachman pulled on the reins, turned the barouche in a wide circle so we faced the San Bartolo hills again and we settled down to a more sedate pace.

'That's put the roses in our cheeks,' said the Baron as he came to ride beside us.

'Can we do it again, Papà?' asked Victorine.

'Another time, perhaps.'

We trotted back towards the harbour and stopped near the cottage.

'Thank you for the outing, Ma'am,' I said. 'The fresh air has certainly blown the cobwebs away.'

'I am refreshed, too,' she said. 'You remind me so much of my dear Charlotte, I knew I should like your company.' She patted my hand.

Victorine hugged the rag doll tightly in her arms. 'Can I keep Peggy for tonight?' she asked as I descended from the coach.

'If you promise to keep her safe for me,' I said.

The barouche's wheels began to turn.

Victorine waved Peggy's calico arm vigorously at me. Smiling, I turned towards the cottage. The Princess might, at first glance, appear a figure of fun, but it seemed to me that she had been poorly treated. I admired her for dismissing her troubles and bravely making the best of things. And, unlikely as it seemed, I felt there was a tentative friendship blossoming between us.

# Chapter 5

'If only you hadn't gone gallivanting off with the Princess yesterday you'd have finished Widow Mancuso's dress earlier,' scolded Ma.

'Well, it's done now,' I said. I finished folding the garment, wrapped it in tissue paper and tied it with a black satin ribbon.

'Come straight home afterwards,' said Ma. 'I need you to help me to finish these shirts before the light goes.'

I went to the post office and collected the parcel of silk for the Princess's ballgown before making my way to Widow Mancuso's. She lived just within the town walls in a once-fine house, now sadly neglected. She received me with old-fashioned courtesy in the gloom of her *salone*, where an almond cake and glasses of wine had been laid out in readiness. I couldn't hurt her feelings by refusing her hospitality.

It was nearly half past five and growing dark when I made my escape back to Harbour Cottage. The full moon was reflected on the water over the harbour wall and the fishing boats were moored in neat rows. Lights were already glowing from behind the shutters in the houses but Ma hadn't yet lit our candles.

I tapped on the front door and hopped from foot to foot, shivering, while I waited for her to let me in. The wind from the sea had sprung up again and I drew my shawl tightly around my shoulders. Perhaps she'd fallen asleep over her sewing. 'Ma,' I called, 'open the door!'

There was no sound from inside. The shutters were folded across the window though not secured, but I couldn't see into the parlour through the tiny gap. I rapped on the casement but still there was no response. I peered at the upstairs windows. Did I imagine it or did the lace curtain move?

Annoyed now, I hurried down the street and into the alley behind the cottages, feeling my way by moonlight. Footsteps came running down the alley. A man blundered towards me out of the shadows. I froze when he loomed over me but I needn't have worried. He didn't say a word but rudely pushed past, thrusting me back against the wall in his haste.

I counted the gates along the wall until I came to the one that gave access to our yard. I found a toehold in the crumbling brick and stretched up to fumble blindly over the top of the wall until cold metal touched my finger. Triumphant, I withdrew the key we'd secreted there. But when I went to unlock the gate, it swung inwards when I touched it. Nonplussed, I ran my hand over the wood and drew in my breath as splinters drove into my palm.

The unease within me flowered into dread when I discovered the kitchen door was wide open. Standing very still, I listened. There was no sound from within.

'Ma?' I whispered.

Nothing.

I fumbled my way to the dresser and lit a candle from the embers of the fire, expecting her to call out at any moment. As candlelight illuminated the room I gasped when I saw the pretty majolicaware had been swept off the dresser and smashed into shards on the tiled floor.

Shielding the candle flame, I tip-toed into the dining room. Fear prickled at the back of my neck. Our sewing boxes were upside-down and the floorboards littered with spools of thread, pins, scissors, thimbles and tailor's chalk. An intruder must have run out of the kitchen door and smashed the lock on the yard gate.

'Ma?' I called.

In the parlour the cushion covers were slashed to ribbons and feathers lay on the rug like a carpet of snow. The bookcase was overturned and volumes lay strewn over the floor.

Frightened now, I sprinted up the stairs, coming to a sudden stop in Ma's bedroom doorway. My stomach lurched. She lay in a crumpled heap on the floor. I fell to my knees, heedless of the wax that seared my wrist in my haste to put the candle down. Blood ran from Ma's temple and her cheek was swollen and turning purple. Her eyes were closed but my own heartbeat steadied a little when I felt her breath against my cheek.

'Ma, wake up!' I shook her gently and she murmured and stirred. 'Thank God!' I lifted her up in my arms and gasped when I saw her wrists and ankles were tied together with coarse twine. Carefully, I placed her on the bed, shocked by the vicious bruising on her arms and throat. I stroked her hair until her eyes opened.

She looked around wildly and tried to sit up. 'He had a stick and he hit me!' She began to shriek with terror and I held her firmly.

'He's gone and you're quite safe.' I rocked her until her cries gave way to hiccoughing sobs. 'I'm going to fetch the scissors and free your hands and feet,' I said.

'Don't leave me!'

'I'll only be a moment.' I ran downstairs, bolted the back door and snatched up my embroidery scissors from the floor.

Ma's wrists and ankles were so swollen and chafed it was hard to sever the twine without slicing her skin. Anger boiled up inside me. 'What happened, Ma?'

'I *told* you they were looking for us,' she said, hysteria rising in her voice again.

My heart sank. This was the worst possible thing that could have happened, when I had so recently persuaded her that her fears were ungrounded.

'Did you see him?' I asked.

She nodded.

'And did you recognise him?'

Her teeth were chattering and I wrapped a blanket around her shoulders. 'No.' She cupped her hands over her eyes. 'It wasn't your father or . . . '

'Sir Frederick?' I asked.

'Who?' Her face was ashen.

'Your mistress's husband.'

She looked perplexed. The head wound must have made her forget temporarily. 'You thought he might be looking for you,' I said.

'Why would he be here?'

'No reason at all. I'm sure your attacker was simply a chance thief,' I said. But, just for a moment, I'd wondered.

'A thief?' She wrinkled her brow and touched the lump on her temple. When she saw the blood on her trembling fingers she frowned.

'A thief,' I said, firmly. 'He ransacked the cottage.'

Ma leaned back against the pillows. 'I didn't tell him about them,' she said.

'About what?'

She closed her eyes and I took water from the ewer and poured it into the basin to wash the blood from her face. She winced as I cleaned the cut on her head. It wasn't deep but, despite my gentle pressure, she groaned as my fingertips explored the swelling around the wound.

'What didn't you tell him?' I asked.

Puzzled, she looked at me with a blank stare. 'Who?'

She was still shaking and confused and I decided any further questions could wait until later. It was too late to report the assault and, besides, I couldn't leave her alone. 'I'll fetch some salve to soothe your wrists and a cold cloth for your head.'

I returned to the kitchen to boil water for tea and sweep up the broken earthenware. Once Ma was asleep I'd tidy up the remaining ravages wrought by the intruder, to spare her further distress in the morning.

We sat together to drink our tea but as soon as I'd taken the empty cup from her she began to retch. I snatched up the slops basin and held it while she vomited.

Afterwards I wiped her face and laid her down against the pillows. 'Go to sleep now, Ma. In the morning everything will be better.'

I sat beside her and waited until she drifted into a doze before going downstairs to tidy up the rest of our jumbled possessions.

My silver thimble glinted on the dining-room floor amongst the scattered contents of my sewing box. I wept when I saw it had been crushed flat. Ma had bought it for me when I was only a little girl and she first taught me to sew. Tears of sadness gave way to rage at the wretch who'd barged into our cottage, frightened my mother nearly to death and despoiled our possessions.

Fuelled by fury, I rushed about like a whirling dervish, righting the furniture, putting away the crumpled books and sweeping up the feathers. I flung the shredded cushion covers into the fire, where they flared and spat as they twisted in the flames.

Once my ire was spent, I was overcome with exhaustion and wearily climbed the stairs to see Ma again. She muttered in her sleep. I lay on the bed beside her and, after a while, I dozed.

Ma was restless for much of the night, her face and limbs twitching while she dreamed. She murmured the name 'Harriet' several times and, once, spoke distinctly. 'Joe,' she said, 'I won't leave Harriet!'

Later, she lay so silent and motionless I touched her in sudden panic. She sighed deeply and opened her eyes, looking at me without recognition, as if she still dreamed.

'Go back to sleep,' I whispered.

She closed her eyes again.

I lay beside her thinking about what she'd said. Harriet . . . There was something about the name that tugged at my memory. I sighed, too tired and upset to think about it anymore.

It was nearly dawn when Ma's sobs awoke me.

She sat beside me, rocking and holding her head in her hands.

'Ma?'

'My head hurts so,' she whimpered.

I got up and dipped a cloth in the ewer.

The cool compress seemed to ease her pain a little. I placed a few drops of lavender oil onto a handkerchief for her to sniff and it was then that I saw the blood trickling from her ear. I wiped it away but Ma looked at the bloodied cloth with horror.

'I'm going to die, aren't I?' she whispered.

'Of course not!' I said, forcing my voice to sound calm, 'but Signor Fiorelli's father is a doctor. I'll fetch him to examine you.'

'Don't go!' She clutched my arm so hard that I knew I'd have bruises later. 'I don't want to die alone!'

I couldn't leave her when she was so upset. 'We'll see how you are later, then.' I changed the subject to calm her. 'You were dreaming last night,' I said. 'Ma, who is Harriet?'

She turned her poor swollen face towards me, her mouth slack with shock. 'Harriet?' she said at last.

'You were mumbling and said, "Joe, I won't leave Harriet."'

She burst into tears. 'God forgive me! I can't take this to my grave.'

'You're *not* going to die!' My curiosity was piqued now. 'Tell me about it and perhaps you'll feel easier?'

Ma's mouth quivered and tears rolled down her cheeks. 'I've kept the secret for so long . . . '

'Then tell me,' I said as persuasively as I could. There was so much she hadn't told me about herself and, sometimes, what she did say conflicted with an earlier story. She'd always been secretive and sometimes that made me feel I didn't know her at all. 'Tell me, Ma!'

Her mouth worked and she brushed tears away. 'I told you about my mistress?'

I nodded encouragingly.

'When Lady Langdon begged me to help her,' said Ma, 'she planned to take her four-year-old daughter Harriet with her.'

'I remember you said she had a baby that died but you didn't mention a daughter.'

'When Sir Frederick came home early that evening, my mistress ordered me to hurry on ahead to the inn with Harriet.'

'You took the child with you?'

'She cried desperately for her mother for days,' said Ma. 'Later, when Joe came to the inn and told me my mistress was dead, I didn't know what to do.' She closed her eyes, tears seeping from under her eyelashes. 'Sir Frederick had already sent out a handbill offering a reward for news of my whereabouts and I'd have been accused of kidnapping as well as theft if I'd gone back. I'd have been hanged.'

'So what did you do?'

'When Joe said we must run to France, I refused to abandon Harriet. I didn't trust Sir Frederick – he had no real affection for his daughter. I couldn't bear for the poor motherless scrap to have no one to love her.'

I stared at her. 'You took Harriet away from her father? To France? How could you do such a terrible thing?'

'Joe wanted to leave her at an orphanage and that's when our troubles began. We quarrelled constantly. He made me sell Lady Langdon's jewels and, when he couldn't find work, spent the money on drink.' She began to rock backwards and forwards. 'I was so alone and I didn't speak French. I was too frightened to take Harriet back to London and leave her on her father's doorstep.'

'So you took her to an orphanage?'

Ma winced as she rubbed her head. 'I couldn't do it.' She began to weep again. 'The little girl's name was Harriet Emilia Langdon,' she sobbed. 'For her safety, I called her Emilia Barton.'

Dry-mouthed, I stared at her. There was a rushing sound in my ears and a sudden heaviness in the pit of my stomach, as if I'd swallowed an iceberg. Black flecks flickered at the edge of my vision and I gripped my hands together to dispel the sudden dizziness. Ma was not my mother. Not only that, but I was not the person I'd always thought I was, either.

'Emilia?'

I opened my mouth but I couldn't speak.

'Marrying Joe was a disaster,' she said. 'He'd imagined we'd be rich if we sold. . . ' She paused and her gaze slid away from me. 'If we sold the contents of Lady Langdon's baggage. He was disappointed by how little her goods fetched, and then he grew angry. He resented being saddled with a child and was so harsh with you I was frightened for your life. So, for your sake, I fled from him to the South of France.'

'How *could* you?' I said. 'You stole me away from my family!' Listening to her story, I didn't know who I really was and nothing would ever be the same again.

Her eyes pleaded with me. 'I did it to protect you.' Tentatively, she reached out to touch my wrist. 'Now you are all that I have in the world.'

I shook my head in disbelief. It felt as if the very ground had crumbled away from under me. I surged to my feet. I had to get away. I rushed down the stairs and out of the front door.

Lifting my skirts, I ran through the echoing streets, not caring where I went. It was raining and I ran and ran until a stitch in my side forced me to lean over with my hands on my knees. Afterwards I sprinted on again until my feet crunched over sand and then splashed into the foaming sea. When the icy water reached my thighs I gasped and stood still.

Tipping up my face to the rain, I wailed and sobbed. A wave surged past me, soaking me up to my waist. The shock of it caused a modicum of sense to return. I couldn't remain standing in the sea.

I would catch my death. Or walk to it. Slowly, I turned towards the shore and ploughed my way through the water, the sand sucking at my shoes. What was I going to do?

I stood on the beach in the grey dawn, dripping wet and shivering in the wind. There really was nothing else to do but to go back to Ma.

Fishing boats were returning to the harbour with their catch by the time I reached the cottage. Gulls screamed overhead as fishwives gutted the catch, tossing the entrails over the harbour wall. Women were already gathering with their baskets, inspecting the mackerel and sardines on display. One or two glanced at me as I passed, my wet dress almost transparent and moulded to my body.

In my mad rush to escape I'd left the front door ajar. Inside, I slipped off my shoes, sand rubbing grittily between my toes. The stairs felt like a mountain to climb and I paused at the top, gathering the strength to face the woman who wasn't my mother after all.

She lay on her side, the pillow stained with fresh blood. There was something about her motionless form that made me pause in the doorway. Then I ran forward, my heart hammering like a blacksmith's anvil. Her sightless eyes were half-open and not a breath stirred the air.

Sarah Barton, the woman I used to call Ma, was dead.

# Chapter 6

Panic-stricken, I ran to the Fiorelli house and hammered on the door, shouting for Dottore Fiorelli. The family were at breakfast and a sea of faces turned to look at me. Tutting and shaking her head, Signora Fiorelli wrapped a blanket around my shoulders while her husband patiently questioned me. I can't remember what I told him but his assured manner was calming.

Alessandro watched anxiously while I told the story, barely able to speak for sobs. Afterwards he and his father left for the cottage and his mother took me upstairs to peel off my sodden clothes. The eldest daughter, Cosima, found me one of her dresses to wear.

Signora Fiorelli sat me by the kitchen fire and wrapped my shaking hands around a cup of soup. 'Drink it quickly, you're chilled to the bone.'

Obediently, I sipped the chicken broth, my teeth chattering against the china rim. I finished the soup and stared into the flames while the sounds of the Fiorelli family talking in hushed tones washed over me. My head spun as I went over and over what Ma had said until a deathly exhaustion overcame me. All I wanted was to climb into bed and pull the covers over my head.

Alessandro and his father returned and Dottore Fiorelli touched me on my shoulder. 'It is with great regret, Signorina Barton,' he said gently, 'that I confirm your mother has passed away.'

'If only I'd come for you last night,' I wept, 'perhaps you might have saved her.'

'Do not reproach yourself,' he said. 'The blow to her head caused an injury to the brain, causing it to bleed. Nothing could have saved her. It was better for her that you stayed close to her during her final hours.'

I couldn't look into his kind brown eyes. 'But I didn't,' I said with a sob. 'We argued and I ran down to the beach. I wasn't there long but when I returned . . . ' I hung my head, recalling her last words to me. *'I did it to protect you.'* I remembered her fingers on my wrist for the very last time. *'Now you are all that I have in the world.'* I buried my face in my hands.

'Hush! She is at peace now,' said Signora Fiorelli. 'And I am going to put you into Cosima's bed and you shall sleep.'

'Swallow this draught,' said the doctor, handing me a small glass. 'When you wake you will feel better.'

I had no energy to protest and allowed myself to be led upstairs. Signora Fiorelli shooed me into bed and closed the shutters. I pulled the covers over my shoulders and, within a few minutes, sank into oblivion.

I stretched and yawned. Peggy lay on the pillow next to me and I stared into her unblinking blue eyes. I clasped her against my chest with my chin on the top of her head as I had done every morning for most of my life. But this wasn't my bed.

I heard a giggle and turned to see Cosima sitting beside me.

'Two redheaded sleeping beauties,' said a voice. Alessandro Fiorelli stood in the doorway, smiling at me. 'How are you?'

Slowly I pushed myself into a sitting position. 'Muzzy,' I said, rubbing my eyes. My mouth was dry. Suddenly the memories of the recent events crowded in. 'I must go home,' I said, throwing back the bedclothes in a panic.

'Not yet,' he said. 'Mamma says dinner is ready. Are you hungry?'

I shook my head but my stomach growled.

Downstairs it was daunting to see so many faces around the table, unused as I was to large families. For as long as I could remember it had only ever been Ma and myself.

Cosima took my hand and urged me forwards. 'You shall sit by Alessandro, who is the eldest of us left at home. Delfina is two years older than Alessandro and lives with her husband and baby Enzo in Fano.' She pointed at two young men who bore a remarkable likeness to their older brother. 'Salvatore and Jacopo are next. I'm the second eldest girl. Then Fabrizio is fourteen and Luca is twelve. Gina is my younger sister and last of all is Alfio.'

'I'm five!' piped up the little boy.

'A late and unexpected gift from God,' said Signora Fiorelli, ruffling his dark curls.

Despite everything, I ate my supper. I spoke only enough to be polite but was comforted by the cheerful chatter of the family. Afterwards, while the women cleared the table, Dottore Fiorelli requested that Alessandro and I join him in his study.

'Signorina Barton,' he said, pushing back the gold-framed spectacles that rested on the end of his imposing nose, 'I have been obliged to report to the police chief, Capitano Bischi, that a thief broke into your cottage and the resulting death of your mother. Alessandro also told them that he saw a man looking in through the downstairs window yesterday afternoon.'

'A man?' I frowned.

'Victorine and I came to see you,' said Alessandro. 'You remember we'd promised to return your doll? Victorine was walking along the top of the harbour wall when I saw a man peering in through your parlour window.'

'When was this?' I said.

'About half past three or four, perhaps. As I approached the cottage I thought he looked . . . ' Alessandro shrugged and turned up his palms. 'He looked furtive. So I called out, "Are you looking for someone?"'

'What did he say?'

'That was the strange thing. He glanced at me and ran off. Great long legs, he had. And a white face.'

'I doubt he'd have run away if he hadn't been up to mischief,' said the doctor.

'I couldn't run after him,' said Alessandro. 'Victorine was in my care. So I knocked on the door and spoke to your mother.'

I opened my mouth to say that she wasn't my mother but the wound was still too raw to talk about that.

'She said you'd gone to deliver a dress not five minutes before but you were expected back within the hour. So Victorine and I walked in that direction, hoping we might meet you and restore Peggy to you.'

'You didn't tell Signora Barton about the man, Alessandro?' queried the doctor.

He shook his head. 'He'd gone so I only said to her that I'd call back.'

The doctor raised one eyebrow. 'It was so urgent to return the doll?'

Signor Fiorelli shrugged but I saw spots of pink flare on his high cheekbones. 'We returned an hour later but the shutters were closed. I presumed Signora Barton was out.'

'Her attacker must have been in the cottage by then,' I said, my stomach churning. 'It was almost dark when I arrived at half past five. It would have taken him some time to search the cottage and then to intimidate . . . ' I pressed my fingers to my mouth.

'What is it?' asked Alessandro.

'When I knocked at the front door I saw the bedroom curtain move. As my mother didn't come, I went into the alley behind the cottage to enter by the kitchen. By then it was dark and a man ran towards me out of the shadows. He nearly knocked me flying.'

'You think your mother's assailant ran out of the back of the cottage when he heard you at the front?' said the doctor.

I nodded and swallowed. If I'd had my door key with me and had let myself in, would he have killed me too?

'Signorina Barton,' said the doctor, 'you must speak to Capitano Bischi and tell him what you saw. There is time enough for that tomorrow. Also there is the matter of . . . ' He hesitated. 'I took the liberty of informing the priest of your mother's death and making arrangements for her burial. Do you have any family or friends I can contact for you?'

'No,' I said. The look of pity on the doctor's face was almost my undoing. I breathed deeply and swallowed back my tears. 'I have no friends or family,' I said. 'None at all.'

The Fiorelli family were extraordinarily kind to me over the following days. Signora Fiorelli insisted I stay with them and Cosima and ten-year-old Gina shared their bed with me. Although I'd often shared a bed with Ma, it was strange to be tumbled together with two friendly girls I barely knew, like a basketful of puppies. At night I listened to their giggles and whispered confidences and wished I had been blessed with sisters.

Dottore Fiorelli escorted me into town to talk to Capitano Bischi. Dark-eyed and dapper, the police chief bade me sit down and I told him my story, omitting Ma's confession. Capitano Bischi shrugged his narrow shoulders and said that her attacker must have been an opportunist thief. Signora Barton's subsequent death had been most regrettable but appeared to be without motive other than theft. He assured me every effort would be made to find the thief but held out little hope of apprehending him. Dottore Fiorelli and I returned to the house in silence.

I owned no mourning clothes but Signora Fiorelli opened the cedarwood chest in her bedroom and took out a crepe mourning gown. 'It was my mother's she said. 'The style is outdated but you may have it if you care to alter it. Sewing will keep you occupied at this difficult time.'

There was a great deal of material in the full skirts and, with careful cutting, I was able to make two dresses and a spencer. It was ironic that I'd only recently finished Widow Mancuso's mourning dress without the slightest inkling that I should be making a mourning wardrobe for myself a few days later.

Ma's body was brought to the Fiorellis' house on the night before the funeral and her coffin laid on trestles in the *salone.* Unable to sleep, I dressed and crept downstairs in the small hours. I peeped into the open coffin, half afraid of what I would see but, despite her bruised cheek, Ma's waxen face was serene with all worry lines erased away. She, at least, was at peace.

I kept vigil beside her body for the rest of the night. I was still angry with her for stealing me away but also nagged by guilt. If I'd taken her fears seriously we'd have moved on and she'd still be alive. Everything about my life had changed and I felt as if I stood on the edge of an abyss, not knowing what terrors awaited me below.

It was still dark when Signora Fiorelli came downstairs. 'I wondered if you might be here, all alone,' she whispered. Leaning over Ma's body, she sighed. 'Too young to die,' she said. She took my hand and I was grateful for her warm touch. We sat quietly together until dawn broke.

Alfio trotted downstairs and Signora Fiorelli firmly led me out of the *salone* and made me sit with the little boy at the kitchen table. I sipped hot coffee and crumbled a piece of bread while I responded to Alfio's childish chatter. One by one the rest of the family joined us.

Alessandro, looking unusually sombre, sat beside me and watched me from under his dark eyelashes. 'You're very brave,' he said. 'But you mustn't be embarrassed to cry. You are amongst friends here.'

At once my eyes began to prickle and I looked down at my folded hands. Ashamed that my feelings towards Ma were so confused, my throat closed up.

Alessandro passed me his handkerchief and patted my arm with great tenderness.

'It is time,' said his father.

Signora Fiorelli led me into the parlour again and gently pushed me towards the coffin. 'You must kiss your mother, Signorina.'

I looked down at Ma's face. She was the only mother I remembered and I was sorry that her life had been so troubled because of me. Leaning over, I touched my lips to her cold forehead for the last time.

Since Ma's coffin was small, Alessandro, his father, Salvatore and Jacopo were all the pallbearers needed.

As they carried the coffin into the church there was a clatter of hooves and I glanced up to see the barouche come to a smart stop outside. The Princess, accompanied by Countess Oldi, descended and hurried after us.

After the burial we returned to the Fiorellis' house for wine, *biscotti* and *panforte*. Cosima and Salvatore were charged with offering the refreshments to the party and Signora Fiorelli whispered to me that it was fortunate she'd set out the best glasses since she hadn't expected to entertain royalty.

The Princess, however, stood on no ceremony and hugged me to her ample bosom. 'My dear little friend,' she said, 'I grieve for you in your loss.' She sighed heavily. 'We are companions in our sorrow.'

'I'm honoured that you came today,' I said.

'I couldn't bear to think you might be alone,' said the Princess, 'but I see your friends are caring for you.'

'The Fiorelli family have been so kind,' I said, imagining how unbearable it would have been if I'd had to arrange everything on my own.

Signora Fiorelli enfolded me in a lavender-scented embrace. 'But now we must decide what you will do next.'

'I've presumed upon your hospitality too long,' I said. 'I'll return to the cottage this afternoon.'

A chorus of protests broke out from her children and only died down once Signora Fiorelli had flapped her hands at them as if they were squawking chickens.

'Whatever would people think?' she said, her expression shocked. 'A young, unmarried lady living alone?'

'Your honour must be protected,' said Alessandro in a firm voice.

I knew they were right. There had been security, or so I'd thought, in two women living together, but my reputation would be at risk if I lived alone. In any case, I doubted I could earn enough by myself to be able to afford the cottage.

'Perhaps I'll find a room to rent,' I said.

Countess Oldi brushed *biscotti* crumbs from her lips. 'Not at all suitable,' she said.

'I doubt you would find one,' said Dottore Fiorelli, shaking his head. 'Not many people would care to be responsible for a young lady on her own.'

'What he means, Signorina Barton,' said Signora Fiorelli, 'is that you are far too pretty for any woman to risk her husband's attention straying if you lived in her household.'

Anxiety fluttered in my breast. Where could I go? Then I had an idea and decided to speak before my courage failed me.

'Ma'am,' I said to the Princess, trying to keep my voice even, 'if you were to find me a place in your household I could undertake any household sewing tasks or alterations to your wardrobe. I speak Italian, French and English fluently and write a clear hand in all three. Perhaps I might be entrusted with your correspondence?' I gripped my hands together while a pulse fluttered in my neck.

The Princess thought for a moment. 'Since I have dismissed my maid, my wardrobe undeniably needs attention. Most of my household are living off the premises, at least until the building works are completed, but there's a little room on the ground floor, if that will suffice?'

I let out my breath in a sigh of relief. 'Indeed it will!'

Later that afternoon I was grateful for the company of Cosima and Signor Fiorelli when I returned to the cottage. I'd dreaded facing the memories of Ma's death on my own.

Alessandro unlocked the door and we followed him inside.

'Oh, no!' said Cosima.

It was immediately apparent that the intruder had returned. Again, the parlour furniture was overturned and even the rug had been rolled up to expose the floorboards.

'Go outside,' ordered Alessandro. 'Hurry now! I shall make sure there's nobody here.'

Wordlessly, Cosima slipped her hand into mine. Her brown eyes were full of pity for me. I allowed her to draw me outside.

The harbour wall was so cold when we sat upon it that it seemed to drain all the warmth out of my body. I shivered and rose to my feet in a panic. What if the intruder were still inside and had harmed Signor Fiorelli? But at that moment he beckoned us back into the house.

'What can he have been looking for to risk coming back?' I said. 'We had nothing of value.'

'Shall I help you to tidy up?' asked Cosima.

I left Signor Fiorelli righting the overturned furniture and Cosima picking up the contents of the sewing boxes while I went upstairs. The mattresses had been sliced open and our clothes tipped out of the chest. My heart sank. I would have to replace the landlady's mattresses, further depleting my purse. One by one I lifted up the scattered clothes and placed them neatly in my travelling bag.

Ma's quilted winter petticoat swung against my ankles as I picked it up and I was surprised by how heavy it was. It was then that I made my discovery. All around the hem, on the inside, was a double row of little cambric pockets. Each one contained a gold coin.

I sank down on the edge of the slashed mattress, my legs suddenly weak. I remembered then that, on her deathbed, Ma had said, *'I didn't tell him about them.'* Perhaps these coins were what the thief had been searching for? I had no idea how Ma could have come by such riches. We'd often gone so short it was hard to believe she'd saved anything much from our meagre earnings.

Agitated, I paced across to the window. However she'd acquired the coins, I needed them now and they'd be a godsend.

Without hesitating I pulled on the petticoat under my skirt and tied the tapes firmly around my waist.

# Chapter 7

Victorine was sitting on the front steps watching the builders when Signor Fiorelli and I arrived at Villa Vittoria. I caught my breath as the guard dogs came bounding up to investigate.

'Down, Titus!' commanded Signor Fiorelli. 'Down, Bruna! Don't worry, Signorina Barton, they're more bark than bite.'

'Nevertheless, I'm pleased you're here to control them,' I said, gripping my bag tightly.

'They'll soon get to know you.' His eyes were full of sympathy for me. 'After all that has happened, coming here must feel very strange. Remember, I am your friend and you may call on me whenever you wish.'

I forced a smile. 'I am a little nervous.'

'Don't be. And come and have dinner with my family again.' He smiled, his teeth white against his olive skin. 'If you can stand the noise and the squabbling.'

Two members of the Papal Guard were standing by the front door, as usual.

'You are coming to live with me, Signorina Barton?' Victorine hopped from foot to foot.

'For a while,' I said.

'Will you play with me?'

I ruffled her dark curls. 'I shall be working for the Princess but when she doesn't need me, then I'll play with you.'

Signor Fiorelli lifted my bags down from the barouche. 'We'll go round to the servants' door,' he said.

Victorine skipped along ahead of us, jumping over heaps of sand and stacked timber boarding.

'I've no experience of the correct etiquette in a royal household,' I admitted.

'Don't be concerned. The Princess is moderate in her demands. She asks only that you are loyal and efficient in your duties.'

'How could I not be loyal to her?' I said. 'I don't know what I should have done if she hadn't offered me a place.'

'Mamma would never have turned you out into the streets,' he assured me.

'I don't intend to be a burden to anyone,' I said.

At the side of the house we paused to look at the stable yard with its rows of horseboxes. The yard echoed to the clang of the farrier's hammer and the acrid scent of burning hooves hung in the air. Lads were busy grooming the horses and mucking out the boxes. Victorine climbed up on the gate and waved at one of the grooms.

'There are forty-eight horses,' said Signor Fiorelli, 'from the smallest piebald ponies to the finest Arabian stallions. And several carriages are housed here for the Princess's use.'

I gazed at him in surprise. 'But this is hardly a palace.'

He laughed. 'No, not yet. The Princess hasn't lived here very long, remember. Before she bought Villa Vittoria she rented the Villa Caprile nearby. That was much grander. Although the Princess and her architect have plans to enlarge Villa Vittoria, she likes to call it her country cottage.'

'So the household is smaller than it was before?'

'Mostly it's the Pergami family who live here,' he said, 'together with Willy Austin, Victorine's nursemaid, two equerries and a few servants. There'll be more space for live-in servants once the building works are finished.'

'And the Baron's family live here also?'

Signor Fiorelli nodded. 'The Baron oversees everything. He hires the staff, which includes his mother, his cousins, brothers and sisters, and controls the household accounts.'

'Victorine's mother doesn't live here then?'

'No,' said Signor Fiorelli. He inclined his head towards me. 'And a word of advice, if I may? You would do well to remember the Baron has almost complete authority over everything at Villa Vittoria.'

Behind the house I glimpsed another avenue of cypresses and a half-completed Italian garden. Soldiers of the Papal Guard patrolled the grounds and two more stood by the back door.

'Are the soldiers always here?' I asked.

'Night and day. Fourteen of them. The Princess is anxious that her husband's spies might infiltrate the villa and poison her,' he added in an undertone. 'The Pope allows her the guard to ensure her safety.'

I glanced at Victorine, who was watching us with bright eyes, and resolved to find out more another time. She grasped hold of her tutor's hand and led the way indoors, past the kitchens.

'I'll take you to Mother Pergami,' he murmured.

I followed him into the servants' dining room. The old lady in black I'd seen when I first visited Villa Vittoria sat at the long table with a younger woman, counting piles of sheets.

'I bring you Signorina Barton,' said Signor Fiorelli.

Signora Pergami's face was deeply wrinkled, like an apple that has sat on a windowsill for too long. Her almost toothless jaw moved from side to side as she looked me up and down. Finally, she nodded in greeting and I dropped a small curtsey.

'Faustina,' she said to the younger woman, 'take her to her room. And then the Baron will see her.'

'I must say goodbye for now, Signorina Barton,' said Signor Fiorelli. 'It's time for your lessons, Victorine.' He held out his hand to the little girl.

Loneliness gripped me for a moment until I saw him glance back at me through the doorway with a grin on his handsome face.

Faustina, who would have been pretty if her features weren't quite so coarse, pushed herself to her feet and picked up my bags.

I followed her along the passage, watching her hips roll from side to side as she sauntered along as if she had all the time in the world. A few locks of black hair had slipped from her hairpins and lay in greasy tresses on her shoulders. She opened a door and dropped my bags on the tiled floor inside. When she unlatched the shutters, light poured in to illuminate a bed covered in a striped cotton coverlet. There was a rag rug on the floor and a washstand with a rough white towel folded on the top. There was no fire in the empty grate.

'There's hooks behind the door for your clothes,' she said. 'A maid'll bring you water in the morning and the privy's out the back beside the stables.'

'Thank you,' I said. The room was whitewashed, with dark timber beams above and terracotta tiles on the floor, simple but adequate.

She gave me the glimmer of a smile. 'Unpack later. My brother's waiting for you now.'

I realised with some surprise that Faustina must be one of the Baron's siblings and therefore Countess Oldi's sister too. Her position at Villa Vittoria was far less elevated than that of the Princess's lady-in-waiting.

We returned along the passage and Faustina knocked on another door. A male voice bade us enter.

The Baron sat at his desk, pen poised to write in the ledger before him. 'Ah, yes, Signorina Barton. You may go, Faustina.'

The Baron beckoned me forward.

A fire crackled in the hearth and the room was very warm. I was close enough to smell his hair oil.

He didn't ask me to sit down. 'The Princess has instructed me to arrange the details of your employment as her sewing woman and, when required, secretary,' he said. 'You shall have your board and lodging and an appropriate salary will be paid half-yearly. You may have an afternoon off each fortnight to visit your mother's grave.'

'Thank you, Baron.' I wasn't brave enough to ask what an appropriate salary might be. In any case, I was in no position to haggle.

He stood up and I was reminded again of how his physical presence dominated a room.

'Your discretion must be absolute,' he said. 'You will not discuss anything you see or hear in the Villa Vittoria with any person outside these walls. Do I make myself clear?' His gaze bored into me.

'Absolutely, Baron.'

'There are those who may wish the Princess ill and it is the duty of every member of this household to protect her. Should you discover any unknown person entering the grounds or the villa, you will inform either myself or one of the guards immediately. Furthermore, the corridors are patrolled all night. You will not leave your room after retiring until the following morning.' He fixed me with an unsmiling gaze that made me faintly uneasy.

I nodded in acquiescence but was startled to learn that the risks to the Princess's well-being were considered so high.

'You will take your dinner now in the servants' hall and the Princess will see you afterwards in her dressing room to outline your duties. Do you have any questions?'

'Not at present,' I said.

The Baron turned his attention back to his ledger and I took it that I was dismissed.

I returned to my room and hung up my other mourning dress and placed my undergarments in the chest. I pushed Ma's travelling bag beneath the bed, along with the package of silk for the Princess's ballgown.

I took Peggy out of my bag and hugged her tightly, trying to dispel the hollow feeling inside me. It distressed me that I had no memories of the time before Ma, or Sarah as I must learn to call her, had stolen me. She'd said I'd cried bitterly for my mother so I must have loved her. And then there was my real father, who was not Sarah's husband Joe as I'd always supposed. I wondered if he could really have been as unkind as Sarah said. She had been fond of her mistress, my real mother, and her judgement of the situation could have been clouded.

I sat on the bed and closed my eyes, letting my mind drift to see if I could remember anything about my long-ago family. My very earliest memories were of hiding behind the furniture with my hands over my ears, listening to a hectoring male voice, but that must have belonged to Joe. The recollection made me feel as forlorn as I had then.

Sighing, I stood up. Sarah's petticoat was heavy but I daren't take it off until I'd found a suitable hiding place for the gold coins. I straightened the bedcover and left the room.

I followed the buzz of conversation to the servants' hall and slipped into a vacant place at the refectory table. The girl beside me pulled her skirt aside to make room.

I smiled. 'I'm Emilia Barton.'

'Mariette,' she said, before continuing her lively banter with another maid. It would take time before I was accepted, I supposed.

Dinner was plain but plentiful, a thick vegetable and bean soup with chunks of crusty bread, but I had little appetite. I glanced at the others, mostly maids but also a number of male servants. Signora Pergami sat at the head of the table beside her daughter Faustina and son Luigi.

There was no formality and one by one the servants began to drift away to resume their duties. Signora Pergami was picking at what were left of her teeth with a knife when I went to ask her how to find the Princess's dressing room.

'The Princess has guests for dinner,' she said. 'You must wait in the hall until she's finished.'

I returned to my room to collect my sewing box and the parcel of silk before sitting down on the hall chair. Gales of laughter emanated from the dining room and a footman went in and out with dishes from the kitchen.

Eventually the door burst open and the Baron and three other men strode across the hall and stood in a noisy group. A maid appeared from the back of the house with coats and hats in her arms and waited by the front door. Willy Austin sloped out of the dining room with his hands in his pockets. Then came the Princess, laughing uproariously at something her companions, a middle-aged couple, had said. Countess Oldi stood silently beside them, nodding and smiling.

The maid helped the couple into their coats and saw all the visitors out of the front door.

The Baron spoke to the Princess in a low voice while she looked up at him with an adoring smile. He rested his hand on her shoulder for a moment before walking away.

I was uncomfortable at witnessing what appeared to have been a private moment. The Princess turned towards the drawing room and I stood up and cleared my throat. She caught sight of me and clapped a hand to her forehead. 'My dear Signorina Barton! I forgot I asked you to attend me. How are you?'

'I'm well, thank you, Ma'am.'

'Still sad, I expect.' She smiled sympathetically. 'But we shall share some rides in the carriage together to take your mind off your wretchedness.'

'You are very kind. If it's not convenient for you to see me now, shall I return later?' I noticed she'd spilled gravy down the front of her bodice.

'Not at all.' She waved at Countess Oldi. 'Come with us, Angelica.'

I followed them upstairs to the Princess's dressing room. It was all I could do not to shudder at the sight that met my eyes. The room was crammed with chests and wardrobes. Garments hung from hooks on the walls and lay in towering piles on the floor. Shoes were tumbled into a corner. Grubby shifts spilled out of drawers and over everything hung the musty odour of unwashed clothing.

'I can never find anything to wear,' complained the Princess. 'Since that traitorous girl left me, I've had a succession of maids to dress me but none has properly managed my wardrobe.'

'I do not profess to be a lady's maid,' I said faintly, 'but perhaps I may see what needs mending, altering or cleaning? I'll arrange everything neatly, though it may be necessary to put some out-of-season items into storage. I fear we must be ruthless and discard items you no longer wear.' I picked up a crumpled dress from the floor. 'Does this fit well?'

The Princess nodded.

'Then I'll sew on this loose button and mend the tear in the hem before sending it to the laundry.'

'Faustina oversees the laundry maids,' said the Princess.

I set to work while she and Countess Oldi sat by the window discussing the Princess's scheme to turn one of the upstairs rooms into a music room.

Some time later I had a pile of clothes ready for mending, a larger one for the laundry, and several items that needed letting out.

'Where would you like me to work, Ma'am?' I asked. 'I shall need a table by a window so that I have sufficient daylight to keep my stitches neat.'

'I'll have a table brought in here for you.'

I glanced at the half-open door that led to the Princess's bedroom and was conscious that it might not always be convenient for me to have free access to the dressing room. I glimpsed a child's bed set at the foot of the large bed, draped with muslin curtains, and a large portrait of the Baron hanging on one wall. 'Are there times you would prefer me not to be working in here?' I asked.

The Princess shrugged. 'If so I shall tell you.'

'Very good.' I unwrapped the bundle of silk I'd brought with me. 'I have brought the material for your new ballgown for you to approve.'

The Princess fell upon the silk with cries of delight. I draped it around her shoulders so that she could see the effect in the looking glass. 'I shall wear this to the Perticaris' ball next month,' she announced.

'Then I will make sure it is ready.'

The Princess smiled. 'I can see I am going to be very happy with my new Mistress of the Wardrobe.'

'Thank you, Ma'am.' I curtseyed, pleased that she had awarded me such a title.

'I have plans for a bathroom with a sunken bath and frescoes on the ceiling,' said the Princess, 'and I have an appointment with my architect now. I'll send for you later, Signorina Barton. I need you to write the invitations for a dinner here on Friday.'

I curtseyed again and the Princess and Countess Oldi returned downstairs.

Gathering up an armful of dirty clothes, I sighed. The organisation of the Princess's wardrobe was going to be more onerous than I had imagined.

# Chapter 8

*March 1819*

I sat by the window with a purple crepe dress spread out before me. The seams had been let out before and there was barely enough material remaining to do so again in order to accommodate the Princess's expanding hips. I made my stitches small and reinforced them at the point of strain, hoping there wouldn't be an embarrassing accident if she bent over suddenly.

The past month had flown by as I settled into the Princess's household. Gradually, I was bringing order to her wardrobe. The fusty smell had gone from the dressing room and, since I'd packed away any items that weren't suitable for the time of year, there was sufficient space to house everything neatly.

Sewing was a quiet pursuit, providing time to reflect, and my thoughts frequently turned to Signor Fiorelli. I smiled to myself at the memory of the way his face lit up whenever we met. Despite that I was still unsettled and confused, plagued by guilt that I had abandoned Sarah to die alone and yet remaining angry with her also for her deception.

A patter of footsteps could be heard along the landing, the door opened and Victorine appeared. 'Haven't you finished?' she demanded.

'Almost,' I said, looking over her shoulder to see her tutor smiling at me.

I added the last stitches to the seam and snipped the thread. 'It's done.'

'It's a beautiful day for a walk,' he said.

Ten minutes later we set off along the avenue of cypresses and into the lane leading down to the town. Victorine walked between her tutor and myself, swinging our hands.

As we reached his home, Signor Fiorelli said, 'I promised Mamma I'd take Alfio with us.' He smiled. 'Victorine likes the company and Mamma enjoys the peace.'

'And how are you, Signorina Barton?' asked Signora Fiorelli when she opened the door with Alfio at her side. 'The Princess is keeping you occupied?'

'She certainly is,' I said. 'She often likes me to sit with her in the *salone* while I'm sewing or writing her letters and she talks to me about Princess Charlotte.'

Cosima came to greet me with a shy smile, while Alfio and Vittoria tugged at Signor Fiorelli's coat, anxious to go out.

'Come and eat with us when you can, Signorina Barton,' said Signora Fiorelli. 'Any friend of Alessandro's is always welcome at my table.' She smiled at me and then gave her son a searching look. I wasn't sure which of us blushed the deepest.

'Can we see the boats now?' asked Alfio.

Signor Fiorelli tickled his little brother's chin. 'Of course we can.'

We said goodbye to Cosima and Signora Fiorelli.

I stopped to buy thread in various colours at the draper's in the town and then, as we passed the grocer's shop, I paused.

'I'd like to call in here for a moment,' I said.

Inside the shadowy shop, the grocer leaned over the counter between a pyramid of cheeses and a bowl of eggs. A row of dried sausages hung from a rack above, their pungent aroma enveloping us.

'I am Signorina Barton,' I said. 'A few weeks ago you displayed a notice in your window regarding our dressmaking services.'

'Indeed I did.' The grocer shook his head sorrowfully. 'My condolences. I was sorry to hear your poor mother has passed on since then.'

'Thank you. Perhaps you remember she came to ask you if there had been any enquiries for us?'

He smoothed down the front of his apron. 'I told her a man had asked me about you. I remember it well because he was a foreigner. And he seemed very curious about you and your mother.'

'Do you remember what nationality he was?'

The grocer shrugged. 'German perhaps. Or English. I thought at first he might have come to ask more questions about the Princess of Wales.' He pulled at his moustache. 'There have been many enquiries of that sort but the Princess is a good customer of mine, with so many mouths to feed up at Villa Vittoria, and I'll not say a word against her.'

Victorine tugged at my skirt. 'I'm hungry,' she whispered.

'Just a moment, sweetheart.' I took her hand. 'What did he look like, Signor?

He turned up his palms and shrugged. 'Foreigners all look the same, don't they?'

I glanced at Signor Fiorelli, who was trying not to laugh. 'Please try to remember.'

The grocer narrowed his eyes while he thought. 'Pale. And very tall. Thin.'

'Then he might be the man I saw outside the cottage,' said Signor Fiorelli.

The grocer took a notepad from under the counter and flipped through the pages. 'This is him,' he said, peering at his notes. 'His name was John Smith and he was staying at the Albergo Duomo. His wife wanted a new dress.'

'My mother enquired but no such persons stayed there,' I said. 'If you see him again, would you be kind enough to let me know? I'm staying at the Villa Vittoria.'

'Ah, working for the Princess?'

'I am.' I glanced down at Victorine. 'And in the meantime we'll have a few slices of your best salami.'

We left the shop, nibbling upon the salami as we walked. When we reached the harbour the children ran ahead of us, squealing with laughter as they chased seagulls.

We passed the cottage where Sarah and I had lived and the door opened. A young woman carrying a baby came out and crossed the street.

'I'm relieved that a new tenant was found so quickly since I had to leave without notice,' I said. 'I loved the cottage and had hoped so much that we'd settle there.'

'Don't look so sad!' said Signor Fiorelli. 'I can't bear you to be unhappy.' He took my hand and lifted it to his lips. 'You must miss your mother very much.'

I didn't answer for a moment. His amber eyes were filled with compassion and I decided to tell him the truth. 'She wasn't . . . ' I closed my eyes, feeling the dull ache of her loss. 'She wasn't my mother,' I confessed.

'Not your mother?'

I shook my head, unable to speak for the sudden constriction in my throat. I'd had no one to talk to about it and I longed to tell him the whole story.

He glanced at the children but they were happy watching the boats. 'She adopted you?' he asked.

'No. She stole me.'

His mouth fell open.

'I didn't know that until the night she died.' All at once the great well of my grief overflowed.

Signor Fiorelli made a small sound of distress and took out his handkerchief. Gently, he blotted my tears. 'Don't cry, Signorina Barton!'

His tenderness made me weep all the more and he put his arm around me. I didn't care what any passer-by might think and buried my face in the comfort of his shoulder, breathing in the scent of clean skin and feeling the rough texture of his coat against my cheek. He rubbed my back and kissed my fingers and at last my tears were spent.

'What did you mean when you said Signora Barton stole you?' he said.

'When I was four years old she took me away from my home and family in London.'

He shook his head as if he didn't believe me. 'But why?'

I related the story Sarah had told me on her deathbed and when I'd finished he sat with his head bowed. At last, he said, 'Family is everything. I cannot imagine what it might be like not to have my family. Of course we fight sometimes but it would be inconceivable for me not to have my parents and siblings beside me, giving their love and support.' He looked up at me. 'They are my life.'

'Sarah and I were never very close,' I said. 'I've been lonely for as long as I can remember. Yes, we travelled together and we relied upon each other, but there was always something missing. We thought so differently about everything.' I shrugged. 'If my real mother hadn't died, then Sarah would have returned me to my father. For most of my life I thought her fears were irrational and it made me angry. I'm still angry.' I swallowed the lump that had risen in my throat. 'She said she took me away because she wanted to keep me safe but all I wanted was a normal life, with a proper family. Like yours.'

'It wasn't her fault your real mother died, though. Did you never wonder why you looked so different from Signora Barton?' he asked. 'I noticed that the first time we met. She was so small, plump and dark while you are tall and slender with hair the colour of a Botticelli angel's.'

I couldn't help smiling, despite my misery. 'Your mother told me you thought I looked like a Botticelli angel. I'm flattered. He's one of my favourite artists.'

He shook his head, his mouth curving into a wry smile. 'Mamma allows me no private thoughts.'

'If I considered it at all, I presumed I looked like Sarah's husband, Joe, the man I thought was my father.' I squinted at the horizon while I tried to picture him. 'His face was always red,' I said. 'Usually he was drunk or angry and his breath smelled rank. Sometimes, in my dreams, I hear his shouts echoing in my head. The memory of him still makes me afraid, even though he left when I was younger than Victorine is now.'

'The children!' Signor Fiorelli reared up and looked around him, one hand clasped to his breast. 'I forgot the children!'

'It's all right,' I said. 'They're over there.' I pointed to where they sat on the wall further down the harbour, playing with a collection of seagull feathers.

'Come,' he said, taking my hand again.

I looked down at his strong brown fingers enfolding my pale ones and felt a flicker of hope or happiness, I wasn't quite sure which.

He rubbed his thumb over my wrist. 'Signorina Barton, will you call me Alessandro?' he said. 'In private, at least. I cannot bear to think of you so very alone.' He glanced up at me, his expression tense. 'I want you to know I am your very good friend.'

A comforting warmth, tinged with elation, blossomed inside my chest. I gripped his hand. 'That means a great deal to me, Alessandro.'

He laughed and kissed my hand. 'Emilia,' he said. 'Such a pretty name, like its owner.'

Upstairs, the banging and hammering reached a crescendo as carpenters constructed bookshelves for the new library, while the Princess paced up and down the *salone* muttering under her breath. Countess Oldi and I bent our heads over our sewing.

At last I could stand it no longer. 'Is there something I can do for you, Ma'am?' I enquired.

The Princess's face was flushed with anger. 'My husband continues to plot against me!' She threw herself down on the sofa beside me.

I put down my needle.

'My lawyer came from Milan to see me this morning,' she said. 'Avvocato Codazzi tells me that snake in the grass Louise Demont was called to testify to the Milan Commission last month. They have employed an architect to make a plan of the Villa d'Este, where I used to live in Como. He went there with Louise and bribed the doorkeeper to let them in. The architect made sketches while Louise informed him of the former occupants of each of the bedrooms.'

'I'm not sure I understand,' I said.

The Princess sighed. 'The Commission hope to provide proof that my relationship with the Baron is adulterous because our rooms were near to one another's.'

'But that doesn't prove anything,' I said. The Baron and the Princess had adjacent rooms at Villa Vittoria but, despite their apparent affection for each other, I'd never seen any evidence of wrongdoing between them.

'The Prince of Wales goes too far!' hissed the Princess. 'He wants to be rid of me but it is impossible for him to prove any adultery, no matter how many lies they tell about me.' Her chin quivered. 'In the beginning, I tried to be a good wife, despite the continuous humiliations and his lack of consideration for my feelings.' She dabbed her eyes, smudging the charcoal on her darkened eyebrows. 'The first time I saw him I was disappointed. He was so fat and not at all like his portrait, but still I smiled and tried to be jolly.' She gave me a wan smile. 'Did you know I was forced to have my husband's mistress, Lady Jersey, as my waiting woman? It was . . . ' She swallowed. 'Intolerable.'

'How humiliating for you.'

'Princesses rarely marry for love,' she said, 'but both parties must make the best of the situation.' The expression in her eyes was inexpressibly sad. 'George never even tried.'

'Is there no way,' I said, treading carefully, 'that you can agree to part amicably?'

'If we divorce it is I who will come out badly, far worse than he will, and my allowance would be cut yet again. I'm already in debt. As a divorced woman I would never be able to return to England.'

'But, forgive me, Ma'am, do you want to return there?'

Countess Oldi put down her sewing and watched the Princess intently with her unfathomable dark gaze.

The Princess stared down at her hands. After a long pause, she sighed. 'Now that my Charlotte is dead there is little reason to return, I admit. I have never been so happy in my life as I am here with my Italian family.'

Countess Oldi dropped her scrutiny of the Princess and lifted up her sewing again, a smile curving her lips.

'Perhaps,' I said, 'you might agree to a divorce if he made you one large payment, enough to live on for the rest of your life? Once you had the funds, there need be no further concern that your allowance might be cut.'

The Princess paced the floor. 'I wish my old adviser, Henry Brougham, were here to discuss this with me.' Her face brightened. 'But my trustees are sending his brother James to see me about my accounts. I will ask him to pass a message to Henry when he returns to England.' She sat down on the arm of the sofa and patted my hand. 'I'm very happy you came to live here, Signora Barton,' she said. 'Not only are we sisters in our sorrow but you always speak such good sense.'

# Chapter 9

The Fiorelli family, including Alessandro's sister Delfina, her husband Franco and their baby Enzo, was gathered around the table. Signora Fiorelli presided over the vast dish of oxtail stew while Dottore Fiorelli poured red wine. Even little Alfio had a splash of it in his water.

'Is your family always as noisy as this?' I said to Alessandro.

He laughed. 'You should hear it when we argue! Since there are so many of us we always have to shout to make ourselves heard.'

The baby began to cry, his wails rising above the hubbub, and Signora Fiorelli lifted him to her shoulder and sang to him, nuzzling into his little neck.

Watching the baby's head nodding against his grandmother's cheek as he fell asleep, I wondered for a moment if my real mother had cared for me so tenderly. The pain of not remembering her was sharp.

I looked around at Alessandro's family and saw how they touched each other all the time: Jacopo playfully punching Fabrizio on his shoulder, Delfina tucking a loose curl into Gina's hairband, Dottore Fiorelli trailing his fingers over his wife's arm as he passed by and tickling Alfio's neck.

'Will you hold him a moment?' Signora Fiorelli handed the baby to me while she fetched cheese and fruit to the table.

The infant was warm and milk-scented, heavy in my arms as he slept.

'Beautiful, isn't he?' said Alessandro, dropping a kiss on his nephew's forehead. 'It doesn't seem five minutes since Alfio was this small.'

'Am I holding him properly?' I asked. 'I've had so little to do with babies.'

Alessandro laughed. 'Don't be nervous. Babies are tough little creatures.'

Did I imagine that Signora Fiorelli had a speculative gleam in her eyes as she looked at us? I bent my head to study Enzo's tiny fingernails and hide my blushes.

After supper, we retired to the *salone* and Cosima played the piano while the rest of us sang.

Later, I glanced at the clock on the mantelpiece and touched Alessandro's sleeve. 'I must go,' I said. 'The Baron locks the doors after dark.'

I said my goodbyes to each member of the family and left with a chorus of good wishes resounding in my ears. Flushed with wine, song and good company, Alessandro and I went out into the darkening evening.

'What a lovely family you have,' I said to him.

'I thank God for them every day.'

Hand in hand, we walked along the tree-lined lane and climbed the hill.

As we walked, I imagined what it would be like to be part of such a family. The safety and security of it would be wonderful but there would be little privacy or space to think your own thoughts.

'You're very quiet,' said Alessandro later, when we reached the avenue of cypress trees leading to Villa Vittoria.

'I was wondering about my lost family,' I said.

He squeezed my hand. 'You'll make a family of your own one day.'

'I hope so.'

He took me by my shoulders. 'You *will*,' he said. And then he kissed me.

His lips were warm and he held me as lightly as thistledown.

A tremor ran through my limbs and I closed my eyes, enjoying the strange sensation.

Then he released me. 'Goodnight, Emilia,' he murmured.

'Goodnight, Alessandro,' I said. I wanted him to kiss me again but he set off home.

Reluctantly, I set off down the avenue. Halfway along, I turned and saw him wave to me before fading into the gloaming.

The servants' dining room was full of gossip about the Princess's visitor.

'Faustina took wine and olives to him while he was talking to the Baron,' Mariette said as we ate our soup. 'James Brougham, he's called,' she said, struggling with the unfamiliar pronunciation. 'He's come to look at the Princess's accounts.'

'The Princess mentioned he was expected,' I said.

Mariette whispered, 'Everyone knows the Princess has big debts in England as well as here. She had to sell the Villa d'Este to pay some of them off and came to Pesaro because she can live more cheaply here.'

'Was the Villa d'Este very grand, then?' I asked.

'Oh, yes!' Mariette opened her brown eyes wide.

'What a pity she had to sell it! Still, Villa Vittoria is being renovated.'

'It's hardly a palace, though, is it?' said Mariette. 'But then, she doesn't live like a princess.' She mopped up the last of her soup with a piece of bread. 'Oh, well, best get on. There's to be a party tonight to welcome Mr Brougham.'

'I wrote the invitation cards,' I said. 'Cardinal Albani will be here, along with the cream of Pesaro society.'

Mariette made a face. 'More work for us mere servants, then.'

I returned to the Princess's dressing room to sew a piece of lace onto the frill of her shift.

Later that afternoon, the Baron pushed open the dressing-room door. 'Signorina Barton,' he said, 'the Princess wants you in the *salone*. You're to write a letter for her.'

Downstairs, a man with a large nose and a determined chin was sitting with the Princess.

The Baron strode over to lean his elbow on the mantelpiece and rest one booted foot on the fender. As usual, he appeared entirely at ease.

The Princess gestured me towards the writing desk. I took out a piece of paper, a pen and the inkwell.

The visitor, James Brougham, glanced at me and then at the Princess with one eyebrow raised.

'Signorina Barton is my Mistress of the Wardrobe and my secretary,' she said. 'You may speak freely before her.'

'I see,' said Mr Brougham, looking me up and down.

I decided I didn't like him.

'I wish to attempt negotiations again with the Prince of Wales,' said the Princess. 'My previous allowance of thirty-five thousand pounds a year is insufficient for my responsibilities now. It cannot be less than fifty thousand.'

I hoped my gasp hadn't been audible and hurriedly dipped the pen in the inkwell.

'The Prince of Wales may not agree,' said Mr Brougham. 'Perhaps a lump sum and a smaller income to follow?'

'I doubt I'd ever receive it,' said the Princess.

I held the pen ready. 'Will you dictate, Ma'am?'

'I never find the right words,' said the Princess, 'especially in English. No, you write it, Signorina Barton, and pass it to Mr Brougham to look at. Address it to Mr Henry Brougham.'

I frowned and the Princess smiled at my confusion. 'Mr Henry Brougham, my adviser, is Mr James Brougham's elder brother.'

'I see,' I said.

'And don't forget to tell the Prince of Wales that if he agrees to my terms, I will promise never to return to England.'

'Yes, Ma'am.' I bent my head over the paper. My previous duties as the Princess's secretary had consisted of writing invitations to dinners or the occasional letter to an acquaintance but this was far more challenging. I took care to write neatly and to set out the proposal as discussed.

As my pen scratched across the paper the Princess and James Brougham continued their conversation while the Baron silently watched them.

'Perhaps it is a good time to finalise a divorce now,' said Mr Brougham, 'before the Milan Commission report their findings to the Prince of Wales. It would be better for all parties if there was a dissolution of the marriage by Parliamentary Bill rather than in open court.'

The Princess shuddered. 'I agree.'

Mr Brougham turned to the Baron. 'I must see the account books to make my report to the trustees. Will you be on hand if required?'

'Most certainly,' said the Baron. 'I shall bring them to you in the morning room.'

'Really, I don't know why my trustees are so anxious about a few trifling debts,' said the Princess. 'If I were to die tomorrow they could all be paid off at once. There are the horses and my jewels, for a start.'

'Unfortunately,' said Mr Brougham, 'your creditors require to be paid now, not at some distant date in the future.'

'I doubt I'd have debts,' grumbled the Princess, 'if my gentlemen and servants before I came to Pesaro hadn't been so incompetent or so determined to cheat me. Thankfully, the Baron is managing my affairs now.' She smiled at him, busy cleaning his nails with a penknife as he leaned against the mantelpiece.

'And you are content to live in Pesaro?' asked Mr Brougham doubtfully. 'It's very provincial and this house does not befit someone of your position.'

'I have rarely felt as settled as I do here amongst the Italians,' said the Princess, 'except for my fear that the Prince of Wales will send his spies to poison or kill me.'

I put down my pen and blotted the ink dry.

'Finished?' asked the Princess.

'Yes, Ma'am.' I handed the letter to Mr Brougham and returned upstairs.

The dressing-room window was left open while I sewed, letting in a gentle spring breeze and the sound of the builders, hammering and sawing. My thoughts began to drift, reliving the warmth and gentleness of Alessandro's kiss. He made my heart beat faster and I hoped, how very much I hoped, that he felt the same about me.

Through the half-open door to the Princess's bedroom, I heard a floorboard creak. The maid had cleaned the room earlier so perhaps the Princess was preparing to take a nap. I tip-toed towards the door, intending to close it, and was surprised to glimpse Mr Brougham's reflection in the mirror, examining the items on the dressing table.

'May I help you?' I asked.

He jumped and dropped a silver-backed hairbrush onto the dressing table with a clatter. 'You startled me!' he said.

I waited, aware that my position in the household did not empower me to accuse him of trespass.

Under my questioning gaze, Mr Brougham turned slightly pink. He cleared his throat. 'You may wonder why I'm here . . . '

I waited.

He coughed again and examined his fingernails. His expression cleared. 'The Princess's trustees have charged me with the responsibility of determining how she lives.'

'How she lives?' I echoed.

‘Indeed,’ he said. ‘My purpose is to discuss ways and means for the Princess to make economies and structure a method of paying her creditors.’

‘And that relates to your presence here in the Princess’s private quarters?’

‘I am required to satisfy the trustees that she is not overly extravagant.’ Mr Brougham glanced around the bedroom until his eyes rested on Victorine’s bed and he frowned. ‘Does the child sleep here, with the Princess?’

‘I believe so.’

‘Every night?’

‘My quarters are downstairs so I couldn’t say.’

‘And the Baron’s bedchamber is adjacent to this room?’

‘The Princess must be protected,’ I said. ‘There have been threats to her life.’

‘I’ve finished here,’ said Mr Brougham. He sounded disappointed.

‘Do let me show you out, then.’

He followed me through the dressing-room door and paused to study the neatly arranged shelves and piles of hatboxes. He ran a finger over a pile of folded nightshifts and lifted the lid of the ottoman to peep inside. ‘Does the Baron keep his clothes in here, too?’

‘Certainly not,’ I said, in as reproachful a tone as I could.

Mr Brougham sighed. ‘Would you say the Princess’s wardrobe is extravagant, Signorina Barton?’

‘Not at all,’ I said, ‘especially for a woman of her rank. Many of the Princess’s clothes have seen better days and I’m in the process of mending and refurbishing them.’

He nodded approvingly. ‘It’s essential to curb any excessive spending.’

I opened the door to the landing. ‘Good day, Mr Brougham.’ I watched him until he had descended the stairs before returning to my sewing.

The encounter left me unsettled. My hackles had risen when I saw him creeping about the Princess's quarters and handling her private possessions. He may have been the Princess's guest but there was something very unsavoury about his line of questioning.

Closing the dressing-room door behind me, I went downstairs and tapped on the door of the Baron's study.

He bade me enter and I found him standing before the window with his back to me, apparently lost in thought. I waited in silence until he turned to face me.

'Was there something, Signorina Barton?'

'Yes, Baron.' I found it awkward to put into words what had to be said. 'When I came to Villa Vittoria you asked me to keep you appraised of any irregularities.'

His eyes narrowed. 'An intruder?'

'No,' I said. 'A guest.'

'Brougham?'

'I found him in the Princess's bedroom looking at her personal items,' I said, finding it impossible to meet his eyes. 'He asked me if you shared her dressing room and about the sleeping arrangements on that floor.'

'And what did you tell him?'

'As little as possible except that Victorine's bed is in the Princess's room and, as far as I was aware, that is where she sleeps every night.'

The Baron nodded his head. 'Which is all perfectly correct. You have done well, Signorina Barton.'

He smiled at me with unusual warmth in his eyes, which made me uncomfortable.

'Then I shall return to my duties,' I said.

He opened the door for me with a flourish. 'I shall tell the Princess you are a good and faithful servant, Signorina Barton. It's good to know you have her welfare at heart.'

'I trust there was never any doubt of that, Baron.' I held my head high and swept from the room.

# Chapter 10

*June 1819*

The sails billowed and cracked above our heads as the Princess's yacht turned into the wind. The breeze teased my hair from the front of my bonnet and I lifted my face to the sun, breathing in the briny air. I had shed my mourning clothes a few days before and was revelling in the cool comfort of lightweight muslin after months of unrelieved black.

'You look happy,' said Alessandro, his eyes bright with enjoyment.

The sunshine touched his hair with bronze lights and I longed to run my fingers through it, as I had the night before when we'd lingered in the avenue of cypresses to kiss each other goodnight. We'd been entranced by the lights of a myriad fireflies illuminating the darkness: a romantic display seemingly provided just for us.

'Emilia?'

I came back to the present, a half-smile playing on my lips. 'It's so exhilarating!' I said. 'We're travelling much faster than we could in any coach.'

A short distance from us the Princess, dressed as an admiral in a naval-style coat and tricorn hat, sat with Willy and the Baron.

She stared out to sea, her tense expression at odds with the jaunty feather in her hat. Countess Oldi, a poor sailor, had retired to the cabin. Victorine peered over the side of the yacht, squealing whenever sea spray splashed her face. Her father kept a tight grip on her skirt.

'It was kind of the Princess to invite us today,' I said to Alessandro. I didn't mention she'd said she liked to see young people enjoying each other's company and that I ought to give handsome Signor Fiorelli more encouragement. Perhaps she'd noticed more than I thought but, unlike most employers, she hadn't forbidden me to have a suitor.

'She looks anxious,' said Alessandro.

I nodded. 'Lord Brougham has written to tell her divorce proceedings against her may start in November.'

'Even if there is a divorce, she can't marry the Baron,' said Alessandro. 'Not while his wife is alive, anyway.'

'They act as if they're a married couple, don't they?' I murmured, watching the Princess lean forwards and touch the Baron's arm. 'However unlikely their relationship, there's a genuine affection between them.'

Alessandro shrugged. 'The Baron knows who butters his bread.'

Victorine prattled to the Princess, making her laugh. Willy was watching the little girl, too. His mouth was twisted in its usual sneer but I remembered Marietta telling me that he'd slept in the Princess's bedroom until the Baron came into her life four years before. Although I didn't much care for Willy, I felt sorry for him if Victorine had displaced him in his adoptive mother's affections.

'What do you mean by "he knows who butters his bread"?' I asked.

'Bartolomeo Pergami wasn't a baron until the Princess bought the title for him,' said Alessandro.

'No!'

'The Princess is surrounded by the Pergami family. They're all living off her wealth. Besides . . . '

'Besides, what, Alessandro?'

'It's gossip,' he said, 'told to me by one of the footmen who was with the Princess before I came to work at Villa Vittoria. I don't know if there's any truth in it but stealing wasn't the only reason Louise Demont was turned off.'

'What then?'

'The Princess discovered the Baron had been, shall we say, rather too *close* to Signorina Demont.'

I caught my breath. 'But if that were the case,' I said, 'wouldn't the Princess have turned off the Baron, too?'

'She's in thrall to him,' said Alessandro. 'He runs every aspect of her household. All her English attendants have gone, the Baron made sure of that, and she can't do without him. I suspect the Princess simply pretends it never happened so all can go on as before. And I wouldn't be surprised if he is making a tidy profit out of managing the household.'

'I don't like to think of her being cheated,' I said. 'She's shown me nothing but kindness.'

'You remind her of her daughter.'

Covertly, I watched the Princess playing a clapping game with Victorine. It wasn't so strange if she looked for a substitute child. 'This morning she said she may go to London to settle the matter of a divorce,' I said.

'London?' Alessandro frowned. 'Surely she wouldn't take the Baron there?'

'That would annoy the Prince of Wales, wouldn't it?' I said, turning my face into the wind.

The jagged coastline slid past as the yacht sliced through the water and I wondered if the Princess really would visit London. I'd never thought much about England but now I was intensely curious. My real father lived there and I'd begun to lie awake at night wondering about him.

Half an hour later the Captain anchored the boat off a small bay.

The crew lashed a sail into a sunshade and placed a table underneath. They spread it with a linen cloth and served a picnic of poached fish and cold meats, crusty bread and bowls of olives with rosemary and lemon. Bottles of white wine cooled in buckets in the shade.

'Sailing always makes me ravenous,' said the Princess, picking up a chicken drumstick and eating it from her fingers. She pulled off a small piece of meat and fed it to Victorine.

Willy piled his plate high and ate without speaking, occasionally throwing a piece of bread to the seagulls who cried overhead.

The Princess, her face flushed by the sun, smiled at the Baron. 'Isn't this perfect, my love?'

I was pleased that the excursion had lifted her spirits. My wine glass was refilled as soon as I finished drinking it and, after we'd eaten the picnic, I grew drowsy in the warmth of the sun. The Princess retired inside the yacht and the Baron sat in the prow with a bottle of wine. Willy stared morosely at the coastline.

Victorine climbed onto my knee and pushed her thumb in her mouth. A few minutes later her eyelashes fluttered and she became warm and heavy in my arms. I yawned and glanced up to see Alessandro was smothering a yawn, too.

He blew me a kiss and smiled lazily at me as I rested my chin on the sleeping child's head. Happiness flowed through my veins. I tipped back my face to watch powder puffs of cloud drift across the cerulean sky. The yacht rocked gently on the swell of the sea while waves slapped rhythmically against the side of the boat.

Unable to resist any longer, I slept.

I put down the pen and waited for the Princess to decide what she wanted me to write in her letter to Lady Charlotte Lindsay. The air in the *salone* was stifling and I looked longingly out of the window. The long, languorous days of summer in Pesaro were so hot that sometimes it was difficult to find the energy to do more than seek out a new patch of shade.

The Princess lay on the sofa, wriggling her bare toes and fanning herself. 'You can write that the report of the Milan Commission is evidence that the Prince Regent is maliciously at work against me,' she said. She poked a finger under her wig and scratched her scalp. 'I desperately want to be rid of him but perhaps separation by agreement has the advantage of avoiding taking the matter to Parliament.' She nodded decisively. 'Tell Lady Charlotte to chivvy Henry Brougham to hurry up and respond to my proposal.'

I dipped the pen in the inkwell and began to write.

The Princess rubbed the heels of her hands in her eyes. 'I must be quit of this blackguard of a husband before I go mad with worry.' She stood up and slipped on her shoes. 'I'm going to sit in the garden. I shan't need you this afternoon if you want to take Victorine swimming. It's too hot for her to walk so take one of the carriages.' She smiled. 'And ask Signor Fiorelli to accompany you.'

'Thank you, Ma'am, I'd like that.' Even the thought of a dip in the sea made me feel cooler.

I found Alessandro sitting in the shade of an oak tree with Victorine. He leaned against the trunk with his sleeves rolled up, listening to her read.

'The Princess has given me the afternoon off to go swimming,' I said.

Alessandro's face broke into a wide smile and the little girl clapped her hands. 'Can we go now?'

'Finish that page first,' he said.

She scowled but picked up her book again, her finger moving beneath each word as she read.

'Very good,' said Alessandro, when she finished. 'There's nothing more fulfilling than seeing a child's mind blossoming,' he said as we walked towards the stables. 'Every child should have the opportunity of an education. I dream of owning my own school one day. One where I can educate poor boys . . . '

'And girls?' I asked.

He smiled. 'And girls.'

'That's a worthy ambition, Alessandro.'

'I've spoken about it to the Princess, since she's so fond of children. Willy, of course, she took into her household but she also adopted several poor orphans in England, fostered them in suitable families and educated them.'

I heard the zeal in his voice. 'You should ask her to support you in such a venture,' I said.

Fifteen minutes later we were in the shell-shaped phaeton jogging along a lane perfumed with the honeyed scent of wild broom. We collected Alfio from the Fiorelli house and soon after we arrived at the wide stretch of golden sand that was the Baia Flaminia. The sea was as calm as a millpond and melded at the horizon with the great dome of blue sky.

Others had come down from their villas and horse-drawn bathing machines were lined up in the surf. I hired one and Victorine and I went inside to change into our swimming shifts.

'Hold on to the bench, Victorine,' I said, tucking her hair into a muslin cap. I rapped on the wall and a moment later the bathing machine began to roll into the sea. Once it came to a standstill I helped her down the steps into the water. I gasped as a wave broke over my thighs.

Victorine squealed. 'It's cold!'

Silver fish swam around our feet as we wriggled our toes on the sandy seabed, avoiding fronds of drifting seaweed. My cotton swimming shift billowed out around me and there was a delicious freedom in the feeling of the water against my naked skin beneath. The sun was hot on my shoulders but the rest of me was blessedly cool.

I heard a shout and saw Alessandro wading towards us with Alfio on his shoulders.

'How are my little mermaids?' he called.

Victorine splashed him and screamed when he splashed her back.

'Isn't it glorious to be cool?' I said, bobbing up and down in the undulating waves.

We stood in a circle holding hands and I sang 'Ring a Ring o' Roses' in English, and told the children the tale of the Plague in London a hundred and fifty years before. 'And then there was a Great Fire,' I said, 'and all the rats and the fleas were burned and the Plague went away.'

We sang the song again and the children shrieked with laughter when we fell down, dipping our heads under the water. My mob cap floated away and my hair tumbled around my shoulders.

'My beautiful Botticelli Venus,' said Alessandro. He picked up one of my curls and lifted it to his lips, never taking his eyes off me. His wet shirt was moulded to his muscular chest and there was a hint of dark hair on the sun-kissed skin exposed by the open-necked collar.

A quiver ran down my spine as I imagined him without his shirt.

'Cold?' he whispered.

I shook my head and he laughed. 'Come on, I'm going to teach you to swim.'

An hour later the children were shivering and their lips turning blue. 'Time to come out of the water,' I said, ignoring their cries of protest.

Victorine and I returned to the bathing machine to change. Afterwards, Alessandro and Alfio were waiting for us, their shirts and breeches already drying in the sun. I rubbed Victorine's hair with a towel and she and Alfio set off along the tideline to collect the mother-of-pearl shells that shimmered in the sand.

Alessandro lay with his arms behind his head and regarded me through heavy-lidded eyes.

I let sand trickle through my fingers whilst my head was bowed so he wouldn't see my blushes. I wanted so much to lie down beside him but propriety forbade it.

'I wonder if your mother was as lovely as you?' he said.

'I wish I could remember her.' I gazed out to sea, listening to the gentle hiss of the waves as they foamed on the shore.

Alessandro touched my mouth with the tip of his finger to still its trembling. 'Your home is here in Italy now.' He turned on his side to face me. 'Emilia, you do know that I love you, don't you?' A muscle flickered in his jaw.

I lifted his hand to my cheek. 'I hoped so,' I said, kissing his palm, 'because I love you, too.'

The tension faded from his face. 'As soon as I saw you,' he said, 'I knew we were destined for each other.'

Joy blossomed in my heart. 'I felt the same. I could hardly look at you in case you guessed.'

Alessandro laughed, his eyes shining. 'Was that why you blushed?'

I leaned towards him to drop a kiss on his salty lips. 'Now you know my innermost feelings,' I said.

'Don't let us keep secrets from each other, Emilia.' His fingers gripped mine. 'Promise me?'

'I promise,' I said, looking into his eyes.

# *Chapter 11*

*October 1819*

**Lyons, France**

I slid the claret silk gown over the Princess's head, careful not to open the loosely tacked seams. We'd purchased the silk from one of the Lyonnais weavers shortly after we arrived here, since the Princess was determined to make an elegant appearance on her arrival in England.

The past weeks had been exhausting. It was ironic that I'd arrived in Pesaro with the firm idea of escaping my nomadic life but no sooner had I found a place to settle than the Princess's restlessness moved us on. Almost overnight her household had been packed into three post-chaises and we'd set off, incognito, for Parma.

The Baron hired Villa San Bono for us in the Piacentini Hills. The Princess received visitors late at night and her emissaries, including her Milanese lawyer, came and went. In the middle of September, a courier arrived with a letter from Henry Brougham warning her not to go to London but to Lyons, where he would meet her. That same night the household left Villa San Bono.

We spent a month in a dilapidated castle in the province of Alessandria, in the King of Sardinia's lands. The Baron and Willy Austin went backwards and forwards to Milan and Como, seeking witnesses to testify for the Princess against the Milan Commission. A month later we departed in a flurry for Lyons. Then we waited for Lord Brougham.

'Be careful of the pins, Ma'am,' I warned as I helped the Princess to remove the emerald gown.

'I cannot bear not knowing what is happening in London,' she said, fretfully. 'I need to know the mood of the people so I'm prepared for the kind of reception I'll receive.' She sighed. 'Write to Lady Hamilton and ask her to travel back to London with me after Henry Brougham's visit. Always assuming, of course, that he does eventually grace me with his presence.'

'There's still no confirmation of his arrival, Ma'am?'

She shook her head. 'It angers me that I came to Lyons at his request instead of going straight to London. Soon the winter will come. I don't care to travel long distances in the cold.'

I sat down at the Princess's writing desk. Once I'd finished her letter I returned to the parlour, where Alessandro was reading with Victorine. Peggy was propped up beside them.

'I see you have two attentive pupils today, Signor Fiorelli,' I said.

'They're working very hard.' Alessandro's smile was warm. 'Perhaps a walk in the garden as a reward?'

Outside, we ambled along the gravel paths while Victorine skipped ahead, clutching Peggy under her arm. Russet leaves twirled down from an oak tree and I smiled as the little girl jumped up to catch them.

'The poor Princess,' I said, shivering in the autumnal breeze. 'Her spies have informed her Louise Demont is in London, ready to provide evidence against her. The sooner Lord Brougham arrives, the better. She'll not regain her equilibrium until she's faced her antagonists in London.'

'Once she's sailed for England, we can return to Pesaro,' said Alessandro. 'I want so much to go home to my family.' A gust of wind loosed one of my curls and he tucked it behind my ear. 'You look strained,' he said. 'What is it?'

I hesitated then said, 'But where is *my* home? Isn't home wherever your family is?'

Alessandro studied my face, his expression suddenly watchful. 'Has all this talk of London unsettled you?'

I shrugged. 'I keep wondering about my father . . . '

'You can't change the past,' he said, kissing the tip of my nose. 'You must look forward to the future.'

I sat up late stitching the Princess's gown and then lay awake while I went over and over in my head what Sarah had told me about my parents. I tried desperately to recall my father but could only hear Joe's cruel voice in my mind. For nearly as long as I could remember there had been just Sarah and me. Alessandro's friendship and love had eased my loneliness but still I wept from a painful yearning to belong, to know that I was connected by blood with another.

In the morning I took out the quilted petticoat and counted the gold coins sewn into the hem. I sat on the edge of the bed for an age deciding what to do.

After breakfast I asked the Baron if it was convenient for the Princess to have a further fitting.

Half an hour later she admitted me to her bedroom and I helped her into the gown. 'It suits you,' I said.

'I shall look very fine once the pearls are sewn onto the skirt,' she said, studying her reflection in the mirror.

I busied myself adjusting a shoulder seam while I plucked up courage. 'Ma'am?' I said. 'There's something I should like to ask you, if I may.'

She waved a hand. 'Ask!'

'Since Henry Brougham has still not arrived,' I said, 'I wondered if it might be helpful if I travelled to England? You wished to know what kind of reception you might receive in London. I could gauge that and write to you. Also, I could visit Lady Hamilton and request she comes to attend you.'

The Princess raised her eyebrows. 'Why would you do this?'

'I have a particular reason for wishing to visit London.'

'Explain yourself!'

My mouth was dry. 'A little while ago I discovered Signora Barton was not my mother.'

The Princess pressed a hand to her mouth, her eyes gleaming. 'How intriguing!'

'When she was dying, she confessed that, after my real mother died when I was four, she stole me from my family,' I said. 'It came as a great shock.'

'I can see that it must have.'

'She was fond of me and worried about my father bringing me up without my mother to guide him and so she took me away. Now that I know I have a father in London, I want to find him.'

'You might not like him,' said the Princess. 'Families can be very strange.' She sighed. 'This I know for a fact.'

I waited, my fists clenched so hard my nails bit into my palms.

'You must finish my gown quickly, Signorina Barton,' the Princess said. 'You will want to travel before the winter storms arrive.'

I was anxious to speak privately with Alessandro but there were ears behind every door in the cramped rented villa where we stayed with the Princess. Later that afternoon, I asked him to accompany me to the remains of the Roman amphitheatre. Hand in hand, we strolled through the cobbled streets of Lyons, my apprehension growing with each step I took.

At last we stood at the brow of the hill looking down to where archaeologists were excavating the outer perimeter of the arena. I could delay no longer and, stomach churning, told Alessandro that I had decided to go to London.

'No!' he said, the colour draining from his cheeks.

'I leave in a few days and the Princess has written to Lady Hamilton telling her to expect me,' I said, trying to keep my voice calm.

'You could be gone for months and you didn't think to discuss this with me?'

The last thing I wanted was to hurt him. 'Alessandro, it was a decision I had to make by myself.' I clenched my fists. 'I cannot be at peace until I find my father.'

'But you don't even remember him.' Alessandro's chin quivered. 'He's nothing to you.'

I looked away from him, unable to face his misery. 'It's true that, even if I find him, it's possible I may feel no sense of connection. But don't you see, I must be sure?'

He gripped my shoulders. 'Please don't go.' His jaw clenched. 'I forbid it!'

One of the archaeologists glanced up from his digging and stared curiously at us.

'I have to know where I come from and you have no right to forbid me to find out,' I said.

He stepped back as if I'd slapped him. 'But I want that right,' he said, quietly. 'I love you and I thought you loved me?'

I couldn't bear seeing pain in his eyes and knowing I was the cause of it. 'You know I do!' I reached out to him. 'But surely you understand how important this is to me? I've spent most of my life wandering with no place to call home. Since Ma died it's as if the ground has been swept from under my feet.'

He grasped my hand. 'If you loved me, you wouldn't leave me.'

I stared at him, wounded he was thinking only of himself and didn't understand my feelings. 'If *you* loved *me*, you'd let me go with your blessing,' I said. I swallowed a wave of nausea and spoke in a more conciliatory fashion. 'I won't be away more than a few months. Don't you want me to be happy, Alessandro?'

He ran his hands through his curls, leaving them standing up like a halo. 'I shall make you happy. I will! I love you, Emilia.' Imprisoning me so tightly in his arms that I could barely breathe, he pressed desperate kisses on my face. 'I can't bear to lose you. Stay here and marry me!'

I struggled free from his clinging embrace. This wasn't how a proposal of marriage should be. Where were the sweet words, the consideration for my needs and the gentle caresses? 'You know I love you,' I said, 'but I could never marry a man who would deprive me of the chance of finding my family.'

Alessandro shook his head. 'Too much time has passed to build the years of memories that bind you into a proper family.' He cupped my chin in his hands and made me look at him. 'My family love you already. Mamma even unwrapped her wedding veil the other day to show me. We'll have children and make a family of our own. Emilia, tell me you'll marry me?'

I shivered with the icy realisation that we had reached an impasse. I'd hoped so much that Alessandro would propose, but not like this.

'Emilia?' he said.

If he truly loved me, he'd wait for me. 'Alessandro . . . ' I said, my voice breaking. 'I *have* to go to England but I'll come back.'

His face crumpled. 'You won't,' he said, his voice leaden with despair. 'You'll find your father and want to stay with him.'

'I will come back!' I said. 'Meanwhile, we can write to each other.' I took a deep breath and gripped his hand. 'Look at me! I love you, Alessandro, and I do want to marry you, but I must come to you with a willing heart. If we marry now I may never have the chance again to look for my father,' I said, 'especially if we have children. I'd always have regrets and that would place an intolerable strain on our love.'

'How long must I wait? I need you here, with me.' He pulled his hand away.

I caught hold of his sleeve. 'Write to me!' I pleaded.

He didn't look at me but after a long moment he gave a brief nod and slipped out of my grasp.

Shaking, I stared after him as he hurried down the hill, unable to believe that Alessandro didn't love me enough to let me leave with his blessing.

# *Chapter 12*

*November 1819*

**London**

It was growing dark when the Lidcomes' carriage drew up outside Lady Hamilton's family home in Portman Street. I glanced through the window at a street of narrow townhouses built of grimy yellow brick and couldn't help thinking how dreary the scene looked by comparison with Italy's golden stone and colour-washed façades.

Mrs Lidcome patted my wrist. 'Are you sure you wouldn't like me to come in with you, Miss Barton?'

I resisted the impulse to accept. I'd set this train of events in motion and must see it through. 'I've trespassed enough on your goodwill,' I said, 'but perhaps you'd wait until I've gone inside?' I'd been worrying ever since we left Dover about what I would do if Lady Hamilton hadn't received the Princess's letter or if she wasn't prepared to allow a complete stranger to stay with her.

'My dear, I wouldn't dream of driving away until I knew you were safe!' said Mrs Lidcome.

The carriage door opened and I sighed to see the rain hissing down and bouncing up off the pavement. It had rained without ceasing ever since I'd gone on deck as the packet approached the white cliffs of Dover. A penetrating wind and sullen sky had done nothing to lift my sadly sunken spirits.

The coachman held an umbrella over my head as I scurried from the carriage. I shivered, contemplating my sodden shoes and damp hem until he returned with my travelling bags, then knocked on the door. I glanced back at Mrs Lidcome, her face ghost-like at the streaming carriage window. She nodded encouragingly.

I'd been lucky to find such a pleasant travelling companion for the sea crossing. Willy Austin had escorted me on the long journey from Lyons to Calais and left me at an inn while he searched for a suitable chaperone. Mr and Mrs Lidcome had been prepared to undertake the task and I'd kept their children amused while their mother suffered the effects of a rough crossing.

The door opened. I swallowed my trepidation and went inside.

Lady Hamilton, dressed in deepest mourning, received me in her elegant drawing room. Unsmiling and dark-haired, she looked me up and down with sharply inquisitive eyes. I dropped a curtsey and she languidly waved her long fingers at a chair.

'I do hope my arrival is not too much of an inconvenience, Lady Hamilton?' The chair was plumply upholstered and I arranged my damp skirt so as not to sully the pristine cream satin.

'The Princess still requires me to join her?' Still she didn't smile or make me feel any less uncomfortable.

'She would be pleased to have your company,' I said. I glanced at the fire crackling in the marble chimneypiece and longed to sit closer to it.

Lady Hamilton sighed as she smoothed her impeccably cut skirt. 'I suppose I must uproot myself to carry out her bidding.' She pursed her lips. 'How does she like her Italian lady-in-waiting?' She leaned her angular body forward and studied me intently.

'The Countess Oldi? I believe the Princess finds her agreeable.'

Lady Hamilton sniffed. 'Pergami's sister, with a failed marriage behind her. And is the Princess still racketing about seeking diversions here, there and everywhere?'

'She's too apprehensive of what the Prince Regent will do next to take a great deal of enjoyment from anything at present,' I said.

'Perhaps that is as well. This would not be a good time to make herself look undignified or foolish. The Prince Regent wouldn't hesitate to use it against her.'

'The Princess would welcome your support. She doesn't care to return to England attended by her Italian household.'

Lady Hamilton fixed me with a gimlet stare. 'And what is your role in her employ, Miss Barton?'

'I suspect you will find it very irregular, Lady Hamilton.' I was absolutely sure that she would. 'The Princess's home in Pesaro is not as imposing as the Villa Caprile and certainly not as magnificent as the Villa d'Este. In fact . . . '

'Yes, yes,' said Lady Hamilton. 'I understand Villa Vittoria is little more than a farmhouse.'

'The Princess calls it a cottage.' I hesitated. 'I'm sure you're aware how informal she can be . . . '

'That may be an extremely polite way of describing her actions, Miss Barton,' said Lady Hamilton with a dry smile. 'Apparently you came to her rescue after she was involved in a carriage accident?'

I nodded. 'Subsequently, she offered me a position as her Mistress of the Wardrobe. And I write her letters for her.'

'How did you come to be in Italy, Miss Barton? What of your family? Surely you were not travelling alone?'

'I was not,' I said. 'But I have come to England to find my father.'

Lady Hamilton's nostrils flared as if she had scented something unpleasant. 'You do not *know* your father, then?'

I shook my head, realising that she probably imagined I was illegitimate. 'My father is Sir Frederick Langdon. He used to live in Grosvenor Street. Perhaps you might have heard of him?'

'Indeed I have.'

Despite my curiosity and excitement, Lady Hamilton's expression discouraged me from asking her about him. 'A little while ago I discovered that I was stolen away from my family by my mother's maid when I was four years old.'

Lady Hamilton drew in her breath. 'I remember the scandal. Sir Frederick's wife drowned herself . . . '

'My mother,' I said.

'Do you have proof that you are who you say you are?'

I shook my head. 'I want nothing from my father, except to meet him.'

'Are you telling the truth? Are you really the lost Langdon girl?' Her tone was sharp.

'Yes.' I looked her firmly in the eye.

'Then at least I may console myself that the Princess has been waited on by the daughter of a member of the landed gentry and not simply another of her waifs and strays.'

'Landed gentry?'

Lady Hamilton gave me a wintry smile. 'Your father owns the Langdon Hall estate in Hampshire.'

'So he isn't still living in Grosvenor Street?' I rubbed my eyes. 'I'd hoped to meet him and then travel back to Italy with you,' I said. 'But if I have to go to Hampshire . . . '

'His London house is still in Grosvenor Street, as far as I am aware,' said Lady Hamilton. 'I daresay he stays there while Parliament is sitting.'

'If he'll receive me, I shall call on him tomorrow.' I sighed. 'All I want is to meet him but I expect, like you, he'll think I'm a fortune hunter.' Suddenly I was so tired I could cry.

Surprisingly, Lady Hamilton's expression softened. 'We shall see what tomorrow brings.' She rang the bell on the side table and a moment later the maid arrived.

'Take Miss Barton to her room and send up some supper on a tray,' said Lady Hamilton.

Relieved that I was to avoid a formal dinner, I dropped a curtsey and followed the maid upstairs.

The guest room was comfortably furnished but chilly, even though a meagre fire burned in the grate. I kicked off my wet shoes and sat by the fire to warm my toes. I must have dozed because the next thing I heard was the maid placing a supper tray on the table beside me.

After I'd eaten the soup and apple pie I began to unpack. Drooping with fatigue, I released the pins from my hair and searched the bag for my hairbrush. It was then I realised with a jolt that Peggy wasn't there. I snatched up the second bag, delved under the folded clothes and then tipped everything out onto the bedroom chair. Nothing. In my haste to pack I must have left the doll behind. I remembered then that I'd last seen her in Victorine's arms.

I missed Alessandro so much and there was a horrible aching void in my chest. My life would be unimaginably bleak if my father wouldn't see me and if Alessandro didn't still want to marry me when I returned to Pesaro. Shivering, I undressed and climbed into the cold bed, close to tears. In Peggy's absence, I hugged one of the pillows, attempting to draw comfort from its downy softness.

The following morning, I awoke when the maid lit the fire. It took me a moment to realise where I was. Grey light filtered in through the curtains and I sat up in bed. The bedroom chair was strewn with possessions and my stomach lurched as I remembered the desperate search for Peggy.

'Did I wake you, miss?' asked the young maid. She wiped her hands on her apron and pulled back the curtains. 'I'll bring you hot chocolate directly.'

'What time is breakfast?' I asked.

'Ten o'clock, miss. Lady Hamilton is always prompt. You'll hear the hall clock chime the hour.'

I'd counted the quarter-hours as the hall clock chimed all through the night while I fretted about Alessandro and imagined how my father might receive me.

After I'd drunk the chocolate I contemplated my tumbled clothes and shook out the blue silk dress I intended to wear while making my visit to Grosvenor Street. I wished I'd laid it flat the night before but natural optimism encouraged me to hope that the creases would fall out from the warmth of my body. The dress had been one of the Princess's cast-offs and, although too short for me, I'd taken it in and added a band of contrasting blue damask to the hem, using the same material to trim the sleeves and bodice.

I took extra trouble with my hair, coiling it up with a ribbon but allowing a few curls to frame my face. As I regarded my reflection in the mirror I remembered how Sarah had stood behind me when I'd looked in another mirror many years ago. She'd shown me how to style my hair to flatter my pointed chin and high cheekbones and taught me the tricks of her trade as a lady's maid. Even though we'd had our differences, she'd always tried to do her best for me. We'd done everything together and I missed her.

I stood by the window looking dejectedly at the rain as it pelted down on the pavement. The sky remained gloweringly grey. The day I'd left Italy there had been blue skies and golden autumn sunshine. I almost wished I'd never come to London.

A carriage rolled along the street, sending up a spray of water and drenching a pedestrian. Turning away from the depressing view I paced up and down, planning, without a great deal of success, what I would say to Sir Frederick. I was uncomfortably aware that he might very well have me thrown out as an imposter.

At last, the hall clock chimed and I slipped on my shoes, still damp, before going downstairs.

Lady Hamilton acknowledged me with a nod as we went into the dining room.

'Good morning,' I said. I hadn't seen her standing up the day before and was surprised to see she was at least six feet tall and certainly six inches taller than myself.

'I trust you were comfortable?'

'Very, thank you.' The pleasantries over, we sat down.

Anxiety at the prospect of meeting my father made my stomach churn but I drank coffee and nibbled a piece of pound cake.

'Your father never remarried,' said Lady Hamilton. 'The death of your mother affected him greatly, I believe. You may not be aware,' continued my hostess, 'that the Prince Regent is a staunch Tory and that Sir Frederick supports him. The Whigs, on the other hand, support the Princess.'

'I know little of English politics,' I said.

'Perhaps you do not take my meaning?' said Lady Hamilton. 'Sir Frederick may not take kindly to discovering that his long-lost daughter has been residing in the enemy camp. Feelings run very high against the Princess in those quarters.'

I placed my coffee cup carefully on the saucer. 'I see. Nevertheless,' I said, 'since I've travelled so far, I shall still visit him, even if he turns me away. I'm very grateful to the Princess for offering me a position in her household when I needed it and my association with her is nothing to be ashamed of.'

Lady Hamilton smiled for the first time. She pushed back her chair. 'If you are ready, I shall send for the carriage.'

I peered out of the carriage window as it progressed along Grosvenor Street. Everything in London was grey and damp. The street was wide and lined with substantial townhouses with slate roofs, not the warm-coloured terracotta tiles I was used to. The brickwork was dusted with soot deposited from the coal-fire smog that loured like a funeral pall over the town. It was all very different from Pesaro. Still, there was a general air of prosperity and elegance in the façades of these buildings, even though I had no memory of ever living in this illustrious area. Lady Hamilton had gone to the trouble of asking her brother, Lord Archibald Hamilton, to find out the exact location of my father's residence and very soon we drew up outside.

I was too nervous to notice more than that the house had a wide frontage with a pillared portico and a freshly painted front door and railings. My knees trembled as the coachman handed me down but I made a show of holding my head high and walking confidently up the stone steps. The lion's head doorknocker was brightly polished and I knocked twice.

A footman opened the door almost immediately.

'I am Miss Barton,' I said. 'I should like to speak to Sir Frederick.'

'Sir Frederick isn't at home.'

I was in such a state of nervous anticipation and false bravado that I could only stare at him. Of course, there was no reason to assume that Sir Frederick would be there.

'I've travelled from Italy,' I said.

'Perhaps you'd care to wait?' said the footman. 'He's expected shortly.'

'I arrived in London only yesterday.'

The footman glanced sideways at me and I knew nervousness was making me talk too much.

The hall was decorated with a bold, almost masculine, paper hanging of purple and gold stripes and the floor was black and white marble. A great number of gilt-framed paintings were displayed on the wall beside the mahogany staircase and a pier glass over a matching console table reflected a magnificent arrangement of hothouse flowers. There was an air of opulence that I had only encountered in the homes of some of our richest clients in Italy and it made me even more nervous.

The footman showed me into a morning room painted in faded pea green and I sat down to wait. The room spoke of ease and comfort, although here the upholstery was sadly worn and the curtains had faded where the sun caught them. All was in such contrast to the sumptuousness of the hall that I suspected the room might not have been redecorated for many years. I indulged a fancy for a moment that my mother had sat by the window at the satinwood desk to write her letters. I wondered if she might have chosen the flower paintings and the china figures in the glazed corner cupboard. I sat quietly, my hands gripped together while I tried, without success, to remember if I had played with my toys at her feet.

I don't know how long I sat there but, when I heard the front door open, my fingers were stiff as I unclenched my hands. There was a murmur of voices and then footsteps echoed across the marble. My heart began to gallop as the footsteps approached. Hurriedly, I pinched colour into my cheeks.

The door opened and a man stood on the threshold. Slightly above average height and a little fleshy around the jowls and stomach, his iron-grey hair was combed back from a noble brow. From the top of his pomaded hair to the tips of his gleaming boots, he looked as sleek and well fed as a pigeon. 'Miss Barton?' he said.

'Yes,' I croaked. 'Sir Frederick, I presume?'

He moved quietly for such a solid man and was inside the room with the door closed behind him in one catlike movement that barely stirred the air. 'My footman said you've travelled from Italy?'

I nodded. Now I was face to face with him I couldn't seem to make my mouth form the words to tell him who I was.

He strode across the room and came to a stop in front of me. His face was impassive but his slate-coloured eyes were watchful.

'I've lived in France and Italy for nearly all my life,' I said. 'But recently I discovered that I was born in England.' It unnerved me to have him study me so intently. 'I've come to London to find my family.'

'Your family?'

My mouth was as dry as sandpaper. 'Before she died, the woman I thought was my mother told me that she was, in fact, my natural mother's maid. She said she'd stolen me and that my name was really Harriet . . . ' I swallowed, unable to look at him. 'Sir, I have reason to believe I am your daughter.'

'Harriet Emilia Langdon.' His voice was low and expressionless.

'I realise this must be a shock to you.'

White-faced, he clenched his jaw. 'Did you really think I wouldn't know my own daughter?'

'It's the truth! I know you may not believe me but . . . '

He lifted a trembling hand to smooth his already smooth hair.

'I'm not a fortune hunter,' I said, lifting my chin. 'I want nothing from you . . . '

'I didn't say I didn't believe you.' Suddenly, he snatched off my hat and tossed it on the carpet.

I gasped and shrank back.

He plucked the ribbon from my hair, allowing it to tumble to my shoulders. Grasping me by my upper arms, he pulled me out of the chair. 'I could never have forgotten the colour of your hair. I'd have picked you out in a crowd.'

His eyes welled with tears and I felt the dawning of hope.

'My dearest child,' he said, 'I searched for you for years all over France and Italy. I advertised a reward for you. And now,' he held me at arm's length to study my face, 'after all these years, you've simply walked back into my life.'

'So you do believe me?'

'Isn't that what I've been telling you, silly girl?' He hugged me to his broad chest again and I felt a sob catch in my throat.

I had found my father.

# Chapter 13

'I never stopped looking for you, Harriet,' said my father, 'even though I thought you might have drowned with your mother. I can't tell you how many false trails I've followed.'

'Perhaps Sarah was right after all,' I mused. 'I always thought she was imagining that people followed us.'

He scowled. 'Don't talk to me of that woman. She's caused me more agony than you can possibly imagine. To lose both my children and my wife so close together was a living hell.' He took my hands. 'You must tell me everything, all the smallest details of your life. We have seventeen missing years to talk about. How did you live? Was it very terrible?' Drawing a chair close to my own, he sat down beside me.

I remembered what Alessandro had said. *'Too much time has passed to build the years of memories that bind you into a proper family.'* But was it really too late for us? Father hadn't rejected me as I'd expected. On the contrary, he'd welcomed me. I began to tell my story.

Some considerable time later I sighed and leaned back in the chair.

'I don't understand how you managed,' said Father, frowning. 'Surely the income earned as travelling dressmakers cannot have been enough?'

'We survived,' I said. 'Sometimes we went hungry but that served to sharpen our appetites to work harder.'

'Did you not have anything to sell that might have brought in some money?'

I saw how his hands gripped the arms of his chair. 'My mother's jewellery, you mean?'

He shrugged. 'Or any other little treasures Sarah might have brought with her?'

'Anything of value was sold while I was still small,' I said. 'Sarah's husband Joe was a drunkard and he drank the proceeds.'

'Nothing remained?' asked my father. A muscle twitched in his cheek.

'Was there something special you're thinking of?' I asked.

'Well,' he said slowly, 'there was a sapphire brooch that had belonged to my mother.' His eyes met mine. 'It was valuable and I should have liked you to have it.'

I shook my head. 'I don't remember ever seeing it. After Sarah died I found gold coins sewn into her petticoat. She'd never mentioned them. When I think of all the times we went hungry . . . '

'She was a wicked thief,' said Father, fists clenched. 'But I'm glad you told me about the coins.'

'Why shouldn't I?' I said.

'I don't want there to be secrets between us.'

'Then will you tell me about my mother now?' I said.

Father stood up and riddled the fire with a poker. When he turned to look at me his face was etched with lines of pain. 'She was beautiful,' he said, 'like you. But when our son Piers died then something in her died, too.'

I hesitated but I had to ask. 'Sarah said you quarrelled with my mother and confined her to her room. She said that's why Mother tried to run away.'

'Of course I confined her!' Father's eyebrows drew together. 'After Piers died she lost her reason and threatened to harm herself.' He looked directly at me. 'And you, too.' He rubbed a hand over his face. 'I tried to keep her safe until she was well again but that interfering maid got her a key. And so Rose ran off and threw herself in the Thames.'

'I wish I could remember her.' I spoke in a low tone, my voice full of regret.

Father curved his hand around mine. 'I cannot give you your mother back and, God knows, I still miss her. But there is someone else you must see. Wait!' He hurried from the room.

I leaned back in the chair and closed my eyes, utterly drained. I'm not sure, perhaps I dozed, but then Father entered the room, steering an elderly lady by the elbow.

'This is Miss Weston, your great-aunt Maude,' he said. 'My mother's sister.'

I bobbed her a curtsey, looking at her with curiosity.

Thin and frail, she leaned upon an ebony stick and blinked watery blue eyes at me. 'Harriet?' she said, her voice quavering. 'You really have returned to us?' Her face was bone-white against her black shawl and a few silvery wisps of hair had escaped from her cap.

'I answer to Emilia,' I said. 'I'm too accustomed to it now to revert to Harriet.'

Father smiled. 'It will take me a little time to become used to that.'

'Frederick told me that your mother's maid stole you away,' said Aunt Maude. 'And you have been abroad with her all this time?'

'She died,' said Father, 'and poor Emilia was forced to live in Caroline of Brunswick's peculiar household. Can you imagine? Still, there's no reason to mention that to anyone. None at all.'

I remembered Lady Hamilton's comments about his support of the Prince Regent.

'I daresay Emilia will have returned abroad before any of your cronies hear where she's been living,' said Aunt Maude.

'Return to Italy? Certainly not!' Father patted my cheek. 'My little girl has come home again and I'm not letting her out of my sight.'

'But I must return . . . ' I began, thinking of Alessandro.

'Emilia, listen to me!' said Father. 'I cannot allow my daughter to return to a life of penury, working as a travelling dressmaker, of all things. Or, even worse, as a servant to that appalling madwoman.' The distaste in his voice made it sound as if I'd been earning my living on the streets.

'The Princess took me in when I was left alone,' I protested. 'Besides, I have friends in Italy.'

'It's perfectly natural for Emilia to wish to return to the life she knows, Frederick,' said Aunt Maude.

'You have *family* in England,' said Father, ignoring his aunt. 'Where are you staying, Emilia?'

'Portman Street,' I said, 'with Lady Hamilton. I shall accompany her back to Italy in a few days' time since she'll be taking up her position as the Princess's lady-in-waiting again.'

'It gets worse!' he said, striding across the room. 'You absolutely cannot stay there. Lady Hamilton's brother is the Radical MP Archibald Hamilton. I'd never hear the end of it! I'll have your luggage collected immediately.'

'Perhaps you should ask Emilia what she wants, Frederick?' Aunt Maud glanced enquiringly at me.

Father came to stand before me. 'My dearest girl, won't you come home? Even if it's only for a few days. I've just found you. Please don't deprive me of that pleasure.'

His expression was so imploring I couldn't have denied him, even if I'd wanted to. 'I'd be very happy to stay with you until I return to Italy,' I said, 'but first I must thank Lady Hamilton for her hospitality.'

'We shall go together,' said Father, picking up my hat from the carpet.

Over his shoulder I saw Aunt Maude watching us. Just for a moment I thought I glimpsed an expression of distress, or was it annoyance, on her face?

'Aunt Maude,' said my father, 'will you instruct the housekeeper to make up the best guest room?'

She bowed her head.

As I left on my father's arm, I glanced back at her. There was no doubt in my mind then that she was looking at me with animosity.

Later, when my father and I returned to Grosvenor Street, the servants were lined up in the hall, waiting for us. The butler, the housekeeper, the cook, two footmen, two housemaids and a scullery maid all bowed or curtseyed as I walked past them. Hot with embarrassment under their curious glances, I was relieved when the housekeeper stepped forward.

'Welcome home, Miss Langdon,' she said.

I glanced behind me and then realised she was talking to me.

'Shall I take Miss Langdon upstairs to the guest room, sir?' Her voice was quietly modulated and her narrow figure as rigidly upright as if she had a steel rod sewn into her stays.

'By all means, Mrs Hope.' Father smiled at me. 'Come down to the drawing room when you're ready, my dear.'

A footman, carrying my travelling bags, followed us upstairs but I couldn't help pausing to look at the paintings that lined the staircase wall.

On the first floor the housekeeper opened one of the panelled mahogany doors. 'The drawing room, Miss Langdon.'

I glimpsed an ornate gilt mirror over a chimneypiece of white marble before she closed the door again. We ascended to the next floor and Mrs Hope led me into the guest room.

It was luxuriously appointed with blue velvet curtains and a silk bedspread heaped with cushions. My feet sank noiselessly into the carpet.

The footman placed my bags on the chest at the end of the bed and silently withdrew.

'If you wish to refresh yourself there's hot water in the ewer,' said Mrs Hope. 'I apologise that we do not have a lady's maid to attend you but I shall send up one of the housemaids to unpack for you.'

'Please don't trouble,' I said. 'I have so little luggage I shall put it away in a trice.'

'It's no trouble at all,' said the housekeeper, bristling. 'We shall make the proper arrangements by tomorrow but we weren't aware of your impending arrival. Is there anything else you require?'

I ran my fingers over the soft towel on the washstand and bent to sniff the hothouse flowers perfuming the air by the dressing table. 'No, thank you,' I said. 'I have everything I need.'

'Very good, Miss Langdon.' Mrs Hope inclined her stately head and silently left the room.

Once the door had closed I toed off my shoes and bounced on the bed. I doubted I'd ever slept on a mattress so soft or in a room so lavishly appointed. Lying on my back and looking at the ceiling, I let out my breath in a sigh of contentment. This would do me very well for the few days before I returned to Italy. Best of all, my father really wanted me to stay. Lady Hamilton, however, had appeared relieved I'd no longer be a burden to her and agreed to send me a note when she was ready to travel. Meanwhile, I had to discover what kind of reception might face the Princess when she arrived on these shores.

There was a soft knock on the door and a housemaid entered. Small and slight, she stopped dead when she saw me on the bed. 'Sorry, miss! Shall I come back later?'

'No, come in, though really there's no need for you to unpack for me.'

Her grey eyes widened and she appeared genuinely shocked. 'You can't do it yourself, miss!'

I opened my mouth to say I'd unpacked my travelling bags more times than she'd had a hot dinner but thought better of it. She'd probably tell Mrs Hope, who would then inform my father that I didn't show proper consideration for my position. 'What is your name?' I asked instead.

'Daisy, miss.'

'Then please unpack my bags for me, Daisy.'

She hurried to oblige before I changed my mind.

After she'd placed the last of my shifts in the drawer and arranged my hairbrush on the dressing table, she said, 'Will that be all, Miss Langdon?'

'I wonder, where is the nursery?' I asked. 'I should like to see the room where I slept as a child.'

Daisy gave me a curious look. 'If you'll follow me, miss?'

I hurried behind her up to the third floor. The stair carpet was of coarse drugget, worn in places, and the walls were painted rather than papered.

Daisy opened a door. 'This was the nursery, miss. It's not been used since . . . ' She looked away.

'Since I left,' I finished. 'Thank you, Daisy. You may go now.'

She hurried away down the stairs.

The nursery was in semi-darkness. I pulled up the blind, exposing safety bars fixed to the windowframe. The walls were painted a faded primrose colour and there was a multicoloured rag rug on the floorboards. I sat in one of the armchairs placed to either side of the mantelpiece. Something hard pressed into my thigh and I put down my hand and retrieved a silver rattle. I shook it gently and wondered if it had belonged to my little brother. It saddened me to have no memory of him. If he'd lived he would have been seventeen or eighteen now.

A brass fireguard was fitted securely over the grate and it was then that I had a fleeting recollection of a stout woman leaning over the flames and remembered the smoky aroma of toasting bread. In my mind I heard the faint echo of a child's voice, my voice, saying, 'Corky, it's burning!' I sat very still, listening, but the memory had gone.

I stood up to spin a globe standing on a nearby cupboard. Opening the cupboard doors, I saw jigsaw puzzles, chalks and a slate, gaily painted wooden bricks and a spinning top. I didn't remember any of them.

There was another door and I pushed it open to find the night nursery. I pulled up the blind and hesitated before opening a chest of drawers and peeping inside. It was filled with neatly folded small items of clothing, from flannel petticoats and a tissue-wrapped christening gown to frilled and tucked cotton dresses. I imagined for a moment how Victorine would squeal with delight at the sight of such treasures. Perhaps my father would allow me to take them back to Italy for her.

My eye was caught by a beautiful doll with a delicately painted wooden face, lying on the white counterpane on the bed. I examined the doll's tiny kid slippers and her old-fashioned dress with panniered skirts. Her blonde hair smelled dusty and was slightly rough under my fingers. Something tugged at my memory. Crying. An all-consuming misery and a tearing sense of loss as I sat in a coach jolting through the dark. Suddenly I was four years old again and sobbing, desolate at having left my beloved Annabelle behind. Now, after all this time, I had found her again and she was a precious link to the life I'd once had.

I kissed the doll's painted face and, still clutching her in my hand, investigated the rest of the night nursery. A curtained alcove contained another bed, a row of wall hooks and a chest of drawers. I wondered if that was where the nanny had slept. There was no sign of a cradle. I pulled down the blinds before returning to the guest room. Carefully, I laid Annabelle on the pillow. Perhaps, one day, Alessandro and I would have a little girl who would love Annabelle as much as I had.

On my way downstairs I stopped on the staircase to take a closer look at the paintings I'd noticed earlier. There were several pastoral landscapes and a number of portraits, some very old, judging by the ruffs, doublets and hose the subjects were wearing.

Father stood at the foot of the stairs. 'There you are!'

'I was admiring the paintings,' I said.

'You like art?'

'I'm not very good at drawing,' I said, 'but I like to look at good paintings.'

Father beamed. 'There! Blood will out. As a young man I had pretensions of becoming a famous artist but, try as I might, the beautiful pictures I saw in my head never materialised on the canvas when I applied my brush to it.'

I laughed. 'Sad, isn't it?'

Father looked up at me, shaking his head sorrowfully but with a smile in his eyes.

'Many of the churches in Italy have wonderful frescoes,' I said. 'It's easy to lose yourself for an afternoon looking at them. One very hot day, when I was no more than ten or twelve, we reached the Scrovegni Chapel in Padua. We went inside to escape the heat and I was entranced by Giotto's fresco cycle of the Life of the Virgin Mary.' I paused. 'That day was the first time I truly believed that God existed.'

'Because no earthly hand could paint anything so glorious without God's guidance?'

'That's it exactly.' I smiled, a warm feeling growing inside me because we'd found common ground. 'Ever since, whenever I had the opportunity, I'd find out if there were any treasures in the town we were visiting and make a point of going to see them.'

'I'll show you my art collection later but we can't stand on the stairs all day talking about Art and God,' said Father. 'Come into the drawing room.'

I'd failed to appreciate the full magnificence of the room I'd glimpsed earlier. Three tall windows flooded it with light, even on such a grey and miserable day. The high ceiling was richly decorated with intricate plasterwork, and softly coloured carpets floated like jewelled islands on a sea of golden parquet flooring.

Aunt Maude sat straight-backed in an armchair. She glanced up at me from her embroidery but didn't return my tentative smile.

Cream damask drapes with muslin under-curtains framed the windows, concealing us from the inquisitive stares of pedestrians in the street below. It was all so sumptuous that I felt completely out of place.

I sat down carefully against the plump cushions on the sofa and Father offered both Aunt Maude and myself a glass of ratafia from a silver tray.

I took a sip and then said, 'Father, who is Corky?'

He started and spilled a few drops of ratafia. Wiping them off the side of his glass with the tip of his finger, he said, 'I thought you remembered nothing from your time here?'

'I don't. Not really. I glanced into the nursery a moment ago and recalled an old lady called Corky burning toast on the fire.' Aunt Maude, I noticed, had put down her embroidery and was watching me with narrowed eyes.

'Miss McCorquordale was your nanny,' said Father, his lips tightening. 'She was an utter disgrace to her profession!'

'How so?''

'On the morning you were discovered to be missing, she was fast asleep in a cloud of gin fumes.'

'What happened to her?'

'Turned off without a reference, of course.'

'Did she usually drink?' I asked. 'Or was she given the gin so that she wouldn't notice when I was taken?'

'Whichever it was,' said Father dismissively, 'she failed in her duties with the most terrible result.'

I sipped my ratafia reflectively. Either my mother or Sarah must have led the poor woman down the path to oblivion.

'Tell me, Emilia,' said Father, 'did you receive any kind of education at all whilst on your travels?'

I bristled at the implication. 'Sarah taught me to read and write . . . '

'I suppose we must be grateful to her for that, at least,' he said.

'I went to school sometimes,' I said, 'but we always moved on just as I began to make friends. We were never in one place for more than a few months. I loved to read and often, when we had work where we lived temporarily on the premises, I'd be allowed to borrow books from the owner's library.'

'What did you read?'

'Poetry, art and philosophy. Botany. Anything I could find on art, architecture and antiquities. I even picked up a little Latin.' I was pleased to note that Father's eyebrows had risen. It appeared that travelling about the continent with a mere maid had allowed me to acquire a surprisingly respectable education.

He nodded. 'What about mathematics?'

'Well, there was the practical side of dressmaking. I learned to estimate quantities of dress material required and do the bookkeeping and place orders. And I can sew, of course, and speak French and Italian fluently.' I didn't mention that I could swear as well as any stable lad in both languages, too. 'I was the one who arranged our lodgings and booked our coach tickets since Sarah found anything other than her native tongue very difficult.'

'Astonishing,' said Father. 'Most girls cannot claim to have learned as much at their schools. Can you dance?'

I shrugged. 'There were village fiestas sometimes.'

'Hmm.' He looked thoughtful. 'There will be deficiencies in your education but nothing, I believe, that we can't remedy. What do you think, Aunt Maude?'

The old lady put down her embroidery. 'It sounds to me as if Harriet, I beg your pardon, *Emilia*, has as much education as she needs for her position in Princess Caroline's household.'

'Ah, well, that's something to discuss in the future,' said Father. He stood up and offered me his arm. 'Shall we look at my paintings, my dear?'

'I should be delighted.'

'Of course, only a part of my collection is here,' he said, as we went into the hall. 'I shall take you to Langdon Hall to see the rest. Now that *is* something I look forward to.'

What a pity, I thought, that there wouldn't be time to visit the family seat in Hampshire before I returned to Lyons and thence to Italy.

# Chapter 14

The candles guttered in the draught as the footmen, James and Edward, brought more dishes to the dining table. Soames the butler watched as the domed covers were lifted with a flourish to reveal a dozen spatchcocked game birds, a great rib of beef and a haunch of venison. My mouth watered at the delicious aroma of roasted meat. After a fortnight in my father's house in Grosvenor Street I was already becoming accustomed to the rich food, which was far more plentiful and of higher quality than in the Princess's household in Pesaro.

'Tell me more about your travels, Emilia,' said my father. 'Where did you go after Florence?' he asked, proffering me a slice of beef.

Every day he asked me to tell him all the places Sarah and I had visited, encouraging me to reminisce about even the smallest villages and towns.

'Once we stayed in a hill-top monastery near Castellina in Tuscany,' I said. 'We mended the altar cloths and turned all the monks' sheets sides-to-middle.'

'A monastery near Castellina?' said Father, putting down his knife and fork. 'And what did you see there?'

I smiled. 'It was more a case of what we didn't see.'

'What do you mean?'

'There was a painting of the Last Supper on a wall in the refectory,' I said. 'It wasn't very good and I asked if one of the monks had painted it. Brother Anselm told us the original had been stolen some years before and the thief had left the poor copy hanging on the wall instead.'

'An audacious thief, indeed,' said Father with a snort of laughter.

'After that we stopped in Siena for a few weeks,' I said as the footman refilled my glass with claret. 'Long enough to find and complete two commissions.'

'Did you see any antiquities or frescoes there?' asked Father. 'And did you by any chance visit an art dealer who has premises just off the square? He writes to me sometimes if he buys a particular treasure he thinks might interest me.'

'You seem to have art dealer acquaintances all over Italy,' I said.

'I buy and sell art and antiquities to my friends and acquaintances. It's my passion,' he said. 'And why not, when I've had no wife or child to spend my fortune on?' He smiled at me. 'Until now, that is. That reminds me . . . I saw a pretty shawl in the window of a shop in Cheapside. We'll go and look at it tomorrow and, if you like it, you shall have it.'

'Father, you spoil me! It was a fan last time, even though I don't have any occasion to use it.'

'Just a few trifles, my dear. And, pretty though it is, you cannot go on wearing that same dress every day. I shall take you shopping.'

Sometimes I awoke in the night, wondering if the last two weeks had all been a dream, but then I touched the finely woven sheets and the silk bed curtains and remembered that I was in my father's house. Such luxury was unimaginable after the penny-pinching life I'd lived with Sarah. I hadn't imagined I'd be so readily welcomed into Father's life. I could only wish that Aunt Maude would accept me more readily.

Sitting opposite me at the table, she ate in silence and rarely engaged me in conversation. She had rejected my tentative advances of friendship ever since I'd arrived. I sighed. Still, she wouldn't be troubled by my presence for a great deal longer. Every day I expected Lady Hamilton to send me a note to say she was travelling to France. And soon after that I'd be back with Alessandro. My thoughts drifted to that summer evening when we'd watched the fireflies dancing in the avenue of cypresses. In the circle of his arms I'd known with absolute certainty that we were meant for each other. But now . . .

'Emilia?'

'Sorry, Father,' I said. 'I was daydreaming.' Alessandro and I *would* make it all right between us again. We had to.

He smiled indulgently and patted my hand. 'And what were you thinking?'

'That I haven't heard from Lady Hamilton.'

'You don't still have it in mind to return to Lyons with her, do you?' The smile had gone from his face. 'I cannot like the thought of you returning to Caroline of Brunswick's household. There's so much still for us to talk about and I haven't yet taken you to see the London sights. And there's a play on at the Adelphi I thought might amuse you.'

I didn't answer him directly. 'Father, have you heard whether Mr Brougham has left for Lyons?' I asked. 'He promised to meet the Princess there before she travelled to England.'

He looked at me sharply. 'Is she definitely coming then?'

I shrugged. 'Mr Brougham led her to believe the divorce case would come to trial this month but soon it will be Christmas. The Princess wants the whole business settled so that she can return to live a quiet life in Pesaro.'

'She wouldn't bring that raggle-taggle band of Italians with her, surely?' Father's expression was incredulous. 'She'd be hounded back across the sea. Especially if she was accompanied by that blackguard Pergami.'

'That's why she wants Lady Hamilton to join her, so that she has an English courtier with her when she arrives.'

'Brougham was in the House of Commons the other day,' said Father, sipping his claret. 'Emilia, if Caroline of Brunswick intends to come to London soon, why the rush for you to return to her side? Will you then return to London with her?'

I shook my head and made a show of cutting up my meat to avoid looking at him. 'I'm no courtier and I have friends in Pesaro,' I said, assailed by a sharp longing for Alessandro. It hurt to remember how we'd parted.

'A special friend?' said Father. When I didn't answer, he said, 'Emilia, we said we wouldn't have secrets from each other. Is there a young man I should know about?'

Reluctantly, I nodded.

'Tell me about him.' He spoke gently.

'His name is Alessandro Fiorelli,' I said, needing little encouragement. 'He's a tutor to the Baron's daughter, Victorine.'

'Tutor?' Father sighed. 'I suppose I shouldn't be surprised . . . you've been working as a dressmaker. It won't have been easy for you to meet suitors appropriate to your birth. What of his family?'

'Alessandro's father is a doctor.' I bristled at the implication that his family wasn't good enough for me.

'Alessandro? Is there a formal understanding between you that you refer to him in such familiar terms?'

'He proposed to me,' I said, 'but we had a disagreement about my coming to London to look for you. Now I don't know . . . ' Reliving the hard words between us before I left Lyons, I knew I couldn't say any more without breaking down.

'Dearest girl, all I want is your happiness,' said Father. 'If this Alessandro Fiorelli is the right husband for you, he'll wait a while and then you can spend the rest of your lives together. But you've only just returned to me and I hope you won't think me selfish if I say how terribly sad I would be to lose you again so soon. Why, I haven't even had time to show you Langdon Hall yet.' He stroked my hand, which was curled into a fist on the tablecloth. 'Are you not happy here?' His tone was imploring.

'How could I not be happy, when you have taken me in and welcomed me?' I said, my thoughts confused. I wanted to return to Alessandro.

'Could you not write to this young man,' said Father, 'and tell him you wish to stay with your newfound family for a few months? Is that too much to ask?'

I rubbed my temples, the quandary making my head ache.

'You'll have to decide soon,' said Father, 'if Lady Hamilton is leaving shortly.'

'I know,' I said, pushing my unfinished food away.

'Shall we take our tea in the drawing room?' said Aunt Maude. She stood up and I had no choice but to follow her lead.

'I shall join you ladies as soon as I've finished my brandy,' said Father.

Aunt Maude and I retired to the drawing room, where the tea tray awaited us. As usual she poured the tea and handed me a cup.

We sat in uneasy silence while I wondered if Alessandro would wait for me if I delayed my return.

'Emilia,' said Aunt Maude, putting down her cup.

I looked at her, surprised. She barely spoke to me except out of necessity.

'Your father can be very persuasive,' she continued. 'But, if you love this young man of yours, you mustn't allow Frederick to keep you apart.'

Her comment surprised me even more. 'I do want to return to Alessandro,' I said, 'but I've longed for a family of my own for so long . . . '

'If you delay your return it may be too late.' She looked intently at me. 'Go to your young man while you can, Emilia.' There was such passion in her voice that I wondered if, in her distant youth perhaps, she'd been parted from a man she loved.

The drawing-room door opened then and Father entered. 'Emilia, shall we play a hand of cards?'

I glanced at Aunt Maude but she was staring at the flickering flames in the hearth, giving no indication that she had spoken to me at all.

The following afternoon, I set off in Father's carriage. He'd refused to let me walk by myself, though I'd protested that I often walked alone through the streets in Italy. Aunt Maude, her lips pursed, commented that no young lady of breeding ever ventured abroad without a companion. So it was that my great-aunt sat opposite me in Father's carriage.

It was bitterly cold and she sat with the collar of her pelisse turned up and her gloved hands folded in her lap. Her face was resolutely turned away from me so she could look out of the window.

The noise in the streets was even worse than in Florence and I caught my breath as a curricle raced past with only a couple of inches to spare. The young blood driving the equipage flicked his whip and surged forward, scattering the pedestrians crossing the road. I was happy to look out of the window as we rattled past an endless array of smart shops. The fashions worn by the stylish London ladies were subtly different from those worn in the Italian cities. I made mental notes as to the cut of a pelisse or the drape of a skirt to take back to Italy for my clients, in due course.

Despite the freezing weather, streams of shoppers ambled along the pavements, girls sold scraps of lace and trinkets from handcarts, here a man juggled oranges and there a beggar in military uniform propped himself up on his wooden leg and proffered his hat for coins. Everything was unfamiliar and therefore interesting to me.

The carriage came to a halt outside the Hamilton house in Portman Street and Aunt Maude didn't object when I took her arm to guide her over the cobbles, which were slippery with ice.

'Are you quite sure you're making the right decision?' she asked as we waited for the door to open.

I wasn't at all sure I was doing the right thing and I'd worried about it most of the previous night, wondering what to do. 'You and Father are the only relatives I have left,' I said.

'That's all very well,' said Aunt Maude, 'but you don't know us at all.'

'But now I have the chance to remedy that,' I said. And it grieved me to realise that if Alessandro didn't understand that I needed time to get to know my father properly, then perhaps he wasn't the right man for me after all.

A footman opened the door and we followed him across the hall to the drawing room.

'Miss Weston and Miss Langdon, ma'am.'

'So, you're not Miss Barton any longer,' said Lady Hamilton, gesturing to us to sit down.

I wasn't sure if I liked being called Miss Langdon. It wasn't who I was. But then, I admitted, sighing, I wasn't Miss Barton either. 'It feels strange,' I said, with a wry smile. 'My father, however, is prepared to compromise and my Christian name remains Emilia.'

'How complicated!'

I didn't say that it was far more complicated for me than it was for her. 'I've brought a letter for the Princess,' I said, 'and should tell you that I've decided not to return to her at present. Now that I'm reunited with my family after so long, I intend to stay here for a while.'

'I daresay the Princess can spare you,' said Lady Hamilton.

'Before I left her, she asked me to ascertain what is happening in London with regard to Mr Brougham's visit to her,' I said. 'It appears he was in the House the other day and not on his way to Lyons as the Princess expected. She believed the matter of a divorce was to come to court this month.'

'I haven't heard any indications of that,' said Lady Hamilton. 'The politicians are far too preoccupied with the massacre at Peterloo.'

'Massacre?' I said.

'Don't you read the papers?' asked Lady Hamilton. 'There was an assembly of sixty thousand Radicals and Whigs in St Peter's Field, near Manchester, to demand the reform of parliamentary representation.'

'I don't see how that influences Mr Brougham's movements,' I said, puzzled.

'The Tory government sent in the cavalry with sabres drawn to dispel the assembly,' said Lady Hamilton. Her expression was grim. 'Fifteen died and seven hundred were injured. The people call it the Peterloo Massacre, after the Battle of Waterloo.'

I resolved to read the newspaper every day after Father had finished with it. Perhaps it would help me to glean some information that could be of use to the Princess. 'That's appalling,' I said, 'but I still don't understand . . . '

'Brougham hasn't time to go gallivanting off to France,' said Lady Hamilton impatiently, 'not while emergency legislation is being debated in Parliament. He's marshalling the rest of the Opposition, Whig and Radical, to argue against the proposed measures.'

'The poor Princess,' I said, 'waiting in vain all this time.'

'She needs careful guidance since she's so prone to impetuousness,' said Lady Hamilton. 'Now is not a good time for her to arrive in London. Much better to wait until next year.'

'I see.' I opened my reticule and took out the letters I'd written that morning. 'This is for the Princess.'

She took the missive from me.

'And then there is another letter for a member of her household,' I said. 'I would be extremely grateful if you would arrange for this to be handed to Signor Fiorelli.'

Lady Hamilton took my letter for Alessandro between the tips of her fingers as if it were something distasteful. 'I am not in the habit of acting as a postman to servants,' she said, 'but I will take your communication this time. Or, if I decide not to travel now, I'll send them on to her.'

My cheeks burned. 'Thank you, Lady Hamilton,' I said. I'd toiled over the letter to Alessandro, writing and rewriting it three times, explaining that I'd decided to stay in England longer than I'd expected. I'd told him I loved him and begged him again to understand and to write to me.

The clock in the hall struck a quarter past three and Lady Hamilton rang the bell on the side table. Our visit was at an end.

'I suppose,' she said, 'since your father is acquainted with the Prince Regent, you may hear of developments the Princess should know about. In such a case you may write to me. Deliver any such letters here and my servants will forward them in my absence.'

I hesitated. 'Would you send a brief note to me now and again when you do return to the Princess? I have no way of knowing how she fares or if there is any small service I can render to her.'

I glanced at Aunt Maude but she sat straight-backed, her gaze fixed on her hands folded in her lap.

Lady Hamilton sighed. 'I will communicate to you any news of particular note. I shall send such letters here to Portman Street and one of my servants will deliver them to you at Grosvenor Street.'

'Then there is nothing for me to say except to wish you a safe journey, Lady Hamilton,' I said.

In the carriage on the way back to Grosvenor Street Aunt Maude sighed. 'Well, you've burned your bridges now, haven't you?'

'I know you don't want me to stay in your home . . . '

'Believe me, it isn't *my* home,' she retorted.

I clasped my hands tightly together. It upset me that she had taken against me. 'I shan't let you stop me forging bonds with my father.'

'I can see that.' She glanced at me, an odd half-smile on her mouth.

I capitulated and asked the question that had troubled me since the first day we met. 'Why do you dislike me so?'

Aunt Maude shook her head. 'I don't dislike you,' she said. 'You were an adorable child. 'You used to call me Auntie Maudie and bring all your little treasures to show me.' She looked out of the carriage window and sighed. 'But that was so long ago. We cannot go back to the past and do things differently with the benefit of hindsight.'

'No,' I said.

Aunt Maude shrugged. 'I didn't want you to have regrets about not returning to your young man, that's all.' She faced me, her pale eyes glittering. 'Once I was considered pretty. I never imagined I'd end up as a spinster, living on my nephew's reluctant charity.'

I studied her fine bone structure and could see that she might have been lovely once. 'I hope I haven't lost my chance with Alessandro,' I said. 'But I need time with my father or I'll always wonder what it would have been like to know him properly.'

'Sometimes the things we want aren't the things that finally bring us happiness,' said Aunt Maude. 'Very well then, since despite my best efforts I haven't managed to frighten you away, we shall make the best of it, my dear. And I sincerely hope we shall regain our earlier friendship.'

Then, to my utmost astonishment, she patted my hand.

# Chapter 15

*Christmas 1819*

**Langdon Hall**

I started awake as Father touched my knee.

'We're here,' he said.

He, Aunt Maude and I had set off early. We'd sought refreshment at the Feathers Inn in Hartley Wintney and afterwards the heavy pastry of the chicken pie had rested uneasily on my stomach. I'd watched the endless countryside roll past the carriage windows until my head began to nod. Aunt Maude dozed and before long I'd succumbed to sleep, too.

'Emilia?' said Father. 'Come and look!' He took my hand as I clambered down, stiff-legged, from the carriage.

We'd come to a halt at the end of a long carriage drive that led through parkland studded with ancient oaks. In the distance was a substantial house built of mellow red brick, with pointed gables and twisted chimneys.

'Langdon Hall,' said Father, unable to keep the note of pride from his voice.

I caught my breath in shocked surprise. 'It's beautiful!' Welcoming lights glimmered behind some of the stone-mullioned windows.

'It was built in the early sixteenth century,' Father said, 'by a Roger Fforbes. It's been in the Langdons' possession for the last two hundred years.'

Although I'd realised Father was wealthy, I'd had no idea that his country house would be so impressive.

Father looked at me intently. 'You have no memories of the Hall?'

I shook my head.

'I can't give you back those years,' he said, echoing my own thoughts. 'If that wretched maid hadn't stolen you away you wouldn't have been deprived of the life you should have had; all the fine dresses, music and dancing lessons, a pony . . . ' He sighed. 'I'll make it up to you.'

'You've spoiled me enough already!' It was true: he'd bought me new clothes, taken me to the theatre and given me books, paints and a gold bracelet.

'Nothing that wasn't due to you.' He reached for my hand. 'Let's go in.'

We returned to the carriage and set off again, the wheels crunching over the gravel. I glimpsed a church tower and a village in the distance through the gathering dusk.

'If the King in his chambers at Windsor Castle is as cold as I am, I doubt he'll live to see the year out,' said Aunt Maude, her teeth chattering.

'I've given orders for fires to be lit in all rooms,' said Father. 'And we must pray for King George to recover from his afflictions.'

If he didn't recover, I reflected, the odious Prince Regent would become King. What would happen then to Princess Caroline?

As we approached Langdon Hall I noticed the glint of light on water immediately in front of the house. 'Is that a lake?' I asked.

Father shook his head. 'A moat. It's no longer used for defensive purposes and the drawbridge is fixed down.'

The carriage rolled onto the bridge, the horses' hooves clattering across the timber. The great gates were open and we proceeded under the gatehouse arch and into a sizeable quadrangle, coming to a halt before the entrance to the Hall. The oak door, framed by carved stone, was flanked by flaming torches to light our way.

Simmering with anticipation, I took Aunt Maude's arm and Father ushered us into the entrance hall. The servants were lined up to greet us and, although I recognised those who had travelled from the London house, there were many more I'd never seen before.

The housekeeper, an imposing figure in black bombazine, curtseyed to us. 'Welcome home to Langdon Hall, Sir Frederick,' she said. 'I trust you had a good journey?'

'Thank you, Mrs Bannister,' said Father. 'Miss Weston and Miss Langdon will wish to refresh themselves but we'll have tea in the library in half an hour.' He turned to the butler. 'Bannister, come with me to discuss the wines for dinner.'

'Very good, sir.'

Aunt Maude's maid, Jane, stepped forward and led her away upstairs.

Mrs Bannister nodded at Daisy.

'This way, if you please, Miss Langdon,' said the girl.

I followed her into the rear hall, up the wide oak staircase and into a long picture gallery overlooking the quadrangle. Marble busts on granite columns were displayed in niches. Halfway along the gallery, Daisy opened a door.

My room was dominated by a carved four-poster bed hung with embroidered claret velvet. Tapestries of hunting scenes adorned the oak-panelled walls and the tall windows were magnificently decorated with leaded lights in a lozenge-shaped pattern. A fire crackled in the hearth and I held out my hands to warm them. Briefly, I remembered the simple whitewashed walls and tiled floor in my little room in Villa Vittoria and marvelled at the contrast.

'Will you change out of your travelling clothes, miss?' said Daisy. 'I wondered if you'd like to wear your new blue gown.'

I nodded. 'And the silk shawl, too, please.' Mrs Hope had protested when I said I wanted Daisy for my maid but I'd been adamant I could train her myself. I liked the girl and she seemed sensible enough. Besides, I wouldn't be in England for more than a few months.

'Shall I fetch your hot water, miss?'

I nodded, luxuriating in the softness of the richly patterned carpet beneath my feet and stroking the silky velvet curtains. There was a window seat heaped with cushions and I couldn't resist sitting in that cosy nest to look out of the window at the garden.

It was almost dark now but I could make out a stone-flagged terrace below with steps leading down to a parterre. To either side of the gardens were rows of substantial yew trees, clipped alternately into balls and cones. Beyond that was the moat.

Excitement made me want to laugh aloud. I could never have imagined that my family owned such a glorious house and I couldn't wait to write and tell Alessandro about it. As usual when I thought of him, a pang of longing lodged somewhere under my breastbone. It was painful to remember how we'd parted and I hoped desperately that, by now, he'd accepted and understood my decision.

When Daisy returned with the hot water, she lit the candles and drew the curtains while I washed and changed.

'I'm impatient to see the rest of the house,' I said.

'It's like going back in time,' said Daisy, 'after the modern conveniences of Grosvenor Street.' She pulled a face. 'There's no indoor necessary house here, for a start.' She giggled and pressed her fingers to her mouth. 'Except for your father's garderobe. His valet showed it to me. It overhangs the moat.'

'It's certainly an ancient house,' I said, standing up. 'Can you tell me where to find the library?'

'Shall I show you?'

I picked up the candlestick. 'I have a suspicion I might lose myself and never be seen again if I roam the galleries alone.'

'You'd have a ghost to keep you company.' There was a hint of laughter in Daisy's grey eyes as we set off down the gallery again.

'I don't believe in ghosts,' I said.

'Perhaps not, miss,' said Daisy, 'but if you wander about in the middle of the night and hear Sir Godfrey Mylton wailing and bemoaning his fate, you might change your mind.'

'And who might Sir Godfrey Mylton be?'

'Roger Fforbes, who built the Hall, was a Catholic and Godfrey Mylton was his priest. After King Henry broke with Rome, the Fforbes family pretended not to be Catholic anymore.'

My candle flickered in a draught as we reached the head of the stairs.

'But Mylton carried on holding services in the secret chapel that Fforbes built,' said Daisy. 'One day the priest was reading the Bible in the library when an unexpected visitor looked through the window and saw him. Mylton ran to warn Roger Fforbes, who shut him in the secret chapel when he saw the soldiers coming over the drawbridge. There was a terrible fight. Roger Fforbes and his sons were killed.'

'What happened to the priest?' I asked.

'After Roger Fforbes died no one knew the priest was in the secret chapel.'

'He starved to death?'

Daisy nodded, her eyes like saucers. 'So his ghost roams Langdon Hall at night.' She came to a stop. 'Here's the library, miss.' She bobbed a curtsey and hurried away.

Father and Aunt Maude were waiting for me, seated on sofas by the blazing fire.

'What a welcoming room!' I said. A candelabra cast a warm glow and the long walls were lined with mahogany bookshelves. The mellow tones of the leather bindings added to the rich appearance of the room quite as much as the Persian carpets, tapestry curtains and comfortable furnishings. Dutch still lifes and interiors were displayed on the two end walls.

'This is our winter drawing room,' said Father. 'In the summer we use the parlour since it has doors to the garden. Do read any of the books that you wish. Some are old and valuable but others are recent and may be more to your taste.'

'Most of my reading has been in Italian and French with some Latin,' I said, 'so I shall enjoy reading in English for a change.'

A footman carried in a laden tea tray and disappeared as silently as he had arrived.

Aunt Maude poured the tea and I handed it around. 'I understand this library caused the downfall of the Fforbes family,' I said.

'Whatever do you mean?' asked Father.

'Wasn't this where Godfrey Mylton was reading the Bible when a chance visitor saw him through the window?'

Father paused in the act of dropping a lump of sugar into his cup of tea. 'I suppose the servants have been gossiping?'

'You can't deny it makes an interesting story.'

'And a story is all it is,' said Father. 'There is no secret chapel. The priest probably escaped through a window. He was never heard of again but, you're right, that was the end for the Fforbeses.'

'Will you take me on a tour of the Hall?' I asked.

'Tomorrow,' said Father. 'You can't see my paintings properly until it's daylight. We'll have some supper in the little parlour this evening and a hand or two of cards before an early night.'

'Travelling is so fatiguing,' said Aunt Maude, smothering a yawn, 'especially over winter roads.'

'Christmas is only a few days away,' said Father. 'We shall have guests. My heir, Adolphus Pemberton, will join us, too.'

'Your heir?' I said.

'My cousin's boy.' Father made an expression of distaste. 'I never cared for my late cousin. I was his fag at Eton and he was a vicious bully. Unfortunately, the Langdon Hall estate is entailed and, since my son died, Adolphus is my closest living male relative. It's a bitter pill to swallow that I cannot leave the house to you, Emilia.'

'I shan't lose any sleep over that,' I said, 'since I wouldn't have the slightest idea how to manage it. Besides, I shall be in Italy.'

Father raised his eyebrows. 'You don't mind?'

'I never expected to inherit anything so I shan't miss it.' This was true but I had to admit to myself how much I was enjoying my new, if temporary, elevated lifestyle.

'I shan't forget you,' said Father. 'You shall have a dowry, if you make a suitable match.'

'I don't expect . . . '

'I know you don't,' he said. 'A dowry, and a trust set up in your name, on condition that I approve of your future husband.'

'I see,' I murmured, hoping he would like Alessandro. There was no doubt a dowry would be welcome but, assuming Alessandro still wanted to marry me, I'd give up any financial expectations to be with the man I loved.

'Still,' said Father, 'I suppose Adolphus Pemberton is marginally better than John Harvey, the next in line to inherit.'

'You don't care for him, either?'

'He's another cousin. His father cut him off without a shilling after he eloped with the daughter of a butcher and ended up going into the trade.' Father shuddered. 'A butcher, imagine that!'

'At least he should have a decent piece of beef for his Sunday dinner,' I said.

'It's no joking matter, Emilia,' said Father. 'It grieves me deeply that I have no son to inherit the estate. I wish I could leave it to you, the only child of my blood.' He sighed heavily.

'I've found you and Aunt Maude,' I said. 'That's all that matters to me.'

'Dear child!' she said.

'Aunt Maude has agreed to bring you out,' said Father.

'What do you mean?' I asked.

'She will prepare you for your presentation at Court in the spring. Afterwards I shall give a ball in your honour and in turn you will be invited to other balls and supper parties, with the opportunities of making friends and meeting a suitable husband.'

'Oh, but . . . '

Father held up his hand. 'You're going to mention Alessandro Fiorelli. Are you engaged to him?'

'Not exactly, though he did propose to me.'

'Then there is no impediment to your having a season. If at the end you haven't met a man you wish to marry, then you shall return to Italy. You will be accomplished and elegant and a suitable bride for anyone.'

'Alessandro is quite happy with me as I am,' I said, bristling.

'And so he should be,' said Father. 'But, as a tutor, he cannot hope to keep you in the comfort you deserve. If you indulge me by spending this time with me first and taking the opportunities that I can give you, then you may return to this young man with a dowry. Once I have approved him, of course.' He reached for my hand. 'Now tell me that you think I'm being fair.'

'More than fair,' I said. 'May I think about it before I agree?'

'We're all tired,' said Father. 'Let's have our supper and we'll talk about it tomorrow.'

After breakfast the next morning, he took me on a tour of the house. We began in the Great Hall, the most impressive room. Six wrought-iron light fittings, each a yard in diameter, hung over the massive refectory table. The beamed ceiling from which they were suspended soared above us. Between the beams the plaster was painted dark blue and embellished with gold stars, like a night sky. It was a great deal more magnificent than the farmhouse dining room at the Villa Vittoria.

'As children,' said Father, 'if it was raining too hard for us to go outside, we used to gallop up and down in here on our hobby horses.'

'Did you have brothers and sisters then?' I asked. It hadn't occurred to me that I might have more relatives.

'There was Cecilia, who was a year older than I, and Margaret, who was two years younger,' said Father. 'Cecilia died of the measles when she was eight and poor Margaret died in childbirth.'

'And my grandparents?'

Father sighed. 'My father died when I was a young man and Mother soon after you were born.'

'So you and I and Aunt Maude are the only remaining members of the family?' I said, sorry not to have known about these other relatives before.

He put his arm around me. 'We have each other now,' he said, kissing the top of my head.

I smiled at him, sad I hadn't grown up secure in the knowledge of his love. It pained me that, despite all his kind attentions, I didn't as yet feel love for him. Alessandro was right, a family was glued together by years of memories and although Father and I were bound by blood, we were not yet bound by love. But if I spent more time with him, as long as Alessandro would wait for me, perhaps we could build those precious memories.

'And here,' said Father, pointing above a heavily carved screen across one end of the room, 'is the minstrels' gallery. We used to hide in there and peep down through the screen when my parents held a party.'

'How wonderful to grow up with this history all around you,' I said. I ran my finger along the polished surface of the table and my footsteps echoed up from the flagstone floor as I went to study the stags' heads adorning the oak-panelled walls.

'Your grandfather was a great huntsman,' said Father, 'but these are of more interest to me.' He showed me a number of gilt-framed landscapes and spent the next twenty minutes teaching me about perspective.

'Forgive me,' he said, at last. 'Once I start talking about my beloved paintings I forget to stop. I've grouped them by type and period in different locations so I'll show them to you as we go from room to room.'

'I've already noticed the Dutch paintings in the library and the Roman busts in the niches in the gallery upstairs,' I said.

'Ah, yes, the Long Gallery,' said Father. 'It's the perfect place to display my treasures and was added by one of our ancestors in the late sixteenth century. Before that there were no corridors and you entered each room through another.' He linked his arm with mine again. 'Shall we see the rest of the house?'

We went all over it, looking into the half-timbered attics where the servants slept, poking our noses into nine bedrooms and several dressing rooms. One door was locked and Father told me it had been my mother's room. His forbidding expression prevented me from asking if I might look inside. We wandered through the hall again, discussed the paintings in Father's study, the morning room and the small parlour. I'd lost count of the various staircases, closets and anterooms.

Father showed me numerous beautiful works of art: Italian Renaissance drawings, Russian icons, English portraits, medieval religious paintings and collections of Roman and Egyptian sculptures and artefacts.

'It's like living in an art gallery,' I said. 'Aren't you ever afraid such a wonderful collection might be stolen?'

Father smiled, as if I'd said something particularly amusing. 'A number of pictures have been stolen from some of my acquaintances over the years but I've taken steps to ensure it won't happen to me. The paintings are very carefully fitted to the walls with a framework of my own devising at the back. A thief can't lift anything down without using a special tool.'

'Perhaps you should raise the drawbridge at night?' I said, only half joking.

Laughing, he said, 'That's not a bad idea but the servants have strict instructions to double lock the doors and to patrol the grounds at regular intervals.'

‘Wise precautions,’ I said.

‘Shall we return to the library now?’

‘I thought you must be lost,’ said Aunt Maude as we joined her by the fire.

‘I might never have been seen again, if Father hadn’t been there to guide me.’

‘What did you think of your father’s art collection?’

‘I’m amazed by it,’ I said, ‘but there’s too much to take in all at once. I shall enjoy revisiting the paintings one by one.’

‘When the weather improves I’ll show you the statues in the garden, Emilia,’ said Father. ‘You’ve seen enough for today and now I have estate business to attend to.’

After he had left us I leaned back on the sofa.

‘You look exhausted, dear,’ said Aunt Maude. ‘I’m afraid your father always forgets the time when he’s talking about art.’

‘But it’s good to feel so passionate about something, isn’t it?’

Aunt Maude pursed her lips. ‘Sometimes you can have too much of a good thing.’

# Chapter 16

On Christmas Eve I awoke to find the inside of my bedroom windows etched with frost flowers. I rubbed the glass with a fingertip to reveal that it had snowed during the night and caught my breath at the beauty of the scene. The sun shone in a blue sky and the skeletons of the leafless trees were crusted as if with glittering diamonds.

When Daisy brought me my morning chocolate she found me curled up on the window seat, watching the birds hopping about in the snow.

'Isn't it lovely?' I said.

'You'll catch your death by the window, miss.'

'I'll have my chocolate in bed,' I said, retreating to the warmth of the four-poster.

Later, dressed in my warmest clothes, I let myself out of the garden door. The terrace was slippery with snow and I trod carefully down the steps to the parterre. Marble statues were placed at intervals around the garden and gravel paths followed the curving patterns of the flowerbeds, all edged with clipped box. In the centre was a pool with a fountain, where icicles clung to the rim of the bowl, as if the water had suddenly been turned to ice by magic. There was something very beautiful about an entirely white garden.

I walked down to the moat and stood on the bank, careful not to slip on the snowy grass. Bulrushes, sparkling with frost, crowded against the bank and a pair of ducks paddled by. I looked back at Langdon Hall, blinking in the bright light reflected off the snow. There were stables to one side of the house and a walled garden to the other with the whole surrounded by the moat. My heart swelled with pride because this lovely place belonged to my father and I was able to claim borrowed ownership of it, even if only for a while. If I'd had a letter from Alessandro the day would have been perfect.

I heard a shout and saw Father hurrying through the parterre.

'You're up early,' he said. His cheeks were ruddy from the cold.

'I couldn't resist coming out to look at the snow,' I said. 'It's a shame our footsteps have spoiled its perfection.'

'There'll be fresh snow tomorrow.' He tucked my arm through the crook of his. 'The servants are bringing in the Yule log and the Christmas greenery this morning. I wondered if you'd like to direct them with decorating the hall?'

'I'd be glad to,' I said.

'Then shall we have an early breakfast before you set to work?'

The Yule log, all wrapped in hazel twigs, was dragged indoors by the gardener and his boys and deposited in the huge stone hearth in the hall. Mrs Bannister clapped her hands for the housemaids to clear up the pieces of brushwood and clumps of mud and moss that lay scattered in its wake. Another maid was sent to fetch the small piece of charred wood retained from the previous year's log and the household watched while the ancient tradition was followed of using it to light the new one. Once the fire was crackling, the servants returned to their duties.

'That should burn for a few days,' said Mrs Bannister. 'Sir Frederick said you're happy to oversee the decorating, Miss Emilia?'

'I look forward to it,' I said.

She bustled away to organise someone else.

The gardener's boys had deposited a large heap of holly and other greenery on the floor and stood by with knives and twine awaiting my instructions. There was much good-natured chatter as I had them running up and down ladders to hang the garlands of greenery.

Two hours later Aunt Maude came to find me. 'It's looking very festive, dear,' she said. 'And the Yule log is blazing well now.'

I wound another piece of ivy around one of the iron light fittings that had been lowered onto the vast refectory table and turned to study the fruits of my labour. The stags' heads had all been given ivy wreaths and a crown of holly. A mistletoe kissing ball decorated with red ribbons hung at each end of the hall and wreaths made from bay, rosemary and holly hung at intervals along the panelling, joined by thick swags of ivy. Scented pomanders were piled in silver bowls and their fragrant perfume mingled with the resinous scent of the Yuletide greenery.

'It's a magnificent room, isn't it?' I said. 'I can imagine the minstrels of a couple of hundred years ago singing in the gallery while the lord of the manor feasted below, throwing chicken bones over his shoulder to the dogs.'

'I'm not sure what Mrs Bannister would have had to say about that,' said Aunt Maude.

I called to one of the gardener's boys and set him to winching the branched light fittings, as large as a waggon wheel, back up to the ceiling before collecting the remaining scraps of greenery into a heap. 'I'll tell Mrs Bannister I've finished.'

'Send a message by one of the servants,' said Aunt Maude. 'Then you must go and change before Adolphus Pemberton arrives. I shall wait for you in the library.'

I added a final sprig of holly to one of the wreaths and then went to find the housekeeper. The buttery was behind the carved wooden screen at the end of the hall and a passage led to the domestic quarters. I glanced through an open door to see a footman polishing an array of silver dishes. A clatter of pans came from the kitchen, together with the sound of frantic chopping and the aroma of boiling ham. I called out to a passing scullery maid.

'Where will I find the housekeeper?'

'Mrs Bannister's room is at the end of the passage, past the still-room,' she said. 'Shall I fetch her for you?'

'That's all right,' I said, 'I'm sure I'll find her.'

The girl gave me an uncertain look and scurried away, the bucket banging against her skinny legs.

I tapped on the housekeeper's door.

She sat at a table and rose to her feet when she saw me. 'Miss Emilia! Is everything all right?'

'I've finished decorating the hall,' I said, 'but a maid will need to sweep up the trimmings.'

Her mouth was set in a disapproving line. 'Will that be all?'

Unsure what I'd done to make her cross, I nodded.

'Robert will escort you back to the hall.'

'I'm sure I shall find the way.'

'Nevertheless,' she said, 'please come with me.'

The footman was called away from the silver cleaning and hastily removed his protective cambric cuffs and apron before leading me back to the hall.

'Thank you, Robert,' I said. 'Will you ask Daisy to attend me?'

He bowed his head and returned to the domestic quarters.

Aunt Maude was walking upstairs as I hurried up to change.

'I found the housekeeper in her parlour and she'll send a maid to sweep up,' I said.

Aunt Maude came to a sudden stop. 'You went to Mrs Bannister's parlour?' Her tone of voice was scandalised.

'Why shouldn't I?' I asked.

'My dear, you have a great deal to learn. *Always* send for the housekeeper and *never* wander about below stairs. The servants will think you're spying on them.'

'I didn't mean any harm.'

'I'm sure you didn't,' said Aunt Maude, 'but you have a position to maintain in your father's house.' She sighed. 'There are so many things I must teach you, to make you fit for society.'

'What kind of things?'

'Not to shrug and wave your hands about when you are speaking, for example.'

'I don't!' I frowned. 'Do I?'

Aunt Maude inclined her head. 'It's an unbecomingly continental habit and will not be well received in an English drawing room. You must never appear excessively animated, Emilia, as you are wont to do.' She patted my cheek. 'Don't look so forlorn, my dear. Your intentions are good and you have natural grace. We'll soon turn you into an English lady.'

We stopped outside her bedroom door. 'I'll see you downstairs in a little while,' I said.

As I walked along the gallery, I heard a horse trot into the quadrangle below. I peered out of the window to see a man in a caped travelling coat dismounting from a grey. I wondered if this was my father's heir. Curious, I watched as a groom came to take the horse but then the man disappeared into the house.

Washed and changed into my blue dress, with my hair brushed and beribboned, I went downstairs to the library. My father was seated by the fire reading a newspaper and Aunt Maude was sorting her embroidery silks.

'You look delightful, my dear,' said Father, standing up as I came in. 'That dress brings out the blue of your eyes.'

'Emilia is so like her mother, isn't she?' said Aunt Maude. 'Have you shown her Rose's portrait, Frederick?'

'There's a picture of my mother?' My heart began to race. 'And you didn't show it to me yesterday?'

Father sighed. 'I thought it might distress you.'

'Why should it?' I asked. 'May I see it?'

'Another time since Adolphus is expected imminently.'

'I saw a man arriving a little while ago. What is he like, this heir of yours?' I asked.

Father steepled his fingers while he thought. 'A well-educated gentleman of fashion,' he said, 'though not interested in art.'

'He's a dandy,' said Aunt Maude, succinctly.

'I can see, then,' I said, 'that he falls far short of being your ideal heir, Father.'

He laughed. 'You're beginning to understand me very well, Emilia.'

Before I could reply the door opened and Robert announced, 'Mr Adolphus Pemberton.'

My father's heir paused in the doorway with a slight smile on his handsome face. His elaborately tied cravat gave him no option but to look down his nose at us. He was taller than average, his long legs encased in buff trousers and his cut-away coat impeccably tailored. His skin was very light against his dark hair, which was carefully arranged in Grecian-style curls.

'Adolphus,' said Father, standing up. 'I trust you had a good journey?'

'Tolerable.' The vision of elegance advanced into the room. 'Aunt Maude,' he said, bowing low before her. 'Always a pleasure.'

'Emilia,' said Father, 'may I present your second cousin, Adolphus Pemberton.'

Adolphus bowed and I inclined my head.

'So,' he drawled, 'it's to be Emilia and not Harriet, is it? And, please, all my friends call me Dolly. Adolphus is terribly formal, don't you think?'

I glanced at Father, who gave me the briefest of nods. 'I'm delighted to meet another member of my family,' I murmured, slightly discomfited by Dolly's gimlet gaze.

'Indeed. Your return to the family fold is the talk of the *ton*, dear coz.' He cleared his throat and sat down on the sofa beside me.

'It was a most' – I paused momentarily – 'unexpected turn of events and I'm only now becoming used to the idea of belonging to a family.'

'Your father called me to meet him at his club,' said Dolly, 'and explained that you'd returned from the dead and how you'd been living.' He shook his head sorrowfully. 'Too dreadful to contemplate.'

'I didn't remember any other life,' I said, 'so I was quite happy.'

'But, in retrospect,' said Dolly, 'you must be very angry with the maid who stole you?'

A momentary picture of Sarah's anxious face formed in my mind and I felt a pang of loss. 'It's exhausting to be angry all the time about something you can't change,' I said. 'Now I must look to the future.'

'And what *does* the future hold for you?' His dark blue eyes were fixed upon me again.

'As to that,' I said, 'it's too early to say.' I knew what he was really asking was if I was going to run off with some of his expected inheritance. 'I'm happy to accept my father's invitation to return to the family home for a while. After that,' I gave him a bright smile, 'who knows?'

'Who knows, indeed?' he murmured.

On Christmas morning it snowed again and, in deference to Aunt Maude's age, we decided not to risk slipping on the ice and took the carriage to St Bartholomew's in the village.

I was uncomfortably aware of the curious gaze of several members of the congregation and of the whispering going on behind gloved hands as we sat in the Langdon pew. I supposed if I had been in their shoes I'd have been curious, too, about the girl who'd returned from the dead. I stared straight ahead, my cheeks burning.

Father squeezed my hand. 'They'll soon lose interest,' he murmured. 'You must admit, it's an extraordinary story.'

Dolly gave me a sideways glance of amusement as he sang the hymns in a light tenor. 'Hold your head up high, coz,' he whispered.

The sermon was interminable and my thoughts drifted to Alessandro, wishing he were beside me. An impatient rustling mounted amongst the worshippers who wanted their Christmas dinner. Prayers were said for the King's recovery from the chill to his lungs.

At last it was over and we milled about with the rest of the congregation to wish the vicar a Merry Christmas. Several churchgoers jostled us, to pass on their festive greetings to Father and to take a good look at me, and I was relieved when we climbed into the carriage and returned to Langdon Hall.

Father's guests, mostly local landowners, a few couples from London, the vicar and several worthy spinsters, began to arrive at five o'clock and were shown into the library. The story of my return had spread like wildfire through the county, and by the time Dolly led me into dinner, I felt like an exhibit in a circus freak show.

In the hall the Yule log crackled and spat in the hearth as the thirty guests found their places. Once we were assembled the vicar said grace and there was a scraping of chairs until everyone was seated.

I looked at the expectant faces as the guests chattered to their neighbours. I sat on Father's right at the head of the table. 'This room was made for feasting,' I said. 'It's wonderful to know that Christmas has been celebrated here for generations and the tradition will continue far into the future.'

'I do hope so,' said Father. 'Although, unfortunately, not with a Langdon at the head of the table.' He turned to Dolly. 'Emilia has made an elegant job of decorating the hall, don't you think?'

'Admirable,' he replied. 'Your daughter has good taste.'

I glanced up at the ivy wreaths wound around the light fittings, the candle flames flickering in the warm rising air. The hall was perfumed with the scent of evergreens, cinnamon, oranges and mulled wine, and I was pleased with my efforts.

Bannister the butler poured the wine and the footmen brought in course after course of dishes. Father explained this was the latest fashion, introduced by the Prince Regent, of serving dinner *à la russe*, instead of having all the dishes placed on the table in two or three removes.

A trio of musicians played in the gallery but the music could barely be heard over the hubbub of conversation.

'Your father keeps a fine table,' said Dolly, as we ate white soup, followed by stuffed pike with oysters and fried sole. 'I believe even the Prince Regent, well known for his extravagant tastes, would find no fault if he were here.'

'Do you know the Prince Regent?' I asked.

'A passing acquaintance,' said Dolly. 'Your father knows him better than I do.' He smiled. 'Prinny admired my cravat.'

'What is he like?' I asked, curious about the Princess's husband.

Dolly cleared his throat and leaned closer. 'Between the two of us, he's rather fat. Although known for his modish style and wit, his extravagant lifestyle has done his figure no favours. When he bends, he creaks.'

'Creaks?'

Dolly's eyes gleamed with malice. 'Whalebone corset,' he said, succinctly.

I stared at my plate and bit the inside of my cheeks in an attempt to stop myself from laughing.

'Did you ever attend any dinners as elegant as this on your travels in Italy?' said Dolly, just as if he hadn't noticed.

'I stayed in some very grand houses.'

'But how unfortunate for you that you were there in your capacity of dressmaker rather than as a guest.'

I sipped my wine and smiled blandly at him. Dolly was amusing and outwardly friendly, but I would take care never to turn my back on him, I decided.

'Your father told me you were living with the Princess of Wales,' he continued.

'I'm surprised he mentioned that,' I said, 'since he forbade me to talk of it.'

Dolly smiled guilelessly. 'But this is all in the family, isn't it? Tell me, how did you find her? Is she as frightful as the Prince Regent says?'

'Not at all!' I said, indignant that he should think so. 'She has her foibles, of course ... '

'You mean she never washes, flirts outrageously, and lives as man and wife with a so-called Italian noble fifteen years younger than herself, who is no more than her servant?'

'If you'd been to Pesaro,' I said, 'you'd know it isn't like that.'

'Ah, but I haven't been to Pesaro,' said Dolly. 'Can't abide foreign travel.' He shuddered. 'Bugs in the bed, garlic in the food, and all that sort of nonsense.'

I paused while I decided how to describe her. 'The Princess of Wales dresses inappropriately sometimes but there's no meanness or pretence about her. Perhaps she grows restless too easily, like a child. I was surprised by how much she dislikes formality, to the extent that she confided in me her distress when she discovered, quite by accident, that her daughter had died. Can you believe the Prince Regent didn't trouble himself to send her the news?'

'I believe he vowed some years ago never again to communicate with his wife.'

'In my opinion, he's behaved badly towards her.'

'In Sir Frederick's household,' said Dolly, 'I'd advise you to keep that opinion to yourself. Your father is vociferous in his condemnation of the Princess, if only to demonstrate his support of the Prince of Wales.'

'Residing in the Princess of Wales's household for several months has brought me to an entirely different, and possibly more informed, judgement of her character,' I said.

'Your father, however, has no patience with any view that doesn't support his own,' said Dolly. 'I heard that the Princess has taken her entourage to Marseilles to overwinter incognito in a small hotel.'

'Marseilles?' This was news to me. 'I expected her to arrive in London soon.'

'The talk in the coffee houses is that she's merely waiting for better weather to sail back to Pesaro.'

I sipped my wine while I reflected on this. So Alessandro was probably in Marseilles. I wondered if Lady Hamilton had travelled to Marseilles, too, and if he had received my letter yet.

The footmen carried in a succession of dishes: roasted goose, venison, a baron of beef, innumerable capons and pheasants decoratively dressed with their tail feathers, and several beautifully ornate Christmas pies.

'I can't eat another thing!' declared Aunt Maude some while later.

The guests raised a cheer as two footmen circled the table, holding aloft a flaming Christmas pudding, while the musicians played a lively march. I recalled Sarah had made an English Christmas pudding one year but I'd found it too rich and preferred *panettone.*

After we'd finished the pudding, bowls of grapes, slices of fresh pineapple, candied apricots and marchpane sweets were placed at intervals along the table. Glasses of dessert wine were poured and then Father rose to his feet and tapped his glass with a spoon.

'Ladies and gentlemen,' he said, 'I wish you all a Merry Christmas and I'm delighted you could be here to share our festive feast today. As you know, this is a very special Christmas since my beloved daughter has been returned to me.' Father looked straight at me. 'Please raise your glasses with me in a toast to Emilia.'

'Hear, hear!' said Dolly.

I blushed as the entire company toasted me, chorusing my name. I pictured Alessandro's face for a moment and heartache and longing washed over me.

# Chapter 17

After breakfast on Boxing Day, a dozen ancient widows and several of the deserving poor of Upper Langdon village shuffled into the hall to collect their annual parcels of surplus food and clothing. Once they had given their grateful thanks, Father and I distributed the Christmas boxes to the servants, who were then granted the rest of the day as a holiday.

Father, Dolly and two other male house guests went off to join the local hunt, leaving Aunt Maude and myself to entertain the wives. We took a turn around the garden and exclaimed how invigorating the fresh air was after the rich food of the previous day, before returning to the comfort of the library fire.

'I remember your mother,' said Mrs Digby, the wife of Father's solicitor. 'You are very like her with the same pale skin, fine features and red hair.'

'I can't picture her at all,' I said. 'I'd hoped coming to Langdon Hall might bring back the memories.'

'And it didn't?' Her hazel eyes were sympathetic.

I shook my head. I only remembered Sarah, who had defended

me from her husband's cruelty and been my companion for most of my life.

'Grief over the loss of a baby sometimes makes a woman lose her reason,' said Mrs Digby. 'I was greatly shocked by the news of your mother's death, though.' She shook her head. 'Rose Langdon always had her feet so firmly on the ground.'

Later, the men returned, full of loud bonhomie and boastful stories of how they'd outwitted the fox. Hunting had sharpened their appetites and they fell on the cold collation laid out for us in the Great Hall.

After dinner, the house guests sent for their carriages and Dolly came to say goodbye, too.

'I understand from your father,' he said, 'that Aunt Maude will be grooming you for your presentation at Court.'

'Mmm,' I said. 'Father believes it's necessary.'

Dolly raised his finely arched eyebrows. 'Why so gloomy? I thought all girls liked to have new clothes and go to balls?'

'He's hoping to find me a suitable husband.'

'Don't you want a husband?'

'Eventually,' I said, 'but I'm not sure about balls and routs.'

He smiled at me down his long nose. 'I shall be on hand to escort you.' He brushed an imaginary speck of dust off his beautifully tailored sleeve. 'I anticipate your return to Grosvenor Street with great pleasure,' he said, bowing.

Once the guests had left, I wasted no time in cornering Father. 'Will you show me my mother's portrait now?' I asked.

He sighed. 'It's upstairs.'

Lighting a candle, he led me up a back staircase to the attics and opened a small door. 'Wait here,' he said.

I peered over his shoulder into a windowless storage room tucked under the eaves. By the wavering light of his candle I caught sight of various discarded pieces of furniture, a broken lamp and assorted trunks. 'Is that a cradle just behind you?' I asked.

'It was yours and then your brother's,' he said. 'I put it away after Piers died, so it wouldn't further distress your mother.' He lifted his candle and pulled back the corner of a dust sheet. 'Here,' he said. Tucking the picture under his arm, he came out of the store, brushing dust off his hands. He carried the portrait to the landing window.

My heart thudded and my mouth was dry as I gazed upon my mother's likeness. Her blue eyes seemed to be looking at me and I couldn't look away. There was a hint of a smile around her full mouth and I touched her painted cheek as if I could bring her to life again.

Father watched me without smiling.

After a moment I sighed. 'Mrs Digby was right. I see a likeness, except that Mother's hair is redder than mine.' There was a hollow ache in my breast, as if I'd been crying for so long that, even when the tears stopped, intense sadness remained. 'I do so wish I could remember her,' I said.

Father looked at the portrait, his face expressionless. 'Some things are best forgotten,' he said.

'May I have the portrait in my room?' Mother's painted face was so lifelike it made me shiver with longing to feel the warmth of her flesh. 'She has such a gentle expression, doesn't she?' I whispered.

Father passed one hand across his face and shielded his eyes for a moment . 'Emilia,' he said, 'it's time I told you what really happened.'

Something in his tone broke the spell cast by the portrait and I stared at him. 'What do you mean?'

'Not here,' he said. 'Come to my study.' He set off down the stairs and I followed him, carrying the portrait carefully in my arms.

In the study he closed the door behind us and lit the candles. Then he took the portrait from me and turned it to face the wall. 'I don't want to look at it,' he said. He sat at his desk, pulled a bottle of brandy out of the drawer and slopped a generous measure into two glasses. He drank deeply from one and then pushed the other towards me.

Tentatively, I sipped it.

'Over the ten or fifteen years before Piers died, there had been a series of notable thefts,' Father began, staring into the candle flame. 'Priceless paintings had been stolen from the homes of the rich and famous and it was assumed the thief was either a member of the aristocracy or had a well-connected accomplice who fed him information.'

'Were those the art thefts you mentioned before?'

He nodded.

'What have they to do with my brother's death?' I asked.

'His birth was difficult,' said Father, 'and Rose was slow to recover. Piers cried a great deal, colic the nurse said, and your mother couldn't love the baby as she should. She spent a great deal of time lying in a darkened room, weeping. Aunt Maude will tell you that. And then one day, Rose found Piers dead in his cradle.' Father drained his brandy glass. 'It may have been a sad but natural event. Or perhaps she saw an opportunity to release herself from her guilt for not loving her son enough.'

I gasped. 'Surely not?'

He shrugged. 'In the light of what happened later I've wondered about that, but we'll never know. Anyway, Rose grieved terribly but then, over the following year, she began to improve. She started to meet her friends again. Sometimes I even heard her laugh. But she became cold towards me.' He sipped his brandy. 'So cold, in fact, that I suspected she was having an affair.'

Shock and distaste rippled through me. That wasn't at all what I'd expected, or wanted, to hear.

'And then one night in Grosvenor Street, I was woken by a noise in the hall. I crept out of my dressing room and went downstairs. The front door was ajar and there was a light in the drawing room. A man was lifting one of my paintings down from the wall.'

'What a shock!'

'It all happened so quickly. He dropped the painting and came at me with a cudgel. I thought my head had split in two when he hit me. I saw him run out of the front door and then I passed out. When I regained consciousness your mother was leaning over me. She'd come out of her room and seen the thief attack me. He had the painting tucked under his arm. She said he barely broke stride, struck me then ran outside.'

'Was he ever caught?'

'No,' said Father. 'The thing was,' he poured himself another brandy, 'when the thief attacked me, the painting was still on the drawing-room carpet, not under the thief's arm.'

'I don't understand,' I said.

'Neither did I, at first,' he said. 'It was the next day that I realised something wasn't right about your mother's story. I went to her room and questioned her. She laughed and said I was imagining things but she kept glancing towards her wardrobe. I threw open the door and rummaged through the contents but there was nothing there. When I climbed onto her dressing-table chair to reach the top of the wardrobe, she stopped laughing.'

My stomach turned over as I realised what he was going to say. 'The picture?' I said.

'Wrapped in one of her shawls.' Father looked at me, his face taut. 'She was defiant. She said she despised me and the thief was her lover. She'd intended to take the painting and run away to Paris with him. I didn't know what to do. The scandal would have been terrible. So I locked her in her room.'

Suddenly I felt like crying. My mother looked so lovely in her portrait and it hurt to know that her sweet smile was merely assumed. I remembered Sarah had told me Mother intended to go to Paris to stay with friends. 'Sarah never mentioned the theft,' I said.

'Of course she didn't!' said Father. 'It was a small but valuable painting and easy to transport. I hung it back on the wall so no one ever realised it had been removed. I begged Rose to tell me who the thief, her lover, was but she set her mouth in a mulish line and refused. It angered me and I told her she'd stay locked up until she did tell me.'

'Was it was someone you knew?'

Father rested his head in his hands for a moment. 'I've tormented myself with that thought for years. But he never tried to contact Rose and she never revealed his identity. She grew pale and thin but remained obstinate.'

'Were there more thefts after that?'

'Not for a couple of years but then they began again so I assumed he'd returned to London and perhaps found another accomplice.' Father looked directly at me. 'I still lie awake at night sometimes, wondering if it's someone I know, one of my friends. It's unnerving to imagine he's been laughing at me all this time.'

'And he used Mother and then abandoned her.'

'If you'd heard the things she said to me,' said Father, his voice bleak, 'things a daughter should never hear, you'd know she was no innocent party.'

I swallowed, sick with the thought of what he had suffered. Mother had used Sarah, too, and it had ruined her life.

'And then,' said Father, 'I came home early from my club one night and I heard Rose's bedroom door close as I came up the stairs. I checked my pocket but I still had my key. I assumed the housekeeper had been careless but, when I tried the handle, the door was still locked. When I unlocked it with my key, I found Rose in her travelling clothes. I realised then that she'd been attempting to escape when she heard me come in and had returned to her room.'

'Sarah made a copy of the housekeeper's key,' I said.

'I was so angry,' Father told me. 'I shouted at Rose and she screamed back at me and then began to laugh so hard she couldn't stop. I had to slap her.' He sat with his head bowed and his fists clenching and unclenching on the desktop.

'What happened then?' I asked.

He heaved a sigh. 'She mocked me and said I couldn't keep her locked up forever. Her lover was waiting for her at Dover. I asked her how she thought she'd survive without money. She taunted me, saying her lover had given her Lord Beaufort's miniatures for safe keeping and they were in her travelling bags, along with her jewels.' He looked up at me. 'Sarah had already gone on ahead with the luggage, you see. Rose and her lover intended to sell the miniatures and live off the proceeds.'

'Miniatures?'

Father opened his desk drawer and took out a wooden box. He lifted out a number of newspaper clippings, yellowed with age. 'These relate to some of the paintings stolen over the past thirty years, mostly in England but also in France and Italy. The press called the culprit the Picture Frame Thief because he usually left a sketch of an empty picture frame in place of the item he stole.'

'How cruel to mock the owner of the painting so!' I said.

'Indeed.' Father picked out one of the clippings. 'This set of three miniatures had been stolen from Lord Beaufort, an acquaintance of mine. We'd been his house guests a couple of weeks before they were stolen and your mother must have fed information back to the thief. As you see, they were extremely valuable.'

I took the clipping from him and read that the three miniatures were of the Infanta Isabella Clara Eugenia, the daughter of King Philip II of Spain and his wife, Elisabeth of Valois. They were painted in 1598 by Sofonisba Anguissola, one of the first female court painters, on the occasion of the Infanta's betrothal to her cousin, Archduke Albert of Austria VII. Shocked, I placed the clipping back on Father's desk.

'Emilia,' he said, leaning forward to look closely into my eyes, 'did you ever see those miniatures?'

'Never!' I said.

He sighed and leaned back. 'They were in the baggage Sarah took away. I hoped you might know what became of them. It's been a terrible burden, knowing that Rose was involved in their theft. I want to return them to Beaufort.'

'Is that why you quizzed me so hard about where Sarah and I had travelled?'

His smile was fleeting. 'You noticed that? I've written to all the art dealers in the towns you mentioned, asking if they might have bought such an item.'

'I wish you'd confided in me before, Father. I'll make a list of the towns we travelled through as far back as I can remember.' I thought of something and pressed my fingers to my lips. 'Those gold coins I found in Sarah's petticoat, do you think they might be the proceeds of such a sale?'

'Almost certainly,' he said. 'I don't believe dressmakers' earnings are high enough to save very much.'

'Indeed not,' I said. If Sarah had sold the miniatures, perhaps we'd been living off the proceeds for years? 'What shall I do with the remaining coins?' I said, in a sudden panic. 'I've already used some of them for my crossing to England.'

'They ought to go to Beaufort,' said Father unhappily. 'But I don't care to explain to him how we came by them. Shall I put them in the bank and use them towards buying back the miniatures, if I can find them?'

I nodded and watched him as he went to stir the embers of the fire. He was still a handsome man. 'Did you never think of marrying again?' I asked.

'I had no heart for another marriage,' he said. 'Anyway, it wasn't possible.'

'Why not?'

He replaced the poker on the hearth. 'I stayed at Rose's side all that night after she tried to escape. By morning she was in despair because she knew her lover would already have sailed to France. She threatened to throw herself in the river off Westminster Bridge. I had an important meeting that morning but I'll never forgive myself for leaving her unattended while she was in such a state.' He paced up and down, running his hands through his hair.

'What happened?' I asked, my stomach churning.

'She escaped and then it was discovered you were missing, too. The servants and I searched for you. After her earlier threat I went to Westminster Bridge.' He covered his eyes with his hand.

'Tell me!' I said, feeling sick.

'We discovered Rose's shoes and shawl but her body was never found.' He swallowed convulsively. 'And that's why I was unable to marry again and lost any chance of a son of mine inheriting Langdon Hall.' He took my hands. 'I thought that you must have drowned with her and everyone I'd loved had left me.' His shoulders began to shake and he clung to me. 'The only good thing to happen out of the whole dreadful business,' he sobbed, 'is that you have come back.'

# *Chapter 18*

## *January 1820*

**London**

Aunt Maude was a hard taskmaster. I walked the length of the Grosvenor Street drawing room with Samuel Johnson's *Lives of the Most Eminent English Poets* balanced on my head. As I made my turn at the end of the room, the book slipped and struck my shoulder before bouncing, painfully, off my foot.

Aunt Maude sighed. 'Really, Emilia, when I think how graceful your mother was . . . '

'I don't want to talk about her,' I said, a tight knot of unhappiness under my breastbone. I had kept her portrait in my room at Langdon Hall but with its face turned to the wall, unable to look into her adulterous eyes.

'That's as may be,' said Aunt Maude, 'but if you will clump about as if you were a stable lad carrying a bale of straw . . . '

'I'm not!' I said, deeply insulted.

'Don't argue with me – it's unbecoming. Glide! You must glide forward like a swan.' She took a few steps to demonstrate. 'Now put the book back and try again. All the other girls will have learned in the schoolroom how to walk gracefully and you don't want to be the only ungainly one.'

Grimacing, I picked up the book.

We'd returned to Grosvenor Street a fortnight before and Aunt Maude had drilled me for several humiliating hours every day on how to walk, to greet new acquaintances, and on which subjects were suitable for conversation. A dancing master had been engaged twice a week to teach me English country dances, Scottish reels, the quadrille and, daringly, the waltz. To my surprise, I discovered I enjoyed dancing.

'That's it!' encouraged Aunt Maude. 'Head up, shoulders down, back straight. Smaller steps, please, Emilia, you are not an elephant.'

As I walked towards the window, I decided Aunt Maude would have been excellent at drilling army recruits. I glanced down at the street and saw a man hurrying away from the house. There was something uncomfortably familiar about his gait that made me pause and stare. I shivered but then the drawing-room door opened. I turned, allowing the book to slide off my head again, and saw Dolly and a young man standing in the doorway.

'Hard at work, Emilia?' said Dolly. 'Aunt Maude, allow me to present Mr Francis Gregory. Francis, Miss Weston.'

Mr Gregory bowed to Aunt Maude as low as his starched neck cloth and tightly fitting coat allowed.

'Emilia, this is Mr Gregory,' said Dolly. 'Francis, Miss Langdon.'

Mr Gregory bowed and brushed back one of the artlessly arranged blond curls that had fallen over his forehead. 'How delightful to meet you!' he said, his pale blue eyes looking me up and down. 'Dolly has told me all about you.' His teeth were rather small in his plump, pink-cheeked face and I was immediately put in mind of the painted cherubs I'd seen in Italian churches.

'I trust we do not interrupt?' said Dolly.

'Not at all,' I said. 'I'm glad to have an excuse to stop walking around with a book on my head.'

Dolly nodded, his mouth twitching. 'I should rather think you might be.' He and his companion sat down, deftly flicking their coat-tails aside in a practised motion.

Aunt Maude tutted under her breath. 'Emilia, please refrain! You must remember to talk only of subjects that will interest our guests.' She turned to Dolly. 'Will you drink tea with us?'

'Alas, I have an appointment with my tailor,' he said, 'but since we were passing I thought we'd drop in for a few moments.'

'I understand your father is holding a ball for you, Miss Langdon?' said Mr Gregory.

'In April,' I said.

'We shall send you an invitation,' said Aunt Maude. 'And I shall be asking your advice, Dolly, for the names of suitable young people to invite.'

The drawing-room door burst open then and Father hurried into the room. 'Have you heard the news?' His eyes gleamed with excitement. 'The King is dead!'

Aunt Maude gasped and pressed her hand to her mouth.

'Well, well,' said Dolly. 'Demented, blind and confined, it can only have been a happy release for Farmer George.'

'At long last,' said Father. He went to the side table, poured out glasses of sherry and handed them to us. 'A toast to King George IV!'

'Long live the King!' said Francis Gregory, rising to his feet.

'The King!' said Dolly.

I put the glass to my lips but did not drink. 'When did it happen?' I asked.

'Last night, at Windsor Castle,' said Father. He rubbed his hands together. 'At last Prinny will be free of the petty restrictions placed upon him by his father.'

It was then I realised that since the Prince Regent was now

the King, the beleaguered Princess of Wales was our Queen. Had anyone, I wondered, taken the trouble to inform her?

That night I awoke suddenly in the grip of a night terror. I sat up and clutched the sheet to my chest while my pulse thudded in my ears. I'd had the same old dream of hiding between the wall and the bed while a man's voice ranted and raved in the next room. I always woke up sweating and sick when I remembered the lash of a belt and a woman's terrified screams.

Of course, it couldn't have been him, but I knew what had sparked my nightmare. Earlier that day I'd seen a man walk away from the house. A man who looked just like Sarah's husband, Joe Barton.

Two weeks later Aunt Maude, Father and I were in the carriage returning to Grosvenor Street after attending church.

'Whatever the Queen may or may not have done,' said Aunt Maude, 'for the King to have taken such a step as to remove her name from the Liturgy is reprehensible. There is no proof of her adultery and therefore no reason to omit her from the customary prayers for the Royal Family.'

'Perhaps it wasn't a wise move on the King's part,' agreed Father, shrugging. 'It's certainly aroused the ire of the general public, thereby increasing her popularity, which, I should imagine, was not at all what he intended.'

'If he wishes to retain his own good standing,' I said, 'surely he must then discontinue his vindictive behaviour towards his wife?'

Father gave me a cold look. 'Emilia, your venturing such an opinion publicly may do us a great deal of harm. Aunt Maude, you clearly need to make greater efforts to teach Emilia what is and is not acceptable in polite society.'

I refused to be cowed. 'I shan't make such observations except in private, Father, but it's obvious to me that the opinion of many is that she's been sorely mistreated. When she arrives in London . . . '

'God forbid!' said Father. 'If she comes to England now, the King will close up the Court, if only to prevent her from presiding over the royal Drawing-Rooms and disgracing the English throne.'

Despite his sharp comment, having been subjected to Aunt Maude's rigorous instruction on exactly what behaviour was suitable for a lady who wished to be accepted by society, I understood that Queen Caroline's conduct might not be considered merely eccentric, but completely unacceptable. I trusted that Lady Hamilton would have an edifying effect upon the Queen, restraining her from her worst excesses and saving her from public censure.

'If there are no Drawing-Rooms,' said Aunt Maude with a frown, 'then the young ladies will not have the opportunity of being presented.'

'I believe that may be the case.'

Aunt Maude sighed. 'Perhaps Emilia should come out at her ball? She may then be presented when the difficulties are resolved. Or even,' she glanced at me, 'upon her marriage.'

I frowned at her.

Father nodded. 'It's possible that the Duchess of Gloucester may hold a Drawing-Room after the Royal Family are out of the second change of mourning in April. I agree however that, in the circumstances, a later presentation is the best course of action. In any case, I can't imagine that Caroline of Brunswick will be Queen for much longer.'

'What do you mean?' I asked.

'The King is determined not to allow her at his Coronation,' said Father. 'She will remain uncrowned. Furthermore, he won't continue to pay her allowance should she come to London.' There was a spiteful gleam in his eyes as he said, 'I imagine that will encourage her to stay overseas.'

I looked thoughtfully out of the carriage window. The Queen would be fretting and, I guessed, no matter what the outcome, she'd still come to London to face her antagonists. The one positive thing she should know was that the ordinary people were on her side while the King's continuing excesses antagonised them. I resolved to write to her again. And, perhaps, I would brave Lady Hamilton's annoyance by including another note to Alessandro.

Mrs Webbe, a sought-after mantua-maker from Pall Mall, had been engaged to provide my new wardrobe for the coming season. It was strange to see her assistant kneeling at my feet as she pinned up the hem of my new spotted muslin dress. I'd lost count of the times I'd crouched at a client's feet performing the same action and, no doubt, I would be doing it again once I'd returned to Italy.

'Miss Langdon has pretty ankles and it would be a shame to hide them when she's ascending the stairs or climbing into a gig,' said Mrs Webbe.

'Not too short!' said Aunt Maude. 'It won't do to attract the wrong kind of attention.'

The assistant paused, unsure as to which instruction she should follow.

'Perhaps just half an inch?' I murmured.

Mrs Webbe nodded in assent and the assistant bent to her task again.

Sometimes I thought that the past few months were only a dream. Finding my father and Aunt Maude and being accepted into their lives was a wonderful thing and there was no doubt that it was very seductive, taking advantage of the many benefits that money brings. Father had even made me an allowance for such fripperies as ribbons and shoe rosettes and it would take some adjustment for me to return to my previous frugal way of life. On the other hand, I wouldn't have to obey the excessively irksome restrictions imposed upon me by propriety and take a chaperone with me everywhere. If all went well, I would have a substantial dowry by the time I was reunited with Alessandro. We might be able to buy a little cottage by the sea . . .

The assistant dressmaker stood up. 'Shall you try on the ballgown next, Miss Langdon?'

We retreated behind the screen in the corner of the drawing room and the dressmaker pinned me into the dress of white satin overlaid with cream gauze. The puffed sleeves were trimmed with lace and I fingered the roses embroidered around the daringly low-cut neckline. I smiled. It was the most extravagant dress I'd ever worn and it made me look entirely different. There's a certain gloss that someone born into money carries with them and, in that dress, I looked like one of them. Remembering Aunt Maude's instructions, I held my head high as I came out from behind the screen and glided across the room.

She clasped her hands and smiled. 'Beautiful, my dear.'

Mrs Webbe tweaked the neckline and smoothed the bodice. 'You need a new pair of stays, Miss Langdon,' she murmured, 'to lift your *décolletage*.'

'No ornaments or feathers in her hair,' instructed Aunt Maude.

'I agree,' said Mrs Webbe. 'Fresh flowers, white of course, are the only acceptable choice for an unmarried girl.'

Feeling as if I were a prize pig being prepared for market, I stood still while the skirt was adjusted. I passed the time daydreaming about Alessandro, imagining him partnering me in the first dance at my ball.

Male voices sounded outside the door and then Father came in, accompanied by Dolly and Francis Gregory.

Father stopped short when he saw me. A muscle flickered in his jaw but then he smiled. 'So this is what I've been spending my money on?'

'Worth every guinea!' said Dolly.

Father took my hands and twirled me around. 'Fine clothes bring out your natural beauty, Emilia.'

'You're so slender,' Dolly observed, 'you could almost be a boy. Don't ever allow yourself to become matronly ... you're so very lovely just as you are.'

'Praise indeed from you,' I said, a faint flush warming my cheeks. Despite his fulsome compliment, it seemed to me that he studied me as if I were a work of art rather than a living, breathing being.

'I came to tell you I was talking to Lord Liverpool this morning,' said Father. 'I mentioned your miraculous return and he expressed an interest in meeting you. It appears he remembers being enchanted by your mother at a musical evening, long before he became Prime Minister. I said you'd invite him to your ball.'

'Aunt Maude and I collected the invitation cards from the printer this week,' I said. A great number of Father's friends were titled but it unnerved me to know that the Prime Minister wished to meet me, if only out of curiosity.

'Aunt Maude,' said Father, 'have you discussed the supper menu for the ball with Cook?'

'Yes, Frederick,' she said. 'And ices and ornamental confectioneries have been ordered from Gunter's. The florist will bring arrangements of white narcissi, early white azaleas and pink and blue hyacinths.'

Father nodded. 'Very good.' He turned to Dolly. 'May I have a word with you in my study?'

'Will it take long?' he asked, glancing at Francis.

'It's time I took my leave,' said his friend. 'I shall see you later, Dolly.'

Before long Father and he were closeted in the study. Mrs Webbe and her assistant soon followed Mr Gregory out of the front door and Aunt Maude and I were left in peace. She sat on the sofa and sighed deeply with her eyes closed.

'You look tired, Aunt Maude,' I said, suddenly guilt-smitten. 'You must let me do more to help with the arrangements.'

'As your father so often tells me, I need to earn my keep.'

'That's awful!' I was shocked that he might say such a thing.

'What is awful is being an elderly spinster with no home of her own,' said Aunt Maude. 'But your arrival here has given my life new meaning,' she said. 'I used to help your dear mother when she was organising a ball but I'm finding I'm too old for it now.' She sighed. 'I miss Rose. She was charming and I loved her quite as much as if she'd been my own flesh and blood.'

'Did you?' Perhaps she'd had two sides to her, the one Aunt Maude liked and then the deceitful side that had made Father suffer so. Still, if Aunt Maude hadn't known that aspect of my mother, I wouldn't spoil her happy memories.

'Your father gave me a home under sufferance,' said Aunt Maude, 'but Rose always made me feel like a valuable addition to the family.'

I felt sorry for her then and hugged her. 'Of course you are, Aunt Maude!' At least my mother had exhibited some human kindness. 'Why don't you go upstairs and rest,' I said, 'while I finish writing out the invitations?'

# Chapter 19

*April 1820*

**London**

I dreamed Alessandro was hurrying along an endless alleyway in the gathering dark while I scurried along behind, desperately trying to catch up with him. I called to him but either he didn't hear me or was pretending not to. I ran faster, shouting his name and reaching out for him, but he was always too far ahead.

Then he stopped suddenly and I ran into the back of him. He caught me by my elbows and held me at arm's length. The dream was so vivid I felt the heat of his hands through the thin muslin of my dress.

'Alessandro, I love you,' I cried. 'Why didn't you wait for me?' I took an involuntary step away when I saw no sign of the usual mischievous humour in his eyes.

'If you really loved me you wouldn't have left,' he said, and walked away without looking back.

I awoke with a hollow ache in my heart and my pillow damp with tears. I'd still received no response to the note I'd sent him, included with my letter to Queen Caroline. Sighing, I turned my thoughts to the ball.

The day had finally arrived. Aunt Maude had worked so hard instructing me on the right way to behave and I didn't want to make mistakes and let her, or Father, down. Dolly and his friend Mr Gregory had called on me regularly to see how I was progressing. Dolly was sometimes waspish but could be most amusing when he set his mind to it. We had spent many an afternoon playacting in the role of guests at the ball, engaging in the art of frivolously polite conversation. Angelic-looking Mr Gregory had made bold attempts at flirting with me while Dolly tried to keep a straight face and coached me on how to give a polite set-down. This had even raised a smile from Aunt Maude, my constant chaperone.

Daisy knocked on the door and came in. 'May I help you dress?' she said.

I swung my feet over the side of the bed, listening to her bright chatter.

'All the furniture's been taken out of the drawing room and it looks like a proper ballroom now,' she said. 'The hired chairs are here. Gilt they are and ever so pretty.'

Once I was dressed I hurried down to the morning room, where Aunt Maude was busy at the writing desk.

'Did you sleep well or were you too excited?' she said, looking up from her book of lists.

'My stomach is full of butterflies,' I said, 'from fear of saying the wrong thing or making a fool of myself.'

'Of course you won't!' said Aunt Maude. 'Your natural grace will carry you through and you don't need to exhibit too much town gloss. After all, you aren't out yet.'

'I have no friends in London. What if no one wishes to dance with me?'

'It's natural to be nervous,' said Aunt Maude, 'but no one would know you haven't had the advantages of being gently brought up.'

‘Sarah was a lady’s maid, not a scullery maid,’ I said indignantly. ‘I wasn’t a complete hoyden.’

‘Then perhaps we have some reason to be grateful to her.’ Aunt Maude smiled. ‘I’m certainly relieved you no longer wave your hands about when you are agitated.’

I sighed. ‘What shall I do to help?’

‘Will you check the latest invitation responses against the guest list, dear? And after breakfast you may inspect the ballroom and the supper room to see that the floral arrangements are in the correct positions. I saw the florist arriving half an hour ago.’

We worked together on the myriad lists, ticking off completed items.

An hour or so later I was crossing the hall to fetch Aunt Maude’s shawl when one of the footmen came out of Father’s study.

‘A letter for you, Miss Langdon.’

I caught my breath. Was it from Alessandro? I recognised the wax seal on the folded paper as Queen Caroline’s. I hurried upstairs to my room and closed the door behind me. I broke the seal with trembling fingers. It was a single, folded sheet of thick paper and I was bitterly disappointed to discover there was no enclosure from Alessandro. The note was written in Italian.

*Dear Miss Langdon,*

*The Queen thanks you for informing her that she has the support of the public in her endeavours to obtain the rights and privileges due to her status.*

*Following the receipt of the news of her exclusion from the Liturgy, she wishes you to know that she will proceed with all haste to England with the intention of arriving on the thirtieth of April.*

*The Queen conveys her kind regards and will be pleased to receive you upon her arrival in London.*

*Caroline*

I remembered so well the sprawling signature that the Princess, now the Queen, had added to the bottom of the letters I wrote on her behalf. I assumed that Baron Pergami now performed that task. Lady Hamilton, still in London, had forwarded my letters to the Queen.

Not only was she coming to England at last, but she was due to arrive in two days! I tucked the letter into the bottom drawer of the chest in my room and went to fetch Aunt Maude's shawl.

She and I were finishing our breakfast when Father came to ask me if he could have a private word. I gulped down the last of my coffee and followed him into his study.

'I have to go out,' he said, 'but I wanted to give you this.' He pulled open his desk drawer and handed me a small oblong box. 'They were your mother's,' he said. 'It seems appropriate you should wear them tonight.'

Curious, I opened the box. Inside nestled a double row of pearls. As I lifted them up I experienced a mixture of pleasure and pain. If my mother had cared more for my happiness than for her own misery, she'd have been here to give me the pearls herself. 'They're beautiful,' I said, 'but will it be distressing for you to see me wear them?'

He looked at me for a long moment and I wondered if it was my mother's face that he saw.

'She must have been wearing them when Sarah took away the rest of her jewellery. You're considerate to think of my feelings, Emilia, but I want you to have them. And the woman your mother was when I married her would have wished it too.' He glanced away from me to gaze out of the window. 'You remind me so much of the young Rose, when she was sweet and innocent, before everything changed.'

'I wish we could go back in time to stop those tragic events,' I said, looking down at the pearls coiled in my palm.

Father sighed heavily. 'We can't go back but I want to give you the best opportunities for a happy and secure future. My wish is that you'll meet a suitable man during the next few months . . . '

'Father, I've already told you that there's a man I love.'

He held up his hand. 'We agreed you would take this time to be sure you're making the right decision – one you must abide by for the rest of your life.' His eyes searched my face. 'And, as far as I'm aware, you've had no communication with Alessandro Fiorelli?'

'No, Father.'

'It's been five months since you saw him last. He'd have found a way to send you a letter if he still cared for you.'

I bowed my head so that Father wouldn't see my distress. Alessandro hadn't written to me in all this time and it worried me that he might not have forgiven me for coming to London. I ran my finger and thumb over every pearl in my mother's necklace as if it were a rosary, saying Alessandro's name in my head as I touched each one.

'I want you to be happy,' said Father gently, 'and I wouldn't want you to refuse a suitable offer of marriage because you're hankering after a man who doesn't love you.'

I stayed silent, afraid I'd cry if I spoke.

'Emilia, you must know Dolly is very fond of you?'

'Dolly?' I looked up, surprised out of my sadness.

'He admires you,' said Father. 'He's sophisticated and good-looking and only six years older than yourself. I've seen how he makes you laugh. It would be a good match.'

I was stunned. 'I thought you didn't like him?'

'Well . . . ' Father shrugged. 'Perhaps I was too hasty. The death of my only son was a devastating blow. I can't break the entail and it's natural I felt antagonistic towards Dolly but since you came home I've had the opportunity to get to know him better. And when he said he'd developed a fondness for you . . . '

I blurted out, 'I don't want to marry a man who is *fond* of me, I want to marry a man who *loves* me!'

'Believe me,' said Father, 'love flies out of the window all too easily. And what then? If you marry Fiorelli, assuming he still wants you, how will you manage when a clutch of children arrive? A tutor's earnings are not large and there will be few servants to help you. You'll become a drudge. There won't be any fine clothes and you'll all be crammed into a small house, wondering where the next meal is coming from.'

'It wouldn't be like that!'

Father raked his fingers through his thick grey hair. 'Emilia, this is the most important decision of your life. You must choose a husband based on who can provide best for you. Dolly will inherit this house and Langdon Hall and the estate. As his wife you would live a life of ease.'

'I don't love Dolly.' The walls of the room felt as if they were closing in and I rose to my feet, my heart thudding.

'Many advantageous marriages begin without love,' said Father, 'but affection follows. And you will have children to love, including a new heir for the estate, carrying Langdon blood.'

'How long ago did you and Dolly cook up this plan between you?' My cheeks burned. 'It's a shame you didn't consult me. I'd have told you then you were wasting your time.'

'Emilia . . . '

'*No*, Father!' Dropping my mother's pearls onto his desk with a clatter, I ran from the room.

Later, Aunt Maude tapped on my bedroom door. She came and sat on the bed beside me.

'Did Father tell you to come and see me?' I asked.

'He didn't mean to upset you,' she said, 'especially today. I'm sure he has no intention of forcing you to marry Dolly.' She placed the box of pearls beside me.

I wiped my eyes with the damp handkerchief balled up in my fist. 'It's not only that I don't want to marry Dolly.' My chin quivered again. 'Alessandro promised to write but he hasn't. I can't bear it if he doesn't still want to marry me.' I leaned back against the pillows.

Aunt Maude stroked my hand. 'Perhaps a letter could have gone astray? Italy is a long way from here.'

'I suppose a letter from him might arrive with Queen Caroline's entourage,' I said. 'She wrote to me saying she expects to arrive at Dover in two days' time.'

'The Queen wrote to you?'

'I expect the Baron wrote it but she signed it. I had so hoped for a note from Alessandro.'

'Try not to worry about that today,' said Aunt Maude. 'I shall send up Daisy with a cool compress for you. We can't have you going to your ball with swollen eyes, can we?'

'The ball! I'd quite forgotten it.' Dread made my shoulders slump. 'How am I going to face Dolly?'

'With your usual good humour,' said Aunt Maude, her tone brisk. 'Remember that no one is compelling you to marry him. And you'll meet any number of suitable men over the coming months so you still have plenty of time to make your choice.' She saw my expression. 'Or not. At the end of the season you shall return to Italy if you wish. Most girls would give their eye teeth for these opportunities.'

I couldn't deny that Father had been very generous but he didn't know how deeply I felt about Alessandro. 'I suppose I ought to apologise,' I said.

'Frederick has gone to an auction to purchase a painting,' said Aunt Maude, 'and won't be back until later.'

I hugged her. 'I don't need a cold compress,' I said, 'and I'll come and help you with the final preparations.'

Aunt Maude patted my hand. 'That's the spirit! Let's go downstairs together.'

# Chapter 20

Aunt Maude, resplendent in purple taffeta with a black lace overskirt, smiled her approval from her seat by my bedroom window. 'You will dazzle them all tonight, Emilia.'

I touched a finger to the lustrous pearls Daisy had clasped around my neck, surprised that they weren't cold. It was almost as if they'd retained the warmth of my mother's skin after she'd last taken them off.

My silken skirts swished around my ankles as I drifted towards the mirror. I stared at my reflection, barely recognising myself. The glorious dress had worked its magic again and new stays had made a great deal more of my *décolletage.* My skin appeared as luminous as my mother's pearls against the cream silk. During the afternoon, the visiting *friseur* had dressed my hair into gleaming copper curls and then entwined them with silk ribbons and lilies. I looked confident and, even to my eyes, beautiful.

'Oh, Miss Langdon!' said Daisy. She clasped her hands to her breast. 'You look so lovely.'

'Thank you, Daisy.'

She handed me the bottle of lily-of-the-valley perfume Dolly had brought me as a gift a few days before. I rubbed a few drops onto my wrists and between my breasts, faltering a little as I wondered how I could face him after what Father had told me.

Aunt Maude stood up. 'Come, Emilia, we shall show your father how beautiful you are.'

'I hope he won't still be angry with me.'

Servants were bustling hither and thither as we went downstairs and the discordant sounds of musicians tuning their instruments drifted from the ballroom. Two men were carrying a harp up the stairs and magnificent arrangements of narcissi and hyacinths perfumed the air.

Aunt Maude tapped on the door of Father's study.

Dressed in formal evening dress of brocade coat, knee breeches and white stockings, he was studying a painting propped up on his desk.

I waited for him to turn around. Would he still be angry with me? Then I forgot my anxious thoughts as I looked at the painting. It was beautiful, a nativity scene showing a stable set against a backdrop of mountains. I peered at the serene face of the Madonna as she smiled down at her babe and took a step closer to scrutinise the brushwork of the flowing drapery of her gown. Frowning, I said, 'It looks like a work by Fra Angelico.'

'It does, doesn't it?' said Father, still gazing at the painting.

'But it can't be,' I said, 'it would be far too expensive.'

'Well done, Emilia,' said Father. 'It seems your travels have not been in vain. This is believed to have been painted by one of Fra Angelico's apprentices.' He laughed. 'And it was still very expensive. But worth it, don't you think?'

'I do,' I said.

'Have you completed that list of places you stayed yet? I want to make a final effort to write to all the dealers to enquire after the miniatures.'

'I'll give it to you tomorrow.'

'My art collection is not subject to the laws of entail,' he said, 'and I'm delighted that it interests you. Perhaps, if you stay in England, one day it will be yours.'

His suggestion astounded me but it wasn't enough to bribe me to marry a man I didn't love.

Aunt Maude coughed to gain Father's attention.

He turned then and stared at me.

Slowly, I twirled around so that he could inspect me from all angles.

He lifted my hand and kissed it. 'I knew you were beautiful,' he said, 'but tonight you are exquisite. I'm very proud to be your father, my dear.'

I looked down at his broad hand with its stubby fingers enclosing my slender white ones. 'Father, I apologise if I was rude earlier today. I'm sure you have the best of intentions in suggesting that I accept Dolly's suit.'

'Not another word!' he said. 'We mustn't spoil your first ball. All I ask is that you don't reject the idea without considering it carefully.'

I sighed, knowing he couldn't change my mind.

'The first guests will arrive soon,' said Aunt Maude.

The drawing room had been transformed into an elegant ballroom, where everything glittered and shimmered. All was brilliantly lit; the crystal chandeliers blazed, gilt torchères gleamed with the glow of beeswax candles, flickering candlelight was reflected into infinity by the faceted mirrors of the wall sconces.

The double doors had been opened wide to the adjacent sitting room, which was set up with card tables for those who did not care to dance. The musicians chatted quietly together while the harpist softly plucked the strings of her instrument and the air was perfumed with beeswax and flowers.

'It's very fine,' said Father, eyeing his reflection in the overmantel and smoothing back his hair.

I was immensely flattered that all this splendour was in my honour. 'A few months ago,' I said, 'I could never have imagined being invited to an event such as this.' I cast a glance around, checking there were sufficient chairs and noting the small tables set out for the chaperones to congregate around. The carpet had been rolled up and removed and the golden parquet floor beneath dusted with powdered chalk to prevent unfortunate slips during the more energetic dances.

'We shall sit down while we can,' said Aunt Maude, lowering herself onto one of the gilt chairs. 'Now, don't forget, Emilia,' she said, 'you may only accept one dance with each partner.'

'But two dances are acceptable with Dolly,' said Father, 'since he's family.'

Nerves fluttered in my stomach. What could I say to Dolly now I knew he wished to propose to me? It would be impossible to meet his eyes. I pulled aside the cream damask drapes to look at the street below. I couldn't see the torches that had been lit on either side of the front door but their welcoming light pooled over the pavement. Then a yellow carriage drew up.

My stomach gave a little lurch. 'The first guests!' I said.

Father nodded at the musicians and they started to play.

Aunt Maude led me to the door and Father joined us. A moment later one of the footmen announced our visitors as they came up the stairs. Then the next ones arrived and there wasn't time to feel nervous as I smiled and curtseyed to a seemingly endless procession.

I found myself looking up at Dolly, who was studying me with a proprietorial air. 'Good evening, Dolly,' I said. My voice was calm but a blush raced up my chest and neck.

'Well,' he drawled, 'the duckling has turned into a veritable swan. Congratulations, Emilia.'

'You look most elegant, too,' I said, regarding his immaculately starched neck cloth and understated but superbly cut coat.

'I shall claim you as my partner for the supper dance,' he said as he moved on.

'Almost all the guests are here now,' said Aunt Maude a while later, 'although we still await Lord Liverpool. Come with me and we'll make conversation until a young man asks you to dance.'

'Supposing nobody does?' I asked, anxious again.

'Of course they will!' said Aunt Maude.

It appeared she was right because in the following five minutes three young men and one elderly one crowded round to ask me to partner them. Father, however, appeared at my side and claimed me for the opening quadrille.

'After that,' he whispered, 'I must do my duty and partner the dowager duchess in the mauve silk.'

Mr Sandys, my dancing master, had agreed to act as Master of Ceremonies and made introductions between the young people. Soon there were enough couples for the first dance.

The strains of the violins, harp and flutes rose above the chatter of conversation. Mr Sandys clapped his hands and organised the dancers in facing rows.

I stood opposite Father at the top of the dance, waiting for it to begin.

My neighbour, a pretty girl with dark curls, smiled at me and said, 'At my last ball I danced until dawn and wore my shoes into holes.'

The violins played the introduction. Mr Sandys nodded and we began. At first I had to take care over the steps but before long my feet remembered what to do as we danced up and down the rows. Father proved to be a competent dancer and I began to enjoy myself. It took half an hour to move all down the set and then to repeat it. At the end I was flushed with enjoyment.

Father led me back to Aunt Maude before going to claim his dowager duchess. I had only a moment to sip a glass of lemonade before my next partner, a Mr Perry, came to find me. He had brown eyes and dark hair that kept falling over his forehead and made me think wistfully of Alessandro.

'I saw you talking to m'sister at the beginning of the first dance,' he said.

'The girl with the dark curls?'

'Araminta. She'll ask you to her ball next month.'

'I should like that,' I said. 'I've been living abroad and don't know many people here.'

'I heard about that,' said Mr Perry. 'M'father knows your father. Quite a mystery, when you disappeared.' He smiled. 'Still, you're home again.'

The music began and, as we readied ourselves, I noticed Dolly and Francis Gregory further up the set with their partners.

'I'd like to hear about your adventures abroad,' said Mr Perry as we began the dance. He wasn't a good dancer but remained cheerful throughout his mistakes. 'Always had two left feet,' he said as he stepped upon my toes again. 'Prefer to be out hunting than indoors practising my steps with some foppish dance teacher.'

Dolly, however, was an accomplished dancer, I noticed. He bowed and twirled with grace and elegance, his cool fingers touching mine briefly as we came together for a moment before we turned to our next partners.

Some two hours and five partners later I was looking forward to the end of a particularly energetic Scottish reel. I'd danced every dance and was hot, breathless and anticipating my supper. Father and a group of older men were watching me and I caught his eye as I spun around. He smiled encouragingly and then I was off again. At last the reel came to an end. I curtseyed to my partner and there was a touch on my arm.

'Come with me, Emilia,' Father said. 'Lord Liverpool is here and has brought a distinguished guest. He was dining with the King when he mentioned he was coming to your ball afterwards. The

King professed a desire to meet you and has accompanied him. It's a very great honour.'

I pressed my hands to my hot cheeks. 'The King! What do I say to him?' I was totally unprepared for such a meeting.

'Curtsey and, if he speaks to you, answer him as best you can.' Father tucked my hand into the crook of his elbow.

I recognised the King immediately from the illustrations I'd seen in the newspapers, except that he was a great deal fatter and much less handsome than his portraits. Flustered, I barely heard the introduction as Father presented me and simply murmured, 'Your Majesty.' I sank into a deep curtsey, my knees shaking, and stared at the King's highly polished shoes for a moment before rising.

'An uncommonly handsome girl, Langdon,' said the King, looking me up and down with protuberant blue eyes. 'You must be happy to have her returned to you after all this time.'

'I had given up hoping for it, sir,' said Father, 'so you may imagine my joy that she's at my side now.'

The King inclined himself slightly towards me. 'Enjoying the dancing, Miss Langdon?'

'I am indeed, sir,' I said. Dolly had been right, the King's corset did creak when he moved, and I suppressed a sudden desire to giggle.

'Used to like dancing myself,' said the King. 'People were astounded by my elegance when dancing the gavotte.'

'I am sure they were,' I murmured.

The King lifted a glass of punch to his small, pink mouth while I wondered what he would say if he knew I had been a member of his wife's household.

Father introduced me then to Lord Liverpool and I curtseyed again. My knees still trembled but I retained my outward composure.

'I knew your mother, Miss Langdon,' said Lord Liverpool. He shook his head sadly. 'A great loss to us all when she passed on. She was beautiful, too.'

'Thank you, sir.'

Over Lord Liverpool's shoulder, I saw Dolly watching us.

'Dancing always made me hungry,' said the King. 'Run along and have your supper.' He waved his hand in dismissal. The presentation was over.

I curtseyed and then Father took hold of my arm and we backed out of the royal presence.

I heard the King say, 'Attractive little thing with her cheeks all flushed and her curls coming loose, don't you think, Liverpool?'

'Very well done, Emilia,' whispered Father. 'What a feather in your cap to be presented to the King tonight!'

I let out my breath slowly and was fanning my hot cheeks when Dolly came to join us.

'The King has an eye for a pretty girl,' he said. 'His approval has sealed your success.'

'I'll warrant the invitations will come in thick and fast now,' said Father with satisfaction.

'It looks as though you've had enough excitement for the moment, Emilia,' said Dolly. 'Shall we take some refreshment?'

He went to fetch our supper and I was grateful to sit down for a while. It had been strange to meet the King; the man Queen Caroline had called a monster. He seemed a surprisingly ordinary, elderly man with broken veins on his nose and cheeks. I wouldn't have looked twice at him if I'd met him in the street. Despite that, knowing he *was* the King had made me nervous.

Dolly returned with supper plates laden with tempting delicacies. Mr Perry and Araminta came to sit beside us as we ate the poached chicken, jellies and sweetmeats.

'What is the King like at close quarters?' asked Araminta, her brown eyes wide.

'Very . . . ' I tried not to picture the King's dissipated complexion. 'Very regal,' I said with as much diplomacy as I could muster.

'But he's terribly fat, isn't he?' she whispered.

‘He called me an “uncommonly handsome girl”,’ I said, trying not to giggle as I avoided answering the question.

Mr Gregory and his supper partner came to sit with us, too, and after a while I felt less agitated and began to enjoy the lively conversation of my companions. By the time we returned to the ballroom the King and Lord Liverpool had left.

The dancing began again and young men were queuing up to partner me. I danced every dance but couldn’t help noticing that Dolly’s gaze was often upon me. Later, he arrived at my side to rescue me when one of my partners, spotty and sandy-haired Mr Fortescue, began to press moist kisses upon my hands.

‘Miss Langdon is my partner for the next dance,’ said Dolly. ‘Let go of her immediately, if you please, and take a turn outside for some air.’ He released the young man’s limpet grip from my fingers. My rescuer’s tone was light but there was an edge of steel to it. ‘And I suggest you don’t imbibe any more punch.’

‘Thank you, Dolly,’ I said, after Mr Fortescue had glowered and taken himself off.

‘Irritating little cub!’ He cleared his throat a couple of times and straightened his perfectly arranged necktie. ‘Emilia,’ he said.

I noticed his usually pale face was even paler than normal.

‘Emilia, I’d hoped for the opportunity to have a quiet word with you tonight. I know that your father has mentioned to you that I spoke to him . . . ’

‘He did.’ I held myself ramrod straight, desperately wishing I were somewhere else and could avoid the forthcoming embarrassing exchange.

‘Perhaps this is not the time to discuss the matter of my proposal.’

‘You haven’t proposed to me,’ I said, ‘only discussed the matter with Father, but I have no intention of marrying *anyone* in the immediate future.’

His lips curved slightly. ‘Neither have I.’

‘Oh!’ I said, taken aback.

'Emilia, marriage is inevitable for both of us but there's no hurry. I do believe, however, that we might make an eminently suitable and convenient match in the fullness of time, should you wish it. My family are constantly pressing me to find a wife and your father wishes to see you settled. Perhaps there's some merit for both of us in suggesting that you are thinking about my proposal but require, say, six months or so to come to a firm decision. What do you say?'

I stared at him but his blue eyes were dark and unfathomable. My thoughts whirled so fast it made me dizzy. I intended to be reunited with Alessandro before six months had passed, unless, of course, he didn't love me anymore. I swallowed hard at the distressing thought. In that case, I might decide handsome, eligible Dolly was a good choice for a husband in the absence of the man I truly loved.

'This proposed mutual arrangement would allow us to make our own decisions in the fullness of time,' said Dolly, 'without familial pressure.'

I fingered the pearls at my neck while I thought. Aunt Maude had impressed upon me that if I passed this season by without accepting an offer, I could find myself on the shelf. I did very much want a family of my own so, however much I loved Alessandro, it would be foolish to slam the door on Dolly's offer.

'Emilia?' A muscle flickered slightly in his jaw.

'I shall consider your suggestion,' I replied.

He glanced at Mr Sandys, who was organising another dance. 'Shall we sit this one out?'

Relieved, I said, 'I must find Aunt Maude.'

She sat with the other chaperones and smiled up at me when we joined her. 'Well,' she said, 'the King's visit was a delightful and unexpected surprise!'

I nodded and smiled, not really listening to the conversations going on around me, and soon my next partner whisked me away. My shoes were pinching and I was relieved it was the final dance.

The guests began calling for their carriages and came to pay their respects before they left. I stood beside Father and Aunt Maude, bidding our guests goodbye. My face ached from constant smiling and trying not to yawn.

Dolly and Francis Gregory were the last to leave.

'The ball was a triumph,' said Dolly.

'An outstanding success!' declared Mr Gregory.

'May I call on you tomorrow, Emilia?' said Dolly.

'We shall expect you,' said Father, giving him a meaningful look.

After the last guest had gone, the musicians left with their instruments and at last the front door was bolted against the night. Yawning, the servants began to move the furniture back into the drawing room and remove the remains of supper to the kitchens.

Father put his arm around me. 'I'm proud of you, Emilia,' he said. 'Now go and get some sleep before the sun rises.'

'Goodnight, Father. And thank you. I shall never forget my first ball.'

He kissed my cheek and I watched him climb the stairs, wishing he'd shown Aunt Maude some sign of appreciation for all her efforts.

The poor lady drooped with exhaustion. 'Come to bed,' I said, gently. 'You've worked so hard to make everything perfect and must rest now.' She looked very frail and I helped her upstairs and handed her into the care of her maid.

Daisy was waiting for me in my room. I was grateful she was there to undress me and take the wilted flowers out of my hair. Afterwards, I lay in bed reliving the evening. I recalled the King's polished shoes and how my knees had trembled when I curtseyed before him; the heady perfume of narcissi and the aroma of beeswax candles. I remembered how Dolly had saved me from unpleasantness with Mr Fortescue and the subsequent extraordinary proposal. I recalled Father's proud smile and heard dear Aunt Maude's words of pleasure at my success.

And then, as dawn light began to creep through the curtains, I thought how I would have been pleased to forgo all of that, if only Alessandro could have been at my side.

# Chapter 21

## *May 1820*

### London

The weeks since the ball had been a constant round of tea and supper parties, balls, routs and excursions, until I barely knew which day of the week it was. I soon discovered that more or less the same guests went to each event and, since I didn't much enjoy gossip, it was becoming tedious. Despite that, I'd made friends with Araminta Perry and her brother and was always pleased to see their cheerful faces.

Dolly and Francis Gregory, both eligible bachelors with prospects, were usually present at these occasions and so I hadn't been surprised to see Dolly at Violet Braithwaite's tea party.

Mr Perry made a point of sitting beside me, as he so often did. I suspected he was half in love with me and was hoping for a sign that I might encourage him. His chatter was generally of a frivolous nature but he aroused my interest when he mentioned the intense speculation in London on the date of the possible arrival in England of Queen Caroline.

'The Lord Chancellor,' said Mr Perry with a speculative gleam in his eye, 'says that bets are being laid, some as large as fifty guineas!'

'The Queen was going to come at the end of April,' I said, 'but then I read in the newspaper that she was indisposed and unable to travel.' I made a point of reading Father's copy of *The Times* every day after he'd finished with it, to keep myself up to date with the Queen's whereabouts and engagements.

'All hell will break loose when she does arrive!' said Mr Perry. 'The King will sue for divorce immediately she sets a foot on the shore at Dover.'

'She has a great deal of support from the people,' I said. 'Perhaps the King will change his mind?' I lowered my voice. 'He's already exceedingly unpopular with the press and if he's perceived to be persecuting her it won't help him rise in the public's esteem.'

'Ah, well!' Mr Perry's eyes gleamed. 'I have another titbit of gossip for you. Did you hear that John Chesterton's father was burgled?'

I shook my head.

'Chesterton told me about it last night,' said Mr Perry. 'One of his father's paintings, a Stubbs, has been stolen. An unusual subject, I'm told, because it wasn't one of Stubbs's usual equine portraits but that of a giraffe. The interesting thing is that it's not the first time.'

'What do you mean?'

'Another Stubbs, a painting of a lion that time, was stolen from the Chestertons five years ago. It seems the impudent rascal bided his time and came back for its companion piece!'

'My father mentioned there'd been a spate of art thefts in the past.'

'What clinched the fact it was the same thief,' said Mr Perry, 'was that he left a sketch of an empty picture frame in its place, glued to the wall.'

'The Picture Frame thief is a scoundrel with a sense of humour it seems!'

It was raining when Aunt Maude and I were ready to leave the party and I suggested that we drive Dolly to his lodgings. No sooner had we all set off than Aunt Maude fell asleep, lulled by the rocking of the carriage and the drumming of the rain on the roof.

'Poor lady,' I said, looking at her papery cheeks all crumpled with sleep. 'She's too old to have to suffer through the season with me. I think we must have a quiet day tomorrow.'

Dolly glanced at her. 'Emilia,' he said in an undertone, 'have you given any more thought to my suggestion?'

'That we tell Father I'm considering your proposal?'

He nodded. 'I've been trying to speak to you privately ever since your ball.'

'I've thought about it because Father keeps pressing me,' I said, unhappily. 'Every time I return from a social engagement he interrogates me about any young men I've met and then lists all your virtues again.'

Dolly smiled. 'I didn't know I had any! That's a marked change from his original opinion of me.'

'He's more than happy now to view you as his future son-in-law,' I admitted.

'I do believe,' said Dolly, studying me through narrowed eyes, 'we might make a good match. Certainly I'd rather marry you than any of those vacuous little misses at the tea party today.'

I thought of Mr Fortescue, who'd pestered me at my ball, and of Mr Chesterton, who was a trifle slow, and of Mr Perry, who was amiable but not blessed with common sense. If I hadn't loved Alessandro, I might have accepted Dolly's proposal. At least then my future would have been secure.

'Well,' he said, 'shall we stop your father from pressuring us by telling him that you will at least consider my offer?'

Aunt Maude stirred and I put a finger to my lips.

There were a great many carriages in the streets, due to the torrential rain, and I watched drops of water racing down the carriage windows. At least I'd been spared the sight of Dolly going down on bended knee to propose to me. I smiled to myself, imagining he'd take great care when kneeling so as not to sully his trousers. There had never been any hint of flirtation between him and myself and any arrangement we reached suggesting a possible attachment between us was purely for Father's benefit. I still hoped Alessandro would write to me and then I'd be certain of his feelings for me.

Soon we arrived at Dolly's lodgings. He thanked us for driving him home before dashing through the rain to his front door.

It was as the carriage splashed down Portman Street that I saw the stork-like figure of Lady Hamilton standing under her portico and directing her servants as they loaded a carriage with travelling valises. I rapped on the roof of our carriage and the coachman pulled up.

Aunt Maude sat bolt upright. 'Are we home?' she murmured.

'Not yet, Aunt Maude dear,' I said. 'I won't be a moment but I want to have a word with Lady Hamilton.' Before she could protest, I snatched up my umbrella and alighted into the street.

'Miss Langdon?' called Lady Hamilton as I hurried towards her.

'I was driving by and saw your servants stowing your travelling bags,' I explained.

'For goodness' sake, come out of the rain!'

I shook my umbrella and sheltered under the portico with her.

'You received the note I sent to you earlier today?' asked Lady Hamilton.

I shook my head. 'I left the house after breakfast.'

'I'm setting off at last to join the Queen.'

'I'd thought she was to arrive at the end of April,' I said, 'but I read in *The Times* that she was too unwell to travel.'

'She decided to return to the Villa Vittoria before coming to England. When she arrived there she had a severe attack of rheumatic fever, which confined her to bed.'

So Alessandro might have remained in Italy or could be part of the Queen's entourage travelling towards Calais.

'Once she'd recovered,' said Lady Hamilton, 'she packed up her jewels and personal possessions and set off for England again. She intends to present herself to the public as their Queen with all attendant ceremony. Unfortunately, she was taken ill with a stomach complaint on the journey and was forced to rest in Geneva. I'm to join her with all speed now and escort her to Calais.'

'You said you wrote me a note, Lady Hamilton?'

'I did,' she replied crisply. 'The Queen has commissioned several new items of clothing from Mrs Webbe in Pall Mall and has sent her favourite dresses to act as patterns. Her Majesty would be pleased if you would visit Mrs Webbe on her behalf to select the colours and materials that will suit her best.'

'Did she give any indication of her preferences?' I said, wondering how I'd be able to arrange this without making Father angry. I'd learned not to mention the Queen in his presence.

'They are all to be in a dignified English style,' said Lady Hamilton, 'since she's determined to present herself as a model Queen of England. She asked me particularly to say she will rely absolutely on your good judgement.'

'I am honoured,' I said.

Lady Hamilton looked up at the sky. 'I sincerely hope the weather doesn't worsen for the Channel crossing.'

I hesitated, unsure if I dared ask a favour, but decided I could lose nothing by it. 'If by any chance Signor Fiorelli is still with the Queen's entourage,' I said, 'would it be possible for you to ask him to write to me?'

Lady Hamilton surveyed me, stony-faced. Then she sighed. 'Love makes the young so reckless, doesn't it? I can't imagine Sir Frederick is happy to have his daughter chasing after a penniless tutor. Besides, I'd heard rumours that you are to marry his heir.'

I looked down at my feet to hide my blush of annoyance. 'Unfounded gossip,' I said.

'*If* I see Signor Fiorelli,' said Lady Hamilton after a moment, 'I will give him your message.'

I looked up at her, unable to disguise my gratitude. Thank you so much, ma'am.'

'Now hurry along, you're delaying me.'

And then I was back in the rain again, skipping through the puddles, heedless of my wet feet.

The following day Aunt Maude and I sat together in the morning room. I'd found *Belinda* by Maria Edgeworth in Father's library and offered to read it aloud. When I opened the first of the three volumes it gave me a little jolt to see that the flyleaf was inscribed with my mother's name in a firm and confident hand, all underlined with a flourish. Tracing the letters with my forefinger I wondered if she had been as self-assured as her signature.

'Rose loved books,' said Aunt Maude, watching me, 'though your father didn't approve of her reading novels. He thought they might inflame her imagination.'

'Perhaps they did,' I said, remembering what he'd told me about her adulterous affair. Maybe reading had encouraged her to seek romance outside her marriage?

'In the time I knew Rose, she was never given to flights of fancy,' said Aunt Maude. 'I always found her to be an eminently sensible young woman, much like yourself.'

The door opened and James the footman entered. 'Miss Langdon,' he said, 'your father requests you attend him in his study.'

I placed the book on the side table. 'I shan't be long, Aunt Maude,' I said.

A moment later James opened the study door and I saw that Dolly was with my father. Both men stood up as I entered.

Father, smiling widely, held out both his hands to me. 'Emilia, my dear child! Dolly has told me the wonderful news.'

I glanced at my *soi-disant* suitor, who looked even paler than usual. A slight tic twitched at the corner of one eye and I was afraid I knew what had happened. 'What wonderful news is that, Father?'

'Why, your engagement, of course!'

I disentangled my fingers from his warm grip. 'I assure you,' I said frostily, 'there is no engagement between us.'

Father waved a hand as if to dismiss my comment. 'As to that, I understand you wish to wait a while before any formal announcement . . . '

'I have made it perfectly clear to Dolly that I have agreed only to *consider* his proposal,' I said. 'I'm in no hurry at all to engage myself to anyone at present.'

Dolly gave his usual nervous cough. 'And I am very happy to give Emilia all the time she needs to make up her mind. Months if necessary. Clearly, there must be no announcement until,' he caught my eye and had the grace to blush, '*unless* Emilia decides to make me the happiest of men.'

He didn't look very happy and it was then that I made up my mind to follow his suggestion. It seemed he wanted his freedom as much as I wanted mine so it was unlikely he'd pressure me into a hasty marriage.

My father breathed heavily and looked at both of us with something akin to anger in his eyes.

'Father,' I said in as conciliatory a tone of voice as I could, 'I am newly arrived in England and everything here is so different for me. I am still learning to know my own family. Allow me a few months more to make what is the most important decision of my life.'

He fixed his gaze on the painting by Fra Angelico's apprentice, which he had hung opposite his desk.

It was curious how calming an effect the sight of it seemed to have on him. His breathing slowed and the tension disappeared from his face as he studied the glowing colours and delicate details.

I caught Dolly's eye and he gave an apologetic grimace.

'Forgive me,' Father said, after a moment. 'It seems such a perfect choice for both of you and I do so want to see you settled. If Dolly is prepared to wait, I suppose I must be content to wait, too.'

'Then, if you will excuse me,' I said, 'I shall return to Aunt Maude.'

Father was spending an increasing amount of time in the House of Commons. At dinner he mentioned the endless speculation regarding the Queen's arrival and the disruption that might ensue. I drew him out on the subject since naturally Her Majesty's affairs were of great interest to me.

'Mr Brougham has suggested advantageous terms on which Caroline of Brunswick might consent to stay abroad,' said Father, 'but the wretched woman has a grossly inflated view of her rights. Furthermore,' he said indignantly, 'he's accused Parliament of being corrupt.'

'I daresay Parliaments over the ages have been accused of that many times,' I observed.

Father ignored my comment. 'The King has sent his old friend Lord Hutchinson off with Brougham to treat with her. His Majesty is determined to get rid of his wife but she'll have to renounce the title of Queen of England if she wants her fifty thousand a year for life.'

'Is the King so concerned about it because the Queen is more popular than he is?' I asked, trying to appear ingenuous.

'Don't ever suggest such a thing outside these four walls, Emilia,' said Father, his brow thunderous. 'I'm working hard to find a place on the King's advisory committee for the Coronation planning and it would destroy all my chances if he thought anyone in my family didn't support him completely.'

Over the next few days Father watched me covertly with a brooding expression on his face and I came to the conclusion that I'd surprised him, not only by my adamant refusal to be propelled into a hasty marriage but also by making that ill-judged reference to the Queen's growing popularity.

When I mentioned at breakfast one morning that I'd received several new invitations, he suggested that I might like some new dresses since the summer was upon us.

‘Thank you, Father!’ I kissed his cheek, wondering if the offer was a bribe for good behaviour.

I now had the perfect excuse for visiting Mrs Webbe, the Queen’s mantua-maker. Aunt Maude, of course, would have to accompany me to Pall Mall and I realised it would be necessary to explain to her my errand on the Queen’s behalf. I was relieved when she brooked no objection.

‘You may count on me to be perfectly discreet,’ she said. ‘And really, I do think that the poor Queen has been very put upon by the King over the years. She may have misbehaved but His Majesty has hardly conducted himself like a gentleman. That she wishes to make a good impression during her visit here shows perfect sense, particularly now she’s the Queen of England and not simply the estranged wife of the Prince Regent.’

‘I have to admit she is a most eccentric lady,’ I confided. ‘Her conduct is not always dignified but, in my experience, she’s good-hearted and very brave in the face of unhappiness.’

‘Between us two,’ whispered Aunt Maude, ‘and whilst certainly not condoning it, I cannot blame her if she sought a little happiness with her Italian steward. After all, the King has had many mistresses.’

I was astonished by Aunt Maude’s attitude. I hadn’t expected it from someone of her age.

‘Sometimes,’ she continued, ‘I’m extremely thankful that I never married. It seems to me that marriage is a risky business.’ She shook her head regretfully. ‘It certainly made your poor dear mother very unhappy.’

‘So Father tells me,’ I said. I didn’t want to talk about my mother now I knew what kind of a person she had been and how much distress she had caused.

We left the mantua-maker after an appointment lasting nearly three hours. Two new dresses were to be made for me, one in pale green muslin with white sprigs and another in the palest primrose yellow with a flounced hem.

I had also made a careful choice of materials for the Queen's dresses, taking into account her florid colouring and rotund figure, and hoped that she would approve. The most useful thing I could do to help the poor woman at the moment was to make her feel elegant and self-assured. While she might be popular with the public, if it came to a confrontation in court, she would be dealing with some of the best legal minds in the kingdom, who all intended to win the King's favour by helping him to destroy her reputation forever.

# *Chapter 22*

*June 1820*

**London**

Father threw down his copy of *The Times* on the breakfast table and sighed. 'The King will be in even more of a temper than yesterday when he reads this. Caroline of Brunswick – I really cannot refer to her as his Queen – set ashore in an open boat, would you believe, rather than wait a few hours until the tide had turned and the *Prince Leopold* could dock at Dover. That's just the kind of undignified and impetuous behaviour that will endear her to the lower orders.'

'It must have taken a great deal of courage to climb down from the *Prince Leopold* into a small boat on the open sea,' I observed.

'For heaven's sake, Emilia! That woman will do anything to play to the gallery,' said Father, scowling. 'The wretched common people had been waiting for her since dawn, milling about on the beach and the cliffs and getting drunk, dressed up as if they were going to a fête. And then the cannon at Dover Castle actually greeted her with the Royal Salute.' His lips twisted in a sneer. 'You can just imagine how cock-a-hoop that would have made her.'

‘Who travels with her?’ asked Aunt Maude.

‘She’s left the Italian rabble behind,’ said Father, ‘except, I believe, for her paramour’s sister.’

Countess Oldi had come to England then but Alessandro had stayed behind. At least the Queen had one of her ‘Italian family’ to support her.

‘Lady Charlotte Lindsay was asked to meet her at Dover,’ said Father, ‘and resume her position as lady-in-waiting, but the King warned her off. Alexander Hamilton, being such a staunch Whig of course, has allowed his sister to wait on her. Hoping to curry favour, no doubt. And that jackass Alderman Wood is with her, and the boy William Austin.’

‘What will happen now?’ I asked.

‘Since Caroline refused to agree terms with the King’s envoys, he’ll sue her for divorce on grounds of her infidelity.’

‘But what about the King’s infidelities?’

‘Emilia! It’s not fitting for a young, unmarried woman to ask such questions.’

‘But . . . ’

‘I will discuss this with you no further! As if there isn’t enough to worry about with all that’s going on in the House, now I have to contend with discord in my own home.’

Still grumbling, Father left for the House of Commons.

I picked up his discarded copy of *The Times* and took it into the garden. A pink rose, just coming into bloom, scrambled over the wall behind an ironwork bench. I sat down in the sunshine and spread the paper over my lap.

*Neither at the landing of William the Conqueror nor that of William III had any arrival in England caused such a sensation*, I read. I wondered what the Queen had felt when she’d landed on these shores after four years abroad. I suspected she’d have chattered away in a cheerful manner, however anxious she might have felt inside. I could picture her waving gaily to the crowds and hoped they’d cheered her on.

*The Times* described the Queen as she walked to the Ship Hotel: *Her blue eyes were shining with peculiar lustre but her cheeks had the appearance of a long intimacy with care and anxiety.*

The anxieties of her situation on top of her enforced parting from the Baron and little Victorine must have been terribly hard for her, especially since she couldn't know how long it would be until she was able to return to Pesaro.

I leaned back against the sun-warmed bench and watched a bee drift lazily past. The sun was hot on my shoulders. I closed my eyes for a moment, imagining I was back in Pesaro sitting on the cliffs of San Bartolo with Alessandro at my side, smelling the honey-scented wild broom and listening to the cries of the gulls over the turquoise sea. I'd had to come to England to find my family but, oh, I did so long to be with Alessandro!

Two nights later, angry shouts from down in the street awoke me. I heard shattering glass and leaped out of bed with my heart hammering. From out of the window I glimpsed a horde of men carrying flaming torches, chanting and waving sticks in the air as they passed by.

'No Queen, no King!' they yelled.

A man dressed in a nightgown leaned out of a second-floor window in the house opposite and a protestor called out, 'Put a candle in the window to show your support for Queen Caroline!'

The observer shook his fist for reply. 'Damned Queenites! Get away with you at once or I'll call the constable!'

The demonstrators roared their displeasure and a hail of stones flew through the air, smashing his window panes. Jagged shards of glass crashed to the pavement below and skittered across the street.

The man shrieked and clasped his head, while the front of his nightgown grew dark with blood.

I jumped at the sound of a thunderous knocking on our front door. Snatching up my wrap, I hurried down to the first-floor landing.

Leaning over the banister, I saw Father and James in the hall below. Their voices were raised in anger as they argued with several men standing on the doorstep.

'Put a light in your windows in support of Queen Caroline!' demanded one of the callers.

'Never!' Father replied vehemently.

'You'll be sorry if you don't!'

Father slammed the front door so hard it reverberated throughout the house. 'Ruffians!'

I hurried back upstairs, lit my candle and placed it on the window-sill. Father wouldn't like it if he knew, even if it did save our windows, but I wanted to support the Queen in whatever way I could.

Downstairs, Father was shouting orders at James and I went to see if any damage had been done.

'Unbelievable!' Father said when he saw me. 'Unruly rabble!' He paced across the hall. 'Queenites, indeed! They ought to be flogged . . . along with that spiteful hoyden Caroline of Brunswick for encouraging them.' He shot the bolts home on the front door. 'Now you can see why any decent and reasonable person wants to be rid of her. God knows how you stood being part of her household, Emilia.'

A firecracker went off in the street and voices chanted, 'Long live Queen Caroline!' and 'No Queen, no King!'

Father turned to James, who was waiting for further orders. 'Don't just stand there, you dolt! Make sure the windows are fast. And bring me brandy!'

'You're not hurt, Father?' I asked.

'It takes more than a few troublemakers to rattle me.' He strode to the front door and checked the bolts again.

'Then I shall say goodnight,' I told him.

Upstairs, I moved the candle to one side of my bedroom window-sill and looked down at the street. Men carrying flaming torches still milled about, waving sticks and chanting their support of the Queen. Glancing along the street, I noticed several of the houses had placed lights in their windows, either out of fear or in genuine support for the Queen.

Eventually the disturbance faded away as the men marched off to break windows elsewhere. I returned to bed and pulled the covers up to my chin.

Somewhere a dog barked, the sound echoing down the deserted street, and, at last, I slept.

In the morning I found the front door open. Father was on the steps surveying the street and a number of servants were sweeping up shards of broken glass. At the house opposite a glazier in a brown apron was boarding up the downstairs windows.

'Disgraceful!' said Father. 'There's broken glass everywhere! The Queenite louts didn't dare damage my house, though.' He planted his hands on his hips and gave me a smug smile. 'They didn't like it when I stood up to them and slammed the door in their grubby faces.'

'We were lucky,' I said.

'No luck about it,' said Father. 'You have to stand firm and show the beggars you simply won't have it.'

'Yes, Father,' I said, reminding myself to take the candlestick off my bedroom windowsill.

'In any case, I heard yesterday the King is set on a divorce and he won't change his mind to please the mob. He's sent Lord Liverpool two Green Bags of evidence against his wife from the Milan Commission. A committee will have to be set up to examine it.'

Father took himself off to his study and I went to the morning room where Aunt Maude was waiting for me.

'Did you manage to sleep through the commotion last night?' I asked.

She smiled. 'I sleep very lightly these days. It was rather exciting, didn't you think? I worry for the horses, though, with broken glass in the street.'

'The Queen has some loyal supporters,' I said.

'It would appear so,' Aunt Maude observed, 'though a mob's good opinion can be fickle.'

'I doubt the so-called evidence against her from the Milan Commission will hold up in court.'

'Don't tell your father,' Aunt Maude whispered, 'but I put my candle in the window last night.'

I laughed. 'So did I!'

Aunt Maude smiled. 'A small victory for Womankind.'

'Father mentioned yesterday that the Queen is staying in Portman Street with Lady Hamilton,' I said. 'I'd like to call upon her. Her household was very informal in Pesaro but it could be very different here. She may not receive me.'

'I trust she will be pleased with the service you rendered to her in advising her mantua- maker.'

'I need to buy ribbons in Bond Street but, afterwards, will you accompany me to Portman Street?' I bit my lip. 'Of course, I haven't asked Father if I may go.'

'While I am not generally in favour of deceit,' said Aunt Maude, 'since Frederick hasn't expressly forbidden you to visit the Queen, I shall make no objection to your plan.' Her blue eyes twinkled. 'In fact, I shall be most interested to see this unusual lady.'

Later that morning, after I had purchased blue silk ribbons in Bond Street, we directed the coachman to drive us to 22 Portman Street. I placed one of my calling cards on a silver salver and a footman in scarlet livery bore it away. Aunt Maude and I waited on the hall chairs.

A short while later the footman returned and asked us to follow him upstairs to the drawing room where the Queen and her ladies sat. Queen Charlotte wore a dress made from material I had selected for her. She had lost a great deal of weight and I supposed this was caused by her recent illness.

I curtseyed deeply to the Queen, Lady Hamilton and Countess Oldi, and helped Aunt Maude to lower herself as far as her poor old knees would allow.

'Pray be seated,' said the Queen. 'How kind of you both to welcome me! So, my Miss Barton is now Miss Langdon?'

'That is so, Your Majesty,' I said. 'I'm pleased to say I have found my father.'

'Is he what you expected?'

'I had no expectations at all, Ma'am,' I said, 'only hopes, so I'm happy he has welcomed me back into the bosom of my family.' I smiled at Aunt Maude. 'Miss Weston is my great- aunt,' I explained.

'How fortunate for you, Miss Weston, to have your great-niece returned to you.'

'I count myself truly blessed, Your Majesty,' replied Aunt Maude. 'It was a terrible thing when Emilia was taken from us as a child. I never expected to see her again.'

'I was sorry to hear that you had been indisposed, Ma'am,' I told the Queen, 'and were unable to travel when you had intended.'

She shook her head and sighed. 'So many delays.'

'I was pleased to read in the newspaper that your arrival was a triumph and that a crowd had gathered to welcome you.'

'It was a great relief to me.' Her blue eyes sparkled. 'We stayed at Canterbury where the cheering of the people kept me awake. It was so all the way to London, with people waving as we went. Men unhitched the horses from my carriage and pulled it themselves!'

'Eventually we requested them to desist,' said Lady Hamilton, 'since otherwise we should not have reached London before nightfall.'

'The rain stopped and the sun shone when we reached Deptford,' the Queen took up the story, clasping her hands to her bosom at the memory. 'We folded back the roof of the carriage so the crowd could see me. They shouted and cheered and said the sunshine was a good omen.'

Smiling, I said, 'I'm sure that it will be.'

'There was a great gathering at Blackheath,' said Lady Hamilton. 'You would have thought it was a midsummer fair. At Shooter's

Hill we could barely pass for the barouches and chaises filled with respectable people come to welcome Her Majesty. And then we progressed to Alderman Wood's house in South Audley Street with all those carriages in procession behind us.'

'I stood on his balcony so my subjects could see me,' said the Queen. 'I waved my handkerchief at the crowds and the cheers were deafening.' She sighed. 'It was wonderful. The King has fled now to Windsor, fearful the people will turn against him. Brougham and Thomas Denman, my Solicitor-General, will meet soon to discuss our strategy.'

There seemed nothing to add to this. Besides, I hoped to turn the conversation so as to ask about Alessandro. 'Was little Victorine well when you saw her last?' I enquired.

'She didn't want me to leave,' said the Queen. 'The Baron had to pull her, weeping, from my arms. It nearly broke my heart. She was sad, too, because her Signor Fiorelli had also left my household.'

'Signor Fiorelli has gone?' I was shocked, knowing how fond he was of Victorine.

The Queen shrugged. 'He is no longer in my employ and I hold you responsible for that, Miss Langdon.'

'Me? But why?'

'When you left my household he suffered. He was miserable and I missed his merry laughter. Why didn't you return to make him a happy man?' The Queen shook her head. 'Anyone could see that you loved each other.'

'I will return to Pesaro,' I said, 'but I wish to spend time with my rediscovered family first.'

'Are you enjoying your time here in London?' asked Lady Hamilton.

I hesitated. 'I promised my father I would do a season before returning to Italy. He wishes me to enjoy any opportunities that may present themselves.'

'And meanwhile you will wear pretty clothes and enjoy all the balls and parties?' said the Queen.

'Yes.' I must have sounded doubtful about that because she smiled.

'I do believe you are finding them tiresome already.'

'I am a little tired of dress-fittings and endless polite conversation with the same people. I miss the freedom I had in Pesaro. Can you imagine what my father might say if I told him I'd been sea-bathing, for instance?'

Lady Hamilton and Aunt Maude looked shocked, while the Queen threw back her head and laughed. 'It is the same for me. I know that as soon as I begin to enjoy myself someone will disapprove and it will probably be written about in the newspapers,' she said. 'Lady Hamilton tells me every day that I must take particular care to restrain myself.'

'Impropriety always comes home to roost in the end,' said Her Ladyship, in a forbidding tone.

'Indeed it does,' said Aunt Maude, looking at me. 'Society can be very unforgiving.'

Lady Hamilton rang the little bell on the table beside her.

'I have enjoyed seeing you again, Miss Langdon,' the Queen told me. 'Call again, won't you? And I haven't thanked you for visiting Mrs Webbe to choose dress materials for me. You have helped to make me look suitably respectable.' She gave me a conspiratorial smile and there was a mischievous glint in her eyes when she continued, 'Even though I might have preferred at least one gown with a low-cut bodice made out of transparent silver gauze.' She turned cordially to Aunt Maude. 'Goodbye, Miss Weston.'

The footman held the door open, waiting for our departure.

Aunt Maude and I both curtseyed and our audience was over.

We ascended the steps to the carriage in silence while I wondered where in Pesaro Alessandro was working now.

'Well!' said Aunt Maude as the wheels began to move. 'The Queen wasn't at all as I expected.'

'What did you expect?' I said.

'She was far more dignified than I could possibly have imagined from all the scandalous tales I've heard about her. And she has a knack of putting everyone she speaks to at ease. I can perfectly well see why the people like her. No wonder the King doesn't want her to return!'

'How do you mean?'

'She forms such a marked contrast with a monarch whose spendthrift and immoral ways have made him justifiably unpopular,' my great-aunt commented. 'I can well imagine why the populace would wish her to take her rightful position as Queen. I suspect Whigs like Alderman Wood and their allies, the Radicals, will do their utmost to use her popularity as a means of bringing down the Tory government.'

'Why, Aunt Maude,' I said, 'I had no idea you knew so much about politics!'

'I may be an old lady no one much notices,' she replied in astringent tones, 'but that doesn't mean to say I don't see what is going on.'

'No,' I said, 'of course not, Aunt Maude.'

A few days later we were returning to Grosvenor Street after meeting Araminta and her brother for an ice at Gunter's. As our carriage rolled up outside, the front door was flung open. The footman pushed a smartly dressed young man out of the hall and onto the front steps.

'Oh, dear!' said Aunt Maude, looking out of the window as she gathered up her reticule and stick, ready to descend. 'An unwelcome caller.'

Protesting volubly, the young man shrugged his shoulders with palms turned upwards in an unmistakably Italian gesture. He gesticulated in the air and my heart nearly leaped out of my chest as I realised who it was. A burst of joy made me laugh aloud. Alessandro had come to England to find me! He must have realised at last how desperately important it had been for me to find my family and come to apologise.

Aunt Maude was blocking the carriage door as she descended one painful step at a time and I couldn't pass without knocking her over.

'Alessandro!' I shouted. Excitement and impatience to be with him again made me fidget as I waited behind Aunt Maude.

She turned to look at me, her mouth pursed reproachfully at my unladylike shout.

I peered between the plumes on her hat and saw Father rush out of the house and barge Alessandro with his shoulder. I gasped as Alessandro teetered on the steps, overbalanced and fell backwards. He landed hard, sprawling on the pavement.

Father threw a small bundle after him and then went inside, slamming the door.

'What a disgraceful scene!' said Aunt Maude.

'Why did Father do that?' I said. Shocked, I watched as Alessandro sprang to his feet, shook his fist at the front door and strode off down the street.

I wriggled past Aunt Maude and hurried after him but he was already a considerable way ahead. Dodging around strolling pedestrians, I ran along the pavement.

Hands tugged at my sleeve and a beggar in a tattered army uniform implored me for alms.

I prised his fingers from my arm. 'Alessandro, wait!' I shouted, but he was too far ahead to hear. I dashed after him, my pulse racing.

At the end of the street I stopped at the junction with Bond Street. Shoppers crowded the pavement and carriages and horses swished past. Frantically, I looked both ways but there was no sign of him. My shoulders drooped and I swallowed bitter disappointment. He'd been so close, almost within touching distance, and I'd lost him! I could only take comfort from the belief that he'd return.

I was nearly home when Father came marching towards me, a scowl on his face. 'Emilia!' He took my arm in a fierce grip. 'Don't ever go running off alone in the street like that again! What will people think?'

'You threw Alessandro down the steps!'

'I certainly did,' he said, his voice cold. 'He's a most objectionable young man.'

'Alessandro is never objectionable!'

'He came marching into my house demanding to see you. He didn't believe me when I said you weren't there. He bellowed your name and had the impudence to start opening doors, looking for you.' He tightened his grasp on my arm.

'You're hurting me!'

'Don't embarrass us both in the street, Emilia.' He spoke through gritted teeth. 'I thought you had acquired a little sophistication but I can see now it's only a thin veneer over your lowly upbringing.'

A red tide of rage rose up in me. 'To treat me so violently clearly demonstrates *your* lack of breeding!' I wrenched myself free from his grip and picked up from the pavement the bundle Father had thrown after Alessandro. A clean napkin tied with silk ribbon enclosed something soft. Untying the bow, I caught my breath when I saw the rag doll. Alessandro had returned my beloved Peggy.

Scarlet-faced with rage, Father snatched the doll from me and dragged me by my wrist until we were inside the front door. He shoved the doll into the footman's hands. 'Get rid of this,' he barked.

'No!' I shouted. I grabbed Peggy and held her tightly to my chest. 'You shan't take her from me. Peggy has been with me on all my travels ever since I was a little girl.' Alessandro knew how much my old doll meant to me and had brought her all this way to return her. I buried my face in Peggy's woollen hair, willing myself not to burst into tears.

Aunt Maude hovered in the entrance hall, her lower lip quivering. 'Thank goodness you're safe, Emilia!'

'No thanks to you,' snarled Father. 'Get out of my sight, old woman!'

Outraged, I stared at him open-mouthed.

Aunt Maude didn't say a word but went upstairs, leaning heavily on her stick.

'How *could* you be so cruel?' I asked.

Father made a visible effort to control himself. 'You were in her care and she failed in her duty. Who knows what might have happened to you had I not intervened?'

I took a calming breath. I wanted to go home to Italy with Alessandro but it would be awful to leave England on bad terms with my father. 'What did Alessandro say?' I asked.

Father rubbed his palm over his face. 'He's in London for a few days.'

'Where is he staying?'

'I didn't ask. He said he wanted to say goodbye.'

'Goodbye?'

Father shrugged. 'I'm going to my study and don't wish to be disturbed.'

I watched his retreating back and heard the study door slam. I didn't understand why Alessandro wanted to say goodbye. My certainty that he must have come to apologise to me suddenly wavered.

I withdrew to my bedroom and placed Peggy on the pillow next to Annabelle. Her embroidered smile looked coarse by comparison to Annabelle's exquisitely painted face but no less precious. Melancholy gripped me. Sarah had taken the trouble to make Peggy for me, no doubt to soothe my tears after my previous doll had been left behind, and I regretted the times I'd been impatient with her.

Meanwhile, poor Aunt Maude, distressed by my father's cruel words, had been banished to her room as a result of my actions.

A few moments later I tapped on her door.

She sat by the window, her knuckles white on the silver handle of her stick.

'I'm sorry for what Father said.' I wrapped my arms around her thin shoulders, upset to see how diminished she looked. 'I've seen a side of him today that I don't like.'

Aunt Maude patted my hand. 'I've endured worse over the years. He's always had a temper if he doesn't get his own way.'

'Shall I fetch my book of poetry to read to you?'

'Thank you, dear. That would be soothing.'

We read and chatted until Aunt Maude's feathers seemed less ruffled but all the time I was conscious that Alessandro was somewhere nearby and I couldn't go to him.

It wasn't until I woke up in a cold sweat during the night that I wondered if he had come to say goodbye because he'd given up waiting for me and found someone else to love.

# Chapter 23

Alessandro didn't try to contact me again and after a month of waiting and hoping, I could only assume he'd returned home. I cried bitter tears into my pillow, whilst nursing resentment against Father for sending him away.

The summer passed in an interminable round of balls, routs and excursions and I became heartily tired of it all. In July I received a proposal from Mr Perry. In his usual jovial fashion, he didn't seem to mind when I refused, which inclined me to believe his heart wasn't deeply engaged. I saw a great deal of Dolly, usually accompanied by Mr Gregory, and was relieved he didn't pressure me to announce a betrothal between us.

In August, Queen Caroline was called before the House of Lords to answer a charge of adultery and it was a nasty, undignified affair. The newspapers were full of salacious and shameful accusations, apparently based on events witnessed by her ex-servants. Father gloatingly read some of the articles aloud to me, as if he hoped to shake my own good opinion of the Queen. Surprisingly, the scandalous behaviour of which she was accused didn't appear to shake the Queenites' good opinion of her. Day after day, cheering supporters from all walks of life lined the streets as she passed by on her way to and from the House of Lords.

In September there was a recess and Father took Aunt Maude and me to Langdon Hall for some country air before the case for the defence began in October.

One morning in early November I knocked on the door of Father's study.

'Look at this on the front page!' he said, shaking his newspaper at me. 'It says, *It's the third anniversary of the death of the late lamented Princess Charlotte, the daughter of our ill-fated Queen.* The newspapers only write in these terms to stir up the rabble,' he commented bitterly. 'The Whigs and the Radicals are behind it, of course. They're using the Queen to curry favour with the common people, to whom she has so aptly allied herself, so as to discredit the King and bring down the Tory government.' He dropped the crumpled newspaper on his desk. 'Was there something you wanted, Emilia?'

'Aunt Maude has woken up with a putrid sore throat,' I said, 'so I must write a note to Millicent Deveraux and cry off from her card party this afternoon. Dolly was to have accompanied us so I must write to him, too. May I send James to deliver the notes?'

Father pursed his lips. 'I think it perfectly proper for Dolly to escort you on this occasion. He is family, after all. Send him to see me before you go, will you?'

'Thank you, Father, and since we need not take the carriage for Aunt Maude's sake, Dolly and I shall be perfectly happy to walk.'

'Absolutely not . . . I insist you take the carriage!' he said. 'I don't want you to be caught up in all the vulgar commotion on the streets. It's likely to be worse than ever after the Lords' final vote on the Queen's trial this afternoon.' He picked up the paper again. 'At least that should see an end to it, though the majority so far has been worryingly small.'

'Fifty-two days now! I can't believe it's gone on for so long,' I said.

'The poor King is completely ground down by worry.'

I refrained from mentioning that the Queen must be in a state of high anxiety, too. 'There was another thing I've been meaning to ask,' I said. 'Have you received any further information on the Infanta's miniatures?'

He shook his head. 'I've written dozens of letters to art dealers in the towns you visited. I haven't received replies from all of them but nothing has come of my enquiries so far. Perhaps you'll rack your brains for any other places Sarah might have sold them?' He sighed heavily. 'You are the only key to solving this puzzle, Emilia.'

'I'll think about it again,' I promised.

Later that afternoon Dolly arrived, immaculately dressed as usual, and went to speak to Father while I fetched my pelisse and bonnet. When I came downstairs I heard voices in the study, raised in argument. I waited in the hall until Dolly emerged. He took my arm with a tight little smile and hurried me out of the front door.

I caught my heel in the hem of my skirt as I tried to keep up with his long stride. I gasped as I teetered on the top step and grasped at his arm.

Dolly caught me and held me tightly against his chest until I regained my equilibrium. But he didn't release me. I looked up and found his gaze fixed on my face. He stared at me for a long moment with unfathomable dark blue eyes and, despite my discomfort, I found myself unable to look away. Then he swiftly kissed the tip of my nose. 'One day, Emilia, you'll make some man very happy,' he said, releasing his grip on me at last.

Flushed with embarrassment, I trotted along beside him to the waiting carriage.

'Your father is becoming extremely pressing on the matter of our engagement,' he said, once we had set off. 'He's determined his grandson will eventually inherit the estate. Perhaps we should bite the bullet?'

I looked at him, aghast. 'I thought you didn't want that, any more than I do?'

He shrugged. 'I'll have to marry someday and you're as congenial as any other girl of my acquaintance.'

'How very flattering!' I said.

He gave a wry smile. 'I apologise, Emilia. That was ungracious. You are a delightfully pretty girl, one whom any man would be delighted to marry. It's simply that I don't feel ready to tie myself down yet.'

'I'm not exactly sought after,' I said, 'since I've received only two proposals. I can't count Mr Fortescue's since he was inebriated at the time.'

'You haven't received more proposals because you don't flirt and flutter your eyelashes like the girls who are desperate to find a husband. That sends a message to those who are actively looking for a wife, rather than simply bent on enjoying female company.'

I sighed. 'Father won't be happy when I tell him I intend to return to Italy next spring.'

Millicent Deveraux, the vivacious daughter of a Whig Member of Parliament, had recently become engaged to the second son of an Earl. She was full of bright chatter about her wedding preparations as we drank tea and nibbled slices of cake.

'I simply can't decide whether to have the wedding next July, before the Coronation, or to wait until August when the fuss is over,' she said. She glanced at Dolly, who was smiling at a comment her mother had made. 'What about you? Have there been any interesting developments yet in that quarter? So many girls have tried and failed. He's very handsome, isn't he?'

'I'm enjoying the season,' I said, heat rising in my cheeks.

'It's nearly at an end,' said Millicent with a pitying smile. 'You must make haste to secure him. Though I did hear a rumour that he's in a little too deep at the gaming tables.'

I was saved further embarrassment when Millicent's mother clapped her hands and shepherded us into fours to play cards.

Two hours passed pleasantly and we were about to leave when Millicent's father rushed into the room. He hadn't removed his coat and his hair was windblown. 'I have an announcement,' he called out over the hum of conversation. His eyes shone and his smile was triumphant. 'The Queen is acquitted! She stood firm and has been vindicated!'

I laughed in relief, delighted that her troubles were at an end. A number of cheers went up and someone called out, 'Long live Queen Caroline!' but others were clearly unhappy about the Lords' decision. Everyone, even the young ladies whose fathers would have tried to divert them from the most sensational reporting as being unfit for their eyes, held vociferous opinions on whether justice had been done.

'Shall we go?' said Dolly over the commotion. 'The streets are bound to be crowded with revellers celebrating the Queen's victory.'

It was dark as we left the Deveraux' house but lights blazed in many of the windows and the streets were nearly as bright as day. The carriage was jostled and rocked from side to side as the hordes ran past, shouting until they were hoarse. Men carried banners painted with slogans such as 'Truth will Prevail', while others waved flaming torches and bottles of ale. There was an air of wild, feverish excitement. Men threw their hats in the air and couples danced to the discordant music of street musicians gathered into makeshift bands. Cannonfire boomed in the distance, women screamed, dogs barked, and all around was the crash of splintering glass.

Dolly remained silent but drummed his fingers on his knee as the carriage forced its way homeward through the press.

The clamour and delight of the crowd exhilarated me and I fervently hoped Queen Caroline was now relieved of her cares and happily celebrating her victory.

A few days later Aunt Maude still suffered from her throat infection and I, too, remained in bed. At noon I told Daisy that I was ill.

'Leave me to sleep until I come downstairs, will you?' I said. 'I don't wish to be disturbed. I'll ring if I need anything.'

Once the door closed behind her I threw back the bedclothes. Father was at the House of Commons and not expected back until late. I dressed warmly since it was foggy outside and locked the bedroom door behind me, placing the key in my reticule. I crept downstairs, heart thudding, and tip-toed across the deserted hall. Thankfully, the front door hinges were well oiled and I was able to slip outside without being noticed. Head down, I hurried away, knowing that I'd be concealed by the fog before I'd gone more than a few steps. I reached Bond Street and hailed a hackney carriage. Once inside, I was in a little world of my own since the fog didn't allow me to see any further than a yard or two ahead.

I'd read in the papers that Queen Caroline was staying in Brandenburgh House in Fulham, lent to her by the Margravine of Ansbach, and very much wished to visit her there to congratulate her on her victory.

By the time the hackney carriage drew up outside Brandenburgh House the breeze off the nearby Thames had dissipated the fog. From somewhere in the distance I could hear the noisy cries and shouts of merrymakers. 'Will you wait, please?' I asked the driver. 'I shan't be long.'

He pulled his scarf up over his ears and settled down.

Brandenburgh House was an elegant, classically styled mansion and a great deal more grand than the narrow-fronted house where I'd visited the Queen on the last occasion. I was pleased to discover she would receive me.

The drawing room was overheated and smelled unpleasantly foetid, an unwashed, greasy smell I recognised from the Queen's wardrobe in Pesaro.

'You have not forgotten me then?' she said, as I rose from my curtsey. Lady Hamilton sat beside the Queen at the fireside but this time there was no sign of the Countess Oldi.

'Of course not, Your Majesty,' I said. 'Since you kindly suggested I might visit you again, I wished to congratulate you on your recent victory.' I noticed that she had lost more weight and the skin on her face and neck sagged.

'I feel no joy,' she said, 'only a great weariness.' Her blue eyes were lacklustre and her hands trembled. She pulled a shawl more tightly around her shoulders and huddled into her chair. 'I am never warm in this miserable country.'

The tumult of noise outside rose to a crescendo and I glanced out of the window at the river. Boats crowded with sightseers waved bottles and handkerchiefs in the air as they called out greetings to Her Majesty. A man stood unsteadily in one of the small boats, singing.

'We have been inundated by the numerous parties that come to ogle Her Majesty every day,' said Lady Hamilton, her face pinched with disapproval. 'The rabble use Brandenburgh House as an opportunity for an excursion.'

'But no one really cares about me,' said the Queen, her chin quivering. 'The whole unpleasant business has been more of a political battle than the championing of a poor, forlorn woman. I have been unwell and am tired of it all ... so very tired.' She picked at a stain on her crumpled skirt.

'I'm sorry to hear that,' I said. I glanced enquiringly at Lady Hamilton, wondering if I should leave, but she gave a slight shake of her head.

'You remind me of happier times in my beloved Villa Vittoria,' the Queen said. 'How I long to return! Do you remember the sea breezes and the sunshine warm on your cheeks, Miss Barton?'

I didn't correct her but simply nodded. 'I miss them, too.'

The Queen sighed heavily. 'But it is not to be until the King has answered my demands. When will my name be restored to the Liturgy?' Her voice rose in agitation. 'Where is my town residence? I should have a palace of my own and a proper household. He cannot expect me to live in borrowed houses while I'm in London.'

'You will return to Pesaro later?' I asked.

'As soon as possible,' she said, closing her eyes. 'After my Coronation. Now that I am confirmed as Queen Caroline, even George cannot deny me that right. I miss my darling Victorine so. She will be bereft without me or Signor Fiorelli.'

'Have you any news of him?' I asked, hope flaring. 'I saw him in London from a distance but wasn't able to speak to him.'

'He wished to improve his English and took a position in London as tutor to two boys,' she said. 'He visited me during the trial recess last September. I always liked that young man.'

'Do you know the name of the family he works for?' I asked, my heart thudding.

The Queen pursed her lips and shook her head.

'It may have been Beacham,' said Lady Hamilton. 'And I believe he said he was staying in Great Marlborough Street.'

I gave a little gasp to realise that Alessandro had been living only a stone's throw from Father's house all that time. 'Thank you very much, Lady Hamilton.'

She surprised me by giving me a conspiratorial smile as she rang the little bell, signifying the audience was over.

The noise on the river increased as the voices swelled in raucous singing.

The Queen covered her eyes.

'I shall take my leave, Ma'am,' I said, 'with my best wishes for your imminent return to good health.'

'Always such a kind girl,' she murmured, 'just like my darling Charlotte.'

# Chapter 24

*November 1820*

**London**

I crept back into Father's house without being caught. I raced up the staircase and was unlocking my bedroom door when Daisy appeared.

'Oh, miss,' she said, 'I was that worried about you!'

I opened the door and pulled her inside. 'Does anyone else know I went out?'

She shook her head. 'I was in the kitchen making you honey and lemon for your throat when I saw you walking past the area steps.'

'I had some business to do,' I said, taking off my bonnet, 'and since Aunt Maude has been ill these past few days . . . '

'I would have come with you,' said Daisy. 'You shouldn't go out unaccompanied, not a young lady like you.'

'Well, no harm done,' I said. 'Daisy, will you do something for me?'

'Of course, miss.'

'I want you to deliver a note.'

'But that's James's job. I'm not sure Mrs Hope will let me . . . '

'If she's not happy you can ask her to speak to me.' I took a pen out of my writing desk and lifted the lid of the inkwell. After a moment's thought I began to write.

*Dear Alessandro*

*I was distressed when my father sent you away in such an unkind manner, especially since you had been so kind as to bring Peggy to me. Afterwards I assumed you had returned to Pesaro but I called upon Her Majesty this morning and Lady Hamilton mentioned you are currently employed in Great Marlborough Street, only a step away.*

*Will you meet me tomorrow at noon by the Cumberland Gate to Hyde Park?*

*In anticipation of seeing you very soon,*

*Emilia*

I sealed the folded paper with a blob of red wax. 'Please take this to Great Marlborough Street, Daisy,' I said.

She took it from me and frowned at the name written on the front.

'It's an Italian name,' I said. 'Signor Fiorelli.'

She echoed the name and I repeated it for her until she said it correctly.

'I don't know which house it is in Great Marlborough Street,' I said, 'so you'll need to make enquiries at the kitchen doors. Signor Fiorelli is employed as a tutor to two small boys.' I fumbled in my reticule and pressed some coins in her hand. 'He's an old friend from when I lived in Italy,' I said, 'but please don't mention this to anyone.'

'No, miss,' said Daisy, giving me a knowing look as she tucked the coins into her bodice.

I sat with Aunt Maude for the rest of the afternoon, simmering with excitement at the thought of seeing Alessandro. Although we'd parted with hard words, in my heart I knew we loved each other.

The Queen said he'd suffered a great deal after my departure. The balls, the presents, the new clothes ... none of that mattered to me as much as Alessandro did. I'd grown to love dear Aunt Maude and was pleased to know, if not love, my father but the endless round of social occasions held no further fascination for me. I'd saved most of my allowance to pay for my passage and was more than ready now to go home to Italy. I hugged myself in pleasurable anticipation of Alessandro's joy when I told him so.

I didn't see Daisy until she came to undress me for bed.

'I had to ask at a great many houses before someone could tell me where Signor Fiorelli was staying,' she said. 'There was a governess at number 12 who knew of him but he wasn't home. I left the note with a maid.'

I sat before the mirror while Daisy combed out my hair, wondering what Alessandro had thought when he received my message. 'I expect to meet Signor Fiorelli tomorrow at noon,' I said. 'I'd like you to accompany me.'

In the morning my reflection was pale and I was full of nervous agitation. At breakfast I was relieved to discover that Father was out so I'd have no need to explain my absence.

When I went to see Aunt Maude, she sat in her shawl by the fireside. 'I'm pleased to see you out of bed,' I said.

'I shall be right as rain in a few days, dear.'

'I'm feeling cooped up and intend to take a walk in the park this morning. Don't worry,' I said, as I saw her expression of consternation, 'I'll take Daisy with me.'

'That's not ideal,' said Aunt Maude, 'but I'm not sufficiently recovered to accompany you. Don't let Daisy stray from your side and come to see me when you return.'

'In Italy I frequently walked alone. No harm ever came to me.'

'No one in London need know what happened in Italy,' said Aunt Maude. 'It's not so much that I fear physical harm, my dear, as long as you hold your reticule tightly ... it's more that I fear malicious gossip might harm your reputation. A young woman can't be too careful.'

'Yes, Aunt Maude.'

Daisy had laid out my clothes and suggested I use a little of my Pears' Liquid Blooms of Roses to brighten my cheeks. I dressed in my favourite walking dress with a matching blue velvet pelisse and plumed bonnet.

'You look a picture, miss,' said Daisy.

I smiled, eager to set off.

Once outside, I was happy that the sun shone, albeit in a pale November fashion. I was relieved it wasn't raining; rain had such a dispiriting effect upon the bonnet plumes. It was a relief, for a change, not to have to meander at Aunt Maude's pace but to be able to walk as fast as was ladylike.

We reached Cumberland Gate but, disappointingly, Alessandro wasn't there.

'We'll walk for a while, Daisy,' I said. I didn't want him to think I was too eager, even though it was hard to contain my excitement.

Ten minutes later, Alessandro still hadn't arrived. It was possible he hadn't received my note, I supposed, but I fervently hoped Father's mistreatment of him hadn't changed his former good opinion of me. Although we'd parted on unhappy terms there had been such easy affection between us before. I knew that some men were not good letter writers but it hurt me that he'd made no attempt to contact me.

'My feet are cold,' I said. 'Let us walk to Stanhope Gate in case he's mistaken the location.' My mood of happy expectancy began to dissipate.

We hurried towards the next gate and I sent Daisy to speak to the flower seller who had a stall there but she returned, shaking her head. I worried we might have missed Alessandro at Cumberland Gate and we dashed back but there was still no sign of him. There we waited a further quarter of an hour while disappointment and humiliation gnawed at me.

'I don't think he's coming, miss,' said Daisy. Her nose was cherry red with cold.

'No,' I said, leaden disappointment in my breast. 'I don't think he is.'

We'd only gone a little way down Park Lane when I heard running footsteps behind us. I whirled around and my spirits soared. Alessandro stood before me. His hair was neat and he wore a well-cut dark coat that could not hide the breadth of his shoulders but his usual smiling demeanour was startling by its absence and that discomfited me.

'I thought you weren't coming,' I said. Two boys stood beside him.

Alessandro made a small bow. 'I apologise,' he said. 'George couldn't find his shoes and then William became tired of walking. May I introduce you?'

The children, wooden boats under their arms, made their bows.

'We shall talk while my charges sail their boats,' said Alessandro, turning away from me.

Once we'd entered the park gates the boys ran ahead towards the Serpentine. Daisy walked a discreet step behind me but, in any case, Alessandro's expression was remote and he didn't speak or offer to take my arm. I hadn't expected him to be so unapproachable and it alarmed me. A bitter wind gusted across the park and pinched my cheeks.

We reached the lake and Alessandro helped George and William to launch their boats and found them each a long stick with which to guide their craft.

'Daisy,' I said, 'will you watch the children while I speak to Signor Fiorelli?'

She nodded and went to stand by the water's edge.

'Alessandro, I can only apologise for Father behaving so badly towards you when you called upon me at Grosvenor Street,' I said, slipping naturally into Italian. 'I ran after you but you disappeared into the crowds.'

'I wanted to return Peggy to you. The doll was on Victorine's bed when you left last year. I thought you might miss it.' His voice was curiously flat and there was no dancing light today in his brown eyes.

'Oh, I did,' I said. 'But what are you doing in London? I thought nothing would induce you to leave Italy and your family?' I knew I was gabbling but Alessandro barely looked at me, gazing over my shoulder at the boys playing with their boats.

'Only my great love for you could have persuaded me to do so,' he said in that same expressionless tone of voice. 'After you left, nothing had any meaning without you. Every day was a torment. And so I came to London to look for you.'

My spirits soared again. He did still love me! 'Why didn't you come and see me? Or at least write to me?'

'You know I tried to see you and I did write to you,' he said. 'It hurt me that you never replied to my letters but I came here to try again to persuade you to marry me. I was dying without you and I wanted to take you home.'

'I promised you I'd return, Alessandro,' I said. 'I know I've been away longer than I expected but . . . '

'You've been away much longer than I expected! When you left you begged me to write to you but after your first letter saying you wished to spend more time with your father, you never wrote to me again,' he said, his voice rising. 'For a whole year you left me believing you still loved me, despite your insistence on staying in London. I hoped every day you would return to me. I was driven mad with misery and in the end I left my home and my work to come to you. But you didn't have the courtesy to tell me you'd become engaged to another.'

'But I haven't!' I said. 'And I didn't receive any letters from you.'

'Don't lie to me, Emilia!' His eyes blazed now.

I stepped back. 'How can you accuse me of that?'

'Your father told me you are engaged to his heir.' Alessandro shrugged. 'It wounded me here,' he pressed his clasped hands to his heart, 'but I understand. These things happen and the life you live now, this rich life, is a dream come true for you. I cannot compete with your future husband, who has everything to offer you.' Shaking his head, he continued, 'But I am deeply disappointed in you, Emilia, for not telling me the truth before I found it out for myself.'

'But I'm *not* engaged to Dolly!' I blazed with anger at Father. How could he have told Alessandro such a lie?

'Emilia, I saw you.'

'Saw me?' I was puzzled.

Alessandro buried his face in his hands for a moment. 'I often waited over the road,' he said, 'praying you'd come out of your father's house alone. I gave notes to the servants, hoping you might come to the window to speak to me. But then I saw you on the steps with your future husband. He held you close, so close, in his arms, and your sweet mouth was waiting for his kiss.'

I gasped. 'It wasn't like that.'

'I *know* what I saw, Emilia. For *you* I have spent months in misery away from my family in a country that is cold and mean-spirited. For *you* I must remain here until my year's contract is finished.' He drew a shuddering breath. 'Enough. I have come to my senses at last and I will not destroy myself any longer by loving a girl who doesn't want me. We shall not see each other again.'

I reached out for him. 'Alessandro! You must believe me!'

He shook off my hand and strode away without looking back.

'You're wrong, Alessandro!' I remained fixed to the spot, trembling with distress. I stared after him as he collected the two boys and set off along the footpath.

A moment later Daisy came and touched my arm. 'Are you all right, miss?'

All right? My heart had broken, not in two but into smithereens. Alessandro and his charges were fast dwindling into the distance. I swallowed. It was too hard for me to speak.

'Come along, Miss Langdon,' said Daisy. Her voice was soft as if she were cajoling a child. 'Let's get you home and I'll ask Cook to make you a nice pot of chocolate. You look fair froze to death.'

By the time Daisy and I reached home we were both chilled to the bone. Daisy brought me the pot of chocolate and I insisted she drank the second cup.

'If you're struck down with a chill, Daisy,' I said, 'it will be my fault.'

She brought me a shawl and rubbed my frozen toes. 'I'll fetch you a hot brick wrapped in flannel,' she said.

An hour later I was warm again though I still shivered each time I relived the terrible conversation I'd had with Alessandro. It made my chest so tight it was hard for me to breathe.

During the afternoon my misery turned to anger. No matter what lies Father had told, Alessandro should have believed me. The deciding factor for him had, of course, been his mistaken belief that he'd seen me about to kiss Dolly. If only I hadn't slipped on the steps!

I was still alternately distraught and seething with rage when the front door slammed and then Father's voice boomed out in the hall. Barely pausing to think, I hurried downstairs, rapped on his study door and went in without waiting for his answer.

'Why did you tell Alessandro I was engaged to Dolly?' I blurted out the question before I'd had a chance to take a calming breath. 'You know it was a lie. And what did you do with the letters and notes he sent to me? I never received any of them.'

The expression in his steel-grey eyes was cold and unflinching. 'I don't care for your tone, Emilia. Do I take it that you have been in contact with that unsuitable foreigner . . . an illicit meeting, in fact?'

'You lied to him!'

'Answer my question, Emilia. At once!' thundered Father. 'Have you been meeting him in secret?'

'Only this morning and I wouldn't have resorted to subterfuge if you'd allowed him to speak to me when he called here.' I was shaking with anger.

'How dare you question my judgement!' Spittle formed at the corners of his mouth. 'Did you think I was going to allow my daughter to throw herself away on a low-born foreigner?'

'You've *always* known I intended to return to Italy and marry Alessandro.'

'And yet you were happy enough to stay here for a year enjoying the benefits I could give you.' His voice was cold. 'What more could I have done for you, Emilia? Have you not been pleased with the ball I gave for you, the new clothes and jewellery and the opportunity to move in the most exalted circles? Why, the King himself came to your ball . . . '

Heat flooded my face. 'You know I'm grateful to you, but my real reason for staying here was in order to get to know you and Aunt Maude.'

Father slumped down at his desk and buried his face in his hands. 'Can't you see, Emilia, I want only the best for you?'

'Alessandro *is* the best for me.' I tried to stop my voice from wavering. 'And you lied to him and now he doesn't trust me.'

My father sighed deeply. 'One day, when you have children of your own, you'll understand how much they can hurt you. You're still young, Emilia, but you will come to learn there is more to happiness than stolen kisses. You must have security.'

'With Dolly, you mean?'

'Be reasonable, Emilia!' He rubbed the bridge of his nose. 'Open your eyes to the advantages of being his wife. You're a beautiful girl and have had every opportunity of meeting a suitable husband during the past year. Many were titled, most were young and several had fortunes, but you gave no encouragement to any of them. It's unlikely you'll meet any suitable men more to your liking in a second season and by then there will be a plethora of younger girls seeking a husband. You may have missed your chance already.'

I shuddered. 'I don't want a second season.'

'A year is a long time to a young man,' said Father. 'Alessandro has probably found another girl by now.'

'He wouldn't!'

'And what then, Emilia,' said Father, 'if you return to Italy and this Alessandro doesn't want you? You could continue to scratch a living as a dressmaker to the bourgeoisie, of course. Dolly, meanwhile, will marry and you'll lose that opportunity. Eventually you'll find life alone in Italy too difficult and come creeping back to me. I'd give you a home, of course, just as I have to Aunt Maude, but she'll be the first to tell you how difficult it is for a spinster living on charity.'

I shivered at the disdain in his voice.

'If you've had a disagreement with Alessandro,' said Father, 'I can only say, however upset you are at present, that you've had a lucky escape. You've returned to the life you were born into and I suggest you reflect on your very good fortune. And remember, Dolly won't wait for you forever.'

Unable to listen to any more of this, I fled.

# *Chapter 25*

*December 1820*

**Langdon Hall**

We arrived at Langdon Hall a week before Christmas. Father had planned a variety of diversions to entertain his house guests over the festive season and, to my dismay, I was expected to act as his hostess.

'I fear I shall say or do something wrong,' I confided to Aunt Maude. I sat listlessly by the fire in the small parlour, planning the guests' table placements for the various dinners.

'I remember your dear mother saying the same thing to me many years ago,' said Aunt Maude. 'She was afraid she'd forget to order the musicians or the flowers and worried about inviting guests who wouldn't find each other congenial.'

'Did she?' I didn't want to think about my traitorous mother, however high she stood in Aunt Maude's esteem. That only proved how double-faced she'd been.

Aunt Maude smiled. 'Dear me, yes! Rose was an anxious hostess as a new bride but soon she forgot her worries and if there was a difficulty she managed to laugh about it.'

I turned back to the scraps of paper, each inscribed with a guest's name, that I was arranging around the perimeter of a larger sheet representing the table. As host, Father had to sit at one end but what was the correct order of placement for the rest of us? There were so many traps for the unwary that could cause offence.

'Rose kept a notebook for entertaining,' said Aunt Maude. 'She included names of the guests, the food and wine served to them, on which date, and other useful information. I wonder if we still have it?'

'It would be terribly old-fashioned now,' I said, distracted by the difficulty of deciding which of two local landowners' wives was the most important.

'Good manners never go out of fashion, dear,' said Aunt Maude with a reproving smile. 'I'll see if I can find it.'

Five minutes later I pushed the placements aside with an expression of disgust and went to look out of the window. There was an area of muddy grass and shrubs below and then the moat. The water was deathly still and very dark today, almost black, reflecting the sullen sky above. A few leaves floated on the surface.

I wondered what my mother had thought about when she drowned herself and if she'd considered at all the needs of the little daughter she was leaving behind. I imagined her being swept along by the current and battered against the banks and bridges. Or perhaps she'd sunk straight down to the river bed and felt the mud squeezing between her toes as she watched the silvery bubbles of her last breath floating to the surface. I closed my eyes, picturing her struggling against the weight of her sodden clothing while her red hair floated around her face like waterweed.

Shivering, I returned to my chair by the morning-room fire. I'd been prone to black moods lately, made worse by succumbing to a feverish chill the previous month after waiting in the cold to meet Alessandro. I couldn't shake off the malaise that ailed me and had given up any hope of receiving an apologetic message from him. It hurt me deeply that he hadn't believed me when I told him I wasn't engaged to Dolly. But Father could be extremely persuasive when it suited his own ends. When I remonstrated with him I'd glimpsed an entirely different side to his character from that of the affectionate parent that he usually presented to me.

The door opened and Aunt Maude returned. 'Dolly has arrived,' she said. 'He's talking to your father but says he'll join us in a while.' Triumphantly, she waved a red morocco-bound notebook at me. 'It was in the library,' she said, 'still on the shelf where Rose kept it.'

I made an attempt to rouse myself from the apathy that gripped me and reached for the notebook with a forced smile. 'How interesting!'

In fact, it was interesting, filled as it was with my mother's bold writing covering all aspects of her life as hostess at both Grosvenor Street and Langdon Hall. There were sketches of flower arrangements, menus, notes on which guests felt an antipathy for each other and mustn't be invited at the same time, addresses of recommended purveyors of hams and cheeses, small orchestras available for hire . . .

I put the notebook down as Dolly opened the door. Since I'd been unwell, I hadn't seen him for more than a few minutes at a time over the past month. 'Dolly, how nice to see you!' I said, noticing his face was even paler than usual and there were deep shadows under his eyes. 'You're a most welcome relief from arranging table placements.'

'I should think so, too,' he said, bowing to Aunt Maude before sitting down beside me. 'Would you care for me to advise you?'

'The very thing! I'm tired of it all.'

The next hour passed pleasantly enough and I even found the energy to laugh at some of Dolly's more acerbic comments about the intended guests. I glanced up at him once or twice to see that he was staring at me with an oddly watchful expression.

Finally, I slipped my lists into the red notebook and closed it. 'Aunt Maude and I shall speak to Cook in the morning about the menus.'

Aunt Maude stood up. 'If you'll excuse me, I'm going to rest before dinner.'

After she'd gone Dolly rubbed his eyes and yawned.

'Tired?' I asked.

He smiled ruefully. 'I've had rather too many late nights recently. And perhaps a little too much wine.'

'You can go to bed early tonight,' I said. 'Straight after supper, if you wish, since there aren't any guests but you. I hope you won't find it too tedious.'

'Indeed not,' he said, 'since you're here to keep me amused.'

'I don't feel very amusing at present.'

He regarded me closely. 'Stand up!' he commanded.

I obliged and he looked me up and down.

'You've lost all your womanly curves.'

'That's not kind!' I sat down again.

'It suits you to be so slender,' he said. 'If I put a laurel wreath on your head and handed you a lyre, I could be looking at a young Apollo.'

I narrowed my eyes at him. 'While you look dissipated enough today to be Dionysus.'

'You surprise me, Emilia, by your knowledge of Greek mythology.'

'I'm not just a pretty face,' I said. 'I've enjoyed the educational benefits of travel. Now, stop appraising me as if I were a piece of horseflesh you were considering at Tattersalls! I've barely shaken off that chill I caught last month.'

'Since neither of us is on top form we shall have to remain cosy by the fireside together, playing cards like an old married couple, shan't we?'

'Which is exactly what Father wants.'

Dolly kept his gaze on his highly polished boots for a moment. Then he cleared his throat and looked at me. 'Perhaps it is time to make both your father and me the happiest of men, Emilia?'

I froze, the light-hearted moment between us gone in an instant.

He placed his long white fingers over mine. 'Neither of us were ready for marriage before. And now your father tells me your Italian lover hasn't come up to expectations.'

I flinched, pulling my hand away, and it was all I could do not to cry out at the memory of my last painful meeting with Alessandro.

'Am I so loathsome to you?' asked Dolly. The expression on his face was hurt.

'I'm sorry,' I said, my voice high and tight. 'Of course you're not loathsome! You're one of the handsomest men in London. All the girls' mothers are hoping to catch you for their daughters.'

He cleared his throat again. 'It's taken me time to be sure I'm ready to settle down,' he said, 'but now I wish, most wholeheartedly and ardently, to make you my wife, Emilia. I've grown exceedingly fond of you. More than fond. In fact, you fill my thoughts so entirely I cannot eat or sleep.'

Struck dumb, I could only stare at him. Was this a declaration of *love*? I noticed once more that tiny tic at the corner of his eye.

'Emilia, say something!' He lifted my hand and pressed it fervently to his mouth. 'Can you not see that I'm desperate for you?' There was a sheen of perspiration on his top lip.

I swallowed. 'I'd no idea you had feelings of that sort for me,' I said, at last.

'I didn't, at the beginning. And when I began to admit to myself that I'd fallen in love with you, I was concerned I'd frighten you away if I began to court you,' said Dolly. 'But now, I can conceal my passion no longer.' He slid off the sofa and onto his knees at my feet. 'Emilia, please will you be my wife?'

I wanted to be somewhere else, anywhere else. I'd become used to considering Dolly as an ally against my father's domineering plans for me. Discovering now that Dolly actually loved me caused such turbulent feelings within me I could not fathom them.

'Emilia?'

'Dolly ...' My mouth was dry. 'You've taken me by surprise.' I forced a smile. 'You gave me no inkling of your feelings.' I closed my eyes briefly, recalling how I'd slipped on the steps and he'd clasped me against him while Alessandro watched us from the other side of the street.

'I know we could make a success of our marriage,' said Dolly, his voice urgent. 'Please, please, say yes.'

'Give me a little time.'

'Time!'

My eyes widened as he raised his voice.

'I apologise,' he said hastily, 'but you've had nearly a year to think about it.'

'Of course I haven't!' I looked down at my hands, twisted together in my lap. Dolly loved me and Alessandro never wanted to see me again. 'The situation is entirely different now.'

'Tomorrow then?'

I looked away from the intensity of his gaze while my heart fluttered against my ribs. 'Give me until Christmas Day, Dolly. There's so much for me to organise here and if you press me too hard, I shall simply say no.'

He let out his breath in a ragged sigh. 'Then I must be patient.'

It was awkward and uncomfortable to look at him and I felt I no longer knew how to speak to him lightly. 'I shall retire to my room for a while,' I said.

Dolly nodded. 'You need time to think. Perhaps I'll rest too.'

We went upstairs together in silence, stopping in the Long Gallery outside the guest room where he was to stay.

'I hope you'll be comfortable,' I said, in formal tones as befitted his hostess. 'Please ring should you require anything.'

'Only you, Emilia,' he murmured, before closing the door behind him.

I avoided being alone with Dolly over the next few days, staying firmly by Aunt Maude's side and busying myself with the forthcoming festivities. At night I lay awake, struggling to decide which course of action to take. The truth was, I didn't know what to do.

It had always been my intention to return to my beloved Italy. Used to a quiet life with Sarah, I'd been overwhelmed sometimes by Alessandro's family, but I knew now that I wanted nothing more than to be a part of it. The cool, polite manners of English society held little charm for me.

Now that Alessandro had cast me aside, there was little point in my returning to Pesaro and my life as an itinerant dressmaker. The time had come for me to decide if I'd be better off staying with my newfound family. Except, of course, that I'd be married to a man I didn't love. The surprising change in our situation was Dolly's unexpected declaration. There were practical advantages in marrying him but could I ever learn to love him?

On the morning of Christmas Eve, I went to supervise the servants while they brought in the Yule log and the festive greenery. The log had stopped smoking and was beginning to spit and crackle in the stone hearth of the Great Hall when Dolly came to find me.

'Shall I help with the decorating?' he offered.

A number of servants were bustling about nearby, bringing me scissors and twine and sweeping up clumps of mud and moss that had fallen off the Yule log so I didn't fear any awkward conversations. I eyed his exquisite coat with its shiny brass buttons. 'It would be a shame,' I said, 'to snag such fine wool with a sprig of holly.'

Five minutes later, Dolly sported one of the linen aprons the footmen usually wore while polishing the silver. 'Thankfully none of my fashionable London friends will see me dressed like this,' he said.

'And you won't upset your tailor by ruining your coat, either.'

He smiled at me. 'What a very understanding girl you are!'

He surprised me by climbing nimbly up a ladder to drape an ivy wreath along the minstrels' gallery and helped me make a kissing ball with the sprigs of mistletoe. We wound ivy and sprigs of holly around the candles on the vast iron light fittings and he stood beside me, watching as the footman hauled the swaying cartwheels of candles up towards the star-painted ceiling again.

'The decorations look splendid,' said Dolly. 'We're a good team, don't you think?' He smiled tentatively.

'Better than I might have imagined,' I said. He had a smudge of dust on his cheek and his cravat was crumpled. I liked him the better for it. To avoid his gaze, I gathered up a holly wreath, climbed onto a chair and placed it on one of the stags' heads. 'How does that look?' I asked.

'Beautiful.' Dolly held out his hand to help me from the chair and his grip was firm and warm as I stepped down. 'Beautiful,' he said again, looking straight into my eyes.

I noticed how long his black eyelashes were but there were little lines of worry etched around his dark blue eyes. He stood so close that his breath fanned my cheek and I smelled his sandalwood cologne. Now that Alessandro didn't want me, would it be so very dreadful to be married to Dolly? I'd be safe, with a home of my own and, in time, children to love.

His hand tightened around my fingers. 'Emilia?' he whispered.

'Yes, Dolly,' I said, suddenly calm as the decision was made. 'Yes, Dolly, I will marry you.'

'Thank God!' he said. He pulled me tight against his chest and pressed dry lips firmly to mine.

# Chapter 26

*July 1821*

**London**

Millicent smiled complacently as the company raised their glasses to the bride and groom.

'She looks like the cat that got the cream, doesn't she?' murmured Araminta. 'As well she might, having secured the second son of an earl. Her grandfather was only a linen-draper, you know.'

'I hope they'll be happy,' I said.

Araminta opened her brown eyes very wide. 'How could she not be, with his fortune?'

I sipped my champagne. Happiness was a fleeting emotion. Once I'd accepted his proposal, seven months later Dolly's declared passion for me appeared to have waned with familiarity. Father, of course, had been delighted when we confirmed our engagement. Almost at once I'd felt as if I were on a runaway wagon careering down a mountainside as he made wedding plans. He decided the ceremony would take place in St Paul's Cathedral in June. I'd demurred, using the excuse that he'd be occupied working with the committee organising the Coronation. Finally, though I'd have preferred to delay even longer, I'd agreed to the end of August, after the London season finished. Dolly and I were to be married at St Bartholomew's in Upper Langdon.

'Your wedding next,' said Araminta, 'and mine in September.' She stretched out her hand and smiled at the diamond on her finger. 'Is your wedding dress finished?' she asked.

I avoided thinking about my wedding. 'Almost,' I said. 'Mrs Webbe was inundated with orders for gowns for the Coronation. I had to beg her to fit me in.' I'd offered to make my wedding dress myself but Father had reeled back in shock. I didn't tell him but it was only my connection with the Queen that finally persuaded Mrs Webbe to accept the commission.

'All the world is being fitted for outfits for the Coronation at present,' said Araminta. 'Oh, look, there's Anna! Excuse me, I must have a word with her.' She drifted away.

I went to find Dolly. He was talking to Francis Gregory, who seemed to appear everywhere we went. It wasn't that I wanted to be alone with my betrothed but I'd become irritated by Mr Gregory's high-pitched laughter and pointed remarks.

'Here comes your little shadow, Dolly,' he said. His smile didn't reach his eyes.

'I thought the boot quite on the other foot,' I replied tartly, 'wherever my fiancé and I go, there you are.'

Mr Gregory's naturally pink cheeks glowed just a little brighter than usual. 'Dolly and I have been friends since our schooldays.'

'And no doubt we'll still be toddling off to play cards together in our dotage, Francis,' murmured Dolly.

'I hear you've been to visit your future mother-in-law, Miss Langdon,' said Mr Gregory. 'Dolly says you were the best of friends by the time the second cup of tea was poured.'

I glanced at Dolly, who refused to meet my eyes and cleared his throat in that annoying way he did when he was nervous. 'It was very pleasant to meet her,' I said. In fact, Mrs Pemberton had such an insipid personality it had been difficult to have any kind of conversation with her.

'I'm so pleased,' said Mr Gregory. 'Warfare with your mother-in-law would be so dispiriting, don't you think?' His blue eyes were guileless. 'Especially when she's going to reside with you.'

I managed to prevent shock from showing in my face. 'That isn't yet decided,' I said evenly. 'Dolly, have you told Mr Gregory that Father is giving us the Dower House on the Langdon Estate?'

'Not yet,' said Dolly. 'But we must also have a place in town. I've no intention of living up to my knees in mud all year round.'

Mr Gregory pressed a hand to his heart as if the very idea pained him.

'I shall be pleased to spend more time in Hampshire,' I said. 'After a while, the social engagements in London all seem to blend into each other.' In fact, I'd begun to loathe them.

Dolly looked down his nose at me. 'You remember what Samuel Johnson said, my dear? "When a man is tired of London, he is tired of life."'

Mr Gregory laughed. 'The perfect solution presents itself! Miss Langdon, you shall live in Hampshire with the delightful Mrs Pemberton for company while Dolly will remain in Mayfair, close to his tailor and his club.'

'No doubt we shall divide our time between town and country,' I said. I had no intention of wasting my energy bickering with Mr Gregory.

'If he'd still been alive you might not have found Dolly's father so pleasant, of course,' continued Mr Gregory. He looked pensive for a moment. 'He had very strong opinions, didn't he, Dolly?'

Dolly shuddered. 'And tried to thrash them into me at every opportunity.'

'Died of an apoplexy while in a rage,' Mr Gregory told me with obvious relish.

I didn't feel it appropriate to comment.

Later the assembled party went outside to wave to the bride and groom as they set off for Scotland in a shiny new landau.

I was relieved when it was time to leave and Father came to find Dolly and me.

'I've been so occupied with the arrangements for the Coronation,' said Father, 'that we haven't yet talked about your honeymoon.'

'Perhaps we might go to Bath?' I said. 'It's very fashionable.' Wherever we went, I hoped there'd be plenty of diversions. The prospect of spending a great deal of time alone with Dolly was daunting.

'Bath is for milksops,' said Father in tones of disgust. 'You shall travel to Italy. You can show Dolly Pesaro and other towns where you stayed.'

'I like the idea of Bath,' said Dolly. 'Travel abroad is so exhausting and the beds are never properly aired.'

'I'd prefer not to return to Italy,' I said. The last place I wanted to go with Dolly was to where I'd been so happy with Alessandro and his family.

'Nonsense!' Father rubbed his hands together with satisfaction. 'It's a perfect idea. It'll give you the chance to retrace your steps and look for the Infanta's miniatures. Despite all my letters I've had no more contact . . . ' He broke off at sight of the expression on my face. 'What is it, Emilia?'

I glanced at Dolly, whose lips were pressed tightly together as he studied his boots.

'I have no secrets from Dolly!' said Father. 'After all, he's my heir as well as your fiancé. He knows all about the missing miniatures.'

'Still,' I said, 'I don't wish to go to Italy.' It would be far too painful to be reminded of what I had lost.

The smile faded from Father's face. 'I thought you were committed to the idea of returning the miniatures to their owner?'

'We need not make final decisions about a honeymoon yet,' said Dolly, clearing his throat again. 'We'll discuss it after the

Coronation. And how are the preparations proceeding, Sir Frederick? I heard that the King's red velvet Coronation robe is to be nine yards long, exquisitely embroidered with gold stars and lined with ermine.'

'Indeed,' said Father, easily deflected to his involvement with the Coronation arrangements. 'His Majesty will be most gloriously attired. A new crown has been made, too, encrusted with over twelve thousand diamonds.'

'And what of the Queen?' I asked.

'It's the King's prerogative,' said Father, 'to decide whether Caroline of Brunswick will be crowned and he'll never agree to it.' He laughed. 'She actually wrote to the Home Secretary declaring it was her right. Her mantua-maker is preparing splendid dresses for her but, mark my words, she'll never wear them in Westminster Abbey.'

'Can the King prevent her attendance?' asked Dolly.

'He won't have her anywhere near him,' said Father. 'He's worried she'll make a scene so he's hired prize-fighters to be dressed as pages. They'll stand by all the doors to the Abbey to stop her.'

Father and Dolly continued to discuss the Coronation while I looked out of the carriage window. I knew better than to defend the Queen in Father's hearing but it saddened me she was not to be crowned. Although declared innocent at the end of her trial, her reputation had been left in tatters. Now that the Radicals and the Whigs no longer found her useful in supporting their cause against the King, her popularity with the people had waned. I'd visited her in March and been shocked by how little care she was taking of her toilette. Unwashed and wrapped in a coarse shawl, she'd laughed about the popular belief that she'd taken to drink but I'd seen tears in her eyes even as she made a joke of it. The next time I went to visit her I was told she was indisposed.

The carriage come to a halt in Grosvenor Street.

'Dolly, there are some matters of estate business to discuss,' said Father. 'Come to my study, will you?'

Dolly handed me down from the carriage and I was pleased to escape upstairs to the solitude of my bedroom.

A few days later, after Aunt Maude had retired early and Father had gone out, I curled up on the chair by the open window in my bedroom and watched the sun setting behind the roofs opposite. I wrinkled my nose as the breeze wafted in the odour of horse dung and drains from the street below. I was bone tired but knew this was the physical manifestation of the sickness of the spirit that troubled me. All those years I'd travelled with Sarah I'd been desperate for a place to belong. Now that I'd returned to my family home, I realised it hadn't brought me fulfilment. I had everything a girl could possibly want but I wasn't happy.

There was a tap at the door and Daisy stood in the doorway, her fingers twisting together. 'Begging your pardon, miss, but may I have a word?' She glanced over her shoulder.

'Come in.' Her expression was so agitated I wondered if she'd come to confess a petty misdemeanour: a scorched lace collar or a snagged stocking.

'It's your Italian gentleman friend,' she whispered.

I sat bolt upright, my stomach twisting into knots. 'He's here?'

'He was in the area outside,' said Daisy. 'I didn't know what to do with him so I hid him in the coal store.'

Flustered, I stood up and smoothed my hair. I wanted so much to see Alessandro but this time I had to face him as Dolly's fiancé. 'Did he say what he wanted?'

Daisy shook her head. 'He was very excitable, miss, you know how foreigners are, and said you're to come at once.'

We hurried downstairs and Daisy led me into the scullery. The day had been hot and it smelled unpleasantly of rotting vegetables. The scullery maid, up to her elbows in greasy water, hastily averted her eyes when I smiled at her.

We stepped into the paved area and stood outside the door to the store room under the pavement.

'I'll go back,' Daisy whispered.

I fumbled with the latch and slipped inside the coal store. The air was heavy with dust and mould. I stood still, listening to the thudding of my heart, until my eyes grew accustomed to the gloom.

'Emilia?' Alessandro's voice was barely a whisper.

Light filtered through the crack around the coal-hole cover in the pavement above. In the dimness I saw him. My mouth was dry and I hardly knew what to say after the way we'd parted last time.

His shoes crunched over scattered coal. It was all I could do to stop myself from running into his arms. Immediately I saw him I knew I still loved him.

'I had to come,' he said in an undertone, 'to warn you.' He spoke in Italian but there was no warmth in his demeanour.

'Warn me?' I'd hoped he'd come to apologise for not believing me when I'd told him I hadn't received any letters from him and that I wasn't, at that time, engaged to Dolly.

'Listen to me.' His voice was urgent. 'You're in great danger.'

Whatever I might have expected, it wasn't that.

'I followed him,' said Alessandro, 'and then I realised who he was.'

'You're not making sense,' I said.

He grabbed me by my upper arms and shook me. 'Listen, will you! That man you're going to marry ...'

'I know what you thought,' I said, 'but I *wasn't* going to marry him when we spoke last.' I had to make him believe me. 'I only became engaged to Dolly *after* you made it so clear you didn't want me.'

'You cannot marry him. He's a murderer.'

Nervous tension made me laugh. 'That's ridiculous!'

He released my arms and stepped back. 'I followed him again last night ...'

'What do you mean, you followed him?'

'I've followed him on many nights,' said Alessandro, shrugging. 'He spends his evenings in gambling dens. *Of course* I wanted to know more about this blackguard who'd stolen the affections of the woman I once loved. But last night I was close enough to hear him cough and then I recalled where I'd seen him before. Do you remember I told you I'd seen a man looking in through your cottage window just before Sarah was murdered?'

'In Pesaro?'

'Exactly! I challenged him then,' said Alessandro, 'but I had Victorine with me and he loped off on those great long legs of his. Don't you understand, Emilia, *he's* the man who killed Sarah? I heard him make that strange cough at the time but thought no more about it.'

I remembered my own fear when a man had loomed up in the darkness in the passageway behind the cottage and almost knocked me over. 'But . . . Dolly has never been to Italy,' I said, bewildered.

'I *heard* him and I *saw* him, Emilia.'

I stared at Alessandro through the gloom. What could have made him imagine such a thing? 'You're jealous!' I said.

He spat on the ground. 'Jealous? I despise him. I tell you again, you must not marry that man.'

'You're jealous of him and you've made up this terrible lie.' My voice was cold. 'I thought better of you, Alessandro.'

'Why can I not make you see the truth?' he hissed, kicking at the coal. 'If you still insist on marrying a murderer, then I can do no more.' Roughly, he pushed past me. He snatched open the door and then he was gone.

I pressed my hands to my burning cheeks and fought back despair. I should have ignored our misunderstandings and made him listen to me. And now there was this dreadful accusation. Alessandro's words rolled around in my head as I fought to regain outward composure.

Eventually I felt calmer and opened the door into the area.

'Emilia?'

I went cold. Father was peering down at me over the railings. Had he seen Alessandro leave?

'What are you doing there?' he said.

I looked back at him, struggling to find an excuse for being in the coal store. Then it came to me. 'I heard a cat yowling,' I said, 'and came to see if it was hurt.'

'Did you find it?'

'It must have run away.'

'For goodness' sake, come up before anyone sees you! Your face is smudged with coal dust.'

'Yes, Father.' I hurried up the steps and followed him inside.

# Chapter 27

Dolly and I had been invited to stay with the Perry family in Northumberland Gardens on the night before the Coronation. Daisy came to wake me at four in the morning, to dress me in my blue satin dress with an embroidered train. She arranged my hair in elaborate curls and topped them with white ostrich plumes. I made my way downstairs, with the feathers bobbing at every step, feeling more than a little self-conscious.

Mr Perry, Araminta, her fiancé Mr Carlton, and Dolly were already in the dining room, dressed in their finery.

'Isn't this exciting?' said Araminta, twirling around to show me her dress of apricot-coloured silk. 'You'd better have some coffee and bread since we don't know when we shall next eat.'

Sir Peter and Mrs Perry came to join us and after a hasty breakfast the ladies and Sir Peter climbed into the waiting carriage, while Mr Perry, Mr Carlton and Dolly set off on foot. We'd barely turned into Whitehall when we came to a stop. Carriages filled the road ahead, making it impassable.

'We'll have to walk,' said Sir Peter. 'If we're separated, go down

Parliament Street and turn left into Palace Yard. Our box is on the second tier, to the front of the first house.'

I held my train over one arm to prevent it being trampled on as we shuffled along in the press of the crowd. All was good-natured, however, with a palpable air of excitement. As we approached Bridge Street we saw the raised and canopied processional route snaking its way from the north door of Westminster Hall, turning down Bridge Street, into King's Road and thence to Westminster Abbey.

It took nearly an hour to walk the short distance to Palace Yard. Already the public had gathered at every window and even on roofs. Some must have camped there all night. Finally, we pushed our way to the tiered pavilions bolted to the front of the houses. An attendant took our tickets and we climbed the rickety steps to a narrow gangway leading to our box. It was disconcerting to look through the gaps between the boards to the ground below but, once seated behind a guard rail, I felt perfectly secure.

'We have a splendid view of Westminster Hall from here,' said Araminta. 'Father had to pull a few strings to secure our seats but it was well worth it, don't you think?'

'It certainly was,' I said. Our box afforded us not only a clear view to the front but also a limited view to our right where the raised walkway turned into Bridge Street.

'The procession will pass directly in front of us after it leaves Westminster Hall on its way to the Abbey,' said Dolly. 'There's nothing to do now but wait until ten o'clock when the King arrives.'

Araminta and I amused ourselves by looking at the ladies' gowns and arguing over which was the prettiest, while Sir Peter and Mrs Perry used their opera glasses to spy out friends and acquaintances. The sun was hot and I was thirsty. An orange seller came along the gangway behind us but we couldn't risk dripping juice on our gowns or staining our white gloves.

Dolly took a silver flask from his coat. 'Brandy?' he asked. I shook my head and he took a sip and put the flask away again.

I watched the crowd, my mind dwelling on the quarrel with Alessandro, wondering what I could have done to make him listen to me. And then there had been his dreadful accusation that Dolly had murdered Sarah . . . I glanced sideways at my fiancé, who toyed languidly with his opera glasses, looking for someone of interest in the sea of faces. It was a ridiculous idea; I'd never seen anyone look less like a murderer than Dolly did. I rubbed my aching temples. Alessandro had never lied to me but in this he must be mistaken.

I was beginning to fidget on my uncomfortable seat when I heard a commotion in the crowd in Bridge Street. There was a series of whistles and then a shout, 'The Queen forever!' I craned my neck to see what was happening. A black carriage, closely followed by a yellow state coach drawn by six bays, forced its way through the crowd.

'Well, well! It seems the Queen has decided to attend the Coronation despite her lack of an invitation,' drawled Dolly as the carriages proceeded slowly down Margaret Street and past Westminster Hall.

I caught a glimpse of the Queen's white face and her feathered headdress at the carriage window. There were cheers and boos in equal measure as the two carriages turned towards the Abbey and disappeared from sight. I was afraid for her. It must have taken a great deal of courage to make that journey, especially since she'd received an uncertain welcome, so different from the ecstasy of the crowd when she was acquitted after her trial.

Nothing seemed to happen for a while but then the noise from the chattering crowd increased.

'Look!' said Mr Perry. He snatched the opera glasses from his mother's hand and trained them towards the Abbey. 'It's the Queen!'

In the distance two figures walked into view and stopped by the West Door of the Abbey. It wasn't possible to see more, even though we all took it in turns to strain our eyes with the opera glasses. Soon the carriages reappeared, stopped before the West Door and the Queen and her escort climbed back in. The carriages then trundled out of sight towards the Poets' Corner door, where the Royal Family were expected to enter the Abbey.

The crowd was full of speculation and word was passed from person to person that the Queen had been barred at both the West and East Doors of the Abbey.

The carriages came around the corner again and stopped outside Westminster Hall. The Queen descended with her escort but the guards shouted and some of the crowd hissed and yelled as she approached the door. Pages clad in scarlet livery and carrying battle-axes surrounded the party. Raised voices drifted towards us on the breeze. There were cries of 'Shame!' and 'Go back to Pergami!' as the guards closed ranks.

Defeated, the Queen climbed into her yellow carriage again and, as the bells of Westminster pealed out, she was driven away.

'That's the last we'll see of her!' said Araminta, unfolding her fan and fluttering it before her face.

'At least the spectacle kept us all amused until the main event of the day,' said Dolly, over the cacophony of the bells.

I watched the crowd in silence, noticing the sun glinting off brilliant jewels and feathers bobbing up and down as people chattered in excited anticipation, while the thought of the poor Queen's utter despair and humiliation almost brought me to tears. I hoped Lady Hamilton would comfort her.

The King was half an hour late. The crowd became restless as the sun grew hotter. Our box was in full sun and my bodice was too tight. Surreptitiously, I eased it under my arms, hoping my face wasn't too unattractively flushed from the heat.

Smart carriages came and went and, at last, the procession emerged from Westminster Hall to loud applause from the crowd. First came the heralds with a joyful burst of sound as they announced their King. Seven herb-women followed, all chosen for their beauty and dressed in flowing white muslin as they strewed lavender, rosemary and flower petals along the processional route.

Dolly nudged me. 'That's the Lord Great Chamberlain holding the mace,' he said, 'and here come the princes of the blood, the Lord Chancellor, the King's gentlemen and other officers of state.'

A collective gasp went up from the assembled company as the magnificent figure of the King appeared, nodding and bowing gracefully to his subjects. I was suffering from the heat but how much worse it must have been for the King in his ceremonial dress. Treading slowly and majestically, he wore a suit made of cloth of silver, lavishly trimmed with gold braid. His vast velvet train of scarlet and gold, trimmed with ermine, was carried by eight pages. He wore a black Spanish hat surmounted by sprays of ostrich feathers and a black heron's plume. Whatever my private opinion of him might be, he looked every inch a King.

A swelling wave of cheers and applause came from the crowd along the processional route.

'What a glorious show!' said Mrs Perry, her eyes shining as the procession finally disappeared from sight.

'And so it should be,' said Sir Peter. 'They say the King's clothes alone cost more than twenty-four thousand pounds!'

Araminta's eyes opened very wide.

'The King wanted his Coronation to be a grander affair than Napoleon's,' said Dolly, 'and I'd say he's achieved that, wouldn't you?'

The spectators settled down to wait while the Coronation took place out of sight inside the Abbey. Araminta and I joined the queue for the ladies' retiring room in the house behind our box and later Sir Peter bought us all a slice of pie. Another trader sold us glasses of negus. Mrs Perry, her nose shining in the relentless sun, complained the negus had been disgracefully watered down but I was grateful to quench my thirst.

After an interminable length of time a roar went up from the crowd as the news filtered out of the Abbey that the King had been crowned. Some threw their hats in the air. A little while later the procession left the Abbey and returned with all pomp and ceremony to Westminster Hall for the Coronation Banquet. Once the dignitaries had disappeared, the crowd began to disperse.

'I'm cramped from sitting so long,' said Araminta as we queued to descend to the street, 'but the celebrations aren't over yet.' The feathers in her headdress had wilted in the sun, giving her a rakish air. 'You will both come with us to the Coronation Fête in Hyde Park this evening, won't you? We'll return to Northumberland Gardens for dinner and then go on to the fête.'

'There's to be a balloon ascent and a pair of elephants pulling a golden carriage in the opening parade,' said Mr Perry.

'How could we miss such an opportunity?' said Dolly.

I gathered my crumpled train over my arm and we were jostled along with the happy horde. All around us was good humour and extravagant praise for the King. I clung to Dolly's arm and reflected soberly on how fickle people were. Only a few months ago there had been fervent support for the Queen but now, humiliated and reviled, she'd been forced to retreat to lick her wounds while the King, puffed up with self-importance, held court amongst his sycophants.

The following morning, I slept very late. I found Aunt Maude waiting for me as I came, yawning, down to breakfast.

'Your father is still abed,' she said.

'He wasn't home when I returned late last night.'

'I've seen the newspaper reports of the Coronation already,' said Aunt Maude, 'but I want to hear all about it from you.' She poured me a cup of strong coffee.

'It was a wonderful spectacle,' I said, 'though it saddened me to see the Queen turned away.'

'A shameful thing!' Aunt Maude shook her head sadly.

I sipped the hot coffee, feeling it revive me. 'I wish she hadn't tried to enter the Abbey. It was demeaning for her to be refused entry.'

'She always was impulsive and strong-willed,' said Aunt Maude. 'She must have known the King wouldn't allow her access and counted on her previous popularity, hoping the crowds would rally behind her. I expect she wanted to provoke him.'

'I imagine you're right.'

Some twenty minutes later I'd described all that I'd seen and Aunt Maude had exclaimed in wonder at it all.

'And then,' I said, 'there was the Coronation Fête in Hyde Park. I shall never forget the elephants in the opening parade. I'd never seen real ones before and I'd no idea they were so big. They wore gold headdresses and the most beautiful, brightly coloured blankets decorated with glittering sequins. And I do wish you could have been there to see the balloon ascent. It made me dizzy to see it rising so high in the air.'

'How wonderful to see the world from a bird's-eye point of view,' said Aunt Maude, 'though I should have been terrified to be so far off the ground.'

'There were Chinese lanterns in the trees,' I said, 'and an illuminated temple topped with a gold crown to commemorate the day. And at the end of the celebration there was a magnificent firework display that went on for at least twenty minutes. I daren't think how much it all cost.'

'I daresay the taxpayer will foot the bill,' said Aunt Maude.

I sipped my coffee. Now the excitement of the Coronation was over there was little to divert my thoughts from my latest distressing meeting with Alessandro. I was sure he was mistaken about Dolly but regretted that my refusal to believe it had prevented us from making up our differences. I sighed deeply, my heart aching for Alessandro.

'You look tired, dear,' said Aunt Maude. 'Shall we have a quiet day reading in the garden?'

'We must hold a ball,' said Father that evening after dinner, 'in continuing celebration of the King's Coronation. Will you arrange it as soon as possible, Emilia? Aunt Maude will assist you.' He smiled. 'And then the next big event will be your wedding, of course.'

My heart sank. 'Hasn't everyone had enough Coronation celebrations?' I said, hoping he'd change his mind. I glanced at Aunt Maude, sitting beside me on the sofa, but her expression remained inscrutable.

'Of course they haven't!' Father leaned back in his chair with a half-smile on his lips. 'The King is delighted that, at last, his people have recognised him for the great man he is. Thankfully Caroline of Brunswick has completely fallen from favour. Turning up at the Coronation, and all that undignified tramping about trying to push her way past the doorkeepers, made her lose any credibility she might once have had. The expectation is that she'll retreat to Italy and her lover with her tail between her legs.'

'She was shamefully treated,' I protested.

'Nonsense! No more than she deserved.' Father smirked. 'She had the gall to write to the King, demanding to be crowned on Monday. Monday! She'll be crowned on no day at all, as far as he is concerned. Anyhow, enough of that wretched woman! I shall inform the King that we're arranging a ball in his honour. He may even grace us with his presence again.'

'Is there a particular advantage to you if he attends the ball?' I asked. I was sure that there must be.

Father pursed his lips. 'He'll be pleased to hear that I continue to be a loyal supporter. And then, as you know, he's a great patron of the arts. He might be interested in my collection.'

I had an inkling of what was in Father's mind. 'And perhaps, if you were his agent, you might have the honour of procuring paintings or sculptures for his residences?'

'I have excellent contacts with art dealers both here and on the continent,' said Father, 'and many satisfied clients.' He smiled winningly. 'You have a good eye yourself, my dear. Should there be something in particular the King has in mind, you would be well placed, while on your honeymoon in Italy, to be my eyes and ears when you visit the dealers to make enquiries.'

'As I said before, Father, I'd prefer to visit Bath. Certainly Dolly doesn't want to go to Italy.' If Alessandro had known how much Dolly disliked foreign travel, he'd never have accused him of being in Pesaro when Sarah was murdered.

'I don't see why not,' said Father, sounding distinctly peevish. 'It's not as if he'll be paying for the honeymoon.'

'Will you give me your guest list for the ball?' I said, changing the subject to something less contentious. 'I'll talk to the printers and the musicians today and see how soon we can make arrangements.'

Father beamed. 'Good girl! Let me know tomorrow so I can be sure to hold it on an evening the King is free.' He stood up. 'I'm going to my club now so don't wait up.'

Aunt Maude sighed heavily after he'd left the room. 'He has no idea how much work is involved in organising a ball.'

Over the following days she and I were kept fully occupied. I threw myself into making the arrangements to divert my thoughts from Alessandro. We visited the printers to find out how soon they could produce the invitations, called on the florists to discuss extravagant arrangements of exotic flowers, hired chairs and ordered a magnificent sugar-paste centrepiece for the supper table from Gunter's. This was to be a Chinese temple, topped with a golden crown, similar to the one in Hyde Park at the Coronation Fête. There would be sugar models of elephants and spectators watching the balloon ascent and the whole arrangement would be set in a sugar-paste landscape representing the park.

Mrs Hope and Cook had discussed with us the elegant dishes to be provided for the supper and embarked upon a frenzied onslaught of additional cleaning and polishing. I interviewed and selected an opera singer to entertain us during the interval in the dancing and booked my dancing master to act as Master of Ceremonies.

At the end of a particularly busy day Aunt Maude and I drooped with exhaustion and could only pick at our supper.

'Whatever's the matter with you both?' said Father, spearing another slice of cold beef. 'Not sickening, I hope?'

I glanced at Aunt Maude's wan face and had an idea. 'We've run ourselves into the ground with preparations for the ball.' I said 'We'd like to visit Langdon Hall for some restorative country air.'

Father chewed contemplatively at his beef. 'Is everything organised? It must be perfect for the King.'

'We can do no more now until the last day or so,' I said. All at once I couldn't wait to escape, not only from the stifling heat in town but also from Father's overbearing ways. 'It would be *such* a pity if I were not well enough to act as your hostess.'

Father sighed. 'I suppose there's no real need for you to remain in town then,' he said.

I glanced at Aunt Maude, demurely contemplating her untouched supper. 'We shall return refreshed and ready to complete the final preparations for your ball.'

# Chapter 28

**Langdon Hall**

The fountain in the centre of the knot garden splashed gently beside us as Aunt Maude and I drank our tea. Bees hummed on the lavender and the sun was warm on my arms. Langdon was a different world from London.

'Take care not to burn, dear,' said Aunt Maude. 'You don't want freckles for your wedding day, do you?'

I reached out and trailed my fingers in the cool water of the fountain. I didn't want to think of my wedding day. Meeting Alessandro again had thrown me into turmoil. I thought endlessly about the dreadful accusation he'd made. Despite our differences I trusted Alessandro, but I was sure he'd come to a mistaken conclusion about Dolly.

'You've been lost in thought for quite five minutes,' said Aunt Maude.

I withdrew my hand from the water. 'Am I making a terrible mistake in marrying Dolly?' I enquired.

She gave me a sharp glance. 'Do *you* think you're making a mistake?'

'I don't love him,' I said, 'but that's not unusual amongst my acquaintances. Father tells me how advantageous the marriage will be . . . '

'But your heart isn't in it?'

I shook my head.

Aunt Maude put down her teacup and squeezed my wrist. 'You mustn't marry Dolly if you dislike him.'

I shrugged. 'He's presentable and well-mannered. I agreed to marry him because he said he loved me. But there's little sign now of the passion he showed for me at Christmas. It seems familiarity breeds complacency.'

'Love may grow once you're married,' said Aunt Maude, but her expression was full of doubt.

'I thought, if I married him, soon I'd have children to love.'

'What if there are no children?' said Aunt Maude.

I pictured Alessandro's big, noisy family and there was a hollow feeling in my heart. 'Then I should be very lonely,' I said.

She sighed. 'Do you still yearn for Alessandro?'

'He doesn't love me anymore,' I said, 'and he thinks I lied to him.'

'Emilia . . . ' Aunt Maude hesitated. 'An unhappy marriage is worse than no marriage at all, however difficult it is to be a spinster. Your parents' marriage was a disaster.'

'I know. He told me about Mother's affair.'

'Affair?' Her expression was outraged. 'There was no affair!'

'Perhaps she hid it from you so you didn't think ill of her?'

'Rose was as honest as the day,' said Aunt Maude, 'and utterly transparent.'

I didn't want to argue with an old lady but I'd seen Father's misery when he told me about my treacherous mother. While I didn't yet love him, his unhappiness had moved me.

A cloud passed over the sun and I shivered. 'Shall we walk?'

We ambled along the gravel paths, stopping to smell the flowers, while all the time I recalled with misery Alessandro's last words to me.

'I wish you remembered Rose,' said Aunt Maude. 'You'd know then how honourable she was.'

'I'm still hurt she abandoned me, a small child, to fend for myself in the world,' I said. 'If she'd really loved me she wouldn't have taken her own life.' My steps faltered as I remembered Alessandro saying, 'If you loved me you wouldn't leave me.'

'I don't know what happened,' said Aunt Maude, 'but I know she loved you above all else.'

Except for her lover. The pain of losing him had made her turn her back on my needs. 'All those years I thought Sarah was my mother,' I said, 'there was something missing. Perhaps I blotted out memories of Mother because they were too painful. Sarah tended to my physical needs but there was never any real connection between us, except that we were bound by our mutual need for survival.'

We walked silently through the avenue of clipped yews and rested on a bench beside the moat. The water level was lower than usual and it was green and turgid, giving off a foetid smell in the summer warmth.

'Frederick usually tells the gardener to open the sluices when the water is so low,' said Aunt Maude. 'The moat refills from the river.' She stared at the water, her brow furrowed.

'Rose kept a diary,' she said. 'I wonder what became of it?'

My interest quickened. Mother's diary might answer my questions, even if it made me unhappy. 'Might it be in the library, where you found her entertaining notebook?'

'Shall we go and see?' said Aunt Maude.

The diary wasn't in plain view in the library.

'She kept it private,' said Aunt Maude, 'and out of your father's way. She wrote in it when we were sitting together sometimes. She often concealed it in her sewing basket but that is long gone.'

We searched the library more thoroughly, taking down each book from the lower shelves in turn. Hours later we brushed dust from our hands and conceded defeat for the day.

The following morning Aunt Maude sat in the wing chair while I used the library steps to investigate the upper shelves. Now that I knew of the diary's existence it had become a matter of vital importance to me to find it.

At last I'd searched all the shelves, rummaged through the library tables, the cupboards and the desk. Frustrated, I came to the conclusion that the diary simply wasn't there.

'Where else might she have hidden it, Aunt Maude?' I asked.

'Your father has kept Rose's bedroom locked ever since she drowned.'

A tiny thrill of anticipation shivered down my back. 'If the room's been undisturbed since she died,' I said, 'then there's a chance it's there.'

'Frederick wouldn't like you going in there,' said Aunt Maude. There was a worried crease between her eyebrows. 'I really don't want to make him angry.'

'I shan't disturb anything.'

'I'd rather not know what you intend to do, in case he questions me. I shall take a turn around the garden.'

I rang for the footman and Aunt Maude left the library by the French windows.

'Robert,' I said, 'will you ask Mrs Bannister for the key to the locked bedroom?'

Five minutes later the housekeeper appeared, her stout figure neat in a black bombazine dress and white collar. 'Robert tells me you requested the key to Lady Langdon's bedroom?'

'I did,' I said.

'Sir Frederick gave express orders for that room to remain locked,' said the housekeeper. 'I'm permitted to enter only twice a year to dust.'

'I shall not require you to enter the room,' I said, 'only to unlock it.' I looked down my nose at her in my best imitation of Lady Hamilton's haughty expression.

After a moment, Mrs Bannister dropped her gaze. 'Very well, Miss Langdon, if that is your instruction.'

My breathing quickened at the thought of what I might discover as we walked upstairs together. Mother's diary might be the key to unlocking the family mysteries that troubled me.

In the Long Gallery the housekeeper stopped outside the locked room, selected a key from the chatelaine around her waist and opened the door.

'That will be all,' I said, outwardly calm. 'I'll send for you to lock it again later.' I stepped over the threshold and closed the door firmly behind me.

It was shadowy dark inside the room and I opened the faded silk curtains and folded back the shutters. Sunshine flooded in, illuminating delicate satinwood furniture, an Aubusson carpet and embroidered bed drapes in soft shades of green, rose pink and violet. The panelled walls were painted the same pale green as the curtains. Something about the room, the restful colour perhaps, was extraordinarily calming. It felt like a safe haven from dangers I didn't even know existed.

I sat on the bed, my fingers stroking the silky coverlet, and watched dust motes drifting lazily in the sunlight. Something teased my memory. Counting. Little hands ... my hands ... clapping. A clear voice singing. I closed my eyes, willing myself to remember as my fingertips caressed the roses and violets embroidered on the coverlet. I gasped as all at once I heard the song in my head.

*Roses are red,*
*Violets are blue*
*Sugar is sweet*
*And so are you!*

Mother's hair silky against my face as she snatched me into her arms and kissed me. The warm, sweet scent of her violet perfume. Giggling and squirming as her lips tickled my cheek. Begging her to sing me the nursery rhyme again. And then the memory was gone.

I reared to my feet but the happy echo from my past had disappeared, taking my mother with it. My pulse raced as I wondered if I'd been wrong about her. Perhaps she'd loved me after all.

Systematically, I began to search the room. The chest of drawers remained full of her clothes; lacy shifts, embroidered nightgowns and silk petticoats, still faintly perfumed with violets. Dresses hung in the wardrobe, too full-skirted for today's fashions but all exquisitely finished. Shoes were arranged in pairs and I slipped on a yellow silk dancing shoe decorated with a tulle rose. It cupped my foot as if it had been moulded to me. Shivering, I took it off and replaced it where it belonged.

I sat at the dressing table and opened the drawers to find a box of fine-milled powder and a half-used bottle of Olympian Dew. A chased silver cachou box held pink lip salve, shrunken and cracked now but still scented with Otto of Roses. I dipped my forefinger into the waxy compound and looked in the mirror while I rubbed the salve on my lips. It made me shiver, with pleasure rather than sadness this time, to think my mother would have been the last person to touch the salve.

I began to look for the diary in earnest. The back of the drawers yielded only a few hairpins and a handkerchief. I searched under the bed, feeling along the frame, I climbed on a chair to reach up to the top of the wardrobe, rolled back the carpet searching for a loose floorboard and, finally, felt inside the chimney.

I stood in the centre of the room. 'Mother,' I whispered. 'Tell me where to find your diary.'

Nothing.

Disappointed and inexpressibly sad, I closed the shutters and drew the curtains. The room was full of shadows again and I tip-toed to the door and left the ghosts to whisper in peace.

The following days passed in outward tranquillity. I retrieved Mother's portrait from where I'd concealed it in the back of my wardrobe and propped it up on the chest of drawers. Her eyes seemed to follow me wherever I went but now that didn't unnerve me. I gazed at her painted face for hours at a time, trying to read her expression. Sometimes I fancied she was trying to speak to me and I sat, motionless, listening for memories of her voice.

Aunt Maude and I visited the pretty Queen Anne dower house on the estate, which was to be my home after the wedding. As I walked through the echoing hall I thought I might be as happy there as anywhere. It had been closed up for years and I'd relish choosing new curtains and wallpaper to my own taste. Living in the house with Dolly as my husband was another prospect entirely and panic fluttered in my breast.

That night I paced up and down while dread made my chest constrict so tightly I could barely breathe. Again and again I wondered about Alessandro's accusation. Soon Dolly would be my husband and England my permanent home. For evermore I would have to forget about Italy and the man I still loved there, despite our difficulties.

In the light of day, I dismissed my night fears and simply avoided thinking about the wedding. Aunt Maude and I drove into the countryside and enjoyed a picnic on the banks of the River Test. I was happy to see her animated when we visited the local churches, explored the churchyards and the tea shops in the nearby towns. She appeared less frail when she was happy.

Mrs Digby, the wife of Father's solicitor, heard that we were staying at the Hall and called upon us. The weather was delightfully warm and we sat on the terrace overlooking the knot garden and the moat.

'It must be eighteen months since we first met, Miss Langdon,' said Mrs Digby. 'And here you are,' she said, 'happy in the bosom of your family and engaged to be married.'

'Two years ago I could never have imagined it,' I said. But then, two years ago I'd been in Italy, the country I loved, anticipating a happy future with Alessandro.

'I'm delighted for you,' said Mrs Digby. Her smile was sincere. 'And, if you won't think me presumptuous, may I say how proud of you your mother would have been?'

'Do you not think Emilia is so very like Rose?' said Aunt Maude.

Mrs Digby nodded. 'The picture of her.'

'I don't recall her,' I said. 'Although, here at Langdon Hall, I've experienced momentary flashes of childhood reminiscences.'

'I'm delighted to hear that, dear,' said Aunt Maude. She patted my wrist. 'I very much want you to remember Rose as she really was.'

'Sometimes I hear her singing a nursery rhyme,' I said. 'And I was standing on the bank of the moat the other day when I recalled her snatching my hand with a warning to be careful.'

'She was always anxious you'd fall in,' said Aunt Maude. 'The water smells dreadful in hot summers like this. Rose worried that if the river flooded, the moat would rise, too, and come into the house.'

'What an unpleasant thought!' I wrinkled my nose, imagining the stinking water seeping under the doors.

'Don't be concerned,' said Mrs Digby, 'the river hasn't flooded in fifty years. And, as far as I remember, Rose was more worried about the thefts of paintings from some of our friends and neighbours.'

'Father mentioned the art thefts to me.'

'We have no valuable paintings since Mr Digby has no interest at all in art.' Mrs Digby smiled. 'So I never had any anxiety that a thief would break in to take them. But Rose fretted that the thief might be someone we knew, someone who might steal Sir Frederick's collection from Langdon Hall. Thankfully, that never happened.'

Mrs Digby gathered up her reticule. 'I've enjoyed meeting you both again and hope I shall see more of you when you're living in the dower house, Miss Langdon.'

'I hope so, too,' I said.

After Mrs Digby had gone, Aunt Maude went inside to rest as was her habit in the afternoons. I wandered through the house, looking again for somewhere Mother might have hidden her diary. Of course, it was possible Father had destroyed it after she died. One place I hadn't searched was his study and I plucked up the courage to enter his private sanctum. I waited until the servants were elsewhere and slipped inside.

The shutters were closed and I dared to open them only a crack to admit some light. One wall was covered with built-in bookshelves but these contained only old ledgers and folios pertaining to estate business and his parliamentary affairs. The remaining walls displayed paintings and the desk was locked. I didn't dare to force it open. I would search the townhouse when I returned to London the next day, I decided. Dejected by my failure to find the diary, I closed the shutters again.

'That's everything packed, miss,' said Daisy. She closed my travelling bags and tightened the straps.

'Will you ask Edward to carry the luggage down?' I said.

'Very good, Miss Langdon.'

I dreaded returning to London. Father's ball was in a couple of days' time and afterwards I'd have to endure making the final arrangements for my impending wedding. Heavy-hearted, I went to look at my mother's portrait for the last time before we left. 'Where is your diary, Mother?' I whispered. I fancied she watched me as I left the room.

I went down to the little parlour to check I hadn't forgotten anything. Sun streamed in through the window and a wood pigeon cooed outside. Suddenly I was four years old again. I remembered kneeling on the window seat watching the pigeons billing and cooing on the lawn outside while Mother worked on her embroidery beside me.

That was it! I raced up the back staircase to the attics. Lighting the candle from the shelf outside the storage room where Father had kept Mother's portrait, I opened the door. Inside was a cradle and various trunks amongst dust-sheeted furniture. I heaved up the lid of the first trunk to find folded curtains and then another packed with children's clothing. I shook out a little muslin dress and there was something so familiar about the sprigs of yellow primroses and the pattern of the lace trimming that I could only believe I had once worn it myself. I fumbled hastily through layers of dresses, vests and tiny nightgowns until I was sure that the trunk contained only clothes.

Throwing back the lid of another trunk, I rummaged through ladies' shoes, a box of paints and a bundle of brushes. My heart nearly stopped when I found a leather book but it was a sketchbook of indifferent landscapes and not a diary at all. Delving deeper into the trunk my fingers scraped against something rough. My spirits soared when I saw it was a wicker basket.

I dragged it out of the trunk with trembling hands. Needle cases, scissors, pins and a small tape measure were arranged in the top tray. A second contained serried ranks of embroidery silks. Underneath that was an embroidery hoop containing a half-finished work of blue tits nestling amongst wildflowers. I burrowed beneath, amongst folded pieces of canvas, and then became very still as I felt smooth leather under my fingertips.

My mouth was dry as I extracted a green leather book with a brass clasp and the name 'Rose Langdon' embossed in gold on the cover. I hugged the book tightly against my breast.

I had found my mother's diary.

# Chapter 29

*August 1821*

**London**

I attempted to read Mother's diary in the carriage on our return to London but after a few tantalising pages the rolling motion of the coach made me feel so queasy I was forced to stop.

'My spectacles are packed,' said Aunt Maude, 'so we'll have to be patient.'

We arrived in Grosvenor Street in time for supper.

'I expected you to return before this,' said Father, his mouth pursed in disapproval. 'My ball is the day after tomorrow. I trust everything is in hand?'

'Absolutely,' I said, 'and I don't anticipate any difficulties.'

'I hope not,' he growled.

When he'd left the supper table for his club, I heaved a sigh of relief and soon I was sitting on the window seat in my bedroom opening Mother's diary. It was strange to read her private thoughts and after an hour or so I began to feel as if I knew her. It brought tears to my eyes to see her comments about me, her *'pretty, clever little girl'*, and her hopes and dreams for me as I grew up. Even if she'd had a lover, it was clear that she'd loved me. She wrote also of her joy when she discovered that she was to have another child and hoped it would make Frederick happy:

*I fear what he will say if I should bear another daughter. Although he is fond of little Harriet he makes sharp comments about me not doing my duty if I don't provide him with an heir.*

I skimmed through the pages, taking little notice of the accounts of parties and social visits, except when she mentioned that Lord Cosgrove's house had been broken into and an important Russian icon stolen.

*So many thefts! Not a month passes without another of our acquaintance being robbed by this impudent thief. There is a deal of speculation that the perpetrator may be one of our own set and that makes me look askance at our friends. The thief has a sense of humour since he leaves a pen-and-ink sketch of an empty picture frame in place of the work he's stolen.*

Presumably, at that point, Mother had not been involved with the thief who later became her lover. I read on.

*I cannot bear to envisage Frederick's rage if one of his paintings were to be stolen. His love for his collection borders on the obsessive and he certainly values it more than he loves his wife or child.*

This distressed me but I could only assume she exaggerated. Then came the entry about Piers's birth.

*Frederick is pleased with me and I am thankful that I have come through the ordeal with no more than the usual difficulties. My darling son, Piers Frederick George Langdon, made his appearance at dawn this morning, two weeks early. He is so tiny but already very dear to me. Sweet Harriet kissed his little cheek and offered to give him her precious Annabelle. In respect of my children I am the luckiest woman alive. They mean everything to me and make up for the rest.*

My heart was very full as I read that entry. But what did she mean by 'the rest'? No matter what might have happened later, Mother had clearly loved her children. Nevertheless, something had happened to drive her into the depths of such despair that she had abandoned me when she drowned herself.

I read the next entries about my little brother's increasingly sickly constitution in the unhappy knowledge that it was to lead to tragedy.

*The poor little mite suffers from colic and screams himself into sobs several times a day. Frederick wants to send him to a wet nurse but I'm determined to continue nursing him myself. I cannot bear to be parted from my poor babe. Bravely, I argued with Frederick. He was very harsh to me and left the house with much ill feeling between us. I am grateful to have dear Aunt Maude to comfort me.*

It was apparent Mother hadn't taken her parental responsibilities lightly. My heart bled for her as I read about Piers's continual colds and his general failure to thrive, despite her best efforts. And then came the terrible day I'd expected to read about.

*My angel has been taken from me. I'm ashamed because I slept well for the first time since his birth since I didn't hear him cry during the night. When I lifted him from his cradle in the morning I knew at once that he had gone. He was already cold and not my Piers anymore. Little Harriet doesn't understand and cries piteously for her baby brother, while Frederick rages and blames me for not sending his son to a wet nurse.*

I touched my cheek and discovered it was wet with tears. The entries for the following months were infrequent but I read of Mother's continuing grief and of my parents' disintegrating marriage. Father coped with his sorrow by blaming her for their son's death. His temper flared at the slightest provocation and I was shocked when I read about the day he'd hit her, splitting open her lip. Two days later he hit her again.

*The floodgates of his rage are opened. Frederick says I disgust him and he cannot bear me in his sight.*

Horrified, I read on, holding the diary with fingers that shook with sorrow and rage.

*I'm rarely able to leave the house since I usually have a black eye, a bruise or a cut cheek. Last night Aunt Maude tried to defend me and was punched for her kindness. I cried and pleaded with him when he threatened to put her out in the street, for she has no income and nowhere to go.*

Sickened, I put the diary down. I couldn't bear to read any more until I'd considered what I'd learned. Father was irascible at times. I'd seen him push Alessandro down the steps but I hadn't imagined he was truly violent.

Confused, I allowed Daisy to undress me for bed, and then lay sleepless in the dark while I tried to make sense of it all.

In the morning, as soon as I was dressed, I reread some of the diary passages, trying to determine if Mother had really been mad or savagely mistreated instead. She appeared to be of sound mind but I'd seen Father's sorrow when he'd told me about their disintegrating marriage. I didn't know what to believe and read on. I gripped the diary tightly when I came to the entry about the theft of Lord Beaufort's miniatures.

*At the Ashworths' rout this evening the assembled company were outraged to hear that the Picture Frame Thief has struck again. Three miniatures of the Infanta Isabella Clara Eugenia, daughter of King Philip II of Spain, have been stolen from Lord Beaufort. Only two weeks before Frederick and I, amongst others, were his house guests at Little Braxton Manor where he had proudly showed us the newly acquired works.*

Puzzled, I rested the diary on my knee. The miniatures had been stolen not long before Mother had drowned. Since she wrote her innermost thoughts in a diary she kept hidden, why had she as yet made no mention of her lover, the Picture Frame Thief? I was still pondering on whether she'd had a lover at all when Aunt Maude tapped on my door.

'Are you ready to go to Gunter's?'

I'd completely forgotten our appointment to inspect the sugar-paste centrepiece we'd ordered for the supper table at the ball. Distractedly, I tucked the diary under my pillow and put on my bonnet.

In the carriage Aunt Maude said, 'Have you read the diary?'

I nodded. 'Not all of it. I'd like you to read it later and tell me if it marries with your recollections.'

The sunshine was hot when we arrived at Gunter's Tea Rooms, where a great number of vehicles were drawn up in the shade of the trees in Berkeley Square. Ladies sat in their carriages eating ices, while their escorts leaned against the railings nearby, chatting to other young bloods. Waiters dodged through the traffic, bringing the ices from the tea rooms to the carriages before they melted.

'I suggest you wait here, Aunt Maude,' I said. The heat didn't suit her and she looked weary. 'Once I've approved the centrepiece, we'll have an ice.'

'That will be delightful, dear.'

I crossed the road to the tea rooms but Mr Gunter was busy with another customer so I sat down to wait.

Two ladies were drinking tea and gossiping at the table beside me.

'My sister was at the Drury Lane Theatre last night,' said a lady in a yellow muslin dress. 'Mr Elliston was performing in a pageant of the Coronation and he impersonated the King so well it was like a portrait. And that wasn't the only interesting event of the evening. The Queen and her party arrived and a whisper went around that she became unwell during the performance.'

My ears pricked up at the mention of her.

'Sick with jealousy, I expect,' said the lady's companion, busy eating a macaroon.

'Still, she stayed to the end and Augusta said she rose and curtseyed to the pit, galleries and boxes. "Positively haggard" was how my sister described her; a complete figure of fun with her wig crooked and her crumpled dress all anyhow.'

My heart bled for the Queen but I didn't overhear any more since Mr Gunter greeted me then. He led me to a workroom where the centrepiece was laid out on a clean cloth. 'As you see,' he said, 'we have yet to make the second elephant and the fountains are under construction.'

'I believe I mentioned that the King intends to honour us with his presence at our ball?'

'We'll work through the night to ensure the piece will be ready for tomorrow,' he assured me.

'It's exquisite,' I said.

Mr Gunter bowed.

'And now I shall purchase two of your excellent ices.'

I returned to the carriage followed by a waiter bringing a violet sorbet for Aunt Maude and a *neige aux pistaches* for me.

We gave our full attention to our ices for a moment. Then I said, 'Aunt Maude, did Father ever treat you violently?'

She paused in the act of catching a dribble of violet sorbet with her tongue. Her eyes filled with tears. 'I tried to stop him from hurting Rose. He was terribly angry and threw me to the ground and kicked and beat me.'

So Mother's account of that event, at least, was true. Softly, I touched my aunt's papery cheek while anger at my father burned inside me like a red-hot ember.

'It wasn't the first time,' she said, 'nor the last. He also threatened to put me in the workhouse.' Her chin quivered while she regained control of herself. 'I've never dared to flout his wishes again.'

'He's been kind to me,' I said, 'and given me so much.'

'Frederick rarely does anything unless it suits him.'

'Do you remember when Lord Beaufort's miniatures were stolen, before Mother died?'

'It was in the papers,' said Aunt Maude. 'There was a terrible furore.'

'Father told me that the so-called Picture Frame Thief was Mother's lover.'

Aunt Maude shook her head decisively. 'Ridiculous! Rose never had a lover.'

'Father and Sarah told conflicting stories. Father said Mother had hidden the miniatures in her luggage. Sarah said there was little of value among Mother's possessions. She ran away with me, taking the luggage, because she was expecting Mother to join her later.'

'I don't understand that at all.' Aunt Maude frowned. 'Rose would have told me what was happening if I'd been there. I'll always regret I wasn't at Grosvenor Street when she died.'

'Where were you?'

'Staying with my cousin's daughter. Her children, all eight of them, had measles, one after the other. She was utterly distracted and needed help to nurse them. Rose insisted I went and told me to stay away until Frederick's anger abated.' Tears spilled down her cheeks. 'When I returned, she was dead.'

# Chapter 30

On the morning of Father's ball I awoke at dawn, propped uncomfortably against my pillows where I'd fallen asleep the previous night still reading the diary. My dreams had been unbearable: visions of Mother's distress at the continuing misery of her marriage and frightening nightmares full of harsh voices and childhood memories of being mistreated by Joe. I remembered being shaken so hard my teeth rattled and then being shut in a cupboard.

I was tired and irritable, particularly as I knew the day would be fully taken up with overseeing the final arrangements for the ball. I groaned at the thought that the last guest probably wouldn't leave until dawn the morning after. I decided to stay in bed for a little longer. I picked up the diary from the pillow beside me and began to read again.

*I must escape from the dread that overcomes me every time Frederick comes home. I am taking Harriet with me to Langdon Hall for a respite.*

And then, three days later:

*I hardly know how to write this. My thoughts are utterly disordered as I grapple with this terrible discovery. I cannot settle to anything for worrying about what to do. I took Harriet out for a long walk to use up my nervous energy but on the way back her poor little legs couldn't carry her any further. I held her in my arms and our tears mingled. I love her so much and I'm very frightened of what I must do.*

There was a tap on the door then and Daisy entered with my morning chocolate.

'Good morning, miss,' she said. 'I guessed you'd be awake early today. The servants have been up since before dawn, busy with the preparations, and the kitchen's already humming.'

Daisy returned downstairs and I scanned the diary as I sipped my chocolate, reading that Mrs Digby had called and, for a short while, Mother had enjoyed her cheerful company. There were accounts of supervising the turning out of the preserves cupboard, discussions with the housekeeper as to what new linen must be purchased from London and a host of other household trivia. But then I came to a passage that made me sit bolt upright.

*I've tried not to think about what I found. I couldn't even write about it before, almost as if I thought it would go away if I pretended it had never happened. But I did find it and there are some things you cannot ignore, no matter what the consequences.*

*It was something Jane Digby mentioned to me that made me curious. One of my servants, a new scullery maid, had been found in a state of collapse after the others below stairs frightened her with tales of a ghostly priest walking the corridors of Langdon Hall at night, rattling his chains. They told her he'd been shut up in a hidden chapel and had died screaming, with no one to hear him. I told the poor girl there was no secret chapel and gave her the afternoon off.*

*When I related the tale to Jane, she shook her head. She said her brother had been Frederick's friend when they were boys and that they had discovered a hidden chapel at Langdon Hall. Frederick had made her brother swear never to reveal the secret but, years later, he had recounted their adventure to Jane.*

A secret chapel! But Father had distinctly told me there was no chapel and the priest had escaped out of the window. Why would he wish to hide it? The days of illicit Catholic priests were long gone. Or had Jane's brother been spinning her a yarn? The answer had to be in the diary.

*Curiosity, and a desire not to dwell on my miserable marriage, set me to the task of searching for the chapel. I poked around in all the obvious places, the attic and the cupboard under the stairs, and spent several hours examining the panelling, searching for a secret door. Then I had the idea of looking for the deeds of Langdon Hall, hoping there might be an old plan showing the chapel's location. Frederick kept folios of legal papers in his study and that was when I discovered his secret.*

I looked up as Daisy brought in a ewer of hot water.

'May I help you dress now, miss?'

'Will you come back in ten minutes?' I said. I couldn't bear to stop reading when a secret was about to be revealed.

Daisy bit her lip. 'Mrs Hope asked me to be quick. She needs me to dust the ladies' retiring room for the ball tonight. And I believe you're supervising the footmen while they clear the drawing-room furniture? They'll start that any minute now.'

Reluctantly, I rose from my bed. 'I'll wear the yellow muslin today, Daisy.'

'Very good, miss. And your white satin ballgown is all ready for you for this evening.' She poured hot water into the wash bowl and laid out the soap and flannel. I barely listened to her chatter as she passed me a towel, stockings and shift while I wondered what it was Mother had found. Might it have been an ancient plan of Langdon Hall or could she have found the chapel itself? I started at a touch on my arm.

'Your shoes, miss.'

I smiled distractedly. 'Sorry, Daisy. I was daydreaming.'

I sat at the dressing table while she brushed my hair. Suppose Mother had discovered the priest's skeleton in the chapel? But although that would have been unpleasant, it wouldn't have inspired such fear.

'Will that be all, miss? Only Mrs Hope is waiting for me.'

I picked up the diary again. Perhaps there was time for another quick look . . .

'Miss Langdon? The footmen will have started moving the furniture out of the drawing room by now.'

I sighed. 'I'll come down.' I hid the diary in the chest and followed her from the room.

Mr Gunter's deliverymen slipped on the freshly washed front steps as they delivered the fragile sugar-paste centrepiece. I nearly cried when I saw that one of the elephants' trunks had snapped, two of the trees had crumbled and the gold crown had toppled off the temple.

'We expect the King this evening,' I said, sounding calmer than I felt. 'Will you take the broken pieces back to your workshop for repair and return them here this afternoon?'

The deliverymen looked relieved I hadn't wept and railed at them, and retreated after promising to make all as good as new.

I had no time to fret since the florists were waiting to be told where to place the enormous urns of sweetly scented roses, lilies and peonies, and Cook wanted me to admire the jellies. I hoped they wouldn't melt in the heat.

I attempted to escape during the afternoon, desperate to read more of the diary and discover what Mother had found. I was halfway up the stairs when a footman called after me.

'The musicians are here, Miss Langdon. They want to discuss where they are to perform.'

I suppressed a sigh of irritation. 'Show them into the drawing room, James.'

'And the men from Gunter's are back and need you to approve the remedial works.'

Later, I was in the dining room, inspecting the glasses to be sure they were properly polished, when I heard Father's voice in the hall.

'The supper table is ready for you to see,' I told him.

Father surveyed the room, his critical gaze taking in the bounteous arrangements of fragrant flowers, the gleaming silverware and starched napery. A vast crystal bowl waited to be filled with punch on a side table and the chandelier, holding the best beeswax candles, was twined with wreaths of rosebuds. He stood silently before the sugar-craft construction representing Hyde Park, running down the length of the table.

'You have surpassed yourself, my clever girl.' He put an arm around my shoulders and hugged me, his eyes shining. 'This is truly a supper table fit for the King.'

It unnerved me that Mother's description of Father's violence was in such contrast to his benevolence to me. I rested awkwardly against his broad chest as he smiled down at me. What should have been comforting made me uneasy instead and I wanted to weep, knowing that either my mother or my father was not what they seemed.

'The King will enjoy the ball since he's in unusually good spirits,' said Father. 'The Queen grows sicker by the minute and he's extremely hopeful he'll be released from any awkwardness in that quarter before long.'

'That's a dreadful thing to say!' I pulled myself out of his arms, unable to bear his touch. The Queen's humiliation at the Coronation must have sent her into a spiral of despair and I pitied her. 'I heard she was taken ill at the theatre.'

'Apparently so,' said Father, rubbing his hands together.

His gloating smile sickened me. 'I'm going to rest now,' I said, 'before it's time to put on my finery.' When the ball was over, I decided I would visit the Queen to offer my commiserations.

I hurried upstairs to my bedroom and curled up on the window seat with the diary.

*Searching the folios on Frederick's bookcase I found the secret staircase almost by mistake. It smelled dank and mouldy and was too dark to explore further without a light. Fortuitously, there was a candlestick on the bookshelves. Perhaps Frederick keeps it there for exactly this purpose? Harriet was resting so I lit the candle and descended the stairwell.*

I gripped the diary tightly, impatient to read on.

*I found myself in an underground chamber. Narrow shafts of daylight entered from above, illuminating a magnificent canvas hanging above an altar. I was drawn to it by the vivid colours and arresting composition.*

*As I passed one of the shafts of light I realised that the whitewashed walls displayed a great number of paintings and drawings. Candle sconces were fixed to the walls at frequent intervals and I lit several. The resulting flickering light illuminated a breathtaking array of works of art. I knew at once that Frederick must have created this secret art gallery hidden in the lost Papist chapel.*

I stopped reading. So there *was* a hidden chapel! But why had Father kept it a secret? He was so proud of his art collection and loved to show it off.

*I came to three tiny oval frames hanging in a row. Holding the candle closer, I saw images of a dark-haired young woman: full face and two opposing profiles. Her expression was grave but there was the smallest curve to her mouth, which made it look as if she were trying not to laugh. I realised with a jolt of recognition that I had seen the little portraits before. They were Lord Beaufort's stolen miniatures.*

I closed my eyes in shock while my heart banged in my chest. I didn't want to believe it but if what Mother had written was the truth, rather than the result of an over-vivid or disordered imagination, the only and inescapable conclusion was that Father was the Picture Frame Thief.

# Chapter 31

I forced myself to smile and greet our guests with equanimity and soon the house was full of chattering people. It was hot and the odour of perspiration mingled sickeningly with the powerfully sweet scent of the flower arrangements, like the stench of decay. I watched Father on the other side of the room, acting the genial host. How had he managed to conceal the fact that he was the Picture Frame Thief for all that time? And what must I do about it now that I knew?

Aunt Maude waved at me from the chaperones' corner. 'Are you quite well, dear?' she asked, her expression full of concern. 'You're very pale.'

I smiled brightly. 'Just a little tired. I can see Dolly has just arrived,' I said, 'and Mr Gregory is with him, as usual.'

'They're late,' said Aunt Maude. 'Dolly's usually so punctual. Indeed, that is one of his better points.'

When they appeared Dolly had a face like thunder and Mr Gregory's eyes were red-rimmed.

'Don't let me come between you two lovebirds,' said Mr Gregory in a waspish tone. 'I shall fetch myself a glass of punch while you bill and coo in a corner.' He marched off.

'Have I upset him?' I asked.

'Of course not,' said Dolly tersely. 'We had an argument. He lost more than he can afford at the gaming tables.'

'It's a shame he didn't stay at home then, instead of bringing his bad temper here.' I heard the vinegar in my voice but had matters of more importance to worry about.

Dolly's mouth tightened and he gave me a look of undisguised loathing before following his friend.

Anger washed over me. Why had I agreed to marry Dolly? Sometimes I didn't even like him, and I certainly didn't like his friend who stuck to Dolly's side like a burr. But, of course, the main reason I'd agreed to marry him was because it pleased Father and, in the present circumstances, that was the worst reason of all. I felt thoroughly out of temper with them both.

Mr Sandys clapped his hands and called out that the first dance would begin shortly.

Father and I were to open the ball and he came to lead me to the top of the set. I flinched momentarily from the touch of his hand and, unable to meet his eyes, looked steadfastly over his shoulder.

The first violin struck a chord and we stood, poised to begin. The music shrieked discordantly in my ears and I wanted nothing more than to escape, to be somewhere quiet far away. Somewhere I could walk along the beach listening to the soothing hiss of the sea. Somewhere far away, like Pesaro.

I moved through the dance like an automaton, bending, pirouetting and jumping, coming together with my father and receding again, while all the time I thought about what Mother had written and what Father had said and what I was to do about it.

We reached the end but there was no escape since I was promised to Dolly for the second dance. He appeared distracted, too, and his smile quite as fixed as my own. We exchanged barely a word and it struck me how joyless our relationship was. Could I stand to go through life with this man at my side? But, of course, if I accused

Father of being a thief, if I reported him to the authorities, Dolly wouldn't marry me after all. And then I would be alone in the world again, but this time penniless and branded the daughter of a thief.

After what felt like a lifetime, the dance ended and Dolly escorted me off the floor.

Almost immediately, one of Father's friends, a portly man whose name I'd forgotten, stood before me.

'Will you do me the honour of partnering me in the next dance?' he asked.

Since I was the hostess, I was obliged to accept.

Half an hour later my lips were fixed into a rictus smile as the dance finished. I lied to my partner for the eighth time that, of course, he hadn't hurt my toe when he stepped on my foot and that the tear in the hem of my ballgown could easily be mended.

'Always had two left feet,' he chortled, drips of sweat falling from his shining brow. 'Well, that was fun! Haven't danced with a pretty girl for ages. Last chance before you're married, eh? Shall we join the next set, too?'

'I've promised to sit with my aunt for a while,' I said in desperation.

Fanning my flushed face, I backed away and instructed a footman to open the windows wider. I went to the card room to collect two glasses of punch. A cluster of young men stood at one side of the room, laughing uproariously, and I noticed with surprise that Mr Gregory, very pink about the cheeks, was at the centre of the group. He was recounting some tale or other with vivacious hand movements while Dolly looked on with a sour expression.

Aunt Maude glanced up anxiously as I went to sit beside her in the chaperones' corner. 'I wish you'd tell me what has disturbed you so,' she murmured as she took the glass of punch from me.

'Not now,' I whispered, still fanning my overheated face. 'Tomorrow perhaps. Here comes Father.'

He hurried up to us. 'The King will be here imminently,' he said. 'You will be ready to greet him, won't you, Emilia?'

I nodded.

He laughed. 'Don't look so overwhelmed!'

I was more overwhelmed by the thought of what the King would say if I told him I believed my own father was the infamous Picture Frame Thief.

I was expected to take part in every dance and the music was shrill to my ears as I moved through the steps whilst an agonising band of apprehension tightened around my head like an iron maiden. Despite the open windows, the stench of overheated bodies packed closely together was unbearable. It was almost a relief when the King's arrival was announced. The music ceased, the dancing came to an abrupt halt and the assembled company bowed and curtseyed in silence.

'Excuse me,' I whispered to my partner. I hurried to join Father as he welcomed the King.

'Perhaps you remember my daughter, sir?'

I sank into a deep curtsey.

'Indeed I do,' said the King. 'Uncommonly pretty girl.'

The music began again and the dancers resumed their positions.

I rose and smiled at the King. Though dressed in the first stare of fashion with jewelled medals glittering upon his chest, I'd forgotten how very fat he was. Once, he might have been handsome. 'We shall shortly be going in to supper, Your Majesty,' I said. 'Perhaps you would care for some refreshment?'

His small pink mouth curved in a smile. 'Delighted,' he said, and offered me his arm.

Followed by Father and the King's equerries, we made a stately progress through the ballroom. The King stopped to bow stiffly, due to his corsets no doubt, and to exchange a word or two with selected guests. I noticed that sweat beaded his forehead and his nose shone greasily in the heat.

We entered the dining room and the footmen sprang to attention.

'I dare to hope you will like the centrepiece for our supper table, sir,' I said. 'Since the Coronation Fête in Hyde Park was such a success, I took the liberty of having the scene recreated in sugar as a tribute to the occasion.'

Amused, the King peered closely at the scene. 'Splendid!' he said, mopping his face with a handkerchief. 'The elephants are particularly fine.'

I held my breath, wondering if he'd notice where the trunk had been mended, but he'd already lost interest.

I stepped away as Father moved in to flatter and encourage our sovereign to taste some of the array of cold meats, soups and jellies. I was relieved to note that the jellies hadn't collapsed in the heat.

The guests began to arrive for their supper and a glimpse of the King. Before long the buzz of conversation and laughter was so loud it was hard to hear what was being said.

Dolly came to sit beside me to eat his supper but was so morose I didn't exchange more than two words with him. His gaze was directed over my shoulder to where Mr Gregory, clearly in his cups, was laughing with an elderly dandy.

After supper we returned to the ballroom where an Italian opera singer waited to entertain us. The King appeared to enjoy the performance, perhaps for the singer's voluptuous bosom quite as much as her musical skills. Whilst it was a pleasure for me to listen to the lyrical cadences of the Italian language again, the singer's high notes sliced into my head like a sharp nail scraped down glass.

The King took his leave after the singer had finished her rendition and the dancing began again. I was much in demand as a partner, not only as the hostess but because my guests were eager to know about my conversation with the King. Several told me how much they were looking forward to the wedding breakfast at Langdon Hall. The band of worry around my head grew even tighter and I counted the minutes until everyone would leave.

Two hours later I was exhausted, not only from the energetic country dances and the excessive heat but from smiling and remaining polite when all I wanted to do was lie down in a darkened room while I decided what to do about Father. At the end of the next set I slipped out of the ballroom and made my way downstairs and into the garden. I paused in the doorway until my eyes became accustomed to the dark. My kid slippers made little noise on the path as I walked past the apple tree to the bench under the rose arbour. Careful of my dress, I brushed leaves off the bench and sat down with a relieved sigh.

The air was still very warm but infinitely cooler than inside the house with the heat generated by sixty guests and hundreds of candles. Music, light and high-pitched laughter spilled from every window. I breathed deeply, inhaling the perfume of the roses that almost overpowered the usual summer reek of city drains.

As I was about to return to the ball, a shaft of light fell across the garden as the door opened. I heard male voices. I looked through the trelliswork of the arbour and saw two men silhouetted against the doorway. They spoke in an angry undertone, too low for me to hear but I realised it was Dolly and Francis continuing their argument.

Dolly placed a hand on his friend's shoulder.

Francis pressed his hands over his ears. 'No!' he said, and ran down the garden towards me.

Dolly called out, 'Francis!' and ran after him, catching him by his coat sleeve as they reached the apple tree not eight feet away from the arbour.

I shrank back in the darkness, peeping at them through the trellis.

Dolly gripped the shorter man by his upper arms and shook him. 'Stop this!' he said. 'It's no good. I don't have any choice.'

Francis muttered something and struggled to free himself.

'I've told you, it's the only way out,' said Dolly, his voice pleading. 'You know I haven't a feather to fly with and this marriage will change all that.'

'I hate her!'

'Don't. She's nothing to me.' Dolly wiped tears from Francis's eyes. He glanced around, then bent to kiss him.

I pressed my fist to my mouth to stifle a gasp. It was absolutely clear to me that this was a passionate kiss between lovers.

Francis wound his arms around Dolly's neck and they fell back against the trunk of the apple tree, their bodies locked together.

When they finally drew apart Dolly tenderly smoothed back a lock of Francis's hair. 'There was only ever you,' he said. 'Nothing and no one will change that. Later, when it's all over, we'll go away together. Just the two of us.'

Shivering with shock, I pulled my feet up onto the bench, accidentally knocking over a plant pot. It tipped and fell with a dull thud onto the path. I curled myself into a ball and my hair snagged painfully on the climbing rose.

'What was that?' said Francis.

I held my breath.

'Nothing of importance,' said Dolly after a moment. 'A fox or a cat, perhaps.'

Francis nodded and the two men kissed again and returned to the house.

Trembling, I untangled my hair from the prickly stem of the rose. So much was explained now. Of course Dolly didn't love me! He never had and never would. His taste lay in another direction entirely. I leaned back against the trellis feeling the tears seeping from beneath my eyelids. How could I bear this terrible discovery so soon after the other?

I don't remember how I endured the last hours of the ball. Dolly partnered me in the final dance and I flinched away from him when he touched my hair.

'Whatever is the matter, Emilia?' he said, frowning. 'You're not going to faint?'

'Of course not,' I said and forced a smile. 'I'm simply a little tired after all the preparation for the ball.'

He didn't speak after that while we danced, his thoughts clearly elsewhere.

I was relieved to be spared having to make polite conversation while I dwelled on what I'd seen in the garden.

Thankfully, the guests finally left. When it came to saying goodnight to Dolly, I became rigid when he pecked my cheek.

At last only Father and Aunt Maude remained.

'You must go straight to bed,' I said to Aunt Maude. 'It's very late for you.'

'I will, dear,' she said. 'We'll talk in the morning,' she whispered. 'Goodnight, Frederick.' She leaned heavily on her stick and looked very frail as she slowly mounted the stairs.

'You will be pleased to hear,' said Father, rubbing his hands together in delight, 'that I managed to have a word with the King regarding my contacts in the art world. He said he'll call on me next time he's looking for a particular piece.'

'So the ball achieved its purpose,' I said, knowing that the King would never buy anything from my Father if I exposed him as a thief.

Benevolent in his triumph, he kissed my cheek. 'You've earned a day of rest tomorrow, my dear.'

I said goodnight and retreated to my bedroom.

Bone-weary, I allowed Daisy to undress me and slip a fresh nightgown over my head. I slid in between the sheets and once Daisy had closed the door behind her, I pinched out the candle. I couldn't bear any more heartache and reading the rest of Mother's diary would have to wait until the morning.

Confused and miserable, I felt under the pillow until my fingers found Peggy's woolly plaits. Hugging her against my chest, just as I had so many times as a child, I curled into a ball and shut out the world.

# *Chapter 32*

Sleep had come quickly to me but I awoke before dawn and relived, over and over again, the passionate embrace between Dolly and Francis. As the sun rose, I tried to understand my disturbed feelings. Occasionally I'd overheard people talk of love affairs between men, in hushed and scandalised whispers, but I'd never seen any evidence of it before. Such a thing upset the natural order of life and bewildered me quite as much as if I'd discovered water flowing uphill.

One thing was certain now, though. I could not and would not marry Dolly. I had never deluded myself that I loved him, it was only his declaration of love that had allowed me to believe there was a chance of making a marriage between us successful. In the light of what I'd seen, Francis would always occupy the first place in his heart and Dolly and I could never have a true marriage.

I pulled myself up against the pillows, thinking about Alessandro's accusation that Dolly had caused Sarah's death. Supposing this was right, what possible reason could he have had? I rubbed my fists against my eyes, trying, and failing, to comprehend. Deep regret for the happiness Alessandro and I might have shared if I'd never left Pesaro nearly choked me.

I slid out of bed and opened the curtains before padding barefoot to the wardrobe to lift the muslin cover from my wedding dress. It was made of the finest ivory silk with a ravishing guipure lace bodice and train but I shuddered when I looked at it. I'd never wear it now. Closing the wardrobe door, I went to the writing desk and took out a fresh sheet of paper to write to Alessandro.

Fifteen minutes later I blotted the ink dry, folded the note and sealed it. Two sheets of crumpled paper lay by my feet, the ink smudged from my tears. I gathered them up and stuffed them into my reticule to dispose of later. I had no intention of allowing them to fall into Father's hands.

My eyes were gritty from lack of sleep and I rubbed them as I pondered again on what I was to do about Father. If he was the Picture Frame Thief and I made this known, disgrace and ruin would fall on us all. I wasn't sure if I was brave enough to face that. If I was, the future for Aunt Maude and myself appeared very bleak.

Sighing, I took Mother's diary out of the chest, hoping it would shed more light on the mystery.

*I've barely slept for worrying about what to do. Harriet climbed onto my knee to kiss away my tears and brought me Annabelle to cuddle. I love my daughter beyond life itself. Whatever I do, I must keep her safe.*

Mother's dilemma had been similar to my own. Her words didn't sound like those of a woman who was prepared to abandon her own child and there was still no indication she'd had a lover. If that were true, I couldn't understand why she had drowned herself.

*Harriet and I return to Grosvenor Street today and I have made a momentous decision. I am going to ask Frederick to agree to a legal separation. He won't like it but he must see that we are both so unhappy this cannot go on. My very existence irritates him and he lashes out at the smallest thing. I rarely dare to initiate conversation and creep about the house trying not to attract his violent attentions. I have written to my old friend Anne-Marie in Paris. When she last came to London, after Piers died, she noticed my bruises and promised that*

*Harriet and I would always have a home with her if we needed it. And now we do.*

I rested the diary on my knee. So Mother's friend in Paris wasn't a lover and Father had lied to me.

*I have taken the miniatures. Frederick is unlikely to visit Langdon Hall for a while and won't know they are missing. I shall return them to Lord Beaufort on my way to Paris and leave it to him to decide what to do about Frederick.*

Mother hadn't intended to steal the miniatures herself, then, but had packed them in the luggage that Sarah later conveyed to the inn at Dover. I turned back to the diary and saw that a week had passed before the next entry.

*My mistake was to believe Frederick would ever be reasonable. He became a madman when I asked for a separation. Only now can I hold a pen again. He has locked me in my room. It would be too shaming for him if anyone saw my terrible bruises. The beating was severe and I heard my fingers and ribs crack. It's still hard for me to breathe and he will not let me see Harriet. We must escape.*

Shocked, I reread the passage. My inclination was to believe Mother's account but I couldn't help remembering Father's distress when he'd told me of her apparent infidelity. I rested my head in my hands, wondering if he could have been so cold-bloodedly deceitful.

The diary was nearly at an end and I gripped it tightly as I read on. There were a few entries about Mother's anguish at not being allowed to see me and then, this final paragraph:

*It will be tonight. Sarah will give Miss McCorquordale a sleeping draught and fetch Harriet to come with us. I pray that by this time tomorrow we will be free.*

The remaining pages were blank. If what Father had told me was true, he'd caught her trying to escape and sat up with her all night. In the morning, after he'd left, she'd escaped and drowned herself. Despair must have driven her to take her own life, knowing Sarah would take me to safety abroad and that Father would never let her go. I ached with pity for my mother. Carefully, I wrapped the diary in a clean shift and hid it again at the bottom of my chest underneath Sarah's quilted petticoat.

Father had abandoned his newspaper on the breakfast table before going out. I read the latest account of the Queen's ill health and was sad to see that she was worse.

While I was reading, Aunt Maude came downstairs. 'What happened last night to upset you?' she asked.

I glanced over my shoulder as James carried in a fresh pot of coffee. 'I wondered if you'd like a ride in the country this morning. To Fulham perhaps?'

Aunt Maude gave me a sharp look. 'That would be very pleasant, dear.'

Half an hour later we were in the carriage. I had asked Dobson the coachman to stop in Great Marlborough Street at the house where Alessandro worked. The butler informed me that Signor Fiorelli was not at home but had taken the young masters out for their morning walk. Reluctantly, I entrusted my note for Alessandro to the butler and returned to the carriage.

'The Queen is very unwell,' I said to Aunt Maude as we set off for Fulham, 'and I want to offer my best wishes for her recovery.'

'I assumed that was the reason for this visit.'

I nodded. 'And I must tell you what I read in Mother's diary. I thought we wouldn't be overheard in the carriage.'

'Perhaps not,' she said, 'but Dobson may report back to your father where we go today.'

'Father has no reason to suspect I've read the diary so I don't believe he'll be spying on me,' I said, 'but I've discovered some terrible things and I don't know what action to take. If the Queen is going to return to Pesaro once she recovers, then I may need to ask her if she'll make room for me in her retinue.'

Aunt Maude pressed a hand to her mouth, her rheumy eyes wide and anxious. 'You're not leaving? But what about your wedding? It's only three weeks away.'

'My discoveries have changed everything,' I said. 'I cannot and will not marry Dolly now. And then there is what I've found out about Father.'

'I *knew* something had happened last night!'

Once I had recounted my story, Aunt Maude shook her head in dismay. I was relieved she'd understood the particular nature of the friendship between Dolly and Francis and I hadn't had to embarrass myself or her by being too explicit.

'You can't possibly marry him now,' she said, her mouth pinched with distaste, 'unless you believe you can use all your womanly wiles to make him suppress his unnatural impulses?'

'Aunt Maude, I saw how tender he was with Francis,' I said. 'I think he really loves him and, even if I wanted to, I don't imagine any woman could seduce Dolly.'

'That sort of behaviour is against the law,' she said, 'but I suppose it does explain why Dolly wasn't more in love with you. I couldn't understand it. You're so pretty and clever, any man should be proud to be your husband.'

'Dolly is very handsome but I didn't fall in love with *him*,' I pointed out.

'I'm bewildered as to why he proposed to you in the first place.'

I gave a wry smile at the thought that he most probably hadn't wanted this wedding either. 'Perhaps because society expects him to have a wife and then he'll be free to continue his association with Francis.'

'When will you break the news that the wedding is not to go ahead?' Aunt Maude asked. 'There will be terrible ructions. Your father . . . '

'I daren't say anything yet,' I said. 'To be absolutely sure of the truth about Father, I must go to Langdon Hall and find the hidden gallery. If I call off the wedding now, after what I've read in Mother's diary, I don't trust Father not to imprison me in my room. Without proof that the chapel exists, I'm sure he would laugh off Mother's diary entry as the wanderings of a disturbed mind. Once I have proof, I can report it to the local magistrate.'

'And what will happen to us afterwards?' said Aunt Maude. Her lower lip trembled.

'I don't know,' I said. I felt as if a deep pit had opened at our feet. I was young and could work as a dressmaker; Aunt Maude had no such reserves to fall back on and I wasn't confident I could earn enough to keep us both. 'If the Queen doesn't take us with her, I suppose I might find someone to marry who would give us both a home.' I rubbed my eyes. 'Or, of course, I could blackmail Father.'

'Emilia!' said Aunt Maude, shocked. 'Although he's my nephew,' she said, 'and I've tried very hard, I cannot care for Frederick. At best, he's two-faced and self-serving.' She leaned towards me. 'It makes perfect sense to me that he is the person who stole all those pictures but, if you did blackmail him, I wouldn't trust him to keep any bargain that is not to his own advantage.'

'Do you believe that what Mother wrote in her diary was true?' I asked.

'I should like to read the diary, if I may, but I have no reason to doubt Rose's account.' My great-aunt looked close to tears. 'When you came to Grosvenor Street it was as if Rose had returned to me. I was so frightened for you and did everything I could to make you feel unwelcome. I wanted to drive you away.'

'I remember,' I said.

'It was to protect you. Frederick is unreliable, perhaps even dangerous if he doesn't get his own way. I believed ... still believe ... he wanted you back only to serve his own ends, whatever they may be.'

I thought about this as the carriage jogged through the streets. 'He had two reasons for encouraging me to stay,' I said. 'Firstly, I'm the most recent lead he has to the miniatures. I'd already guessed that but thought his reasons were altruistic and he wished to return them to Lord Beaufort. Having read Mother's diary, now I think he wants to keep them himself.'

'Surely, if they are found, he must know you would want them to be returned to their rightful owner?'

I shrugged. 'Perhaps he thinks he can tell me they have been returned but then keep them in his secret gallery. Or maybe, because he knows I'm interested in art, he believes he can persuade me to share his illicit pleasure in them.'

Aunt Maude sighed. 'And what do you think is his other reason for wanting you to stay?'

'He's never liked Dolly very much but, since it's not possible to change the entail, at least Langdon Hall would stay within his direct bloodline if I had children with Dolly.' I smiled briefly. 'Perhaps he hasn't realised that his heir might not care to give him a grandson since his proclivities lie elsewhere.'

Aunt Maude reached for my hand and we spoke no more until the carriage drew up outside Brandenburgh House.

The steward told us that the Queen was indisposed and I asked if we might speak instead to Lady Hamilton. A few moments later she came downstairs to receive us.

'We came to pay our respects to the Queen, Lady Hamilton,' I told her, 'but we understand that she is still unwell?'

'She is,' said Lady Hamilton. Her eyes were deeply shadowed and her fingers plucked anxiously at her skirt. 'Dr Holland is with her now. She has a great deal of pain in her stomach and, despite being bled and taking heavy doses of magnesia and laudanum, can find no relief.'

'I am sorry to hear that.'

'Her spirits are very low,' said Lady Hamilton, 'and I'm terribly afraid she has lost the will to recover.'

'I was outside Westminster Abbey when she was so shamefully turned away,' I said. 'I daresay that affected her greatly?'

Tears made Lady Hamilton's eyes glitter. 'She shut herself in her bedroom for four hours afterwards and would speak to no one. Then, at supper, she put on the semblance of unusual gaiety but, however hard she tried to hide her distress from her friends, she deceived only herself. Tears of anguish rolled down her face, even as she laughed and joked.' Lady Hamilton was unable to say more while she struggled to regain her customary control.

'She has a good friend in you, Lady Hamilton,' I said, 'and she'll be relieved that you are at her side to support her.'

'The Queen has made her will and I fear ... ' Lady Hamilton broke down again. 'I don't know how to ease her grief!'

'Please convey to her our sincere best wishes,' I said, at a loss for what else to say.

'Pray for her, won't you?'

'We will.'

Lady Hamilton nodded without saying more and Aunt Maude and I left.

The following days passed quietly. Aunt Maude and I sat in the morning room while she read Mother's diary. Frequently she would stop to wipe her eyes and we discussed particular passages in an undertone.

'Aunt Maude,' I said, 'I've been wondering about something.'

She closed the diary, marking the place with a piece of silk ribbon. 'What is it, my dear?'

'Something occurred to me and I can't get it out of my head.' I swallowed. 'Do you think it possible that Father drowned my mother?'

Aunt Maude bowed her head over the diary and then looked up at me, her expression bleak. 'I'm very much afraid I do.'

I nodded and returned to my confused and wretched thoughts. All the while I listened for the doorknocker, hoping there might be a response from Alessandro to my note. I sat by the morning-room window for hours at a time, in case he was outside, trying to catch my attention, but as the days passed without word from him, I finally lost hope of any reconciliation.

At breakfast one morning Father picked up the newspaper James had placed beside him. Then he banged his cup down on the table, slopping coffee onto the starched tablecloth, and laughed. 'At last!' he said.

I mopped the stain with my napkin. 'What is it, Father?'

'She's dead! The King is free at last!'

The jubilation in his expression made my stomach turn over. 'The Queen is dead?'

Aunt Maude gasped.

'Isn't that what I said? Listen to this!' Father rustled the paper and began to read aloud.

*'The Queen had suffered terrible pain all the previous day. Dr Holland, who was attending her, felt her pulse at twenty-five minutes past ten at night and then closed her eyelids. He declared, "All is over."'*

'Thank God,' said Father, his face wreathed in smiles. 'Now there'll be celebrations in certain quarters.'

'How can you be so cruel?' I said.

'For goodness' sake, Emilia!' He frowned at me and rose to his feet. 'It's time you saw the truth about that woman.'

I stood up, clenching my fists in fury. 'You don't know the first thing about her.'

He strode from the room and slammed the door behind him.

I shed tears for the Queen's turbulent and misunderstood life, reliving those moments when our paths had crossed in Pesaro. I remembered her kindness when Sarah died and how she'd taken a fancy to me when I'd comforted her after she spoke of her daughter's death. I remembered her romping on the floor with Victorine and that perfect, sunny day when Alessandro and I had joined her for a picnic on her yacht. She may have been anathema to her husband and much of society but I had seen another side of her character. All she'd wanted was to be loved for herself. I supposed that was no different from what most people wanted, myself included.

# Chapter 33

On the morning of Queen Caroline's funeral a week later I looked out of the window to see grey skies and rain. I still hadn't heard anything from Alessandro. I'd hoped desperately we might have made a fresh start together and I tried not to think about how empty my life would be as the last vestiges of that dream faded.

Daisy came to help me dress and I defiantly chose to put on my old mourning gown as a mark of respect for the Queen, despite what Father would say.

'Mr Soames says half the shops and businesses will be shut today,' said Daisy as she buttoned my cuffs. 'And a crowd's already gathered at Hyde Park Corner, waiting in the rain to follow the funeral procession.'

'I understood the Prime Minister had insisted on sending the procession north of the city?'

'That's as may be,' said Daisy, 'but the people want to say goodbye before her body is returned to Brunswick. They won't put up with the procession being hurried out of sight.' She nodded with satisfaction. 'They'll block the road if Lord Liverpool tries to send the procession north.'

I reflected on how fickle the populace was. They had abandoned the Queen during the Coronation and, now she was dead, wanted to show their support again.

After my maid had gone, I sank down on the window seat and miserably watched raindrops running down the glass. Aunt Maude had finished reading Mother's diary and agreed with me that nothing could be done until we knew for sure that the secret gallery existed and that it contained stolen paintings. We would all travel to Langdon Hall the day after tomorrow since the wedding was set for ten days' time. It made me queasy with anxiety to contemplate not only calling that off but also telling the authorities about Father. And then what would Aunt Maude and I do? I had some money saved from my allowance but it wouldn't keep us for long.

One good thing had come out of the bad. I was now sure my mother had loved me. I took the diary from its hiding place under Sarah's quilted petticoat, to comfort myself by rereading Mother's loving words.

A while later, I tucked the diary under the petticoat again. I'd given the residue of the gold coins to Father to keep in his safe. I wondered how long Sarah had kept them concealed. If I knew that, I might be able to work out in which town she'd sold the miniatures and then be able to trace them. Of course, she might have sold them one by one as we had need of the money. Hurriedly I ran my fingers all over the petticoat, double checking each coin pocket. Nothing. Sighing, I replaced the petticoat in the chest and went downstairs.

Father was in the dining room finishing his breakfast. 'I shall be out until early evening, Emilia,' he said. 'I'm going to Kingston to view an auction of Dutch paintings. I expect to purchase an addition to my collection and we shall take it with us when we go to Langdon Hall.' He rubbed his hands together. 'And the happy day of your wedding is fast approaching now. You will have final arrangements to make.'

'I plan to do all that is necessary,' I said, non-committally.

'I must leave now if I'm not to be late. The roads are likely to be congested with that wretched woman's funeral procession.'

'My maid tells me that crowds have turned out to pay their respects to the Queen.'

'You see, she's still causing trouble even after her death,' said Father. 'Is that why you're all dressed up in black like a crow this morning?'

I lifted my chin and gave him a defiant look. 'She was kind to me,' I said.

He frowned. 'You know nothing about what she was really like.'

'On the contrary,' I said, 'I lived in her household and probably know her a great deal better than most, including you.'

Father pushed back his chair abruptly. 'I'm not going to argue with you and make myself late.'

I sighed in relief when the front door banged behind him. Since Aunt Maude hadn't come downstairs I went to tap on her bedroom door.

She lay back against the pillows with her white hair in a thin plait over one shoulder. 'Are you unwell?' I asked, distressed to see her like this.

She shrugged. 'I feel very old today,' she said. Her bony fingers plucked at the sheet. 'My heart flutters so.'

'May I bring you something?'

'I was thinking about dear Rose,' she said. 'I keep wondering if there was something I could have done to save her.'

'You cannot blame yourself for what happened.'

She turned towards me, her milky blue eyes full of anxiety. 'You must be careful, Emilia. Your father is unpredictable.'

I enfolded her trembling hand in mine. 'Having read Mother's diary I'm forewarned of how quickly his mood can turn,' I said. 'But I must find out the truth about the hidden gallery.'

'And what then?'

'Once I tell Father I have no intention of proceeding with the wedding I suspect he'll attempt to coerce me into marrying Dolly.

As for what action to take over the stolen paintings . . . ' I sighed. 'I don't know yet. Will you be well enough to travel to Langdon?'

Aunt Maude nodded.

'There's something I must do today,' I said. 'I shall pay my last respects to the Queen when her funeral procession passes on its way to Harwich. Her body is to be conveyed to Brunswick for burial.'

'You can't go out alone, Emilia!'

'Perhaps I'll take Daisy.' I had no intention of doing any such thing. 'I shouldn't be longer than an hour or two. In my ancient travelling cloak, no one will take me for a lady.'

'If you really must go, come and see me as soon as you return.'

'I will.' I kissed her forehead. 'Rest now.'

I went to my room to fetch my cloak. It was at the back of the wardrobe as I hadn't used it since Father had provided me with clothes suitable to my position as his daughter. My fingers brushed against my old travelling bag and something occurred to me. I dragged the bag out of the back of the wardrobe and then delved in it again to find Sarah's bag. One after the other I turned them inside out. I fetched the embroidery scissors from my work basket and ripped open the seams in the linings. Quickly, I searched between the lining and the scarred and worn leather of each bag, hoping there might be a hidden pocket for the miniatures, but there was nothing. Sighing, I returned them to the wardrobe.

As I was tying the ribbons of my cloak, I paused. Where else might Sarah have hidden the miniatures? It must have been in something we always carried with us, somewhere that no one would think of looking. Slowly, I turned to face the bed. Peggy, faithful companion of my childhood, lay on my pillow, her familiar woollen smile still slightly crooked. I squeezed her but she was so well stuffed that her body was hard. I unbuttoned her dress and reached for my scissors.

I hesitated, finding it difficult to cut open my old friend, then carefully snipped the stitches down her calico back. Pulling out the stuffing I recognised the shredded fabrics, each one reminding me of one of our commissions: Signora Donati's afternoon dress, Maria Lagorio's first Holy Communion dress, my own lawn shift, the Conti bride's wedding dress ... I fingered the white silk gauze. I hadn't restuffed Peggy for years and the Conti wedding had been in Florence shortly before we arrived in Pesaro. Could Sarah have put those scraps inside Peggy more recently? I dug my fingers more deeply inside the cavity. And then I found it. A small silk bag. I teased apart the drawstring. Catching my breath, I extracted an exquisitely painted miniature portrait in an oval gold frame. The Spanish Infanta.

I stared at it in horrified fascination. Had I carried the priceless miniatures around with me everywhere I went? I remembered Sarah being sharp with me once or twice when I was a child when I'd nearly left Peggy behind. And, more recently, I'd forgotten all about the doll and left her with Victorine when I came to London. Sarah had used me, an innocent child, to conceal stolen goods. I could only assume she had sold the other two miniatures and that accounted for the gold coins in her petticoat. A bubble of hysteria welled up inside me and I hugged the remnants of the doll against me and laughed and laughed until I cried.

Once I'd recovered my equilibrium, I replaced the miniature, restuffed Peggy and sewed up the seam. I left the doll in her usual place on my pillow and went to see Aunt Maude again.

When I told her of my discovery, she pressed her fingers to her breast, her expression horrified.

'I wanted you to know where the miniature is hidden,' I said. 'It's a precautionary measure, to give you some proof to take to the authorities if anything should happen to me.'

Aunt Maude gripped my wrist. 'We cannot let it! I couldn't bear that.'

'Don't worry,' I said. 'I'm going to see the funeral procession now before it's too late but I shan't be very long.'

She nodded, her eyes wide and troubled.

I kissed her cheek and hurried downstairs.

The hall clock struck eleven. I waited until there were no servants in the hall and then let myself out of the front door. It was raining, that fine but persistent drizzle that always finds a way to trickle down your neck inside your collar. I pulled down the brim of my bonnet and scurried down Grosvenor Street towards Hyde Park.

A moment later I heard running footsteps from behind and then someone caught my sleeve. I gasped and whirled around, thinking it was a pickpocket. Shock made me stumble. 'Alessandro!'

He gripped hold of my hands. 'I didn't mean to frighten you.' The shoulders of his coat were dark, saturated with the rain, and his curly hair plastered to his head.

The flash of joy I'd felt on seeing him was mixed with disbelief. 'But what are you doing here?'

'You didn't answer my letter,' he said, 'and I've been loitering outside all morning hoping to catch you. I thought you might want to see the Queen's funeral procession.'

'I do.' I frowned. '*Your* letter? I wrote you a letter but you didn't reply.'

'Yes, I did.' His expression was grim. 'I gave it to your footman. It must have been intercepted. And it's not the first time, either.'

Anger seethed in my breast. 'My father must have taken them. He admitted he'd taken your earlier letters.'

A group of pedestrians jostled us as they hurried past.

'There's no time to talk now,' said Alessandro. 'Let's walk towards Hyde Park Corner, we should see the procession there.' He turned up his coat collar against the downpour, took my arm and set off at a brisk pace.

I splashed along beside him, my shoes sodden from the puddles.

He didn't slow his pace as we crossed Grosvenor Square and turned into Upper Grosvenor Street. Already we could hear the noisy crowd ahead.

'Hurry or we'll miss the procession. Afterwards we'll talk.'

As we neared Hyde Park the streets were crammed with pedestrians. We turned into Park Lane and Alessandro took a fierce grip on my wrist as we were swept along with the boisterous crowd making its way towards Hyde Park Corner.

People of all classes had turned out to follow the procession. Many wore mourning dress whilst others had tied on black armbands or waved batons with a flutter of black crepe at the tip. Men on horseback and numerous carts and carriages edged forwards amongst the horde, bringing howls of abuse from those who had their toes run over.

Wagons had been set across the street to form a barricade and the hubbub of the crowd echoed all around. The throng pressed so close around us I was fearful I'd lose my footing. I was relieved to have Alessandro hold my hand so firmly and didn't want ever to let him go. Ahead, men carried banners and shouted out repetitively but the hum of the multitude was too great for me to hear what they were chanting.

I turned to a man beside me, tightly pressed against my arm as we shuffled along. 'What's happening?' I asked. 'Is the funeral procession nearby?'

'It was stopped at Kensington about half past nine,' he said, wiping rain off his face. 'Lord Liverpool wanted it to go past the gravel pits and up northwards.' He grinned. 'But the people weren't having it and blockaded the road with wagons. Carts and carriages piled up behind and the procession couldn't go neither forward nor back.'

The horde ahead of us in the park began to chant, 'Through the city! Through the city!'

'Sounds as if the procession's on the move again,' said the man. 'There was talk last night of bolting the park gates so it couldn't cut through there and go up New Road.'

After a while we reached Hyde Park Corner and discovered a contingent of mounted Life Guards waiting there. Some of the mob shouted insults at them and an apple core sailed over their heads to resounding laughter from the spectators. All the while the Life Guards, resplendent in their scarlet and gold uniforms, stared straight ahead, ignoring the rain and the affronts to their dignity.

'I can't see that it's possible for the procession to make its way through so many people or the barricades,' said Alessandro.

By now we were near the gates to the park and pressed flat against the railings. 'The poor Queen!' I said. 'Even in death nothing is straightforward for her, but she would have been pleased the people turned out to support her at the end.'

A boy perched on his father's shoulders shouted, 'They're coming! They're coming along Knightsbridge.'

I peered along the road and through the haze of rain saw only the crush of people, men on horseback and a line of gigs and barouches along the side of the carriageway.

There was a muttering in the crowd and a few men climbed onto the gates for a better view. A ragged cheer rose, swelling to a roar as a dozen soldiers on horseback appeared, riding two by two, along Knightsbridge. All the while, rain continued to fall from the leaden sky.

The progress of the cavalcade was slow. At last the soldiers drew level with us and then passed, followed by three mourning coaches, each drawn by six black horses and interspersed with eight marshals riding in pairs, a troop of mounted soldiers and a dozen pages in black cloaks and headbands.

'Here comes the hearse,' said Alessandro. Like most men around us, he took off his hat and bowed his head.

The people fell silent as the Queen's hearse, decorated with the Royal Arms and drawn by eight matched horses with black plumes on their heads, trundled slowly forward, led by black-clad mutes.

Tears sprang to my eyes as I imagined the Queen inside her coffin, cold, lifeless and alone, wrapped in a shroud, hands folded across her breast. I preferred to picture her slightly crooked smile and the love in her eyes as she'd dandled Victorine on her knee.

A woman shouted out, 'God bless Queen Caroline!' and the cry was taken up by others until it became a resounding chorus. I joined in wholeheartedly, my tears mingling with the rain.

Alessandro reached for my hand and I grasped his as another four mourning coaches, a trumpeter blowing a mournful salute, various sodden dignitaries on foot and eighteen mounted soldiers passed by. I wasn't sure, but I believed I saw Lady Hamilton and Willy Austin at the window of one of the coaches.

'The Queen always faced her enemies so bravely,' said Alessandro, wiping his eyes. 'I shall miss her. Little Victorine will be inconsolable.'

The funeral procession jerked to a sudden halt. Angry shouts came from the front.

'What's happening?' I asked.

'I can't see,' said Alessandro as the muttering of the crowd grew louder.

A man balanced on top of the park gates yelled out, 'They're coming back!'

The cavalcade slowly turned around, carriage by carriage, and headed towards the gates again.

'Perhaps there were too many people for the procession to push through without injuring someone,' said Alessandro. He stood on tip-toe, peering ahead, while I glanced at the familiar lines of his firm jaw and high cheekbones. I longed to reach up and touch his face.

The crowd surged through the gates before the cavalcade reached it. There were yells as the press of people from behind propelled us forward and I struggled to stay upright, frightened we'd be crushed underfoot.

Alessandro's hand, wet with rain, slipped from mine as I was dragged away from him. A moment of pure panic washed over me when I was lifted up as if by a tidal wave. Just as suddenly I was deposited on the ground again as the crowd shoved through the park gates and we came out on the other side like a cork popping from a bottle. I hurried to a patch of open grass. Men and women ran past and a child shrieked without stopping. I looked wildly around for Alessandro.

'Emilia!'

I heard his shout and spun around to see him hurrying towards me.

He caught me to his chest, his face full of consternation. 'Are you hurt?'

I shook my head, showering him with raindrops from my bonnet.

'This country!' he said, making a face of disgust. 'Does it ever stop raining? Let's move out of the way until the crowd thins and then I'll escort you home.'

We stood back, seeking shelter from the rain under an oak tree, though I was already wet through to my shift. The procession moved slowly by, going north along the carriageway running parallel to Park Lane.

A mass of people ran past and a man on horseback cantered by, yelling, 'To Cumberland Gate! Stop the procession at Cumberland Gate!'

The rain abated a little and I pulled at Alessandro's hand. 'Let's see what's going on.' We walked briskly alongside the procession.

Hoarse shouts of, 'Through the city!' and 'Shut Cumberland Gate!' came from all sides as the seething multitude rushed towards the north side of the park.

The funeral cavalcade began to move faster and we ran to keep abreast as it proceeded at an unseemly pace towards Cumberland Gate.

The Guards galloped past, their horses' hooves kicking up clods of turf behind them and spattering the spectators with mud. In turn spectators threw stones at the soldiers' backs. A horse squealed and reared up, its forelegs flailing, terrifying those who couldn't move away fast enough.

With their feet slipping on the muddy ground, some men laboured to grip hold of the soldiers' mounts and wrestled with their bridles to make them halt. Others clung on to the Queen's hearse, knocking down the mutes and turning the terrified horses around yet again so that the cavalcade was facing in the direction of the city.

There was a crack of thunder and the heavens yawned. All around us bellowing men and soldiers brawled in the torrential rain. The mob rocked the mourning coaches and dragged the soldiers off their whinnying horses.

A shot rang out.

The crowd went silent for a fraction of a second and then roared in anger.

More soldiers galloped into the park from the nearby barracks, shouting, 'Clear the way!'

I caught my breath in fright as a great brute of a grey brushed my arm when it raced past. The smell of sweating horseflesh hung in the air after it had gone.

The soldiers let out another round of shots and Alessandro shouted, 'Let's go!'

Women screamed and then all descended into chaos. People ran in all directions, slipping and falling in the mud, desperate to get away as more shots echoed through the air. A man yelled and I was knocked down to the muddy grass and trampled in the rush.

Winded, I pulled myself into a sitting position as booted feet raced past. A horse pranced on the spot, inches away from my face. Terrified, I looked up at the underside of its belly while I heaved for breath. A deafening volley of shots rang out overhead. The horse galloped away and people all around me screamed and shrieked.

I crawled on all fours through the mud to find a place of safety but my cape and skirt were tangled around my legs. My bonnet was lost somewhere and my hair fell over my face in sodden strands. Yelling soldiers on horseback milled amongst the scattered crowd, carbines and pistols at the ready. I looked around for Alessandro amongst the confusion but couldn't see him.

Horrified, I saw a man on the ground thrashing about in agony as his companions attempted to staunch the blood gushing from his thigh. A little way off a woman sat on the ground comforting a fallen man. Blood ran from his head.

My breath was catching in my breast but fear gave me the strength to push myself to my feet. Shaking, I leaned against a tree and scanned the crowd in desperation while I prayed under my breath. And then I saw him.

'Alessandro!' I screamed. Pushing my way through the milling crowd, I fell to my knees at his side. He was slumped on the ground against the park railings, his eyes closed and his coat front scarlet. Sobbing, I cradled him in my arms while terror froze the blood in my veins.

# Chapter 34

The kitchen in Great Marlborough Street was clean and dry and smelled of the comforting aroma of the apple pie cooling on the scrubbed pine table. The cook, a monstrously fat woman stuffed into a clean apron, stood watching me with her hands on her hips. 'He'll do,' she said, as I finished winding the bandage around Alessandro's upper arm and tied the end in a knot.

The wound would leave an ugly scar and it made me want to weep to see Alessandro's beautiful smooth skin damaged. But it might have been so much worse.

'It's only a graze,' he murmured, though his face was alarmingly white.

'A deep one,' I said.

Pain and blood loss had made him faint several times and it had taken over an hour for us to reach his employer's house. Now that he was safe, I trembled with the delayed shock.

'I've never heard of such a disgraceful thing, soldiers firing into the crowd like that,' said the cook. 'Sounds as if you got away lightly by all accounts, but you've both had a bad fright. I'm going to make you a nice pot of tea. And you, miss, can go in the scullery and have a bit of a wash. Lizzie will bring you some hot water.' She looked me up and down. 'You could be one of them mudlarks after a day spent scavenging by the river.'

I looked down at my skirt but didn't see how 'a bit of a wash' could possibly make me look presentable. 'Thank you,' I said. 'Tea would be most welcome.'

'We need to get on with preparing the dinner and you'll be in our way here,' she said brusquely. 'You can have a sit in the kitchen office while you drink your tea.'

Ten minutes later I had mopped the front of my dress, washed my face and combed my tangled hair with my fingers. The cook, Mrs Bowker, had sat us down in her tiny office and the kitchen maid brought us a tea tray, leaving the door ajar behind her.

My hands shook so much the cup rattled in the saucer.

Alessandro took it from me with his left hand. 'Emilia,' he said, 'it's all right. We're both safe.'

'Will your employers be angry with you?'

He shook his head. 'Since I'd been in the Queen's employ they gave me the day off to observe the funeral procession. I must be fit to undertake my duties again tomorrow. What about you? Will your father be angry?' He smiled, and for a moment his face was lit by mischievous humour just as it used to be. 'Mrs Bowker was right – you do look like a mudlark.'

'Father is out of town today so I hope to return before he does.' I sipped the scalding tea and burned my tongue. 'Alessandro . . . '

He reached out and laid his hand on my wrist. 'I know. There are many things to discuss.'

Little shivers ran through me as he caressed the delicate skin with his thumb.

'You've grown so thin, Emilia,' he said. 'You look as fragile as a sparrow. Has finding your family not brought you the contentment you expected?'

'I wrote to tell you what I'd discovered about Dolly, though I still cannot understand how he might have been the one who hurt Sarah. But there are worse things that I haven't told you,' I said. 'Things I've discovered about my father.'

'I don't understand – what things?'

My mouth trembled. 'Alessandro, I'm frightened!'

'Tell me.'

His sudden gentleness was more than I could bear and my face crumpled.

'Don't, Emilia! I can't bear to see you cry.' He stroked my cheek.

'I hardly know where to start,' I said. 'Of course I'll break off my engagement to Dolly . . . you see, I found out that he loves someone else . . . but before that I must go to Langdon Hall, my father's house in Hampshire.' The jumbled facts of what my mother had discovered came tumbling out and Alessandro patiently made me stop and asked me questions until he had the whole story straight.

'You're right to be afraid, Emilia,' he said at last, pulling me close. 'You cannot return to that house.'

'But I must! There's Aunt Maude, you see.' I leaned against his chest, inhaling the scent of his skin, my pulse skipping to feel him so close. 'And I have to know if the secret gallery exists. If it does, Father must be brought to justice.'

'And what kind of life will you have here once people know you are the daughter of a thief? Your father's actions will ruin you!' He paced away from me to look out of the small window at the teeming rain. 'I've been so unhappy in this city,' he said. 'I came here only to find you and make you come home. I never intended to stay. My contract is almost at an end and I shall return to my family gladly. But . . . '

'But what, Alessandro?'

'My life has no meaning without you in it.' His voice was quiet and full of hurt. He sat down close beside me. 'I have regretted every day since you left me that I didn't let you go with my blessing.

It was selfish and wrong of me to try and stop you. I apologise unreservedly. Emilia, I shall ask you one more time and, if you refuse, I'll never mention it again.' He cupped my face in his hands so that I had to look at him.

The pupils of his amber eyes were very black and as I met his gaze it seemed that I looked deep into his soul. I yearned to tell him that I loved him, that I'd never stopped loving him, but I wasn't quite brave enough.

'Emilia,' he said, 'you have my heart in your hands.'

I sat very still, hardly breathing, while my own heart somersaulted under my ribs.

'You are as necessary to me as water and without your presence I die a little more each day,' he said. 'I want nothing more than for us to spend the rest of our lives together. Emilia, my dearest Emilia, please will you marry me?'

Alessandro still loved me, even after all that had happened between us. All my past fears that he would abandon me, as others had in my past, if I allowed myself to love him entirely, evaporated like mist in sunshine. I had no doubts now; we would put things right between us. I let out a sob of pure joy. 'Yes, Alessandro,' I whispered, 'yes, I will marry you.'

He released his breath on a long sigh and caught me against his chest before burying his face in my hair. Then his beautiful mouth was warm against my lips and his hungry kiss was so full of passion that I felt as if I was liquefying into a river of molten sunlight.

At last we drew apart and I was left trembling with wanting him.

'I cannot give you the life of privilege you have become used to,' he said, 'but I will do all in my power to bring you happiness.'

'I haven't been truly happy since I left Italy,' I said. 'I certainly haven't been happy since I left you. And as for my life of privilege: I came to London with nothing and I shall leave with nothing, but if I have your love I shall consider myself the richest girl in the world.'

'But I have such ambitions, Emilia!' His face glowed. 'I've had so much time to think lately and I want to set up a school to educate poor children. You know how Princess Caroline loved little children? I spoke to her about it on many occasions and she promised to provide funds to help me set it up, but now . . . ' He sighed.

'Together, somehow, we shall find a way to build your school.'

He kissed me again, more gently this time.

As I slid my arms around his neck, the door creaked open.

'Ho! It's like that, is it?' said Mrs Bowker with a broad smile.

A scarlet flush rose in my cheeks.

'Mrs Bowker,' said Alessandro, 'may I introduce you to the future Signora Fiorelli?'

She gave a shout of laughter. 'Well then, congratulations are in order, I do believe.'

Alessandro kept a tight hold on my hand while we thanked her.

'I must go home,' I said.

'I shall walk you,' said Alessandro.

Mrs Bowker and I spoke in unison. 'Oh, no, you won't!'

'You must rest now, Alessandro.' I squeezed his hand.

'I'll send the second footman with you, miss,' said Mrs Bowker. 'I'll call him.' She winked at Alessandro and left us alone together.

'I expect to go to Hampshire the day after tomorrow,' I said. 'I'll look for the secret gallery and decide what to do about bringing Father to justice. And then I shall tell Dolly I have no intention of marrying him. All the guests will need to be informed that the wedding will not proceed . . . ' I swallowed at the prospect of the fury and recriminations to come.

'Emilia, I'm worried for you,' said Alessandro. 'If you find the proof you're looking for, you must not challenge your father outright. If what your mother wrote in her diary is true, he may be violent. You must leave Langdon Hall and come straight to me and we'll speak to the authorities together. Will you promise to do that?'

I nodded, relieved I wouldn't have to face Father alone.

'I want to come with you but I'm bound to remain here until the end of the month.' He chewed his lip and looked at me with worry in his eyes.

'I shall be quite all right,' I said. 'Besides, you must rest and let your arm heal. I'll come to you as soon as I can.'

Footsteps sounded in the passage outside.

Alessandro kissed me swiftly. 'Take great care, *cara mia*.'

'*A presto, amore mio*,' I replied.

I entered the house by the area door, hoping to return to my room unobserved. The scullery maid, however, gasped at the sight of me and dropped a pan with a clatter.

'It's all right, Annie,' I said. 'I slipped over in the mud. Will you ask Daisy to attend me with some hot water?'

I crept past the kitchen door with my shoes squelching and hurried upstairs to my room. Glancing at my reflection in the mirror, I saw why I'd frightened Annie. My hair hung in dripping rat's tails, there was blood on my face and my sodden cloak and mourning dress were encrusted with mud. I laughed, thinking how much worse it would have been if I'd been wearing white muslin. I looked into the mirror again. My eyes shone and my cheeks were flushed. Despite my torn and filthy clothes, I hadn't looked so happy or pretty since I'd arrived in England. I hugged my arms around myself, hardly able to believe that Alessandro loved me and that, no matter what, soon we would be together.

Daisy entered with a jug of hot water and her eyes widened. 'Annie said you'd had a mishap in the mud, miss.'

'I'm afraid my clothes are beyond redemption, Daisy.'

'There's blood on your face. You're not hurt, miss?'

I shook my head. 'Though others were. I went to see the funeral procession and there was a riot. Soldiers fired on the troublemakers.'

'I heard that,' said Daisy. 'If I'd known you were caught up in it I'd have been that worried . . . ' She shook her head. 'You'd better take everything off, miss.' She poured clean water into the basin and set out towels and soap.

Once I was clean I hurried to see Aunt Maude and found her sitting in a chair by her bedroom window.

'Thank goodness you're back!' she said. 'I was worried.' She looked me up and down. 'I saw your clothes were in a dreadful state when I spied you creeping in by the area door.'

I smiled. 'You don't miss anything, do you?'

Aunt Maude looked at me with her head on one side, like a robin. 'You seem surprisingly cheerful considering you've been to a funeral.'

'I have so much to tell you,' I said. 'The procession turned into a riot and I was caught up in it.'

She pressed a hand to her chest. 'You're not harmed?'

I shook my head. 'But Alessandro was with me and he was wounded . . . '

Aunt Maude gasped.

'Thankfully, it's not serious, though others were badly hurt. But then something wonderful happened.' My heart was nearly bursting with joy as I told her that Alessandro had proposed and I had accepted.

'My dear Emilia!' Aunt Maude reached for both my hands. 'I'm so very happy for you.'

'It makes me angry to think that if Father hadn't intercepted Alessandro's letters it might have happened before,' I said. 'This is a bitter-sweet day. I'm very sad I shan't see the Queen again.'

'Poor lady,' sighed Aunt Maude. 'And there will be difficult times ahead for you. Frederick will take the news of the cancellation of your wedding very badly, I fear.' She twisted the end of her plait round and round her fingers. 'You know, in the beginning he was passionately in love with Rose, or perhaps in love with the person he wanted her to be. It wasn't until she didn't conform to his ideal, when she voiced opinions that weren't the same as his own, that he changed so towards her.'

'And you fear his feelings towards me might change in a similar way?'

'Frederick is a formidable enemy,' she said, 'and you must be extremely careful, Emilia.'

# *Chapter 35*

*August 1821*

## Langdon Hall

Father, Aunt Maude and I travelled together to Langdon Hall, with our ladies' maids, Father's valet and the head footman following in the smaller carriage. The rain drummed down on the roof all the way, making our nerves taut as we jolted over muddy roads full of pot holes.

'I suggested that Dolly drive down in the carriage with us,' said Father, 'but it appears he has more important matters to attend to in town today. You'd have thought an appointment with his tailor or his wine merchant would be less pressing than the company of his fiancée and learning to run the estate that will one day be his.' He smiled in grim satisfaction. 'I gave him a piece of my mind ... and serve him right if he nearly drowns when he rides down on horseback tomorrow.'

I shifted my feet, which were pressed uncomfortably against the box containing my wedding gown, with the canvas-wrapped painting

Father had bought at the auction tucked in behind. Daisy hadn't allowed Dobson to put the dress box with the other luggage on the roof, not even under a tarpaulin, in case the rain penetrated it and spoiled the silk. Perhaps I'd give the gown to her later. I certainly wouldn't wear it when I married Alessandro. I smiled to myself, imagining how wonderful that day would be, filled with happiness and laughter. I wouldn't care if I were dressed in rags so long as Alessandro loved me.

'What are you smiling about, Emilia?' asked Father. 'Dreaming of your wedding day?'

I laughed. 'Yes,' I said, 'I was.'

At last we arrived at Langdon Hall. It was raining still and as the carriage clattered over the drawbridge the water in the moat appeared black, reflecting the bruised sky. Dobson brought the carriage to a stop close to the front door and servants ran out with sheltering umbrellas before we dashed into the hall.

Soon we were drinking tea in the library, while the servants lit candles to dispel the gathering gloom.

Father unwrapped the new painting and propped it against the wall beneath his other Dutch interiors. 'Come and tell me what you think of this, Emilia,' he said. 'As soon as I saw it I thought how appropriate it was.'

The painting depicted a couple, holding hands and standing on a tiled floor in a darkly panelled room. Sunlight poured in through a large window with diamond-shaped leaded lights, illuminating the damask of the young woman's claret gown and the white linen of the man's shirt.

'It always strikes me how calm and peaceful these Dutch interiors are,' I said. 'There's a beautiful underwater quality to the light, similar to when sun falls onto a stream and you catch a flash of the secret world existing beneath.'

'It's called The Proposal,' said Father. 'It looks as if the portrait has captured the young woman just as she's about to speak.'

'It makes me curious,' I said, 'about whether she said yes.'

'Of course she did,' said Father. 'They would never have been painted together if she hadn't.'

'Well, I hope they had a happy marriage.' Sipping my tea, I wondered how soon I was going to have an opportunity to search his study for the staircase to the secret gallery.

The following morning, after breakfast, Father retired to his study. 'I have a considerable amount of estate business to attend to,' he said, 'and on no account wish to be disturbed.'

My heart sank since that meant he wouldn't be going out.

After he'd left the room, Aunt Maude said, 'Frederick will be expecting us to be making the final arrangements for the wedding.'

'That would be a waste of our time,' I said. My stomach knotted again at the prospect of breaking off the engagement. 'Instead, perhaps we should write to the guests to tell them it's cancelled? We can post the letters after I've broken the news.'

Aunt Maude bit her lip. 'Supposing Frederick were to find them?'

'Hide them in your work basket,' I said.

We sat in the library and worked steadily on the letters for the next hour until Samuel came to inform us that Mr Cole, the tenant of Little Langdon Farm, had called to speak to Father.

'Sir Frederick is in his study,' I said.

'I did knock,' said Samuel, 'but there was no response. I've looked everywhere for him since Mr Cole is so anxious to speak to him. I wondered if Sir Frederick might have gone out?'

I frowned. 'Not so far as I am aware. I shall speak to Mr Cole.'

He stood in the hall, as solid as an oak tree, with his hat grasped in his meaty hands. He wore a rough tweed coat and old-fashioned breeches on his sturdy legs. 'Ah, Miss Langdon! A pleasure to see you again but it's Sir Frederick I need to speak to on a matter of some importance.'

'Come into the little parlour, Mr Cole,' I said. 'Father is in his study and has asked the servants not to disturb him. I'll see if he'll come and speak to you for a moment.'

'That's kind of you, Miss Langdon.'

He perched his bulky frame incongruously on a delicately carved chair and I hurried across the hall to tap on the study door. There was no reply so I turned the handle. The door was locked and I returned to explain to Mr Cole that my father must have gone out after all.

Some two hours later Aunt Maude and I had finished the letters and stowed them away in her work basket ready for posting. I was sitting on the window seat watching the rain pock-mark the moat when Father appeared in the library with his newspaper rolled up under his arm.

'There you are!' I said. 'It's nearly time for dinner and I thought you'd gone out.'

'Not in this perpetual rain,' he said. 'Besides, I told you I was working.'

'So you did,' I said, 'but Mr Cole came to speak to you and I tapped on the door to no response.'

Father shrugged. 'Been there all morning. I didn't hear you. Probably concentrating too hard. What did Cole want?'

I looked at my father thoughtfully. 'He asked you to call on him at Little Langdon Farm. The rain is damaging the harvest and Lower Meadow is half flooded.'

'I'm not sure what he expects me to do about it,' grumbled Father.

'He asked if you'd visit him this afternoon. He's very anxious.'

Father sighed. 'I'll go after we've had our dinner. I need to talk to him about a rent increase anyway.' He opened up his newspaper and settled down to read about the shameful story of the people's rebellion during the funeral procession of the late Queen.

I made an excuse to slip away and hurried upstairs to my bedroom. Mother's diary was wrapped in a scarf and concealed behind the lining of my old travelling bag. I hastily reread the passage where she explained how she'd found the staircase to the gallery.

*Searching the folios on Frederick's bookcase I found the secret staircase almost by mistake. It smelled dank and mouldy and was too dark to explore further without a light. Fortuitously, there was a candlestick on the bookshelves. Perhaps Frederick keeps it there for exactly this purpose? Harriet was resting so I lit the candle and descended the stairwell.*

I closed my eyes and pictured Father's study in as much detail as I could remember. The folios were the clue. If Mother had been examining the folios when she found the access to the staircase, surely it must be behind the bookshelves? Father hadn't been in the study when I knocked on his door but I guessed he might have been out of earshot in his hidden gallery.

I hid the diary again and went downstairs as Robert announced dinner was ready.

I was in an agony of impatience as we ate and could hardly force down a morsel while Father had a second helping of beef pudding and seemed inclined to linger over his claret and cheese.

Aunt Maude watched me as I crumbled a piece of bread and tried to make light conversation about the continuing bad weather.

At last Father pushed back his chair. 'I suppose it's no use delaying any longer,' he said, glancing at the rain hammering against the window panes.

'Will you take the carriage?' asked Aunt Maude.

Father shook his head. 'Cole wants us to look at the Lower Meadow so I'll ride Shadow. Don't forget Dolly will probably arrive before I return.'

Aunt Maude and I retired to the library and a short while later I heard hooves clattering across the courtyard.

'This is my opportunity to look for the hidden chapel,' I said, 'before Father returns and Dolly arrives. Of course, even if I find it, the stolen paintings may no longer be there. Then I can't prove

Father is the Picture Frame Thief. You go and have your rest, Aunt Maude. Later, I'll come and tell you what I've found.'

I accompanied her to the bottom of the staircase and waited until she'd turned onto the first-floor landing. Glancing around the hall to make sure no servants were watching, I slipped inside Father's study.

I stood in the centre of the room, hearing only my own heartbeat. There was a faint smell of tobacco smoke. The leather desktop was clear except for a brass ink pot and a blotter. Five paintings hung on the walls and, as far as I could see, there was no concealed jib door. The bookcase, built in three sections across one entire wall, was stacked with buff leather-bound folios, books and a few artefacts: a Chinese vase, a small bronze figure of a boy throwing a discus, a candlestick and a stone urn full of marble chips.

I went to look more closely at the first bookcase, pulling at it to see if it would move forward, but it was firmly fixed to the wall. I tried the same with the middle section and then the final one. A flash of excitement left my pulse racing when it moved a fraction. Perhaps this was it? I pushed aside some of the folios on the shelves, not quite sure what I was looking for, and it wasn't until I reached the lowest shelf that I saw the brass hinges in its back corner.

I searched the shelves above more carefully until I found another set of hinges in the centre and the last set higher up. My fingers scrabbled behind the folios at the opposite end of the shelves and in only a moment I had located two small bolts. I lifted them up and hinged the bookcase out towards me. I drew in my breath as a door was revealed in the wall behind. Grasping the ring handle, I paused for a second and looked at my fingers, imagining my mother's hand in the same place all those years ago when she, too, had found this door.

It wouldn't open. There was a small keyhole under the handle but no key. Frustration boiled up in me and I could have screamed. I took a deep breath. Think! What was it Mother had written?

*Fortuitously, there was a candlestick on the bookshelves. Or perhaps Frederick keeps it there for exactly this purpose?*

I looked at the adjacent set of shelves and there was the candlestick. A tinderbox and a used spill lay beside it. Where would Father keep the key? Somewhere nearby ... I rummaged through the marble chips in the stone urn without success and then my gaze fell on the Chinese vase. I tipped it towards me and heard a slight chink of metal against the porcelain. Peering inside, my heart leaped when I saw a key. My trembling fingers pushed it into the door lock and it turned with a satisfying click. I opened the door and shivered when I saw the rickety stairs disappearing down into blackness. The air smelled like that of a crypt.

It took me several attempts with the tinder box to light the candle but finally I stepped, dry-mouthed, onto a winding staircase. The sound of my footsteps was muffled as I felt my way down and my elbows brushed against the dusty, cobwebbed walls to either side. I tried not to think how suffocating the blackness would be if I dropped the candle. At the foot of the steps was a rusty gate with a key in the lock. I turned this and pushed the gate inwards.

Then I stepped through.

# Chapter 36

The air was very still. It almost felt as if I was suspended in time in the cool, mould-scented gloom. Light filtered in through four narrow clerestory windows set high up along one wall, dimly illuminating a sizeable underground chamber. I had found the hidden chapel!

My eyes began to make out shapes in the shadows and I caught my breath at the sight of a group of people, all dressed in white, watching me. I froze until I realised they were marble statues, similar to those in the garden. I held my candle high to study the unseeing eyes in finely modelled faces and the sculpted drapes that barely concealed the perfect proportions of the figures. One of the women had a chipped nose but this did nothing to render her any less beautiful.

Reluctantly, I turned away from the statues. Four pews in the centre of the space were arranged back to back in a rectangle, facing the walls, and there were a number of marble plinths displaying artefacts such as an urn, a marble bust or stone figurine. Bronze sconces in the sinuous shapes of water serpents, each supporting half a dozen candles, were fixed to the walls at regular intervals. I lit them from my candlestick and the decorative mirror behind magnified the flames into brilliance. Only once the chapel was ablaze with candlelight did I allow myself to look properly at the paintings.

My eyes were immediately drawn to the glorious oil-painted panel of The Last Supper hanging above the altar. I lit the torchères to either side of it and the candlelight made the gold of the saints' haloes gleam and the rich colours on the canvas glow as if they were alive. The painting was in the style of Botticelli. Each of the apostles had a face so full of character that the artist must have modelled them on persons known to him. The central figure of Christ stared directly back at the viewer and there was something about his benign expression that made it almost impossible for me to look away. Enchanted, I stood in front of the painting for I knew not how long, wondering why it seemed familiar.

I had to find the proof I needed before Father returned. One by one I studied the other paintings, discovering that each was a treasure. I'd gleaned some knowledge of Italian art during my travels and was astonished by the number of religious works in the chapel that were beautiful enough to rival any I'd seen in Florence, Siena, Arezzo, and the many ancient hilltop churches and monasteries in the area surrounding them.

The religious paintings were hung at the altar end of the chapel but as I worked my way back towards the entrance I discovered secular subjects too, grouped together by type: portraits, landscapes, architecture, even some studies of exotic animals. Hurrying now, I glanced at the remaining pictures but wasn't sufficiently knowledgeable to recognise the artists.

Then I saw something that made me gasp. Two tiny portraits in oval gilt frames hung on the wall, spaced about a foot apart. Each showed the profile of a dark-haired girl, one facing to the left and one to the right. I knew without a shadow of doubt that the portrait that completed the trio was in my bedroom, safely sewn inside the rag doll I'd carried everywhere with me since I was a child. But how had these two miniatures come to be here? Father had told me he was still searching for them.

Confused, I returned to the altar, intending to blow out the candles on the torchères, but was seduced again by the beauty of The Last Supper. I sank down onto the pew placed in the perfect position to appreciate it and, for a moment longer, allowed the peace imparted by Christ's calm gaze to wash over me.

'It is magnificent, isn't it?' said my father's voice behind me.

Immobilised by shock, I was unable even to turn my head. Astonishingly, Father didn't sound angry and so I took my lead from him and behaved as if my presence here was nothing out of the ordinary. 'It has me utterly bewitched,' I said. That at least was true.

'I didn't know when I acquired it that it was by Sandro Botticelli,' said Father, 'but, like you, I fell in love with it as soon as I saw it. I knew immediately I had to have it.'

'Botticelli?' I said, shock making my knees weak. I tensed as Father sat down on the pew beside me.

'One of his early works from the time he was apprenticed to Fra Filippo Lippi. Imagine having such exceptional talent at only twenty years old!' Father sighed. 'It was this painting that made me abandon my own dreams of being an artist. I knew I could never achieve half of what he had, even if I spent every day of my life working at it.' He shook his head. 'That was when I decided to collect art instead and this was the first item in my collection.'

'It's an extraordinarily fine one,' I said. I glanced at the iron gate at the other end of the chapel. It stood open still and I took a calming breath and forced myself to smile at Father.

'I'm delighted you like it!' he said. 'It's clear you are the child of my blood, even if you didn't grow up at my side. This collection is infinitely precious to me and I've wanted so much to share the pleasure of it with the right person. Shall I show you the rest?'

I nodded and tried not to flinch as he slipped his arm under mine. Aunt Maude had said he was unpredictable and, if I ran, I might provoke him to violence. All I could do was humour him until it was safe for me to leave. It wasn't hard to look fascinated as he took me from one extraordinarily beautiful work of art to another. My mouth was dry and my mind whirled as he attributed works to Raphael, Titian, Donatello and Caravaggio as well as a number of other artists I hadn't heard of. Some were cartoons or sketches, others oil on canvas, but there was an indefinable magic about them all that left me in no doubt that they were priceless.

'What do you think of them?' asked Father.

He must have stolen them or he'd have exhibited them where he could boast about them. I couldn't begin to imagine how he had managed to steal so many treasures without being caught. 'I'm almost speechless,' I said at last. 'It's overwhelming.'

'Come and look at these,' he said, face glowing with excitement. 'Although the works of the Italian Renaissance are closest to my heart, I enjoy the best of every kind of art, be it Roman sculptures and artefacts or something more contemporary, like these exotic animal studies by Stubbs.'

I peered closer to look at paintings of a giraffe, a lion, a rhinoceros, a tiger and a monkey. Something teased my memory.

'The especially interesting thing about these,' said Father, 'is that Stubbs usually paints horses.'

Then I recalled a comment that Araminta Perry's brother had made to me. A Stubbs painting of a giraffe had been stolen and the thief had left behind a tiny sketch of an empty picture frame. A companion painting by the same artist had been stolen five years previously from the same owner. My stomach clenched. This was all the proof necessary to confirm Father was indeed the Picture Frame Thief.

'Fascinating,' I said.

'I brought Dolly down here,' Father said, 'but he wasn't interested in the paintings except to ask about their value. Philistine! He thinks they'll be part of his inheritance but I'm damned if he'll have them! You appreciate them properly and shall have them after I'm gone. I wish now that I'd shown you my collection before.'

'Why didn't you?' I asked. Would he admit to me that he'd stolen the paintings?

'This is my secret place,' he murmured. 'When the world treats me badly, this is where I come to drink in the beauty of some of the finest art mankind has to offer. It's a place of healing. After Piers and your mother died, whole days passed by without my noticing while I sat here.'

'And you didn't want to share it with anyone before?'

'It is a question of trust,' he said. 'I made an error of judgement when I showed it to Dolly.' He grasped me by my upper arms so that I had to face him. 'Now I've seen the way you look at these paintings with such awe, I believe I can trust you.' He looked straight into my eyes and it took a great effort of will for me not to recoil. 'I can trust you, can't I?' he said.

I met his eyes with a guileless gaze. 'I couldn't bear it if any harm came to these priceless pieces.' I was sincere about that, at least, but I had to persuade him to tell me more. 'Where did you find The Last Supper?' I asked, freeing myself from his grip to look at the painting again.

'Where else but Italy?' said Father, smiling. 'I was twenty-one and near the end of my Grand Tour, learning the language, painting and sketching the scenery and visiting the churches to see the frescoes. I stayed at a monastery in the hills of Tuscany. This magnificent panel was hanging in the refectory.'

I caught my breath. That was why it had seemed familiar!

'As soon as I saw it I was lost,' said Father. 'It made me angry that none of the monks appeared to appreciate its beauty.'

'So you took it?' I hoped he wouldn't notice the tremor in my voice. I had to convince him I condoned his thefts. If I didn't, well, I knew from Mother's diary how violent he could be.

He crowed with laughter. 'I was in that place for three weeks! The monks believed I was very devout, spending hours alone in my cell reading the Bible. During that time, I secretly painted a replica. As soon as it was finished I took down the original while the monks were at their prayers, replaced it with my copy and escaped into the night.'

I made an attempt to look surprised. 'That was you?'

'It made me smile when you mentioned you'd seen my painting,' he said, 'even though you were unkind enough to say it was poorly rendered.'

'But the panel is at least five feet long,' I said. 'However did you bring it home without damaging it?'

'With great difficulty! I wrapped it in a blanket and strapped it to the side of my horse. I travelled overnight. When I reached Florence I had a stout box made for it, wrapped the painting in muslin and packed it with straw. I hired a carriage and conveyed it to Livorno from whence I sailed for London.'

I felt sick to know he'd stolen something so precious from his hosts and now was boasting about his betrayal of their trust. 'A very clever move,' I said, unable to look at him.

He laughed again. 'Wasn't it? You cannot imagine my exultation when the boat sailed out of the harbour. The painting was mine!'

He had no sense of right and wrong at all. 'And I daresay,' I said, 'you'd have liked to see the monks' faces when they were eating their bean stew the next day and noticed their painting had been replaced by a copy?'

'I confess, I've often been amused by that thought.'

It made me want to cry that this man, who had appeared to be the affectionate father I'd wanted all my life, had turned out to be utterly despicable.

'Collecting became a compulsion for me,' he continued, voice bubbling with enthusiasm. 'Once I came into my inheritance, I made this hidden chapel into a gallery worthy of my growing collection. I returned to Italy as often as I could. These marble statues came from a site near the Colosseum in Rome.' He caressed the hair and cheek of the lovely girl with the chipped nose. 'Venus,' he said. 'Isn't she perfection? She reminded me of your mother before she became such a shrew. The bribe I paid to the guard was only a fraction of the value of the statues. It was surprising how often fine sketches and paintings were displayed in perfectly ordinary churches, with little to prevent me simply lifting them off the walls when the priest's back was turned.'

His elation as he described his vile trickery revolted me and I decided to change the subject. 'There's something else I wanted to ask you, Father,' I said, beckoning him towards the small oval frames. 'Are these the miniatures of the Spanish Infanta?'

'How sharp of you to guess.'

'Where did you find them? I've spent so much time thinking about where Sarah might have sold them.'

'Ah, well.' Father rubbed his nose. 'Perhaps I wasn't entirely straight with you about that but I hoped so much you'd lead me to the missing one. After your mother's maid stole you away I chased after her but lost the trail by the time I reached Lyons. I promised all the art dealers I knew a reward for information leading to the miniatures and, of course, to Sarah.'

It hurt that the miniatures appeared to have been of more importance to him than his own daughter was. 'You thought that was the best chance of finding me?' I asked.

He rubbed his nose again and his gaze slid away from me. 'Yes, of course. If Sarah sold one of the miniatures, I thought I'd be able to pick up her trail again and find the other two.'

'And me?' I asked, although by now I knew what his priority had always been.

He waved his hand dismissively. 'I thought it unlikely Sarah would sell all three at once because they'd be a good nest egg for her future. And then, a year or so later, I had a stroke of luck. Her husband, Joe Barton, came to London to find me. He and Sarah had separated and he sold me information about her whereabouts. I gave him half what he asked for, with the promise of the rest if the information was verified. He told me Sarah remained in Milan with the child.'

The child. Me. Shadowy images of those dark days returned along with memories of always being hungry and scared. 'What happened then?' I asked.

'I set off for Milan straight away. I put up reward notices and visited all the local art dealers. You can imagine my joy when I found one of the miniatures. I bought it and was able to glean enough information from the dealer to trace Sarah to Verona. But then the trail went cold again.'

'Hunger and desperation must have forced her to sell it,' I said. 'She was always frightened because she thought someone was searching for us but she didn't know if it was you or Joe. That's why we moved so often.'

'Very annoying it was, too,' said Father, a nerve twitching in his jaw. 'Since Barton knew her best of all, I paid him to find Sarah and discover what she'd done with the remaining miniatures. Several times he nearly caught up with her before she did a moonlight flit. It infuriated me that she kept getting away with it. Still, Barton has proved a useful employee over the years.'

'What do you mean?'

He shrugged. 'I don't like getting my hands dirty. I locate the paintings I wish to acquire and spy out the lie of the land. Barton then brings them to me for a consideration.'

'So I *wasn't* imagining it,' I said, 'when I thought I saw him leaving the house in Grosvenor Street?'

'I told him never to come to my house again,' said Father, scowling. 'Such careless disregard for my instructions could have caused untold trouble.'

'When did you find the second miniature?'

'Two years ago. I had a letter from a contact in Florence saying he'd bought a fine miniature from an Englishwoman. It was impossible for me to leave the country at that time as Parliament was sitting and I was involved in negotiations for several paintings for one of my clients.'

'So it wasn't you Sarah thought was chasing us?' I said.

He shook his head. 'Not recently. I decided it was time Dolly earned his inheritance.'

'Dolly?' Puzzled, I shook my head.

'He'd landed himself in severe debt with his gambling habit and his tailor was threatening to have him taken up and sent to debtors' prison. He went to Florence in my place. I said I'd settle his bills if he collected the second miniature, and promised him half the value of the third if he found Sarah and, shall we say, persuaded her to tell him where she'd hidden it. My patience had worn very thin by then.'

I pressed a hand to my mouth. So Alessandro *hadn't* been mistaken about Dolly being in Pesaro.

'Unfortunately,' said Father, his lips pursing in annoyance, 'when he found Sarah, she obdurately refused to tell him where she'd hidden the miniature and he was disturbed before he was able to beat the information out of her. Stupid fool lost his nerve, returned to England and said he'd never go back to Italy again.'

I stared at my father, stunned into silence. I remembered again how that man, a very tall man, in the alley behind the cottage had nearly knocked me over. It must have been Dolly, exactly as Alessandro had said. The true horror of it was that Father condoned, had even suggested, the beating Dolly had given to Sarah, the beating that caused her death. Would he also have condoned it if Dolly had beaten me, his own daughter, in search of the information he sought?

'Emilia?' Father was looking at me with a puzzled air.

There was something very wrong with him. He had no conscience at all. 'I'm surprised that Dolly found himself able to beat Sarah so severely that she died,' I said. 'He isn't usually so . . . ' I floundered.

'What?'

'What I mean is, I'd have expected him to be far too anxious that he might risk dirtying his coat.'

Father hooted with laughter. 'How true! The threat of debtors' prison, however, was a bit of a stiffener. He had little choice but to do as I told him.'

I had to know. 'When you decided you wanted me to marry Dolly, did you know he didn't like women?'

Father narrowed his eyes. 'What do you mean?'

'I mean,' I said, 'that Dolly prefers men. Do you care so little for me that you are prepared to condemn me to a hollow marriage, possibly without children?'

'Without children?' He frowned. 'You *must* have children, Emilia. A son is essential to maintain two centuries of the same bloodline at Langdon Hall.'

'Well, you're barking up the wrong tree if you believe it's a foregone conclusion that Dolly will be able to bring himself to give you a grandson,' I told him.

A dry cough came from behind us.

Father and I spun around to see Dolly emerging from the shadows at the foot of the stairs.

# Chapter 37

Dolly's face was pale and strained as he confronted us. 'So I was right. I suspected you'd seen me with Francis in the garden.'

'How did you know?'

He gave a tight little smile. 'I was your partner in the next dance and you couldn't look at me. I'd heard a noise in the garden and there were rose petals in your hair.'

'Rose petals?' said Father.

'It was oppressively hot on the night of the ball and I had a headache,' I said. 'I went into the garden for some air. It was dark. The petals must have fallen on my hair when I hid in the arbour. I was so shocked when I saw Dolly and Francis together that I barely noticed my hair was tangled in the climbing rose.'

'Together?' said Father, frowning.

'Kissing,' I said. The expression of revulsion on Father's face was almost comical but at least it showed that, however else he'd tried to bend my will to further his own aspirations, he hadn't been aware of Dolly's particular inclinations.

Father pinched the bridge of his nose. 'I'm sure whatever Dolly may choose to do in his private life, Emilia, it won't prevent him from doing his family duty.'

White-hot rage rose up in me then and I threw caution to the winds. 'It's clear to me now, Father, how very little you've ever cared for my happiness. If you think I'm still going to marry him, you're quite mistaken.'

'Of course you'll marry him!' said Father through gritted teeth. 'You can't cry off so close to the wedding.'

Dolly coughed again. 'I rather think it's too late for that, Sir Frederick. You've told Emilia that I assaulted Sarah Barton, an assault that led to her death.' He shuddered. 'I still have nightmares about the sound of her skin splitting when I hit her. And now I'm sure that Emilia knows about my illegal relationship with Francis, I must take measures to protect us.' He took a step back. 'I will not risk my reputation and my inheritance, not to mention my liberty, should either of you decide to inform the authorities.' Spinning on his heel, he returned through the metal gate at the bottom of the stairs.

'It's hardly in my interest to denounce you,' Father called after him.

'Perhaps not,' said Dolly, his hand on the gate. 'But I believe Emilia is made of sterner moral fibre than you and that places me in a very difficult position.'

I realised with a jolt what he was about to do and launched myself at him.

Too late!

Dolly slammed the gate, turned the key and slipped it in his pocket.

'Don't be ridiculous, Dolly.' Father thrust his hand through the iron bars. 'Give me that key!'

'Let us out, Dolly.' My voice was calm but my fury was supplanted by fear. A flicker of panic at the thought of being locked in an underground chamber made my chest tighten.

Ignoring me, he said, 'As you may have noticed, it's raining in torrents.'

'What does that have to do with anything?' asked Father, rattling the gate.

'Have you noticed how high the river is?'

'Of course I have! I've been this very afternoon to see Cole at Little Langdon Farm. Lower Meadow is under water.'

'Opening the sluices to the moat will drain off some of the excess from the river.'

'You can't do that!' Father rattled the gate again. 'Idiot! The moat is already lapping over its banks. It'll overflow.'

'And when it does,' said Dolly, 'it will flood the cellars, including this chapel.'

I let out an involuntary moan.

'I'm afraid you've left me no other choice, Sir Frederick,' said Dolly. He rubbed his eyes as if he were deathly tired. 'I've grown quite fond of you, Emilia, and it grieves me that I'm forced to take this course of action.'

'Then don't,' I said.

He shook his head. 'I must. If I let you go now, even upon your solemn promise to say nothing, your conscience would eventually make you disclose who was responsible for Sarah's death.'

'Open this gate right now, Adolphus,' said Father, 'and we'll say no more about it.'

Dolly sighed. 'I'm sorry it had to be like this,' he said, 'but I understand drowning is quick and peaceful, Emilia, if you don't struggle.' His mouth twitched in a ghastly smile and he lifted his hand in farewell.

'Adolphus!' shouted Father. 'Come back at once and let me out! Don't you understand? If you flood the chapel, you'll destroy some of the finest art in Western civilisation.'

The sound of Dolly's footsteps faded as he climbed the staircase. And then the door to the study slammed shut.

I swallowed, my mouth suddenly dry. Whether the chapel flooded or not, we were incarcerated underground and I didn't like it. I didn't like it at all.

Father lost his temper. He bellowed and raged, calling down curses upon Dolly. He lifted an urn off one of the marble plinths and used this as a battering ram against the gate. Rusty or not, the gate stood firm. At last the paroxysm of rage dwindled and he sank down to the floor with his head in his hands, breathing heavily.

I started at the sound of a crash and then a large stone bounced into the chapel. Daggers of broken glass fell from one of the windows and skittered across the floor. A shadow moved across the casement.

Father looked up, his eyes red-rimmed. His usually smooth hair stood up in iron-grey tufts. 'Adolphus, let me out!' His voice was hoarse from shouting.

I winced as one by one the other three windows shattered, too. Tip-toeing over the broken glass, I sat down on one of the benches, my hands clenched together in my lap to stop them from shaking.

After a minute, Father pushed himself to his feet with a grunt and came to sit a few feet away.

'I always knew Dolly was a rotten egg,' he said. 'Just like his father. I can't bear to think of him getting his hands on Langdon Hall.'

I turned away from him. A damp draught drifted in through the windows and the torrential rain hissing down outside could now be clearly heard through the broken panes. My pulse began to skip as I studied the clerestory windows thoughtfully. At the top of the walls against the ceiling, beginning perhaps eight feet up, they were two feet wide and one foot high. The frames were set within the apertures, narrowing the access space to little more than an arrow slit turned sideways. Black depression descended upon me as I lost my last hope. I couldn't possibly wriggle through such a tiny gap.

A long time passed.

I wept a little as I pictured Alessandro's face. We'd come so close to being free to enjoy the rest of our lives together. I consoled myself by thinking that at least I'd die knowing he loved me.

'Emilia?' Father's voice breaking the silence made me jump. 'How did you find the chapel?'

We were going to drown anyway so what did it matter now if I told him? 'I found Mother's diary.'

He frowned at me. 'Where?' His voice sounded like a bark. 'I searched for it everywhere.'

I shrugged. 'In her work basket. It was in the attic with her portrait.'

'What else did you discover?'

'That you used to beat her so violently she was in fear for her life. That she wasn't unfaithful to you. And that you are the Picture Frame Thief.'

Father sighed. 'I loved her, you know. In the beginning anyway.'

'Did you drown her?' I asked. I had to know the truth.

He looked at me with bloodshot eyes. 'No,' he said.

I had no idea if I could believe him and still didn't understand why Mother might have drowned herself, however cruel he had been to her. 'You don't love anything other than yourself and your paintings,' I said. 'Or should I say, other people's paintings that you have stolen?' Then something caught my eye and icy fingers of fear ran down my spine. 'Look!' I said, staring at the wall in horror.

Water lapped through the broken window panes and trickled down the whitewashed wall, streaking it with muddy brown. Puddles grew on the floor and, as we watched, joined together into a pool.

Father moaned. 'I had those windows put in to air the chapel. I hid them behind bushes on the narrow strip of ground between the Hall and the moat. I was concerned damp might spoil the canvases.'

I laughed mirthlessly. 'Damp is certainly going to spoil them now,' I said. 'Did you never think about the moat flooding?'

'Of course I considered it!' he snapped. 'It hasn't flooded for generations. In any case, this is Dolly's fault for opening the sluices.' He gave me a baleful stare, his grey eyes as cold as slate. 'You sound just like Rose, always criticising and nagging at me. I couldn't bear the way she used to look at me with those great blue eyes of hers brimming with tears of disappointment.'

I didn't answer him. He clearly believed he was without fault and I wasn't going to waste my last breaths arguing with him.

The water began to flow faster. It gushed through the windows now in a stinking stream, filling the chapel with the reek of decay.

Father ran to the altar and snatched off the heavily embroidered cloth. He dragged a pew over to one of the windows and climbed onto it to stuff the cloth into the broken glass. The flow of water slowed to a dribble.

'There!' he said, jumping off the pew. 'We must find something to put in the other windows.'

'If we had a hammer and nails we could use some of the paintings to board them up.'

He gave me a look of outrage and at that moment the sodden altar cloth burst out of the windowframe and thumped to the floor. A great surge of water followed.

After a while, water eddied around my shins and I sat on a pew with my feet up. I was frightened to see how fast it was pouring in now that the level in the moat was higher than the windows. It made a terrible rushing noise and it was all I could do not to sob with terror.

Father sat on the altar with his arms wrapped around his knees and his gaze focused on The Last Supper.

Some of the candles had burned out, leaving nothing but drifts of acrid smoke. I splashed around the gallery and pinched out half of the remaining flames. I would relight them later when the others had burned down. I didn't want sit in the dark, waiting to drown, any longer than I had to.

It wasn't long before the water in the chapel was knee-high. My muslin skirt was sodden as it lapped over the seat of the pew and my panic was rising as fast as the water. I had to do something! Unable to sit still, I waded back and forth to try and keep warm as the cold began to seep into my bones. I looked up at one of the windows again and had an idea.

'Father!'

Slowly, he turned his head to look at me.

'Bring me that bronze urn,' I said. I climbed up onto the pew he'd left beneath the window and stood beside the cascading water.

He stared at me sullenly.

'Quickly!'

Moving as slowly as if he were walking through treacle, he brought me the urn. 'What are you going to do?'

I didn't answer but snatched the urn from him with both hands and raised it above my head. I brought it down to thump against the timber framework of the window. I gasped as my arms went into the torrent of cold water and it diverted a stream onto my face. I lifted the urn and held my breath as I crashed it into the frame again. Several minutes later I was wet to my skin but the windowframe had splintered.

'Don't just stand there watching!' I cried.

Father stepped up onto the pew and we began to prise away the broken wood. It wasn't easy since we couldn't see properly through the tumbling water.

Eventually I ran my fingers around the frame removing any remaining sharp fragments as best as I could.

'Now help me bring the altar,' I said.

We heaved the heavy table across the floor with the water swirling around our waists and placed it under the window. It took a great deal of grunting and swearing on Father's part and straining on mine, but we managed to lift the pew onto the altar table. I leaned against the wall to catch my breath.

'It's a waste of time,' said Father flatly. 'The window is too small.'

'I'd rather die trying to escape than by sitting here until I drown,' I said. 'You must shove me through as far as you can so I'm not forced back by the pressure of the water.'

Father chewed at his lip, water dripping from his hair. 'It's a pointless exercise but you're courageous to try.'

'Then put your back into lifting me up and pushing me out,' I said, not feeling brave at all. 'If I survive I'll fetch help. Then there's a chance we can save the paintings.'

'Save the paintings?' Hope lit up his eyes again. 'By God, we'll give it our best shot, Emilia.'

My teeth chattered as I hauled myself out of the swirling water onto the altar and then the pew.

Father clambered up beside me. The stinking waterfall poured down between us, carrying clumps of weed and twigs.

'Pray Dolly isn't outside watching,' I said, toeing off my shoes. I composed myself, taking several deep breaths and sending up a prayer of thanks that Alessandro had taught me how to swim. 'I'm ready,' I said. I took another breath, stepped into the cascade and gripped the window aperture. The powerful force of the water made me stagger.

Father grabbed me firmly around my thighs and hoisted me up.

I propelled myself through the window, coming to a painful halt as my hips wedged in the narrow opening. Water filled my ears and all sounds were muffled. Wriggling, I kicked my legs but it was hopeless.

Father was still thrusting me forwards and I fought frantically against him, trying to slide backwards. I was running out of breath and stars danced before my eyes. My hands flailed desperately in the muddy water and my hair wound itself around my face like tentacles.

If Father didn't pull me back, I'd drown. A furious rage rose up in me and gave me the strength to kick back at him, hoping that with my last breath at least I'd knock his teeth out. That last vicious kick twisted my wedged torso and at the same time Father gave me a violent shove.

There was a wrenching pain in my hip and I shot through the opening into the churning waters of the moat.

# Chapter 38

My chest felt as if it would burst. I watched the bubbles of my last breath floating away as I drifted in the murky depths. Inside I was screaming *I don't want to die!* but at the very second I believed all hope had gone, I saw light above. I scissored my legs, all tangled up in my skirt, and then my face surfaced. A half-submerged shrub clawed at my cheek as I heaved in damp air.

I sank again. A current thrust me along and I thumped against something hard. Terror made me thrash my feet and I came, coughing, to the surface again. Langdon Hall loomed above and an expanse of black water lay ahead. The sky was darkening and still it poured with rain.

The water pushed me inexorably backwards and I grazed my heels on rough brickwork. The moat sucked me down and, when my foot disappeared into a void, I realised with horror that the current had carried me back to the very window I'd escaped from. I pushed against the wall with my feet, using the very last of my strength.

The water boomed and echoed in my ears as I bumped against the wall while muddy water swirled before my eyes. I was tired. So tired.

I floated face down. Something snagged painfully at my hair, jerking my head up. Water foamed around me and then I was on my back and rain pattered on my face. I breathed in great harsh gulps of air.

'Emilia!'

My eyes opened. I blinked at the rain-filled sky above as I was towed through the water by a pressure around my neck. Sleep ... I closed my eyes again.

My arms were hauled upwards and my legs slid along muddy ground.

'Emilia!'

Alessandro's face came into view above me. He smoothed the tangled hair off my face and covered my cheeks, my eyes, my nose with kisses while tears and rain dripped off his chin.

'Is it really you?' I whispered.

He scooped me up against his chest and buried his face in my neck. 'I thought I'd lost you!' His voice cracked as he rocked me against him.

I slid my arms around his shoulders.

Alessandro kissed my forehead again. 'I've been so angry with myself that I let you come here on your own.'

I leaned my forehead against his, my strength slowly seeping back. We were on the grass on the opposite bank of the moat. My hip ached. There was a long tear in my skirt and blood blossomed on the filthy muslin from a throbbing gash on my thigh.

'How did you come to be in the water, Emilia?'

I drew in my breath sharply. 'Father!' I tried to stand but Alessandro held me back.

'You must rest,' he said, 'and then I shall take you away from this place.'

I shook my head. 'You don't understand! Father will drown if we don't save him.'

Alessandro's shocked gaze never left my face until I finished recounting the story that tumbled out of me.

'So we must hurry and break through the gate into the chapel!'

'It's magnanimous of you to want to save your father after all he's done,' said Alessandro through gritted teeth.

'I must save the paintings,' I said, 'and Father must face justice.'

'What about Dolly?' Alessandro's lip curled contemptuously.

'I want him punished, too ... but we must send the gardener to close the sluices immediately.'

Alessandro pulled me to my feet and, hand in hand, we hurried off to seek assistance.

Mrs Bannister, after her first shocked reaction to us dripping on her polished floors, moved into action and set the scullery maid to boiling water in as many pans as would fit on the fire.

Mr Bannister sent for the gardener to turn off the sluices and Samuel and Robert went to fetch a crowbar to force open the gate to the chapel.

'Have you seen Mr Pemberton?' I asked Mr Bannister.

'Not since he arrived.'

'Will you find out where he is now? Take care, though. He may be dangerous.'

Mr Bannister sighed. 'A pretty pass that Sir Frederick's heir should conduct himself in such a fashion.'

I knew how shocked he'd be once he was appraised of the full extent of Father's perfidy.

'Go upstairs now, Miss Langdon,' said Mrs Bannister. 'Daisy will help you to wash and change and dress your wounds.'

'Later,' I said. 'Send for the parish constable, will you? Tell him to bring men with him. And take my aunt a message to say I'm perfectly all right and I'll come and see her when I can. We'll need candles to light the way.'

Robert reappeared, brandishing a crowbar.

Alessandro and I ran to the study with Samuel and Robert in tow. The door was locked.

'Look!' said Alessandro. He bent down and picked up a key from the floor.

'Dolly must have dropped it in his hurry to open the sluice gates,' I said.

Alessandro pushed the key into the lock. 'It's not the right one,' he said, examining the key and trying again. 'It's too small.'

There was no time to waste. 'Robert, force the door open!' I said.

His eyes widened.

'If you don't, I will.'

He grinned and forced the crowbar into the frame. The door swung open.

The bookcase was back in place and the three men stared blankly around the room.

I hurried to release the hidden bolts.

Alessandro drew in his breath as the bookcase hinged forward and exposed the door leading down to the chapel.

'Try the small key,' I said and smiled as the narrow door opened with a click. I started down the stairs. 'Father, we're coming!'

The river stench rose up to meet me and I gasped at the cold as I stepped into the chest-high water at the foot of the stairs.

Robert followed close behind, grimacing at the stink as the filthy water swirled around him. I held up my candle and he started to prise open the gate lock under the water.

I peered through the iron bars into the chapel. Several more of the candles had burned away.

Father waded through the water towards us holding a canvas above his head. 'Thank God!' he said. 'I thought you'd never come to let me out, Emilia.'

I noted he didn't profess any concern for me.

'Take this to safety,' he said, 'and I'll fetch the others.' He slid the painting through the bars into my hands.

'I'd have drowned if Alessandro hadn't pulled me out of the moat,' I told him.

'Never mind that now,' said Father. 'Get that gate open and help me!'

Robert forced open the lock with a cry of triumph.

Alessandro took the canvas from me.

'Careful!' said Father. 'It's a cartoon by Titian. Whatever you do, don't let it get wet.'

'Emilia, you've been in the water too long already,' said Alessandro. 'You'll catch a chill. Go upstairs now.'

'Don't you dare!' said Father. 'We need all hands.'

Alessandro balled his fists. 'Have you no care for your daughter? She nearly drowned!'

I caught Alessandro's sleeve and pulled him back. 'Later,' I said. I'd no intention of leaving until the precious paintings were safe but it warmed my heart to know that Alessandro, at least, cared for my health.

First of all came The Last Supper. Father and Alessandro carried it between them, resting it on their heads as they waded through the rising water. It was too heavy for me to handle safely on my own and Samuel helped me convey it upstairs. We propped it against the wall in the hall and Mrs Bannister stood by with clean muslin cloths to blot any wet fingerprints off the rest of the paintings as they arrived.

One by one, Alessandro and Father lifted the paintings off the walls, starting with the largest and the most valuable, and bringing them to me at the staircase. I carried them up to the doorway of the study and passed them to Mr Bannister. He carried them to the next person in the chain. All the while the water grew deeper in the chapel.

The candles began to flicker as they burned down. Soon there was so little light I feared for Alessandro and Father's safety. I called for more candles and waded into the chapel to replenish the sconces. Water lapped under my chin now and it was difficult for me to walk through it. Once the new candles were lit, I swam back to the stairs.

Alessandro handed me the Stubbs painting of a giraffe. 'The water is deep and your father and I are both tiring,' he said, wearily pushing a lock of hair from his eyes.

'Don't take any risks, Alessandro,' I said.

'Don't stop to talk!' shouted Father.

We brought out another five paintings but by then the water was so high that Alessandro and Father had to swim on their backs with the paintings held over their chests. The movement of the water as they splashed past the sconces extinguished yet more candles.

I shuddered as I peered into the increasing gloom of the chapel. The stinking water appeared black as Alessandro swam slowly back to me. I reached out to catch his collar and pulled him to the stairs. 'You're exhausted,' I said, noticing the wound on his shoulder was bleeding again. 'Come and rest.'

'Your father's bringing the last one,' he said, handing me the two miniatures of the Spanish Infanta.

I placed them on the stair above us as Alessandro dragged himself out of the water.

Several more of the candles flickered out and I squinted into the darkness to see Father swimming towards us. His breath came in rasping gasps as he clambered onto the stairs, his knees shaking.

I took the canvas from him. It was soaked.

Father collapsed in a half sitting, half lying position. 'That's the last of them.'

'Come upstairs,' I said. Now that the immediate crisis was over and the paintings were safe I was full of dread at what was to come next. I would have to ask the footmen to confine Father until the magistrate arrived.

'There are still the statues and the urns and the bronzes . . . '

I stared at him. 'But they're already underwater. They'll be safe until the flood goes down.'

'I'm not risking any harm coming to them!' He pushed himself off the steps and slid into the water.

'Father . . . come back!'

'No!' said Alessandro, gripping my wrist as I made to follow. 'It's too dangerous.'

Almost all the candles were submerged and the black water was only a foot below the ceiling. I shivered. Father didn't answer but I saw his dark head bobbing up and down like a seal's as he swam away.

At the far end of the chapel he disappeared under the water. A moment later I caught a glimpse of a white face coming towards us in the deepening gloom. The statue of Venus. There was a great deal of splashing as Father struggled to drag the marble statue along.

'He's lost his wits,' said Alessandro. 'It's far too heavy.'

As we watched, the statue slipped and sank. The surface of the water undulated and slapped against the walls. I waited for Father's head to reappear. It didn't.

'Where is he?' asked Alessandro, his voice tight.

My teeth chattered with cold and fear.

The water became still, all secrets hidden below its dark depths.

I couldn't bear it. I launched myself into the water and struck out towards the other end of the chapel. Alessandro's voice calling my name rang in my ears. I concentrated on reaching the place where I last saw my father. My head bumped against the ceiling and I couldn't touch the floor. And then I banged my knee on one of the statues and knew I must be in the right place. I took a breath and dived.

It was pitch black and I moved my hands before me. I touched several of the statues but each one was upright. I went to the surface to gulp another breath and dived again. This time I went deeper and my questing hands found Venus, lying on her side. Blindly, I ran my palms along the cold marble and paused as my fingers touched cloth. Scrabbling at it, I felt flesh underneath. Frantically, I tugged at the cloth but it was trapped under the statue. My chest was tight. I had to breathe!

Pushing myself upwards, I burst through the surface and gasped for air. I screamed and nearly choked when Alessandro swam into me. He tried to pull me away but I fought him off. 'He's trapped! I have to go back.'

I dived again and felt Alessandro beside me in the black water. I found his hand and guided him back to where Father lay trapped. Together we heaved at the statue and rolled it to one side. We hauled at Father's clothing and towed him back towards the light.

Mr Bannister waded towards us and dragged Father out of the water.

I knew it was too late as, coughing and sobbing, I watched Mr Bannister and Alessandro pressing in vain on Father's chest.

At last Mr Bannister shook his head. 'I regret to say that life is extinct,' he murmured.

Alessandro carried me upstairs to the study. He placed me on Father's desk chair, chafed my hands and kissed me, murmuring words of endearment while I wept.

# *Chapter 39*

It was still raining as the three of us stood by my father's open grave with the vicar. There were no other mourners. The story of Father's secret life as the Picture Frame Thief had spread faster than the plague and his neighbours and acquaintances had stayed away. I was glad of it since I didn't have to face them.

After Father drowned, once I'd stopped weeping from the shock, Alessandro took Shadow from the stables and galloped off after Dolly. From London he followed Dolly and Francis's trail as far as Dover, where he ascertained they'd taken the packet to France. Alessandro then returned to Langdon Hall to answer the magistrate's questions.

The vicar's voice droned on and Alessandro held my hand tightly as Father's coffin was lowered into the ground. Aunt Maude offered me a handkerchief. My tears were not for my father the man, but for the loss of my dream of having a parent to love and respect.

As soon as the service was over the vicar nodded to us without comment and took himself off, his duty done.

'Shall we go?' said Aunt Maude.

I took her arm and the three of us walked through the churchyard towards the waiting carriage.

Mr Digby, my father's lawyer, sat at his desk and sipped a glass of ratafia.

Gripping the arms of my chair, I waited for him to enlighten me on my financial situation.

He replaced the glass on the desk and smiled. 'My wife sends her good wishes and would be pleased if you and your aunt would call on her.'

A flush warmed my cheeks. Aunt Maude and I were tarnished by our connection with Father and there hadn't been a single caller at Langdon Hall since his death. 'Please tell her she is very kind,' I said. I liked Mrs Digby and was grateful to her for being prepared to support us but I wouldn't place her in an awkward position.

'On to matters of business,' said Mr Digby. 'In consultation with the magistrate, I employed an investigator, on behalf of the Langdon Hall Estate, to confirm Mr Fiorelli's account of events. Your father's heir, Mr Adolphus Pemberton, together with Mr Gregory, did indeed flee to France.'

'Did the investigator find them?' I asked.

Mr Digby shook his head. 'It can only be assumed Mr Pemberton intends to evade the justice that awaits him here. He would be extremely unwise to return.'

Dolly's actions had resulted in Sarah's and my father's deaths and he should pay the price. I pictured him squabbling with Francis as they were forced to lead an impoverished, vagabond life. That would be some small punishment but it angered me that, so far, he had escaped retribution. 'I have now been in communication with your father's second cousin, Mr Harvey,' continued Mr Digby. 'He's more than delighted to live at Langdon Hall and to manage the estate, an unthought of possibility for him. He is aware that he cannot inherit until seven years after Mr Pemberton's last sighting, when he can be declared legally dead. Mr Harvey intends to sell his trade premises and move his family here in two months' time.'

'I hope he'll be happy at Langdon Hall,' I said.

Mr Digby peered over his gold-rimmed glasses. 'I believe he and his wife and their eight children will find the Hall a considerable improvement on three rooms above a butcher's shop in Fetter Lane.'

'I'm sure they will,' I said, smiling slightly at the thought of Father turning in his grave.

'Mr Harvey asked me to send you his reassurances that if you and Miss Weston wish to remain at Langdon Hall or the townhouse in Grosvenor Street until you have finalised your plans, he will be pleased to allow it. He intends to retain the servants, too, should they wish to stay.'

I was relieved, knowing how much consternation there had been in the servants' hall. 'My aunt and I will be travelling to Italy with my fiancé as soon as I've finished sorting through my father's papers.'

Mr Digby nodded approvingly. 'And have you discovered any further information amongst Sir Frederick's effects regarding the origins of the stolen paintings?'

I shook my head. 'Not yet.' The paintings and sculptures had been removed to a place of safekeeping by the magistrate and the complicated job of returning them to their rightful owners had begun. I had personally undergone a humiliating meeting with Lord Beaufort to return all three miniatures.

'Your own situation is not entirely bleak,' said Mr Digby. 'Sir Frederick came to see me following your engagement ...' He paused. 'Perhaps I should say your *first* engagement. He made a new will, which still stands.'

'But the estate is subject to the entail?'

'Indeed. However, Sir Frederick's art collection ...' He coughed. 'That is, the collection he purchased *legally* over the years, with funds inherited from your grandmother and from his own dealings in art, was entirely his to gift to you.'

'I see.'

'The collection will be valued but, should you wish to sell it, there is likely to be sufficient to provide an income for the remainder of your life.'

I stared at him, dumbfounded.

'Furthermore,' said Mr Digby, leaning his elbows on the desk and steepling his fingers, 'there is your mother's jointure. This settlement was arranged on your mother's and father's marriage to provide an income for your mother, *or her surviving children*, should your father die first.' Mr Digby smiled. 'This sum, quite separate from the Langdon Hall Estate, is now due to you.'

A short while later Mr Digby ushered me out of his office.

Alessandro and Aunt Maude were laughing together as I joined them in the waiting room.

'There you are, dear,' said Aunt Maude with a smile. 'Your young man was telling me such an amusing story . . . ' She frowned. 'Is there something wrong?'

'No,' I said. 'Quite the contrary.'

Alessandro took my arm. 'Let's return to the carriage and you can tell us all about it.'

Dazed, I explained my good fortune as the carriage rolled away.

'It's no more than you deserve,' said Alessandro.

'But nothing will bring back your dear mother,' said Aunt Maude.

We sat in contemplative silence during the rest of the drive.

'I'm going to sort through Father's papers,' I said. 'I'm sick of the task and want to finish it today.'

'Let me help,' said Aunt Maude.

Alessandro settled down in the library to write a letter to his family and Aunt Maude and I went into the study. I opened the window to let out the smell of dank decay that still drifted up from below. The water had been pumped out of the chapel but it would take months to dry properly.

'Aunt Maude,' I enquired, while opening one of the desk drawers and lifting out a pile of loose papers, 'will you mind leaving England? Be honest now.'

'Mind?' She laughed. 'You cannot imagine how flattered I am that Alessandro has invited me to come to Pesaro with you. There's nothing for me here, without you.'

'Now I can afford to give you an income of your own,' I said. 'It won't be a great deal but enough for you to be independent. So if you don't want to go so far from home ... '

Her cheeks went bright pink and her mouth worked. 'Wherever you are *is* home to me, Emilia,' she said, her eyes glinting with unshed tears. 'But I thank you from the bottom of my heart.'

I hugged her, so grateful an aunt had come into my life. 'I suppose we'd better finish the task in hand.' I pushed a handful of papers towards her. 'We need to put unpaid bills in one pile, anything relating to paintings or art in another, and glance at the rest.'

An hour later I heard Aunt Maude take a gasping breath.

'What is it?'

Her face was bone white and she pressed a hand to her heart.

I sprang to my feet. 'Aunt Maude? Are you ill?'

She opened her mouth but didn't speak and then held out a paper to me with violently shaking fingers.

I took it from her and saw that it was a bill, dated July 1820. It read:

*To the final quarterly sum for the continuing care and confinement of Rose Langdon.*

*To defraying the funeral costs of Rose Langdon.*

I sank down on the desk chair and read it again while my pulse raced. 'I don't understand,' I said. 'My mother drowned herself ... Didn't she?'

'Frederick found her clothes by the river but her body was never found,' whispered Aunt Maude.

'But if she didn't drown herself ... ' I crumpled the paper in my fist, suddenly breathless with rage. 'Father had her put away! He shut her up for *eighteen years* because she threatened to expose him as a thief. And he let me think she abandoned me because she didn't love me enough. If Father weren't already dead, I'd kill him with my bare hands.' Searing hatred for him ripped through me. He had lied to me, told me Mother was a thief and an adulteress, but far worse than that, he had deprived me of her love.

Alessandro pushed open the study door. 'I'm going to the post office …' His voice trailed off when he saw me hunched over Father's desk. 'What is it, Emilia?'

I couldn't speak and Aunt Maude quickly explained what we had discovered.

Alessandro enfolded me in his arms with a muttered curse and I buried my face in his neck. He held me tightly and rubbed my back as I wept for the loss of my mother, knowing now for certain that she had always loved me.

# Chapter 40

*September 1821*

**Italy**

Alessandro and I waited in the *salone* of the Villa Vittoria. Late-September sunshine cast lozenges of light across the floor and I smiled, remembering my first visit here when I'd found Princess Caroline laughing uproariously while she played a boisterous game with Victorine. Even though she'd often been unhappy, the Princess always grasped life with both hands, taking small pleasures where she could find them.

Footsteps, light and quick, sounded in the hall and then the door burst open.

'Signor Fiorelli!' Victorine let go of Countess Oldi's hand and launched herself into Alessandro's arms, shrieking with delight as he whirled her around.

'I hope we find you well, Countess?' I enquired.

She shrugged. 'Life is very quiet at Villa Vittoria these days.'

Alessandro put Victorine down and bowed to the Countess.

'The Baron returned from hunting a moment ago,' she said, 'and wishes to speak to you, Signor Fiorelli.'

Victorine tugged at Alessandro's hand to gain his attention again.

'Look at you!' he said. 'You've grown up into a fine young lady since I saw you last.'

The child giggled. 'Why have you been away so long?'

'I went all the way to England to fetch Signorina Barton.'

'Mamma died in England,' said Victorine, her happy smile fading.

'We met her there,' I said, 'and she told me how much she missed you.'

'I miss her, too.' The corners of her mouth turned down.

Alessandro stroked her hair. 'But I have some happy news for you. Signorina Barton and I are to be married next week and we've brought you an invitation to our wedding.'

'We'd be delighted if you and the Baron would honour us with your presence also,' I said to Countess Oldi.

'I should like that. And will you be living in Pesaro?'

'I've been offered a position teaching at the University of Bologna but haven't yet decided to accept it,' Alessandro told her.

'Bologna is so far from your family!' the Countess exclaimed.

'I know,' he said unhappily.

The whole Fiorelli family had welcomed Aunt Maude and me with such warmth that it felt as if we were being wrapped in a soft blanket. I didn't want us to make our home too far away from them, either.

Footsteps rang out across the hall and the Baron strode in, carrying the pungent odour of hot horseflesh with him.

I'd forgotten how tall and handsome he was, with his glossy black hair and curled moustache, but he looked older and I fancied there were worry lines around his eyes.

'Papà!' cried Victorine. 'Signor Fiorelli and Signorina Barton are to be married and I'm invited to the wedding!'

'Then you shall have a new dress,' said the Baron. 'But now I wish to talk business to Signor Fiorelli. Go to the kitchen, my sweet. Faustina has baked almond cakes.'

After Victorine had pulled the Countess away with her, the Baron said, 'Congratulations on your engagement, Fiorelli.'

'Thank you,' said Alessandro, smiling at me.

'It's very strange to be here without the Princess,' I said.

A shadow passed over the Baron's face. 'At least her unhappiness is at an end,' he said. 'I heard you were back in Pesaro, Fiorelli, and I was going to ask you to come and see me.'

'Oh?' said Alessandro.

'I'm happy to tell you that Victorine's future is secure as she has inherited the Villa Vittoria.'

'The Princess loved her,' I said.

The Baron nodded. 'She always showed great compassion for children. Did you know that some years ago she adopted a number of orphans in London and had them trained in useful trades?'

'She spoke of that to me many times,' said Alessandro. 'She wanted to help the poor children of Pesaro, too.'

The Baron nodded. 'Although she made no provision in her will, I want her wishes to be honoured.'

I held my breath, hoping I'd correctly anticipated the drift of his conversation.

'As the Princess's steward,' the Baron continued, 'I managed her household finances. Her spending was sometimes impulsive. As a result of this I kept aside a sum for contingencies. I believe you, Signor Fiorelli, may be the right person to implement the Princess's wishes.'

'A school?' asked Alessandro breathlessly.

'Exactly!' said the Baron.

'She intended to give poor children a good education,' said Alessandro, his voice bubbling with enthusiasm, 'to help them make their way successfully in the world.'

'The sum available is enough to buy a house to convert into a school,' said the Baron, 'and to provide a home for the headmaster. The remaining funds, carefully invested, will allow salaries for two or three teachers. Is this project something you would be prepared to undertake?'

'Yes! Yes, it is!' Alessandro's eyes glowed. 'At least,' he glanced apologetically at me, 'I should like this very much if my fiancée is agreeable.'

I reached out to grasp his hand. 'It's a marvellous idea,' I said. 'But I have one proviso. The school must honour Princess Caroline's name.'

The Baron smiled. 'She would have liked that,' he said.

A week later afternoon sunshine bathed the wedding party in liquid gold. Mamma Fiorelli and her daughters had been cooking all week to prepare the sumptuous feast laid before us on tables decorated with wild flowers. There were vast platters of *frutti di mare*, chicken with lemon and garlic, bowls of glistening olives and salads, pyramids of crusty bread, and purple figs stuffed with mascarpone and drizzled with flower-scented honey.

We sat in the Fiorellis' garden listening to the happy chatter of the guests and I kept glancing at Alessandro, hardly able to believe my good fortune that he was now my husband. He squeezed my hand and light danced in his eyes as he gave me one of his infectious smiles.

Dear Aunt Maude sat beside me, her pinched, careworn expression replaced by a contented smile. All the people I loved were close by and I couldn't remember ever having felt so at peace.

Papà Fiorelli raised his wine glass for yet another toast. *'Per cent'anni!'* For a hundred years! The guests echoed him. 'If my son's marriage is only half as happy as mine,' he said, 'then he will be a very happy man indeed!'

I glimpsed Alessandro's mother, who was blushing like a bride herself.

Salvatore, one of my new brothers-in-law, called out, 'A kiss for the bride!'

My veil, lent by my mother-in-law, fluttered in the breeze and Alessandro laughed as it tickled his face. He kissed me and a roar of approval and clapping burst out around us.

'Hurrah for the newlyweds!'

Alessandro whispered in my ear, 'You're so beautiful you take my breath away.'

I touched Mother's pearls at my throat. I wore them with the lovely dress embroidered with roses that I'd worn for my first ball. How long ago it seemed!

'Let there be music!'

The musicians struck up on a mandolin and a violin.

'We must lead the dancing, Signora Fiorelli,' said Alessandro with a mischievous smile. He offered me his arm.

I touched Aunt Maude's cheek. 'Is it too noisy for you?'

Smiling, she shook her head. 'Music makes me feel alive again. Go and dance with your husband.'

Alessandro and I opened the dance and every touch of his hand on mine sent shivers up my spine.

'You must dance with every man who asks you,' he said, his gaze fixed on my mouth, 'and be aware they'll all kiss you to make me jealous.'

'I'll never give you cause to be jealous of another man,' I promised.

He smiled and kissed me slowly and thoroughly while the guests whistled.

Other couples joined us in the dance and the pace of the music quickened. Alessandro's father came to claim me and whirled me around as fast as a man half his age would have done.

The Baron took his place and was light on his feet as he moved expertly through the steps. 'How the Princess loved to dance!' he exclaimed.

I fancied there was a wistful tone to his voice. I glanced over his shoulder and missed a beat when I saw Aunt Maude dancing sedately with Papà Fiorelli. Her face was flushed and her eyes bright. Breathless, I danced with all Alessandro's brothers, even little Alfio, and then, one by one, the rest of the male guests.

Then Alessandro came to claim me again and we laughed as Victorine and Alfio spun in circles around us, showering us with rose petals.

The sun was slipping down behind the hills when, at last, we returned to the table, where Cosima sat with Aunt Maude, teaching her to speak Italian.

As it grew dark, Salvatore and Jacopo went around the garden with burning tapers and lit the myriad candles hanging from the trees in glass jars. The flickering flames reminded me of the fireflies that had enchanted us in the avenue of cypresses at Villa Vittoria so long ago.

Alessandro and I exchanged a few words with every guest, receiving their good wishes.

Later, as the garden became illuminated by moonlight, Mamma Fiorelli came to see us with her little grandson Enzo asleep on her shoulder. 'It's time for you to go,' she said. 'And don't worry, Emilia, I will look after your aunt. Tonight is for your husband.'

I hugged her. 'We'll come and take Aunt Maude home with us tomorrow.'

Mamma Fiorelli kissed Alessandro and he hugged her tightly.

The young men in the party began to whoop with exuberant high spirits.

Alessandro and I kissed Aunt Maude and said our goodbyes to the older guests.

A violinist, playing a lively jig, led the noisy procession out of the Fiorelli house and down the lane. Alessandro's friends sang at the tops of their voices, banging wooden spoons upon saucepans and setting off firecrackers as we danced along until we reached our temporary home on the cliff overlooking the sea.

Alessandro lifted me in his arms and carried me over the threshold to the accompaniment of loud cheers and whistles. Laughing, he pushed a group of revellers out of the door when they tried to follow us inside.

'Goodnight, my friends,' he said. 'Go back to the party and finish the wine.'

One by one they went off unsteadily down the lane, still singing. As the merry racket faded away, Alessandro closed the front door.

All at once it was quiet.

'Alone at last!' he said. Gently, he took me in his arms and rested his forehead against mine. 'Did I tell you how beautiful you are today?'

'Only about a hundred times.' I smiled into the dark.

'Not enough then,' he said, nuzzling my neck.

Shivers of desire and apprehension ran down my back.

He lifted off my wedding veil and pulled the pins from my hair, allowing my curls to tumble around my shoulders. 'My Botticelli angel.' His breath stirred against my cheek. Sliding his hands under my hair, he kissed me and when we drew apart, my heart was singing.

'Let's go upstairs,' he whispered.

We stopped to kiss again on the top stair, a long slow kiss that made me melt inside.

In the bedroom the shutters were open and the bed was bathed in a shaft of silvery moonlight. I hung back, excited but also a little afraid.

Alessandro sensed my nervousness. He drew me past the bed to the window and opened the casements. A fresh, salty breeze stirred my hair as we stood, hand in hand, listening to the soothing ebb and flow of the sea on the sand below.

'If it hadn't been for Princess Caroline, we might never have met,' said Alessandro quietly. 'And a year ago I thought I'd lost you. I've never been so miserable.' He pulled me closer to his side. 'But, tonight, here we are at the beginning of the rest of our lives together and I'm filled with such joy and thankfulness I think my heart might burst.'

I watched him wipe away a tear and loved him all the more for his display of emotion. 'I loved you from the beginning,' I said. 'But it frightened me. I thought I'd been abandoned by those I loved and that perhaps I wasn't worthy of lasting love.'

'I promise to love you forever,' he said, cupping my face. 'Even when we argue, as married couples do, I'll never abandon you. Never, never!'

As I looked deep into his eyes something shifted inside me, as if a key had unlocked my heart. All tension drained away and I was filled with exultation. *This* was how it felt to trust and love another so absolutely. *This* was how it felt to be whole.

'Emilia?'

'Yes,' I said. I kissed him, gently at first, but then with rising passion. I slid my hands inside his shirt and shivered as I touched his firm, warm skin.

His lips were hot and urgent as he fumbled with my buttons and ribbons and finally freed me from my dress, shift and stockings. He lifted me in his arms and carried me to the bed.

Afterwards, Alessandro caressed my hair as we lay entwined together in the moonlight. 'My wife,' he murmured sleepily. 'Forever.'

'Forever,' I whispered.

He curled himself protectively around my back. And then he was asleep.

I nestled into the curve of his body, revelling in the silken touch of our naked skin. My husband's sleeping breaths echoed the peaceful murmur of the sea outside and very soon my own breathing rose and fell in the same rhythm.

# Historical Note

## *Caroline of Brunswick 1768 – 1821*

Whilst researching *The Dressmaker's Secret*, I went to Pesaro and walked up Monte San Bartolo to look for Villa Vittoria, visited the harbour, dipped my fingers in the fountain in the Piazza del Popolo and swam in the Baia Flaminia. All the while I had the feeling Caroline was looking over my shoulder and smiling at the places she had loved.

In 1794 Princess Caroline of Brunswick was twenty-six years old and longing for marriage and children. In London, George, the Prince of Wales, enjoyed an extravagant lifestyle that had plunged him into severe debt amounting to £630,000 (equivalent to fifty-nine million pounds today). His father, George III, urged the Prince to marry, since Parliament would then increase his annual allowance to £65,000. Reluctantly, he agreed. The King approved of the match to Caroline since she was not only his sister's daughter but also a Protestant.

The cousins had never met and Lord Malmesbury was sent to Brunswick as the Prince's envoy to arrange the marriage treaty. He reported that blonde and blue-eyed Caroline had '*a pretty face and tolerable teeth*'. Her short figure was '*not graceful*' though she had a '*good bust*'. Lord Malmesbury was, however, perturbed by her free and easy manner and over the following months made attempts to prepare her for the cool formality of the English court. It was a challenging task. The Princess was impulsive, lively and indiscreet, careless about her toilette, laughed too loudly and enjoyed, perhaps too much, the company of men.

The Prince of Wales appointed his current favourite mistress, Lady Jersey, as Caroline's lady-in-waiting. Lady Jersey was to meet her on her arrival in Greenwich but, in the first of many such snubs to the Princess, kept her waiting most of the day.

The Prince and Princess met for the first time on 5 April 1795, three days before their wedding. Caroline wore an unflattering dress that Lady Jersey provided for her, after persuading her that it was more suitable for the presentation than her own had been. Caroline kneeled to the Prince, who then formally embraced her. He immediately recoiled and called for his equerry to bring him a glass of brandy. Without another word, he left the room. Caroline, affronted by the Prince's behaviour, commented that he was very fat and not as handsome as his picture. Later, it transpired that Lord Malmesbury's attempts to teach the Princess to be particular about her personal hygiene had failed.

The wedding was as disastrous as the first meeting. The Prince of Wales was agitated and almost unable, or unwilling, to utter his responses during the ceremony, and by the evening had consumed so much brandy he collapsed on the floor by the fireplace in the bridal chamber. Despite this, nine months later Caroline gave birth to Princess Charlotte.

The unhappy marriage was made worse by the malicious gossip spread by Lady Jersey, who poisoned the Royal Family against Caroline. Regardless of his own infidelities, the Prince was desperate to divorce his wife but the government wouldn't sanction it without proof of her adultery. Three days after Princess Charlotte's birth the

Prince drew up a will leaving all his property to Roman Catholic commoner Maria Fitzherbert, whom he had married illegally and without the essential royal permission ten years before. He referred to Maria as '*the wife of my heart and soul*', and to '*the woman who is call'd the Princess of Wales*', he bequeathed one shilling.

The humiliations continued and, deeply hurt, Caroline reacted with increasingly reckless behaviour. In 1796 the papers carried reports that Lady Jersey had intercepted and opened letters written by Caroline to her mother, in which she made rude comments about the Royal Family and referred to the Queen as '*Old Stuffy*'. The press took Caroline's side against the already unpopular Prince of Wales. They called for Lady Jersey's dismissal as lady-in-waiting and were critical of the Prince's spiteful attitude to his wife. The couple separated but the cruelties continued when Caroline was denied proper access to her daughter or an allowance or home suitable to her position.

Caroline rented a house in middle-class Blackheath, where she held some kind of a court of her own, encouraging politicians and society figures to visit. She hosted eccentric and sometimes wild parties, entertaining visitors while she sat on the ground eating raw onions or romped on her knees on the carpet with Princess Charlotte. She indulged in flirtations with a number of her guests, though her behaviour was somewhat sobered by the knowledge that adultery, in her case, was a treasonable offence. Her manner was extraordinarily open and confiding and she formed many intense, but often short-lived, friendships without regard for whether or not they were appropriate.

During the following years she became the protector of seven or eight orphan children, finding them good foster homes and supervising their education. Caroline's fondness for children and for irresponsible jokes led to a national scandal. She adopted a baby, William Austin, but teased her friend Lady Douglas by pretending that she had given birth to the boy herself. Later, in 1806 and following a quarrel between the two women, Lady Douglas spread vindictive rumours about the purportedly illegitimate baby. At once a secret committee was set up by the King, the so-called Delicate Investigation, but, to the Prince of Wales's chagrin, it was proved William Austin was not Caroline's natural child. She, however, was condemned for her 'loose behaviour with men' and left with a stain upon her character.

In 1814, with Napoleon apparently a spent force, a service of thanksgiving was held at St Paul's, followed by celebrations hosted by the Prince Regent and attended by royalty, peers of the realm and ministers of state. Caroline was incensed at being excluded and the press considered such treatment of her as shameful. The populace cheered her in the streets and at the opera. Exhausted by the continuing humiliations and political and financial wrangling of life in England, Caroline, now forty-six years old, decided to leave the country. She visited her family in Brunswick and thence travelled to Italy, appointing a lady's maid, Louise Demont, in Geneva on the way. Upon leaving Milan she hired a courier, thirty-year-old Bartolomeo Pergami. In Florence, in an attempt to assimilate the Italians, whom she had grown to love, Caroline bought a black wig and darkened her eyebrows.

Pergami became indispensable to her and was always attentively by her side. Following a sojourn at Villa d'Este by Lake Como, they set off on extensive travels. Caroline elevated Pergami to the position of chamberlain and purchased land and a title for him before they travelled back to Italy, eventually arriving in Pesaro in 1818. The facts of her life from 1819 to her death on 7 August 1821 are outlined in *The Dressmaker's Secret* except for her interactions with Emilia, whose fictional story is woven throughout.

Following her death, Caroline's body was conveyed to Brunswick Cathedral, where one hundred maidens carrying lighted candles lined the aisles. As her coffin was placed in its vault a prayer was said and then the maidens extinguished the candles.

Caroline was her own worst enemy and impossibly unsuited to life as a royal princess. Although her common touch was popular with the people, her sometimes outrageous behaviour made her a painful burr in the side of the Royal Family. Despite that, she was courageous and loving. She tried to live for the moment, squeezing as much joy as possible into a life led under difficult and often extremely unfair circumstances.

# Further Reading

If you wish to find out more about Caroline of Brunswick, I suggest the following books:

*The Unruly Queen* by Flora Fraser
*Rebel Queen* by Jane Robins
*Caroline* by Thea Holme